CHEMISTRY
An Investigative Approach
Revised Edition

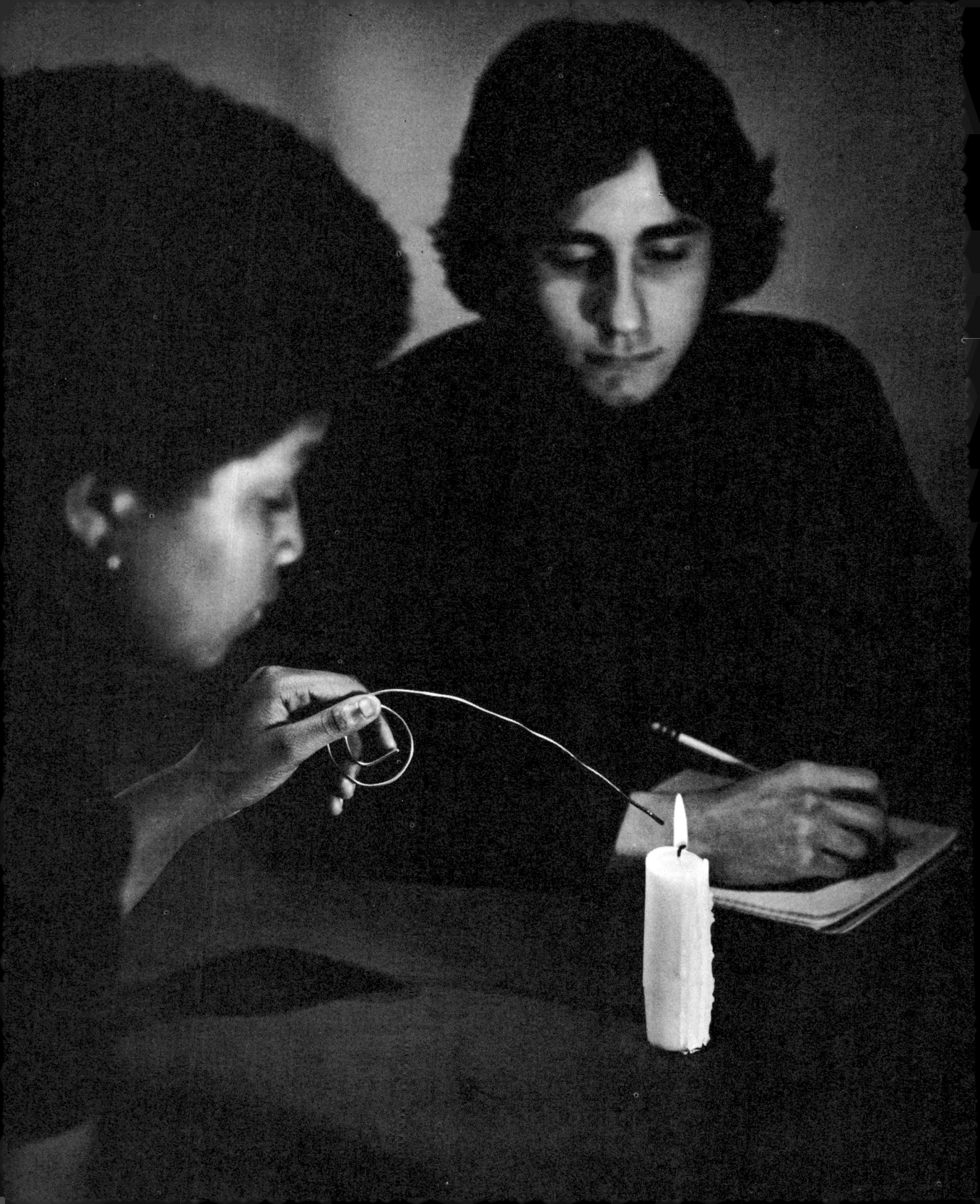

CHEMISTRY

An Investigative Approach
Revised Edition

F. ALBERT COTTON
Texas A & M University
College Station, Texas

C. LEROY DARLINGTON
Montclair High School
Montclair, New Jersey

LAWRENCE D. LYNCH
Palo Alto Senior High School
Palo Alto, California

Houghton Mifflin Company   Boston

ATLANTA   DALLAS   GENEVA, ILLINOIS   HOPEWELL, NEW JERSEY   PALO ALTO   TORONTO

Dr. F. Albert Cotton is the Robert A. Welch Professor of Chemistry at the Texas A & M University, where he conducts an extensive research program. He is the author of over 300 research papers and of several widely read textbooks for college and graduate level courses. In addition, he has lectured throughout the world, and is a member of the National Academy of Sciences. Dr. Cotton has a thorough knowledge of the requirements at all levels of chemical education and has shown a consistent interest in upgrading educational materials and techniques.

Mr. C. LeRoy Darlington teaches chemistry at Montclair High School in Montclair, New Jersey. He has taught science and mathematics courses in both private and public secondary schools for more than twenty years. Earlier, he worked on the Manhattan Project as a member of E.I. Du Pont de Nemours and Company. He has continued to do industrial research in chemistry and to take many courses in chemistry and other sciences during his summers.

Mr. Lawrence D. Lynch is principal of the Palo Alto Senior High School in Palo Alto, California. He has taught high school chemistry for more than 15 years. He taught first in Denver, Colorado and later in Beverly Hills, California. Earlier, Mr. Lynch was a research chemist with Union Carbide and Carbon Chemical Corporation at the Oak Ridge Laboratory. Mr. Lynch has worked in the area of curriculum development for a number of years, and has been an active member of the CHEM Study writing team.

Copyright © 1980, 1976, 1973 by Houghton Mifflin Company

All rights reserved. No part of this work may be reproduced or transmitted in any form or by any means, electronic or mechanical, including photocopying and recording, or by any information storage or retrieval system, without permission in writing from the publisher.

Printed in the U.S.A.

ISBN: 0-395-27839-2

PREFACE

Chemistry is one of the important branches of science. Nearly every phase of our lives is affected by the results of scientific activity. In the last two centuries, during which man has taken a scientific approach to the understanding and control of his environment, the conditions of his life have changed more than in all of the preceding hundreds of thousands of years. The influence of science on man's life will become even greater in the future. In view of this, it is certain that all of us must have some understanding of what scientific activity is, how it is carried on, and how it will affect mankind. The study of chemistry is one way to become familiar with these things.

Chemistry deals with all of the substances making up our environment and with the changes these substances undergo. These changes may be such seemingly simple and commonplace ones as the burning of a candle, or such complicated and delicately balanced ones as the processes that sustain life in our own bodies. The substances themselves may be such hard and strong materials as rocks and steel, or such apparently insubstantial ones as the gases of the air.

This book presents chemistry as it is today. Great emphasis is placed on experimentation and observation as the basis for all knowledge of chemistry or, indeed, of any other physical science. The unifying principles of the subject are developed in a logical way, with laboratory work providing a basis for this development. There are many more facts known than any person can hope to remember, but science is not simply a collection of facts. The true goal of science is the understanding and interpretation of those facts in a consistent way, that is, a grasp of the principles upon which the facts ultimately rest.

In this book we emphasize these principles and the way they have been, and can again be, derived from experimental observation. You will be engaged in scientific activity in fundamentally the same way as any scientist is engaged. In other words, to some extent you will become a scientist as you study chemistry. In addition, you will learn about some of the harmful effects that technology has had on our environment. You will be encouraged to consider how we may deal with these problems and use a knowledge of chemistry to avoid creating further environmental problems.

As a supplement to this book we also offer a small volume of *Supplementary Readings and Investigations*. The chemistry presented in it can be understood using the principles developed in this book.

At the end of this course you may be acquainted with only a small fraction of all that is known about the science of chemistry. You will realize, however, that all of the things that you do not know (some of which, as yet, nobody knows) are not unknowable. On the contrary, they can become understandable when approached with the proper logical scientific methods. We hope further that you will realize that the exploration of new frontiers in chemistry—or any science—is an activity full of excitement and satisfaction.

The Authors

TABLE OF CONTENTS

1 Chemistry: The Science of Materials and Energy

1 What Chemistry Is Like — 2

WHY AND HOW TO STUDY CHEMISTRY *1-1 What Is Chemistry? 1-2 Chemists Do Experiments* 1-3 Scientific Observation and Description 1-4 Behavior of Solids on Warming 1-5 Melting Temperature SCIENTIFIC PROCEDURE *1-6 The Method of Science 1-7 Observation 1-8 Recording the Facts 1-9 Organization of Observations 1-10 Searching for Regularities and Logical Conclusions 1-11 Wondering Why 1-12 Communication of Results* 1-13 What Controls Combustion? What is Produced?

How To Use a Burner 9 *Chemistry and Our Environment* 18
Chapter Summary 25 *Questions and Problems* 25

2 Matter and Mass — 28

WHAT IS MATTER? *2-1 The Fundamental Property of Matter 2-2 The Law of Gravitation 2-3 Units of Mass* 2-4 Measuring Mass 2-5 Determining the Mass of an Object WEIGHT AND MASS *2-6 What Do We Mean by Weight and Weighing? 2-7 Air Buoyancy 2-8 The Buoyancy Effect 2-9 How to Correct for Buoyancy 2-10 Some Other Sources of Error in Weighing★* DOES MASS CHANGE MEASURABLY IN CHEMICAL REACTIONS? 2-11 Investigating Some Reactions for Possible Changes in Mass *2-12 Mass is Conserved in Chemical Reactions*

The Care and Use of Balances 36 *Joseph Weber* 40 *Roy J. Plunkett* 52
Chapter Summary 51 *Questions and Problems* 53

3 Energy — 54

RECOGNIZING AND MEASURING ENERGY *3-1 Energy Has Many Forms 3-2 Potential Energy 3-3 Thermal Energy and Temperature 3-4 Units of Energy and Temperature* 3-5 The Specific Heat of Metals LAWS GOVERN ENERGY CHANGES *3-6 Conservation of Energy 3-7 Energy and Work 3-8 Energy Tends to be Dissipated Spontaneously 3-9 Randomness or Disorder Is Also Important 3-10 Conservation of Mass and Energy★*

Benjamin Thompson 59 *Pioneers in the Measurement of Temperature* 60 *How to Use a Graduated Cylinder* 63 *Thermal Pollution* 68
Chapter Summary 72 *Questions and Problems* 72

NOTE: Numbered sections that appear in roman type are Investigations. Starred sections are supplementary.

4 Measurements, Units, and Accuracy — 74

UNITS OF MEASUREMENT *4-1 The Metric System   4-2 Comparing Metric and More Familiar Units   4-3 Density* 4-4 Measurement of Density NO MEASUREMENTS ARE ABSOLUTELY ACCURATE *4-5 Uncertainty in Measurement   4-6 Uncertainty in a Derived Quantity: Addition and Subtraction   4-7 Uncertainty in a Derived Quantity: Multiplication and Division* 4-8 Measuring Heat Effects During Burning and During a Phase Change *4-9 Calculating Uncertainty in a Derived Result   4-10 Calculating Uncertainty Using Percentages   4-11 Significant Figures: Approximating Uncertainty   4-12 Large and Small Numbers   4-13 Random and Systematic Errors*★

James Prescott Joule 89
Chapter Summary 102   *Questions and Problems* 102

2 Atoms and Molecules: Building Blocks of Matter

5 Atoms and Molecules in Gases — 106

PRESSURES AND VOLUMES OF GASES ARE RELATED *5-1 From Everyday Experience to a Scientific Model   5-2 Defining and Measuring Pressure   5-3 Pressure Measuring Instruments   5-4 How Pressure and Volume Are Related: Boyle's Law   5-5 Another Look at the Particle Model of a Gas* RELATIVE MOLECULAR MASSES—ATOMS, MOLECULES, AND MOLES *5-6 The Masses of Equal Volumes of Gases   5-7 Pressure-Volume Relationships of Different Gases   5-8 Some Other Properties of Gases   5-9 Avogadro's Hypothesis   5-10 Atoms and Molecules   5-11 The Mole* GASEOUS REACTIONS *5-12 Applying Avogadro's Hypothesis to Gaseous Reactions   5-13 Chemists' Shorthand: Formulas and Equations   5-14 Gaseous Reactions Can Give Information About Molecular Formulas   5-15 Proving the Formula of Nitric Oxide*★

Robert Boyle 117
Chapter Summary 135   *Questions and Problems* 136

6 More About the Molecular Model for Gases — 138

ATOMS AND MOLECULES HAVE DEFINITE MASSES *6-1 The Definition of Molar Masses   6-2 Pressure of a Mixture of Gases: Partial Pressures* TEMPERATURE CHANGES AFFECT PRESSURE AND VOLUME *6-3 The Volume Occupied by a Mole of Gas at Different*

Temperatures: Charles' Law **6-4** *The Absolute Temperature Scale*
6-5 *A Quantitative Investigation of the Reaction of a Metal with Hydrochloric Acid* **6-6** *Combining Boyle's Law, Charles' Law, and Avogadro's Hypothesis* **6-7** *The Molecular Model Accounts in Detail for the Ideal Gas Law*★

Smog 144
Chapter Summary 158   *Questions and Problems* 159

7 Elements, Compounds, and Chemical Reactions — 162

BASIC PARTICLES OF MATTER **7-1** *Atoms and Molecules in Liquids and Solids* **7-2** *All Gases Can Be Liquefied*★ **7-3** *Elements and Compounds* **7-4** *Analyzing Compounds* **7-5** *Decomposing a Common Substance* **7-6** *Chemical Analysis* **7-7** *Chemical Synthesis* **7-8** *Synthesis of a Metal Oxide* **7-9** *Definite Proportions* THE ELEMENTS: THEIR NAMES AND SYMBOLS **7-10** *The Elements* **7-11** *Symbols for the Elements* **7-12** *Molecular Formulas* REPRESENTING MOLECULES AND CHEMICAL REACTIONS **7-13** *Molecular Structures and Models* **7-14** *Equations for Chemical Reactions* **7-15** *What a Chemical Equation Tells Us* **7-16** *Writing and Balancing Some Equations*

John Dalton 177
Chapter Summary 189   *Questions and Problems* 189

8 Patterns of Chemical Behavior — 194

SOME IMPORTANT TYPES OF REACTIONS **8-1** *Reactions with Oxygen* **8-2** *Active Metals React with Water* **8-3** *Comparing Reactions of Metals with Water* SOME ELEMENTS HAVE SIMILAR PROPERTIES **8-4** *The Alkali Metals: A Chemical Family* **8-5** *Another Chemical Family: The Halogens* A SEARCH FOR REGULARITIES LEADS TO THE PERIODIC TABLE **8-6** *Other Elements Have Similar Properties* **8-7** *The First Periodic Table* **8-8** *The Modern Periodic Table* **8-9** *Properties Vary Smoothly within Groups* **8-10** *The Noble Gases Are Unreactive but Useful*★

Iron and Aluminum Oxides 198   *Dmitri Ivanovich Mendeleev* 209
Ida Tacke Noddack 214
Chapter Summary 216   *Questions and Problems* 217

9 The Electrical Nature of the Atom — 218

MATTER HAS ELECTRICAL PROPERTIES **9-1** *Basic Characteristics of Electrical Charges* **9-2** *Factors Determining the Amount of Attraction or Repulsion* **9-3** *Neutralization of Charges* **9-4** *The Nature and Origin of Electrical Charges and Forces* ATOMS ARE MADE OF POSITIVE AND NEGATIVE CHARGES **9-5** *More Detailed Evidence for Charged Particles* **9-6** *The Charge and Mass of the Electron* **9-7** *How Thomson Measured the e/m Ratio*★ **9-8** *Ions in Gases* **9-9** *Ions in Solution* **9-10** *The Relation Between the Moles*

of Electrons and the Moles of Metals Used in Electrolysis
9-11 *Ions, Atoms, and Electrons*

Robert Andrews Millikan 231   *Michael Faraday* 240
Chapter Summary 245   *Questions and Problems* 246

3 The Nature of the Atom

10   The Structure of the Atom — 250

ATOMS HAVE A DEFINITE DESIGN   10-1 *Rutherford's Scattering Experiments*   10-2 *Rutherford's Atomic Model*   10-3 *The Nuclear Atom, a Successful Scientific Model*   10-4 *Atomic Size*   10-5 *Measurement of the Size of a Molecule*   10-6 *Neutrons*   10-7 *Basic Characteristics of the Nuclear Atom*   10-8 *Isotopes*   ATOMIC NUCLEI MAY DECOMPOSE   10-9 *Radioactivity*   10-10 *Nuclear Reactions and Nuclear Energy*   10-11 *Calculating the Energy of Nuclear Reactions*★

Ernest Rutherford 256   *Harold C. Urey* 266   *Nuclear Energy* 270
The Golden Age of Atomic Discoveries 273
Chapter Summary 276   *Questions and Problems* 276

11   Interactions Between Light and Atoms — 278

THE NATURE OF LIGHT   11-1 *Light as a Wave Motion*   11-2 *Does Light Behave Like a Series of Waves?*   11-3 *Light Is Made Up of Different Frequencies*   11-4 *Light Is a Form of Energy*   11-5 *Photons and the Photoelectric Effect*   ATOMS EMIT ONLY CERTAIN FREQUENCIES OF LIGHT   11-6 *Atomic Spectra*   11-7 *The Hydrogen Spectrum*   11-8 *A Model for Explaining the Hydrogen Spectrum*   11-9 *The Energy Levels of a Hydrogen Atom*   11-10 *Bohr's Model of the Hydrogen Atom*   11-11 *Continuous Spectra*★   11-12 *Molecular Spectra*★

Max Planck 292   *Solar Energy* 294   *Niels Bohr* 308
Chapter Summary 316   *Questions and Problems* 316

12   The Atom According to Quantum Mechanics — 318

FROM THE BOHR ATOM TO THE ATOM OF QUANTUM MECHANICS   12-1 *Bohr's Model Fails for Other Atoms*   12-2 *Matter Is Found To Have Wave Properties*   QUANTUM MECHANICS EXPLAINS ATOMIC STRUCTURE   12-3 *Probability*   12-4 *Orbitals and the Principal Quantum Number*   12-5 *The Uncertainty Principle*★   12-6 *Shapes of s Orbitals*   12-7 *p Orbitals*   12-8 *d and f Orbitals*   12-9 *More About the Shapes of Orbitals*

Louis Victor de Broglie 322
Chapter Summary 335   *Questions and Problems* 336

13 The Atomic Basis for Periodic Properties — 338

BUILDING UP COMPLEX ATOMS *13-1 The Helium Atom, and Electron Configurations 13-2 The Lithium Atom, and the Exclusion Principle 13-3 Atoms with Atomic Numbers Four through Ten 13-4 Atoms Beyond Neon* THE PERIODIC PATTERN OF ELECTRONIC CONFIGURATIONS *13-5 The Recurrence of Closed Shells 13-6 Properties Vary Periodically with the Atomic Number* 13-7 The Solubility of Salts: A Search for Patterns *13-8 Electron Configurations Tie In with Properties 13-9 How Moseley Determined Atomic Numbers* ★

Henry G. J. Moseley 354
Chapter Summary 358   *Questions and Problems* 358

Chemical Bonds and Molecular Structure

14 Chemical Bonds — 362

ELECTRICAL FORCES HOLD ATOMS TOGETHER *14-1 The Hydrogen Molecule 14-2 The Chemical Bond as an Overlap of Half-filled Orbitals 14-3 Interaction Between Helium Atoms 14-4 Representations of Chemical Bonding 14-5 The Bonding of Fluorine* HOW BONDING OCCURS IN SECOND ROW ELEMENTS *14-6 The Bonding Capacity of Oxygen Atoms 14-7 The Bonding Capacity of Nitrogen Atoms 14-8 The Bonding Capacity of Carbon Atoms 14-9 The Bonding Capacity of Boron Atoms 14-10 The Bonding Capacity of Beryllium Atoms 14-11 The Bonding Capacity of Lithium Atoms 14-12 A Word of Caution about the Word "Valence"* SOME BONDS HAVE IONIC CHARACTER *14-13 The Bonding in Gaseous Lithium Fluoride 14-14 Ionic Character in Bonds to Fluorine 14-15 Ionic Character in Bonds to Hydrogen* 14-16 Covalent and Ionic Compounds *14-17 Electron Density Maps*★

Linus C. Pauling 381   *Disposal and Recycling of Wastes* 386
Chapter Summary 391   *Questions and Problems* 392

15 The Shapes of Molecules — 394

ELECTRON PAIRS FORM BONDS IN DEFINITE DIRECTIONS *15-1 Experiments Tell Us about Molecular Architecture 15-2 The Shapes of Some Molecules Formed by Oxygen and Nitrogen 15-3 The Tendency for Electrons To Form Pairs 15-4 The Electron-Pair Repulsion Theory of Molecular Shapes 15-5 Hybrid Orbitals* DOUBLE AND TRIPLE BONDS *15-6 Double Bonds 15-7 Triple Bonds 15-8 Resonance 15-9 Molecular Orbitals*★

MOLECULAR SHAPE AND DIPOLE MOMENTS *15-10 Definition of a Dipole Moment   15-11 How Molecules Can Have Dipole Moments   15-12 Structural Isomers   15-13 Geometric Isomers*
15-14 Molecules and Their Shapes   *15-15 Another Way of Describing Double and Triple Bonds★*

Robert Sanderson Mulliken 413   *Peter Joseph Wilhelm Debye* 420
Chapter Summary 433   *Questions and Problems* 434

16 Molecules Formed by Carbon 436

ORGANIC COMPOUNDS ARE MADE OF CARBON ATOMS *16-1 Sources of Carbon Compounds   16-2 Composition and Structure of Organic Compounds   16-3 Determining Empirical and Molecular Formulas*
HYDROCARBONS PROVIDE THE FRAMEWORK OF ORGANIC COMPOUNDS *16-4 Nomenclature   16-5 Saturated Hydrocarbons   16-6 Some Properties and Reactions of Hydrocarbons   16-7 Unsaturated Hydrocarbons   16-8 Systematic Names for Noncyclic Hydrocarbons   16-9 Aromatic Hydrocarbons: Benzene and Its Derivatives*
FUNCTIONAL GROUPS DETERMINE THE BEHAVIOR OF ORGANIC MOLECULES *16-10 General Characteristics of Functional Groups   16-11 Alcohols   16-12 Some Properties and Reactions of Alcohols   16-13 Aldehydes   16-14 Carboxylic Acids   16-15 Ketones   16-16 Esters   16-17 Soap and Detergents   16-18 Functional Groups Containing Nitrogen: Amines and Amides   16-19 Right- and Left-Handed Molecules★*

Friedrich A. Kekulé von Stradowitz 454   *Percy Lavon Julian* 458   *Robert Burns Woodward* 470   *Persistent Insecticides* 474
Chapter Summary 481   *Questions and Problems* 481

17 Solids, Liquids, and Solutions 484

SOLIDS, LIQUIDS, AND SOLUTIONS HAVE SPECIAL PROPERTIES *17-1 The Attractive Forces Causing Condensation of Gases   17-2 The Nature of Crystalline Solids   17-3 The Nature of Pure Liquids   17-4 The Nature of Solutions*   17-5 The Packing of Atoms or Ions in Crystals FORCES BETWEEN ATOMS VARY IN STRENGTH *17-6 Interatomic Forces   17-7 Covalent Bonds and Network Solids   17-8 Metallic Bonding* COMPLEX FORCES HOLD COMPOUNDS TOGETHER *17-9 Additional Forces in Compounds   17-10 Van der Waals Forces and Molecular Compounds   17-11 Covalent Bonds and Network Solid Compounds* IONIC SOLIDS ARE SIMPLY EXPLAINED *17-12 Factors Favoring Ion Formation   17-13 Structures of Ionic Compounds   17-14 Properties of Ionic Compounds   17-15 Some Quantitative Ideas about Ionic Crystals★* MOLECULES INTERMINGLE TO FORM SOLUTIONS *17-16 Solvent Properties   17-17 Molecular Dipoles   17-18 Solubility of Electrolytes in Water   17-19 Boiling and Freezing Points of Solutions   17-20 Units for Expressing Concentration: Molarity* HYDROGEN HAS UNIQUE BONDING

xii   Table of Contents

PROPERTIES *17-21 Hydrogen Bonds 17-22 Energy of Hydrogen Bonds 17-23 Where Hydrogen Bonds are Found 17-24 The Nature of the Hydrogen Bond 17-25 The Significance of the Hydrogen Bond*

Johannes Diderick van der Waals 495
Chapter Summary 524 Questions and Problems 524

18 Giant Molecules   528

GIANT MOLECULES ARE BUILT OF SMALLER UNITS *18-1 Giant Molecules Are Usually Polymers 18-2 Condensation Polymerization 18-3 Addition Polymerization 18-4 The Preparation of Some Polymers* THE GIANT MOLECULES OF NATURE *18-5 Proteins 18-6 Starches and Cellulose 18-7 Rubber 18-8 The Nucleic Acids, DNA and RNA 18-9 The Double Helix and the Template Idea 18-10 RNA* SILICATES ARE INORGANIC GIANT MOLECULES *18-11 Silicates★ 18-12 Asbestos, Mica, and Glass★*

Choh Hao Li 539 Paper 542 Wallace Hume Carothers 545
Chapter Summary 554 Questions and Problems 555

5 The Dynamics of Chemistry

19 Chemical Reactions and Calculations   558

CHEMICAL FORMULAS AND EQUATIONS *19-1 Valence, the Combining Capacity of Elements 19-2 Using Valence Rules to Write Formulas 19-3 Writing the Names of Compounds 19-4 Behavior of Solid Iron in a Water Solution of Copper Sulfate 19-5 How Chemical Equations Are Written* CHEMICAL CALCULATIONS 19-6 Mass Relationships in Chemical Changes *19-7 Stoichiometry 19-8 The Mole Method 19-9 Conversion Factors 19-10 The Manufacture of Sulfuric Acid: Mass-Mass Chemical Calculations 19-11 The Manufacture of H_2SO_4: Mass-Volume Calculations 19-12 The Manufacture of H_2SO_4: Volume-Volume Calculations 19-13 Stoichiometry in an Organic Reaction★*

Automotive Smog 582
Chapter Summary 584 Questions and Problems 584

20 Energy Effects in Chemical Reactions   588

HEAT IS ABSORBED AND RELEASED IN CHEMICAL REACTIONS *20-1 A Practical Illustration 20-2 Heat Content of a Substance 20-3 Measuring the Heat of Reaction 20-4 Additivity of Reaction Heats 20-5 Heats of Formation 20-6 Predicting the Heat of a*

Chemical Reaction   20-7 Calorimetry   MOLECULES STORE ENERGY
20-8 The Energy of a Molecule★   20-9 Energy Changes on Warming★

Germain Henri Hess 596   Can Our Activities Change the Climate? 600
Chapter Summary 605   Questions and Problems 605

21   The Rates of Chemical Reactions   608

MANY FACTORS DETERMINE REACTION RATES   *21-1 Definition of the Rate of Reaction   21-2 The Nature of the Reacting Substances*
21-3 *A Study of Reaction Rates   21-4 Effect of Concentration on Reaction Rate   21-5 Reaction Mechanism   21-6 Effect of Temperature on Reaction Rate*   ENERGY PLAYS AN IMPORTANT ROLE IN REACTION RATES   *21-7 Distribution of Kinetic Energies   21-8 Activation Energy   21-9 Activation Energies and Heats of Reaction   21-10 Catalysts   21-11 Enzymes, the Catalysts in Living Things★*

Henry Eyring 629   Edward Calvin Kendall 636
Chapter Summary 640   Questions and Problems 641

22   Chemical Equilibrium   644

EXAMINING SOME QUALITATIVE ASPECTS OF EQUILIBRIUM
22-1 Equilibrium in Chemical Reactions   22-2 Equilibrium in a Phase Change   22-3 The Boiling Point   22-4 Solubility Equilibrium   22-5 Recognizing Equilibrium   EQUILIBRIUM IS DYNAMIC   *22-6 The Dynamic Nature of Equilibrium in Chemical Reactions   22-7 The Dynamic Nature of a Phase Change Equilibrium   22-8 The Dynamic Nature of Solubility Equilibrium   22-9 Direct Proof of the Dynamic Nature of Equilibrium*   THE STATE OF EQUILIBRIUM CAN BE ALTERED   *22-10 The State of Equilibrium   22-11 Altering the State of Equilibrium   22-12 Predicting New Equilibrium Concentrations: Le Chatelier's Principle   22-13 Application of Equilibrium Principles: The Haber Process*   INVESTIGATION OF THE QUANTITATIVE ASPECTS OF EQUILIBRIUM   *22-14 A Quantitative Approach to Chemical Equilibrium   22-15 The Equilibrium Constant for a Reaction   22-16 The Law of Chemical Equilibrium   22-17 Determination of the Solubility Product of Silver Acetate   22-18 Quantitative Treatment of Solubility Equilibrium   22-19 Calculation of the Solubility from the Solubility Product   22-20 Will a Precipitate Form?   22-21 The Law of Chemical Equilibrium Derived from Rates of Opposing Processes★   22-22 Factors That Determine Equilibrium★*

Norbert Rillieux 648   Ozone, Life on Earth, and the SST 660
Harold S. Johnston 662   Josiah Willard Gibbs 672
Chapter Summary 686   Questions and Problems 687

23 Acids and Bases — 690

WATER ITSELF IS AN ELECTROLYTE *23-1 Strong and Weak Electrolytes* *23-2 Water As a Weak Electrolyte* *23-3 Changing the Concentrations of H^+ and OH^- Ions* *23-4 Acidic, Basic, and Neutral Solutions; the Definition of pH* *23-5 The Change of K_w with Temperature*★ ACIDS AND BASES HAVE CHARACTERISTIC PROPERTIES *23-6 Properties of Aqueous Solutions of Acids* *23-7 Properties of Aqueous Solutions of Bases* *23-8 An Explanation of the Properties of Bases* FINDING AMOUNTS OF ACID AND BASE 23-9 Determining the Hydrogen Ion Concentration of Solutions, Using Indicators *23-10 HCl and NaOH in the Same Solution* *23-11 The Process of Titration in Theory* 23-12 The Process of Titration in Practice ACIDS VARY IN STRENGTH *23-13 Weak Acids* *23-14 Determination of K_A* *23-15 Calculation of $[H^+]$* *23-16 Competition for H^+ Among Weak Acids*★ *23-17 Hydrogen Ion in the Proton Transfer Theory of Acids*★

Acid Rain and Snow 706 *Care of Burets* 710 *Gilbert N. Lewis* 720 *Chapter Summary* 722 *Questions and Problems* 723

24 Oxidation-Reduction Reactions — 726

UNDERSTANDING REDOX REACTIONS 24-1 The Chemistry of an Electrochemical Cell *24-2 Examination of the Electrochemical Experiment* *24-3 Oxidation-Reduction Reactions in a Beaker* *24-4 The Function of the Salt Bridge* *24-5 Some Other Redox Reactions* *24-6 Competition for Electrons* HOW OXIDATION-REDUCTION REACTIONS CAN BE PREDICTED *24-7 Measuring Cell Potentials* *24-8 Measuring Half-Cell Potentials* *24-9 Using Reduction-Potential Tables* BALANCING EQUATIONS FOR OXIDATION-REDUCTION REACTIONS *24-10 Oxidation Numbers* *24-11 Using Oxidation Numbers* *24-12 Using Half-Cell Reactions to Balance Equations* USEFUL ELECTROCHEMISTRY *24-13 Batteries*★ *24-14 Using a Voltage to Drive a Reaction*★ *24-15 The Process of Electroplating*★

Svante August Arrhenius 750
Chapter Summary 751 *Questions and Problems* 752

Appendixes

1. RELATIVE STRENGTHS OF ACIDS, 754
2. STANDARD REDUCTION POTENTIALS, 755
3. MOLAR MASSES OF THE ELEMENTS, 756
4. ELECTRONIC CONFIGURATIONS OF THE ELEMENTS, 758
5. NAMES, FORMULAS, AND CHARGES OF SOME COMMON IONS, 760
6. SOLUBILITY OF COMMON COMPOUNDS IN WATER, 761
7. UNITS, 762
8. INTRODUCTION TO LABORATORY WORK, 763

Glossary 768   **Acknowledgements** 779   **Index** 782

STUDY AIDS

A number of features are included in this book to help make your study and learning more effective and pleasant:

Important words and terms are printed in boldface type when they are introduced in the text: **model, noble gases**.

Each section of a chapter is followed by a short summary and review questions to help you check your comprehension of the section.

At the end of each chapter there is a summary of the important understandings to be gained from the chapter, and a longer list of questions and problems.

Appendixes 1–7 provide various charts and tables of data you will need to refer to from time to time.

Appendix 8 outlines procedures and general instructions for safe and efficient laboratory work, and includes photographs of common laboratory apparatus.

The Glossary at the end of the book provides concise definitions of important words and phrases.

Pronunciations are given for some of the more difficult words: Geiger (GY-ger). The syllable that bears the greatest emphasis when the word is spoken appears in capital letters (GY). The guide below will enable you to work out the pronunciations.

GUIDE TO PRONUNCIATION

a, like *a* in c*a*t	alphabet	AL-fuh-bet
ah, like *a* in f*a*ther	farming	FAHR-ming
ai, like *ai* in f*ai*r	share	shair
ay, like *a* in c*a*ke	Asia	AY-zhuh
aw, like *aw* in s*aw*	always	AWL-wayz
e, eh, like *e* in b*e*d	test	tehst
ea, like *ea* in f*ea*r	here	hear
ee, like *ee* in f*ee*d	leaf	leef
ew, like *ew* in n*ew*	neuter	NEW-ter
i, ih, like *i* in f*i*n	system	SIHS-tum
y, ye, eye, like *i* in f*i*nd	alive	uh-LYV
o, like *o* in h*o*t	otter	O-ter
oo, like *oo* in f*oo*d	lunatic	LOON-uh-tihk
oh, like *o* in h*o*me	rainbow	RAYN-boh
oi, like *oi* in b*oi*l	poison	POI-zuhn
ow, like *ow* in f*ow*l	fountain	FOWN-tuhn
u, uh, like *u* in f*u*n	blood	bluhd
er, like *er* in fin*er*	treasure	TREHZH-er
ur, like *ure* in fut*ure*	mercury	MER-cur-ee
g, like *g* in *g*as	regret	ree-GREHT
j, like *j* in *j*ump	magic	MA-jik
z, like *s* in hi*s*	busy	BIHZ-ee
zh, like *z* in a*z*ure	pleasure	PLEHZH-er

INTRODUCTION

The textbook you are about to use—*CHEMISTRY: An Investigative Approach*—has a very distinguished ancestry. The beginning was in 1959, when a steering committee, headed by Professor Glenn T. Seaborg, a Nobel Prize laureate in chemistry, determined to provide the high schools with a truly new and contemporary approach to the teaching and learning of chemistry. A team of teachers, some from universities and some from high schools, was assembled to plan and write such a "new wave" textbook. The first version, together with a laboratory manual, was written in the summer of 1960 and used in 23 high schools during the school year 1960-61 by about 1300 students. The book was called *Chemistry—An Experimental Science,* and that title, like the title of the book you are now reading, summarizes the basic theme: that chemistry is a science which must be approached through experiment and investigation.

The original book was warmly received by teachers and students. A second and then a third edition were prepared, each new edition embodying changes based on the experience of teachers and students in using the preceding one. By 1966, the course had been used by over 700,000 students in high schools in the United States, and by many students abroad.

The basic idea in working out the original text was to show the steps by which scientists proceed to get at the fundamental facts and key principles governing the behavior of the world around us. That basic idea has been retained in this book. Insofar as it appears practical to do so, the concepts and facts, as well as the feel and flavor, of chemistry are introduced by way of experiments. These experiments are performed by the student just when the text discusses the ideas that they illustrate. Thus, the notion that scientific ideas and facts arise from experiment is emphasized.

Of course, the idea that science is more than just a collection of facts—however accurate, striking, and numerous the facts may be—is also emphasized. The patterns and regularities to be found in experimental observations and the models or theories that can be constructed to account for such patterns are a most important part of the human enterprise called science. In this book, the correlation of experimental data so as to reveal its patterns and the development and testing of models and theories are given full emphasis along with the laboratory work. Science at its best involves the fruitful interplay of experiments and theories, and this book endeavors to convey to the student an appreciation of this interplay.

This book is written with the belief that the best way to present chemistry is to present all of its most fundamental principles, including concepts of energy, rate, and equilibrium in chemical reactions, the three-dimensional structure of molecules, chemical bonds as they occur in liquids, solids, and gases, and the ideas of atomic structure and chemical periodicity. Naturally, all of these principles and concepts cannot be presented completely, but they can be—and are—presented in an honest, though simplified, form. They are presented so as to form a foundation on which further study may build, without requiring that anything learned here need later be "unlearned."

Finally, in this book we hope to convey an awareness of the significance and potentialities of scientific activities in shaping the future of man and his world. Even the student whose future occupation may have no direct dependence on science will find his career indirectly influenced by the activities of chemists and other scientists. We hope that the study of this book will help all future citizens to understand and wisely judge the growing impact of technological advances on their physical and social environment.

Just in the past few years it has become widely recognized that many of our activities, based on technology, are having profoundly harmful effects on our environment. We are polluting the air and water, and are introducing substances that seriously disturb the established ecological relationships. Many of these environmental problems are essentially chemical in nature. Therefore, it is appropriate—and even urgent—that in this book, which emphasizes the principles of chemistry as a pure science, some attention be paid to these important environmental problems. We have therefore included a series of brief articles, called *Chemistry and Our Environment,* explaining the chemical aspects of some of our main environmental problems. We hope thereby to show that chemistry is not only a great intellectual discipline, but a vital tool in shaping the future of mankind intelligently.

We should like to acknowledge our great debt to others in the preparation of *CHEMISTRY: An Investigative Approach*. It is an authorized revision of *Chemistry—An Experimental Science,* created by the Chemical Education Material Study with the support of the National Science Foundation. For permission to prepare this revision, the authors hereby express their appreciation. This permission is not an endorsement by Chemical Educational Material Study of the present volume.

Knowing that this book, like all human endeavors, must be imperfect in many ways, we earnestly encourage all users, teachers and students, to give us their criticisms and suggestions.

The Authors

1

Chemistry: The Science of Materials and Energy

Chemistry is the science of materials—their composition, properties, interactions, and transformations. Since chemical transformations normally involve energy changes, chemists must concern themselves with energy, too. Chemistry is the study of all the elements found in nature, as well as a few that have been made by man, and of the hundreds of thousands of compounds that they form.

Chemical knowledge has resulted from years of work by many chemists. Each chemist was started on his search by questions such as the following ones. "What will this be like if I heat it?" "How can I purify this?" "What will cause this color to change?" Most important of all, perhaps, is the question: "Why do these things happen?"

In order to find answers to his questions, a chemist must often devise entirely new experiments, and new equipment, as well. He usually makes many false starts and fails many times. The fact that you read only of successes may make science seem cut and dried and very easy. Nothing could be further from the truth.

In this unit you will learn how chemists ask their questions, carry out experiments, and record and analyze the results. You will learn this not merely by reading, but by thinking about certain questions and by doing your own experiments. For chemistry, like all science, is a matter of thinking and doing. You will also become familiar with the basic concepts of matter and energy and with the ways in which measurements can be made and expressed.

1 What Chemistry Is Like

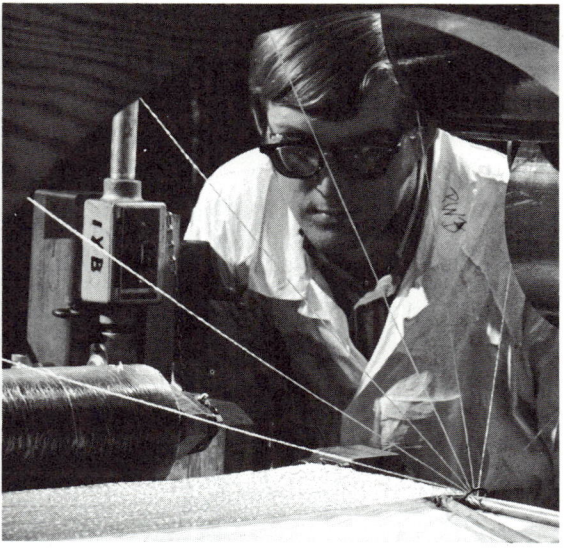
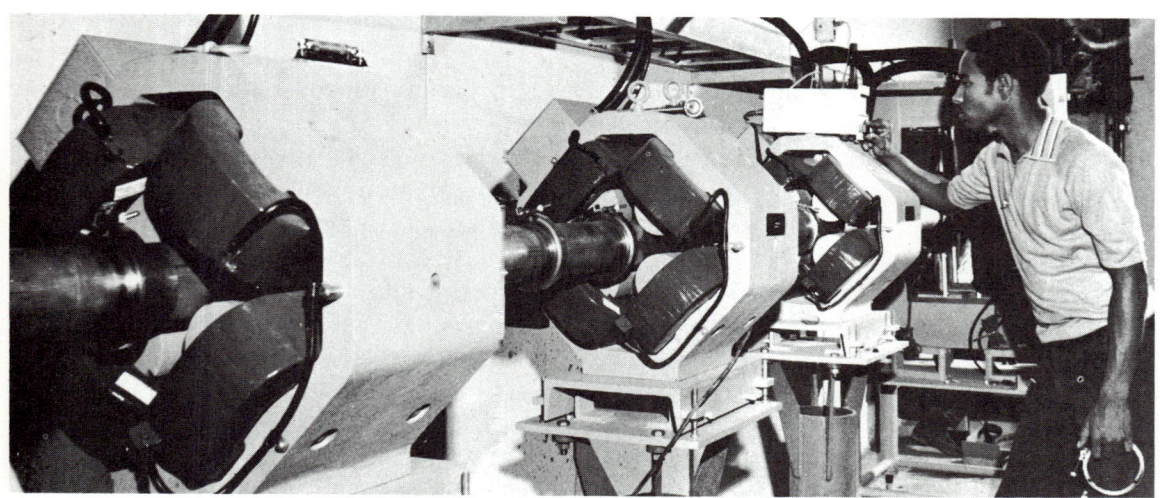

Clockwise, from the upper left, the scientists are engaging in cancer research, producing fibers for use in purifying sea water, and preparing for high-energy experiments with the fundamental particles of matter.

Much of what you see in your surroundings every day owes something to the work of chemists. For this reason alone, it is important to find out just what chemists do. While no two chemists go about things in exactly the same way, there is a discernible pattern in their approach.

Perhaps more than any other science, chemistry relies on experiment. One reason for this is that the chemist is dealing with things that he cannot see and count easily: atoms and molecules.

Another reason lies in the fact that although matter follows general laws, many of which are already known, there are frequently new factors involved in a chemical idea or a chemical problem. As a result, the only way to find out "if it works" is to experiment.

It is especially appropriate, then, that your study of chemistry will involve experimentation from the start. Only by doing experiments can you truly learn what chemistry is.

Why and How To Study Chemistry

1-1 *What Is Chemistry?*

Those of us living in an advanced technological society enjoy certain comforts and conveniences that we usually take for granted. We know, however, that the quality of the life we enjoy is quite different from that experienced only a few generations ago. It is also quite different from what exists even now in large sections of the world.

How do advanced technological societies like ours differ from less advanced ones? For one thing, we have available a great many *materials*, designed for a wide variety of specific uses. One may be just right for use in a space satellite, another for repairing a valve in the human heart. Many have less dramatic, but nevertheless important, uses. Permanent-press fabrics, rust-resistant metals, and artificial turf are just a few of many possible examples. See, for example, Figure 1-1.

Another thing that a society like ours must have in abundance is *energy*. It must be available on demand, in suitable forms. The

Figure 1-1 *An artificial heart valve. Most of the materials of which it is composed did not exist 30 years ago.*

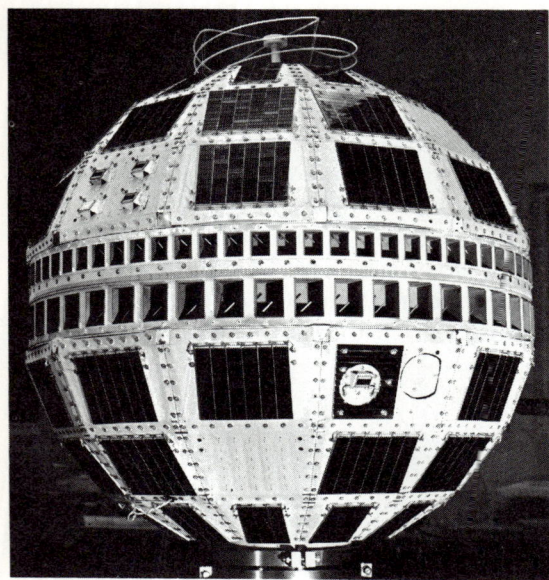

Figure 1-2 *A model of a Telstar communication satellite. The dark squares on the surface are solar batteries that convert sunlight into electrical energy.*

energy. Those chemical processes that produce a lot of energy are often carried out for just that purpose. The burning of fuels is a good example.

A chemist's work, then, involves both energy and materials. Or, to use a more general term than "materials," both energy and *matter*. Think of matter in the ordinary sense of the word, for now. The scientific meaning of the term will be made clear in the next chapter.

It is not enough to say that chemists work with matter and energy. A football player could be said to work with both matter and energy, too. When blocking or tackling, he is applying plenty of energy in moving a lot of matter. But he is not called a "chemist" as a result, certainly not by his opponent. It is the way in which a chemist uses matter and energy that is distinctive.

Probably the most accurate definition of chemistry is simply that it is "what chemists do." The spirit of that definition, the spirit of *doing*, is the keynote of this book.

forms that are most suitable for powering a satellite, as illustrated in Figure 1-2, may be quite different from those that help to regulate a human heart or light a city or propel the family car.

What has chemistry to do with all of these things? The answer is, quite a lot. *Chemistry is the science of materials.* Chemists put simpler materials together to make more complex ones, and break more complex ones down into simpler ones. They change some materials into other ones that will better meet a specific need.

What about energy? Energy is essential to the study of materials. All of the processes by which chemists manipulate materials involve energy. Chemical processes sometimes require energy, and sometimes produce

1-2 Chemists Do Experiments

In order to learn how to play the chemistry game you must do experiments. Leonardo da Vinci (1452-1519) stated that "those sciences are vain and full of errors which are not born from experiment, the mother of all certainty." The words of that remarkable Italian genius are no less true today.

All that chemists have learned is based on the results of experiments. Chemists are still experimenting, because what they know does not begin to approach what they would like to know. An experiment might even turn up something that no chemist has yet realized he does not know. In addition, chemists are experimenting in order to help solve such obvious problems as those involved in maintaining a livable environment.

Much of what you will learn in this course will derive from your experiments. You cannot, of course, learn everything by experiment. There is not enough time, for one thing. We will also have to rely on language, in order to transmit to you the fruits of others' experiments.

Even if time were not a factor, one would hardly expect to have chemistry classes running huge atom-smashers in order to find out for themselves what atoms are. (Figure 1-3 shows one large atom-smasher.) Many experiments can be done in regular school laboratories, however. By doing such experiments, you can have the necessary experience of confronting some facts and seeing what you can make of them. If you learn something about chemistry in this way, your appreciation of all the rest will be far more penetrating and realistic. Besides, doing experiments can be fun!

One of the interesting and even exciting things about chemistry is that very simple experiments can often be quite fruitful. They may be full of unexpected facts and things to think about. Careful observation may turn a seemingly ordinary occurrence into an informative experiment. In order to illustrate this, you will perform some experiments involving a candle. It may surprise

Figure 1-3 *The National Accelerator Laboratory in Illinois. The "main ring," partially shown, is four miles in circumference.*

you to find how little you really know about what a candle does. Some very distinguished scientists have been working on the problem. Even today they are still discovering new and important things about what happens inside the hottest part of the flame. You should not suppose that the humble candle is beneath your notice.

INVESTIGATION

1-3 Scientific Observation and Description

The work of many scientists is not confined to their laboratories. An oceanographer may travel thousands of miles to collect and analyze samples of ocean water from many parts of the globe. Biologists may attach tiny radio transmitters to the necks of grizzly bears in Yellowstone Park to find out how widely the bears roam and where they spend their winters. Other biologists spend years in the Antarctic or in tropical jungles to study organisms in their natural habitats.

Most chemists and physicists, however, spend the greatest part of their time in their laboratories, where they can control the conditions under which they carry out their experiments. Even scientists who study such complex problems as water and air pollution must return to their laboratories, where they can vary a limited number of conditions at one time.

The key to all successful scientific research is the ability to make accurate observations, regardless of whether the work is done in the field or in the laboratory. From good observations made on preliminary experiments the scientist can then determine what conditions he may want to control in later, more refined experiments.

PURPOSE

The main reason for doing this experiment is to discover how many observations you can make of a candle and its burning process. By comparing your observations with those of your classmates after you have finished, you may discover observations that you had not thought important enough to record.

MATERIALS

Lab notebook Ruler
Matches   Candle, mounted on a "drip catcher"

PROCEDURE

1. Examine the candle carefully. Record whatever characteristics of the candle seem important to you.

2. Light the candle. Record in your notebook as many observations as you can while the candle burns for 10 to 15 minutes. Leave space in your notebook to list additional observations made by other class members. You may fill in this list during the postlab discussion of the experiment.

RELATED PROBLEMS

1. Make as complete a list as you can of the conditions that existed in your laboratory during the burning of the candle.

2. Which of these conditions may have had an important effect on the way in which the candle burned? Which conditions may have had little or no effect on the burning candle?

3. If you now wanted to make a careful study of the rate at which a candle burns, what conditions would you want to control?

4. As a result of your observations, make a list of unanswered questions that this experiment has raised in your mind.

INVESTIGATION

1-4 Behavior of Solids on Warming

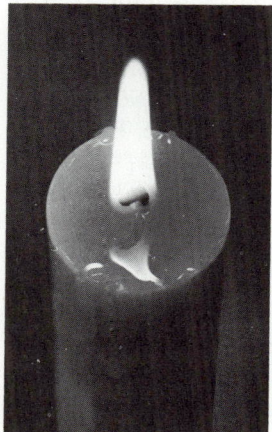

Figure 1-4

If you observed the candle with care, you probably noticed several things about it that were new to you. As a result of your observations, you may have asked such questions as: "What is the colorless liquid in the hollow at the top of the burning candle?" (Figure 1-4), or "Is this liquid merely melted wax or has something else been formed in the burning process?"

Perhaps the liquid is just melted wax. But how do you know this? What kind of experiment would help you decide whether it is wax or something else? Is it possible that wax might behave differently than another substance when it is heated?

In order to answer these questions, it may be necessary for you to find out how various other substances behave when they are heated.

PURPOSE

In this experiment you will compare the behavior of several different substances when they are heated. As a result of your observations, you will try to find a clue to the identity of the material at the top of the candle.

MATERIALS

Ringstand, 2 rings
Wire gauze
Burner
Candle
7 test tubes, 16- × 125-mm
Beaker, 250-ml
Crucible tongs
Knife or spatula (for cutting off piece of candle)
Scissors or pliers (to cut wire)
Small pieces of sulfur, steel wool, tin, lead, and copper (one of each)

40-50 cm copper wire, #16
Silver chloride powder
Metric ruler
Steel wool or sandpaper
Gummed labels
Paper towels

PROCEDURE

Part A. Comparison of Different Substances

1. Use a piece of #16 copper wire to tie together seven 16- × 125-mm test tubes into one bundle. (The millimeter, abbreviated mm, is a unit of length and the milliliter, ml, used below, is a unit of volume. These units will be discussed in Chapter 4.) Twist the ends of the copper wire together tightly enough to prevent the test tubes from slipping out of the bundle. (See Figure 1-5.)

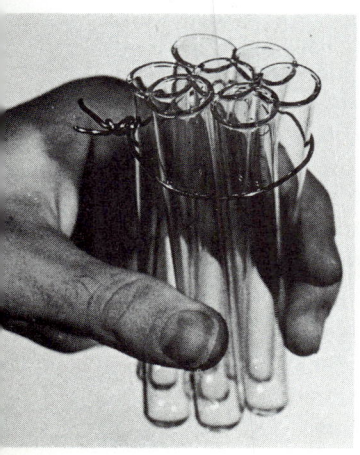

Figure 1-5

2. Distribute the following materials among the seven test tubes: a piece of sulfur, a piece of candle wax, and a ball of steel wool, each about the size of the small eraser on the end of a pencil; strips of sheet metal or metal foil approximately 10×40 mm (about $\frac{1}{2} \times 1\frac{1}{2}$ in) made of each of the metals, lead, tin, and copper; enough silver chloride powder to cover the bottom of a test tube. Polish the metal strips with a piece of steel wool or fine sandpaper before dropping them into the test tubes.

3. Place a burner on the base of a ringstand. Attach an iron ring to the ringstand about 50 mm (2 inches) above the top of the burner. Place a wire gauze on the ring and a 250-ml beaker on the gauze.

4. Attach another iron ring to the upper part of the ringstand. Make two hooks from pieces of #16 copper wire about 100 mm (4 inches) long. Hang the bundle of test tubes by means of the hooks from the upper iron ring. Carefully lower the bundle of test tubes into the 250-ml beaker below so that the bottoms of the test tubes are about 25 mm (an inch) from the bottom of the beaker, as in Figure 1-6.

   Be sure to wear your safety glasses throughout the rest of the experiment!

5. Light the burner and heat the bottom of the beaker as strongly as possible. (See the accompanying directions on how to use a burner.) Watch the materials in the test tubes carefully. Record your observations immediately, in your laboratory notebook, as changes take place. Pay particular attention to the melting behavior, but be sure to record all other changes in color and condition as they occur.

6. If continued heating fails to melt some of the samples, remove the burner. Protect your fingers with a piece of folded paper towel and loosen the screw on the lower iron ring. (*Be careful. It is hot.*) Lower the ring holding the hot beaker and turn it to one side, out of the way, where it can cool. Immediately replace the burner under the bundle of test tubes and continue to heat them as strongly as possible. (See Figure 1-7.) Record all further changes, even after the samples melt. Finally, when nothing else happens, turn off the burner and let the tubes cool.

7. If any of the materials did not melt when they were heated in the test tubes, obtain fresh samples of these materials. Hold a piece of the material in a pair of crucible tongs and heat it in the hottest part of the burner flame (as shown in Figure 1-8). Record all changes.

What Chemistry Is Like 9

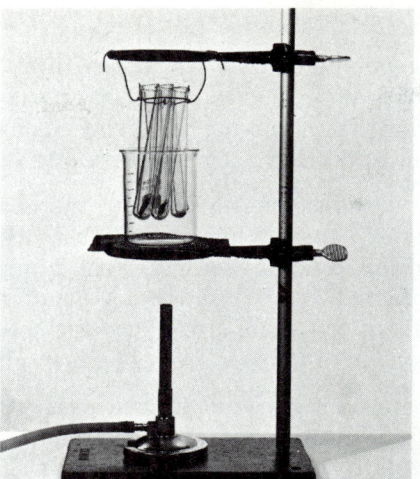

Figure 1-6

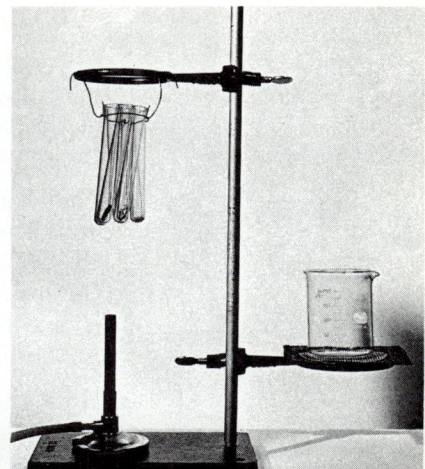

Figure 1-7

How To Use a Burner

To light your burner, turn the gas supply valve halfway on and hold a lighted match slightly above the edge of the burner top. If the resulting flame is yellow, turn the air-adjustment sleeve (Figure 1-8a) until the yellow color disappears. A perfectly adjusted flame either is nearly invisible or has a blue-violet color. If the base of the flame rises above the top of the burner, turn the gas pressure down. This can be done either by adjusting the burner gas valve or the main gas valve if your burner has no valve. Make all adjustments gradually, not suddenly.

Burner flames have different temperatures in different parts of the flame. The hottest part is indicated in Figure 1-8b. *Caution:* If your burner produces a nearly invisible flame, be careful. You can be severely burned if you are not alerted by *seeing* the flame. It is very easy to forget the burner is lighted as you concentrate on other work.

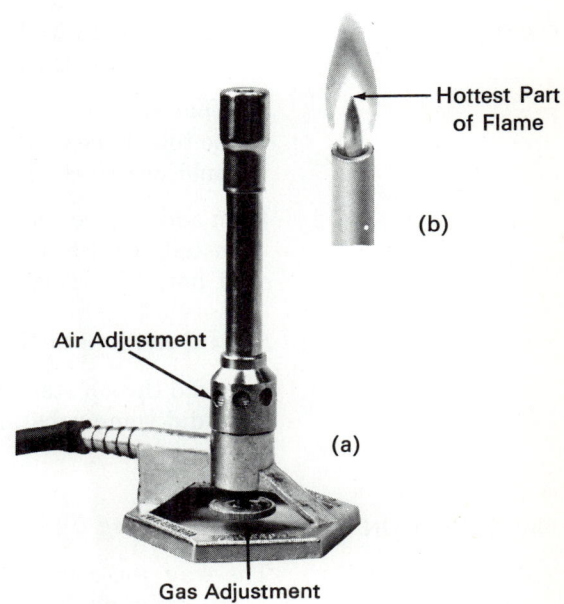

Figure 1-8

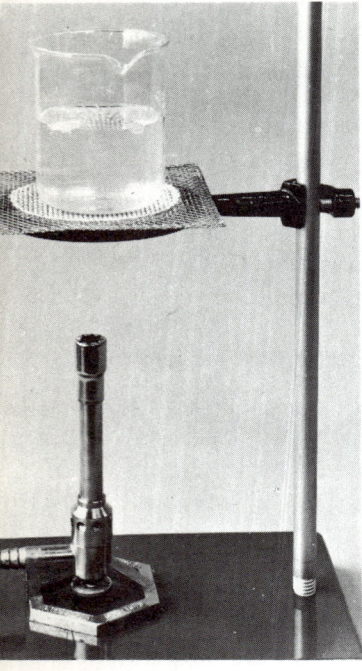

Figure 1-9

Part B. Comparison of Materials from the Candle

1. Remove the bundle of cool test tubes from the ringstand. Place a wire gauze on an iron ring about 50 mm (2 inches) above the top of the burner. Fill a 250-ml beaker or a small can about one-third full of water and place it on the wire gauze, as in Figure 1-9.

2. Pour a few drops of the liquid from the bowl of a burning candle onto a piece of paper. When it solidifies, break off a small piece of the solid and place it on the water contained in the beaker or can. Cut a chip of candle wax from the bottom of your candle. Both pieces should be about the same size. Place the second piece on the water in the beaker or can, apart from the first piece.

3. Heat the beaker or can and its contents with a burner flame and note when each substance starts to melt. Allow the melted substances to solidify, then discard them in the waste jar. Do not pour the liquid substance into the sink.

RELATED PROBLEMS

1. How does your observed order of melting for the substances tested compare with that observed by other members of the class?

2. What generalized statement, if any, can you make based on the combined observations?

3. What can you now conclude about the identity of the material in the top of the candle? Can you think of any arguments that someone could use to challenge your conclusions?

4. In addition to the differences in the melting order of the substances tested, what other changes did you observe? Can you explain them? If not, what further questions do they raise?

Some Questions to Wonder About

Why did the substances tested begin to melt at different temperatures? Why did some substances not melt?

INVESTIGATION

1-5 Melting Temperature

The way you compared melting temperatures in the previous experiment was good enough for a few samples. But with hundreds of chemicals to test, that sort of direct comparison of one substance with another is not practical. A more practical method is to measure the melting point of each substance separately. Then these characteristic temperatures can be filed for comparison with the measured melting point of any other substance.

In this experiment you will examine a common solid, the moth repellent, *para*dichlorobenzene (*pa*-ra-dy-*kloh*-roh-BEN-zeen).

PURPOSE

The purpose of this experiment is to study closely how *para*dichlorobenzene behaves, first as it is cooled from a temperature slightly above its melting point, and then as it is heated slowly from a temperature below its melting point. You will find out how the temperature of the *para*dichlorobenzene changes within certain regular time intervals during the cooling and heating processes.

MATERIALS

Utility clamp Burner
Ringstand, ring Beaker, 400-ml
Wire gauze
2 thermometers ($-10°C$ to $110°C$ range)
Clock or watch with sweep second hand
*Para*dichlorobenzene, in an 18- × 150-mm pyrex test tube

PROCEDURE

Part A. Cooling Behavior

1. To make the recording of data in this experiment easier, you should work with a partner. One partner should prepare a table in his notebook that will allow him to record the temperature, time, and cooling behavior. An additional copy of the data should be made for the other partner. (See the sample table following Part B.)

2. Fill a 400-ml beaker three-fourths full of tap water at about room temperature (below $30°C$). Place the beaker on the base of your ringstand.

3. Slowly heat a large test tube, $\frac{1}{3}$ to $\frac{1}{2}$ full of the moth repellent *para*dichlorobenzene, over a low burner flame until it melts, as shown in Figure 1-10. Place a thermometer in the liquid and continue heating until the temperature is between $65°C$ and $70°C$. Continuously move the test tube back and forth through the flame.

4. Clamp the test tube containing the melted *para*dichlorobenzene into position above the beaker of water.

5. Now clearly divide your labor. Place a clock or watch in such a way

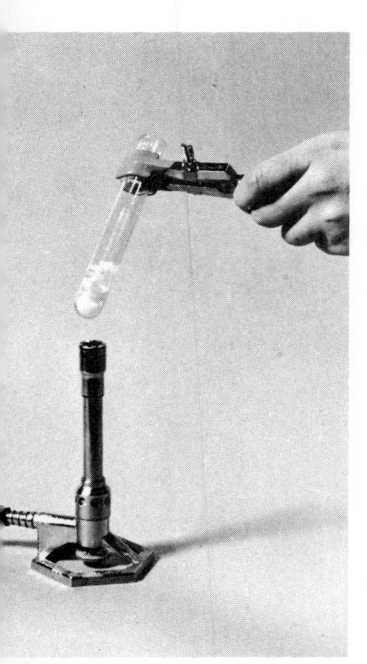

Figure 1-10

that one partner, the recorder, can read the time every 30 seconds. This partner should also record all the observations noted by the other.

6. When all is ready, check the temperature of the *para*dichlorobenzene and record it to the nearest 0.2°C. (Be sure that your eye is at the same level as the top of the mercury column in the thermometer. Otherwise, your reading will be in error.) On signal by the recorder, immediately immerse the lower half of the test tube into the water bath and clamp the test tube in position. Hold the thermometer against the side and just off the bottom of the test tube so that it will become embedded in that position when the *para*dichlorobenzene solidifies, as in Figure 1-11. Record the temperature each 30 seconds until a temperature in the upper thirties is reached. Note also when solidification starts and when it is complete.

Part B. Warming Behavior

The observer and recorder should exchange duties at this point.

1. Raise the test tube out of the water bath and turn the clamp around so that the test tube is on the other side of the ringstand, opposite the water bath, as shown in Figure 1-12. Place the beaker, three-fourths full of water, on an iron ring and wire gauze. Heat the water bath to approximately 70°C. (Start with hot tap water, if available, to save time.) Then turn off the burner, but leave it in place.

2. Record the temperature of the solidified *para*dichlorobenzene to the nearest 0.2°C. With a second thermometer measure the temperature

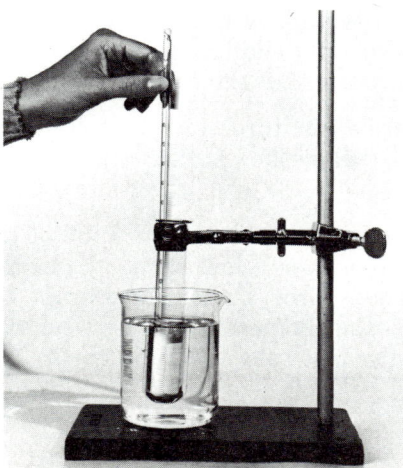

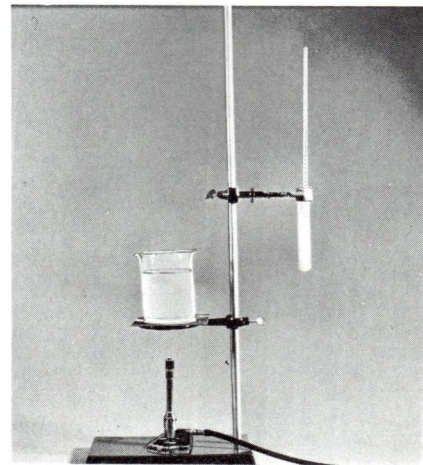

Figure 1–11   **Figure 1–12**

of the hot water bath to the nearest 1°C. On signal by the recorder, immerse the lower half of the test tube into the water bath until the water level is above the level of the *para*dichlorobenzene. Record the temperature of the *para*dichlorobenzene each 30 seconds, noting also when melting starts and when it is complete.

3. As soon as the solid becomes free from the walls of the test tube, move the thermometer gently up and down. Occasionally note the temperature of the water bath with the second thermometer. If the water temperature drops below 60°C before the solid is melted, be prepared to warm it with the burner flame to keep the temperature between 60°C and 65°C. Continue to move the thermometer up and down in the test tube. Record temperatures until the temperature of the *para*dichlorobenzene is about 60°C.

DATA AND RESULTS

The following type of table should be reproduced in your notebook and used for recording the data obtained on cooling. A similar table should be made for the data obtained on heating.

Time (minutes)	Temperature (°C)	Observations and Remarks
start		
0.5		
1.0		
1.5		

(Continue every half minute for about 8 minutes)

To understand the results of this experiment, it helps to record the data in graphical form. A *graph* is a kind of picture that shows visually how one numerical property (say, the temperature of the substance) varies as some other numerical quantity (say, the time that has elapsed) changes. A set of numbers, such as those you have recorded from your experimental observations, contains the information. But a graph of these numbers, one set against the other, helps you to *see* how the two measured quantities are related to each other. In the physical sciences, especially physics and chemistry, many quantitative experiments are done. Graphical presentation of data is very often used. It is therefore important that you learn to plot and interpret graphs.

Use a full page of your notebook for making the graph, representing time along the horizontal axis and using two spaces for one minute.

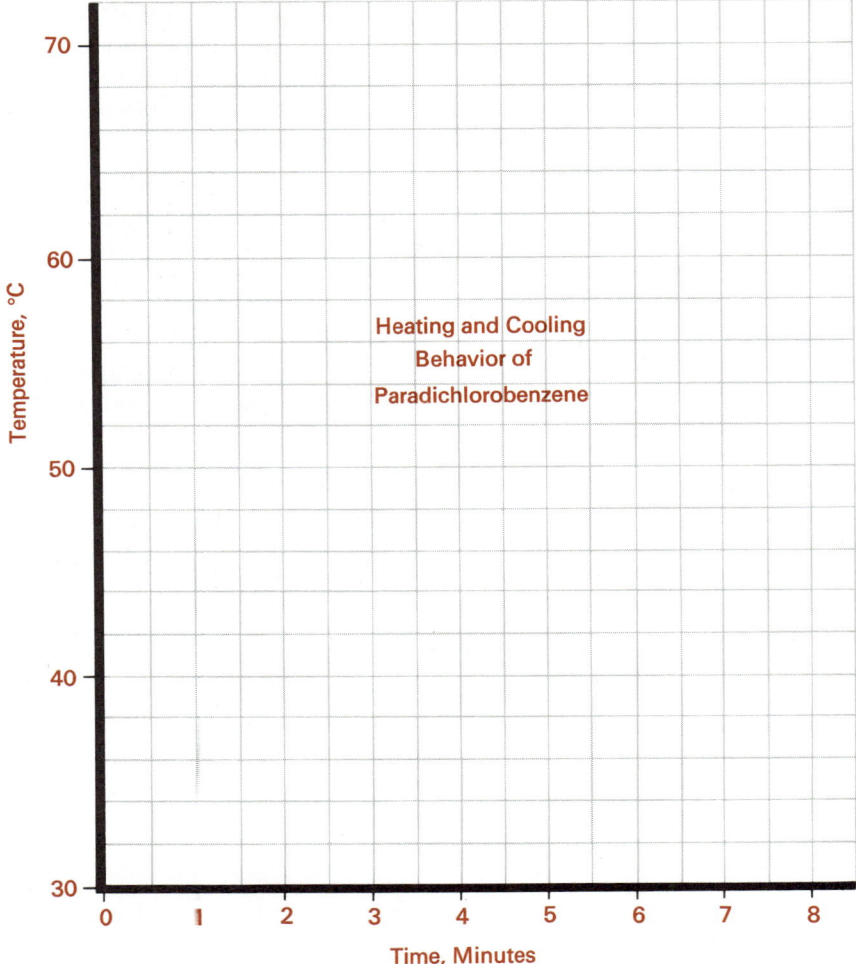

Figure 1–13

Represent temperature on the vertical axis, allowing one space for each two degrees, starting at 30°C. This scale will make the range of data plotted on each axis about the same. (See Figure 1-13.)

Plot heating and cooling temperatures on the same graph. For each cooling temperature and its corresponding time, make a small cross. For the heating data use small circles. In this way you can distinguish between the two series of plotted points.

Using a red pencil, draw a smooth curve to represent the heating behavior of the *para*dichlorobenzene. Using a black pencil, do the same for the cooling behavior. Be sure the graph has a title and a label on each axis. At the top of the graph sheet place your name, your partner's name, the date, and the name of your experiment.

Study your experimental results as shown in the graph. Write a summary report on the experiment. In the first paragraph describe your graph. In the second paragraph give your interpretation of the shapes of the heating and cooling curves. (To what does each part of the curve correspond?) Include the graph with your report.

RELATED PROBLEMS

1. What effect would increasing the amount of *para*dichlorobenzene have on the shape of the melting or cooling curve?
2. Based on your data, what is the melting temperature and what is the freezing temperature of *para*dichlorobenzene?

A Question To Wonder About

Why do the heating curve and the cooling curve have the characteristic shapes shown by your graph?

ADDITIONAL INVESTIGATIONS

To be undertaken as extracurricular experiments. Consult your teacher before proceeding.

1. Devise an experiment in which you can obtain data to plot a heating curve for the evaporation of a liquid substance.
2. Find cooling and warming curves for another solid substance.

Summary

Chemistry is the science that deals with materials and how they are changed from one form into another. These changes generally involve the absorption or production of energy. The chemist, like all scientists, relies on careful observation to gather facts. Experiments are carefully planned and carried out under controlled conditions, usually in a laboratory, so that observations can be made accurately and systematically. The observations you can make on the burning of a candle, the behavior of various substances when heated, and the process of melting all raise important questions. The questions relate to the nature of various forms of matter and how they behave.

Self Check

1. State briefly what you think the science of chemistry is all about.
2. Why do chemists perform experiments?
3. Summarize as briefly as you can what you learned in each of the three experiments you have just performed.

Scientific Procedure

1-6 The Method of Science

Now that you have had some experience in doing experiments, see if you can discover some method in your efforts. First, you started each experiment by making careful observations, remembering to account for *all* conditions that could affect the results. You organized your data or observations in some logical way, in the form of a table or a graph, for example. At this point perhaps you were able to spot a trend or pattern, such as the long, flat portion of your heating or cooling curve for *para*dichlorobenzene. And, finally, perhaps you were able to draw some conclusions. For example, to what process does the flat portion of each curve correspond?

You were given some "questions to wonder about" following your experiments. They represent a further stage of scientific procedure, "wondering why." This is a stage that many scientists find the most challenging. It requires an exercise of the imagination like that which artists enjoy.

This summary of how you did your experiments, together with the final step, "wondering why," is also a rather good description of what scientists do. The processes described are often presented as the "scientific method." Of course, we have omitted one very important step in your work, the lab report. A scientist must be able to communicate the results of his efforts if they are to benefit anyone else.

As you may have guessed from the foregoing discussion, the "scientific method" is not a set of rules that scientists follow literally. But it does describe pretty accurately the logical, systematic way in which scientists go about their work. You must remember that science is the result of human efforts, and, as such, subject to human limitations. The scientific method is perhaps the best effort possible for determining objective scientific truth. Such truth must be independent of human strengths and weaknesses, free of vanity, self-interest, prejudice, and just plain stupidity.

You should not try to memorize and follow a set of rules for playing the chemistry game. Instead you should try to develop a feeling for what is involved in each step. Then you will be able to apply your understanding to any situation you might wish to investigate. In order to help you to develop such an understanding, each of the steps will be examined in greater detail.

1-7 Observation

Careful observation is the foundation of all successful experimentation. Perhaps the greatest pitfall to beware of in this step is the tendency to look for desired or expected results. Frequently, a preconceived idea about what should happen will get in the way of careful, impartial observation. One of the most interesting things about scientific investigation is that many important discoveries have come from the observation of things that previously had seemed trivial or unrelated. Then some careful observer noted an overlooked detail or saw relationships that others had missed, and a new breakthrough occurred.

Undoubtedly, when you compared your observations of a burning candle with those that a careful, trained observer could make, this point became clear. You probably were surprised by the large number of items you could have reported, but overlooked. The observation phase of the scientific method, that is, getting all the facts and getting them accurately, is by no means easy or trivial. It is a challenge and it is important. It is the basis for all the later steps.

Another problem faces any competent observer. It is necessary to consider all factors that could affect the outcome of the experiment. For this reason, it is essential to set up a situation in which outside conditions that might affect the experiment can be regulated. It is important to be aware of all conditions, whether or not you think they affect the experiment.

1-8 *Recording the Facts*

An accurate record of the things you observe in your experiment is of primary importance. This record must be made at the time the work is being performed, as illustrated in Figure 1-14. It must be carefully checked for both accuracy and completeness as you proceed. In nearly all cases it should contain such information as the date, time, and conditions of temperature and humidity in the laboratory, plus any other factors that might conceivably be important.

A widely told story illustrates the need to account for all relevant factors. A certain chemist had done important research on the chemical composition of the earth's crust. Other scientists, however, were having difficulty in reproducing some of his results.

Later, a young assistant noted that the chemist smoked cigars constantly, and seldom worked without a cigar in his mouth. Could this fact account for the different results? Indeed, it was found that the differences could be accounted for by the small amounts of chemicals in the ash of a cigar. Some of the ash must have found its way into the earth samples that the chemist had been analyzing! The lesson is obvious: All conditions surrounding an experiment are important.

Accuracy is of utmost importance also. The laboratory investigator should always double-check his figures, instrument readings, and other measurements.

1-9 *Organization of Observations*

Many experiments fail because the facts, no matter how carefully observed and accurately recorded, are not organized so that they reveal a trend or pattern, that is, a **regularity.** Sometimes regularities are most readily seen by arranging the data in a carefully prepared table. Another arrangement is graphical presentation of the result. You have used both of these methods in your most recent experiment. Later in this course you will find that scientists have developed other tools to help in the organization and presentation of data. Some of these include the use of symbols, equations, and mathematical expressions. Any method is permissible, however new, that presents data clearly to everyone.

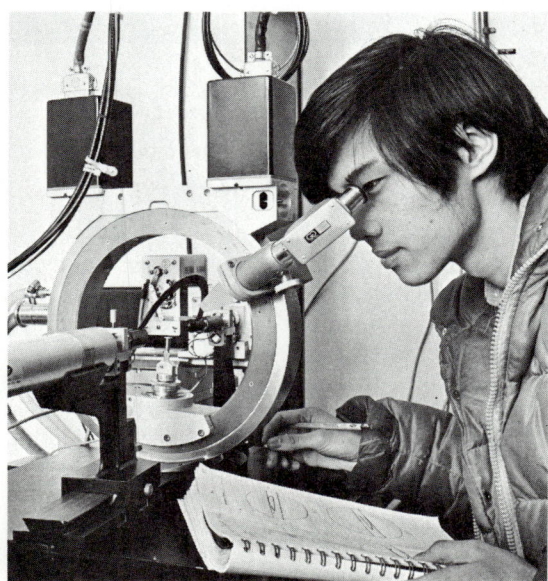

Figure 1-14 *A graduate student in chemistry carefully records all the facts in conducting his experiment.*

Chemistry and Our Environment

One of the main reasons why you and every other citizen should know something about chemistry is because of our shared responsibility for our environment. Our way of life today depends heavily on the use of chemical processes. We extract chemical raw materials from the earth—metal ores, petroleum, natural gas, coal, salt, sulfur—and even oxygen and nitrogen from the air. We put back into the environment not only the various finished products that are made from these raw materials but a host of unwanted by-products as well. Many of our other activities, which are not really a part of the chemical industry, also have effects on the chemistry of our environment. For example, the burning of coal to generate electricity puts several gases into the atmosphere, at the rate of thousands of tons per day. Similarly, the burning of gasoline in automobiles and other vehicles using internal combustion engines puts many undesirable gases into the air. Some of these can be very harmful unless they are kept to a low level. Other environmental problems, which have a chemical aspect to them, include the disposal of solid wastes, the treatment of sewage, the control of insect pests, and the use of chemicals to preserve food and enhance its nutritive value. Also, supersonic aircraft, flying in the upper atmosphere, may affect the amount of the sun's energy that will reach us here on the surface of the earth.

Science, like art or music, can be studied and appreciated for its own sake, and there is certainly nothing wrong with such a purely intellectual approach. However, it is of vital importance today, when man's impact upon his environment has become so great, for all responsible citizens, both scientists and nonscientists, to learn all they can about the scientific principles upon which man's relationship to his environment is based. We can no longer go on abusing our home, the earth, as we do now. Neither is it realistic to think that we can turn back the clock and revert to a more primitive way of life, lacking in all the materials and machines to which we have become accustomed. We must learn to control our activities so that we live in harmony with our environment. This can only be done by making thoughtful and intelligent decisions based on knowledge.

Throughout this book, you will find special articles on chemistry and the environment. These will occur wherever the discussion of chemistry brings us to a topic that has to do

Mid-Manhattan in the grip of highly polluted air, on November 25, 1966.

with an environmental problem. This book will present the science of chemistry as a science. Chemistry will be presented and developed in a logical manner so that the basic concepts and principles may be appreciated. The presentation will not be shaped to the purpose of discussing the chemistry of our environment, for that would only make the fundamental principles more difficult to grasp. On the other hand, the articles entitled *Chemistry and Our Environment* will help and encourage you to stop and think about the chemical nature of our environmental problems and how, with knowledge and self-discipline, you might solve them. You may be surprised at how soon your knowledge of chemistry will increase your ability to understand some of our environmental problems.

BARRY COMMONER *is a scientist who has long been alerting people to environmental problems. He is widely regarded as one of America's best informed and most articulate spokesmen on the subject. If we are able to solve the environmental problems that now face us, much of the credit will be due to such pioneers as Barry Commoner.*

1-10 Searching for Regularities and Logical Conclusions

At a casual glance, the great wealth of material that nature presents for investigation may seem to have no pattern, no rhyme or reason, in it. But as careful systematic study goes forward, some patterns do emerge. These regularities provide clues to the way nature behaves. Knowledge of the way nature behaves is the goal of science. To this end, the chemist searches for some consistency, or trend, in the information he has collected. If a regularity is found, the chemist will try to express it in a concise, accurate way. This statement will be the conclusion he draws from his observations. It is very important, however, not to draw too broad a conclusion from the observations made. The regularity seen may be perfectly real and yet the conclusion may be false. Let us consider the experience of a hypothetical chemistry student in order to illustrate the point.

Wilbur Bright was a beginning chemistry student. His imagination had been stimulated by the statement that fuels are a source of energy. He tried to think of all the different fuels that he had encountered in everyday life. There was the heating oil that his family used to heat their house, for one. Then there were gasoline for the family car, kerosene for the cooking stove that he took on camping trips, the natural gas that his mother used for cooking, and the wood that they burned in their fireplace. He thought of coal, too. He had seen a huge pile of coal at a nearby plant once and asked his father what it was. His father had mentioned, among other things, that coal is virtually pure carbon. (Carbon is one of those substances that chemists call "elements." Wilbur had read about them somewhere. If you want to find out what they are, you can look ahead a few chapters, but it is not necessary for this story.)

Then Wilbur remembered something else. He had read somewhere that gasoline contains carbon. Did all fuels contain carbon? Wilbur went to the library to investigate. He was elated to find that all of the fuels he had thought of *did* contain carbon. He had found a regularity!

Wilbur immediately began to wonder why carbon should be found in all fuels. He started to look for a book that might have the answer. A book about rocket fuels was the first one to catch his eye. Wilbur had hardly got beyond the first page when he noticed something quite surprising. Some important rocket fuels contain no carbon at all!

Wilbur's conclusion had been based on too limited a list of fuels. The regularity that he had thought he had found did not hold for *all* fuels. He had generalized beyond the limits of his data.

1-11 Wondering Why

When your experiments have been carefully performed, regularities discovered and evaluated, and logical conclusions reached, then you are ready for the most creative activity of all: "wondering why." How does your new discovery relate to other things that you know? How does it fit into, or alter, your view of the universe? Jacob Bronowski has tried to explain the satisfaction that scientists find in such an endeavor, in his book, *The Common Sense of Science*. "It is the unity of nature, living and dead, for which our thought reaches . . . We seek to find nature one, a coherent unity. This gives to scientists their sense of mission . . . of aesthetic fulfillment: that every research carries the sense of drawing together the threads of the world

Figure 1–15 *Communicating results, an important part of science, can be done directly in seminars (top) or through scientific journals (bottom).*

into a patterned web." Later, in examining some regularities, you will see how fruitful it can be to "wonder why" they exist.

1-12 Communication of Results

Suppose a chemist, or other scientist, has done an experiment that has provided new information. When he or she has carefully evaluated the information, has identified any regularities, and has, perhaps, reached some conclusions, it is important that others working in the same field be informed of this work.

What Chemistry Is Like 21

One of the biggest problems in science today is that of having the results of research clearly written and widely circulated. Many scientific journals and digests have been developed to publish research findings, and meetings devoted to certain areas of research are becoming increasingly important. (See Figure 1-15.) A clear, understandable, written report of a research project is the most critical part of scientific communication. The results of your experiments will be of little value unless you can communicate them. To do so requires a knowledge of language and the ability to use it.

INVESTIGATION

1-13 What Controls Combustion? What Is Produced?

Once again, you will experiment with a burning candle. Try to apply what you have learned about the scientific method to arrive at some conclusions about the actual process of burning.

In your description of a burning candle some of your recorded observations might have been: the candle decreases in length as it burns; candle material is consumed; heat and light are produced. Did you stop, however, and ask what is really happening to the candle and what is causing it to happen? You know the candle is burning, but what does this word "burning" actually mean? Today we know a great deal, but not everything, about the burning process. Only two hundred years ago, the world's foremost scientists were almost entirely ignorant concerning changes that take place when something burns.

PURPOSE

In this experiment you will attempt to find out why a candle burns and what products are formed during the burning process.

MATERIALS

Candle
Cobalt chloride paper
200 ml limewater
4 wide-mouth bottles, 8-oz (or 1000-ml beakers)
Wide-mouth bottle, 4-oz (or 500-ml beaker)
Small cylinders of carbon dioxide, oxygen, and nitrogen gas
4 Erlenmeyer flasks, 250-ml
Glass tube or straw

PROCEDURE

This experiment is designed to be performed as a demonstration. Everyone should observe and record what happens.

22   Chapter One

Part A.

1. Invert an eight-ounce, wide-mouth bottle over a burning candle as in Figure 1-16. Test the thin liquid film that forms on the inside of the bottle with a strip of cobalt chloride test paper.

2. Moisten a second piece of cobalt chloride test paper with a drop of distilled or demineralized water. (Such water has been treated to remove its impurities.)

Part B.

1. Determine the length of time in seconds that a candle continues to burn when an eight-ounce bottle is placed over it as shown in Figure 1-17.

2. Relight the candle and repeat procedure B-1, using a four-ounce bottle in place of the larger one. Note: 1000-ml and 500-ml beakers may be used in place of bottles.

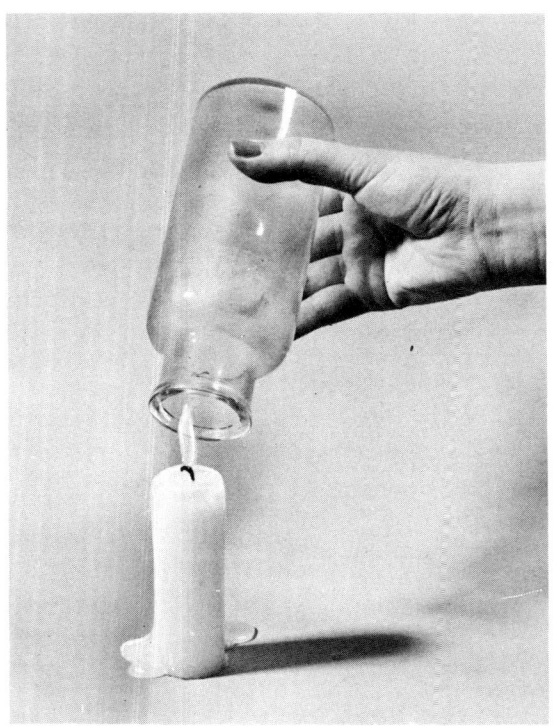

Figure 1-16

Figure 1-17

As often happens in science, the attempt to answer one question raises others. What causes the candle flame to be extinguished when confined for a short time? Two possible answers follow.

a. The burning process produces a gaseous material that somehow puts out the flame.
b. The burning process consumes a gaseous material present in air. When this material is gone, burning stops.

Part C.

1. Place a 250-ml Erlenmeyer flask over a burning candle as in Part B-1. After the candle flame has gone out, place the flask upright on the table. Obtain a second, clean 250-ml Erlenmeyer flask to use as a "control."

    Scientists use the term **control** to designate an experiment identical to the one of interest in all respects *except one*. With the use of a control it will be seen whether any result is due to the one factor in question or to something else. In this case, if the control flask shows the same result as does the flask that was over the candle nothing has been proved. If only the flask that was over the candle gives a reaction with limewater, however, the candle must have had something to do with it.

    Add about 25 ml of limewater solution to each of the flasks. Swirl the solution in each flask simultaneously until a change occurs.

2. Using a glass tube or a straw, blow into a third flask containing some limewater solution.

3. Briefly bubble some carbon dioxide gas from a cylinder or small generator into a fourth flask containing some limewater solution. It appears that combustion does indeed produce a new gaseous substance that causes limewater to become cloudy. Which of the two possible answers proposed at the end of Part B is correct? If a new gaseous material *is* produced, answer a. could be correct. Has it been proved to be correct? Part D may help you decide.

Part D.

1. Hold an eight-ounce bottle upside down and blow a stream of nitrogen gas into it. When all the air has been replaced with pure nitrogen, quickly place the bottle over the lighted candle.

2. Hold an eight-ounce bottle upside down and blow a little oxygen gas into it. *Do not allow pure oxygen to replace all the air—only about one third of it.* Place the bottle over a lighted candle. Note any change in the appearance of the candle and record how long it burns.

RELATED PROBLEMS

1. What conclusions can be drawn from this experiment?

2. Does the evidence eliminate the possibility that something other than water caused the observed changes in the cobalt chloride test paper? Explain your answer.

3. If the liquid film is water, where does it come from?

4. What are the two products of combustion that you have identified?

5. Finally, on the basis of your observations, give the most logical explanation for why the candle goes out when confined. Is there any reason to suppose that carbon dioxide or nitrogen "put it out"?

ADDITIONAL INVESTIGATIONS

1. In order to get a more satisfactory answer for Related Problem 5, you might try the following experiment, if time permits.

    Fill two eight-ounce bottles with water. Cover them with a glass plate and invert them in a pan, or trough, containing at least one inch of water. Attach a rubber hose to a cylinder of oxygen and bubble the oxygen gas into each of the two bottles until about half of the water has been displaced from each. Now bubble nitrogen gas into one of the bottles until all of the water has been displaced. Bubble carbon dioxide gas into the other one until all the water has been displaced.

    Remove the bottles from the pan of water one at a time and lower each bottle over a burning candle. Measure the length of time each candle burns.

    Can you now say what really causes the candle to burn or not to burn?

2. Since your experimental results suggest that your breath contains carbon dioxide, try to find out if you are also exhaling any oxygen. Fill an eight-ounce bottle with water and invert it in a pan as you did above. Take a deep breath and bubble your exhaled air through a rubber tube into the bottle until all the water has been displaced. Invert the bottle over a candle and see if the candle burns. If it does burn, see how long it will burn, as before. What does this prove?

Summary

The method of science can be summarized in the following series of steps. (1) Careful observations are made. (2) The observations are recorded promptly and accurately. (3) The recorded facts are organized, often in the form of a graph. (4) The facts are examined to see if they follow any regular pattern. (5) The scientist then attempts to think why any detected pattern might exist. (6) The results of the scientist's work are communicated, as clearly as possible, to others. Of course, the results and conclusions from one experiment usually raise further questions. New experiments and observations are then made, in an effort to get a better understanding of nature, and the whole process is repeated.

Self Check

1. Why is it important for scientists to communicate their work to others?
2. Why should observations be recorded promptly?
3. What kinds of arrangement can help to reveal a regularity in lab results?
4. When a candle burns, what substances are consumed? What substances are produced?

Chapter Summary

Chemistry is the science that deals with materials. The properties and composition of materials, and how one material can be transformed into another are of chief concern to a chemist. The energy changes that take place during the transformations are also studied by chemists. What we know about chemistry today, and what will be discovered in the future, is, and must be, derived from experiments and observations. An experiment is a planned and controlled series of observations. The gathering of experimental facts is only the first step in the scientific method. The facts must be recorded, organized, and analyzed, with the object of finding regularities. Finally, there must be a search for explanations. All of this generally leads to further experiments, in order to answer new questions that have arisen. It is important for scientists to communicate the results and conclusions of their experiments to others.

Questions and Problems

1. During recent years the world has been experiencing a population explosion. In many countries the home building industry has been unable to keep up with this great increase in population. Make a list of building materials that have been used for many years in the construction of houses. Now make a list of recently developed materials that either are being used in house construction or might possibly be used for this purpose.
2. Many people are finding or suspecting that pollutants seriously affect their lives, including their ability to earn a living. The following is a practical problem that could very well occur.

    A dairy farm is located about fifteen miles from a large chemical plant. All water for the farm is obtained from deep artesian wells. After many years of careful breeding and attention to the nutritional requirements of his cows, the farmer has been able to develop a healthy herd of cows that produces an extremely high yield of milk. Now, however, he notices that his cows are less active than usual. In addition, for the first time in years their milk production has begun to fall off significantly.

Even though your understanding of the scientific method is somewhat limited at this early stage of the course, make a list of all things that you think a scientist should do in trying to find an answer to this farmer's problems.

3. In this chapter it was said that the scientific method is an effort to make scientific truth objective, that is, independent of human strength and weakness, of vanity, self-interest, prejudice, and just plain stupidity. How can you explain the fact that so many members of your class made observations of the burning candle that you did not notice or bother to record?

4. Suppose you took a tray of ice cubes out of the refrigerator and found that the temperature of the ice cubes was about $-15°C$. If you know that ice melts at $0°C$, make a sketch of the curve you might get on a graph prepared by plotting the temperature of the ice cubes against time as these ice cubes gradually turned into liquid water.

5. Why did you draw the shape of the curve as you did in Question 4? Does it have any similarity to the curve you obtained for *para*dichlorobenzene? If it does, what limited generality did you have to make before drawing your curve? Without actually measuring the temperature change of melting ice cubes, what other information do you need to be sure that the curve you predicted is reasonably accurate?

6. Although you can cultivate the habit of making careful observations through the study of chemistry, this habit is useful in many other areas. Give some examples of how the ability to make accurate observations might be extremely valuable in such widely separated areas as law and automotive mechanics.

7. What do you understand to be the meaning of a controlled experiment? How many conditions of an experiment should be varied at one time to get the best results from the experiment?

8. The following data were obtained in the laboratory when a volume of gas was measured at various temperatures. (The temperatures were measured in kelvins, K. The kelvin is a unit of temperature used by scientists. It is defined in Chapter 6.)

Volume of Gas (ml)	Temperature (K)
300	273
350	318
400	364
450	409
500	455
550	500
600	546

(a) Write a statement expressing the relationship between gas volume and temperature.
(b) Prepare a graph of the data representing the relationship.
(c) Write a mathematical expression for the relationship.

9. Discuss the advantages of recording in some concise, orderly manner data concerning changes that occur in an experiment. How would you organize a data sheet for an experiment in which you were not sure what you might find?
10. What common techniques are employed to present experimental data visually to someone else?
11. Why is it so necessary to control the conditions of an experiment as carefully as possible?
12. Why do you believe that it is necessary (or not necessary) to record *all* observations made during an experiment?

2 Matter and Mass

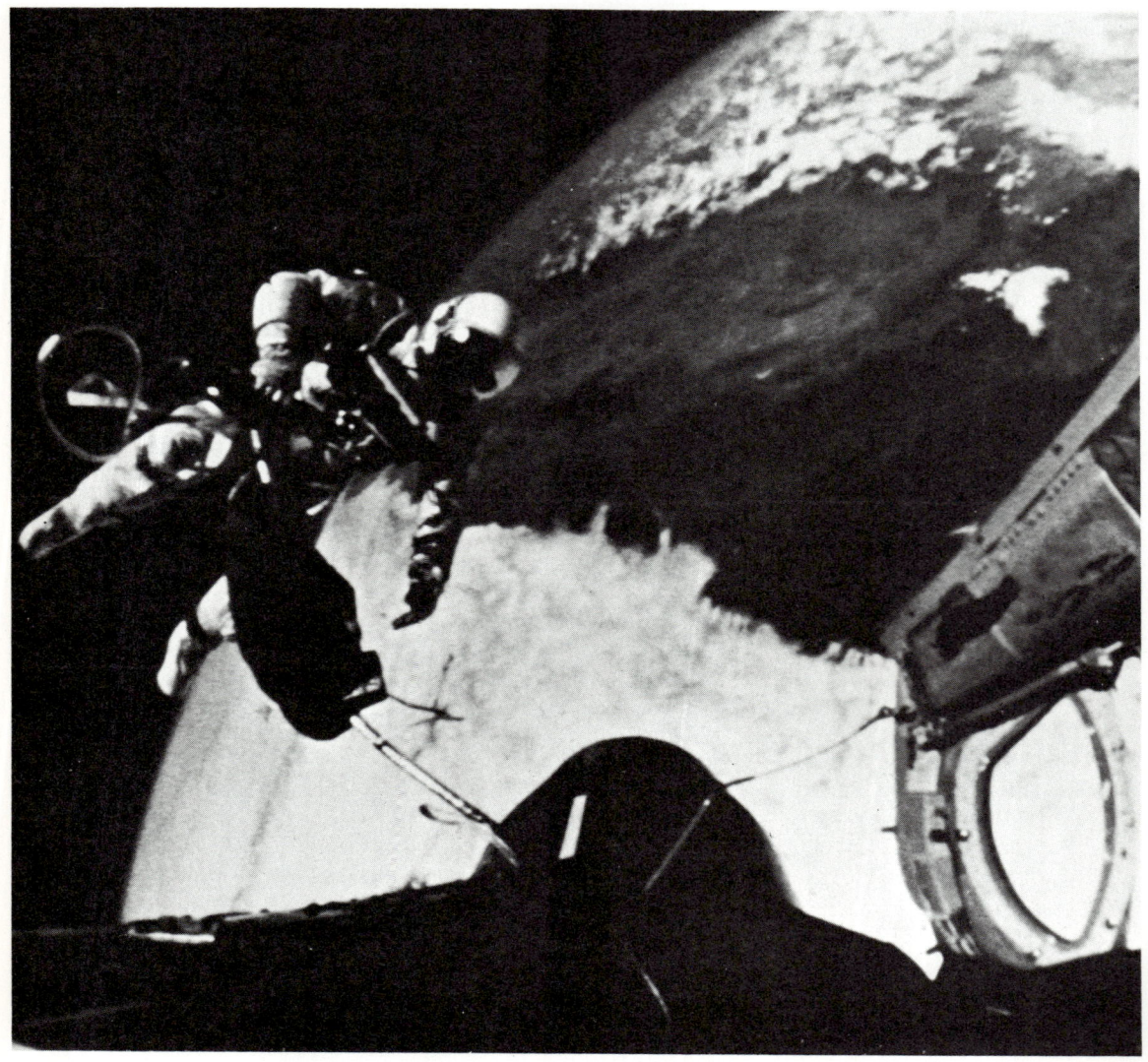

Astronaut Edward White maneuvering in a weightless environment outside the Gemini 4 spacecraft.

Since chemistry deals with the many forms of matter, and the changes that they can undergo, it is important for you to know how matter is defined. What is there about matter, in all its many forms, that makes it matter? How do we go about measuring the amount of matter present in any particular substance? When a substance is changed into one or more others, does the total amount of matter present remain the same? Before you can understand chemistry as a science, you must have some basic ideas about the answers to such questions as these.

What Is Matter?

2-1 The Fundamental Property of Matter

In Chapter 1 you read that chemistry is the "science of materials." Neither "materials" nor the related word, "matter," was defined at that time. Both words should be examined carefully at this point.

In your experiments you worked with several materials. They were obviously different in many ways. Candle wax is softer than copper and melts much more easily. Copper is shiny and has a characteristic color, while pure candle wax is dull white. Sulfur is yellow but turns dark brown when it melts. Copper stays the same color when it melts, but turns black when heated more strongly in air. Sulfur, when heated strongly in air, burns with a blue flame, and, like wax, eventually disappears.

The three materials copper, sulfur, and candle wax have quite a variety of properties. You can no doubt think of many other materials, with many more properties. Nevertheless, they *all* have one thing in common. They all consist of matter. Even the invisible gases given off when wax or sulfur burns are made of matter. The gas that burning sulfur gives off does not escape notice, because of its pungent odor. Nor do the invisible and odorless gases that are normally in the air around us go unnoticed when they are moving rapidly, as in a windstorm.

Chemists try to understand how matter can exhibit so many different properties. But what do they mean by "matter"? What fundamental property does all matter have? Think of the answers to a few other questions first. Why does the air stay around the earth instead of drifting off into space? Why is London Bridge falling down, rather than up, or sideways? Do your answers suggest a property common to all matter? Of course. All material bodies are attracted to the earth. All matter has *weight*.

2-2 The Law of Gravitation

The observation that all forms of matter have weight, that is, are attracted toward the earth, can be developed into a fundamental definition of matter. Sir Isaac Newton did that over 300 years ago, when he proposed his Law of Gravitation. Newton reasoned that the force of attraction between the earth and some other object, say an apple hanging from a tree, is really a two-way business. It is a *mutual* attraction, of the earth to the apple as well as of the apple to the earth. It is the tree that keeps them apart. When the apple gets disconnected from the tree, the apple and the earth move toward each other! Because the earth is so large and the apple so small, it is the apple that does practically all (not absolutely all) this moving, and we say that the apple falls to the earth.

Newton wrote the following equation to describe the force, F, acting between two bodies.

$$F = \frac{GM_1 M_2}{d^2} \quad (1)$$

The symbols M_1 and M_2 represent the masses of the two bodies, and d is the distance between them. (The letter G is a "proportionality constant." The constant makes it possible to calculate the magnitude of F, given the other values.)

The **mass** of a body is by definition the amount of matter it contains. This is taken to be true whatever the particular kind of matter, whether hard or soft, liquid or solid, visible or invisible, or anything else. Equation (1) used the shorthand of algebra to state several things compactly. It makes the following important statements.

1. The force between two bodies will increase in direct proportion to the amount of matter in each one. If we double the amount of matter in one, the force is doubled. If we double the amount of matter in each of them, the force is multiplied by 2×2, or 4.
2. As the bodies are moved apart, the force between them decreases with the *square* of the distance. Thus, when the distance between two bodies is doubled, the force is reduced to one-fourth its initial value.

The Law of Gravitation was proposed by Newton on the basis of observations astronomers had made of the motions of the moon, the earth, and other planets. Its correctness has since been checked and confirmed again and again by experiments of many kinds, both terrestrial (those performed on earth) and astronomical. In short, the Law of Gravitation was discovered, checked, and proved by those procedures we call the scientific method. It is now accepted as a law of nature.

We present this law without reviewing the evidence for it. Scientists often find it necessary to accept such laws without testing, for themselves, the validity of the laws. They assume that if other scientists have worked carefully and reasoned correctly, their conclusions are likely (though not absolutely certain) to be right and can be accepted. Their conclusions are accepted *not* because some brilliant and famous person (like Sir Isaac Newton) said so, but because proper scientific methods were used to reach them. The validity of the conclusions can always be checked again, if necessary. It is worth noting that, even though Newton was probably the most brilliant and productive scientist in history, he actually made a few mistakes, which were caught by lesser men using the scientific method.

The Law of Gravitation can help you to understand the idea of weight, and the relation between weight and mass. Suppose you wish to determine the mass (M_1) of a small body on the earth. Call the earth's mass M_2. If the two bodies do not interpenetrate, the distance (d) between them, in equation (1), is the distance between their centers. (See Figure 2-1.) More exactly, it is the distance between their "centers of mass." A center of mass is the point at which all of the mass of a body could be concentrated and still give the same gravitational effect to an outside body. Assume, further, that the earth is a perfect sphere. (It is not, exactly. The effect of its actual shape will be discussed later in the chapter.) Then, if your measurements are made on the earth's surface, the distance to be used in equation (1) will always be the same, as Figure 2-2 shows.

Using the foregoing assumptions, equation (1) can be rewritten as shown in equation (2). M represents the earth's mass, R represents its radius, and m represents the

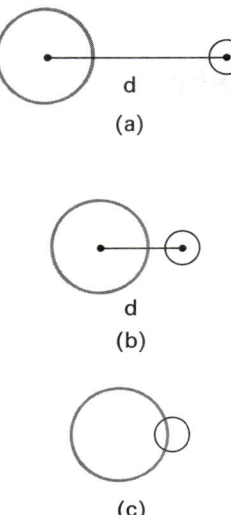

Figure 2–1 *How the distance, d, between two bodies is defined (a,b), provided the two bodies do not interpenetrate (c).*

mass of the object being studied in a laboratory at the earth's surface. Since G, R, and M are constant, GM/R^2 is also constant. Let us call its constant value C.

$$F = m \times \frac{GM}{R^2}$$
$$= m \times C \qquad (2)$$

A small body's weight, F, at a given point on the earth's surface, then, is directly proportional to the mass of the body.

Dividing equation (2) through by C gives the following equation.

$$m = \frac{F}{C} \qquad (3)$$

Equation (3) shows that the mass, m, of an object can be determined by measuring the force of attraction between it and the earth and dividing by the constant C. As you have seen, C is based on the properties (mass and radius) of the earth.

2-3 Units of Mass

In order to state the mass of an object it is convenient to have a unit of mass. A **unit** is a standard amount of something in terms of which all other amounts of the same thing may be stated. Butter is sold by the pound, eggs by the dozen, and milk by the quart. The pound, the dozen, and the quart are the usual units for these foods.

Similarly, in the United States, the dollar is the unit of currency. When you want to state how much money you have, you state how many dollars, and fractions of dollars, you have. Of course, if you go to Canada, you find that the unit of currency there is also the dollar. But the Canadian dollar is not usually worth exactly the same amount as the United States dollar. It is sometimes

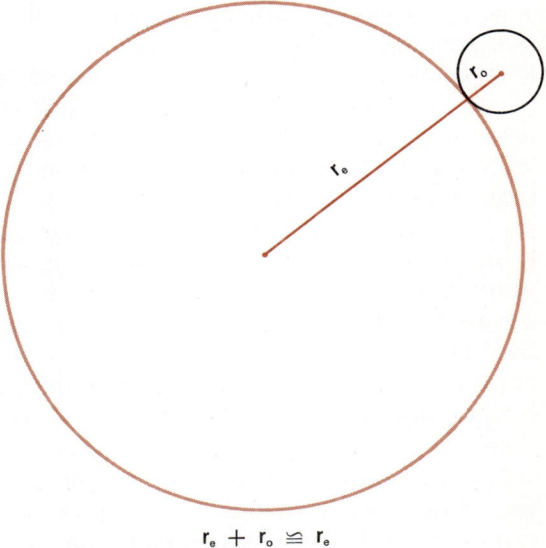

$r_e + r_o \cong r_e$

Figure 2–2 *For objects on the earth's surface, the center-to-center distance equals the earth's radius, r_e, plus the object's radius, r_o, which practically equals r_e.*

Figure 2-3 *This copy of the standard kilogram is kept at the U.S. National Bureau of Standards.*

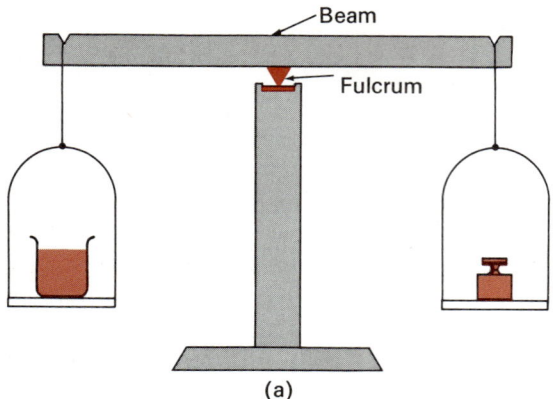

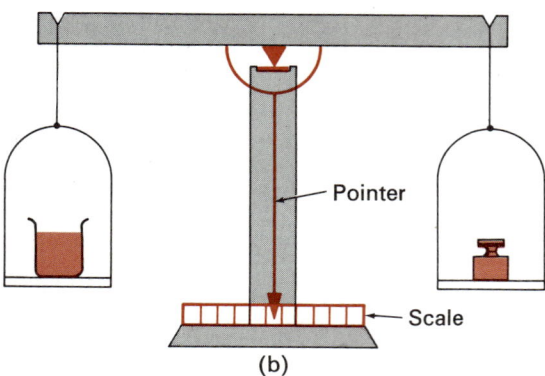

Figure 2-4 *Two versions of a simple balance that can be used to measure the mass of an object by direct comparison with a standard. The added pointer and scale in (b) make the balance easier to use.*

worth more and sometimes less, depending on economic conditions in the two countries. If you go to Mexico, you find that the unit of currency has a completely different name (the *peso*) and a distinctly different value.

There may be good reasons for tolerating a wide variety of units in commercial life. But scientists would have no reason for allowing such a situation to exist for the units in which basic scientific measurements are made. It is only sensible to have one set of units for all the basic quantities that scientists measure, and to have all scientists in all countries use this set. Scientists agreed to do just this over a hundred and eighty years ago, and it will be necessary for us to discuss the present form of the system that they adopted. We shall consider only mass units at this point, however, postponing a further discussion of units until the next chapter.

The international scientific unit of mass is the **kilogram.** It is a *defined* unit, which means that it has been set equal to the mass of an agreed standard object. The standard kilogram is a block of platinum metal kept in the town of Sèvres, a suburb of Paris. A number of exact copies have been made. One copy, shown in Figure 2-3, is in the United States National Bureau of Standards.

The kilogram is equivalent to about 2.2 American pounds. Since this unit is inconveniently large for most laboratory work, the **gram,** by definition one thousandth of a kilogram, is more frequently used by chemists.

1 kilogram = 1000 grams

1 gram   = 0.001 kilogram

2-4 Measuring Mass

There are two basic ways of measuring mass. One is by *direct comparison*. A device called a **balance**, which consists of a beam on a fulcrum, is used. It is illustrated in Figure 2-4. The beam must first be adjusted so that it is in perfect balance. Suppose that an object of known mass is placed on one arm of the beam at a precisely known distance from the fulcrum and another object is placed on the other arm at precisely the same distance from the fulcrum. If the beam remains balanced, we can conclude that the two objects have the same mass. If we make the comparison in air, our conclusion may be slightly in error, as explained in Section 2-7.

The balance shown in Figure 2-4 can be made more flexible. Suppose, for example, that the object of unknown mass is placed at a distance d from the fulcrum and that, in order to balance it, the known mass has to be placed a distance $d/2$ from the fulcrum. Then the unknown object must have a mass just half as great as the object of known mass. Further experimentation would show that balance is always achieved when the product of its mass and distance from the fulcrum is the same for each object. Thus, for any two masses, m_1 and m_2, at distances d_1 and d_2, respectively, from the fulcrum, we have the following relationship.

$$m_1 d_1 = m_2 d_2$$

If the mass of object two, m_2, is known, the mass of object one, m_1, can be calculated as follows.

$$m_1 = \frac{d_2}{d_1} m_2$$

This principle is used in the balance shown in Figure 2-5, in which the *rider* on the scale

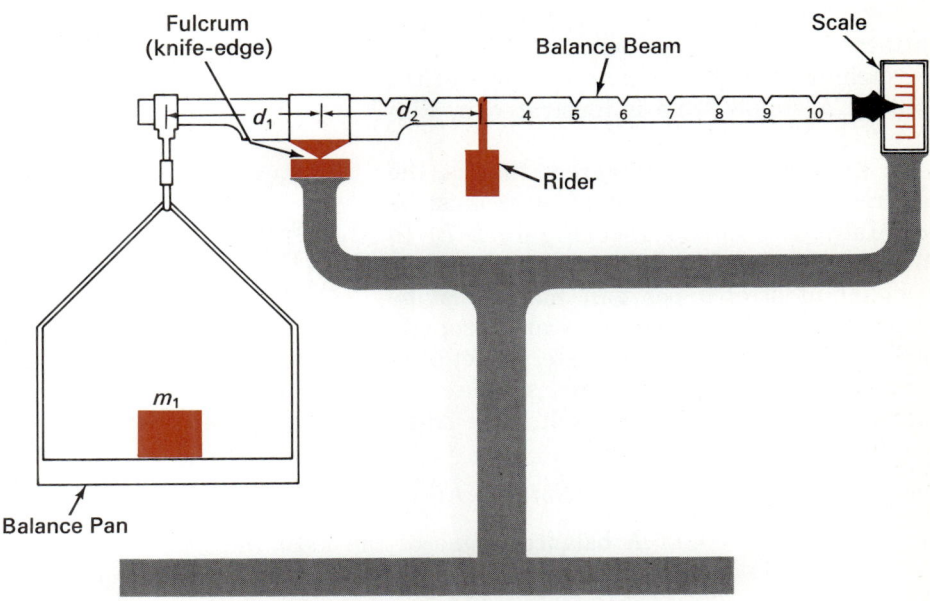

Figure 2-5 *A further modification of a simple balance.*

is an object of known mass. The object of unknown mass (m_1) is placed on the balance pan, and the distance (d_2) of the rider from the fulcrum is adjusted until balance is achieved. The greater the mass of the unknown object, the greater the distance through which the rider must be moved along the scale. For convenience, the scale is marked directly in mass units, and it is notched to make accurate placement of the rider more certain. The triple-beam balance shown in Figure 2-8b is commonly used in scientific laboratories. As you can see, it is basically the same as the one discussed in this paragraph.

With the balance shown in Figure 2-5, the idea of direct comparison of unknown with standard masses has been lost somewhat. There are other devices in which direct comparison is eliminated altogether. Such devices are known as **scales,** and their use is another basic method of measuring mass. Scales usually contain a spring of some kind, which stretches or twists when an object is attached to it. The amount of twisting or stretching is proportional to the mass of the object. A simple scale is illustrated in Figure 2-6.

A scale must be **calibrated,** that is, the degree of stretch per unit of mass must be carefully determined. (See Figure 2-7.) In order for the calibration to be correct, the amount of stretch per unit mass must be reproducible. A well-made scale, properly calibrated, can be quite accurate. Direct comparison, however, is more simple and straightforward, and, as a result, more trust-

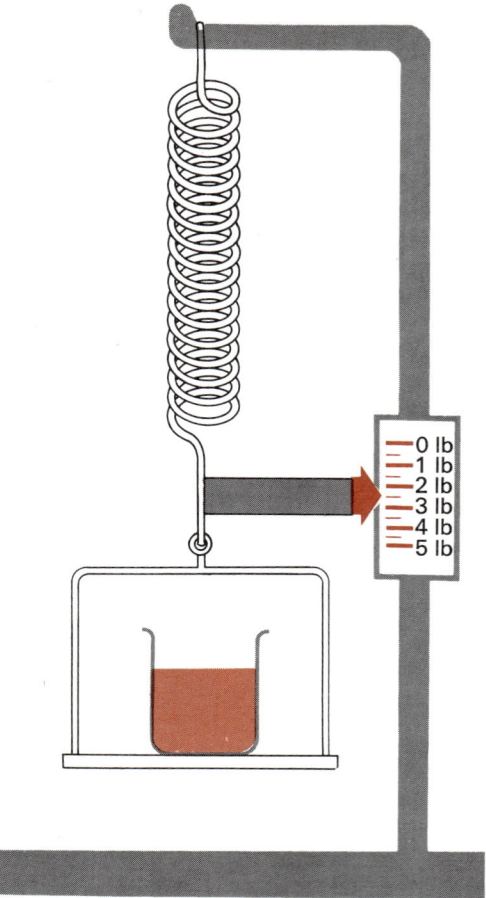

Figure 2-6 *A scale that operates on the principle of stretching a spring.*

worthy. Most scientific measurements of mass are made with some sort of direct comparison device.

INVESTIGATION *2-5 Determining the Mass of an Object*

A balance is one of the most important instruments in a chemistry laboratory. A number of different types of balances are available. The choice of a balance depends on the requirements of the particular experiments that are being carried out. Many balances commonly used in high school laboratories can determine the mass of an object to the nearest 0.01 gram. A good analytical balance can be used to find the mass to the nearest 0.0001 gram.

In using a given balance you must know more than just how to operate it. You must also understand the limitations of the instrument and the care that must be observed in order that the balance may continue to perform satisfactorily. Balances are delicate and can be damaged if they are not used properly. Be sure to read carefully the accompanying directions for care and use of balances.

In performing this experiment you will gather data that may be used and analyzed later, so it is particularly important for you to keep a neat and accurate record of your observations.

PURPOSE

The main reason for performing this experiment is to find out how dependable your results are when you use a balance to determine the mass of an object.

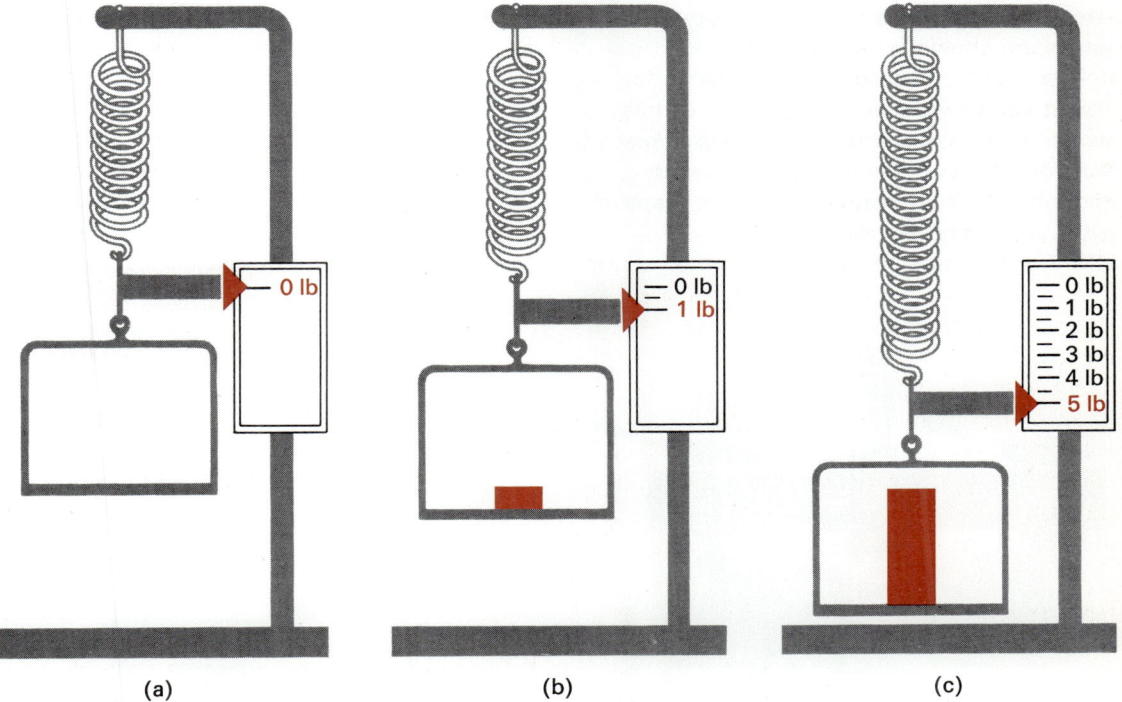

Figure 2-7 *A spring scale is calibrated by finding the amount of stretch caused by objects of known mass.*

MATERIALS

Balance, centigram   Objects of unknown mass, labeled

PROCEDURE

You will work with one or two of your classmates. An object with an unknown mass will be given to each of the groups in the class.

The Care and Use of Balances

A balance is one of the most important instruments in the chemical laboratory. In order for a balance to provide accurate results for a long period of time, it must be used with care. Two common laboratory balances are shown in Figure 2–8.

The following procedures should be followed *each time* you use any type of balance. Never assume that a balance is ready for use until you have checked it, even though only a short time may have elapsed after you checked it before.

1. The balance should be located on a firm, level surface in an area that is free from vibrations, drafts, or nearby sources of heat. Avoid areas heated by direct sunlight.
2. If a balance is equipped with a leveling bubble, check to see if the balance is level.
3. Before you place anything on the balance pan, be sure the pan is clean. Move all the balance riders to their zero positions and make certain that the beam moves freely. Be sure that the pointer swings an equal distance to each side of the zero mark. If the balance has a magnetic damper, check to see if the pointer comes to rest on the zero mark.

(a)

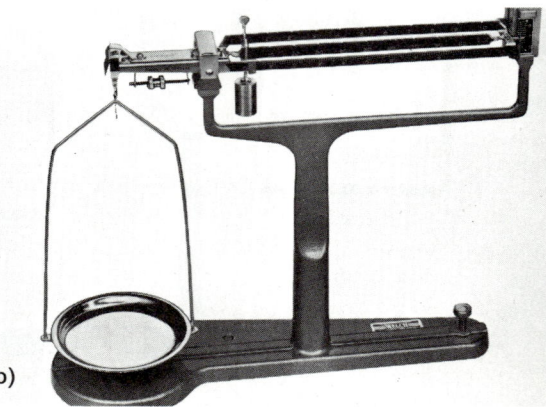

(b)

Figure 2–8 *It is possible to weigh accurately to the nearest 0.5g on the platform balance (a) and to the nearest centigram, 0.01g, on the triple-beam balance (b).*

1. Using the balance that has been assigned to your group, measure the mass of your unknown object as accurately and carefully as you can. Do not attempt to change the adjustment of your balance during the experiment. Record your results in your notebook at once and pass your unknown object to the next group. Determine the mass of another unknown object and continue until you have determined the mass of all the unknown objects available for this experiment.

Most balances have a threaded nut that can be used to bring the pointer to zero. Check with your teacher before trying to make any adjustments.

4. If your balance is equipped with a beam-arrest lever, use it to raise the beam before placing any object on the pan. Whether it has such a lever or not, place any object on the pan *gently* in order to prevent damage to the knife-edges or other parts of the balance.
5. *Never* weigh any type of solid or liquid chemicals directly on a balance pan. Use *dry* glass, porcelain, plastic, or metal containers to protect the pan. Some dry solids can be placed on smooth weighing papers. If you spill any chemical on a balance pan or in the area around the balance, clean it up at once.
6. Do not place any hot objects on a balance pan. Such objects may damage the pan. Also, the heating of the balance by these objects will make accurate readings impossible.
7. To determine the mass of an object, move the largest rider first to a position slightly less than the mass of the object being measured. Use the smallest rider only after you have approached the mass of the object as closely as possible with the larger riders. The small rider will be in the proper position when the balance pointer swings an equal distance to each side of the zero mark or stops on the zero mark when a magnetic damper is being used. If your balance has a beam with notches on it, the riders must be in the notches when you make your readings. If they are not, your results will be meaningless.
8. The mass of the object being measured is equal to the sum of the readings of all the riders being used. Record the mass at once in your laboratory notebook or an appropriate data sheet. Do not write your measurements on scraps of paper. Pieces of paper are easily lost and mistakes can be made in copying the results over into a more permanent record.
9. After you have recorded your measurements, lift your object carefully from the balance pan. If the balance has a beam arrest lever, use it to raise the beam from its knife-edges *before* you remove your object. The riders on a balance without a beam arrest lever should *not* be returned to zero after the balance has been used. This will prevent the beam from moving continuously while the balance is not being used and will lessen the wear on the knife-edges.
10. Before leaving, check both the balance and the area around it to make sure that you have not spilled anything. Leave everything ready for the next person who will use the equipment.

2. A table similar to the one following this procedure will be prepared on the chalkboard. Locate the number of your balance in the table. In the row opposite this number write the mass of each of the unknown objects you measured, in its proper column. When all of the spaces have been filled in by other class members, copy this data in your own notebook.

3. Examine the data reported by the class. Did any one balance produce results that were consistently higher or lower than the results obtained on most of the other balances?

4. Four or five groups should now measure the mass of unknown Sample A on the balance that gave consistently high results. Four or five other groups should measure the mass of the same sample on the balance that gave consistently low results. These results should be copied on the chalkboard under columns headed "Sample A, High Balance" and "Sample A, Low Balance." Copy this data in your notebook also.

DATA AND RESULTS

Balance #	Sample A	Sample B	Sample C	etc.
1				
2				
etc.				

RELATED PROBLEMS

1. Several balances were used to determine the masses of the various unknown objects in the first part of the experiment. What was the difference between the highest mass recorded and the lowest mass recorded for each object? Why does one object appear to have different masses when measured on different balances? List as many reasons as possible for these apparent differences.

2. Examine the original data table again. If you do not consider the highest mass recorded and lowest mass recorded for each unknown, what is the difference between the remaining high and low results for each unknown? How do these differences compare with the differences between masses recorded under the columns headed "Sample A, High Balance" and "Sample A, Low Balance"?

3. How can you explain the fact that one balance gave high results and one gave low results?

4. With all the data you now have for Sample A, what would you report as the mass of Sample A? Why?

5. The teacher will list on the chalkboard the weight in ounces of several of the unknown samples used in the first part of the experiment. From your original data table find the average mass in grams of one of these unknowns. Using this result, calculate the values of conversion factors, f and F, that will make it possible to convert ounces into grams and kilograms into pounds. These factors are determined by the following equations.

$$f \times \text{(weight in ounces)} = \text{mass in grams}$$
$$F \times \text{(mass in kilograms)} = \text{weight in pounds}$$

Using these conversion factors, calculate how many grams of butter are in one pound of butter.

Knowing your own weight in pounds, use the conversion factors to find your mass in kilograms.

Summary

While different forms of matter vary greatly in their properties, they all have one thing in common. They have mass. As a result of having mass, bodies of matter attract one another. The greater the mass of either body, the greater the attraction. The attraction diminishes with the square of the distance between the bodies. When you measure the force by which a body is attracted to the earth, you are measuring its weight. If this is done under proper conditions, you can determine the mass of the body. The units in which scientists express mass are the kilogram (about 2.2 pounds) and the gram. A gram is one thousandth of a kilogram. Mass can be measured with a balance, in which a direct comparison of one mass with another can be made, or with a scale, in which the comparison is indirect.

Self Check

1. State in words what the Law of Gravitation says about how the force between two objects depends on their masses and on the distance between them.
2. About how many grams are there in a pound?
3. What two quantities must be equal when a balance is "balanced"?

Weight and Mass

2-6 What Do We Mean by Weight and Weighing?

The processes just described for determining mass are not totally unfamiliar to you. The use of scales, at least, is certainly familiar to all. When you step on a bathroom scale, for example, you say that you are weighing yourself, or finding out what your weight is. But have you determined your mass? Is there

Joseph Weber (1919–)

NEARLY THREE HUNDRED YEARS have passed since Isaac Newton proposed an equation that correctly predicts the effects of gravitational forces. It has been about sixty years since Albert Einstein considered gravity and some of its properties in his General Theory of Relativity. No one, however, has yet been able to explain how gravity works.

Joseph Weber, Professor of Physics at the University of Maryland, is a modern scientist who hopes to unlock some of the secrets of this common but mysterious force. Since 1958 he has been trying to detect the extremely faint gravitational waves predicted by Einstein. Since the amount of matter required to produce waves strong enough to detect must be enormous, Weber decided to focus his attention on some of the stars and galaxies in outer space. He calculated that the energy from incoming gravitational waves should be able to set up vibrations in other large bodies. For a while he considered using the entire earth as a detector, but discarded this idea when he realized that earthquakes and other disturbances deep within the earth would interfere with the vibrations he was looking for.

In 1968 Weber constructed two large aluminum cylinders, weighing three and a half tons each, to act as his detectors. Sensitive crystals on the sides of the cylinders generate minute amounts of electric current when the cylinders vibrate. To eliminate the possibility of local disturbances affecting the results, he placed one cylinder in Maryland and the other about six hundred miles away near Chicago. If the two detectors record vibrations simultaneously at times when the earth itself is not vibrating, the energy must be coming from some source not associated with our own planet.

When the instruments began to operate in late 1968, a number of signals began to be recorded. By scanning various parts of the sky, Weber soon discovered that the waves appeared to originate from the center of the great galaxy of stars of which the solar system is only a tiny part.

Scientists still do not know how to explain the origin of the waves, but one more step has been made toward a better understanding of gravity.

a difference? Or are "mass" and "weight" just two different terms, one familiar and the other more sophisticated, for the same thing?

There are definite differences, both theoretical and practical, between weight and mass, and between weighing something and determining its mass. To begin with, weighing something, that is, measuring its attraction to the earth (or to some other body) is only one way of determining its mass. The expression "determine the mass of" is far more general than the term "weigh." Secondly, the mass of an object is a precisely defined quantity, whereas its weight is simply the result obtained in one particular type of measurement. Thirdly, there are many sources of error in weighing. The weight of an object, as compared to its actual mass, may vary quite a lot, depending on where, when, and how the measurement is made. (For example, see Figure 2-9.) Later in the chapter, several of the key sources of error in weighing are discussed.

It should be clear to anyone living in the space age that the words "weight" and "mass" cannot possibly mean the same thing. An object continues to have mass, even when it has no weight! As astronauts move away from the earth, their weight decreases. When the gravitational pulls of earth and moon are just balanced, their weight is zero. The condition of weightlessness can be achieved in other ways as well, yet the objects in question always have their proper *masses*.

Mass is fundamental and inherent. Weight is merely the force exerted on an object by a body such as the earth, and it varies with local conditions. Mass and weight are not the same.

2-7 Air Buoyancy

You have seen that there is a direct relationship between the weight of a body and

Figure 2-9 *Astronaut Edwin Aldrin walking on the moon. Is his weight the same as on the earth? Is his mass the same?*

its mass. You have also seen that a simple way to find the mass of an object is to place it on one arm of a balanced beam and then place objects of known mass on the other arm, until the beam is balanced again. The mass of the body is exactly equal to the sum of the masses of the objects that just balance it, however, only if the entire procedure is carried out in a vacuum. A vacuum is literally empty space. In order to weigh in a vacuum, it is necessary to have the balance and the objects of known and unknown mass inside an airtight box, from which all the air has been pumped. Normally we do not want to contend with the inconvenience of weighing things in a vacuum. So it is necessary to find out why the presence of the atmosphere makes weighing inexact and, if possible, how to make a correction for it. The correction we need is called a *buoyancy correction*.

The following experiment, which your teacher will demonstrate for you, will help you to understand the buoyancy effect. It will also help to make clear the discussion, which follows the experiment, on how to apply a buoyancy correction.

INVESTIGATION

2-8 The Buoyancy Effect

Even if we had the equipment that would make it possible for us to weigh an object first in air and then in a vacuum, the differences that we would be able to determine would be relatively small. In order to measure larger differences we will weigh objects in air and also in water. Since the mass of a given volume of water is considerably greater than the mass of an equal volume of air, its buoyancy effect should be much larger than that of air.

PURPOSE

The purpose of this experiment is to determine how much difference there is in the weight of an object when weighed in air and its weight when it is weighed in water. It will help you in learning how to make corrections for the buoyancy effect. With the use of such corrections you will be able to determine the true mass of any object.

MATERIALS

Lead, 1 piece
Aluminum, 1 piece
Beaker, 100-ml
Balance, single-pan or double-pan
Support for double-pan balance, if such a balance is used
Hook for supporting objects on balance pan
60-80 cm of thread or very fine wire
Paper towel, several pieces
Detergent solution
Medicine dropper

PROCEDURE

1. Obtain a piece of lead and a piece of aluminum, each weighing the same amount to the nearest 0.1 gram. Tie a piece of thread or very fine wire around the piece of aluminum and hang it from the hook on your balance. You may use either a single-pan or double-pan balance. If you use a double-pan balance, you should place the balance on a box or other suitable support that will raise the base of the balance five or six inches above the bench top, as in Figure 2-10a. If you use a single-pan balance, use the arrangement shown in Figure 2-11a. Find out the apparent weight of the piece of aluminum and record the result in your notebook under the heading, "Weight of aluminum in air."

2. Now raise a small beaker of water beneath the piece of aluminum in such a way that the entire piece of metal is completely immersed in the water, as in Figure 2-10b or 2-11b. A wooden block may be

Matter and Mass 43

Figure 2–10a

Figure 2–10b

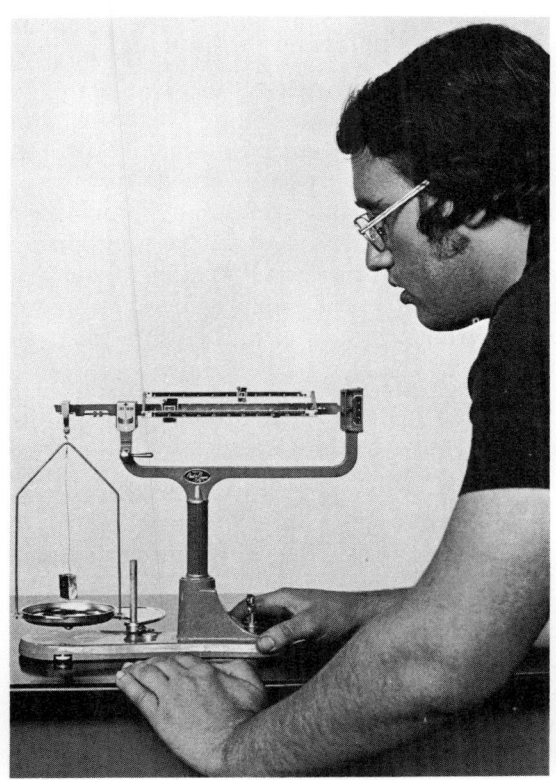

Figure 2–11a

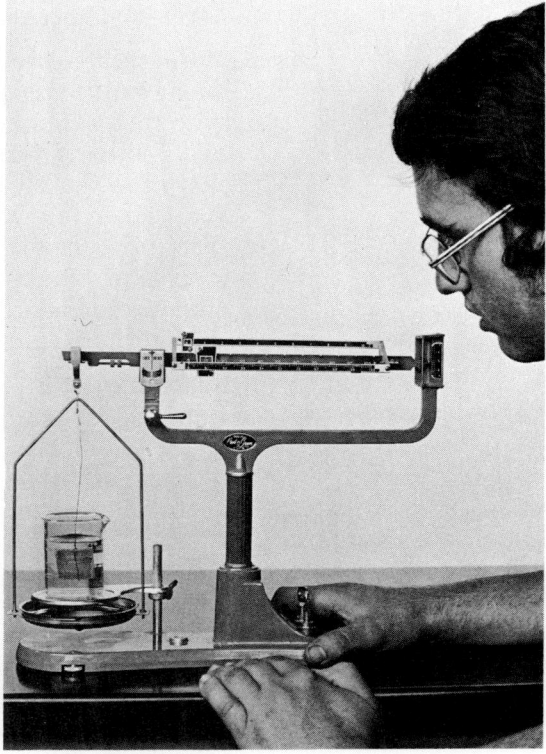

Figure 2–11b

used to support the beaker, as in Figure 2-10b. Record the weight of the aluminum again under the heading, "Weight of aluminum in water."

3. Tie some thread or fine wire around the piece of lead. Hang the lead from the hook on the balance and weigh it first in air and then in water. Record the weights in your notebook under the headings, "Weight of lead in air" and "Weight of lead in water."

4. Place a small beaker on a towel on the desk top. Add 4-5 drops of detergent solution to the beaker. (This solution gives the water some characteristics that are desirable in this experiment.) Add water to the beaker until it is completely full and a little water runs over the top. Hold the piece of aluminum that you used in steps 1 and 2 by its thread and lower it slowly into the water, allowing the water that has been displaced to run out of the beaker onto the towel. Remove the aluminum from the beaker of water. Carefully dry off the outside of the beaker. Place the beaker with its remaining water on a balance and weigh it. Record the result in the space marked "Weight of beaker and water after immersing aluminum."

5. With the beaker still on the balance carefully add water to the beaker until it is almost full. Use a dropper to fill it just to the brim with water to avoid running it over. If a drop or two should overflow, use a piece of paper towel to blot up the excess, but do not try to move the beaker, because you will probably spill some more. Weigh the beaker full of water and record your result in the appropriate place in your table. Subtract the weight recorded in step 4 from this weight to determine the "Weight of water displaced by aluminum." Remove the beaker of water from the balance and be sure that you dry off the balance pan if you spill any water on it.

6. Repeat steps 4 and 5 with lead instead of aluminum. Record the difference in weights as "Weight of water displaced by lead."

DATA AND RESULTS

Leave space in your notebook for each of the following. Make all measurements in grams.

Weight of aluminum in air
Weight of aluminum in water
    Loss of weight of aluminum
Weight of full beaker of water
Weight of beaker and water
    after immersing aluminum
        Weight of water displaced
           by aluminum

Weight of lead in air
Weight of lead in water
    Loss of weight of lead
Weight of full beaker of water
Weight of beaker and water
    after immersing lead
        Weight of water displaced
           by lead

RELATED PROBLEMS

1. How did the weight that the aluminum lost when it was immersed in water compare with the weight of water displaced by the aluminum?

2. How did the weight that the lead lost when it was immersed in water compare with the weight of water displaced by the lead?

3. Although your piece of aluminum and your piece of lead weighed almost the same amount at the start of the experiment, how can you explain the fact that they lost different amounts when they were immersed in water?

4. When any object is immersed in water, what must the surrounding water be doing to the object?

5. If the weight of water displaced by some object is less than the weight of the object in air, what will the object do when it is dropped into water?

6. Suppose that an object is completely immersed in water. Suppose, further, that the weight of the water it displaced was greater than or equal to the weight of the object in air. What would the object do if it were released in the water?

7. When you have inflated a balloon with your breath and tied off its end, you have found that it would drop to the ground if you let go of it. On the other hand, you have observed that a similar size balloon filled with helium would rise into the air. If you assume that *all* matter, including helium, has mass, how can you explain the fact that the helium balloon rises into the air?

2-9 How To Correct for Buoyancy

We shall now apply some of the knowledge gained in the preceding experiment in explaining how to correct weighings for the buoyancy effect. You have seen that if an object is weighed while it is immersed in water, it weighs less than when it is weighed in the usual way. The amount of weight it appears to lose is equal to the mass of the water that occupies the same volume as the object itself. By adding the mass of water displaced to the balance reading obtained when the object is immersed in water, we get the same result that we would get if we simply weighed the object in air.

It might seem to you that all of this is an unnecessary exercise. If we want to know the weight of an object in air, we should just weigh it in air. Why weigh it in water and then make a correction for the buoyancy effect of the water? The experiment was not done so that you would be able to weigh objects in water and then correct the weights for the buoyancy effect. The purpose was to show, in a way that is easily seen and measured, just what the buoyancy effect is. The buoyancy effect due to air is not so readily apparent. When we weigh an object in air, a buoyancy correction also must be applied, if we wish to find the actual mass of the object from the weight read on the balance.

Air has a buoyancy effect on any object in it. An object will weigh less in air than it would in a vacuum. Its weight will be less by the mass of the air that would occupy the same volume as the object. Since it is very inconvenient to weigh things in a vacuum, we weigh them in air and then make the required buoyancy correction to get a number that is closer to the actual mass. The actual mass is what the balance would read if we made the weighing in a vacuum, and if there were no other sources of error.

The buoyancy corrections that must be applied to weights obtained in air are much smaller than the corresponding ones for weights obtained in water. A given volume of air weighs only about $\frac{1}{800}$ as much as the same volume of water. The piece of aluminum required a buoyancy correction of 1.5 grams when it was weighed in water. The buoyancy correction would be only about

$$\frac{1.5}{800} = 0.0019 \text{ gram}$$

if it were weighed in air. This correction, when added to its weight in air, gives the weight that the aluminum object would have in a vacuum, that is, its true mass. In most of your laboratory work in this course you will be using balances that can only be read

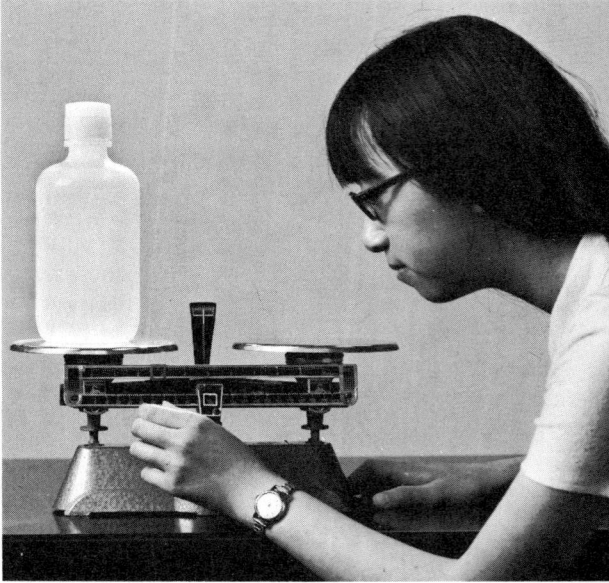

Figure 2–12 *When weighing a large, light object in air, it is necessary to make a buoyancy correction.*

to the nearest 0.01 gram. As a result, so small an error is not important.

There is one way in which the buoyancy correction for air can become important, even when you are only reading weights to 0.01 gram. What if you were weighing a very large object? Suppose, for instance, that you were weighing a vessel that can contain one quart, and so displaces a quart of air, as in Figure 2–12. A quart of air has a mass of about 1.2 grams under normal conditions of temperature and atmospheric pressure. A correction of that size would have to be made in order to get an accurate estimate of the true mass of the object in question. Later you will perform an experiment in which the mass of an object of roughly this size has to be determined from its weight in air.

In the meantime, the important thing to be understood about the buoyancy effect is that it causes a difference between the reading you make on the balance, that is, the weight of the object, and the mass of the object. The existence of this effect is one important reason why weight and mass are different.

2-10 Some Other Sources of Error in Weighing ★

You have just seen that the mass of an object, as determined by weighing it on the earth's surface, will be in error if we do not take account of the buoyancy effect of the earth's atmosphere. The importance of this effect is greatest for objects that are both light and large. There are other errors that can occur in weighings. A few of them will be mentioned here to emphasize that the weight of an object can only be considered equal to its mass if the weighing is done under strictly controlled conditions. Naturally, the instrument used to make the weighings must itself be of high precision, in good working order, and carefully checked against standards of known mass. Let us consider some of the more subtle errors that may come into weighing.

The earth is not smooth, is not uniform in composition, and is not exactly round. All of these facts have an influence on the accuracy of mass determinations. The reasons are rather obvious once you stop to think about them. Because the earth is not smooth, we have to consider the altitude at which weighings are made. The radius of the earth is about 4000 miles. If we weigh an object at a point on the surface that is 4000 miles from the center of the earth and then weigh it again on a mountain that is 4 miles high, the results will differ. As shown in the equation expressing Newton's Law of Gravitation, given in Section 2-2, the greater the distance, the less the force. The weight obtained on top of the mountain will be less than the weight at the lower altitude by a factor of

$$\left(\frac{4000}{4004}\right)^2 = \text{about } 0.998.$$

This small error would never be noticed in the experiments to be done in this course. It can be detected in the very precise measurements required in some research.

A similar effect results from the fact that the earth is not a perfect sphere, but is instead slightly flattened, as Figure 2-13 shows. At sea level on the equator you are about 26 miles farther from the center than at sea level near one of the poles. Thus the pull of the earth on an object will be less at the equator than at a pole by a factor of roughly

$$\left(\frac{4000}{4026}\right)^2 = \text{about } 0.987.$$

Figure 2-13 *This photograph, taken from an altitude of 22,300 miles, shows that the earth is not a perfect sphere.*

The fact that the composition of the earth is not uniform may affect weight in a way that can be put to practical use. On the average the earth has a mass per unit volume of about 5.5 grams per milliliter. However, the mass per unit volume for oil is only about 0.9 gram per milliliter. If a weighing is made at a point on the earth that is directly over a large volume of oil, the pull of gravity will be a little less than if the volume occupied by the oil were occupied by matter of average density. There is less matter in a given volume of oil than in the same volume of stone. Hence, there is less pull exerted on the object being weighed. This is an extremely small effect, but instruments have been invented to detect it, and in this way some large deposits of oil have been discovered.

If weighings are made with the type of device called a scale, the various errors just mentioned will affect the results with full force. When balances are used, these errors are eliminated. Let us go back to the general mathematical relationship on which the balance is based.

$$m_1 d_1 = m_2 d_2$$

As before, m_1 represents the unknown mass that we want to measure, and m_2 is the known, or standard, mass used for comparison. The distance of mass from the fulcrum is d. We can then rearrange the equation to read as follows.

$$\frac{m_1}{m_2} = \frac{d_2}{d_1}$$

Any of the three previously-described effects, which depend on the earth's form and composition, will decrease (or increase) the earth's pull on both the object of unknown mass and the reference object by the same fraction. As a result, the ratio of the two masses will be unchanged. Therefore, the ratio of the two distances will not change and the weight of the object being measured will still be given correctly by the balance. This is the great advantage of a balance over a scale. Many possible errors in determining mass can be eliminated because of the direct comparison principle involved in a balance.

None of the three errors in weighing discussed in this section will show up in the experiments to be done in this course. They have been mentioned to stress a point. *Weight is not mass.* It is only an approximate measure of it.

Summary

Weight and mass are not the same thing. Weight is proportional to the pull exerted by the earth (or some other body) on an object. The magnitude of this pull may be changed by many conditions even though the actual mass of the object (that is, the amount of matter it contains) does not change. One of the conditions that influences the weight of an object, making it different from the mass, is the buoyancy effect of the medium in which it is immersed when weighed. The weight of the immersed object will be less than its mass by a definite amount. This amount is equal to the mass of the medium that has the same volume as that of the object being weighed. In order for weight to be equal to mass, even if there are no other sources of error, the object must be weighed in a vacuum. For practical reasons, we usually weigh objects in air. A correction for the buoyancy effect of the air may then be necessary in order to give a result with the desired degree of accuracy.

Self Check

1. Why was there a greater decrease in the apparent mass of the piece of aluminum than in that of the lead, when the two were weighed in water?

2. Why is the buoyancy correction that is necessary when weighing in air smaller than the one required when weighing in water?

Does Mass Change Measurably in Chemical Reactions?

INVESTIGATION

2-11 Investigating Some Reactions for Possible Changes in Mass

You saw in an earlier experiment that when a candle burns, some of its substance, wax, is lost and two new substances, water and the gas called carbon dioxide, are formed. You also found that the presence of the gas called oxygen is necessary and that the oxygen is consumed in the process. In short, you found that the burning of a candle is a process that involves many changes of substances into other substances.

PURPOSE

You will try to find out if there is any change in the *total* amount of matter involved in several different types of changes.

MATERIALS

Copper wire, #16 (about 100 cm)
Balance, platform
Candle
100 ml dilute sulfuric acid
Ice cube or several pieces of crushed ice

100 ml dilute $BaCl_2$ in water
2 flasks, 250-ml
Flask, 500-ml
Gallon jar or bottle

PROCEDURE

Part A.

1. Wrap one end of a piece of #16 copper wire tightly around the lower end of a candle that is approximately 2 cm in diameter and 5-6 cm long. Use the wire as a handle to lower the candle to the bottom of a one-gallon, wide-mouth mayonnaise jar with a screw cap or a one-gallon plastic milk bottle with a tight-fitting lid or other suitable stopper. Cut off the copper wire so that it will fit inside the jar

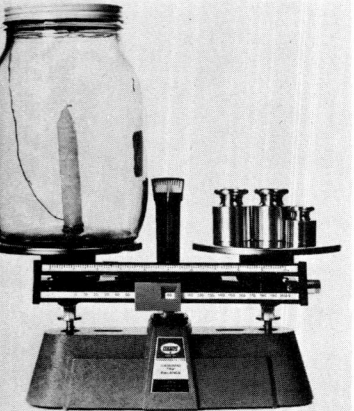

Figure 2-14

or bottle without interfering with the lid, as shown in Figure 2-14. Place the jar or bottle with the candle, the wire, and the lid on a suitable balance. Adjust the riders to give equilibrium as you would when making a weighing.

2. Use the wire handle to raise the candle out of the jar. Light the candle and quickly lower it to the bottom of the jar. Immediately seal the jar and observe what happens to the equilibrium of the balance as the candle burns.

3. After the flame has gone out, open the jar. Either blow a stream of air into the jar to sweep out the combustion products or turn the jar upside down for several minutes to let the combustion products escape. Now place the jar containing the candle on the balance pan with the lid beside it. Adjust the balance to equilibrium if necessary. Relight the candle and observe any change in the balance.

Part B.

1. Place a 500-ml flask, half full of water at 30–35°C, on a balance.
2. Drop in some pieces of ice and stopper the flask.
3. Adjust the balance to equilibrium and observe its behavior as the ice melts. Is there any change in total mass?

Part C.

1. Place two 250-ml flasks on the same pan of a balance. Add about 100 ml of a dilute barium chloride solution to one of the flasks. Add about 100 ml of a dilute sulfuric acid solution to the other flask. Adjust the weights to equilibrium.

2. Pour the contents of one flask into the other. Note any changes that take place. With both flasks, one now empty and the other containing a new substance, on the balance, note any change in mass.

RELATED PROBLEMS

1. How and why did your observations in Part A-3 differ from those in Part A-2?

2. What can you conclude about the changes that take place in the *total* amount of matter involved when matter is changed from one form into another form?

A Question To Wonder About

How would you compare the type of change that took place in Part B with the changes that took place in Parts A and C?

2-12 Mass is Conserved in Chemical Reactions

Perhaps the most basic law of chemistry is the **Law of Conservation of Mass.** This is a generalization based on the experience of thousands of chemists in experiments of many kinds. As you saw earlier, when a generalization has been so long and so drastically tested that there seems to be no possibility of its being incorrect, it is customary to call it a law. This law states that:

Matter may be changed from one form to another but cannot be created or destroyed.

This law is of tremendous importance in chemistry because chemistry is primarily concerned with the processes by which one form of matter is converted into another. In his efforts to make sense of these changes the chemist finds it extremely valuable to know that, however different the properties of the starting materials and the products may be, the total mass will be the same.

It was in great part due to the work on combustion by the French chemist Lavoisier (lah-vwah-zee AY) that the Law of Conservation of Mass was established as a basic rule in chemistry. It was such a crucial contribution that for this alone Lavoisier deserves to be called the father of scientific chemistry. (He did many other important experiments, too.) Lavoisier showed by direct and careful experiments that when chemical changes such as burning occur, the total mass of all substances before and after remains unchanged. In the preceding experiment you saw the kind of evidence that Lavoisier used in proving this. All such experiments show that there is no measurable change of mass in a chemical reaction.

Summary

In the kinds of processes chemists deal with, the total mass does not change measurably. Matter changes its form, but the total quantity of matter remains constant. This is called the Law of Conservation of Mass. It is one of the most fundamental and valuable rules the chemist uses in attempting to understand and interpret his experiments.

Self Check

1. Why do we call the Law of Conservation of Mass a "law"?
2. When a candle burns, it gets smaller. Yet the Law of Conservation of Mass says that when all the products of this burning process are taken into account, there is no decrease (or increase) in mass. How do you reconcile these two statements?

Chapter Summary

The basic property of all forms of matter is that they have mass. Mass can be defined as the amount of matter present, whatever its form or appearance. Mass can be measured by using Newton's Law of Gravitation. This states that two objects attract each other in direct proportion to the product of their masses, and in inverse proportion to the square of the distance between their centers of mass. The most usual way to measure mass is to take the earth as one of the objects and measure how strongly another object is attracted to it. We call this the weight of the object. Weight will be equal to mass only if the measurement is made under precisely controlled conditions. Mass is measured in

Roy J. Plunkett (1910–)

IN 1938 A RESEARCH CHEMIST named Roy J. Plunkett made a very important discovery because he knew and believed the Law of Conservation of Mass. What he discovered was the plastic called Teflon. One of the many useful products now made with this plastic is the "nonstick" cooking utensil.

Plunkett was doing some research that required a gas called tetrafluoroethylene (tet-rah-*floo*-roh-ETH-il-een). He obtained this gas in a metal tank that, although small, contained a great deal of the gas because the gas had been put into it under pressure. The quantity of gas in the tank was measured by weighing the tank before and after filling it.

As Plunkett's work proceeded, he would release small amounts of the gas for use in his experiments. One day, he found that he could get no more gas from the tank and assumed that he must have used it all up. However, as he lifted the tank from the bench, he had the feeling that it was heavier than one might expect an empty tank to be. Following up on this, he weighed it and found it was indeed heavier than the weight given for the tank alone.

He drew the only possible conclusion based on the Law of Conservation of Mass. There must still be some matter in the tank, even if it was no longer the gas that had been put in. He had the tank cut open and found that it contained a white powder. In some way, part of the gas tetrafluorethylene had been converted into this new substance as it stayed in the tank under high pressure. The substance was Teflon, and deliberate experiments soon showed that it could be made regularly and efficiently from tetrafluoroethylene.

---

units called kilograms, or, for smaller masses, in grams. A gram is equal to a kilogram divided by 1000. One of the conditions required for weight to be equal to mass is that the weight be measured in a vacuum, in order to avoid the buoyancy effect of the air. If an object is weighed immersed in water, there is an even larger buoyancy effect. A correction for buoyancy can be made by adding to the apparent weight the mass of the medium (such as air or water) that would occupy the same volume as that of the object being weighed. Weight and mass are not the same thing.

A fundamental law of chemistry is the Law of Conservation of Mass. This states that in any chemical process the total mass must remain the same.

Questions and Problems

1. The distance from the earth to the sun is approximately 90 million miles. If another planet had a mass twice as large as that of the earth and was located 180 million miles from the sun, how would the gravitational force between the sun and the earth compare with the gravitational force between the sun and the other planet?
2. An object was weighed on the earth with a spring scale and was found to weigh 100 grams. If this object could be weighed with the same spring scale on a planet that had twice the mass of the earth but the same diameter as the earth, what would its weight appear to be there?
3. If the same object that was weighed on another planet with a spring scale (Question 2) were then weighed on the same planet with a balance and weights having known masses, what would its weight appear to be?
4. Recall your weighing experiment, described in Section 2-5. You probably found that the mass of a piece of metal did not always appear to be the same when it was measured on a number of different balances. Make a list of all the factors you can think of that could affect the accuracy of measurements made on a balance, in addition to those mentioned in the chapter.
5. A block of balsa wood was placed on one pan of a balance and was exactly balanced with metal weights having a known mass of 99.95 grams. Measurements showed that the balsa wood displaced 0.19 gram of air and that the metal weights displaced 0.02 gram of air. Find the mass of the balsa wood block to the nearest 0.01 gram.
6. Using the concept of the buoyancy effect, explain why a metal ship floats, but will sink if it gets a hole in its hull.
7. The buoyancy experiment (Section 2-8) involved the use of two pieces of metal that had the same masses but were much different in size. How might you explain this difference?
8. You have found out that weight is a measure of gravitational forces between two bodies. From your general knowledge of gravity, would you describe it as a short-range force or a long-range force? Explain briefly.
9. What other natural forces can make two bodies attract each other? Would you describe these as short-range or long-range forces?
10. Silver oxide is a solid compound that will produce metallic silver and gaseous oxygen if it is heated to a high temperature. Describe briefly an experiment that could be used to determine whether the mass of the silver oxide originally present is equal to the sum of the masses of the products.

3 Energy

Pioneer 10 starting off on its trip to the vicinity of Jupiter, and beyond. Ample chemical energy is provided by the rocket fuel.

Whenever one form of matter is changed into another, energy is either produced or consumed. This may happen very spectacularly, as when a bomb explodes, or in a fireworks display, or more quietly as in the burning of a candle. Each produces energy, giving off heat and light. In addition to heat and light there are many other forms of energy, such as electricity. Because energy changes go hand in hand with chemical changes, it is just as necessary for you to familiarize yourself with some basic ideas about energy as it was to learn about the basic properties of matter. In this chapter, you will learn some things about energy that you will need in your study of chemistry.

Recognizing and Measuring Energy

3-1 Energy Has Many Forms

Energy is the second basic ingredient of chemistry. Like matter, energy is not easy to define in the abstract, and such a definition will not be attempted. Instead, let us begin by asking, and trying to answer, some practical questions about energy. What forms does energy take? Can these forms be interconverted? How and in what units is energy measured?

One thing that energy in some of its forms can do quite directly is to make matter move. Matter in motion is said to have **kinetic energy**. (The word "kinetic" comes from the Greek word *kinetikos*, which means "motion.") For example, when a pool cue strikes a ball, the ball rolls. The rolling ball has kinetic energy. If that ball strikes another and both of them move away from each other, the kinetic energy has then been partly transferred to the second ball. An automobile, a battery-powered golf cart, and a steam locomotive are examples, when in motion, of objects possessing kinetic energy. The kinetic energy is imparted to them by some other type of energy, in each case by a different one.

The automobile is driven by *chemical* energy. There is energy locked up in the gasoline molecules. It is released when the gasoline burns. This chemical energy is partly converted to the kinetic energy possessed by the moving pistons. This energy is, in turn, converted by mechanical means into the motion, or kinetic energy, of the car.

The battery-driven golf cart is moved by *electrical* energy. The steam engine is driven by *thermal* energy, that is, heat. Light is still another form of energy. Let us now examine several important forms of energy a bit more closely.

3-2 Potential Energy

Consider the situation shown in Figure 3-1. A marble on the upper level at the left is at rest (position 1). It is then pushed gently to the brink of the slope (position 2), and it proceeds to roll down the slope (position 3). It continues along the lower level (positions 4 and 5) and then rolls up the slope on the right (position 6), finally coming to rest on the upper level at the right (position 7). Between position 2 and position 7 the marble had kinetic energy. Yet it started with none and it ended with none.

We say that at positions 1, 2, and 7 the marble has **potential energy.** It has this potential energy because of its height. As it begins to roll down, it begins to convert potential energy into kinetic energy. When it is rolling along the lower level (positions 4 and 5) potential energy has become kinetic energy. As the marble rolls up the other slope, its kinetic energy is being converted back into potential energy. When it finally comes to rest, it has only potential energy. The particular kind of potential energy involved here can be called gravitational potential energy, because it was acquired by an original upward movement against the force of gravity.

Another very simple example of potential energy is furnished by a stretched rubber band, Figure 3-2a. This can be called mechanical potential energy. As shown in Figure 3-2b, it is transformed into kinetic energy when the rubber band is released.

Finally, let us look at one more example showing changes in kinetic and potential energy. This one introduces a new concept that will be of importance later. Figure 3-3 shows a cart approaching a spring at constant speed. From the instant the cart touches the spring it begins to slow down. It is losing kinetic energy. What becomes of this energy? It is converted into the potential energy of the compressed spring. There

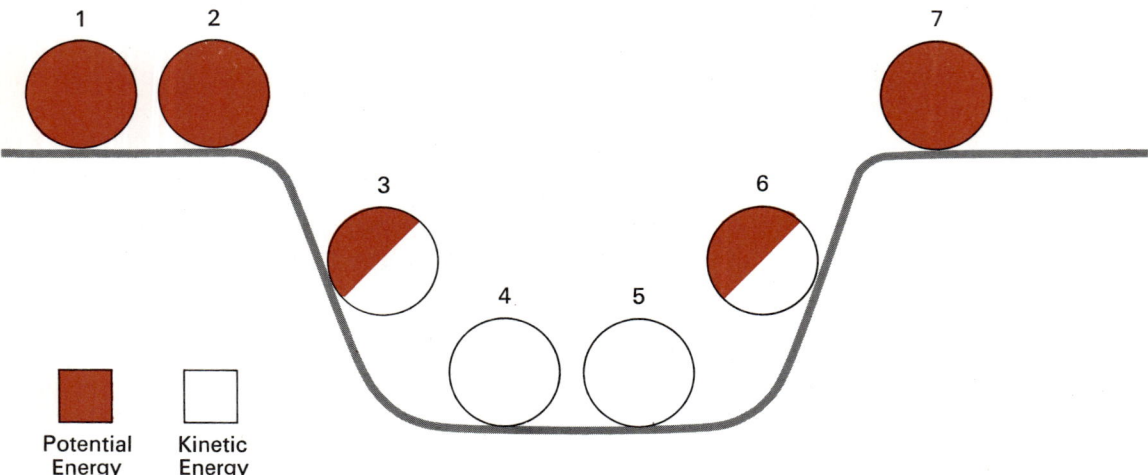

Figure 3–1 *A marble rolling into and out of a trough converts potential energy to kinetic energy and back to potential energy again.*

comes an instant when the cart is at a dead stop. All of its kinetic energy has become potential energy. The cart then begins to pick up speed in the opposite direction as the spring expands, and finally the cart moves away at uniform speed. All potential energy has been converted back into kinetic energy.

The process you have just considered is called an **elastic** collision. There is no net change in the amount of kinetic energy of the cart. Only the direction of its motion is changed. The spring is left as it originally was, with no change in potential energy. The concept of an elastic collision is one that you will have use for later.

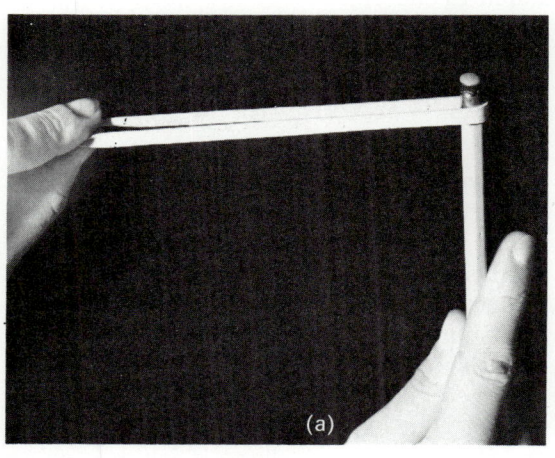

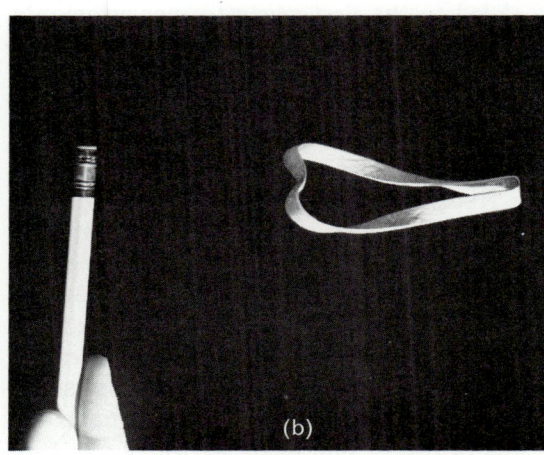

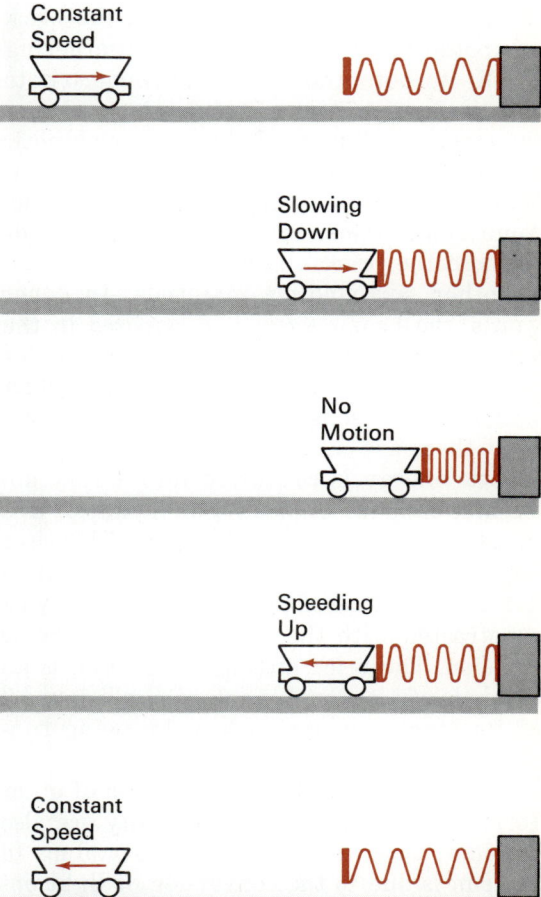

Figure 3–2 *The work done in stretching the rubber band in (a) gives it potential energy. In (b) the potential energy has been converted into kinetic energy.*

Figure 3–3 *An example of an elastic collision. Kinetic energy is converted to potential energy and again to kinetic energy with only a reversal of direction.*

3-3 Thermal Energy and Temperature

One of the most familiar, and most important, forms of energy is heat, or thermal energy. Nearly every physical and chemical process consumes or produces heat. Thermal energy is often the final form into which other types of energy are converted. For example, suppose a brick on a shelf falls to the ground. At first, on the shelf, the brick has potential energy. As it falls, gathering speed, its potential energy is converted to kinetic energy. Then, suddenly, the brick hits the ground and stops. It has neither kinetic energy nor its original potential energy. What has become of the energy? It has become heat, which, in turn, rapidly spreads through the brick, the ground, the air, and all other surrounding materials. In other words, the heat energy is *dissipated* in the surroundings. In this case the quantity of heat would be too small for you to detect without very sensitive instruments. However, there are cases in which mechanical or kinetic energy is converted into a very noticeable amount of heat. For instance, if a motorist has to "slam on the brakes" his tires and brake linings become very hot. Incidentally, the case of the falling brick may be constrasted with the example of the elastic collision. The brick hitting the ground is an example of a perfectly **inelastic** collision. All of its kinetic energy is lost, by being converted into heat.

Examples of the direct conversion of other forms of energy to thermal energy are also familiar. The warmth you feel standing in the sun is due to the conversion of light energy into heat energy. Electric lights and electric heaters both illustrate the conversion of electrical energy into heat. The electric light is designed to transform as much of the electrical energy as possible into the form of light, but some heat is also produced. In the case of the heater, most of the electrical energy is converted to heat but some light is also generated.

The amount of thermal energy a body contains determines its temperature. The more heat a body contains, the higher its temperature. A rise in the temperature of an object is an indication that thermal energy is being added to the object.

3-4 Units of Energy and Temperature

A convenient way of measuring energy is to convert the energy to heat, if it is not already being produced in that form, and let the heat be absorbed by water. The rise in temperature of the water can be measured, and from this it is possible to calculate the quantity of thermal energy that the water absorbed. You will measure energy in this way in your experiments. The use of water and its temperature changes for measuring heat led to the first accurate measurements of energy. Scientific units of energy and temperature were therefore first based on the properties of water.

In order to obtain a convenient unit of temperature, the temperature range from the freezing point to the boiling point of water, under very exact conditions, was divided into 100 parts. Each of these was called a Celsius (or centigrade) degree. When we also assign to the freezing point of water a temperature of zero, we have what is called the Celsius, or centigrade, temperature scale.

The amount of heat required to raise the temperature of one gram of water one Celsius degree is reasonably constant between 8°C and 80°C. This amount of energy has been defined, again using precise conditions of measurement, as the **calorie.**

Benjamin Thompson, Count Rumford (1753–1814)

BENJAMIN THOMPSON, who took the name of Count Rumford when he was named a Count of the Holy Roman Empire in 1791, is chiefly known for his work on the relationship between heat and mechanical energy, but was a man with many other scientific accomplishments. Morever, he led a varied and colorful life.

Born in Massachusetts in 1753, he decided to support the British cause in the American Revolution, but left America for England when the British were driven out of the Boston area. Spending most of his life in London, Munich, and Paris, Rumford divided his time between government service and scientific research on diverse subjects. These subjects included the explosion of gunpowder, better methods of signaling at sea, the development of smokeless chimneys, the measurement of light intensity, and many problems involving heat.

For a number of years Rumford held several high positions in the Bavarian government at Munich. As Inspector General of Artillery for the Bavarian Army, he was responsible for the manufacture of cannons and guns. He frequently observed the boring of brass cannons at the Munich arsenal and was amazed at the great amount of heat acquired by the guns during the boring process. Because the heat generated by the boring appeared to be inexhaustible as long as the boring continued, Rumford began to suspect that heat was not some material fluid that could flow in and out of bodies as many eminent scientists believed. Lavoiser had even named this fluid the *caloric* in 1787.

In order to confirm his suspicions, Rumford first used one of the best balances in Europe to prove that ice did not gain in weight when it melted to form water. Following this, he set up a lathe driven by two horses. The lathe was used to turn a blunt tool inside a metal cylinder surrounded by a known amount of water. After $2\frac{1}{2}$ hours of continuous turning, the heat produced by the friction caused the water to boil, demonstrating a clear relationship between mechanical energy and heat. From his experiments he concluded that heat could not possibly be a material substance but must be "nothing but a vibratory motion among the particles of the body."

Since, as you have learned, all forms of energy are interchangeable, use of the calorie is not restricted to the measurement of heat energy alone. It can be used to measure the amount of energy in any form, and scientists have long used it for that purpose. Recently, in order to aid the communication of scientific results, a new system of units has been adopted by the international scientific community. The International System of Units (SI) includes neither the calorie nor the Celsius scale.

The standard international unit of energy is the **joule** (JOWL), abbreviated **J**. Its basic definition, which we need not discuss here, is in terms of certain electrical units. One calorie equals 4.184 joules.

The standard international unit of temperature is the **kelvin,** abbreviated **K**. A kelvin is equal in magnitude to a Celsius degree, but the temperature scale (that is, where one starts with zero) is different, for a reason that you will learn in Chapter 6. Because the kelvin and the Celsius degree are equal in size, scientists still find it convenient to use the Celsius scale to state temperatures in many cases, and we shall use Celsius temperatures most of the time in this book. Your laboratory thermometers are all calibrated on the Celsius scale.

Similarly, although the joule is the official international unit of energy, scientists still use the calorie as a practical unit of measurement. It is especially convenient when using the temperature rise of water to determine the amount of thermal energy that it has taken up. In this book we shall use both the calorie and the joule as energy units. The calorie will be used in several experiments and in many discussions of the energies of chemical reactions. The joule will be used in some discussions in order that you do not forget about it.

Pioneers in the Measurement of Temperature

THE FIRST SERIOUS ATTEMPT to develop a device for measuring temperature was made in 1593 by the famous Italian astronomer and scientist Galileo, who invented an instrument called a thermoscope. The thermoscope was not very accurate, but its creation led other scientists to explore better ways of doing the job. A fairly accurate thermometer using alcohol was made in 1641.

The most outstanding work, however, was performed by Gabriel D. Fahrenheit in the early 1700's. Fahrenheit, who had been born in Poland, spent most of his life in the Netherlands, where he established a precision mechanics shop for constructing and repairing scientific instruments. In experimenting with both alcohol and mercury thermometers, he recognized the need for establishing fixed points that could be readily reproduced for developing a satisfactory temperature scale. He found that a mixture of equal parts of ice and salt produced the lowest temperature that he could achieve in his laboratory. After assuming that the temperature of human blood was the same for all people, he developed a scale that divided the range between the temperature of his salt-ice mixture and that of human blood into ninety-six equal steps. On this scale, he also found that water boiled at 212 degrees and froze at 32 degrees. Since these two points were much easier to reproduce than his earlier points, he decided to base his scale on the freezing and boiling points of water. The fact that his temperature for human blood was

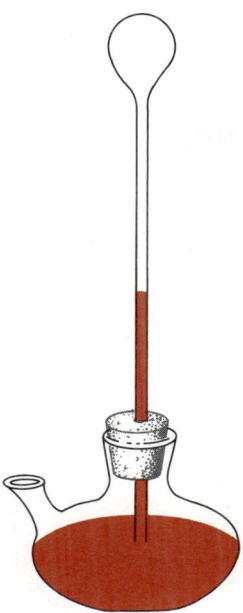

Galileo's thermoscope. The glass bulb was warmed in the hand and the open end of the tube was then placed in the water. As the bulb cooled, water rose in the tube.

William Thomson, Lord Kelvin, first proposed the internationally recognized temperature scale that bears his name.

lower than today's accepted value was probably due to a slight irregularity in the bore of his thermometer tubes.

About thirty years later, Anders Celsius, a Swedish astronomer who had built the observatory at Uppsala University, suggested that it would be simpler to use a temperature scale divided into one hundred degrees between the freezing and boiling points of water. For many years his scale was called the centigrade scale until an international conference of scientists in 1948 named it the Celsius scale in honor of its inventor. The Kelvin scale, named in honor of William Thomson, Lord Kelvin, uses the same size degrees as the Celsius scale but assigns a value of zero to a point 273 degrees below the freezing point of water.

INVESTIGATION

3-5 The Specific Heat of Metals

The **specific heat** of a substance is the amount of heat needed to change the temperature of one gram of that substance one Celsius degree. You have seen that the amount of heat needed to change the temperature of one gram of water one Celsius degree is one calorie. The specific heat of water is one calorie per gram per Celsius degree.

You may reasonably wonder whether other substances will also require one calorie of heat to change the temperature of one gram one Celsius degree or whether their specific heats may vary widely.

There probably have been times when you have cooled off a hot object by putting it into cool water. You have been satisfied that the hot object became cool, but were not concerned with the temperature of the water that was used to do the cooling. In this experiment you will place a hot piece of metal of known mass and temperature into some cool water of known mass and temperature. As the metal cools, the water will become warmer. Knowing the mass of the water, and that one calorie will raise the temperature of one gram of water one Celsius degree, it will be easy for you to find how many calories of heat passed from the metal into the water.

PURPOSE

In this experiment you will measure and compare the specific heats of several metallic elements.

MATERIALS

Ringstand, ring
Wire gauze
Burner
Graduated cylinder, 100-ml
2 beakers, 250-ml
Aluminum, copper, iron, and lead samples

Thermometer
Utility clamp
Balance
Heavy-duty string or thread
Paper towel

PROCEDURE

You will measure the specific heats of four metals: aluminum, copper, iron, and lead. The method that you will use will be the same for each of the metals.

1. Determine the mass of a piece of metal to the nearest gram.

2. Tie a *strong* piece of thread or string around the piece of metal. Hang the piece of metal from a clamp above a 250-ml beaker about half full of water. Lower the clamp until the metal is completely

Figure 3-4

submerged in the water but does not touch the bottom. The same clamp can be used to hold a thermometer. (Figure 3-4 shows the experimental set-up.) Bring the water to a boil and keep it boiling until you are ready to remove the metal. Read the temperature of the boiling water.

3. Use a graduated cylinder to measure accurately 100 ml of water. (Read "How To Use a Graduated Cylinder.") Pour this water into another 250-ml beaker, placed on a piece of folded paper towel to insulate the beaker from the desk top. Since one milliliter of water has a mass very close to one gram, you can assume for this experiment that the mass of 100 ml of water is 100 grams.

4. Measure the temperature of the 100 grams of water. Try to estimate the temperature to the nearest 0.1 Celsius degree. For best results, the water temperature should be about 3-4 degrees below room temperature.

5. Remove the piece of metal from the boiling water and quickly lower it into the 100 grams of cool water.

How To Use a Graduated Cylinder

You should use a graduated cylinder to measure the volume of a liquid when your experiment requires a moderate degree of accuracy. If more accurate measurements are needed, you should use a *buret* or a *pipet*. By comparing the graduations on several different size cylinders, you will find that it is possible to measure volumes to the nearest 0.1 ml with most 10-ml cylinders, while you can read a 100-ml cylinder only to the nearest 1 ml. To insure the most precise measurements, you should select a graduated cylinder with a capacity not much different from the volume you wish to measure. It would be unwise to use a 100-ml cylinder to measure a volume of 9 ml, if a smaller cylinder is available.

Most liquids form a curved surface at the top when they are poured into a cylinder, as shown in Figure 3-5. You will notice that

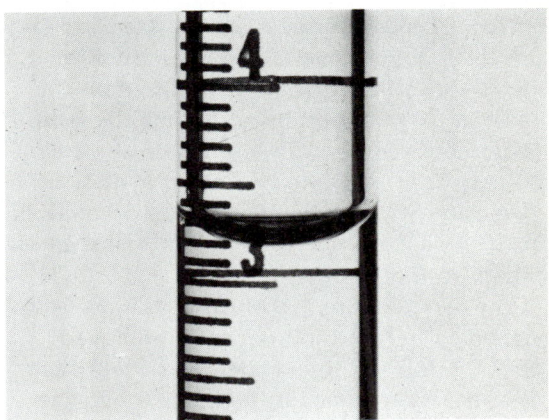

Figure 3-5

6. Carefully stir the water around the piece of metal and read the thermometer to the nearest 0.1 degree when the temperature stops rising.

7. Remove the metal from the beaker and pour out the water. Refill the beaker with a fresh 100 ml of water and repeat the experiment with another piece of metal.

DATA AND RESULTS

Tables for recording the following data and results should be written in your notebook before you come to the lab.

---

the outer edges of the surface are slightly higher than the center of the surface. This curved surface is called the **meniscus** and is caused by the attraction between the liquid and the walls of the container.

When you measure the volume of a liquid in a cylinder, you should be sure that your eye is at the same level as the meniscus. You then read the volume by noting the line on the cylinder that is directly opposite the *bottom* of the meniscus as illustrated in Figure 3-6. If your eye is above or below this level, your reading will not be accurate. In cases where the bottom of the meniscus appears between two of the lines on the cylinder, you can estimate how far it is above the lower line. With a cylinder that is graduated in one-milliliter intervals you can make an estimate to the nearest 0.1 ml.

The temperature for which a cylinder is designed to be used is frequently engraved near the top of the cylinder. The most accurate measurements can be made when the temperature of the liquid in the cylinder is close to this temperature.

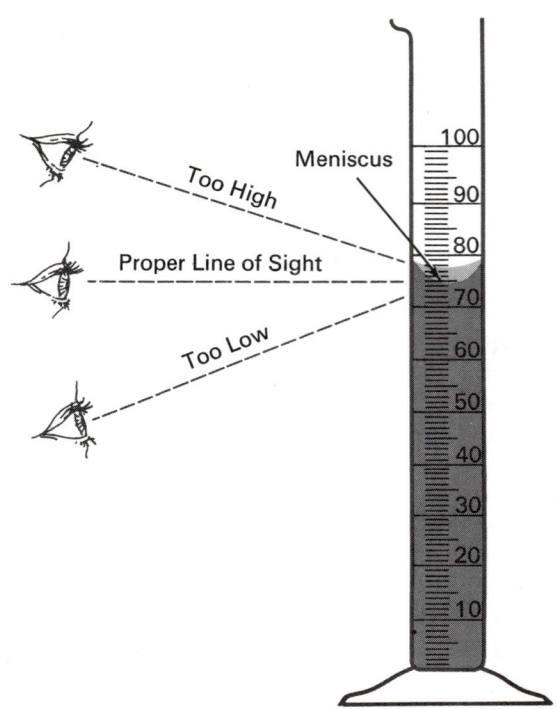

Figure 3-6

Type of metal	Mass of cool water (g)
Mass of metal (g)	Final temperature of cool water and metal (°C)
Temperature of boiling water (°C)	Starting temperature of cool water (°C)
Final temperature of cool water and metal (°C)	Temperature gain of cool water (°C)
Temperature loss of metal (°C)	

Calculations

1. A gain of one calorie of heat raises the temperature of *one* gram of water one Celsius degree. Hence, to find the heat gained by m grams of water, undergoing a temperature change Δt, simply multiply $m \times \Delta t$. (The Greek letter *delta*, Δ, is often used to mean "a change in" something.)

2. The heat lost by the metal is the same as that gained by the water, assuming that there are no other significant heat transfers during the experiment. Hence, to find the heat loss per *gram* of metal per *degree* (its specific heat) divide the answer from step one by the mass of the metal and by its temperature change.

3. Record your specific heat for each metal as well as the class average specific heat for each metal.

RELATED PROBLEMS

1. Try to propose a theory that would explain how heat is transferred from a hot piece of metal to cool water.

2. In this experiment we did not take into account the heat needed to warm the beaker containing the cool water, since it was relatively low in comparison to the heat absorbed by the water. Where else might there be sources of error in the experiment?

Summary

There are many forms of energy. One of the most important is heat, or thermal energy. Other important forms of energy are light and electrical energy. Kinetic energy is the energy of a moving object due to its motion. Potential energy is a kind of energy that may not be noticeable at a certain time, but which can appear in some other form. A coiled spring, a stretched rubber band, or an object resting in a high place all have potential energy that can be converted into kinetic energy. The temperature of an object is a measure of how much thermal energy it contains. When a body gains thermal energy, its temperature rises, unless the body is undergoing some type of transformation, such as a change from liquid to solid. An important scientific temperature scale is the centigrade or Celsius scale, in which the temperature difference between the freezing and boiling points of water is divided into 100 degrees and the freezing point is called 0°C. The official international unit of thermal and other forms of energy is the joule. A useful unit for laboratory work is the calorie, which is approximately the amount of heat required to raise the temperature of one gram of water one Celsius degree at room temperature.

Self Check

1. For any three forms of energy, think of a sequence of events that might occur in everyday life, involving the conversion of one of these forms into each of the others.
2. How many calories would be required to raise the temperature of ten grams of water from 15°C to 25°C?

Laws Govern Energy Changes

3-6 *Conservation of Energy*

You have seen that energy can take many forms and that these forms are interconvertible. For nearly 150 years chemists and physicists have carried out experiments in which they have measured the amount of one kind of energy that could be obtained from another. They have consistently found that the total amount of energy involved stays the same. One form of energy, such as electrical energy, may be completely converted into another form, such as light energy. Or a single form of energy may produce more than one new form. Remember that the electrical energy used in a light bulb, for example, produces heat as well as light energy. Whatever the energy change, or changes, the total amount of energy does not change. It changes in form, but not in amount. Furthermore, energy has never been observed to appear spontaneously, that is, to come out of nowhere.

Scientists summarize their experience with energy in a law, called the **Law of Conservation of Energy.** (It is also known as the First Law of Thermodynamics.) This law states that:

Energy can be neither created nor destroyed. When energy changes form, the total amount of energy is conserved.

This law is one of the most important discoveries of science. It is believed to be true because every attempt to prove it false has failed.

The First Law of Thermodynamics, like the Law of Conservation of Mass, is a generalization. You should recall the experience of the student who generalized about fuels. He found that a generalization may be incorrect if the range of observations on which it is based is too limited. The First Law of Thermodynamics has been checked against such an enormous range of observations, so often, by so many people, that we have great confidence in it. Nevertheless, it does not apply in conditions extremely different from those in which it was developed and tested. Neither does the Law of Conservation of Mass. Certain processes, chiefly involving atomic nuclei, are governed by an even more basic law, which combines these two. It is called the Law of Conservation of Mass *and* Energy, and we will explain and discuss it later. It should be stressed that, in all chemical processes, mass and energy are each conserved separately, as stated in the two separate laws.

3-7 Energy and Work

We all know what work means in the ordinary sense of the word. It means effort put forth to do a task. Energy is expended, and we are tired when we finish. The exact scientific definition of the word is not very different in meaning, but it is more precise. In science, the amount of work done is defined as the magnitude of a force times the distance over which it is applied. **Work** is force times distance. The units of work are the same as those of energy, calories or joules. Work and energy are closely related, but they are not exactly the same thing. The relationship is that it takes energy to do work, and work can generate energy.

Suppose a boulder lies at the bottom of a hill and a bulldozer is used to push it to the top. If the dozer puts a certain continuous force, F, on the boulder to keep it moving up the slope, and moves it a distance d, the work done by the dozer is equal to $F \times d$. The dozer is able to do this work because its engine burns gasoline, creating thermal energy. The thermal energy is converted into the kinetic energy of the moving dozer and the boulder in front of it. Some of this energy is converted into heat (and noise, also a form of energy). Some is converted into the gravitational potential energy that the dozer and the boulder have gained in going to the top of the slope. If the boulder is allowed to roll back down the slope again, its potential energy will be converted partly into kinetic energy and partly into heat. The heat is produced by friction as the boulder rolls. Eventually the boulder will come to a stop, when all of its kinetic energy has been converted into heat. It leaves a trail of heat that is soaked up in the surroundings.

The most common reason for our interest in energy is a very practical one. We can use it to do work instead of having to do it with our own muscles. We use a source of energy, such as an internal combustion engine, an electric generator or battery, or water falling from a high place to a lower one, and try, directly or indirectly, to do some useful work with the energy.

A hundred years ago the potential energy of water approaching a waterfall was used fairly directly. The kinetic energy of the falling water was converted into the kinetic energy of a water wheel, which, in turn, drove some piece of machinery. Today, natural waterfalls, or those created by dams, drive electrical generators. The electrical energy is sent to other places, where it activates electric appliances, doing work.

3-8 Energy Tends To Be Dissipated Spontaneously

Whenever there is a large amount of energy concentrated in one place, it tends to spread into the surroundings. An obvious example of this is that any body that is warmer than its surroundings spontaneously gives off heat to the surroundings. It continues doing this until its temperature has become equal to that of its environment. Similarly, a body that has high potential energy because it is at a high altitude will fall spontaneously if not supported, losing potential energy in the process.

All moving objects are subject to frictional forces. Unless they are continually supplied with propulsion energy, they gradually slow down. All of their kinetic energy is eventually given off to the surroundings in the form of heat.

When the positive and negative terminals of a battery are connected by a conductor of electric current, a spontaneous flow of current takes place. Eventually, all of the energy that was originally concentrated into the small volume of the battery is dissipated into the environment in the form of heat. The battery itself and the conductor connecting the terminals both get warm as current passes, and this warmth is given off to the surroundings.

In chemistry, the tendency for energy to be given off is an extremely important factor. Nearly all chemical processes that occur spontaneously give off energy. You have seen a typical example as you watched the candle burn. It is true that this is not a spontaneous process in the sense that it will start by itself. The candle must first be lighted. But once lighted, and kept supplied with oxygen, the candle continues to burn with no further stimulation. The process of combustion is, in that sense, spontaneous.

3-9 Randomness or Disorder Is Also Important

Besides the tendency to lose energy, there is a second factor that helps to determine whether something will happen of its own accord. This factor is the natural tendency for disorder, or randomness, to increase.

Chemistry and Our Environment
THERMAL POLLUTION

The production of energy is necessary to sustain an advanced technological society. Whatever its initial form, all of the energy that is produced finally ends up as thermal energy, which must be absorbed by the environment.

Until very recently, the amount of thermal energy resulting from industrial energy production was much less than that already in the earth and that being constantly received from the sun. Today, however, such large amounts of thermal energy are added to the environment that the balance of nature is upset. This effect is known as thermal pollution.

The most conspicuous sources of thermal pollution are plants that generate electricity. Such plants either burn coal or oil or use nuclear processes (which you will learn about in Chapter 10) to generate heat. The heat energy is then used to power a motor or turbine that drives an electric generator, thereby producing electricity. It is impossible to convert all of the heat energy into electrical energy, however. Scientists know that even the most efficient machine that could be designed could not completely convert thermal energy into some other form of energy, such as electricity. At least some of the thermal energy *must* leak away as heat, even under the most ideal circumstances. In fact, the type of machinery that exists today converts less than half of the thermal energy into electrical energy. The rest is given off as heat in the immediate vicinity of the power plant. Eventually, *all* of the electrical energy also becomes converted into heat energy. The heat produced from the electricity creates no local impact because it is distributed over a wide area.

Two types of problems can arise as a result of thermal pollution. The first one occurs in

These twin hyperbolic cooling towers at the Rancho Seco Nuclear Station in California dissipate waste heat by recycling the cooling water at a rate of a half-million gallons per minute.

the immediate region of the generating plant. Usually water from a lake, a river, or an ocean bay is taken in to absorb the excess heat and then discharged back into its source. As a result, the water temperature near the plant can rise significantly, especially in lakes of small to medium size. Many fish and other forms of aquatic life can live within only a relatively narrow range of temperatures. Even small temperature changes can be as devastating as the introduction of a poisonous material. Thus, thermal pollution can alter the relationships between the organisms and their environments, that is, upset the ecological balance. The location of new power plants and the plans for cooling them must always be considered very carefully. Almost any body of water might be good for a power plant, but the power plant might not be good for the ecology of the body of water.

The second type of problem created by the tremendous, and still rapidly growing, rate of power generation involves the whole earth. All the energy produced in all the power plants must be absorbed by the surface of the earth or by the atmosphere, eventually. At what point will this thermal energy begin to affect the climate? There is not enough information to answer this question or to plan future control of our energy use. We need to know much more about how the earth functions, including the chemical and physical processes involved. We must be able to predict future power needs and to assess their effect on the environment.

Thermal pollution is both a short-range and a long-range problem. Young people interested in science and in their environment will find interesting and important work to do in helping to solve this problem.

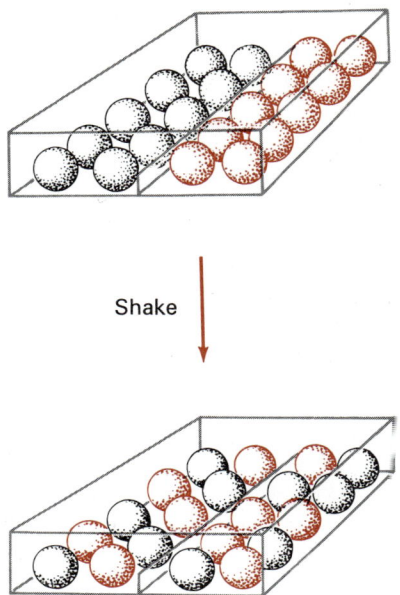

Figure 3-7 *When a box of highly ordered marbles is shaken, order is destroyed, and one of about one million possible disordered arrangements results.*

Suppose you had a box, like the one shown in Figure 3-7, containing 10 white marbles on one side of the partition and 10 black ones on the other. The arrangement of the marbles in terms of color is a very orderly one. It is not a random one. Suppose you now pick up the box and shake it for a while, and then place it on the table again. Soon all of the kinetic energy imparted to the marbles in shaking will have been lost, and they will be at the same height and temperature as before. Thus there cannot have been any net change in energy resulting from the shaking process.

There has been a tremendous change in the orderliness of the arrangement, however. Order has been lost! Randomness is now the characteristic feature of the arrangement.

Each time the box is shaken a different disordered arrangement will result. But there is only one chance in about a million that the completely ordered arrangement with which we started will ever be recovered. The natural tendency is for the originally ordered arrangement to become, and to stay, disordered.

In most chemical processes that proceed spontaneously, there is both a loss of energy *and* an increase of randomness in the way in which the matter exists. That is, both of the factors that favor the occurrence of a spontaneous process are operating in the same direction, mutually supporting each other.

For a case in which both factors are favorable, there can be no question that the process will be spontaneous. For instance, as a candle burns, energy is given off as heat and light. At the same time, the matter that was arranged in an orderly pattern in the candle becomes entirely converted into gases, which spread out, in a disorderly fashion, all over the room. Prediction of results is more difficult when one factor is favorable and the other is not.

3-10 Conservation of Mass and Energy ★

In Section 3-6 it was noted briefly that in processes involving atomic nuclei and other particles that are smaller than atoms, the laws of conservation of mass and energy are not always obeyed separately. Instead, they are obeyed in a kind of combined form. Let us go into that a little more deeply here.

The great theoretical physicist Albert Einstein first recognized that such a thing could happen. He was able to derive from his theory of relativity an exact mathematical relationship between mass and energy. It is the well-known equation: $E = mc^2$.

E represents energy, m mass, and c the speed of light.

Only when enormous amounts of energy are involved is this relationship significant. To see why, let us substitute some numbers into the equation. The speed of light in a vacuum is 3.00×10^8 meters/sec. (Its speed in air is virtually the same.) Let us assume that 1.00 g of matter is to be converted into energy.

$$E = (1.00 \times 10^{-3} \text{ kg}) \times (3.00 \times 10^8 \text{ m/sec})^2$$
$$= 9.00 \times 10^{13} \text{ kg-m}^2/\text{sec}^2$$

Within the system of units we are using, one joule is, by definition, one kg-m²/sec². Hence:

$$E = 9.00 \times 10^{13} \text{ joules.}$$

This is a fantastically large amount of energy. It could supply the electrical energy normally required by about 5000 homes for a year!

The point is that very little matter corresponds to an enormous amount of energy. Or, from the opposite point of view, unless an enormous amount of energy is released, no balance could ever detect the corresponding decrease in mass. For example, when one ton of coal is burned, the amount of energy released is about 3×10^{10} joules. Using the above numbers, you can readily calculate the corresponding loss of mass.

$$m = \frac{E}{c^2} = \frac{3 \times 10^{10} \text{ kg-m}^2/\text{sec}^2}{9 \times 10^{16} \text{ m}^2/\text{sec}^2}$$
$$= 3 \times 10^{-7} \text{ kg} = 3 \times 10^{-4} \text{ g}$$

There is no known instrument that could detect a loss of about a ten-thousandth of a gram in a total mass of about 1000 kg. This is one part in about 10^{10}. A balance capable of detecting such a difference is not a practical possibility.

We say that coal (and the oxygen used to burn it) contain potential energy, which is converted to heat energy when the coal is burned. It may be that this potential energy is really in the form of mass, but we have no way of proving it. Therefore we do not say it. As far as experiments can tell us, the mass remains unchanged in all ordinary chemical and physical processes. So does the sum total of all the forms of energy, provided we use the idea of potential energy. Hence, we are justified in using separate conservation laws for mass and energy.

In certain processes involving the atomic nucleus, the amounts of energy released per gram of matter are so great that instruments *can* detect the change of mass. In Chapter 10 we shall examine such processes.

Summary

One of the most important laws of chemistry states that energy can be neither created nor destroyed. It can only be changed from one form into another. This is called the Law of Conservation of Energy. When a force acts on an object and moves it through a distance, work is done. Work may be converted into energy, and vice versa. Therefore work can be measured in the same units, joules or calories, as energy. There is a natural tendency for energy to be dissipated spontaneously. There is also a natural tendency for any orderly arrangement, left to itself, to lose order. Processes in which either or both of these things happen can occur spontaneously.

Self Check

1. State in your own words the Law of Conservation of Energy.

2. Why is this called a law? Why do scientists believe it to be true?

3. What two things are characteristic of spontaneous processes?

Chapter Summary

Energy may take a variety of forms. It may be changed from one form to another without loss. In many cases energy is at least partially converted into thermal energy, which is dissipated in the surroundings. Important forms of energy are heat (thermal energy), light, electrical energy, kinetic energy, and various forms of potential energy. Work is defined by scientists as force times distance. Energy can do work, or work may be done to build up energy. Therefore, work and all forms of energy are measured in the same units. These units can be defined in terms of heat and temperature.

The temperature of a body increases as its content of thermal energy increases. A useful scientific scale of temperature puts the freezing point of water at 0 degrees and its boiling point at 100 degrees. This is the Celsius, or centigrade, scale, and the degrees are called Celsius degrees, or centigrade degrees. Between 8°C and 80°C it takes approximately the same amount of heat to raise the temperature of one gram of water one °C. This quantity of heat, or any other form of energy or work, is called one calorie. Another unit of energy, used internationally, is the joule. There are 4.184 joules per calorie.

Questions and Problems

1. If you held an ordinary rubber ball in your hand at a height of three feet above a smooth concrete floor and dropped it, at what point would the ball possess the maximum potential energy? the maximum kinetic energy?
2. Would you expect the rubber ball (even if it were an extra bouncy "superball") to return to the exact height from which it was dropped? Explain your answer.
3. Water coming over the American side of Niagara Falls drops 167 feet before hitting the rocks at the bottom. Would you expect to find any difference between the temperature of the water at the top of the falls and its temperature after hitting the rocks? Explain.
4. Consider a large moving van and a small compact car moving in parallel lanes at high speed on a highway. What factors do you think determine the amount of kinetic energy that each vehicle has? Which one would have the greater amount of kinetic energy? In what disastrous way might this be demonstrated?
5. Suppose you were given a small, four-wheeled cart, 6-8 inches long, together with a balance, a series of weights of known mass, an inclined ramp about three feet long, and a wooden block that

could be placed on a table top a short distance from the lower end of the ramp. Describe a series of experiments that you might perform to make a rough estimate of the effects of mass and velocity on the kinetic energy of a moving body.

6. An automobile is moving at a speed of 60 miles an hour on a long, flat, straight stretch of highway. Explain what would happen if the engine were put into neutral by disengaging the clutch while the car was traveling at that speed.

7. Consider an electric motor connected to the drive shaft of an electric generator. If the motor is started with electricity from another source, do you think that the electricity from the generator could be used to keep the motor going once the first source is disconnected? Explain.

8. On the basis of your answers to Questions 2, 6, and 7 what would seem to be the biggest obstacle to the complete utilization of mechanical energy?

9. Throughout the centuries many scientists and amateur experimenters have tried to design "perpetual motion" machines that would keep going forever once they were started. Why have none ever been successful? What are the possibilities that such a machine may be invented in the future?

10. Fuel oil produced in a petroleum refinery is frequently pumped long distances through pipes to the location where it is used. Describe all of the energy changes involved in pumping fuel oil two hundred miles from a refinery to a power plant, and then producing electricity that is conducted to a house about twenty miles outside the city limits. Also describe how energy would be lost along the way.

11. How would you explain the difference between heat and temperature?

12. Assume you are given a piece of lead, a piece of aluminum, a piece of iron, and a piece of copper, each having the same mass. Assume, further, that you then heat them to 100°C in a beaker of boiling water. Using a thermometer and four beakers, each containing an equal amount of water, how would you find out whether the pieces of metal contained the same amount or different amounts of heat?

13. In an early experiment to find a relationship between mechanical energy and heat, Count Rumford used two horses to drive a lathe. At a later date, James Watt, a Scottish engineer known for his development of the steam engine, measured the amount of work a single horse could do in one hour. If work is defined as force times distance, how would you find out how much work a horse could do in one hour? (Watt's work eventually led to the definition of one horsepower.)

14. Suppose you had three coins all showing heads. If you shook up the coins and dropped them on the floor, would you expect them to land in a different arrangement? How many different arrangements is it possible to get with three coins? Why does shaking them up, in this case at least, tend to produce a more random array?

15. A certain high school has a student body of about 1500 students. During what times of a typical school day is the motion of the students most random? When may they be found in the most ordered arrangement? Is there any time when the motion of the students is completely random? Explain.

16. You have read that many sources of energy produce thermal pollution in our environment. Describe other forms of pollution created by these same energy sources.

4 Measurements, Units, and Accuracy

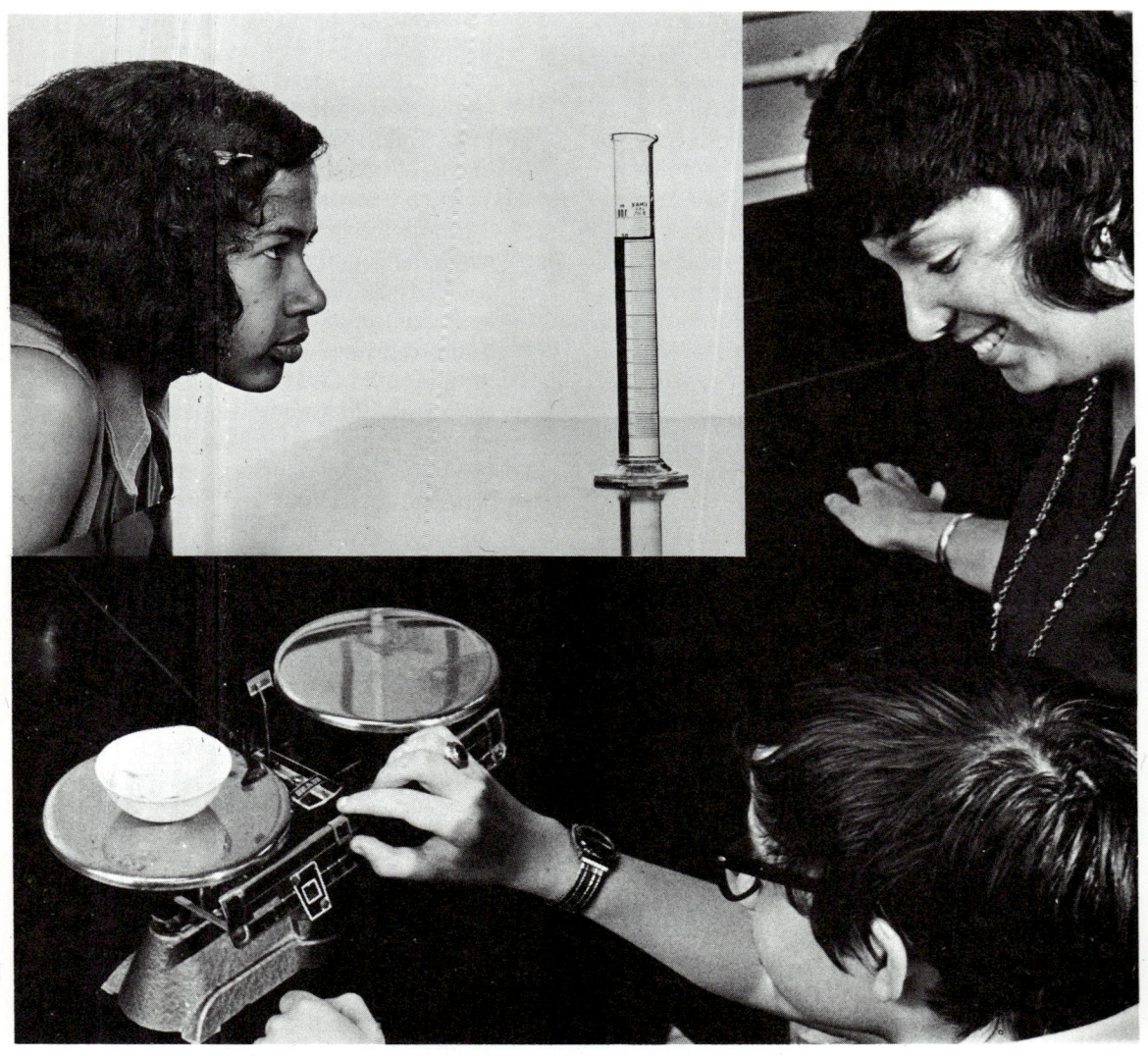

Quantitative measurements are an essential part of chemistry.

Chemists observe the behavior of matter. They study the properties of substances in order to be able to purify them, identify them, and learn how one substance may be transformed into another. Many of a chemist's observations are of a quantitative nature. That is, the magnitude of some property such as temperature, length, or volume is measured and expressed as a number. Careful measurements of mass and energy are often necessary if a chemical process is to be fully understood. In order to make such measurements, the chemist must have a system of units in which to express his data. He must understand how to estimate the accuracy of his measurements and how to express his results so that their degree of accuracy is unmistakable. Finally, he must know how to determine the limit of accuracy of a quantity that is calculated from other, measured, numbers.

Appropriately, experiments have already formed a part of your study of chemistry. It is also appropriate that some attention be given at this early time to the problems connected with measurements, their accuracy, and the units in which they should be expressed.

Units of Measurement

4-1 The Metric System

You will not proceed very far in chemistry before you discover that many investigations involve measurement. You will, for example, have to record changes in length, volume, and temperature. The masses of materials before and after a reaction and the length of time involved in a reaction will also have to be measured.

Obviously then, it is important for you to know how these measurements are usually expressed. You know something about the metric system already. You were introduced to the kilogram and the gram as units of mass and the joule as a unit of energy or work in previous chapters. Now we must discuss the other principal units in the metric system.

The metric system is used exclusively in many countries today. Even in the United States, where it is not in commercial use, the metric system is used in scientific research.

One great advantage of metric units of measurement is that they are based on a decimal system. This makes conversion to many different units simply a matter of shifting the decimal point. A second advantage is that, among scientists, these units are standard in every part of the world.

Let us consider the metric units of length in order to show the advantages of a decimal system. The kilometer is used for the large distances that would be given in miles in the United States. Similarly, the meter is the metric unit corresponding to the foot or yard. The relationship of meters to kilometers is neat and simple.

$$1000 \text{ meters} = 1 \text{ kilometer}$$

Contrast this relationship with the following much more complicated relationships, which are an unnecessary burden to your memory.

$$5280 \text{ feet} = 1 \text{ mile}$$
$$1760 \text{ yards} = 1 \text{ mile}$$

The meter is the basic unit of length in the metric system. It was originally intended to be one ten-millionth of the distance from the equator to the north pole. This may seem to be an arbitrary choice, but it is not nearly so arbitrary as are most nonmetric units. They were defined in terms of such variable quantities as the size of someone's foot, a man's pace (the yard), and the distance from a man's elbow to the tip of his middle finger (the cubit).

Even so, the very precise measurement of the meter in terms of a fraction of the earth's surface presents practical problems. Therefore, the standard of length first employed was a metal bar. This standard meter bar was made by French scientists and kept, in France, under the most scrupulous care regarding temperature and environment. A number of exact copies of the bar have been carefully made. One of them, shown in Figure 4-1, is kept in the United States by the National Bureau of Standards. On the basis of this standard, the meter is a defined unit, like the kilogram. In 1960, scientists switched to a *natural* definition of the unit of length, based on the light emitted by the gas called krypton.

4-2 Comparing Metric and More Familiar Units

A meter (m) is slightly longer than a yard, being 39.37 inches. For many measurements in the laboratory that require a much smaller unit, the centimeter (cm) is used. The centimeter is $\frac{1}{100}$ of a meter or 0.39 of an inch. When an even smaller unit is required, the millimeter (mm), $\frac{1}{1000}$ of a meter, can be used. A millimeter is, of course, $\frac{1}{10}$ of a centimeter.

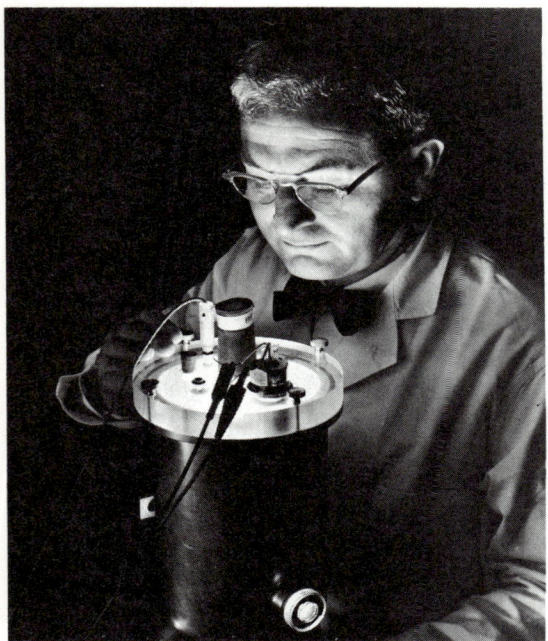

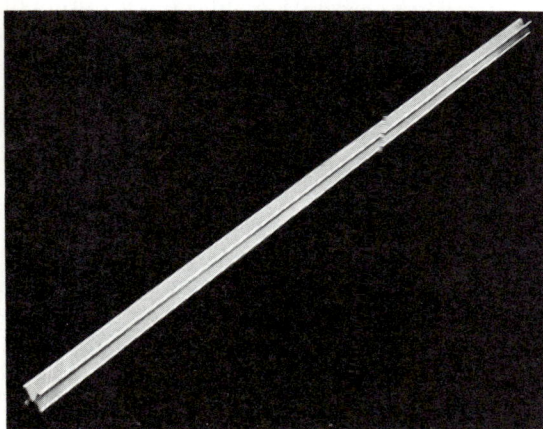

Figure 4-1 *The scientist in the top photo is observing light given off by the gas krypton. The wavelength of this light is the new standard of length. The meter bar (bottom photo) is an exact copy of the former length standard.*

Centimeters, millimeters, and inches are contrasted in Figure 4-2.

There are other units of length, but the three just given are sufficient for your purposes. It is not important to remember the exact relationship between the metric units and the familiar inches, feet, and yards. Until you acquire from experience a feeling for the size of each metric unit, it will help if you remember roughly how the two sets of units compare. (Some familiar measurements are contrasted with their metric equivalents in Figure 4-3.)

For practice, you should measure the widths and lengths of several small objects using both the metric system and the units

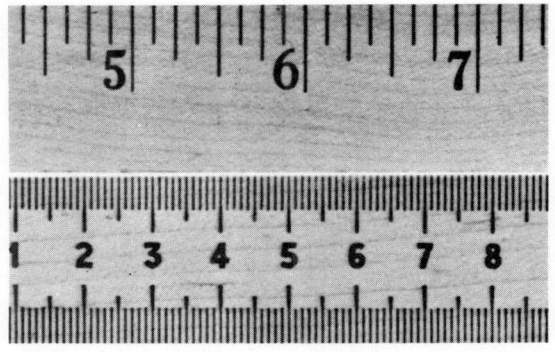

Figure 4-2 *The numbered divisions on the meter stick are centimeters, each of which is divided into millimeters. The numbered divisions on the yardstick are inches.*

Figure 4-3 *The record for the long jump (left photo) in 1970 was 8.90 meters, or 29 feet, 2½ inches. The 100-yard dash (right photo) is nearly 9½ yards shorter than the 100-meter dash.*

used in the United States. In doing so, you will notice the ease with which metric units, expressed as decimals, can be used in a variety of computations. This is especially evident when decimals are contrasted with fractional notation.

Suppose you measured the dimensions of a small rectangular block with an ordinary ruler and got the following results.

$$\text{Length} = 4\tfrac{3}{4} \text{ inches}$$
$$\text{Width} = 3\tfrac{7}{8} \text{ inches}$$
$$\text{Depth} = 2\tfrac{1}{2} \text{ inches}$$

Calculate (1) the perimeter of the largest face, and (2) the volume of the block. The answers are: (1) $17\tfrac{1}{4}$ inches, (2) $46\tfrac{1}{64}$ cubic inches.

The dimensions of the same block, expressed in metric units, follow.

$$\text{Length} = 12.1 \text{ cm}$$
$$\text{Width} = 9.8 \text{ cm}$$
$$\text{Depth} = 6.4 \text{ cm}$$

Now calculate (1) the perimeter of the largest face, and (2) the volume of the block, both in metric units. The answers are: (1) 43.8 cm, and (2) 759 cubic cm.

Which set of calculations was easier to do?

The most commonly used unit of volume in the metric system is the liter (LEE-ter). It is equal to 1.057 quarts. For smaller volumes, the milliliter (ml), is used. It is equal to $\tfrac{1}{1000}$ of a liter. In other scientific books you may see small volumes expressed in cubic centimeters, cm^3 (or cc). It was originally intended, when the metric system was set up, that one ml would be identical with one cm^3. There is, in fact, a slight discrepancy. For all but very precise work, the difference is unimportant. The milliliter is the small unit of volume used in this book.

In order to acquire a direct sense of how large these units are, you should measure the capacity of a few familiar vessels in liters and milliliters. An average coffee cup, for example, will be found to hold about 250 ml, or 0.25 l. (See also Appendix 7.)

You have already seen that the units of mass are the kilogram and the gram. A gram (g) is a fairly small unit. (One ounce is equal to 28.35 g.) It is a convenient size for laboratory work and will often be used in this book. In many scientific articles you will find reference made to an even smaller unit, the milligram (abbreviated mg), which is $\tfrac{1}{1000}$ of a gram.

4-3 Density

A characteristic property of every pure substance is its density. **Density** is the quantity of matter in a given unit of volume. In the metric system, density is most often expressed as the number of grams per milliliter, abbreviated g/ml. This combination, g/ml, may itself be considered a unit, a unit of density. This particular unit of density is extremely common. It is used by scientists for liquids and solids without exception. For gases, which are much less dense, the unit grams per liter, g/l, is often used.

The unit of density is our first *derived*, or *secondary*, metric unit. A secondary unit is one made up of two (or more) of the primary or basic units. In everyday life one of the most common derived or secondary units is that used to measure speed. In the United States it is miles per hour. In the metric system speed is stated in meters per second, or kilometers per hour.

Since density is expressed in terms of two primary units, grams and milliliters, two measurements must be made in order to determine it. You should have a chance to do some measurements from which densities can be determined. The following section provides suitable ways for doing this.

INVESTIGATION

4-4 Measurement of Density

Density is an important property of matter. Since different types of matter usually have different densities, the measurement of the density of an unknown substance can help in determining the identity of the material. In addition, the density of a solution depends on how much material is dissolved in the solution. It is frequently possible to measure the density of a solution and find the amount of dissolved material in it by referring to tables showing the densities of mixtures of known composition.

Density may be determined in several different ways. In one method, involving regularly shaped objects, you can first find the mass of an object by using a balance. To find the volume of the object you can measure its dimensions and calculate its volume using standard formulas. For example, the volume of a rectangular solid is equal to length times width times height. The volume of a cylindrical solid equals π times the square of the radius times the height. Using the general formula, density $= \dfrac{\text{mass}}{\text{volume}}$, or $d = \dfrac{m}{V}$, you can then calculate the density.

For an irregularly shaped solid that will not dissolve in or react with a liquid, it is possible to find the volume of the solid by immersing it in the liquid. Since a solid object will displace its own volume of a liquid, you can lower the object into a graduated cylinder containing a measured volume of liquid. The difference between the new level of the liquid in the cylinder and the original level will equal the volume of the solid.

In your buoyancy effect experiment you discovered that the apparent loss in mass of an object immersed in a liquid is equal to the mass of the liquid that it displaces. This is frequently referred to as Archimedes' principle. The same principle can be used in finding the density of an object. The mass of an object is first found by weighing it in air on a balance. The mass of the liquid it displaces can then be found by weighing the object while it is immersed in a liquid. The difference between the two weighings represents the mass of the liquid displaced. To find the volume of the displaced liquid, which is the same as the volume of the immersed object, you start with the same formula for density that is shown above, density $= \dfrac{\text{mass}}{\text{volume}}$. This formula can be rewritten as follows:

$$\text{volume} = \dfrac{\text{mass}}{\text{density}}, \text{ or volume of liquid displaced} = \dfrac{\text{mass of liquid displaced}}{\text{density of liquid displaced}}.$$

Figure 4-4a

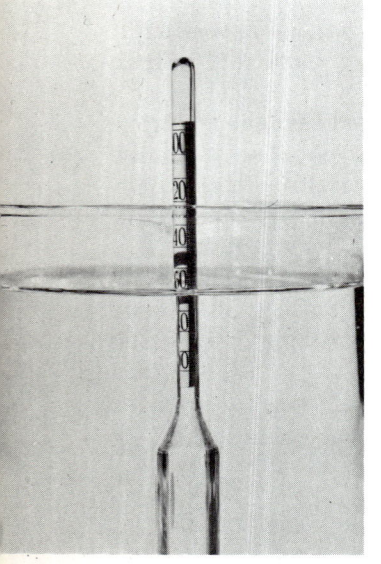

Figure 4-4b

Knowing that the volume of the liquid displaced is equal to the volume of the solid, you can now calculate the density of the solid.

In each of the above illustrations a method for finding the density of a solid has been described. There are several ways to find the density of a liquid. It is quite easy to find the mass of a known volume of a liquid by weighing the liquid in a graduated cylinder of known mass and calculating the density directly from your measurements of the mass and the volume.

A very simple device, known as a hydrometer, can also be used to determine the density of a liquid. A **hydrometer** is a sealed tube with a weight in the bottom, as shown in Figure 4-4a. The upper part of a hydrometer contains a scale. A hydrometer will float higher in a dense solution than it will in a less dense solution. The density of a solution can be found by reading the hydrometer scale where the surface of the liquid crosses it as in Figure 4-4b.

A hydrometer does not measure the density directly, but is designed to measure a value known as the "specific gravity" of the liquid. **Specific gravity** is defined as the ratio of the mass of an object to the mass of an equal volume of water, or as the ratio of the density of a substance to the density of water.

It is important to realize that the volumes and densities of all substances change with temperature. At 4°C one milliliter of water weighs 1.0 gram. At 25°C one milliliter of water weighs 0.997 gram. If the hydrometer that is being used is designed to measure the specific gravity of a solution based on the density of water at 4°C, then the specific gravity reading is numerically equal to the density.

PURPOSE

In this experiment you will measure the density of a solid and of a liquid, using direct and indirect methods.

MATERIALS

Metric ruler (or calipers)
Graduated cylinder, 100-ml
Graduated cylinder, 25-ml
Beaker, 250-ml
Ringstand, ring
Rectangular or cylindrical piece of metal
Balance, centigram (platform balance if necessary)
Wire gauze
Thread
Salt or alcohol solution
Hydrometer

PROCEDURE

Section I. Density of a Solid

Part A. Direct Measurement of Mass and Volume

1. Before starting this experiment prepare data tables like those suggested in "Data and Results."

2. You will be given a rectangular or cylindrical piece of metal. Find its mass by weighing it on a balance to the nearest 0.01 gram. Record the result in your table.

3. Measure the dimensions of your solid. A metric ruler will be satisfactory, although other more precise instruments such as calipers may be used if they are available. Try to estimate your measurements to the nearest 0.1 millimeter, if possible. If you have a rectangular solid, measure its length, width, and height. If the solid is a cylinder, measure its height and diameter. Record all measurements.

4. Use the proper formula to calculate the volume of the solid.

5. From your measurements of mass and volume find the density of the solid and record your result in your table.

Part B. Determination of Volume and Density by Displacement of Water

1. Use the same piece of metal that you started with in Part A.

2. To find the volume of your piece of metal, first fill a graduated cylinder about half full of water. Record the initial volume of water in the cylinder.

3. Tie a piece of thread around the metal and carefully lower the metal into the cylinder of water. Record the final volume occupied by the water and the metal in the cylinder.

4. Find the volume of the piece of metal, which is the same as the volume of water displaced, by subtracting the initial volume in the cylinder from the final volume.

5. Using the volume found in the previous step, and the mass of the metal found in Part A, calculate the density of the solid. Record the result in your table.

Part C. Density by the Buoyancy Effect

1. Using the thread tied to your piece of metal, hang the metal from the hook above the pan of a triple beam balance, as in Figure 2-11a.

(Be sure that the metal has been dried off after Part B.) Determine the mass of the metal by weighing it in air. This should be the same mass you found in Part A.

2. Use either the platform support on the balance or a ring on a ringstand to raise a beaker of water under the piece of metal in such a way that the metal is completely immersed in the water. (See Figure 2-11b.) Record the apparent mass of the metal immersed in water.

3. Subtract the apparent mass of the metal immersed in water from its mass in air. This gives you the mass of the water displaced.

4. The volume of water displaced equals the mass of water displaced divided by the density of the water. The density of water at room temperature is 1.0 gram per milliliter. Thus, the volume of water displaced, in milliliters, is numerically equal to the mass displaced, in grams. This, of course, is equal to the volume of the metal. Record this result in your table.

5. From the mass of the object in air and the volume of the object calculate its density.

Section II.   Density of a Liquid

Part A.   Direct Measurement of Volume and Mass

1. Determine the mass of a clean, dry 25-ml graduated cylinder to the nearest 0.01 gram by weighing it on a balance.

2. Pour about 15 ml of a solution assigned to you by your teacher into the graduated cylinder. Read and record the volume of the liquid as accurately as you can. Try to estimate your readings to the nearest 0.1 ml.

3. Find the total mass of the cylinder and the liquid it contains by weighing it on the same balance you used in step 1. Record this value.

4. Subtract the mass of the empty cylinder from the mass of the cylinder and the liquid to find the mass of the liquid.

5. Calculate the density of the liquid.

6. Dispose of the liquid as instructed by your teacher.

Part B.   Use of a Hydrometer

1. Carefully lower a clean hydrometer into the solution on the reagent table. If the solution is not deep enough for the hydrometer to float without touching the bottom, pour some of the solution into a tall

cylinder. Read the specific gravity shown on the scale of the hydrometer and record it in your table.

2. Compare this reading with the result obtained in Part A of this section.

DATA AND RESULTS

As you have learned, the orderly arrangement of observations and measurements is one of the most important parts of any experiment. Having appropriate tables for recording data in your notebook before you do an experiment is an excellent way to insure that you will have such an arrangement.

You should copy the following tables in your notebook before you do this experiment, and record all your measurements in these tables as you do the experiment. It is especially important to do so in this experiment, in order to prevent the possibility of confusing data from one part of the experiment with that from another.

Calculate all densities to the nearest 0.01 g/ml.

Section I. Density of a Solid

Part A.

Mass of solid (g)
Length (cm)
Width or diameter (cm)
Height (cm)
   Volume (ml)
   Density of solid (g/ml)

Part B.

Initial volume in cylinder (ml)
Final volume in cylinder (ml)
   Volume of water displaced (ml)
   Volume of solid (ml)
Mass of solid (g)
   Density of solid (g/ml)

Part C.

Mass of solid in air (g)
Apparent mass of solid in water (g)
   Mass of water displaced (g)
Density of water (g/ml)
   Volume of water displaced (ml)
   Volume of solid (ml)
   Density of solid (g/ml)

Section II. Density of a Liquid

Part A.

Mass of cylinder and liquid (g)
Mass of empty cylinder (g)
   Mass of liquid (g)
Volume of liquid (ml)
   Density of liquid (g/ml)

Part B.

Hydrometer reading of specific gravity

RELATED PROBLEMS

1. Your teacher will tell you the known density of the piece of metal that you used in your measurements. How close were your measurements of the density of your piece of metal to the known density? How can you account for any differences?

2. Where do you think there might be errors in each of the methods you used?

3. In making your measurements in this experiment you were told to find the mass to the nearest 0.01 gram. In measuring length you tried to read your metric ruler to the nearest 0.1 millimeter. Similarly, you attempted to read the volume of the water in the graduated cylinder to the nearest 0.1 milliliter. Do you think that these were reasonable limits for the particular measuring devices that you used? If you were to repeat your measurements several times, how closely do you think it would be possible to duplicate the same measurements you got the first time you made them, assuming you did not remember what the original measurements had been? What do you think would be reasonable limits to expect for your various measuring devices?

4. If you could actually measure 15.0 ml of water to the nearest 0.1 ml, what percentage error would 0.1 ml make? What percentage error would 0.2 ml make?

Percent error
$$= \frac{\text{Difference between actual value and experimental value}}{\text{Actual value}} \times 100$$

5. What percentage error would you expect to have in trying to measure 1.0 ml of water to the nearest 0.1 ml?

6. At 32°C a 1000-ml flask held 995 grams of water. At the same temperature the same flask held 800 grams of another liquid. What was the density of the water in the flask at that temperature? What was the density of the other liquid? What was the specific gravity of the other liquid compared to water at 32°C? How many grams of water would the 1000-ml flask hold at 4°C?

A Question to Wonder About

Why do you think different substances have different densities?

Summary

The metric system is a set of units agreed upon by scientists in all countries for making and reporting scientific measurements. In most countries other than the United States the metric system is also used for everyday measurements as well. For mass the defined unit is the kilogram and commonly used smaller units are the gram ($\frac{1}{1000}$ of a kilogram) and the milligram ($\frac{1}{1000}$ of a gram). For length, the unit is the meter, and common smaller units are the centimeter ($\frac{1}{100}$ of a meter) and the millimeter ($\frac{1}{1000}$ of a meter). A metric unit of volume is the liter, which is about equal to a quart. Smaller volumes are measured in milliliters ($\frac{1}{1000}$ of a liter).

The density of a substance is the mass of it contained in a unit volume. A useful unit for expressing the density of a liquid or a solid is grams per milliliter.

Self Check

1. What are the basic units of mass, distance, and volume in the metric system?
2. What do the prefixes kilo-, centi-, and milli- signify? Illustrate each, using units you have learned in this chapter.
3. How could you determine the density of a small, irregularly-shaped piece of a solid substance?

No Measurements Are Absolutely Accurate

4-5 Uncertainty in Measurement

As you have learned, chemists devote a lot of time to making measurements. In making measurements, calculating results, and reporting them to others, chemists face certain problems. One such problem is the uncertainty involved in all measurement. Uncertainty has two main causes: (1) the measuring limitations of the equipment used, and (2) the skill and accuracy of the observer himself.

A related problem arises when chemists deal with rather large or small numbers. These numbers must be readily usable and, at the same time, express only the degree of accuracy warranted by the equipment and the conditions of the experiment. By concentrating on a few basic points, you can understand these problems and get on with further work in chemistry.

Consider the following statements about the melting behavior of *para*dichlorobenzene, which you studied earlier (Section 1-5).
1. The melting point is 53°C.
2. The melting point is 53.2°C.
3. The melting point is 53.203°C.

You might believe that the third statement says more than the second, and that the second says more than the first. Actually, for the experiment you performed, the second statement is the most informative one.

Any scientific statement conveys knowledge about an occurrence or an object. The statement is careless if it asserts less than is known. It is misleading if it asserts more than is known. The most accurate statement conveys just what is known and no more. A scientist must decide whether to list the melting point of *para*dichlorobenzene as 53°C, 53.2°C, or 53.203°C by considering which value reports the measurement most accurately.

Recall how you measured the melting point of *para*dichlorobenzene. (A similar

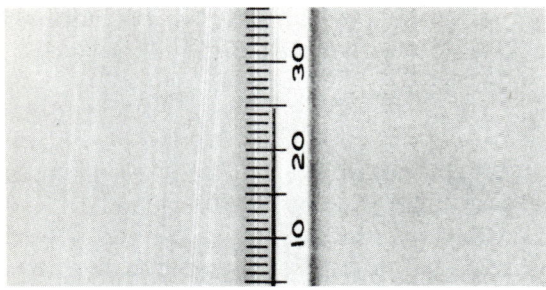

Figure 4–5 *An ordinary laboratory thermometer. What temperature do you read?*

thermometer is shown in Figure 4-5, to refresh your memory.) Did your temperature measurements permit you to say that the melting point is 53°C, not 54°C? Probably, yes. It is not difficult to read a thermometer to that degree of certainty. Can you read the thermometer so as to distinguish 53.0°C and 53.2°C? This is more difficult. It depends on the thermometer and your skill in its use. It also depends on whether the temperature of the solid during melting is uniform throughout. Still, a magnifying glass permits more certainty in the reading of the scale, and slower heating increases the uniformity of the sample temperature. Using this type of extra care, you would find that the melting point of *para*dicholorobenzene is 53.2°C, not 53.0°C. Consider, however, the chance of reading between 53.200°C and 53.203°C. With the equipment available to you, it is not possible.

A careful measurement, then, could establish the melting point as 53.2°C. Therefore, it is the second statement (m.p. = 53.2°C) that tells just what is known. The first statement (m.p. = 53°C) does not tell all that is known since only two figures are given, though three were measured. The third statement (m.p. = 53.203°C) tells far more than is known since the last two figures could

not be determined. The most accurate statement, then, gives three figures. The number 53.2°C is said to have three "significant" figures.

4-6 Uncertainty in a Derived Quantity: Addition and Subtraction

Results of scientific observations are often combined. For example, in your next experiment you will determine the change of water temperature during the combustion of a candle (or during the solidification of candle wax). The change in temperature, which is written Δt, is the result of two measurements, not just one. Thus it is, like density, a derived quantity. It is obtained not from direct measurement but from some mathematical combination of direct measurements.

Consider, for example, the following data that one student got. It shows the change in temperature of some water that was heated by a candle.

Temperature after heating	38.5°C
Temperature before heating	9.3°C
Difference (temperature change), Δt	29.2°C

In accordance with good scientific practice, you should express the temperature change as a number that tells just what is known, but no more or less. To do this, let us see how the uncertainties in the two temperature readings fix the uncertainty in the difference, Δt.

Suppose the "temperature after heating" is measured by a second student who reads it as 38.3°C, and that a third student reads it as 38.7°C. Apparently, different students making the same measurement may differ in their readings by a few tenths of a degree. According to the students' readings, the temperature recorded as 38.5°C could be as high as 38.7°C (0.2° higher) or as low as 38.3°C (0.2° lower). This can be expressed as follows.

Temperature after heating = 38.5 ± 0.2°C

(The symbol ± is read "plus or minus.")

Let us assume that the three students' results accurately represent the uncertainty normally present in this type of measurement. Then the same uncertainty is present in the first temperature measurement, and the following calculation results.

Temperature after heating	38.5 ± 0.2°C
Temperature before heating	9.3 ± 0.2°C
Difference (temperature change), Δt	29.2 ± ?? °C

To decide what uncertainty to give to 29.2, let us first determine what combination of temperatures gives the largest value of Δt. The first temperature could be as low as 9.1°C and the final temperature could be as high as 38.7°C. Then the difference would be 29.6°C. This particular combination causes an uncertainty in the difference equal to the sum of the uncertainties in the parts, 0.2 + 0.2 = 0.4.

What is the smallest possible value for Δt that you can get from the given data? If you say 28.8°C, you are correct. Again the uncertainty in the result is 0.4°C. Thus, 0.4°C is the maximum uncertainty that can result from the subtraction of the two temperatures. The value of Δt must lie between the limits 28.8°C and 29.6°C. The temperature difference, including limits of maximum uncertainty, should be expressed as follows:

$$\Delta t = 29.2 \pm 0.4°C.$$

The foregoing example shows that the uncertainty in a derived quantity is determined by the uncertainties in the measurements that are used to calculate it. Similarly, for *any* addition or subtraction, the range of uncertainty in the result is simply the sum of the ranges of uncertainty in the numbers used. The following expression illustrates this rule, for our example.

$$(\pm 0.2) + (\pm 0.2) = \pm 0.4$$

Let us try another problem involving the same rule. The mass of a sample of water is determined by subtracting the mass of the empty can from the mass of the can when it contains water.

(mass water)
  = (mass can + water) − (mass empty can)

Suppose the mass of the can is 61 ± 1 grams and the mass of the can plus water is 406 ± 1 grams. Show, by calculating maximum and minimum values, that the correct result is the one predicted by the rule for addition and subtraction, that is, 345 ± 2 grams.

4-7 *Uncertainty in a Derived Quantity: Multiplication and Division*

When numbers must be multiplied or divided to obtain a derived quantity, it is again necessary to carry through the individual uncertainties in order to determine the uncertainty in the final answer. That is, maximum and minimum values again must be calculated. The resultant uncertainties do *not* turn out to be the sum of the uncertainties in the individual measurements, however, as was the case with addition and subtraction. The following discussion of an uncertainty calculation for a result obtained by multiplication shows this difference.

Suppose we wish to calculate the area of one face of a rectangular block. The edge lengths, which we can measure to ± 0.1 cm, are found to be 12.1 cm and 9.9 cm. The area of the face is:

$$12.1 \text{ cm} \times 9.9 \text{ cm} = 119.79 \text{ cm}^2.$$

This result *looks* very accurate. But how accurate is it? How many of those five digits are significant? That is, how many of them can we really trust?

Since each measurement is certain to ± 0.1 cm, these lengths *could* be as great as 12.2 and 10.0 cm, or as small as 12.0 and 9.8 cm. Once again we must consider the worst possible combinations of uncertainties in the measured quantities. In this way, we find that the area could be as large as

$$12.2 \text{ cm} \times 10.0 \text{ cm} = 122.00 \text{ cm}^2$$

or as small as

$$12.0 \text{ cm} \times 9.8 \text{ cm} = 117.60 \text{ cm}^2.$$

These two answers differ by 4.40 cm². Thus, the area of the face is 119.79 ± 2.20 cm². Chemists would consider it misleading to report the area as 119.79 since this implies that the answer is known to the nearest hundredth of a cm², and you have just seen that the answer is known only within a range of a few cm², specifically ± 2.20 cm². Although this indicates that there is uncertainty in three digits, the major contribution to the uncertainty is in the first digit to the left of the decimal point. In other words, if you are uncertain as to the nearest whole cm², there is no use in worrying about the tenths of a cm². For these reasons chemists would round off the answer to the ones place. The answer would be written as 120 ± 2 cm², showing only one uncertain digit, the other uncertain digits being small enough, in comparison, to disregard.

In order to compare uncertainty in division with uncertainty in multiplication, consider the following example. Using maximum and minimum values, it summarizes a calculation of the uncertainty in the density of an iron meteorite fragment found in the United States.

Example

$$m = 950.3 \pm 0.2 \text{ g}; \quad V = 120 \pm 1 \text{ ml}$$
$$d = m/V$$
$$d = 950.3 \text{ g}/120 \text{ ml} = 7.919 \text{ g/ml}$$
$$d_{\max} = 950.5 \text{ g}/119 \text{ ml} = 7.987 \text{ g/ml}$$
$$d_{\min} = 950.1 \text{ g}/121 \text{ ml} = 7.852 \text{ g/ml}$$
$$\frac{d_{\max} - d_{\min}}{2} = 0.068 \text{ g/ml}$$
$$d = 7.919 \pm 0.068 \text{ g/ml}$$

The error term in the last line indicates that the major contribution to the uncertainty is in the second digit to the right of the decimal point. Because all digits to the right of this digit are negligibly small compared to it, they are disregarded. For this reason a chemist would write the result in rounded-off form as

$$d = 7.92 \pm 0.07 \text{ g/ml}.$$

The answer is rounded off to 7.92 because 7.919 is closer to 7.92 than it is to 7.91.

James Prescott Joule (1818–1889)

WHEN JAMES JOULE WAS seventeen years old, he was fortunate to have John Dalton, the scientist who proposed the Atomic Theory of Matter, as a teacher. Within five years Joule had embarked on a series of studies that he continued for almost half a century. Having read how Count Rumford had established a relationship between mechanical energy and heat in 1798, he decided to determine whether the amount of heat produced by a given amount of work was always the same, regardless of what method was used to produce the heat.

In his first experiments he measured the amount of mechanical energy needed to drive an electric generator. He then determined the amount of heat produced by the electricity flowing through a wire. Following this, he pumped water through thin pipes and compared the heat produced by the friction of the water in the pipes with the amount of work needed to operate the pump. In later experiments he measured the heat produced by compressing a gas, by stirring various liquids in insulated containers, and by rubbing two iron disks together under water. He even studied Rumford's drilling experiments and calculated the relationship between the mechanical energy expended by the horses that drove Rumford's lathe and the heat produced by the drilling. This was a calculation that Rumford himself had never attempted.

After eight years of work, Joule summarized all of his results and announced a value for what we refer to today as the "mechanical equivalent of heat." Expressed in modern units named in honor of this brilliant scientist, his results of 4.150 joules of mechanical energy for each calorie of heat were not very different from today's accepted value of 4.184 joules per calorie.

Joule's work with heat soon led him to propose a kinetic theory of matter that related all of the effects of heat on matter with the motion of the particles in matter and the attraction between them.

Joule was so enthusiastic about his work that he even took time during his honeymoon in Switzerland to measure the rise in temperature of water in a waterfall when it reached the bottom of its descent. History does not record whether his bride shared his excitement in his work at this point.

INVESTIGATION

4-8 Measuring Heat Effects During Burning and During a Phase Change

Temperature measurements like those discussed in the example on uncertainty in addition and subtraction (Section 4-6) enable you to calculate the quantity of heat liberated when a known mass of candle wax is burned. Heat, as a form of energy, is measured in joules or calories. A convenient way to measure heat is to let it be absorbed by a known amount of water and measure how much the temperature of the water rises. It has been shown by accurate measurements that for water in the temperature range from 8° to 80°C, the temperature will rise 1.00°C when 1.00 gram of water absorbs 4.18 joules or 1 calorie of heat. Conversely, 1 calorie of heat is released when 1.00 gram of water is cooled 1.00°C. (The specific heat of water varies slightly from 1.00 cal/g°C, depending on the temperature of the water. The difference between the actual value and 1.00 is small in the above temperature range. The error that is introduced into calculations when the specific heat of water is taken to be 1.00 cal/g°C is small compared to the errors in the other measured quantities. For this reason any error in the specific heat of water may be disregarded in your calculations.)

As stated before, the official international unit of energy or heat is the joule. However, the calorie makes a very convenient *working* unit because of its simple relationship to the properties of water. To calculate the amount of heat, q, that has been absorbed by m grams of water when its temperature has risen Δt°C, you need only carry out the multiplication: $m \times \Delta t$. The answer comes out directly in calories, as the following calculation shows.

mass (g) × temperature change (°C) × specific heat (cal/g°C)

= heat (cal)

Because the specific heat of water is 1 cal/g°C, the equation can be written as follows.

$$m \times \Delta t = q$$

The change that occurs when a solid melts or a liquid freezes is an example of a **phase change.** In Section 1-5 you found that the temperature of a melting solid or freezing liquid remains constant as long as the two states of matter, liquid and solid, are present. The heat energy absorbed in changing the solid to a liquid is released when the substance changes from a liquid to a solid.

The type of change in which new substances are formed is called a **chemical change.** When you examined the combustion products of a candle in Section 1-13, you found that the candle wax, on burning, is converted into at least two substances: carbon dioxide and water. Since these products have a different composition from the starting materials, this is an example of a chemical change.

PURPOSE

In this experiment you will find the amount of heat produced by the burning of a candle (Part A) and compare this with the amount of heat released as wax changes from a liquid to a solid (Part B). You will also determine the uncertainty produced by multiplying and dividing measured quantities.

MATERIALS

Candle and lid	Wire gauze
Balance	15-20 cm glass rod
Ice cubes	Test tube, 18- × 150-ml, with wax
Burner	Test tube, 18- × 150-ml, empty
Thermometer	2 beakers, 250-ml
Ringstand, ring	Test tube holder

Large fruit juice can with openings around base, both ends removed
Small can with two holes at opposite sides, only one end removed

PROCEDURE

Part A. Heat of Combustion

In this experiment you will warm a known mass of water with the heat given off by a burning candle, and find the resulting rise in the temperature of the water. Knowing the mass of water and the rise in temperature, you can calculate the amount of heat given off by the candle, if you assume that all heat from the burning candle was used only to heat the water.

You will also find the mass of the candle before and after burning. Knowing the amount of candle wax that burns, together with the amount of heat given off in the process, you can calculate the heat produced per unit mass of candle wax.

Before beginning, read the procedure through carefully and draw up a suitable data table. Entries that should appear in your table are listed under "Data and Results."

1. Stick a candle to a tin can lid and determine their combined mass to the nearest 0.01 g, as shown in Figure 4-6. Record the mass of the candle and lid in your data table. Note the balance number so

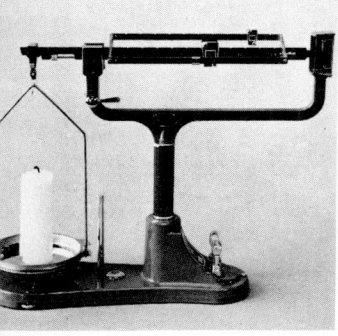

Figure 4-6

Figure 4-7

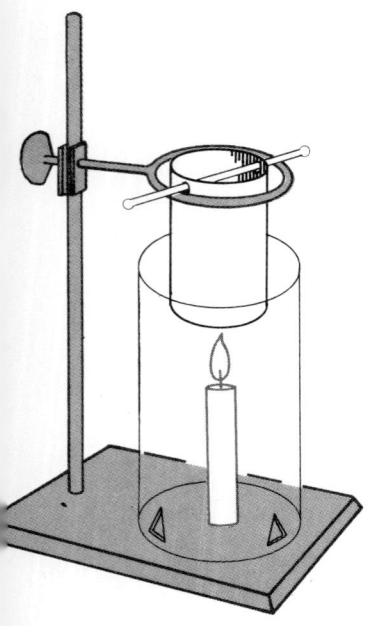

Figure 4-8

that when you again determine the mass of the candle you can use the same balance. (This eliminates any error you might introduce due to differences between balances.)

2. Place an empty can with one end removed on a platform balance, as in Figure 4-7, and determine its mass to the nearest 1 g. Record the mass in your data table.

3. Set up the apparatus as in Figure 4-8 so that the flame of the candle when lit (do not light it yet) will almost, but not quite, touch the bottom of the can. For the chimney use a large can open at both ends.

4. Fill the weighed can about two-thirds full of cold tap water. Do not measure the mass or volume of the water at this time.

5. Cool the water with ice, if necessary, so that its temperature is about 10 to 15°C below room temperature. Add the ice directly to the water. Remove any remaining ice when the desired temperature has been reached.

6. Read and record the temperature of the water to the nearest 0.2°C. Light the candle and heat the water, stirring it gently, until it reaches a temperature about as much above room temperature as it was below at the start. Carefully blow out the candle flame. Continue to stir the water, while watching the thermometer reading, until the highest temperature has been reached. Record the highest temperature to the nearest 0.2°C, as before.

7. Determine the mass of the candle and lid on the same balance that was used before. Make certain that any drippings from the candle are included when determining the new mass. Record the mass.

8. Determine the mass of the can and water on the same platform balance that was used before, and record this mass. Repeat the experiment if time permits.

Part B.   Heat of Solidification

In this experiment you will use some melted wax, at its melting point, to produce a change in the temperature of a known mass of water. You can assume that all of the heat exchanged between the wax and the water is produced by solidification of the wax. The heat required to change the temperature of the beaker is small enough to neglect in this experiment.

Prepare a data table containing the entries given in "Data and Results."

1. Obtain an 18- × 150-mm test tube partially filled with wax (about

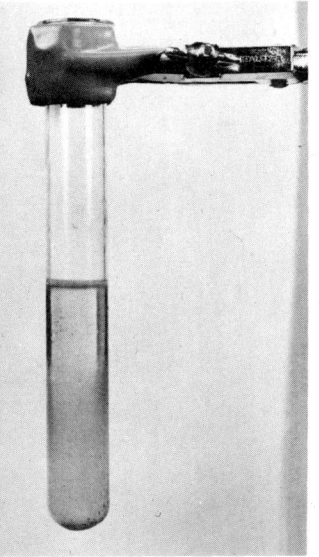

Figure 4-9

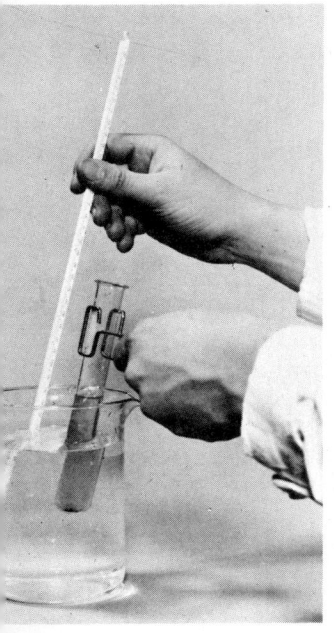

Figure 4-10

10 g of wax). Determine the mass of the test tube with the wax to the nearest 0.1 g. Record the mass.

2. On the same balance, determine the mass of a similar clean, dry test tube, provided by your teacher, in order to estimate the approximate mass of an empty test tube. Record its mass to the nearest 0.1 g.

3. On a platform balance, determine the mass of an empty 250-ml beaker. This beaker will contain the cool water to be warmed by the solidifying wax. Record the mass in your table, to the nearest gram.

4. Fill the beaker about half-full with cold tap water. Use ice, if necessary, to adjust the temperature of the tap water to 1 or 2°C below room temperature. Determine the mass of the beaker and water using the same balance. Record the mass.

5. Add water to a second 250-ml beaker, or a small tin can, until it is about half full. Then heat the water to boiling. Place the test tube that is partially filled with wax in the boiling water bath. Keep it in the bath until the wax is *just* melted. Be careful to avoid overheating.

6. Using a test tube holder, remove the test tube containing the melted wax from the hot water. Allow the wax to cool until the first sign of cloudiness (solidification) is evident, as in Figure 4-9. While the wax is cooling, measure and record the temperature of the cool water bath to the nearest 0.2°C. Quickly place the test tube containing the slightly cloudy wax into the cool water and stir the water gently with the test tube. (See Figure 4-10.) Note the temperature of the water as you stir and continue to observe the temperature until no further increase is noted. (When reading the thermometer, immerse it in the water midway between the test tube and the wall of the beaker.) Record this maximum temperature to the nearest 0.2°C. Repeat the experiment if time permits.

DATA AND RESULTS

Part A.

Balance number(s)
Room temperature (°C)
Mass of candle and lid before burning (g)
Mass of candle and lid after burning (g)
   Mass of candle burned (g)
Mass of can and water (g)

Mass of empty can (g)
  Mass of water heated (g)
Temperature of water before heating (°C)
Temperature of water after heating (°C)
  Temperature change of water, Δt (°C)
  Quantity of heat required to warm the water, q (cal)
  Heat of combustion of the candle material (cal/g)

Part B.

Balance number(s)
Mass of test tube and wax (g)
Mass of empty test tube (g)
  Mass of wax (g)
Mass of beaker and water (g)
Mass of empty beaker (g)
  Mass of water (g)
Temperature of water before (°C)
Temperature of water after (°C)
  Temperature change of water, Δt (°C)
  Quantity of heat absorbed by the water, q (cal)
  Heat of solidification of the wax (cal/g)

Calculations

Using your data and the rules about uncertainty that you have studied, make the calculations indicated in the foregoing tables.

RELATED PROBLEMS

1. Considering the assumptions made in either the "heat of solidification" or "heat of combustion" calculation, would you expect your results to be lower or higher than the accepted value? Explain.

2. Compare the class averages for heat of combustion and heat of solidification. How much larger is one than the other?

A Question to Wonder About

Why is the heat of combustion so different from the heat accompanying the phase change?

ADDITIONAL INVESTIGATIONS

1. Devise an experiment to measure the heat effect for the phase change—gas to liquid or liquid to gas—for some suitable substance.

4-9 Calculating Uncertainty in a Derived Result

In the preceding experiment you calculated the maximum uncertainties in your results for heat of combustion and heat of solidification of candle wax. The following example summarizes a calculation of the uncertainty in a typical value of q, using maximum and minimum values. It will be useful to compare the following result with those obtained using other methods of determining uncertainty that will be discussed in following sections. In each case, we shall use the results calculated in Section 4-6 for the mass of water heated by a candle and the resulting temperature change of the water.

Example

$$m = 345 \pm 2 \text{ g}; \Delta t = 29.2 \pm 0.4°C$$
$$q = m \times \Delta t = (345 \pm 2)(29.2 \pm 0.4)$$
$$= 10{,}074.0 \pm \text{? cal}$$
$$q_{max} = (347)(29.6) = 10{,}271.2 \text{ cal}$$
$$q_{min} = (343)(28.8) = 9{,}878.4 \text{ cal}$$
$$\frac{q_{max} - q_{min}}{2} = \frac{392.8}{2} = 196.4$$
$$q = 10{,}074.0 \pm 196.4 \text{ cal}$$

The result shows uncertainty in the third digit to the left of the decimal point. Thus, digits to the right of this digit are "insignificant," both in the quantity calculated *and* in the estimate of uncertainty. A chemist would report the following result, in which both numbers are rounded off to the third digit to the left of the decimal point.

$$q = 10{,}100 \pm 200 \text{ cal}$$

4-10 Calculating Uncertainty Using Percentages

In Section 4-9, one way of calculating the uncertainty in a derived result has been summarized. It is the same method you used to determine the uncertainty in your experimental result. You may have wondered whether there was an easier way, since it requires a lot of time and effort to calculate maximum and minimum values. Scientists do have a more convenient way of calculating uncertainties. The method that you have used up to now was best for helping you to understand what uncertainty is all about. From now on you will be able to use a more convenient method.

Scientists use the following rule for calculating uncertainties in results obtained by *multiplication and division*.

The **percentage uncertainty** in the result is equal to the sum of the percentage uncertainties in all of the factors, divisors, or dividends.

In order to understand how the "percentage rule" is used, study the following example. So that you can readily compare the percentage method of calculating uncertainty with the method used in the preceding section, the same figures are used.

Example

$$m = 345 \pm 2 \text{ g}; \text{ uncertainty}$$
$$= \left(\frac{2}{345}\right)(100) = 0.58\%$$
$$\Delta t = 29.2 \pm 0.4°C; \text{ uncertainty}$$
$$= \left(\frac{0.4}{29.2}\right)(100) = 1.37\%$$

sum of % uncertainties

$$= 1.37\% + 0.58\% = 1.95\%$$

$$m \times \Delta t = 10{,}074 \pm \left(\frac{1.95}{100} \times 10{,}074\right) \text{ cal}$$

$$= 10{,}074 \pm 196 \text{ cal} = 10{,}100 \pm 200 \text{ cal}$$

The calculated uncertainty, 196 cal, does not differ from that obtained previously by a much more laborious method.

4-11 Significant Figures: Approximating Uncertainty

You have seen that, while the quantity 10,074.0 cal can be calculated from the measured values of m, Δt, and the specific heat of water, the uncertainty in this number is ± 196 cal. (Remember: The error in the specific heat of water is negligible and so it was disregarded.) Consequently, none of the numbers beyond the third one in 10,074.0 has any significance. For this reason we round off both 10,074 and 196 to the nearest hundred and write $10{,}100 \pm 200$. The number 10,100 contains only three significant figures. The first three figures are reliable within rather narrow limits, whereas all figures farther to the right have no reliability whatsoever.

As a general rule, **significant figures** are digits that are completely certain and one more. The "one more" has some uncertainty in it. For example, in $10{,}100 \pm 200$, the first two figures are quite certain. The last significant figure is most likely a 1, but it could differ from this by ± 2.

With the definition of significant figures established, a convenient rule can be given for stating the maximum uncertainty in a derived quantity. *A derived quantity can have only as many significant figures as there are in the least precise quantity from which it was calculated.*

In the example we have been using, there are only three significant figures in each of the quantities used and thus there can be only three significant figures in the calculated number of calories, according to the preceding rule. In this case, the rule is consistent with our earlier calculations that showed the third digit to be uncertain.

Let us consider another example. Suppose that you were determining the density of a solid aluminum block of regular dimensions. Suppose, further, that you were able to determine its mass to ± 0.01 g and each edge length to ± 0.02 cm. Using the data given in the following example, together with the previously-discussed rule, you can easily calculate the density to the correct number of significant figures. You might also wish to show that your answer is consistent with the result that is obtained by adding up the individual percentage errors.

Example

$$m = 41.93 \pm 0.01 \text{ g}; \; l = 4.02 \pm 0.02 \text{ cm};$$

$$w = 2.80 \pm 0.02 \text{ cm}; \; h = 1.38 \pm 0.02 \text{ cm}.$$

$$d = \frac{m}{l \times w \times h} = \frac{41.93 \text{ g}}{(4.02)(2.80)(1.38) \text{ cm}^3}$$

$$= 2.70 \text{ g/cm}^3 = 2.70 \text{ g/ml}$$

The answer is given to 3 significant figures, since that is the number of significant figures in the least accurate measurement(s).

What result is obtained by summing the individual percentage uncertainties?

$$l = 4.02 \pm 0.02; \; (0.02/4.02)(100) = 0.5\%$$

$$w = 2.80 \pm 0.02; \; (0.02/2.80)(100) = 0.7\%$$

$$h = 1.38 \pm 0.02; \; (0.02/1.38)(100) = 1.4\%$$

$$m = 41.93 \pm 0.01; \; (0.01/41.93)(100) = 0.02\%$$

$$\text{sum of \% uncertainties} = 2.6\%$$

$$d = 2.70 \pm (0.026)(2.70) = 2.70 \pm 0.07 \text{ g/ml}$$

This result shows that the zero is the uncertain figure, as was predicted by our simpler rule.

Note, for comparison, that the accepted value for the density of aluminum, determined to 4 significant figures, at 20°C, is 2.698 g/ml. Does this value lie within the range of the result just calculated?

4-12 Large and Small Numbers

In chemistry there is frequently a need to use very large and very small numbers. For example, one dimension of a molecule might be 0.000000021 cm, while the number of molecules in a milliliter of gas at room temperature is approximately equal to 20,000,000,000,000,000,000. Obviously, it is impractical to write such numbers in the way they have just been written. It is possible to write very small and very large numbers, and more ordinary numbers as well, in a more compact way. Numbers written in such a way are said to be in "exponential form."

Numbers written in exponential form have powers of ten to indicate where the decimal point is located. For example, all numbers greater than $100 \,(10 \times 10 = 10^2)$ but less than 1000 are expressed as a number from 1 to 10 times 10^2. Thus, 200 is 2×10^2, 635 is 6.35×10^2, 527.28 is 5.2728×10^2, and so on.

A tabulation of some powers of ten, expressed as ordinary decimal numbers and as exponentials follows. Note that numbers less than one involve negative powers of ten.

$$1{,}000{,}000 = 10^6$$
$$100{,}000 = 10^5$$
$$10{,}000 = 10^4$$
$$1{,}000 = 10^3$$
$$100 = 10^2$$
$$10 = 10^1$$
$$1 = 10^0$$
$$0.1 = 10^{-1}$$
$$0.01 = 10^{-2}$$
$$0.001 = 10^{-3}$$
$$0.0001 = 10^{-4}$$
$$0.00001 = 10^{-5}$$
$$0.000001 = 10^{-6}$$

When multiplying exponential numbers, add exponents, as in the following examples.

$$(3 \times 10^4)(2 \times 10^5) = 6 \times 10^9$$
$$(4 \times 10^{-5})(1 \times 10^{-7}) = 4 \times 10^{-12}$$
$$(6 \times 10^6)(4 \times 10^{-8}) = 24 \times 10^{-2} = 2.4 \times 10^{-1}$$

When dividing, subtract exponents.

$$(8 \times 10^3)/(4 \times 10^6) = 2 \times 10^{-3}$$
$$(6 \times 10^{-4})/(2 \times 10^{-7}) = 3 \times 10^3$$
$$(2 \times 10^{-8})/(4 \times 10^{-5}) = 5 \times 10^{-4}$$

When such numbers are squared, the exponents are multiplied by two.

The square of 2×10^4 is 4×10^8.
The square of 5×10^{-6} is 2.5×10^{-11}.

When obtaining the square root, divide the exponent by two. If it is not divisible by two, shift the decimal point until the exponent is a multiple of two.

The square root of 9×10^{-8} is 3×10^{-4}.
The square root of 2.5×10^5 is 5×10^2
$(2.5 \times 10^5 = 25 \times 10^4)$.
The square root of 3.6×10^{-7} is 6×10^{-4}
$(3.6 \times 10^{-7} = 36 \times 10^{-8})$.

Let us return to the two unwieldy numbers with which this section began. They are expressed in exponential form as 2.1×10^{-8} cm

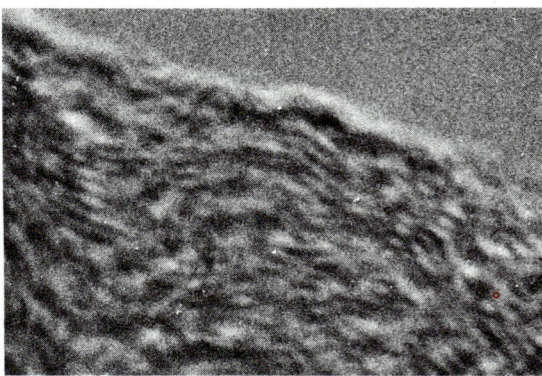

Figure 4-11 *The Milky Way (above photo) is approximately 10^{25} cm in diameter. The photo on the right, taken with an electron microscope, shows layers of atoms in a sample of carbon, 5×10^6 times its actual size.*

and 2×10^{19} molecules per ml. (The applicability of exponential form is further illustrated in Figure 4-11.)

The exponential form has an additional advantage besides providing a compact way to write very large and very small numbers. It provides a way to indicate the number of significant figures in a very large number. This can be illustrated using the number $10{,}100 \pm 200$ that we obtained earlier. To indicate that only the first three digits are significant, without indicating that the uncertainty in the third one is ± 2, you should write the number in exponential form as 1.01×10^4. When you write 10,100 it is not clear whether the last two zeros are significant or whether they are just there to show the decimal place. When you write 1.01×10^4 it is perfectly clear that there are only three significant figures. If there were five significant figures, you could show this unambiguously by writing 1.0100×10^4. In exponential form, there is no other reason for including the last two zeros, except to show that they are significant figures.

4-13 *Random and Systematic Errors* ★

You have learned that no measurement can be made with absolute accuracy. Your laboratory experiments showed you this directly. There is always some uncertainty in a measurement, no matter how carefully it is made. A scientist cannot hope to make perfectly accurate measurements. What he does hope, and try, to do is to make sure that his measurements are accurate enough for the purposes to which he intends to put the results. In order to plan his experiments to give the desired accuracy, he must understand the sources of error. He must understand what kinds of errors may occur in his measurements.

The most obvious thing to consider is the instrument that he is using. Is it capable of measuring what it is designed to measure to the required accuracy? For example, if it is necessary to determine the mass of an object to the nearest 0.1 gram, it is necessary to use a balance that is sensitive to 0.1 gram or less. A change of 0.1 gram must be noticeable, and the scale that is read or adjusted must have markings (graduations) of this size, or smaller.

In Section 2-5 your class performed an experiment in which the mass of an object was determined by a number of people using several different balances. All of these balances were supposedly sensitive to the desired level of accuracy, that is to 0.1 gram. You should have in your notebook the combined results obtained by the entire class. At this point you should examine those results. You will see that even though each balance is supposed to be sensitive to 0.1 gram, the measurements do not all agree to the nearest 0.1 gram. The many results differ from one another in two ways.

First, the results obtained by *different* teams using the *same* balance are not all exactly the same, but they do not differ by much.

Second, one balance consistently gave high or low results, *no matter who used it.*

The two patterns, or regularities, just pointed out arise from the fact that there are two kinds of errors. In the first case you are dealing with **random errors.** These are errors that occur because you are working at the limit of sensitivity of the instrument. The readings made by different people, or different teams, may differ by one or two of the smallest units being read. You experienced this same problem in reading your thermometer to the nearest 0.1°C. If several people try to read an instrument as accurately as it can be read, the various readings will tend to differ a little. Such a result also occurs when one person makes several readings on different occasions. If a large number of readings are made, some will be above and some will be below the "true" value. That is why this sort of error is called random. Any single reading is as likely to be high as it is to be low.

It is easy to minimize the effect of random errors, just because of the fact that they are random. If you take a large number of readings and average them, the average is sure to be nearer the "true" value than are many of the individual readings. This is because, with a large number of readings, the positive errors and the negative errors will tend to cancel out.

You should have observed a second pattern in the class results. A certain balance always gave a higher (or lower) result than the other balances. This is an example of a **systematic error.** A systematic error is one that always goes in the *same* direction. No matter how many times, or by how many people, the measurement is made, if there is a systematic error in the instrument, or in the method used to read it, the results will be off in the *same* direction. In the example we are considering, the source of the systematic error is the result of a misadjustment, perhaps due to damage, of a particular balance. That same source of error is there every time the balance is used and no amount of repetition

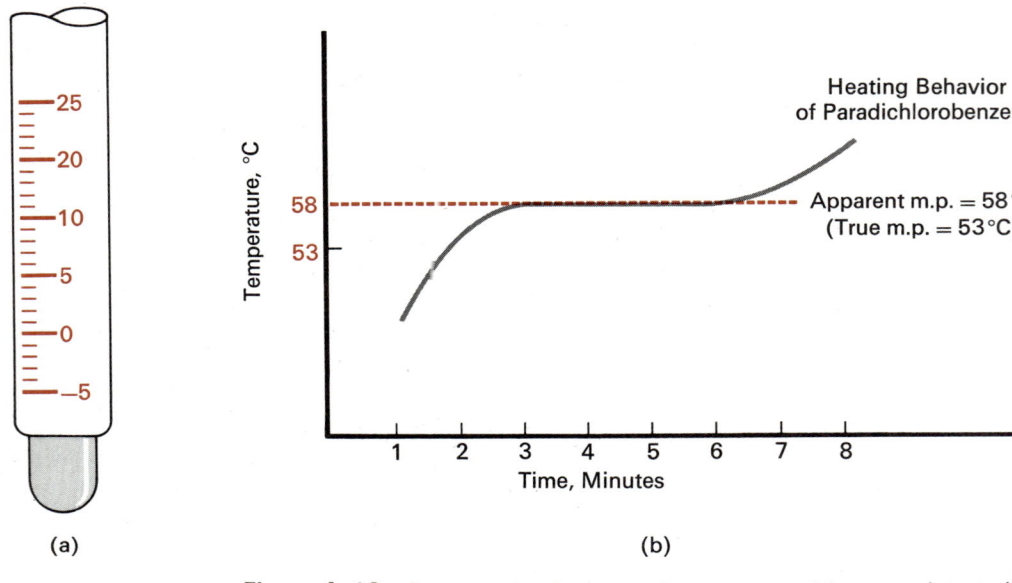

Figure 4–12 *An example of systematic error caused by an undetected mistake in the labeling of a thermometer.*

can ever "average out" such an error. Other examples of this type of error are shown in Figures 4-12 and 4-13.

Systematic errors can result from sources other than the intrinsic incorrectness of an instrument. They can arise from the use of impure materials, for example. Suppose you had determined the density of the metal aluminum in the following manner. You took a block of aluminum and cut several rectangular pieces from it. For each piece you measured the mass ten times and averaged the results. For each piece you measured the edges ten times, calculated the volumes from the measurements, and then averaged the results. You then calculated the density of each piece. Finally you averaged the results for each piece. If you had done all this carefully, you would certainly expect to have a value for the density that is as accurate as your measurements of mass and of length could give. But suppose that, unknown to you, your sample of aluminum was impure. Suppose you did not have pure aluminum, but an alloy of aluminum that contained a few percent of iron. Iron has a higher density than aluminum. Therefore, despite all your care to eliminate *random* error, there would still be a *systematic* error in your results. Your final answer would be a little too high.

Systematic error often results because some factor that influences the result is overlooked. Suppose, for example, that you have a balance on which you can determine the mass of one liter of water (which is close to 1000 g at 25°C) to the nearest 0.1 g. Suppose, further, that the assumed volume is actually correct to the nearest 0.1 ml. You might then suppose that you could calculate the density to an accuracy of ±0.0002 g/ml.

Measurements, Units, and Accuracy 101

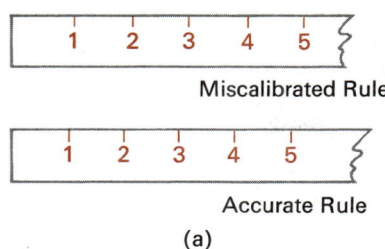

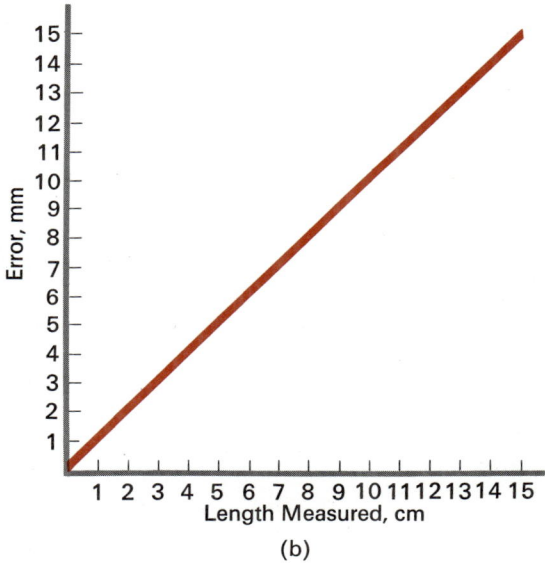

Figure 4-13 *An example of systematic error caused by miscalibration of a rule.*

(This accuracy can be readily calculated with the use of percentage uncertainties.) Unless you had made a buoyancy correction to the mass measurement, you would be wrong. A liter of air, under normal conditions, has a mass of 1.20 grams. If you did not apply a buoyancy correction, you would be using a mass for the water that is 1.2 grams too low in calculating the density. This would cause your density determination to have a systematic error of about 0.0012 g/ml instead of merely a small random error of 0.0002 g/ml. No matter how often you might repeat your measurements, if you continued to forget the buoyancy correction, your answer would always be about 0.0012 g/ml below the correct value.

In planning experiments that involve quantitative measurements, a chemist must be alert to three things. First, he must be sure to have at his disposal instruments sensitive enough to make measurements as accurate as he requires. Second, he must repeat his measurements several times and average the results, in order to minimize the effect of random errors. Third, he must try to anticipate the various sources of systematic error (incorrect instruments, impure materials, the overlooking of factors that affect his measurements) and eliminate them.

Summary

All measurements are subject to errors. There is thus an uncertainty in the value that we assign to any measured quantity. It is important to state the uncertainties in measured quantities correctly, and to allow for the effect of these uncertainties on the uncertainties in all quantities calculated from them. A quantity obtained by addition or subtraction of other quantities has an uncertainty equal to the sum of the uncertainties in all the quantities used to calculate it. A quantity obtained by multiplication or division has a percentage uncertainty equal to the sum of the percentage uncertainties in the quantities used to calculate it. In any quantity, the significant figures are the digits that are certain plus one more. A derived quantity should have only as many significant figures as there are in the least precise quantity used in calculating it. Very large and very small numbers can be most conveniently expressed in exponential form.

Self Check

1. In stating the value of some quantity, how many digits should be given?
2. How many significant figures can a calculated quantity have?
3. How would you express 0.00045 in exponential notation?

Chapter Summary

You learned in Chapter 1 that experiment is the key to chemistry. This chapter has shown that accuracy is the key to experiment.

The degree of accuracy that can be achieved is, in part, determined by the limitations of the measuring instruments being used and the skill of the observer. Because of these limitations, any measured quantities have some uncertainty in them. These uncertainties are further compounded when the measured values are combined mathematically in order to calculate derived quantities.

With the use of significant figures, scientists can conveniently express the degree of uncertainty in their results. Exponential notation can be useful in indicating the number of significant figures, as well as in handling very large and very small numbers.

Scientists throughout the world use the metric system, both in their own work and in the communication of their results.

Density is a derived unit for expressing the amount of mass in a unit of volume. The densities of solids and liquids are usually expressed in grams per milliliter.

Questions and Problems

1. How many meters are there in a kilometer? How many grams in a kilogram?
2. How many milliliters are there in a liter? How many milligrams in a gram?
3. Is a liter larger or smaller than a quart? Which is larger, a meter or a yard?
4. In track and field competition 1500 meters is sometimes referred to as the "metric mile." Which race is longer, the one-mile run or the 1500-meter run?
5. How is density defined? Why is density described as a "derived" quantity?
6. How is the Celsius temperature scale defined?
7. What does the symbol Δ (delta) usually mean?
8. What is the definition of a calorie? Is the calorie a primary or a derived unit?
9. How are joules and calories related?
10. What part of the earth's circumference was originally chosen to represent the distance of one meter? Could any other distance have been used to represent a meter?
11. What causes the uncertainty in the measurement of temperature with a thermometer? In the measurement of mass with a balance?

12. How do you determine the uncertainty in a result derived by adding or subtracting measured quantities?
13. How do you determine the uncertainty in a result derived by multiplying or dividing measured quantities?
14. How do you determine the percent error in a number with an expressed uncertainty?
15. How many significant figures are represented in the number 1,253.7? If this number represents a measured quantity, which figure contains the greatest possible error?
16. Express the following numbers in exponential form: 14,729; 368; 0.0059.
17. What is the meaning of a negative exponent in an exponential number form?
18. How many milligrams are there in a kilogram? (Use exponential notation.)
19. In the number 6240 is the zero significant? What about in the number 6.240×10^{37}?
20. If an error of two parts in a hundred is a 2% error, what percent is an error of four parts in 400?
21. Assume that you have a balance that can find the mass of an object to the nearest 0.01 gram. If a certain experiment requires results with an error no larger than 0.1%, should you use a sample having a mass of 1.00 gram, 10.00 grams, or 100.00 grams? Explain.
22. What is the smallest size sample you should use with a balance that can find the mass to the nearest 0.0001 gram, if your allowable error is 0.1%?
23. If 2.54 cm equal an inch, how many centimeters are there in $6\frac{7}{8}$ inches?
24. How many inches are there in 6.5 meters?
25. What is the maximum value the number 68.5 ± 0.4 can have?
26. The sum of 16 ± 3 and 14 ± 2 has what uncertainty?
27. What is the smallest number that could result from subtracting 16 ± 4 from 23 ± 2?
28. Solve and express in exponential form:

$$\frac{625 \times 5200}{0.0013 \times 0.025}$$

29. Round off the number 26,060 to three significant figures.
30. What is the quotient of 6.25×10^2 divided by 25?
31. Determine the square root of 1,440,000 using exponential notation.
32. A measured density is 2.36 g/ml. The best official value is 2.41 g/ml. What is the percent error?
33. How many significant figures are in the quotient: 122/31.29?
34. Change the distance of one mile to meters.
35. Three kilometers equals how many miles?
36. What is the uncertainty in the product $(314 \pm 3)(696 \pm 4)$?
37. How many significant figures are there in the number 26,400?
38. The number 0.0140 has how many significant figures?
39. What is the quotient of 1.50×10^4 divided by 3.75×10^{-2}?
40. Using exponential forms, multiply 62,800 by 0.005.
41. Express in exponential form and solve:

$$\frac{1500 \times \sqrt{16900}}{225,000 \times 0.00013}$$

42. A group measures a quantity sixty-five times and the result becomes 25.6 ± 0.3. What is the percent error in the measurement?

2

Atoms and Molecules: Building Blocks of Matter

To define matter as that which has mass is a good start, but it does not really answer the question, Of what is matter composed?

For thousands of years, men have asked this question. In trying to answer it, they have created many theories and tried many experiments. The problem became one of the most challenging in all of science, for scientists were trying to understand things so small that they could not be seen or manipulated individually. Great ingenuity and skill were required to gather evidence of atoms and molecules. Single atoms are on the order of 10^{-8} cm in diameter and the smallest amount of matter detectable by ordinary means contains several billion billion molecules. (An exception is shown in the field ion microscope photograph at the beginning of Chapter 7.)

By means of logical steps, based on experiments that can be repeated and verified anywhere in the world, it is now possible to show that atoms exist. Furthermore, it can be shown that there are different kinds of atoms, all built from even smaller particles. It is possible to show how atoms combine to form molecules, and to measure the forces holding them together. The hundreds of thousands of known chemical compounds can be understood in terms of the molecules of which they are composed.

This unit will help you to understand how the existence of atoms and molecules can be inferred from experimental results.

5 Atoms and Molecules in Gases

Gaseous nebula in the constellation Serpens, photographed in red light.

How can you visualize something you cannot see? One way is to think about it in terms of something that you can see. If there are sufficient similarities between the two things, you can assume that the visible one must at least approximate the invisible one. In such a case, a scientist would call the visible one a *model* for the invisible one.

In using the terms "atom" and "molecule," chemists have for years been dealing with models. Atoms and molecules are much too small to have been seen by anyone. Chemists have been able to describe the behavior of such small particles quite accurately, nevertheless.

Observations of gas behavior have provided much of the evidence for the existence of atoms and molecules. In solids or liquids atoms and molecules are more constrained than in gases. Since those in gases are freer to display their individual characteristics, a study of gases is a logical first step in the study of atoms. In this chapter, properties of gases are presented and related to atomic models.

Pressures and Volumes of Gases Are Related

5-1 *From Everyday Experience to a Scientific Model*

Scientists often use a familiar system to help them understand an unfamiliar one. The familiar system is called a model for the new one. Usually, though not always, the model is something that has been observed directly with the senses. It is *macroscopic*. The phenomenon under study is often produced by very small, or *microscopic* things, that cannot be seen directly. A model, in this scientific sense, is not to be confused with a simple scaled-down (or scaled-up) replica like a model airplane or a model train. It is an analogy, rather than any sort of copy or imitation (as a "train of thought" is a series of thoughts, one following the other like cars of a railroad train). A model does not even have to be a physical thing. It may sometimes be a mathematical model, that is, a group of equations. When this is so it is called a theoretical model, but the basic idea remains the same.

A **model** explains something new and unfamiliar, or something invisible, in terms of what is familiar or visible.

Of course, no model can be completely faithful to the real phenomenon it represents, but it may still be a great help in interpreting what has been observed. If you follow the analogy of the model, you often think of new experiments to perform. The results of these experiments may either support the model further or they may show its limitations. In either case, you will have acquired more information about the real system.

108 Chapter Five

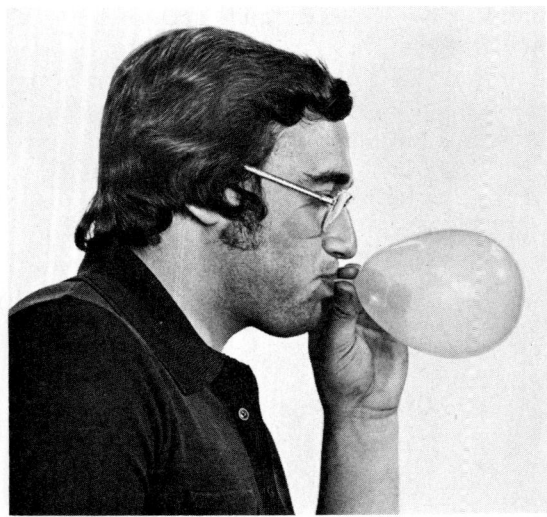

Figure 5-1 *Air blown into a balloon inflates it. How can this familiar process be explained?*

Consider the process of inflating a balloon, shown in Figure 5-1. As you blow into the balloon again and again, it expands and becomes harder. Apparently the air is pushing on the inside of the balloon, stretching its elastic surface. Why does the gas push outward more and more on the inner surface of the inflating balloon? Why does the gas continue to push outward, preventing a collapse of the balloon, even after you have stopped blowing into the balloon and tied it off?

The very simple observations that can be made, and the questions that naturally arise in your mind as you observe the inflation of a balloon, might stimulate you to think of a model in order to explain the observations and answer the questions. What sort of a model might that be?

To devise a good model, you should make use not only of direct observations, but also of all other knowledge that might have a bearing on the problem. In this case you could use the idea that air, which is being blown into the balloon, is believed to be made up of small particles that are never still. That is, these particles never slow down and come to a full stop. (Otherwise, air in a box, for example, would eventually become a handful of particles lying on the bottom of the box, and that never happens.)

The air particles enclosed in a container could be likened to billiard balls moving on a table, such as the one shown in Figure 5-2. Suppose a billiard ball is set in motion and allowed to strike one edge of the billiard table. It makes an elastic collision (as described in Section 3-2), rebounds, and moves until it strikes another edge. Then it moves off in still another direction and strikes another edge. If the ball continues to roll, it will go on striking the edges of the table at different places. If more billiard balls are put

Figure 5-2 *Billiard balls striking the edges of a table suggest a model to explain the inflation of a balloon.*

on the table and set in motion, they do the same thing, while also striking each other and rebounding. The greater the number of billiard balls added, the greater the number of collisions between the billiard balls and the table edges.

The edges of a billiard table are not elastic like the rubber balloon. If they were, and many billiard balls were striking them, the edges of the table would stretch farther and farther apart, just as the balloon does when more air particles are blown into it. Although the idea of billiard balls on a billiard table is not identical in all respects to that of air particles in a balloon, there are common features. It is these common features that help us answer the question: Why does a balloon expand as air is pushed into it?

With this model, you can reason that perhaps as air particles are put into the balloon, they push on it like billiard balls striking the edges of a billiard table. As more air goes into the balloon, there are more particles striking its inner surface, causing the elastic balloon to expand.

This is how a model is developed. The process begins with a question about an occurrence that is not well understood. An answer is proposed in terms of an occurrence that is well understood.

In our example, the origin of gas pressure in the balloon is what we wanted to explain. But air particles are invisible, and it is difficult even to sense the presence of the air. It cannot ordinarily be seen, tasted, or smelled. It cannot be heard or felt, if there is no wind. But if air particles behave the same way that billiard balls do, then billiard balls can be used as a model. Billiard balls are readily seen and felt. Their behavior has been thoroughly studied and is well understood.

A model is successful if it meets the following requirements.

1. The model is well understood. Regularities in the behavior of the model system have been carefully established.
2. There are close similarities between the system of interest and the model system.

The model that we have developed in our example is successful for the following two major reasons.

1. The way in which a billiard ball rebounds is well understood. The amount of push that a billiard ball exerts on the table edge at each bounce can be calculated mathematically.
2. There is a very strong similarity between the behavior of billiard balls and that of gas particles. This is shown by the fact that the billiard ball calculations correctly describe the pressure behavior of gases.

5-2 *Defining and Measuring Pressure*

We have just developed some simple ideas about the nature of gases and why they exert pressure and take up volume. This model is basically correct, but it needs to be made more detailed and more quantitative. A quantitative model must be consistent with the results of experimental measurements and any calculated results obtained from them. Let us begin with the notion of gas pressure and see how we can define it so as to make it an exactly measurable quantity.

Scientists have agreed to define pressure as the amount of force or "push" on a unit of area. The pressure that a particular object creates depends not only on the force it exerts, but also on the area over which that force is spread. For example, an object weighing 1000 pounds resting on a level surface 10 square feet in area exerts a pressure of 100 lb/ft^2. If another object of the same weight covers 100 square feet of surface, the

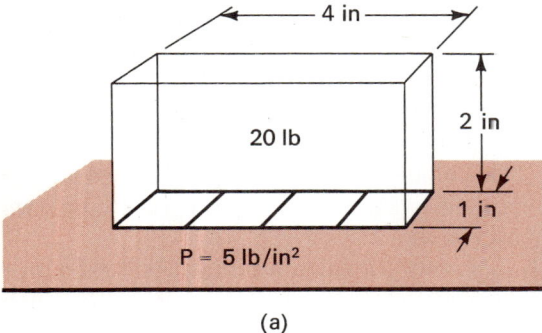

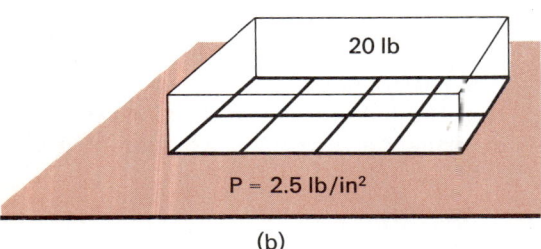

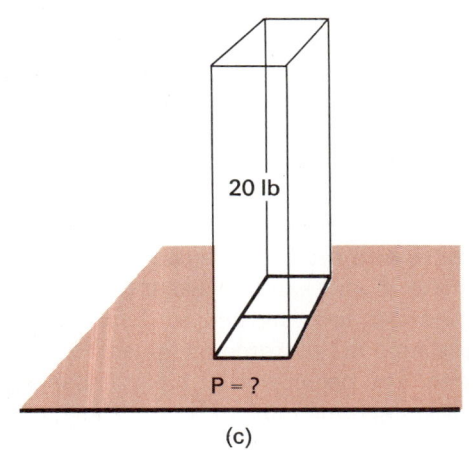

Figure 5-3 *The 20-lb block in (a) exerts a pressure of 5 lb/in². Placed as in (b), the same block exerts half as much pressure. What pressure does it exert in (c)?*

pressure is only 10 lb/ft². Figure 5-3 shows a block that weighs 20 pounds. Placed as in (a), it exerts a force of 20 lb over 4 in². The resultant pressure is 5 lb/in². In (b) the same force is spread over 8 in², and the pressure is 2.5 lb/in². What is the pressure exerted in (c), with the same force acting on only 2 in² of surface?

In many chemical experiments a unit of pressure called the *atmosphere* is used. As illustrated in Figure 5-4, this is the amount of force exerted by the earth's atmosphere at sea level on a unit area. (It amounts to 14.7 pounds per square inch.) In scientific research this quantity, one atmosphere, is used as a unit. The pressure of the earth's atmosphere varies, however, depending on the humidity, altitude, and latitude. In order to avoid such variability, the pressure unit one atmosphere (abbreviated **atm**) is precisely defined. An **atmosphere** is the pressure exerted by a column of pure mercury exactly 760 mm high, measured at 0°C.

It is not necessary to specify the shape or cross-sectional area of the column of mercury. This is because each unit of area can be thought of as bearing the mass of the column of mercury that stands directly over it. Changes in size or shape of the entire column are only changes in the number and arrangement of the component columns, each standing on its own unit area of base. Thus, when the height of the column is fixed, the pressure it exerts is fixed, because pressure is not governed by the total mass, but only by the mass over each unit of area.

Sometimes the atmosphere is too large a unit to be convenient. It is then common practice to use as a unit the pressure exerted by a column of mercury only one millimeter high. The new unit is exactly $\frac{1}{760}$ of an atmosphere. Because this small unit is defined as the pressure due to one mm of mercury, it was for a long time called (and often still is)

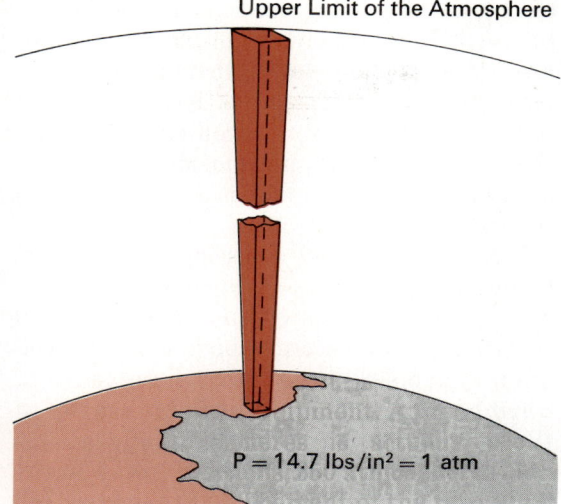

Figure 5–4 *A column of air over a unit area of the earth's surface exerts a pressure of one atmosphere. The earth's atmosphere does not actually have a sharp boundary.*

of the mercury column there is a vacuum, that is, there is no air or other gas in it. As a result nothing is pushing down on the surface of the mercury on that side. On the other side, the manometer tube is connected to a bulb that contains some gas. Notice that the mercury column is higher on one side than the other. A line has been drawn horizontally at the level of the mercury column on the left side. All of the mercury below this line on one side is balanced by the mercury below this line on the other side. Then the downward pressure exerted by the additional mercury on the right side must counterbalance the pressure of the gas on the left side. If the gas pressure is to be expressed in Torr, this pressure is directly given by the extra length of the mercury column on the right side. In the example shown, this pressure is 105 Torr.

simply "a millimeter of mercury." This designation is objectionable for various reasons, and today the name **Torr** has been widely adopted for the unit. Torr comes from the name of the Italian scientist Evangelista Torricelli (1608–1647), who advanced the idea that we live in a sea of air that presses on us. He did pioneer work on the measurement of pressure and invented the barometer. The barometer is a pressure-measuring instrument, described in the following section.

5-3 Pressure Measuring Instruments

In order to measure gas pressures directly in Torr, a device called a **manometer** can be used. The simplest form of manometer is the closed-end manometer shown in Figure 5-5a. In the space above the right side

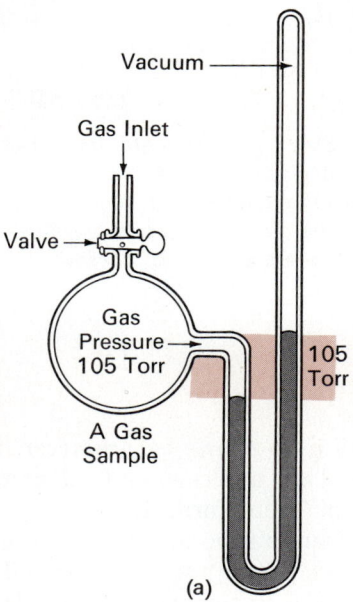

Figure 5–5a *A closed-end manometer. It shows a gas pressure of 105 Torr.*

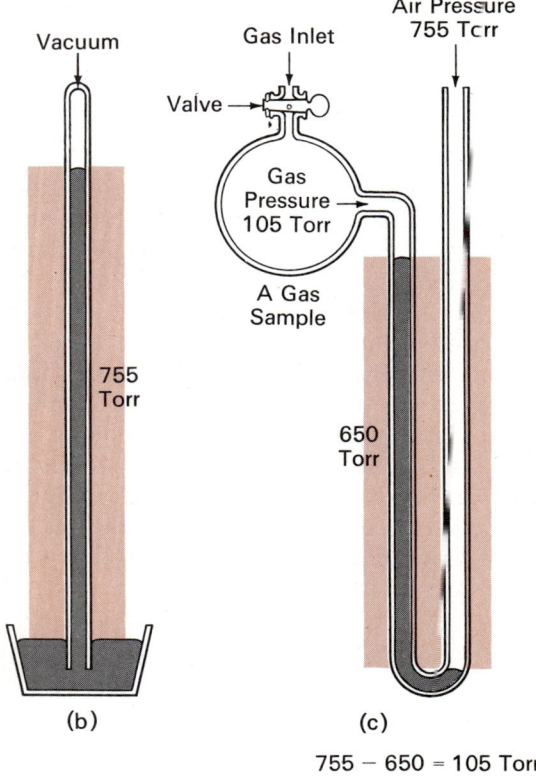

Figure 5–5b,c *The barometer (b) shows an atmospheric pressure of 755 Torr. The open-end manometer (c) registers 650 Torr, the difference between atmospheric pressure and the gas pressure.*

Figure 5-5b shows a barometer, which is actually just a closed-end manometer in slightly different form. In this case, there is only one glass tube and it has no gas in it. The mercury column is supported by the pressure of the atmosphere on the surface of the mercury in the dish. Hence, the height of this column is the pressure of the atmosphere, in Torr. In general, the term **barometer** is used for any device that measures the pressure of the atmosphere. The type shown in Figure 5-5b is the most common one. Note that in the illustration atmospheric pressure is shown as 755 Torr, that is, 5 Torr less than the defined unit, one atm.

Figure 5-5c shows a second type of manometer, called an open-end manometer. It differs from the closed-end manometer in that there is no vacuum above the right column. It is open to the atmosphere and that column, therefore, has a pressure on top of the mercury equal to the prevailing atmospheric pressure. Hence, the difference in the heights of the two sides of the mercury column shows the difference between the pressure of gas in the bulb and the prevailing pressure of the atmosphere. In order to use an open-end manometer it is necessary also to have a barometer to measure the prevailing atmospheric pressure.

In the example shown, the two sides of the mercury column differ in height by 650 mm. This means the pressure difference between the gas in the bulb and the atmosphere is 650 Torr. Since the barometer shows that the pressure of the atmosphere is 755 Torr, the pressure in the bulb must be 755 − 650 = 105 Torr. The same result was obtained by direct reading of the closed-end manometer.

5-4 How Pressure and Volume Are Related: Boyle's Law

Your own experience with the gaseous substance air has probably taught you something about the pressure-volume relationship of a gas. In blowing up a balloon or an automobile tire, for example, you have found that the pressure increases as the amount of air

Atoms and Molecules in Gases   113

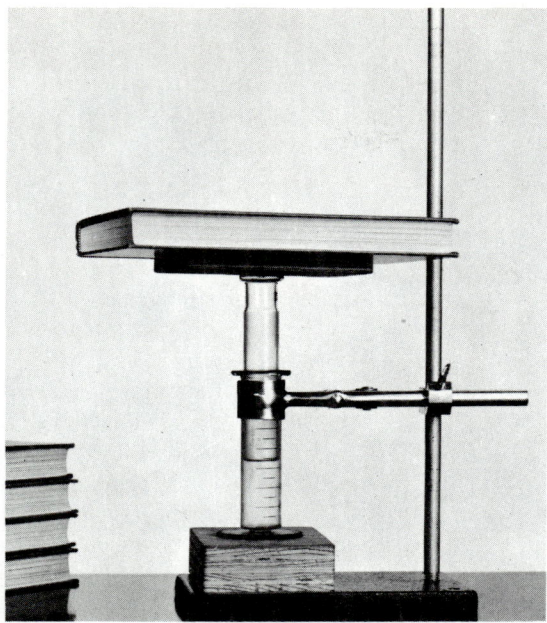

Figure 5-6 *A simple apparatus to demonstrate Boyle's law. The pressure acting on the air trapped within the syringe can be varied, and the resulting volumes of air can readily be noted.*

put into the balloon or tire increases. This is the same as saying that when you take a certain quantity of air from a large space (the earth's atmosphere) at low pressure and push it into a much smaller space, the pressure it exerts becomes greater. In general, when a given amount of gas is in a container of large volume, its pressure on the container walls is low. When it is put into a smaller container it exerts more pressure on the walls.

This idea can be expressed *quantitatively*. That is, there is an exact, measurable relationship between the volume occupied and the pressure exerted by a fixed amount of any gas. This relationship was discovered about 300 years ago by an Irish scientist, Robert Boyle. It is called Boyle's law.

With the apparatus shown in Figure 5-6, the relationship that Boyle discovered can readily be shown. Your teacher may be able to demonstrate it for you. If not, the data given in Table 5-1 will show you what happens. A certain quantity of air has been trapped in the syringe, to which a special

TABLE 5-1 Pressure-Volume Relationships

Observations		Calculated Relationships	
Number of Books N	Volume (ml) V	Pressure (g/cm²) P	$P \times V$
1	34.2	1.43×10^3	4.89×10^4
2	27.4	1.82×10^3	4.99×10^4
3	21.4	2.22×10^3	4.75×10^4
4	18.1	2.61×10^3	4.72×10^4
5	16.3	3.01×10^3	4.91×10^4

Inside diameter of syringe = 2.0 cm
Atmospheric pressure = 1025 g/cm²

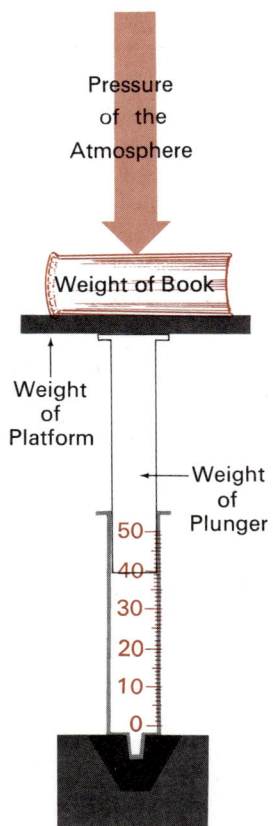

Figure 5–7 *The pressure on the air in the syringe is equal to the sum of the weights of the plunger, platform, and book divided by the area at the end of the plunger, plus the pressure of the atmosphere.*

rubber seal has been added to prevent leakage. (See Figure 5-7.) Since all books placed on the top platform are supported by the column of air in the syringe, which has a constant cross-sectional area, the addition of each book increases the pressure by the same amount. This is assuming that each book has the same mass, which is approximately true.

It can be seen in Figure 5-8 that, as books are added, the volume of air decreases. The actual volumes, as read from the graduations on the syringe in a typical experiment, are listed in Table 5-1.

Now that we have some numbers we may begin to think quantitatively. We may ask the question: Is there any simple mathematical relationship between the pressure and the volume? Before we can answer that we must know the actual pressure corresponding to each volume.

We shall calculate the pressures in grams per square centimeter. Strictly speaking, of course, the gram is a unit of mass, rather than of force. As such, it is not the appropriate unit to use in expressing pressure (force per unit area). However, as you learned in Chapter 2, force is directly proportional to mass: $F = mC$. (C is a constant, explained in Section 2-2.) All of the measurements in our present experiment are made under the same conditions. Thus, it is legitimate to use grams, rather than the corresponding unit of force in the metric system, and it is more convenient for us to do so.

Figure 5-8 *As books are added to the apparatus, the volume of the air trapped within the syringe decreases.*

The total pressure on the confined sample of air is the sum of three contributions. The push due to the book or books is an obvious one. If we have chosen a syringe with an inside diameter of 2 cm, its cross-sectional area is calculated as follows.

$$\text{Area} = \pi \times r^2 = 3.14 \times (1 \text{ cm})^2 = 3.14 \text{ cm}^2$$

The average mass of a book can be determined. Suppose it is 1240 grams. Then each book on the platform contributes

$$\frac{1240 \text{ grams}}{3.14 \text{ cm}^2} = 395 \text{ g/cm}^2$$

to the total pressure on the confined gas.

Another contribution to the total pressure is the mass of the inner (solid) part of the syringe and the platform atop it. They also rest on the 3.14 cm² cross-sectional area. Let us suppose that the mass of these objects is 16.0 grams. Then their contribution to the pressure is

$$\frac{16.0 \text{ g}}{3.14 \text{ cm}^2} = 5.10 \text{ g/cm}^2.$$

There is one more contribution. It is not as obvious as the others but it is a very important one, nevertheless. The earth's atmosphere is constantly pressing down on the whole apparatus, including the 3.14 cm² cross-sectional area in the syringe. Let us suppose that we read a barometer and find that atmospheric pressure at the time the experiment is being performed is 754 Torr. This reading is equal to 1025 g/cm², since mercury has a density of 13.6 g/cm³ at room temperature. A column of mercury 75.4 cm high exerts a pressure of 1025 g/cm².

$$75.4 \text{ cm} \times 13.6 \text{ g/cm}^3 = 1025 \text{ g/cm}^2$$

We can now calculate the total pressure on the confined air at each volume reading. For instance, when three books are on the platform, we have the following total pressure.

Contribution to Pressure	Pressure (g/cm²)
three books	$3 \times 395 = 1185$
platform, etc.	5
earth's atmosphere	1025
	2215

This can be rounded off to three significant figures and written as 2.22×10^3 g/cm². The other pressures listed in Table 5-1 are calculated in the same way. Only the pressure due to the books is different in each case.

Boyle observed and recorded similar data and then looked for any regularity that might exist. He found a very simple regularity; the pressure multiplied by the volume at that pressure always gave the *same* number, within experimental error. In order to obtain this regularity, he had to make all of his measurements at the same temperature. For the data in Table 5-1, these products, $P \times V$, are listed in the last column.

Boyle's law can be stated as follows: At a fixed temperature, the product of the pressure and the volume of a given mass of gas is constant. In algebraic form, with P representing the pressure, V the volume, and C a constant, Boyle's law is as follows.

$$P \times V = C$$

Another way of stating Boyle's law is to say that for a fixed temperature and mass of gas, the pressure is inversely proportional to the volume. (Or, alternatively, the volume is inversely proportional to the pressure). This means that if one of these quantities changes by a factor of n (say, 2) the other changes by the inverse factor, $1/n$ (say, ½). In other words, if you double the pressure, you halve the volume.

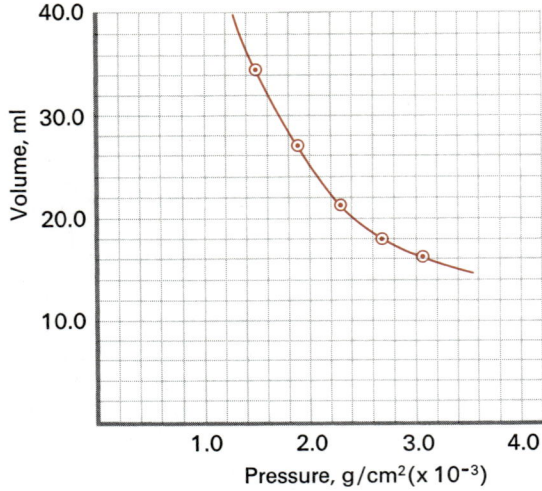

Figure 5-9 *This graph of the* P *and* V *data illustrates the inverse relationship between* P *and* V, *as stated in Boyle's law.*

There is another important way of showing the Boyle's law relationship. When values of P and V are plotted against one another, a curve like the one in Figure 5-9 is obtained. This particular curve was drawn from a plot of the data in Table 5-1. Its shape is characteristic of an inverse proportion.

5-5 Another Look at the Particle Model of a Gas

We began this chapter by devising a very simple model of what a gas might be like, in order to explain some simple observations. We now have established a quantitative relationship, Boyle's law. Using this law, we can test the particle model more thoroughly.

Can the particle model account for other information about gases? An acceptable model must fit all the facts. Fitting all but one will not do. You already know some other facts about gases. You know that they are

Robert Boyle (1627–1691)

Robert Boyle, right, discussing some of his laboratory equipment.

ROBERT BOYLE, who was born in Ireland, was the fourteenth son of the Earl of Cork and received his formal education in England, Switzerland, and Italy. He became interested in medicine, agriculture, chemistry, astronomy, and physics at an early age. Boyle is best known, however, for his work in chemistry and for his pioneering experiments with gases and their properties.

In the early 1650's Otto von Guericke (GAI-ri-kuh), a German scientist, discovered how to build a pump that could produce a vacuum. After many spectacular experiments and demonstrations, he published the results of his work in 1654. Boyle read about von Guericke's work in 1657 and immediately began to construct a pump that was more efficient than the one he had read about. The main feature of Boyle's apparatus was a glass chamber in which he could observe the behavior of various objects as the air was pumped from the chamber. It did not take him long to discover that small animals in the chamber died and candles were extinguished when the air was removed from it. He demonstrated that sound would not pass through a vacuum and also confirmed that air pressure was responsible for the operation of the barometer invented by Torricelli in Italy 27 years earlier. In his first book, published in 1660, Boyle also reported that air appeared to be elastic.

Boyle continued to experiment with his pump and vacuum chamber, and two years later reported that the pressure and the volume of air were inversely proportional if the temperature was held constant. This is the relationship that is known today as "Boyle's law." Boyle never did arrive at an adequate explanation for the behavior of air, although he did disagree with other scientists who thought that air must be somewhat like a pile of coiled wool that springs back when you push down on it.

In reporting many of his later experiments in the fields of heat and temperature, light and color, and in the analysis of many common substances, Boyle described his experimental methods in great detail. He was one of the first scientists to emphasize the importance of orderly procedure and refined technique in carrying out scientific experiments. Although he lived about 300 years ago, Robert Boyle's work still serves as an excellent model of good scientific procedure.

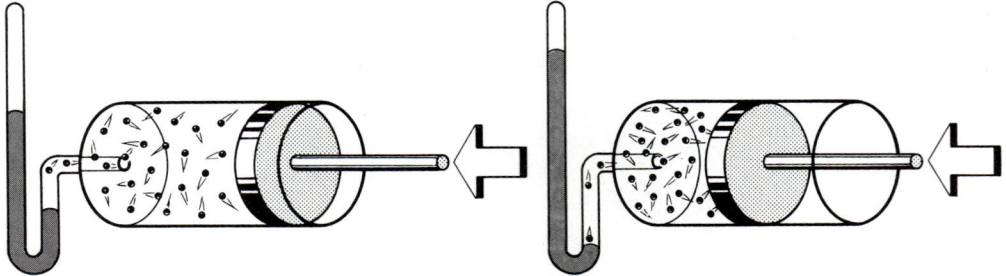

Figure 5-10 *A fixed number of gas molecules in a cylinder (left) exert pressure due to collisions with the walls. If the volume is halved (right) while the temperature (and, hence, the kinetic energy of the molecules) remains constant, the number of collisions per unit time doubles. Thus, the pressure doubles.*

relatively light, since gas bubbles rise in liquids such as water. You also know that gases flow, or diffuse, rapidly through very small holes. For example, even the smallest crack or puncture will allow air to escape from a tire.

The model to be tested against these facts can be stated in detail as follows. A gas consists of a large number of very small particles that move very rapidly in random directions. These particles are, on the average, very far from one another in comparison with their own size.

Both the ease with which gases diffuse and their low densities are accounted for by this model. Moreover, the qualitative facts about the pressure of gases are also explained. Each time a particle strikes the wall of the container and bounces off, it imparts a small "push" to the wall. Because there are an enormous number of particles in an average-sized container and their directions of travel are entirely random, the net effect of all the individual collisions is to exert a continuous and uniform pressure.

Now let us test the model to see whether it can explain the quantitative regularity called Boyle's law. Does it follow from this model that if you take a given quantity of gas and decrease its volume by a factor $1/n$, that the pressure should increase by exactly the factor n?

If you reduce the volume to $1/n$ times the original volume, you have n times as many particles in each unit of volume. Clearly, if the number of particles per unit of volume increases by n, the number of collisions with the walls becomes n times as great in any period of time. Thus the pressure, which is proportional to the number of collisions on each unit area of the wall per unit of time, increases by n. Halving the volume, for example, should double the pressure, as shown in Figure 5-10.

Summary

In the study of gases, the concept of pressure is important. Pressure is the force exerted per unit of area. Its units can be grams per square centimeter, or, as is more general in chemistry, atmospheres. One atmosphere is equal to 14.7 lbs/in^2, or 1034 g/cm^2. One atmosphere is also equal to the pressure of a 760 mm column of mercury. A smaller unit is the Torr, which is $\frac{1}{760}$ of an atmosphere, or the pressure of a 1 mm column of mercury.

Pressure is measured with open or closed manometers or with a barometer, a glass tube filled with mercury and inverted in a small pool of mercury. The surface of the pool is open to the atmosphere. Air pressure on this surface supports a column of mercury 760 mm high at standard atmospheric pressure.

Boyle's law relates the effect of a change of volume with a change of pressure or vice versa, for a given mass of gas at constant temperature. Mathematically it is: $PV = C$. In other words, pressure and volume are inversely related. If pressure goes up, volume goes down.

The particle model fits what is known about gases. According to the model, gases are made up of many minute, rapidly moving particles that are relatively far apart in comparison to their size. This is consistent with the observations that gases are light and can diffuse through small openings. The model is also consistent with Boyle's law. If pressure is due to particles hitting the sides of the container, then when the container is made smaller, the same number of particles have more collisions with the walls per unit of time. Hence, the pressure goes up when the volume is reduced.

Self Check

1. How many atmospheres are in 1 Torr?
2. What is the difference between an open and a closed manometer?
3. Describe a mercury barometer.
4. What are the units of the constant in Boyle's law, $PV = C$?
5. If the pressure of a liter of gas is increased by a factor of 25, what is the resulting volume, assuming constant temperature?

Relative Molecular Masses—Atoms, Molecules, and Moles

INVESTIGATION *5-6 The Masses of Equal Volumes of Gases*

Gases seem to be weightless. Yet if a gas is composed of molecules, the molecules must have substance, hence they must have mass. If it is possible that a gas may have a measurable mass, you may very possibly wonder whether all gases have the same mass or whether they may differ in their masses.

PURPOSE

You will compare the masses of equal volumes of several gases under the same conditions of temperature and pressure.

MATERIALS

Plastic bag, one-quart size Thermometer
Rubber band Barometer

Medicine dropper
Balance
Graduated cylinder, 2-liter
One-hole rubber stopper, size 5 or 6
Sources of oxygen, carbon dioxide, and possibly another gas
Bottle or jar, half-gallon (or $2\frac{1}{2}$-liter acid bottle)
50-60 cm rubber tubing
Water container or trough

PROCEDURE

Since the volume of a gas varies with its temperature and pressure (conditions that matter), precautions must be taken to keep these variables the same when the masses are determined. Great care must be taken to avoid errors in weighing, since the masses of the gas samples will be at most only a few grams. A few greasy finger marks or a moisture droplet can provide enough mass to interfere with the experiment.

Before coming to the laboratory, read the experimental directions and organize a table for the data and your calculated results. See the suggested tables at the end of the experiment.

Part A.   Oxygen Gas

Figure 5-11

1. Obtain a one-hole rubber stopper (size 5 or 6) with a deep groove cut around it about 1 cm from the large end. Fold, in small pleats, the open end of a plastic bag (one-quart size) around the large end of the stopper. Hold it firmly in place with a rubber band that goes around it in the groove.

2. Remove the rubber cap from a medicine dropper. Hold the glass portion with a towel and carefully twist the tapered end of the dropper part of the way into the small end of the stopper until it is firmly held. (You should now have an assembly like the one in Figure 5-11.)

3. Press out any air in the bag by smoothing it flat. Replace the rubber cap on the medicine dropper and weigh this assembly to the nearest 0.01 g. Record the uncertainty as ±0.01 g.

Figure 5-12

4. Remove the rubber cap and connect the assembly to an oxygen gas source supplied by your teacher. Allow the bag to become fully inflated, as shown in Figure 5-12. Hold the bag assembly by the stopper, and disconnect the rubber tubing from the medicine dropper. Allow any excess gas to escape so that the gas in the bag will be at room pressure, but do not squeeze the bag. Then replace the rubber cap.

5. Place the bag assembly containing the gas at room temperature and

pressure on the balance and record the mass to the nearest 0.01 g. Record the uncertainty as before. *Optional:* If directed to by your teacher, repeat steps 4 and 5 to check your work.

Part B.   Carbon Dioxide Gas

1. Fill the bag with carbon dioxide gas, using the source supplied by your teacher. Make sure that the bag-and-stopper assembly is empty and dry.

2. Repeat steps 4 and 5 of Part A. Try to have the same volume of gas at the same temperature and pressure as before.

3. Repeat the filling and mass determination with carbon dioxide, if instructed by your teacher to do so.

Part C.   Obtaining the Volume of the Bag

If the bag contains a gas other than oxygen, remove the gas. Then fill the bag with air by blowing into it. Try to have the same volume of gas as you used in each of the previous experiments.

1. Measure the volume of air or oxygen in the bag by the method shown in Figure 5-13. Completely fill a large bottle or jug (about $\frac{1}{2}$-gallon size) with tap water. Fit it with a stopper and invert it in a large container of water. Remove the stopper under the water.

2. Remove the cap from the medicine dropper and in its place attach a length of rubber tubing. Put the other end of the hose into the neck of the inverted bottle.

3. Gently press on the bag so that all of the gas will displace an equal volume of water in the bottle. Finally, smooth out the bag to remove all the gas.

4. Pinch the tubing to close it and remove it from the water container. Place a solid stopper in the neck of the bottle under water. Remove the stoppered bottle from the water and set it upright on the table.

5. Measure the amount of water required to refill the bottle, using the largest graduated cylinder available to you. Record the volume of the gas displaced and the uncertainty of this measurement.

6. Record the room temperature and pressure.

DATA AND RESULTS

Tables for recording the following information should help you to organize your data and results. The next section tells how the results can be calculated.

Figure 5-13

Part A. Oxygen Gas

(All measurements in grams.)
Mass of bag and stopper with oxygen, O_2
Mass of empty bag and stopper
    Apparent mass of O_2 gas
        Mass of air displaced by gas-filled bag
        Actual mass of O_2 gas

Part B. Carbon Dioxide Gas

(All measurements in grams.)
Mass of bag and stopper with carbon dioxide, CO_2
Mass of empty bag and stopper
    Apparent mass of CO_2 gas
        Mass of air displaced by gas-filled bag
        Actual mass of CO_2 gas
Ratio of this mass to that of O_2

Part C. Volume of the Bag

Volume of water displaced by a bagful of air (l)
Room temperature (°C)
Atmospheric pressure (Torr)
    Density of air at above temperature and pressure (g/l), from Table 5-2.

Calculations

In order to obtain the true masses of the gases you weighed in this experiment, you must apply a buoyancy correction. As you learned earlier, the apparent mass of an object weighed in air is less than its true mass by the mass of the air that it displaces. Usually when an object is weighed in air, the mass of air that is displaced is so small in comparison to the mass of the object that it can be ignored. However, when the object weighed is large and its density is not much greater than that of air itself, the apparent mass must be given a sizeable buoyancy correction. Only with such a correction can a good approximation to the true mass be obtained.

    The necessary calculations for this experiment can be summarized in the following steps.

1. Find the apparent mass of oxygen in the bag by subtracting the mass of the empty bag from the apparent mass of the bag full of oxygen.

2. Calculate the mass of air displaced by the gas-filled bag. The volume of water displaced by a bagful of air is equal to the volume of air

TABLE 5-2 Density of Dry Air in Grams per Liter, ±0.01 g/l, at Various Temperatures and Pressures

Pressure (Torr)	Temperature			
	15°C	20°C	25°C	30°C
600	0.97 g/l	0.95 g/l	0.94 g/l	0.92 g/l
610	0.98	0.97	0.95	0.93
620	1.00	0.98	0.97	0.95
630	1.02	1.00	0.98	0.97
640	1.03	1.01	1.00	0.98
650	1.05	1.03	1.01	1.00
660	1.06	1.05	1.03	1.01
670	1.08	1.06	1.04	1.03
680	1.10	1.08	1.06	1.04
690	1.11	1.09	1.07	1.06
700	1.13	1.11	1.09	1.07
710	1.14	1.12	1.10	1.09
720	1.16	1.14	1.12	1.10
730	1.18	1.16	1.14	1.12
740	1.19	1.17	1.15	1.13
750	1.22	1.19	1.17	1.15
760	1.23	1.21	1.19	1.16
770	1.24	1.22	1.20	1.18

that the bag, full of gas, must have displaced. The mass of air displaced by the bagful of gas, then, can be found by multiplying the volume of air displaced (in liters) by the density of air (in grams/liter), from Table 5-2. (Include the limits of maximum uncertainty in your result.)

3. Find the actual mass of oxygen in the bag by adding the mass of air displaced to the apparent mass obtained in step 1.

4. In a similar way determine the mass of carbon dioxide gas in the bag and the mass of any other gas used.

5. Compare the mass of each gas measured to that of oxygen by dividing each mass by the mass of the comparable volume of oxygen. Express each ratio as a decimal fraction. Include the uncertainty in this derived result.

A Question to Wonder About

Is there any relation between the comparative masses of equal volumes of gases and the relative masses of molecules?

ADDITIONAL INVESTIGATIONS

How do the masses of equal volumes of gases collected at atmospheric pressure, but at higher than ordinary temperatures, compare with those at room temperature?

5-7 Pressure-Volume Relationships of Different Gases

Boyle's law was deduced from pressure-volume data for air, such as those in Table 5-1. Similar experiments with other gases show that Boyle's law applies to all of them. That is, for all gases, $PV =$ a constant. The numerical value of the constant for each gas is proportional to the mass of gas used. If the mass is doubled, the magnitude of PV doubles. So if the pressure-volume behavior of one gas is to be compared accurately with that of another, the masses of the gases used must be specified. It would seem that the simplest thing to do would be to use the same mass of each gas. (Care must be taken to measure the PV products of all the gases at the same temperature. Temperature, as you will see later, also influences the magnitude of PV.)

Let us imagine that an experiment like the one with the syringe and books was done as follows.

1. The confined gas was a sample of pure oxygen (instead of air, which is a mixture of several gases).
2. The quantity of oxygen confined was 32.0 grams.
3. The temperature was 0°C. (Obviously this would be difficult to do with our simple apparatus.)
4. The measurements were all made accurately enough to give three significant figures.
5. The pressures were expressed in atmospheres and the volumes in liters.

The preceding experimental conditions were not chosen arbitrarily. (The reasons for these choices will become clear to you later.) The result obtained under such conditions is always: $P \times V = 22.4$ liter-atm. (The unit obtained when a volume in liters is multiplied by a pressure in atmospheres is conventionally written "liter-atm".)

Let us now suppose that the same experiment, under the same conditions, is done with two other gases. One is called hydrogen chloride, the other ammonia. Using 32 g of each gas at 0°C, the following results are obtained.

Oxygen	$PV = 22.4$ liter-atm
Hydrogen Chloride	$PV = 19.6$ liter-atm
Ammonia	$PV = 42.1$ liter-atm

Each gas follows Boyle's law, but a 32-g sample of each one gives a different "constant."

There is another way to express our results. It follows naturally from a pertinent question. What mass of each gas must we use to get the same constant as that for 32 g of oxygen (22.4 liter-atm)? It is an easy question to answer. The size of the constant is proportional to the mass of gas used. Thus, for ammonia, if 32 g give a constant that is too large, less gas must be used. To reduce the constant from 42.1 to 22.4, it will be necessary to use only 22.4/42.1 times as much gas.

$$\frac{22.4}{42.1} \times 32.0 = 17.0 \text{ g}$$

Conversely, for hydrogen chloride, more than 32.0 g are needed, the exact amount being:

$$\frac{22.4}{19.6} \times 32.0 = 36.5 \text{ g.}$$

5-8 Some Other Properties of Gases

The pressure-volume relationships that we have just considered in detail for three gases have led us to two equivalent statements. One is that equal masses of different gases, measured at the same temperature, give different PV products. The second statement is that equal volumes of gases, measured at the same temperature and pressure, have different masses. (See Figure 5-14.) Our particle, or billiard ball, model of gases does not provide an explanation for the behavior summarized in these statements. Must we abandon our model? If necessary, we would do so, since a model is only suitable as long as it is compatible with all experimental re-

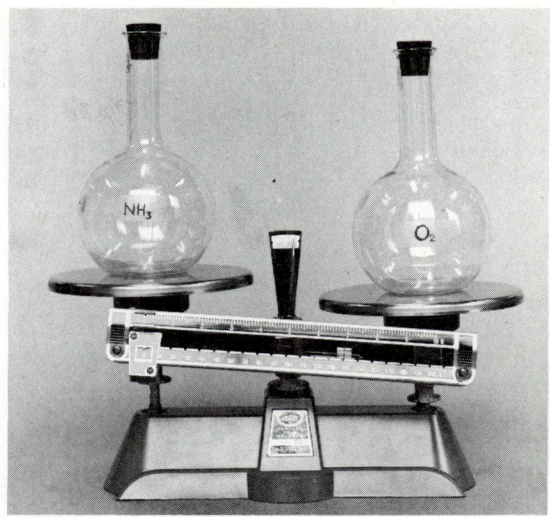

Figure 5-14 *Equal volumes of gases, measured at the same temperature and pressure, have different masses.*

sults. It often happens, however, that a model can be modified in some way so that it explains new results, as well as the older ones for which it was originally designed. In order to decide what sort of a modification might be suitable for our gas model, let us consider some other properties of gases.

Ammonia, oxygen, and hydrogen chloride are quite readily prepared in the laboratory and so their properties are easily observed. We can, of course, make some observations about oxygen simply by looking around us. Since pure air, which contains oxygen, is colorless and odorless, it is reasonable to assume that oxygen must be a colorless and odorless gas. Like oxygen, but unlike several other gases, both ammonia and hydrogen chloride are colorless. They are by no means odorless, however. Anyone who has experienced the characteristic odor of ammonia, in smelling salts or in an ammonia-containing cleaning solution, for example, can believe

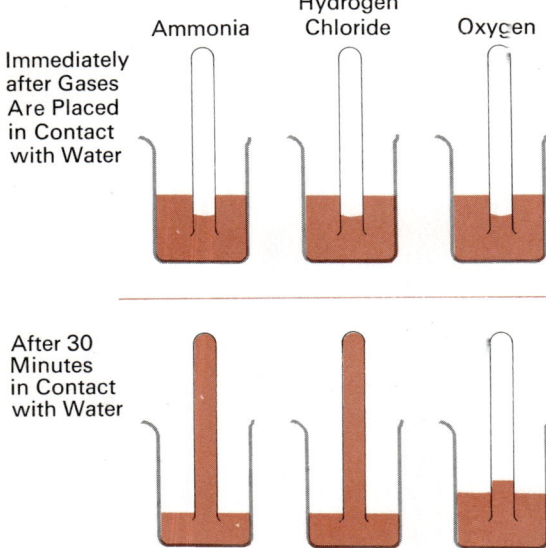

Figure 5-15 *An experiment showing different solubilities of several gases in water. Ammonia and hydrogen chloride are highly soluble, dissolving completely.*

that. The odor of hydrogen chloride is no less irritating than that of ammonia, but it has its own distinctive "sour" quality.

Dissolved oxygen is necessary for aquatic life, but only a relatively small amount of oxygen will dissolve in water (31.6 ml per liter of water at 25°C). Both ammonia and hydrogen chloride dissolve in water to a much greater extent than does oxygen. At least thirty times as great a volume of either gas will dissolve in water at 25°C. Figure 5-15 depicts a simple experiment that shows these dramatic differences in solubility.

How can we explain why these three gases are so different, and yet all follow Boyle's law? We were able to explain Boyle's law by assuming that gases consist of very small (invisible) particles, in rapid motion. But maybe the particles themselves are different.

If that is the case, then such differences could account for the observed differences in gas properties, such as solubility, color, and odor. And if the particles differ in ways that produce such different macroscopic properties, it would be rather incredible if they did not have different masses, too. So our assumption that all gas particles are not the same leads to a possible explanation for the different masses of equal gas volumes. Let us pursue this possibility further.

5-9 Avogadro's Hypothesis

We noted earlier that the volume of a balloon increases as we blow more air into it. This simple observation suggests that the volume of a gas depends on the number of gas particles it contains. So it is reasonable to suppose that gases having the same volume have the same number of particles, if the gas volumes are measured at the same temperature and pressure.

Suppose that an ammonia particle has a mass less than that of an oxygen particle by a factor of 17.0/32.0, while a hydrogen chloride particle has a mass greater than that of an oxygen particle by a factor of 36.5/32.0. In Section 5-7 you learned that the PV product for 32.0 g of oxygen at 0°C is 22.4 liter-atm. If the pressure were one atmosphere, then the volume occupied by the oxygen gas would be 22.4 l. If the assumption that equal volumes of gases contain equal numbers of particles is true, what would be the masses of equal volumes of ammonia and hydrogen chloride, measured under the same conditions? If there are n particles of oxygen in 22.4 liters, the mass of each particle is $32.0/n$ grams. From this you can see that the mass of an ammonia particle is $17.0/n$ grams and the mass of a hydrogen chloride particle is $36.5/n$ grams. Obviously, n particles of ammonia would have a mass of 17.0 grams and

n particles of hydrogen chloride would have a mass of 36.5 grams. These are the results that we calculated from pressure-volume data in Section 5-7. (See Figure 5-16.)

The assumption that equal volumes of gases, measured at the same temperature and pressure, contain equal numbers of particles is consistent with all earlier assumptions about gases and with all of the experimental results we have discussed. This assumption, or *hypothesis*, was first suggested in 1811 by the Italian scientist Amadeo Avogadro, and is known as **Avogadro's hypothesis.** The correctness of this idea has been shown by many experiments and is accepted beyond any doubt. It could, therefore, be called a law, but, curiously, it is still generally referred to as Avogadro's "hypothesis."

5-10 Atoms and Molecules

In preceding sections of this chapter we have discussed some of the facts and reasoning that support our model of a gas. Before discussing other properties of gases, it will be helpful to develop our picture of gas particles a little further. In doing so, certain facts and conclusions will be presented that chemists have learned from experiments, although the experiments themselves will not be discussed until later. Some new terminology will also be introduced. The process of learning any science includes learning the names of things as well as the facts and the concepts. The facts and ideas are truly the science. The specialized vocabulary makes possible discussion and communication of facts and concepts.

The particles making up nearly all known gases (all but six) are called **molecules.** These molecules themselves consist of two or more smaller particles called **atoms.** The six gases that do not consist of molecules are

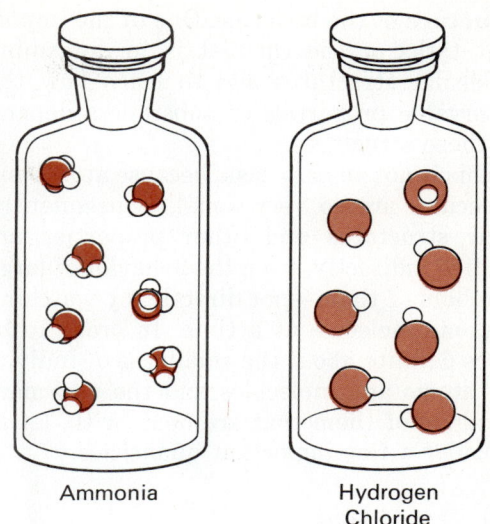

Figure 5-16 *Equal volumes of gases at the same pressure and temperature have the same number of molecules, but the molecules are not identical.*

those in which the particles are themselves single atoms. They constitute a particular group of chemical substances called the **noble gases.** The word "noble" indicates that their atoms do not combine with one another to form molecules.

It is important to remember that the *general* properties of gases that depend on their consisting of small, high-speed particles are not dependent on whether the particles are single atoms or the groups of atoms called molecules. The *specific* characteristics of a given gas, such as its color, odor, solubility in water, and the mass of its particles, do depend on what kinds of atoms comprise the molecules.

Some properties depend on the exact way in which the atoms comprising the molecule are arranged. This arrangement is called the

structure of the molecule. One of the important tasks of the chemist is to determine molecular structures and to learn how the observable properties of substances depend on these structures.

This is not an easy task, because atoms and molecules are so very small. Consequently, their structures and other properties are studied indirectly, from the behavior of large numbers of them—not directly, by watching any one molecule at a time. In order to be more definite about the smallness of individual atoms and molecules and the enormous numbers of them that are dealt with, let us consider a few important quantities.

5-11 The Mole

Samples of matter that are ordinarily dealt with, even very tiny specks barely visible to the naked eye, contain a huge number of atoms or molecules. For this reason it is convenient to count molecules in very large groups. These groups are called *moles*.

It is not uncommon to measure objects that come in large numbers by groups. For example, oranges are often sold by twelves, called dozens, and you count pennies by the hundred, the unit 100 being called a dollar. The mole is a much larger group of objects, but the principle is the same. It is perfectly possible to talk about a mole of oranges or pennies, but the mole is more appropriately used for very much smaller objects. For the time being, the only such small objects with which you will be concerned are atoms and molecules. Later, you will encounter other objects for which the mole is an appropriate unit.

Just how large a group of objects is a mole? The number of objects in a mole has been found by experiment to be 6.02×10^{23}. A **mole** is 6.02×10^{23} atoms, molecules, or other objects.

It is probably not surprising to you that

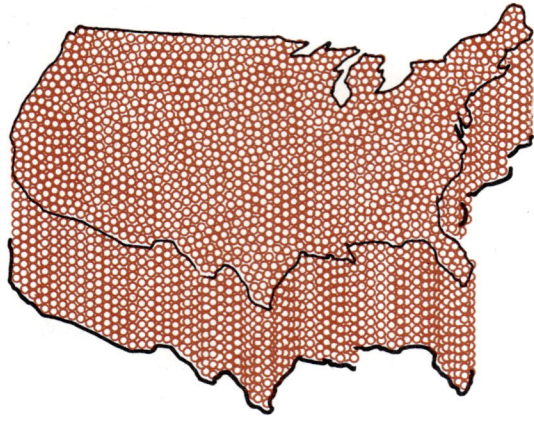

Figure 5-17 *If the entire surface of the continental United States were covered with layers of beads 1 cm in diameter, it would take about one million layers to use up one mole of beads.*

the unit for molecules is very large. (Figure 5-17 may give you a better idea of just how large it is.) You may find it surprising that the mole is such a peculiar number, 6.02×10^{23}. (More precisely, it is 6.02252×10^{23}.) Why this number instead of some exact "round" number such as 10^{23} or 10^{20} or one billion (10^9)?

The reason is that this rather odd number came from experimental measurement. It was defined in terms of a measurable quantity. The mole was first defined as the number of molecules in a sample of oxygen having a mass of exactly 32 grams. (See Figure 5-18.) In 1961 the definition was changed, basing the mole on measurements that can be made with still greater accuracy. The mole is now defined as the number of atoms in exactly 12 grams of a form of carbon known as carbon-12. This form of carbon is known as an *isotope* of carbon. Isotopes are discussed in Chapter 10. The new definition resulted

Figure 5-18 *A mole (32 g) of oxygen gas is a readily measurable quantity. At 1 atm and 0°C, a mole of oxygen would fill slightly more than three basketballs.*

volved will be far smaller than we can measure in our own laboratories.

You were introduced, in Section 5-9, to Avogadro's hypothesis. That is, equal volumes of gases, measured at the same temperature and pressure, contain equal numbers of molecules. It follows that the number of molecules in 32.0 g of oxygen is the same as that in 17.0 g of ammonia, or 36.5 g of hydrogen chloride, or in any substance that, in the gaseous state, occupies the same volume as that occupied by 32.0 g of oxygen. Thus, a mole of any gas consists of 6.02×10^{23} molecules (or atoms). The number, 6.02×10^{23}, is known as Avogadro's number, in honor of that scientist's contribution to its eventual determination.

It is important to know precisely what objects make up a given mole. When chemists speak of a mole of the gas oxygen, they mean 6.02×10^{23} *molecules*. When they refer to a mole of the noble gas neon, they mean 6.02×10^{23} *atoms*. In some cases, it may be helpful to specify just what objects you have in mind. For example, a mole of water *molecules* contains 6.02×10^{23} oxygen *atoms* and 12.0×10^{23} hydrogen *atoms*.

in a change of only -0.0045% in the number of particles in a mole, which is unimportant in all but a few highly specialized aspects of chemistry. The original definition of the mole is valid for all practical purposes. If we consider the mole to be the number of molecules in 32.0 grams of oxygen, any error in-

Summary

It is found by experiment that for 32.0 grams of oxygen at 0°C the product of the pressure, P (in atmospheres), multiplied by the volume, V (in liters), is equal to 22.4. When the product P times V, of 32.0 grams of other gases is measured at 0°C, it is found that, while PV is a constant, it is not equal to 22.4. To make P times V equal to 22.4, a different mass of each gas must be used. This is because the particles making up different gases have different masses.

At the same temperature and pressure, equal volumes of all gases contain the same number of molecules. This statement is called Avogadro's hypothesis, although it is no longer a hypothesis but a well-established fact. In 22.4 liters of any gas at 0°C and one atmosphere pressure—which for oxygen is 32.0 grams—there are 6.02×10^{23} particles. This number is called Avogadro's number and is, by definition, one mole of the gas.

The particles making up a gas are, in general, called molecules. Molecules are made up of atoms. A mole is 6.02×10^{23} atoms, molecules, or other objects.

Self Check

1. The PV product of 40 g of gas A gave a constant of 10, while the PV product of 40 g of gas B gave a constant of 20. What mass of gas B would it be necessary to use to get the same constant as that for gas A?
2. What is Avogadro's number?
3. Define a mole of any substance.

Gaseous Reactions

5-12 Applying Avogadro's Hypothesis to Gaseous Reactions

Certain pairs of gases, when mixed together, combine with one another to form new gases, or to form liquids or solids. Not all pairs of gases do so by any means. For example, if oxygen is mixed with either ammonia or hydrogen chloride, no interaction takes place at room temperature. The molecules mingle together, but nothing further occurs. If ammonia and hydrogen chloride are mixed, however, a white powder is formed immediately.

Suppose that a vessel containing hydrogen chloride at a particular pressure is connected with another vessel of the same volume containing ammonia, at the same pressure, and the connecting tube is then opened so that the gases can mix. The white solid forms immediately and the gas pressure drops to zero.

According to Avogadro's hypothesis, the two vessels of gas must have contained the same number of molecules, since they had the same volume and pressure and both were at room temperature. The experiment showed that the number of ammonia molecules was just sufficient to combine with all of the hydrogen chloride molecules, since both gases were completely used up. The obvious conclusion is that the white solid product is composed of ammonia molecules and hydrogen chloride molecules in exactly equal proportions, that is, a 1:1 ratio.

There are many other reactions between gases, or between a gas and a liquid or solid, in which gases are consumed or formed in simple volume ratios. Some examples follow. They are illustrated in Figure 5-19.

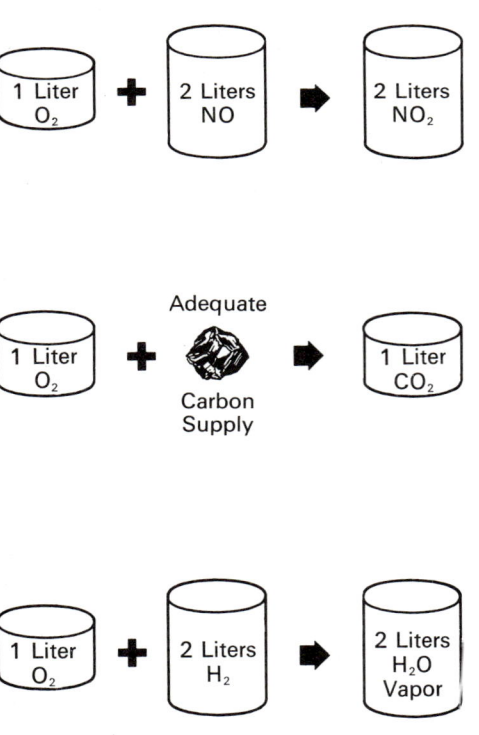

Figure 5-19 *Measurable volumes of gases can be combined chemically to form measurable volumes of products.*

1 liter of the gas oxygen
plus
2 liters of the gas nitric oxide
form
2 liters of the gas nitrogen dioxide

1 liter of the gas oxygen
burns
coal (mainly carbon)
forming
1 liter of the gas carbon dioxide

1 liter of the gas oxygen (at 100°C)
plus
2 liters of the gas hydrogen (at 100°C)
form
2 liters of water (vapor at 100°C)

In view of Avogadro's hypothesis, all these results give us knowledge about how many molecules of one kind react with how many molecules of a second kind to form how many molecules of the product, insofar as gaseous materials are involved. From the first example, it follows that:

1 molecule of oxygen
combines with
2 molecules of nitric oxide
forming
2 molecules of nitrogen dioxide.

Similarly, from the other two examples, you can deduce the following relationships.

1 molecule of oxygen
combines with
coal (mainly carbon)
forming
1 molecule of carbon dioxide

1 molecule of oxygen
combines with
2 molecules of hydrogen
forming
2 molecules of water

The preceding examples involve an extremely important line of reasoning. Without having seen any molecules, or even having tried to see any, you were able to figure out how many molecules of various kinds must be involved in forming molecules of other kinds.

In order to carry out this important line of reasoning, it is, of course, necessary to know the results of the appropriate experiments. Experiments that give such results involve three basic steps. First, the volumes of the gaseous substances that are to be reacted are measured, at the same temperature and pressure. Then these measured volumes are mixed together, so that they can react. Finally, the volumes of the gaseous products must be determined, at the same temperature and pressure as that of the reacting gases.

5-13 Chemists' Shorthand: Formulas and Equations

In Figure 5-19, you probably noted that the containers were labeled in a code, or shorthand. For example, in one case, you saw something that we can write here on the page, omitting the drawings of the containers, as

$$O_2 + 2H_2 \longrightarrow 2H_2O.$$

This says the same thing as the statement that we made earlier, in words.

1 molecule of oxygen
combines with
2 molecules of hydrogen
forming
2 molecules of water

It is obvious that the statement in words, even if written out in continuous lines, is long and cumbersome. The *compactness* of the code or shorthand is its most obvious advantage, although it has others.

Chemical symbols, formulas, and equations will be treated in more detail a little later, but it is worthwhile to start getting used to them right now. Their use will be of help in the following section.

In the "code statement" at the beginning of this section, the letters H and O, numbers on the same line with the letters, and numbers that are subscripts to the letters are used. The letters H and O represent atoms of hydrogen and oxygen respectively. (Other atom symbols seen in Figure 5-19 are N for nitrogen and C for carbon.)

A subscript number following a letter tells how many such atoms, if there are more than one, are present in each molecule. Thus the code word, or **formula,** for a water molecule is H_2O. It states that this molecule contains two hydrogen atoms and one oxygen atom. Similarly, both oxygen molecules, O_2, and hydrogen molecules, H_2, contain two atoms of the appropriate type.

The number in front of the formula for a molecule tells how many molecules are involved. When there is only one, the number 1 is not written.

Finally, the entire set of formulas, written with plus signs between some and an arrow (or maybe an equals sign), makes what we call an **equation.** It tells us that some molecules combine with others, in *certain proportions*, to produce still others, in *certain proportions*.

5-14 *Gaseous Reactions Can Give Information About Molecular Formulas*

It is often possible to gain information on how many atoms make up certain molecules by carefully analyzing the data for reactions involving gases. For instance, the formula O_2 for the oxygen molecule (rather than, say, O or O_3) can be justified by such reasoning. Consider the reaction of nitric oxide, a colorless gas, with oxygen, another colorless gas, to produce nitrogen dioxide, a reddish-brown gas. In this reaction, two molecules of nitric oxide react with one molecule of oxygen to form two molecules of nitrogen dioxide.

The reasoning from this point on is very simple. All of the atoms in the one oxygen molecule have been divided among two molecules of nitric oxide. Since the two nitric oxide molecules are identical, each must have taken an equal number of atoms from the oxygen molecule. Conclusion: The oxygen molecule cannot contain less than two identical atoms. It may contain any other *even* number of atoms, 4, 6, 8, etc., but it cannot contain 1, 3, 5 or any *odd* number of atoms. Thus, the simplest acceptable structure for an oxygen molecule is one with two atoms joined together. There are other kinds of data that show this simplest, two-atom model to be correct.

Once we have learned that molecules of hydrogen and oxygen are H_2 and O_2, respectively, it is possible to derive from the reaction by which these two gases form water the molecular formula of the water molecule. Let us now consider how to do that.

First, exactly two volumes of hydrogen and one volume of oxygen are mixed together. Both volume measurements must, of course, be made at the same temperature and pressure. Then an electric spark is introduced into the mixture. The two gases do not react when they are simply mixed, but react violently when the slightest disturbance, such as a spark, acts upon them. The reaction is instantaneous, and all of the hydrogen and oxygen are consumed.

The next step is to measure the volume of water produced by the reaction. It must be heated above its boiling point, 100°C, in order to convert it all to a gas. Then its volume is measured and compared with the volume of hydrogen measured earlier. (Both volumes must be compared at the same temperature and pressure.) The volume of water vapor produced is found to be equal to the volume of hydrogen that reacted.

These facts, interpreted in accordance with Avogadro's hypothesis, mean that two molecules of hydrogen combine with one molecule of oxygen to form two molecules of water. The only possible interpretation of this is that each of the water molecules consists of two hydrogen atoms and one oxygen atom. Their molecular formula is H_2O. The entire process of combining hydrogen (molecules) with oxygen (molecules) to produce water (molecules) can then be expressed in symbols by the chemical equation that is already familiar from the earlier discussion.

$$2H_2 + O_2 = 2H_2O$$

Figure 5-20 shows molecular representations of the reactions discussed in this

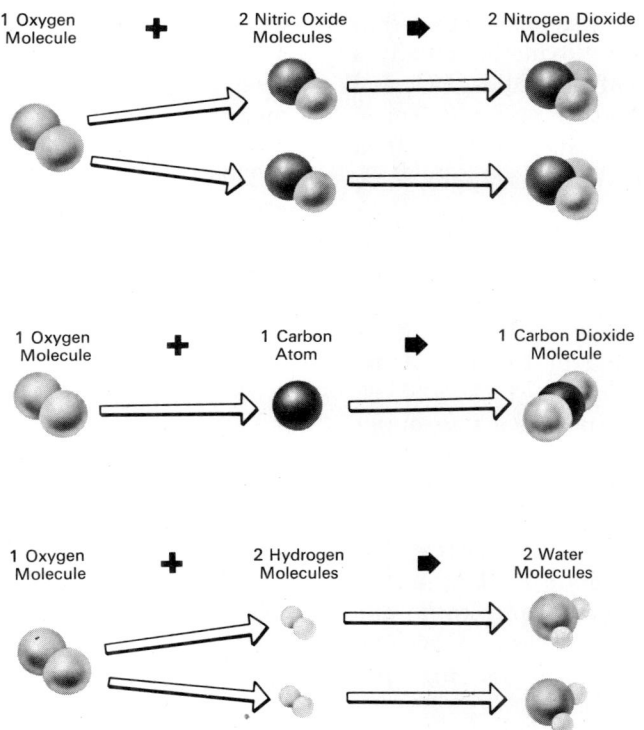

Figure 5-20 *With molecular formulas known, interactions among individual atoms and molecules can be stated. Those shown above illustrate the logic that has led to the creation of chemical equations.*

section, and of the carbon-oxygen reaction discussed earlier. You have just seen how it is possible to reach some conclusions about molecular formulas from experimental results. Although we have not yet considered how the *shapes* of molecules are determined, molecular models are presented in this chapter to help you visualize what is going on. Chapter 15 is devoted to a discussion of molecular shapes.

5-15 Proving the Formula of Nitric Oxide ★

Nitrogen and oxygen combine to form several different compounds, called oxides of nitrogen. Two of these are NO and NO_2, which are named nitric oxide and nitrogen dioxide, respectively. It has been found that the gas called nitrogen dioxide is actually a mixture of two kinds of molecules, NO_2 and N_2O_4. At room temperature and one atmosphere pressure, most of the molecules in a sample of nitrogen dioxide are actually N_2O_4 molecules.

The question then arises: Does a similar situation prevail for NO, nitric oxide? Does nitric oxide really consist entirely of NO molecules, or does it consist of N_2O_2 molecules? Or does it contain a mixture of NO and N_2O_2 molecules?

A hundred years ago, for reasons no longer of interest to us, some chemists thought that nitric oxide molecules should have the formula N_2O_2, and not NO.

The question of which is the correct formula did not have to remain for long a subject of argument, provided one knew and accepted Avogadro's hypothesis. However, while Avogadro's hypothesis was first put forward in 1811, many chemists did not accept it, even 50 years later. As late as 1899 there were still chemistry books, written by well-known chemists, in which nitric oxide was given the formula N_2O_2. With Avogadro's hypothesis, a simple experiment can show conclusively that this cannot be so, and that the correct formula must be NO. Let us see how this is done.

The experiment can be done quite easily because of the following facts. (1) Nitric oxide and oxygen are only slightly soluble in water. Thus, these gases may be collected and stored over water. (2) Nitric oxide and oxygen react quickly to produce nitrogen dioxide ($NO_2 + N_2O_4$). (3) Nitrogen dioxide is very soluble in water.

The experiment consists of allowing a known volume of nitric oxide to react with a larger known volume of oxygen. The volume of the remaining oxygen can then be measured and the volume of oxygen that reacted can be calculated. (See Figure 5-21.)

This experiment shows that a given volume of nitric oxide reacts with *half* its own volume of oxygen. Consider now the two possibilities, depending on the correct formula for nitric oxide.

Case 1: $2NO + O_2 = 2NO_2$ (dissolved)
Case 2: $N_2O_2 + O_2 = 2NO_2$ (dissolved)

In Case 1, two ml of nitric oxide will react with only one ml of oxygen. In Case 2, two ml of nitric oxide will react with two ml of oxygen. The experiment clearly shows that Case 1 is correct. The true molecular formula of nitric oxide is NO.

Summary

By measuring the relative volumes of gases involved in chemical reactions, it is often possible to tell how many molecules of each kind are involved in the reaction. It is also

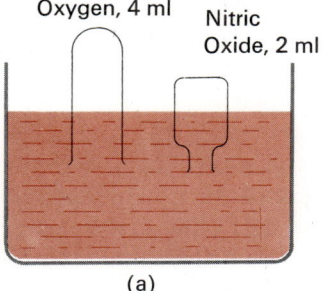 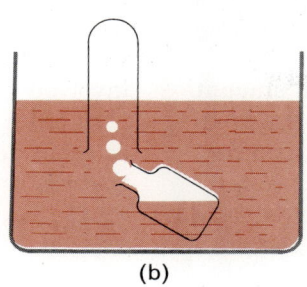 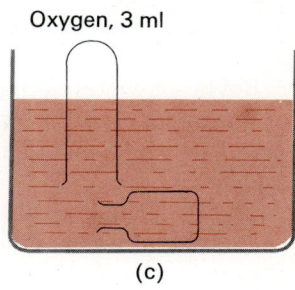

Figure 5-21 (a) At the same temperature and pressure, 4 ml of oxygen and 2 ml of nitric oxide are collected over water. (b) The nitric oxide is added to the larger tube, where it reacts with the oxygen. (c) After reaction, 3 ml of oxygen remain.

possible to gain information on how many atoms there are in each type of molecule.

Molecules may be represented in a shorthand notation, such as H_2O for the water molecule. Reactions may be represented by equations such as

$$2H_2 + O_2 \longrightarrow 2H_2O.$$

This shows that two molecules of hydrogen, H_2, react with one molecule of oxygen, O_2, to form two molecules of water, H_2O.

Self Check

1. How do you know that the simplest formula for oxygen is O_2?
2. Write the chemical equation for the formation of water from the gases hydrogen and oxygen.
3. The ammonia molecule is composed of one nitrogen atom, N, and three hydrogen atoms, H. Write the molecular formula for ammonia.

Chapter Summary

Very simple, everyday observations, such as the process of blowing up a balloon and the behavior of billiard balls on a billiard table, suggest that the properties of gases can be explained with the use of a model. According to the model, gases consist of minute, invisible particles called molecules. The molecules move rapidly, making elastic collisions with the walls of their container. These collisions cause the pressure exerted by the gas.

Pressure is defined as the force per unit area. Useful units for pressure are the atmosphere and the Torr. These are the pressures exerted by columns of mercury 760 mm and 1 mm high, respectively. Gas pressure is measured with devices called manometers and barometers.

Boyle's law states that, for a fixed amount of gas and a constant temperature, the product of pressure and volume, $P \times V$, is a constant.

The molecular model fits all known facts concerning gases and helps explain the pressure-volume regularity.

Avogadro's hypothesis states that equal volumes of gases at the same pressure and temperature contain equal numbers of molecules. One mole of a gas contains 6.02×10^{23} atoms or molecules. This number is known as Avogadro's number.

From the behavior of gases, the conclusion is reached that molecules of different gases have different masses. By studying the relative volumes of gas consumed (or produced) in a chemical reaction, chemists can determine how many of each kind of molecule are involved in the reaction.

In order to write the composition of molecules in terms of the atoms they contain, there is a shorthand notation, as illustrated by O_2, H_2, H_2O, and NH_3, for molecules of oxygen, hydrogen, water, and ammonia, respectively. The reactions by which gases combine to produce new molecules can be compactly expressed by using such symbols to write chemical equations. For example, the reaction of nitric oxide, NO, with oxygen, O_2, to give nitrogen dioxide, NO_2, is written as follows.

$$2NO + O_2 \longrightarrow 2NO_2$$

Questions and Problems

1. The balloons that are used for weather study are quite large. When they are released at the surface of the earth they contain a relatively small volume of gas compared to the volume they acquire when aloft. Explain.
2. Consider two identical balloons. At sea level, both balloons are inflated with hydrogen. One balloon is inflated to three times the volume of the other. How will the volumes of these two balloons compare when they reach 20,000 feet? Explain.
3. Methane, CH_4, and oxygen, O_2, are made to react in a sealed metal container. Under the conditions used, the products are hydrogen, H_2, and carbon dioxide, CO_2. Write the equation for the reaction.
4. Using the equation in question 3, state whether the pressure in the vessel changes and, if so, how.
5. Consider two closed containers of the same volume. One is filled with hydrogen gas, the other with carbon dioxide gas, both at room temperature and one atmosphere pressure.
   (a) How do the number of moles of the two gases compare?
   (b) How do the number of molecules of the two gases compare?
6. Gas is slowly added to the empty chamber of a closed-end manometer like the one in Figure 5-5a. Draw a picture of the manometer, showing, in millimeters, the difference in height of the two mercury levels:
   (a) before any gas has been added to the empty gas chamber.
   (b) when the gas pressure in the chamber is 300 Torr.
   (c) when the gas pressure in the chamber is 760 Torr.

(d) when the gas pressure in the chamber is 865 Torr.
7. Repeat question 6(b), but with an open-end manometer like the one in Figure 5-5c. Assume that the atmospheric pressure is 760 Torr.
8. The gas sulfur dioxide combines with oxygen to form the gas sulfur trioxide as follows.

$$2SO_2(gas) + O_2(gas) \longrightarrow 2SO_3(gas)$$

What ratio would you expect for each of the following?

(a) $\dfrac{\text{number of } SO_3 \text{ molecules produced}}{\text{number of } O_2 \text{ molecules consumed}}$

(b) $\dfrac{\text{volume of } SO_3 \text{ gas produced}}{\text{volume of } O_2 \text{ gas consumed}}$

9. Nitrogen gas combines with hydrogen gas to form ammonia gas as follows.

$$N_2(gas) + 3H_2(gas) \longrightarrow 2NH_3(gas)$$

What ratio would you expect for each of the following?

(a) $\dfrac{\text{number of } NH_3 \text{ molecules produced}}{\text{number of } H_2 \text{ molecules consumed}}$

(b) $\dfrac{\text{volume of } NH_3 \text{ gas produced}}{\text{volume of } H_2 \text{ gas consumed}}$

10. One liter of hydrogen gas reacts with one liter of chlorine gas to produce two liters of hydrogen chloride gas, all measured at the same temperature and pressure. From this information show how many atoms are in a molecule of hydrogen and a molecule of chlorine.
11. Four liters of a gas are stored in a cylinder at a temperature of 15°C and a pressure of 300 Torr. The temperature remains constant while a piston moves in the cylinder until the volume of the cylinder is 3 liters. What is the pressure at this new volume?
12. Suppose that the total pressure in an automobile tire is 30 lbs/in² and you want to increase the pressure to 40 lbs/in². What change in the amount of air in the tire must take place? Assume that the temperature and volume of the tire remain constant.
13. At room temperature the density of mercury is 13.6 g/cm³, and that of water is 1.0 g/cm³. How high a column of water is required to exert a pressure of one atmosphere? Give your answer in millimeters and in feet.
14. If 80 g of a substance with a density of 10 g/cm³ were made into a cube, what pressure, in g/cm², would that cube exert on the surface supporting it?
15. Suppose that you have both an open-end and a closed-end manometer connected to the same gas bulb. The closed-end manometer shows a pressure in the bulb of 430 Torr. If the atmospheric pressure is 750 Torr, what will be the difference between the two columns of mercury in the open-end manometer?
16. An 11.2-liter sample of unknown gas, at 0°C and 1 atm pressure, has the same mass as 22.4 liters of oxygen under the same conditions.
    (a) What is the ratio of the masses of the molecules in the unknown gas to the masses of O_2 molecules?
    (b) How many moles of the unknown gas are present?
17. A certain gas is represented by the symbol, X_n, where n stands for the number of atoms in a molecule of element X. It is known that 6.02×10^{23} atoms of element X have a mass of 35.5 grams. If 11.2 liters of X_n at 0°C and 1 atm pressure also have a mass of 35.5 grams, what is the value of n in X_n?

6 More About the Molecular Model for Gases

Scientists can learn about the upper atmosphere by sending instrument-carrying balloons aloft.

The molecular model of a gas has proved helpful in understanding how the pressure and volume of a given sample of gas are related. The regularity called Boyle's law, which expresses a quantitative relationship between pressure and volume, is consistent with the model. You now know that, with the use of Avogadro's hypothesis, the relative masses of molecules can be determined from pressure and volume measurements. Nevertheless, there are still some basic questions that can easily be thought of concerning the behavior of gases and the ability of the molecular model to explain their behavior. Nothing has yet been said about the actual masses of molecules, how pressure and volume are related for a mixture of gases, and how the pressure and volume of a gas are influenced by temperature. These fundamental problems will be considered in this chapter.

Atoms and Molecules Have Definite Masses

6-1 The Definition of Molar Masses

In Chapter 5 the concept of the mole was introduced. It was defined as 6.02×10^{23} molecules, atoms, or other small objects. It was also stated that a mole of oxygen is 32.0 grams of oxygen. Using these two facts, it is easy to calculate the mass of an oxygen molecule.

$$\frac{32.0 \text{ g/mole } O_2}{6.02 \times 10^{23} \text{ molecules/mole } O_2}$$
$$= 5.32 \times 10^{-23} \text{ g/molecule}$$

For most practical purposes, this number is neither convenient nor useful. The actual mass of one molecule is far too small to measure directly. It is much more useful to the chemist to have a unit of mass that can be measured on a laboratory balance, and which corresponds to the vast numbers of molecules he ordinarily deals with in performing experiments. The chemist, then, finds it convenient to deal with *molar* masses, rather than the masses of individual atoms or molecules.

The mass of one mole of a substance is its **molar mass.**

The molar mass of a substance is the mass of 6.02×10^{23} atoms, molecules, or other objects. For example, the molar mass of a substance consisting of atoms (such as argon gas) is the mass of 6.02×10^{23} atoms. Similarly, the molar mass of a substance consisting of molecules (such as oxygen gas) is the mass of 6.02×10^{23} molecules.

We should note that chemists often refer to the mass of a mole of atoms as the "atomic mass," or even the "atomic weight," of the substance. Likewise, the mass of a mole of molecules is often referred to as the "molecular mass" or "molecular weight" of the substance. These terms, taken literally, suggest the masses of individual atoms and molecules, although that is not what chemists usually mean when using the terms. *Molar mass* expresses the desired meaning more precisely, and can be used for either atoms or molecules. We shall use that term in this book.

The mass of a mole of oxygen gas is 32.0 g; that is its molar mass. The mass of any other gas that occupies the same volume as does 32.0 g of oxygen, at the same temperature and pressure, is the molar mass of that gas.

Since a volume of hydrogen chloride gas equal to the volume of 32.0 g of oxygen gas at the same temperature and pressure has a mass of 36.5 g, the molar mass of hydrogen chloride must be 36.5 g. Similarly, the molar mass of ammonia is 17.0 g.

By experiment it can be shown that a mole of hydrogen gas has a mass of 2.0 g. Since the hydrogen molecule contains two atoms, the molar mass of hydrogen atoms is 2.0/2 = 1.0. Similarly, the molar mass of oxygen atoms must be half of the molar mass of its molecules, 32.0/2 = 16.0.

The molar mass of water can be calculated simply by adding the molar masses of the atoms that make up the molecule, as follows.

$$2 \times \text{molar mass of H atoms} = 2 \times 1.0 = 2.0$$
$$1 \times \text{molar mass of O atoms} = 1 \times 16.0 = \underline{16.0}$$
$$18.0$$

This means that a mole of water is 18.0 g of water.

The masses of all the known kinds of atoms have been measured and expressed in the unit of grams per mole. In many cases these masses are known with great accuracy. For example, the molar mass of the average oxygen atom as found in nature is known to be 15.9994! For our purposes, in this book, the rounded-off value of 16.0 is accurate enough. A table giving the molar masses of all atoms to the highest available accuracy is found at the end of this book and also in Appendix 3. You can find there, for example, that the average molar mass of carbon atoms as they are found in nature is 12.01. Rounded off to one decimal place, for our use, this becomes 12.0.

The following example shows how to use the table for determining the molar mass of a compound, given its molecular formula. The compound in question is the gas carbon dioxide, which you detected as a combustion product in Section 1-13.

Example

Carbon dioxide has been found to consist of CO_2 molecules. That is, each molecule contains one atom of carbon and two atoms of oxygen. What is its molar mass?

$$1 \times \text{molar mass of C} = 1 \times 12.0 = 12.0$$
$$2 \times \text{molar mass of O} = 2 \times 16.0 = \underline{32.0}$$
$$\text{Molar mass of } CO_2 \qquad\qquad\quad 44.0$$

6-2 Pressure of a Mixture of Gases: Partial Pressures

So far you have studied the behavior of pure gases, those in which all molecules are identical. How would a mixture of gases behave? Consider the experimental situation depicted in Figure 6-1, which shows three one-liter bulbs at 25°C. The first bulb contains 0.0050 mole of air and has a pressure, indicated on its manometer, of 93 Torr. The second bulb contains 0.0011 mole of water vapor, which exerts a pressure of 20 Torr. The third bulb contains both 0.0050 mole of air and 0.0011 mole of water vapor. The manometer on this bulb registers a pressure of 113 Torr.

This situation shows that the pressure exerted by the mixture of gases is the sum of

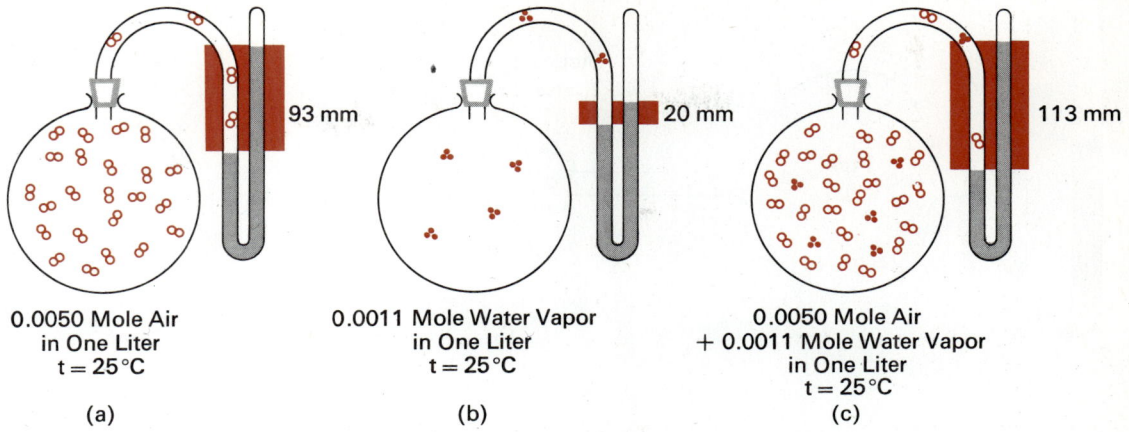

Figure 6-1 *An experiment showing the additivity of partial pressures. The total pressure is the sum of the partial pressures.*

the pressures exerted by each of the gases when it is alone in the vessel. This can be stated as follows, with P_T representing the total pressure. P_1 and P_2 represent the pressures exerted by air and water, respectively, when each is alone in the bulb.

$$P_T = P_1 + P_2$$
$$113 \text{ Torr} = 93 \text{ Torr} + 20 \text{ Torr}$$

The contribution to the pressure made by each of the gases in a mixture (P_1 and P_2 in the preceding example) is called the **partial pressure** of that gas. It is the same as the pressure that the gas would exert if it were alone in the same vessel, at the same temperature. In this example, the total pressure is 113 Torr, the partial pressure of air is 93 Torr, and the partial pressure of water vapor is 20 Torr.

The pressure behavior of a mixture of gases, each exerting its pressure independently of the others, is consistent with the kinetic theory of gases. As the different kinds of molecules move rapidly about in the container, they occasionally collide. The collisions are elastic and the particles continue on their way, eventually striking the wall and thereby making a contribution to the pressure. The average pressure, then, is the same as it would have been if there were no collisions between the different kinds of molecules. Of course, only gases that do not combine or otherwise react with one another can be expected to behave in this simple way.

The concept of partial pressures is an important one. To test, and perhaps improve,

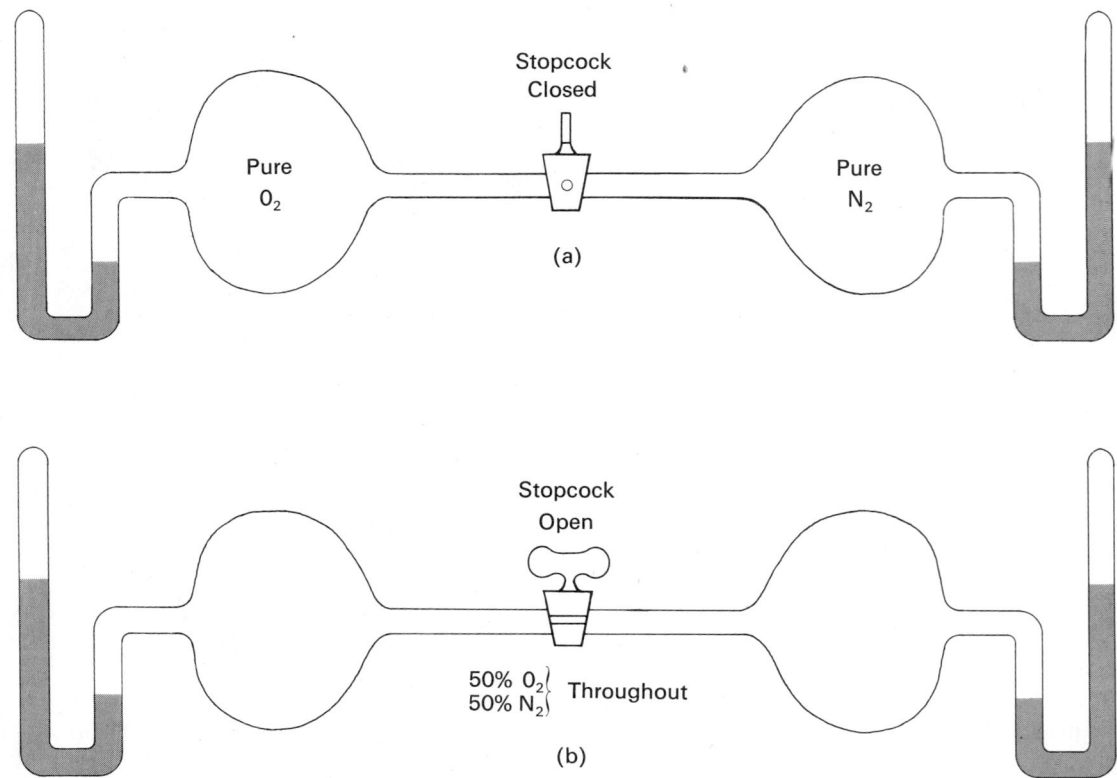

Figure 6-2 In the above experiment, the total pressure remains the same when the stopcock is opened, since each gas exerts a partial pressure in the mixture equal to one-half of its original pressure.

your understanding of it, consider the experiment shown in Figure 6-2. At the start there is pure oxygen, O_2, in one bulb and pure nitrogen, N_2, in the other. The bulbs have the same volume. The pressures, as shown, are the same, and the whole apparatus is at the same uniform temperature. It follows from Avogadro's hypothesis that the number of O_2 molecules is equal to the number of N_2 molecules. What will happen to the pressure when the stopcock is opened?

As Figure 6-2b shows, each manometer continues to indicate the same pressure. This is exactly what you should have expected. When the stopcock is opened, all of the molecules can move throughout the interior and they do so. Very soon the entire apparatus is filled with a uniform 50-50 mixture of O_2 and N_2 molecules. The O_2 molecules are now occupying twice the volume they occupied at first. The same is true of the N_2 molecules. By Boyle's law, each gas exerts only half the pressure it did at first. Both contribute equally to the total pressure.

Let us examine these relationships in another way.

1. Original pressure of O_2
   = original pressure of N_2.
2. Final pressure of O_2
   = $\frac{1}{2}$ original pressure of O_2.
3. Final pressure of N_2
   = $\frac{1}{2}$ original pressure of N_2.
4. Final total pressure
   = final pressure of O_2
   + final pressure of N_2
   = $\frac{1}{2}$ original pressure of O_2
   + $\frac{1}{2}$ original pressure of N_2.
5. Therefore,
   final total pressure
   = original pressure of oxygen
   = original pressure of nitrogen.

Undoubtedly, the most important mixture of gases we ever meet is the air around us. You saw earlier (Section 1-13) that one component of air is oxygen. About one-fifth of the molecules in air are O_2 molecules. Practically all the rest of the molecules are N_2 molecules. If the pressure of the atmosphere at a certain time is 750 Torr, the partial pressures of oxygen and nitrogen are as follows.

$$P_{O_2} = 1/5 \times 750 \text{ Torr} = 150 \text{ Torr}$$
$$P_{N_2} = 4/5 \times 750 \text{ Torr} = 600 \text{ Torr}$$

In addition to O_2 and N_2 molecules, air also contains small amounts of several other gases, including carbon dioxide, CO_2, produced by burning fuels and in other ways, and the vapor of water. There are also atoms of the only gases that consist of single atoms. These gases, called the noble gases, include argon, helium, krypton, and neon. The com-

Figure 6-3 *Portable devices enable scientists to determine the nature and concentration of air pollutants.*

ponent of air that varies the most from time to time and place to place is water vapor. Over desert areas there are only tiny traces of water vapor, while over tropical oceans as much as 5 percent of the air may be water vapor.

Unfortunately, the air everywhere, especially in and near urban areas, contains small but harmful amounts of many other gases. (See Figure 6-3.) These are pollutants, the "ignoble" gases. Three of the chief ones are nitrogen dioxide, NO_2, carbon monoxide, CO, and sulfur dioxide, SO_2. The first two come chiefly from automobile engines. Most of the SO_2 comes from the burning of coal and oil, which contain small amounts of sulfur.

Chemistry and Our Environment
SMOG

The term "air pollution" simply means the presence in the lower atmosphere of undesirable materials. The undesirable materials are called *pollutants*. Practically anything that is not normally in the air due to natural processes is undesirable and must be considered a pollutant. Some pollutants are gases, but others are tiny airborne particles. In this and a later article we shall look at the nature, the sources, and the dangers of some of the major air pollutants.

London Smog. The word *smog* is derived from a combination of the words smoke and fog. It is a well chosen word, since it emphasizes that the kind of visible pollution to which it refers results from a combination of smoke (and associated combustion products) with moisture in the air.

The major cause of this type of smog is the burning of coal or oil that contains sulfur. The sulfur reacts with oxygen to form the gas sulfur dioxide, SO_2. Once the sulfur dioxide is in the air, it is gradually oxidized further to form sulfur trioxide, as shown in the following equation.

$$2SO_2 + O_2 \longrightarrow 2SO_3$$

When the air is moist, the sulfur trioxide reacts with the water vapor to form sulfuric acid.

$$SO_3 + H_2O \longrightarrow \underset{\text{sulfuric acid}}{H_2SO_4}$$

The sulfuric acid molecules, in turn, attract additional water molecules until the final product is a haze of droplets of sulfuric acid, a deadly, corrosive, poisonous substance.

Trafalgar Square, London, at mid-morning, May 12, 1952.

This 18th century marble statue at the Palazzo Giusti in Venice clearly shows the ravages of pollution.

of beauty and history, can be seen today in the city of Venice, in Italy. The level of sulfuric acid is sufficiently high in the air of Venice, due to nearby factories, that statues and building ornamentation are being "erased." The stone of which these statues and buildings are made is mainly composed of calcium carbonate. If you drop a lump of calcium carbonate into a beaker of dilute sulfuric acid, it dissolves quickly, giving off carbon dioxide.

$$\underset{\text{calcium carbonate}}{CaCO_3} + H_2SO_4 \longrightarrow \underset{\text{calcium sulfate}}{CaSO_4} + H_2O + \underset{\text{carbon dioxide}}{CO_2}$$

That is exactly what the air of Venice is doing to its buildings and sculpture!

In addition to oxides of sulfur and the resulting sulfuric acid, other pollutants enter the air as a result of burning coal and oil. For one thing a certain amount of finely divided carbon—soot—is produced. This is why buildings in cities turn black. Needless to say, the inhalation of this soot is not good for the lungs. Worse yet, there are a number of carbon-containing compounds that are also emitted in small quantities. These substances are *carcinogenic,* that is, they cause cancer.

Sulfuric acid in the air is very toxic to all living things. In animals it attacks bronchial and lung tissue and its effects can be fatal. In London in 1952 the sulfurous smog became so severe, due to a static weather pattern, that it led directly to the deaths of at least 4000 people in 5 days. Long-term continuous exposure to lower levels of this type of pollution must surely be injurious to health, even if disability and death do not result quickly. As a result of the fatal smog of 1952, the open burning of coal in home fireplaces in London was banned. Since this had been the major form of home heating in that city, the smog was greatly reduced. There is not much chance of another killing fog in London or any other city at the present time. However, the burning of coal and oil to heat buildings, to supply certain forms of industrial power, and, most particularly, to run electric generators still goes on, at an ever increasing rate.

A very dramatic and sad consequence of this type of pollution, to anyone with a sense

As urban areas continue to get larger and our demands for energy continue to increase, air pollution of the type we have just been discussing will become more and more of a problem. What is the answer? One obvious answer is that coal and oil of the lowest possible sulfur content will have to be used. This will add to the cost very substantially, since the removal of sulfur from these materials on a large scale is expensive. Another possibility is to devise efficient methods for removal of sulfur oxides and other pollutants from the chimney stacks of the power plants. In the long run we may have to level off our demands for power.

TABLE 6–1 The Approximate Composition of Dry Air on a Typical Day in a Large City

Substance	Formula	% of Molecules
Nitrogen	N_2	78.1
Oxygen	O_2	20.9
Argon and other noble gases	Ar (Ne, Kr, He, Xe, Rn)	0.9
Carbon dioxide	CO_2	0.03
Hydrogen	H_2	0.07
Pollutants	NO_2, SO_2, and others	

Table 6-1 gives a list of the chief components of "dry air," that is, air from which the water vapor has been removed, on a typical day in a large city.

Summary

Chemists usually concern themselves with molar masses, rather than the actual masses of individual atoms and molecules. To illustrate, the molar masses of H_2 and N_2 are 2.0 and 28.0 (grams per mole), respectively. From these numbers we can say that the molar mass of ammonia, NH_3, is 14.0 g + (3 × 1.0 g) = 17.0 g.

In a mixture of gases, the total pressure of the mixture is made up of the partial pressures of the individual gases. The partial pressure is the pressure that would be exerted by the molecules of a gas if they were the only molecules present, under the same conditions of volume and temperature.

The partial pressure of a given gas in a mixture is the same fraction of the total pressure as the fraction of its molecules in the total number of molecules in the mixture.

Dry air is a mixture of gases, consisting of about one-fifth O_2, about four-fifths N_2, and about one percent of other gases, mainly CO_2, argon, and, unfortunately, pollutants.

Self Check

1. What are the units for molar masses?
2. Using the table of molar masses at the end of this book, calculate the molar mass of a molecule with the formula CH_4O.
3. If a mixture of hydrogen and nitrogen with a total pressure of 2 atm contains 3 moles of H_2 and 1 mole of N_2, what is the partial pressure of each gas?
4. What are the two main constituents of air?
5. What general term describes the gas SO_2 when it is found in the air?

Temperature Changes Affect Pressure and Volume

6-3 The Volume Occupied by a Mole of Gas at Different Temperatures: Charles' Law

In Chapter 5 you learned that 32.0 grams of oxygen at 0°C and one atmosphere of pressure occupy a volume of 22.4 liters. This is an experimental fact. The reason for choosing that particular mass of oxygen was not given. Now you realize the significance of using 32.0 grams. This is, *by definition*, a mole of oxygen. You also know that on the basis of Avogadro's hypothesis, a mole of any other gas will occupy this same volume at the same temperature and pressure.

It is customary to take the conditions

$$\text{pressure} = 1 \text{ atmosphere}$$
$$\text{temperature} = 0° \text{ Celsius}$$

as standard temperature and pressure, abbreviated as STP. The general rule then is: *One mole of any gas occupies 22.4 liters at STP.*

Boyle's law can be written as

$$P \times V = 22.4 \frac{\text{liters} \times \text{atm}}{\text{mole}}$$

for one mole of gas at 0°C. With the use of this expression, it is easy to predict how the volume will change as the pressure is changed. Suppose, however, that the temperature of the gas is changed, while the pressure is kept constant. How will the volume change?

This question was studied and answered in 1787 by a French scientist, Jacques Charles. He discovered an extraordinary regularity, which we call **Charles' law.** This states that *all gases expand by the same fraction of their original volumes when they are heated over the same temperature range*. It can be illustrated by a simple experiment.

Into a small-bore glass tube, about half a meter in length and closed at one end, a drop of mercury is placed. The mercury falls and eventually traps a sample of air in the bottom of the tube, as shown in Figure 6-4a. Since the tube has a uniform bore, the length of the air column can be used as a measure of its volume. The mercury plug can move up and down as the gas expands or contracts and thus maintain a constant pressure.

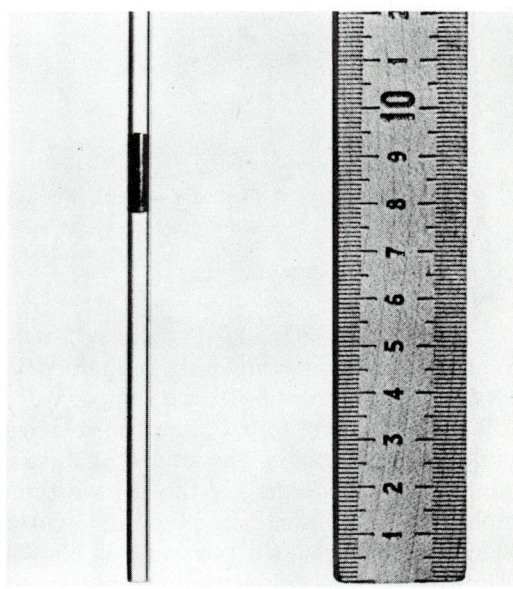

Figure 6-4a *An apparatus for demonstrating Charles' law. A measurable volume of air has been trapped in the glass tubing by the mercury plug.*

 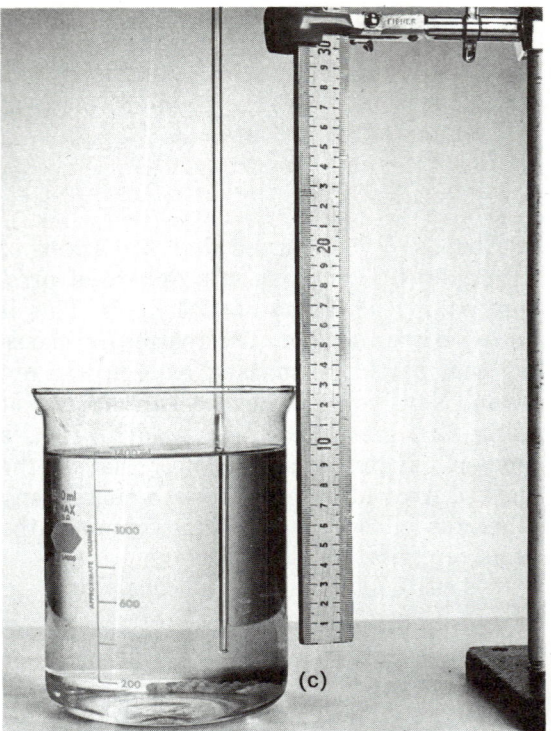

Figure 6–4b,c *The volume of air trapped in the glass tubing undergoes changes when the tubing is placed in an ice-water mixture (b) and in hot (nearly boiling) water (c). These changes demonstrate Charles' law.*

When the tube is completely immersed in ice water, as in Figure 6-4b, the temperature of the gas will come to 0°C, and the relative volume (as measured by the length of the air column) can be recorded. If the tube is then completely immersed in boiling water (100°C) the volume will increase, as shown in Figure 6-4c. The volume may be measured at intermediate temperatures and at still higher temperatures. An oil bath is used in place of the water bath at temperatures greater than 100°C (since the boiling point of oil is well above that of water). In this way the kind of data shown in Table 6-2 can be

TABLE 6–2 Change in Air Volume with Change in Air Temperature

Temperature (°C)	Relative Volume (Compared to Volume at 0°C)
200	1.73
150	1.55
100	1.37
50	1.18
0	1.00

collected. It is important that the measurements of volume be made with the air sample completely immersed in the bath. Otherwise, serious errors will result.

The data in Table 6-2 are for air, but Charles' experiments showed that the same results are obtained for all gases.

Throughout these measurements the pressure on the confined column of air remains the same. It is equal to atmospheric pressure plus the pressure of the small column of mercury. You have seen that at constant pressure the volume increases as the temperature increases. Since at any given temperature the product of $P \times V$ is a constant, the increase in V at constant P as the temperature rises means that the product $P \times V$ itself increases with temperature. It then follows that if the temperature of a given mass of gas is increased while its volume is held constant, its pressure must increase as the temperature increases. It is natural and appropriate to ask whether these facts can be accounted for by the particle model of a gas that was discussed earlier.

The pressure of the gas must be determined by how hard and how often the molecules strike the walls of the container. This, in turn, depends on how much energy of motion, *kinetic energy*, the molecules have. As heat is added to a gas in order to raise its temperature, the energy of the molecules is increased. The molecules absorb the heat and move faster. They strike the walls of the container harder and more often. The molecular model of a gas can thus explain the increase of pressure at constant volume as the temperature is raised.

6-4 The Absolute Temperature Scale

If the results recorded in Table 6-2 are considered carefully, a very interesting idea emerges. As you cool a gas at constant pressure, its volume gets smaller and smaller. Will there be a temperature sufficiently low that the volume of the gas will become zero?

As a matter of fact, such a thing will not ever really happen because all real gases eventually condense to liquids. But suppose condensation did not occur. What is the temperature at which the volume would become zero? The easiest way to answer this question is to take the data in Table 6-2 and make a plot of temperature against relative volume. This is shown in Figure 6-5. The five

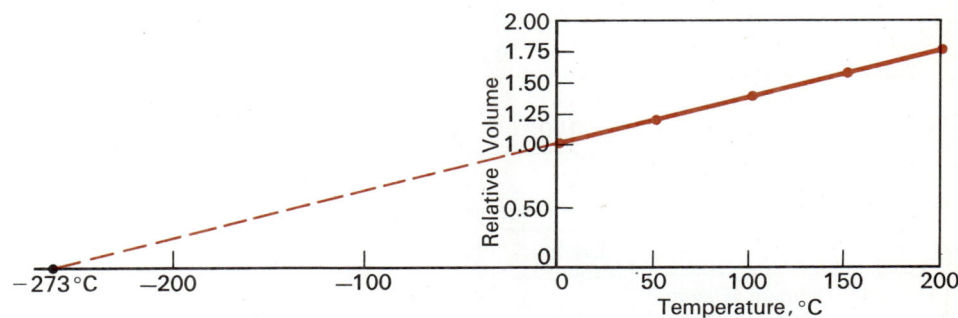

Figure 6–5. *The straight line that results when data from Table 6–2 are plotted is extrapolated to zero volume. The corresponding temperature is −273°C, or 0 K.*

points that were measured define a straight line. If this line is extended to temperatures below 0°C to a volume of zero, the corresponding temperature is −273°C.

This temperature, −273°C, is of fundamental importance in science. It is the temperature at which the molecules have no kinetic energy. They cease to fly around and hence they no longer cause a pressure that can support the mercury drop. The mercury drop would fall to the bottom of the tube, and the volume of the gas would be zero.

The temperature at which gas molecules would have no kinetic energy is an absolute minimum and thus is called the absolute zero of temperature. The "zero" of the Celsius scale has no such absolute significance. It is simply a convenient point (the freezing point of water) to use in calibrating thermometers.

You can now understand the reason why the fundamental international temperature scale is the Kelvin or absolute scale and the unit of temperature is the kelvin. These standard units were introduced in Chapter 3 with a promise that you would later find out why they are used. We can now make good on that promise.

Since −273°C is the lowest temperature that there can be, it is the zero to end all zeros. It is the absolute zero from which to start a scale of absolute temperatures. There can be no negative temperatures with such a scale. A unit is needed for stating temperatures on this Kelvin or absolute temperature scale. For practical reasons scientists decided to make this unit exactly equal to the Celsius degree, but it is called a **kelvin** (*not* a kelvin degree).

In absolute, or Kelvin, temperatures, water freezes at 273 K and boils at 373 K. Figure 6-6 shows some other melting and boiling points, in both Celsius and Kelvin temperatures. As you can see, to convert a Celsius temperature to a Kelvin temperature, simply add 273. The most accurate measurements show that the freezing point of water is 273.150 K, but for our purposes 273 K will be accurate enough.

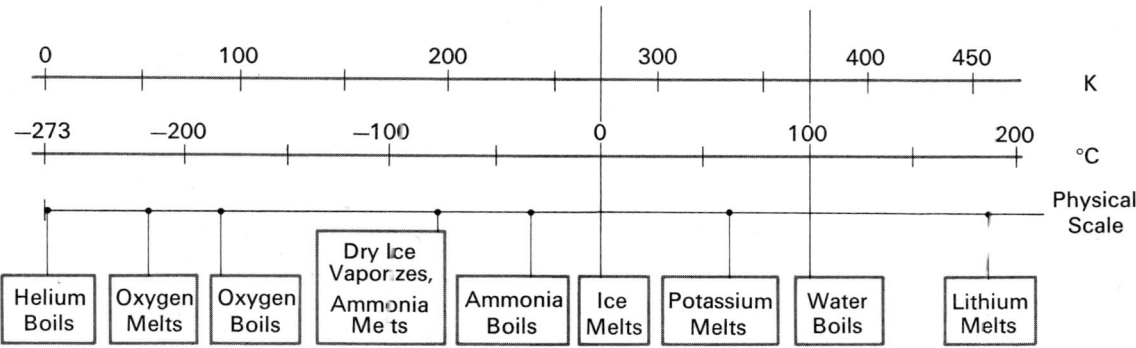

Figure 6-6 *A comparison of the Kelvin and Celsius temperature scales, showing the melting and boiling points of some substances on these scales.*

INVESTIGATION

6-5 A Quantitative Investigation of the Reaction of a Metal with Hydrochloric Acid

In Chapter 5 you found that one mole of any substance is 6.02×10^{23} objects. You learned later that one mole of any gas occupies 22.4 liters at STP. When you react various substances with each other, you might wonder whether there is any relationship between the number of moles of reacting materials and the number of moles or volumes of the products that are formed. In this experiment you will examine some of these relationships. You will react magnesium metal and a solution of hydrogen chloride in water, producing hydrogen gas.

PURPOSE

You will find out how many liters of dry hydrogen gas, measured at room temperature and one atmosphere pressure, can be produced by one mole of magnesium metal. Then you will see whether there is any relationship between the number of moles of magnesium metal used and the number of moles of hydrogen gas formed.

MATERIALS

Gas-measuring tube, 50-ml
Thermometer
Barometer
Ringstand and clamp
Beaker, 400-ml
Graduated cylinder, 10- or 25-ml
Metric ruler
Magnesium ribbon, 5 to 6 cm long
Copper wire, 22 gauge, 15 cm long
One-hole stopper, size 0 or 00 to fit measuring tube
Battery jar or large beaker, 2- or 3-liter
Hydrochloric acid, 10 ml of dilute acid, labeled 6 M

PROCEDURE

1. Obtain a piece of magnesium, Mg, ribbon approximately 5 cm long. (Use a shorter piece at higher altitudes.) Measure the length of the ribbon carefully and record it to the nearest 0.05 cm. Your teacher will give you the mass of one meter of the ribbon. Since the magnesium ribbon is uniform in thickness, you can calculate the mass of your piece.

2. Fold your piece of magnesium ribbon so that it can be encased in a small spiral cage made of fine copper wire. Leave about 5 cm of copper wire to serve as a handle. (See Figure 6-7.)

3. Set up a ring and utility clamp in position to hold a 50-ml gas-measuring tube that has been fitted with a one- or two-hole rubber stopper.

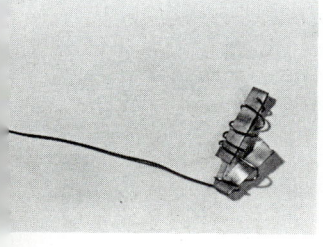

Figure 6-7

4. Take out the stopper temporarily. Incline the gas-measuring tube slightly from an upright position and pour in about 10 ml of hydrochloric acid, labeled 6 M HCl.

5. With the tube in the same position, slowly fill it with tap water from a beaker. While pouring, rinse any acid that may be on the sides of the tube so that the liquid in the top of the tube will contain very little acid. Try to avoid stirring up the acid layer in the bottom of the tube. Bubbles clinging to the sides of the tube can be dislodged by tapping the tube gently.

6. Holding the copper coil by the handle, insert the metal about 3 cm down into the tube. Hook the copper wire over the edge of the tube and clamp it there by inserting the rubber stopper. The tube should be completely filled so that the stopper displaces a little water when put in place.

7. Cover the hole in the stopper with your finger and invert the tube in the container of water. Clamp it in place. The acid, being more dense than water, will diffuse down through it and eventually react with the magnesium metal, as shown in Figure 6-8a. (Copper will not react with the acid solution.)

8. After the reaction stops, wait for about 5 minutes to allow the tube to come to room temperature. (See Figure 6-8b.) Dislodge any bubbles clinging to the sides of the tube.

9. Cover the hole in the stopper with your finger and transfer the tube to a large cylinder or battery jar that is almost filled with water at room temperature. Raise or lower the tube until the level of the liquid inside the tube is the same as the level outside the tube, as indicated in Figure 6-9. This permits you to measure the

Figure 6-8a

Figure 6-8b

volume of the gases in the tube (hydrogen and water vapor) at room pressure. Remember to read the volume with your eye at the same level as the bottom of the meniscus, as you did when using a graduated cylinder. Record your measurement of the gas volume to the nearest 0.05 ml.

10. Remove the gas measuring tube from the water and pour the acid solution it contains down the sink. Rinse the tube with tap water.

11. Record the room temperature. Your teacher will give you the room pressure or will assist you in reading the barometer.

   The experiment may be repeated with another sample of magnesium to check your results, if time permits.

DATA AND RESULTS

Your data table should include the following.
Mass per unit length of magnesium ribbon, (g/m), from the teacher
Length of your piece of magnesium ribbon (cm)
Volume of hydrogen, saturated with water vapor (ml)
Temperature of the water (°C)
Temperature of the room (°C)
Barometer reading, room pressure (Torr)
Vapor pressure of water at the above temperature (Torr), from Table 6-3

Calculations

1. Determine the mass of the magnesium you used from the grams per meter relationship and the length of the ribbon.

2. Determine the number of moles of magnesium used. The molar mass of magnesium is 24.3. (That is, there are 24.3 grams per mole of magnesium atoms.)

3. Determine the partial pressure of the hydrogen gas, as explained in the following discussion.

   Since the hydrogen gas was collected over water, the gas in the tube consists of a mixture of hydrogen gas and water vapor. The total pressure caused by these two gases is equal to the room pressure. (See the hypothetical case discussed in Section 6-2.) Mathematically this can be expressed as follows.

$$P_{H_2} + P_{H_2O} = P_{room}$$

Figure 6-9

TABLE 6–3 Vapor Pressure of Water at Various Temperatures

Temperature (°C)	Pressure (Torr)	Temperature (°C)	Pressure (Torr)
15	12.8	23	21.0
16	13.6	24	22.4
17	14.5	25	23.8
18	15.5	26	25.2
19	16.5	27	26.7
20	17.5	28	28.3
21	18.6	29	30.0
22	19.8	30	31.8

The pressure in the room is given by the barometer reading. The pressure of the water vapor, P_{H_2O}, can be determined from Table 6-3. The values in the table were obtained by measuring the pressure of water vapor over the surface of liquid water at various temperatures. The partial pressure of the hydrogen can then be calculated as follows.

$$P_{H_2} = P_{room} - P_{H_2O}$$

4. Determine the volume of the hydrogen gas at one atmosphere pressure (760 Torr).

   You have learned that for a given temperature, the product of the pressure and volume of a gas is a constant: $PV = C$. To calculate the new volume, V_{new}, at 760 Torr, the following mathematical relationship can be used. P_{H_2} is the partial pressure calculated in step 3.

$$V_{measured} \times P_{H_2} = V_{new} \times 760$$

$$V_{new} = V_{measured} \times \frac{P_{H_2}}{760}$$

5. Calculate the volume of dry hydrogen that would be produced by one mole of magnesium at room temperature and one atmosphere pressure.

RELATED PROBLEMS

1. Given that one mole of Mg produces one mole of hydrogen, H_2, what is the volume of one mole of hydrogen at room temperature and one atmosphere pressure? (Calculate the answer from the results of your experiment.)

2. If one mole of hydrogen has a mass of 2.0 g, what is the density of hydrogen, in grams per liter, at room temperature and one atmosphere pressure? (Calculate the density using your experimental results.)

ADDITIONAL INVESTIGATION

Determine the volume of hydrogen gas produced when a mole of another metal reacts with an acid.

6-6 Combining Boyle's Law, Charles' Law, and Avogadro's Hypothesis

Three basic laws concerning how gases behave have now been discussed. One of these is Boyle's law. It states that for a given mass of gas at a constant temperature (that is, when mass and temperature are not allowed to change):

$$P \times V = \text{a constant.} \qquad (1)$$

Charles' law states that for a fixed mass of a gas, the product $P \times V$ varies directly with the temperature, provided that temperature is measured on the absolute scale, that is, in kelvins. Charles' law says, in effect, that

$$P \times V = \text{another constant} \times T. \qquad (2)$$

T is the temperature in kelvins. Equation (2) states both Boyle's and Charles' laws at the same time.

Avogadro's hypothesis states that for a particular set of values of P, V, and T all gases contain the same number of molecules. This means that they contain the same number of moles. If at constant temperature the number of moles of gas is increased, the product $P \times V$ must increase in direct proportion. This can be built into the equation as follows.

$$P \times V = R \times N \times T \qquad (3)$$

N represents the number of moles and R is a constant, called the gas constant. Because of Avogadro's hypothesis, it has the same value for all gases. Its numerical value, of course, depends on the units used for P and V. The usual ones are atmospheres for pressure and liters for volume.

The numerical value of R is obtained from experimental measurements. You have already learned that one mole of a gas at 0°C (273 K) and a pressure of one atmosphere (760 Torr) occupies a volume of 22.4 liters. This fact can be used to find the value of R. First, equation (3) is rearranged as follows.

$$R = \frac{P \times V}{N \times T} \qquad (3a)$$

Then the quantities just mentioned are substituted into it.

$$R = \frac{1 \text{ atm} \times 22.4 \text{ liters}}{1 \text{ mole} \times 273 \text{ K}}$$

$$= 0.0820 \frac{\text{liter} \cdot \text{atm}}{\text{mole} \cdot \text{K}} \quad (3\text{b})$$

It is now possible to write the **ideal gas law:**

$$PV = (0.0820) \times NT. \quad (4)$$

If any three of the four quantities, P, V, N, or T, are known, the other one can be calculated. This is called the *ideal* gas law because real gases deviate slightly from it. Very accurate measurements are required to detect the deviations under most conditions.

The ideal law is very useful for calculations that involve gases. Suppose, for example, that you wanted to design a vessel to hold 3 moles of nitrogen at 25°C and one atmosphere pressure. What volume should the vessel have? The following example shows how you could find the answer.

Example

First, list the three known quantities, in the proper units.

$$P = 1 \text{ atm}$$
$$N = 3 \text{ moles}$$
$$T = 25°\text{C} = 298 \text{ K}$$

Next, rearrange the equation so that the quantity you need to calculate is on the left and all the known quantities are on the right.

$$V = \frac{(0.0820) \times NT}{P}$$

Then substitute the values of the known quantities and do the arithmetic.

$$V = \frac{(0.0820)(3)(298)}{1}$$

$$= 73.3 \text{ liters}$$

6-7 The Molecular Model Accounts in Detail for the Ideal Gas Law ★

Using some physics and some algebra, we can show that the molecular model explains the ideal gas law. In other words, we can *derive* the gas law, equation (4), from the model.

Let us assume that we have a cubic container, l centimeters on an edge, containing n molecules, each of which has a mass m. The temperature is T kelvins.

According to the model, the gas molecules fly around in the box in random directions, making elastic collisions with the walls. The model does *not* say that all the molecules must have the same velocity, and as you will see later in this book, they do not. We can, however, work with their *average* velocity, v. A particle of mass m moving with velocity v has two properties: momentum, M, and kinetic energy, KE. Elementary physics defines these properties as follows.

$$M = mv$$
$$KE = \tfrac{1}{2} mv^2$$

When a molecule makes an elastic collision with the walls of the container it transfers momentum to the wall, as shown in Figure 6-10. If it approaches the wall with a velocity v it must leave with a velocity of $-v$ since its direction is reversed. Thus, the momentum transferred is equal to the difference between the initial and final momentum of the molecule.

Momentum transferred per collision

$$= mv - (-mv) = mv + mv = 2\,mv. \quad (5)$$

How many collisions will occur per second? On the average, one third of the molecules, $n/3$, are bouncing back and forth between each pair of parallel walls of the container.

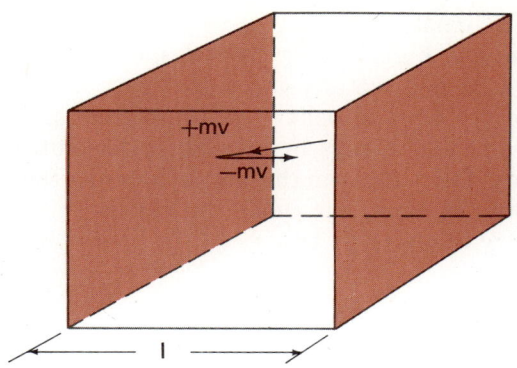

Figure 6-10 *A molecule strikes a container wall with momentum +mv, leaves with momentum −mv, and thus transfers 2mv to the wall.*

The time required to travel from one wall to the other and back is $(2l)/v$. Thus, the number of collisions that one wall will receive per second can be expressed as follows.

$$\frac{\text{collisions}}{\text{second}}$$

$$= \frac{\text{number of molecules bouncing back and forth}}{\text{time required per round trip}}$$

$$= \frac{n/3}{(2l)/v} = \left(\frac{n}{3}\right)\left(\frac{v}{2l}\right) = \frac{nv}{6l} \quad (6)$$

Pressure may be expressed as the amount of momentum per second that is transferred to a unit area of wall. Thus, the pressure on one wall is given as follows.

$$P = \left(\frac{\text{collisions}}{\text{second}}\right)\left(\frac{\text{momentum}}{\text{collision}}\right)\left(\frac{1}{\text{area}}\right) \quad (7)$$

Substitution of the detailed expressions for each of these factors gives the following expression.

$$P = \left(\frac{nv}{6l}\right)(2mv)\left(\frac{1}{l^2}\right)$$

$$= \frac{1}{3}\left(\frac{n}{l^3}\right)(mv^2) \quad (8)$$

Notice the quantity l^3 in the denominator. It is, of course, the volume of the container. Let us replace l^3 by V.

$$P = \frac{1}{3}\left(\frac{n}{V}\right)(mv^2)$$

$$PV = \frac{1}{3}nmv^2 \quad (9)$$

The quantity mv^2 is simply twice the average kinetic energy of a molecule. When heat energy goes into a gas, two things happen. The temperature of the gas rises and the heat energy becomes kinetic energy of the molecules. The relationship between T and kinetic energy is a direct one.

$$T = (\text{a constant}) \times KE = C \cdot mv^2$$

With rearrangement, and the use of a new constant $C' = \dfrac{1}{C}$, the following expression results.

$$mv^2 = C'T \quad (10)$$

On substitution of the preceding expression into equation (9), the following equation results.

$$PV = \frac{1}{3}n(C'T) \quad (11)$$

Finally, instead of stating the number of molecules as such we can use moles. That is, we can replace n by $(6.02 \times 10^{23})N$, where N is the number of moles.

$$PV = \left[\left(\frac{C'}{3}\right)(6.02 \times 10^{23})\right]NT \quad (12)$$

All of the quantities in the square brackets are constants. We may use R to represent the entire collection and write:

$$PV = RNT. \quad (13)$$

The molecular model has thus given us exactly the same equation that was derived from experiment. Remember that Boyle's law and Charles' law are both purely experimental. They simply state what can be observed. These, combined with Avogadro's hypothesis, give the ideal gas law.

Since our model accounts so well for experimental results, it can be taken very seriously. We must begin to believe that molecules really exist and that gases do consist of minute, high-velocity particles, even though we do not see any individual particles. We do not *have* to believe all these things, and we should always be on a sharp lookout for any facts that might argue against them. But, by now, our model seems useful and very likely to be true.

Summary

The term STP means standard temperature and pressure, 0°C and one atmosphere. Boyle's law ($PV = C$) is true only if the temperature does not change. Jacques Charles discovered that all gases expand by the same fraction of their original volume when heated over the same temperature range. This is known as Charles' law.

The molecular model accounts for the effect of temperature on gas properties. When gas in a container is heated, it absorbs energy. The gas molecules move faster as they gain kinetic energy. As a result, they strike the walls of their container more often and with more force, and so the pressure increases. If not contained, these faster-moving molecules would fly farther apart, causing the volume of the gas to increase.

If a gas is cooled, kinetic energy is withdrawn from the molecules and they slow down. Therefore, they exert less pressure on the walls of the container. Is there a temperature at which the volume becomes zero? Physically, no—because gases condense and form liquids. But a plot of volume versus temperature can be drawn showing at what temperature the volume would be zero if condensation did not occur. This temperature, approximately −273°C, is called absolute zero. Absolute zero is defined (for your purposes) as that temperature at which the gas molecules have no kinetic energy.

On the absolute, or Kelvin, scale, the freezing point of water is 273 K, the boiling point, 373 K. It is clear from this that the unit on the scale (the kelvin) is the same size as the familiar Celsius degree. To change a Celsius temperature to the corresponding Kelvin temperature, add 273 to the Celsius reading. For instance 22°C (a normal indoor temperature) is 295 K.

Self Check

1. What is STP?
2. State Charles' law.
3. Describe the kind of experiment that supports Charles' law.
4. Explain how the molecular model accounts for pressure and volume changes when the temperature changes.
5. What is absolute zero in words? In numbers?
6. Convert 50°C to the Kelvin scale.

Chapter Summary

The molar mass of a substance is defined as the mass in grams of a mole of the substance. Thus, 32 grams is the molar mass of O_2.

When two or more gases that do not react with one another are placed in the same container, each exerts pressure as though the other gas(es) were not present. Thus, the total pressure is simply the sum of the individual or partial pressures.

Temperature changes affect pressures and volumes in gases. Charles' law states that all gases expand by the same fraction of their volume when heated through the same temperature range. If a gas is heated in a closed container so that its volume cannot increase, the pressure will rise. The temperature scale used in measuring gas temperature changes is called the absolute, or Kelvin, scale. It is based on the theoretical extension of a volume-temperature graph that reduces temperature to a point at which the volume of a gas becomes zero. This occurs at $-273°C$. The Kelvin scale uses this temperature as absolute zero and uses the same size unit as the Celsius degree. On this scale the freezing point of water ($0°C$) is 273 K and the boiling point of water ($100°C$) is 373 K. To convert Celsius readings to Kelvin, add 273 to the Celsius reading.

The three separate laws, Boyle's law, Charles' law, and Avogadro's hypothesis can be expressed by one equation: $PV = RNT$. This is called the ideal gas law. R is a constant, called the gas constant.

Questions and Problems

1. How many molecules are there in 22.4 liters of a gas at $0°C$? In 11.2 liters?
2. What relationship exists between temperature and the number of molecules in a molar volume of a gas?
3. Why does the pressure build up in a tire on a hot day? Explain in terms of the kinetic theory.
4. Why is it desirable to express all temperatures in kelvins when dealing with gases?
5. How does a kelvin compare in size with a Celsius degree? How does a Kelvin temperature compare with a Celsius temperature?
6. Consider two closed containers of the same volume. One is filled with hydrogen gas, the other with carbon dioxide gas. Both gases are at room temperature and one atmosphere pressure. If the temperature of the hydrogen is now raised, how do the two gases compare in each of the following?
    (a) pressure
    (b) volume
    (c) number of moles
    (d) average molecular kinetic energy
7. Two identical cylinders are fitted with movable pistons, thus allowing the volumes of the enclosures to vary. One cylinder contains a certain amount of nitrogen gas and the second cylinder contains an equal volume of oxygen gas, both at room temperature and pressure.
    (a) How do the number of moles of the two gases compare?
    (b) How do the number of molecules of the two gases compare?
    (c) If the mass of the nitrogen is 7.00 g, what is the mass of the oxygen?
    (d) If the temperature of the oxygen is lowered, how do the two gases compare with respect to pressure, volume, number of moles, and average molecular kinetic energy?
8. The boiling points and freezing points of certain liquids are listed below in degrees Celsius. Express these temperatures on the absolute temperature (Kelvin) scale.

Liquid hydrogen: boiling point = $-253°C$; freezing point = $-259°C$
Liquid oxygen: boiling point = $-183°C$; freezing point = $-218°C$
Liquid helium: boiling point = $-269°C$
Liquid nitrogen: freezing point = $-210°C$

9. The boiling points and freezing points in kelvins of certain liquids are listed below. Express these temperatures on the Celsius scale.

   Liquid neon: boiling point = 27 K;
   freezing point = 24 K
   Liquid chlorine: boiling point = 238 K;
   freezing point = 166 K
   Liquid acetylene: boiling point = 201 K;
   freezing point = 192 K

10. A glass bulb has a mass of 108.11 grams after all the air has been removed from it. When filled with oxygen gas at atmospheric pressure and room temperature the bulb has a mass of 109.56 grams. When filled at atmospheric pressure and room temperature with a gas sample taken from the mouth of a volcano, the bulb has a mass of 111.01 grams. Which of the following possibilities could account for the data?

    A pure gas: CO_2 OCS Si_2H_6 SO_2
    NF_3 SO_3 S_8
    A mixture: half CO_2, half Kr

    (Use the table of molar masses in Appendix 3.)

11. When the bulb in Question 10 is filled with nitrogen gas at standard conditions, it has a mass of 109.93 grams. When filled at STP with one of the gases listed below its mass is 109.15 grams. Which is the molecular formula of the unknown gas?

    HCl CH_4 NH_3 Cl_2 C_2H_2 CO SO_2 CO_2

12. What is the molar mass of a gas if, at 0°C and one atmosphere pressure, 1.00 liter of the gas has a mass of 2.00 grams?

13. If a 2.50-g sample of a gas has a volume of 500 ml at STP, what is the molar mass of that gas?

14. If exactly 100 ml of a gas at 10°C are heated to 20°C (pressure and number of molecules remaining constant), the resulting volume of the gas will be which of the following? (a) 50 ml; (b) 1000 ml; (c) 100 ml; (d) 375 ml; (e) 103 ml

15. If the volume of a gas exerting 800 Torr of pressure is changed from 50 ml to 250 ml (temperature and number of molecules remaining constant), the resulting pressure of the gas will be which of the following? (a) 50 Torr; (b) 4000 Torr; (c) 16 Torr; (d) 160 Torr; (e) 250 Torr

16. A carbon dioxide fire extinguisher of 3 liters volume contains about 10 pounds (4.55 kg) of CO_2. What volume of gas could this extinguisher deliver at standard conditions?

17. If you wanted a carbon dioxide fire extinguisher of 10 liters volume to deliver approximately 10,000 liters of CO_2 gas, how many pounds of CO_2 would you need?

18. Compressed oxygen gas is sold at a pressure of 130 atm in steel cylinders of 40 liters volume.
    (a) How many moles of oxygen does such a filled cylinder hold, if the pressure is measured at 0°C?
    (b) How many kilograms of oxygen are in the cylinder?

19. How many moles of nitrogen gas are contained, at 25°C, in a 25-liter cylinder at a pressure of 200 atm? How many kilograms of nitrogen are in the cylinder?

20. A 1.50-liter sample of dry air in a cylinder exerts a pressure of 3.00 atm at a temperature of 25°C. Without change in temperature, a piston is moved in the cylinder until the pressure in the cylinder is reduced to 1.00 atm. What is the volume of the air now?

21. After traveling some distance on a hot day, an automobile tire shows an absolute air pressure of 40 lbs/in² and a tem-

perature of 30°C. A few days later the temperature of the tire is 5°C. Assuming that the volume of the tire and number of molecules of air in the tire do not change, determine the air pressure in the tire at 5°C.

22. The density of liquid carbon dioxide at room temperature is 0.80 g/ml. How large a cartridge of liquid CO_2 must be provided to inflate a life jacket of 4.0 liters capacity at STP?

23. The density of liquid carbon dioxide is 0.80 g/ml at room temperature. How large a raft could be inflated with a 25-ml cartridge of liquid CO_2? (Assume 1 atm pressure and 25°C.)

24. A student collects a volume of hydrogen over water. He determines that 2.00×10^{-3} mole of hydrogen and 6.0×10^{-5} mole of water vapor are present. If the total pressure inside the collecting tube is 760 Torr, what is the partial pressure of each gas?

25. Nitrogen gas is collected over water, and it is determined that 4×10^{-3} mole of N_2 gas and 8×10^{-5} mole of water vapor are in the collecting tube. If the nitrogen gas exerts a pressure of 800 Torr, what is the total pressure exerted inside the collecting tube?

26. A candle is burned under a beaker until it extinguishes itself. A sample of the gaseous mixture in the beaker contains 6.08×10^{20} nitrogen molecules and 0.76×10^{20} carbon dioxide molecules. The total pressure is 764 Torr. What is the partial pressure of each gas?

27. A mixture of gases contains 4.04×10^{12} molecules of nitrogen, 0.98×10^{12} molecules of oxygen, and 0.40×10^{12} molecules of carbon dioxide. Determine the pressure contributed by each gas if the pressure of the mixture is 840 Torr.

28. A sample of nitrogen is collected over water at 18.5°C. The vapor pressure of water at 18.5°C is 16 Torr. When the pressure of the sample has been equalized against atmospheric pressure, 756 Torr, what is the partial pressure of nitrogen? What will be the partial pressure of nitrogen if the volume is reduced by the factor 740/760?

29. A sample of hydrogen collected over water at 25°C and a pressure of 760 Torr occupies 100 ml. What would the volume occupied by the gas be at the same temperature and pressure if it were dry? (Vapor pressure of water at 25°C is 23.8 Torr.)

30. A cylinder contains nitrogen gas and a small amount of liquid water at a temperature of 25°C. (The vapor pressure of water at 25°C is 23.8 Torr.) The total pressure is 600 Torr. A piston is pushed into the cylinder until the volume is halved. What is the final total pressure, if the temperature is kept at 25°C?

31. A cylinder is fitted with a piston, thus allowing the enclosed volume to vary. A sample of gas plus a small amount of liquid water are introduced into the cylinder at a temperature of 20°C. (Vapor pressure of water at 20°C = 17.5 Torr.) The total pressure is 400 Torr. The piston is moved, changing the volume of the enclosure from 2 liters to 5 liters. What is the final total pressure at 20°C?

32. Suppose the tube used to gather the data in Table 6-2 were placed in a block of dry ice (solid CO_2) which has a temperature of -78°C. What would the relative volume of the gas be? What temperature is this on the absolute scale?

33. Suppose an experiment similar to that shown in Figure 6-2 were set up, except that the O_2 pressure was 20 Torr while the N_2 pressure was 10 Torr. What is the final total pressure when the stopcock is opened? What is the partial pressure of each gas?

7 Elements, Compounds, and Chemical Reactions

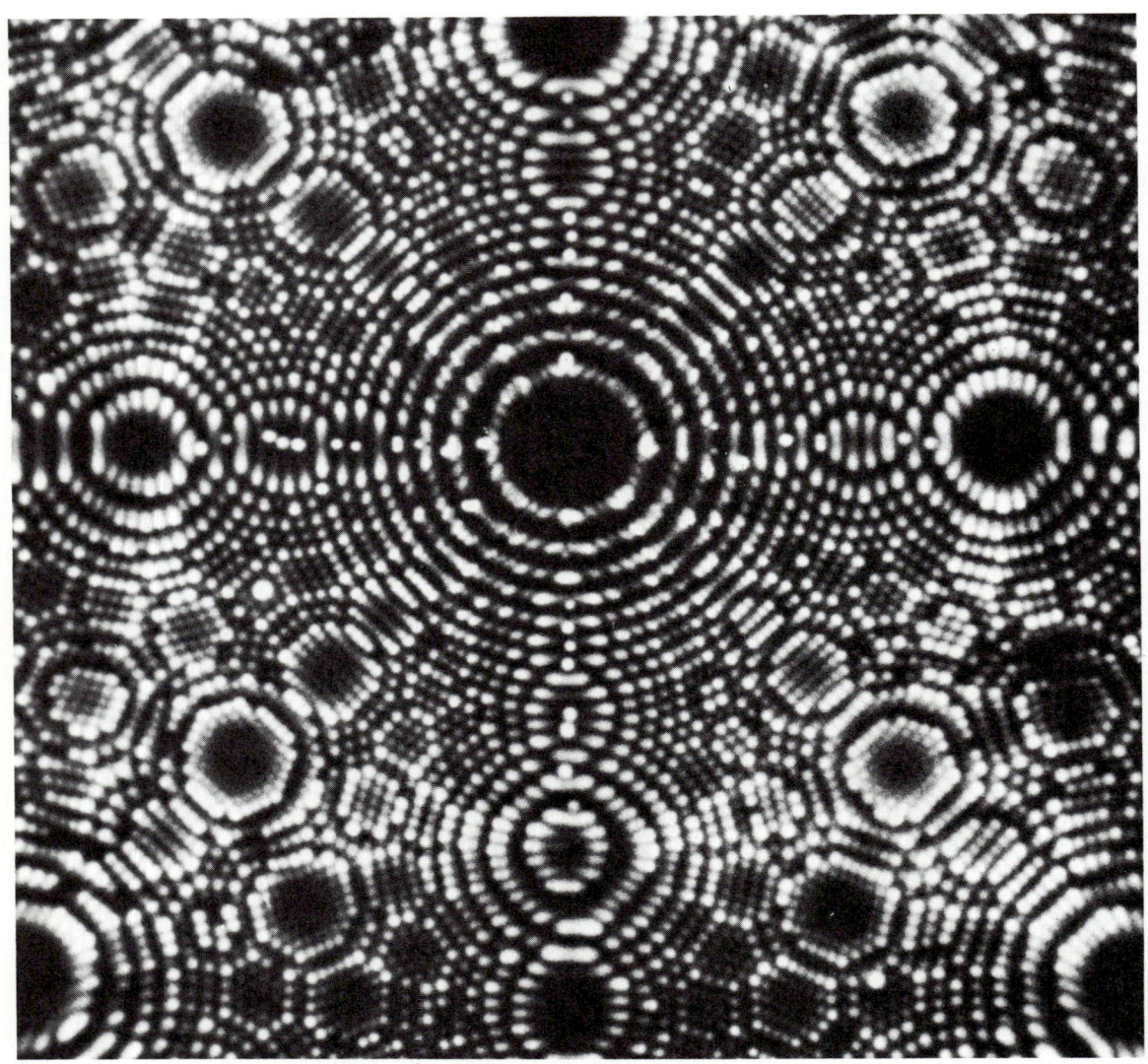

The pattern of atoms on the fine tip of some iridium metal at 21 K. The image was produced by a field ion microscope, one instrument with which individual atoms can actually be shown to exist.

You have seen how the basic properties of gases can be accounted for in terms of a model. In that model a gas is made up of a vast number of tiny particles called molecules. This model works so well, and there is so much other evidence for the actual existence of molecules, that chemists are sure that molecules are real.

In this chapter you will learn more about molecules. You will learn how chemists can break them down in order to find out what atoms compose them. You will also see how molecules can be built up.

You will also become acquainted with the different kinds of atoms that nature provides. Less than a hundred of these different kinds of atoms are combined in various ways to build up the millions of different substances in the universe.

Basic Particles of Matter

7-1 Atoms and Molecules in Liquids and Solids

The particle model for gases was developed in preceding chapters. It was shown that many properties of gases can be understood and correlated by assuming that a gas consists of rapidly moving particles called molecules (or, for a very few gases, atoms). All gases, however, can be made to condense to liquids, and all liquids can then be made to freeze, forming solids. From the opposite point of view, nearly all of the substances normally met as solids or liquids can be converted into gases by raising the temperature enough. In many cases it may not be possible to produce a very high pressure of gas without reaching temperatures at which the molecules break down. Nonetheless, the substance is at least partially or slightly converted into a gas or vapor form.

Substances can, in general, be transformed back and forth from the gas phase to condensed phases (that is, the liquid and solid phases). This fact is most easily explained by assuming that, in general, the same molecules are present in all phases. Gases condense at low temperatures because there are very weak forces of attraction among the molecules. As the temperature is lowered, the molecules move more and more slowly until eventually their speed (or kinetic energy) is no longer able to overcome the forces tending to draw them together. The molecules then fall together into a heap.

The nature of the attractive forces between molecules, the question of how the molecules arrange themselves in the condensed phases, and some other related topics will be discussed more fully later. The object of this brief discussion has been simply to point out that liquids and solids as well as gases are made up of atoms and molecules. The only difference is that in the condensed

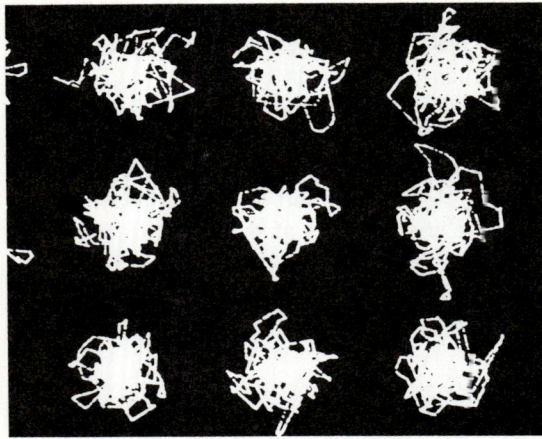

Figure 7–1 *Molecules move only around well-defined positions in the solid state. This computer-generated representation shows the paths resulting from the motions of molecules in the solid state. Each cluster represents two molecules, one behind the other.*

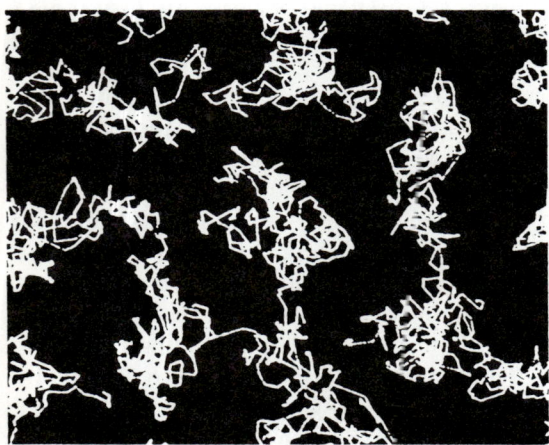

Figure 7–2 *Molecules move more freely in the liquid state than in the solid state, but the total number of molecules per unit volume is not much different. Compare the paths of these molecules in the liquid state with those in the solid state.*

phases the molecules are very close together and have much less kinetic energy. They still move, as indicated in Figures 7-1 and 7-2, but not nearly so far or so freely as they do in a gas.

7-2 All Gases Can Be Liquefied ★

You have seen how the gas law, $PV = RNT$, was developed as a combination of the simpler, experimental laws of Boyle, Charles, and Avogadro. This law can also be derived mathematically if gas molecules are assumed to be particles of negligibly small volume, lacking any attractive or repulsive forces between themselves.

According to this gas law, if the product PV is plotted against increasing pressure, for a fixed mass of gas at a constant temperature, a horizontal line will be obtained. Nearly all real gases behave in just this way at pressures up to several atmospheres. Above about 10 atmospheres pressure, however, most gases begin to deviate from this behavior. The deviations become more pronounced the lower the temperature and the higher the pressure.

A gas that strictly follows the law $PV = RNT$ at all temperatures and pressures is called an ideal gas. The law itself is called the ideal gas law. No truly ideal gas actually exists, although most common gases, such as oxygen and nitrogen, behave in practically ideal fashion at ordinary temperatures and pressures.

Real gases are all more or less nonideal. That is, they fail to follow exactly the behavior required by the ideal gas law. For nitrogen, the form of the deviation shown in Figure 7-3 is typical. Note that the deviation is more pronounced at the lower temperature. However, the general shape of the curves is the same at both temperatures. At each temperature the product PV at first dips

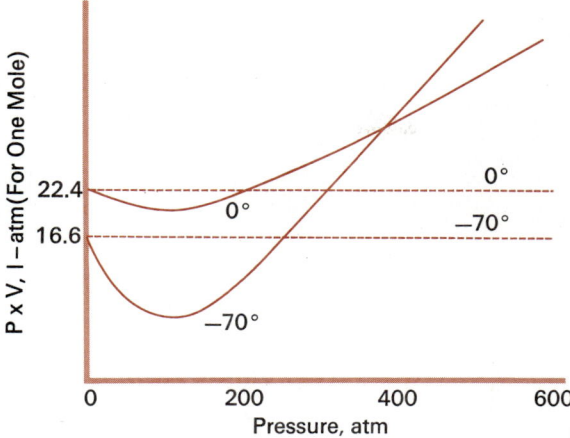

Figure 7-3 The solid curves show the nonideal behavior of N_2, since PV varies with temperature and pressure. The dashed lines show that for an ideal gas PV is constant at all temperatures and pressures.

You can see that, with increasing pressure, eventually a further increase in pressure causes the product PV to increase. PV finally becomes greater than the value required by the ideal gas law. The reason for this is quite simple. The ideal gas law takes no account of the fact that each molecule has a small but finite volume. For oxygen at STP, the molecules themselves occupy only about 0.1% of the total volume. Thus, neglecting the volume of the molecules themselves is a valid approximation under these conditions. However, as the pressure is increased, the total volume of the gas eventually becomes so small that the portion occupied by the molecules themselves constitutes an appreciable fraction of it. Each molecule is no longer free to move around the entire volume of the container. Instead the volume in which it can move is less than that of the container by a considerable fraction. For air at 200 atm and 0°C, the molecules themselves occupy about one-fourth of the total volume. Therefore, further increases in pressure become less and less effective in reducing the volume. V does not fall as rapidly as it should, and so PV increases.

The attractive forces between molecules of real gases are important in liquefying gases for two reasons. First, and most obvious, these are the forces that keep the molecules close together in the liquid state. Oxygen and nitrogen are gases at room temperature because the thermal energies of the molecules are much greater than the small energies that these attractive forces can provide. The molecules have too much kinetic energy to stick together. However, at sufficiently low temperatures, when the thermal, or kinetic, energies of the molecules are small, the attractive forces predominate. The molecules cling together, and the substances become liquids.

The temperature required to liquefy a gas depends on the pressure. When a gas is

below the value required by the ideal gas law. At extremely high pressures, PV is too large.

The cause of the low values for PV at pressures up to about 200 atm is the existence of small attractive forces between the molecules. As pressure is increased, volume is decreased. The average distance between molecules becomes shorter. At shorter distances, attractive forces, which vary inversely with the sixth power of the distance, begin to contribute to drawing the molecules together. These forces work with the pressure to reduce the volume. The volume decreases more quickly with increasing pressure than it should according to the ideal gas law. Therefore the product, PV, has a value less than that predicted by the ideal gas law. These intermolecular attractive forces are extremely important in the liquefaction of gases. We shall discuss this process later in this section.

highly compressed, the molecules are kept closer together. The attractive forces can therefore overcome the kinetic energy more easily. Gases under pressure condense to liquids at higher temperatures than uncompressed gases.

The other way in which the attractive forces between molecules assist in their liquefaction is more subtle, but very important. Ask yourself this question. How can you cause a gas to become cool enough to liquefy? For nitrogen and oxygen at 1 atm pressure, temperatures as low as $-190°C$ are required. The real answer cannot be "put a sample of the gas into a superrefrigerator (technically called a *cryostat*) which is at or below the required temperature." That only raises the question of how you get the cryostat down to such a low temperature in the first place. The real question is this: Starting with a container full of gas at 25°C, how do you get at least some of this gas cool enough to make it liquefy? Two main processes, often used together, can accomplish this.

First, if a gas initially contained in a certain small volume V_1 at a pressure P_1 is allowed to expand quickly to a new volume and pressure, V_2 and P_2, it will cool off. The process is illustrated in Figure 7-4. This is because the gas does work in moving the piston. Even an ideal gas will be cooled in this way. The energy required to do this work must be taken from the internal energy, that is, the thermal (kinetic) energy, of the gas. Since the gas has less thermal energy after its sudden expansion, it is cooler. Thus, by first slowly compressing a gas, such as air, at room temperature and then allowing it to expand suddenly, it can be cooled. This cool gas can then be used to cool another sample of gas that has already been compressed at room temperature. This second sample of gas is then allowed to expand suddenly and it cools still further. This process can be repeated over and over. However, at each step,

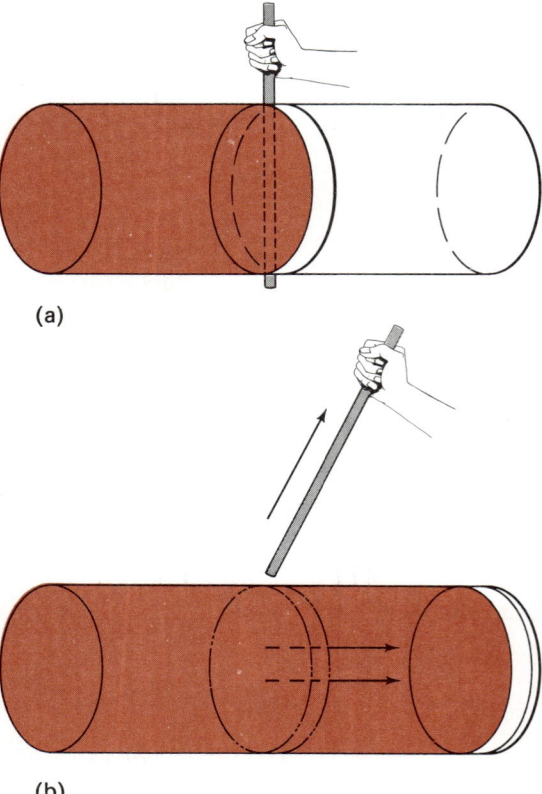

Figure 7-4 When a gas is allowed to expand as shown, it cools because it does work in moving the piston.

less gas must be used and the additional cooling effect is smaller. Still, this simple process, in which thermal energy is removed from a gas by letting it do work on its surroundings, enabled the French scientist Claude to liquefy air in 1902.

There is a second process that can help greatly in cooling gases. It depends on the nonideality of gases. Suppose we again have a gas at initial conditions T_1, V_1, and P_1. Now, instead of letting the gas push on a piston and do work as it expands, we simply punch a hole in the separator and let the gas expand

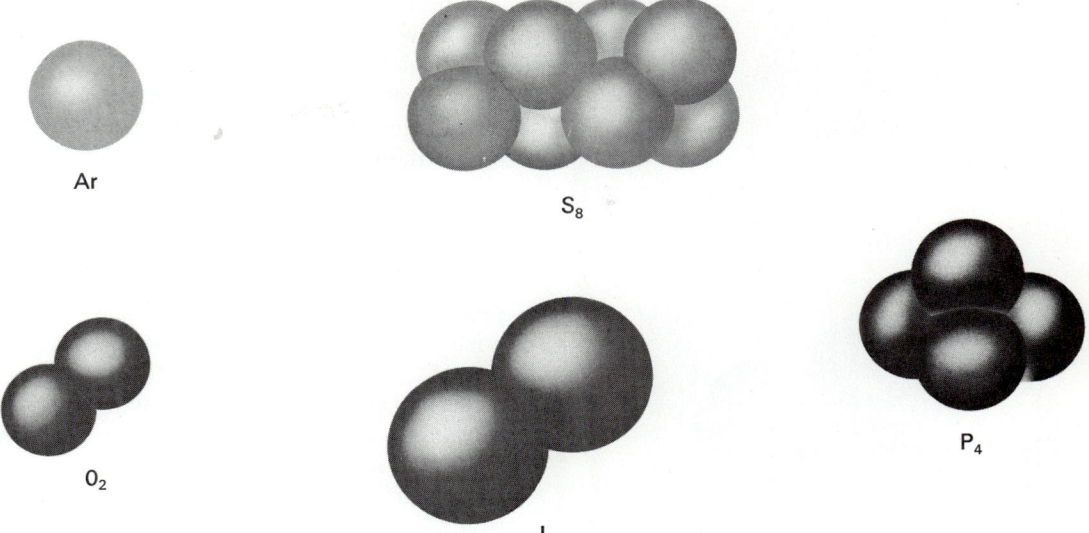

Figure 7-5 *The molecules of the elements differ in complexity. Shown here are models of some that range from a single atom to as many as eight atoms. They are argon, Ar, oxygen, O_2, iodine, I_2, phosphorus, P_4, and sulfur, S_8.*

into a vacuum. None of the internal energy of the gas is consumed by doing work on its surroundings. Nevertheless, the temperature of the gas drops. Why? Remember the attractive forces between gas molecules, which cause deviations from ideal behavior. When a gas expands, the molecules move away from each other. Thermal (kinetic) energy is used up in order to work against the attractive forces between the gas molecules. The temperature of the gas becomes lower. This effect is called the *Joule-Thomson effect*.

In present-day technology, expansion against a piston (the effect used by Claude) and the Joule-Thomson effect are often combined to liquefy air and other gases efficiently. In household refrigerators the Joule-Thomson effect alone is used. The purpose of the motor in a refrigerator is to compress a gas (gases such as CF_2Cl_2 are usually used). The process of compression generates heat, which is given off into the room. The compressed gas, at room temperature, is then allowed to expand suddenly through a pinhole into an evacuated space. The Joule-Thomson effect cools the gas. The cool gas then absorbs heat from the interior of the refrigerator, and the cycle is repeated.

7-3 Elements and Compounds

A molecule is a group of atoms bound together in a definite way. Two types of molecules are possible, depending on the kinds of atoms they contain.

1. A substance having molecules that contain only one kind of atom is called an **element.** Several such molecules are shown in Figure 7-5. As we have already noted, in the noble gases the "molecules" are

Figure 7-6 Models of molecules of the compounds ammonia, NH_3, nitrogen dioxide, NO_2, and carbon monoxide, CO.

When you think about these definitions, and how to apply them in practice, it becomes clear that the *purity* of a substance is of great importance. You cannot determine whether a given substance is an impure element or a compound, since both contain more than one type of atom. Nor can you find out what kinds of atoms, and how many of them make up the molecules of a compound, if the compound is contaminated. First you must separate the one substance that you wish to analyze from all others, that is, purify it.

A pure substance is one in which all of the molecules are the same. Its composition is uniform and thus it has certain constant and characteristic properties. The purity of any given sample of a substance is determined by how close its properties are to those of the truly pure substance. Of course no sample of any substance is ever absolutely and completely pure. What we seek in practice is a sample in which the level of impurities is too low to have a measurable or observable effect upon its characteristic properties. To illustrate these points, some familiar substances can serve as examples.

7-4 *Analyzing Compounds*

Water is probably the most familiar substance in our environment, and it is readily recognized. At least a rough judgment about the purity of a given water sample can be made by comparing its properties with those known to be characteristic of "pure" water. Characteristic properties of water include its appearance and feel, its density, the way it flows, the temperatures at which it freezes and boils, and its ability to dissolve solids such as sugar and salt.

The presence of impurities often causes easily observed changes in one or more of these properties. Any colored impurity is

single atoms. These are included among the elements, since there is obviously only one kind of atom in the "molecule." The metallic elements, such as iron, tin, gold, and silver consist of single atoms packed together closely but without the formation of any individual groups that chemists can identify as molecules.

2. A substance having molecules that contain different kinds of atoms is called a **compound**. Figure 7-6 shows some molecules of this type.

certainly obvious, since color is not a characteristic property of pure water. Salt dissolved in water lowers its freezing point, an effect which is put to use in melting winter ice.

Assuming that a sample of pure water can be obtained and that you can verify that it is pure by examining its properties, you may ask: "Is it an element or a compound?"

It is possible to show by a simple experiment that water is a compound. This experiment is one that your teacher may be able to demonstrate for you, using an experimental setup similar to that shown in Figure 7-7. While a complete explanation of this experiment will have to be postponed until later, the main result can easily be seen. The passage of an electric current through slightly acidified water causes the water to break down into two gases. One of these gases is a pure substance called hydrogen and the other is a pure substance called oxygen.

The reasoning is as follows. Since water can be decomposed into two other substances, it must consist of at least two kinds of atoms. Hence, water must be a compound, not an element.

It happens that the two substances into which water can be decomposed are elements. Compounds are not always broken down completely to elements. Quite often simpler compounds are formed, as you will see in the following experiment.

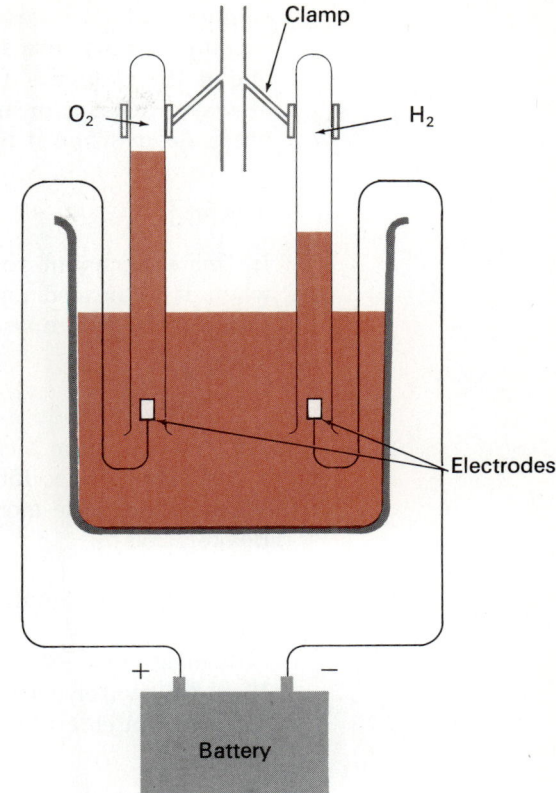

Figure 7-7 *Electrolysis of (acidified) water yields H_2 and O_2. The volume of H_2 produced is twice that of O_2.*

INVESTIGATION

7-5 *Decomposing a Common Substance*

Relatively few substances require electrical energy for their decomposition. Most everyday materials will decompose on sufficient heating. As heat is absorbed, the substance becomes increasingly unstable. Finally the forces holding the molecule together are overcome. The actual temperatures at which substances break down vary depending on the stability of the molecules. Molecules composed of more than two elements do not necessarily break down into those elements. They may decompose only into simpler molecules, or into a simple molecule and one of the elements. Complex molecules can break apart in a number of ways. Sodium bicarbonate, commonly known as baking soda, will break down when it absorbs moderate amounts of heat.

PURPOSE

In this experiment you will find out how sodium bicarbonate changes when it is heated and examine the products formed when sodium bicarbonate decomposes.

MATERIALS

1-2 grams sodium bicarbonate Utility clamp
3 test tubes, 18- × 150-mm Wooden splints
Glass bottle, wide-mouth, 4-oz pHydrion test paper
Beaker, 250-ml Cobalt chloride test paper
Ringstand Balance, centigram
Glass tubing, 6 mm diameter, 35 cm long with 90° bend 5 cm from one end
Rubber stopper, one hole, #1 or size to fit test tube
10 ml limewater (saturated calcium hydroxide solution)
1 ml barium chloride solution (2% barium chloride in water)

PROCEDURE

1. Find the mass of an 18- × 150-mm test tube to the closest 0.01 gram.

2. Add enough powdered sodium bicarbonate to the weighed test tube to give a height of 3-4 cm of powder in the tube. Then find the mass of the test tube and its contents.

3. Set up the apparatus as shown in Figure 7-8. Place limewater solution to a depth of 2-3 cm in the test tube. Place the test tube with the limewater in a 250-ml beaker, which serves as a support. Put the beaker on the desk top near the base of the ringstand and lower the bent glass tubing into the test tube so that its lower end is below the surface of the limewater.

4. Heat the test tube containing the sodium bicarbonate with your burner. Heat it gently at first and then more strongly. Observe any changes that occur inside the test tube of sodium bicarbonate and in the limewater solution.

5. After one or two minutes of heating, remove the heat. Loosen the

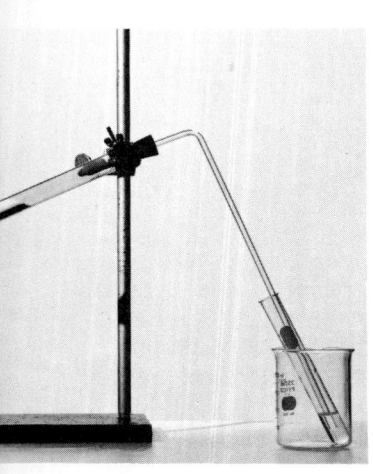

Figure 7-8

utility clamp. Raise the clamp and the test tube to a higher position on the ringstand to make it possible to remove the bent glass tubing from the limewater solution. Avoid touching the test tube because it is hot.

6. Wipe off any limewater remaining on the bottom of the bent glass tubing. Wipe out any solution inside the lower end of the tube with a small piece of paper towel.

7. Place a wide-mouth bottle on the desk top near the ringstand. Lower the test tube so that the bent glass tubing nearly reaches the bottom of the bottle as in Figure 7-9.

8. Again heat the test tube of sodium bicarbonate for several minutes. Light a wooden splint and lower it into the bottle. Observe results. Light another splint and lower it into a bottle full of air. Compare the results.

9. Remove the heat. Raise the test tube. Remove the bottle. Place an empty test tube in a 250-ml beaker that is about half full of cold water. Lower the test tube of sodium bicarbonate until the glass tubing reaches the bottom of the empty test tube as in Figure 7-10. Heat the sodium bicarbonate strongly for several more minutes and observe changes that take place in the empty test tube. Remove the heat. Raise the test tube of sodium bicarbonate and allow it to cool.

10. When the test tube has cooled, remove the rubber stopper with the glass tubing in it and determine the mass of the test tube and its contents to the nearest 0.01 gram.

11. Examine the test tube in the beaker of cold water. If there is any liquid inside, test it with a piece of cobalt chloride paper. Note any change in the color of the paper. Refer to the results you got in Section 1-13 to help interpret your observations.

12. Pour the solid product remaining in the original test tube onto a piece of paper. Moisten a piece of pHydrion test paper and touch it to the product. Note the color of the paper.

13. Put a small amount of sodium bicarbonate on another piece of paper and touch it with a piece of moist pHydrion paper. Note the color of the paper. (This test paper is used for finding the amount of acid or base in a substance. Acids and bases will be discussed later in the course. In this experiment it is enough to note whether the two powders tested give the same colors or different colors to the paper. This can provide some evidence as to possible changes that may or may not occur during the heating of sodium bicarbonate.)

Figure 7-9

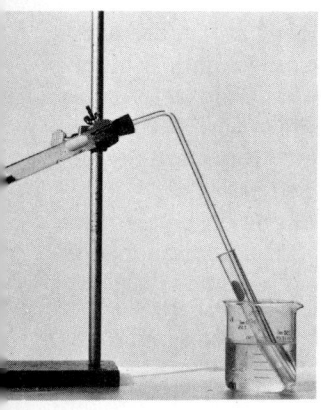

Figure 7-10

14. Dissolve the remaining solid product in about one-half test tube of demineralized or distilled water. Add 3-4 drops of 2% barium chloride solution and note any changes that take place.

15. Dissolve some sodium bicarbonate powder in about one-half test tube of demineralized water. Add 3-4 drops of 2% barium chloride solution and note any change that takes place.

RELATED PROBLEMS

1. Make a list of all evidence that indicates that sodium bicarbonate was decomposing into some new substances while it was being heated.

2. Refer to Section 1-13 to interpret the meaning of the test with limewater. From this test and the test with cobalt chloride paper what two new substances could you identify with a reasonable degree of certainty?

3. The chemical formula for sodium bicarbonate is $NaHCO_3$. What would be the smallest number of molecules of $NaHCO_3$ needed to produce one molecule of each of the two substances you identified in Problem 2? After you have subtracted each of these molecules from the original sodium bicarbonate molecules needed for their formation, what would appear to be the formula of the remaining product?

4. What additional tests could you make to confirm your answer to Problem 3?

7-6 Chemical Analysis

You have seen that the pure substance sodium bicarbonate can readily be broken down into three other substances. From this it is clear that sodium bicarbonate is a compound. Earlier you learned that water, too, is a compound, since it can be decomposed into simpler substances. Water, unlike sodium bicarbonate, can readily be broken down directly into its elements. The products, hydrogen and oxygen, cannot be broken down further chemically, since hydrogen contains only hydrogen atoms and oxygen contains only oxygen atoms. The products of your sodium bicarbonate decomposition would have to be broken down further to produce separately the elements of which sodium bicarbonate is composed.

Finding out what kinds of atoms are present in various pure substances is an important activity in chemistry. This activity is known as *analysis* or *analytical chemistry*. When only the kinds of atoms are determined, the process is called *qualitative* analysis. When the amount of one or more of the different kinds of atoms is determined, the process is called *quantitative* analysis.

7-7 Chemical Synthesis

You have seen that compounds can be broken down, or decomposed, into simpler ones, or even into elements. The reverse process, that is, building up more complex molecules from simpler ones, or from atoms, is also an essential part of chemistry. This process is known as *synthesis*.

Chemists synthesize a wide variety of compounds, some of which find uses in every area of our daily lives. Some compounds are so important to our economy that they are produced on a very large scale. For example, the amounts of ammonia, NH_3, and sulfuric acid, H_2SO_4, that are produced are usually measured in tons.

Ammonia molecules consist of hydrogen and nitrogen atoms. It is possible to make ammonia in a single step by combining hydrogen, H_2, and nitrogen, N_2, molecules in giant reactors at high temperature and pressure. This process is discussed in greater detail in Chapter 22.

Sulfuric acid is made in three steps. First sulfur, S_8, is burned. In this process it combines with oxygen to form sulfur dioxide, SO_2. The sulfur dioxide is then made to react with more oxygen to form sulfur trioxide, SO_3. Finally, the sulfur trioxide is combined with water, H_2O, to give sulfuric acid, H_2SO_4. The synthesis of sulfuric acid is discussed in greater detail in Chapter 19.

In the following experiment you will synthesize a compound simply by heating a metal in the presence of air. The results of such syntheses, together with those obtained from analyses, have led to some of the most fundamental laws and concepts in chemistry.

INVESTIGATION

7-8 *Synthesis of a Metal Oxide*

Many of the elements known as metals will combine with the oxygen in the air when they are heated. The resulting substance is a type of compound called an *oxide*. Copper is a metal that combines quite rapidly with the oxygen of the air when it is heated, although it does not do so at ordinary temperatures and is thus often used as a protective coating. Iron is an element that, in the presence of a trace of water, will combine with oxygen even at room temperature.

PURPOSE

You will attempt to learn something about the nature of the oxide that forms when copper is heated in air. By comparing the ratio of the mass of copper to the mass of oxygen in your experiment with the ratios found by other class members using different masses of copper you will find out whether any regularities exist.

MATERIALS

1.75–2.25 grams of powdered copper   Burner
Pipestem triangle   Balance, centigram
Ringstand, ring
Porcelain evaporating dish, 75-mm diameter

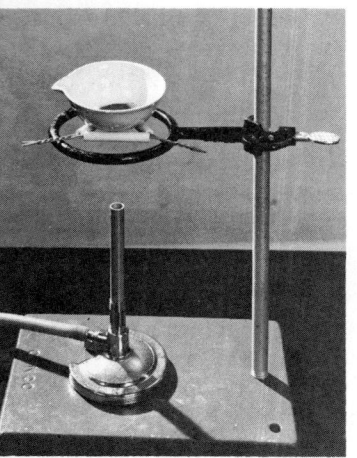

Figure 7-11

PROCEDURE

1. Determine the mass of a 75-mm porcelain evaporating dish to the nearest 0.01 gram.

2. Sprinkle a thin layer of powdered copper around the inside of the dish until you have added between 1.75 and 2.25 grams of the metal. Find the combined mass of the dish and the powdered copper.

3. Place the evaporating dish on the triangle on the ringstand as shown in Figure 7-11. Light the burner and heat the evaporating dish as strongly as possible for 20-25 minutes. If two burners are available, use them both to heat the dish. The dish should glow red if it is being heated sufficiently.

4. Turn off the burner and allow the dish to cool completely to room temperature. Again determine the mass of the dish and its contents.

DATA AND RESULTS

Record the following data and results in your notebook. All measurements are in grams.

Mass of dish and copper
Mass of empty dish
  Mass of copper

Mass of dish and copper oxide
Mass of empty dish
  Mass of copper oxide

RELATED PROBLEMS

1. Calculate the mass of oxygen that combined with the copper.

2. Calculate the ratio: $\dfrac{\text{Mass of copper}}{\text{Mass of oxygen}}$.

3. Why was it necessary to cool the evaporating dish completely before the final determination of its mass?

4. Each student, or group, doing this experiment will have begun with a different mass of copper before heating and will have found a different increase in mass after heating. How did the ratios of copper to oxygen, calculated by different groups in Problem 2, compare? Were they all different or was there any similarity in them?

5. What was the average ratio in all the experiments just done by the class?

6. Calculate the percentage of copper and oxygen in your sample of copper oxide. Compare your results with those obtained by other class members.

7. To calculate the average uncertainty in this experiment, first find the difference between your ratio and the average ratio for the class. Then, after everyone in the class has done the same thing, find the average of these differences. Your answer is the average uncertainty in the class result. It is also known as the average difference or average deviation. (Do not distinguish between high and low results in doing your calculations. Use absolute values for the differences, rather than positive and negative ones.)

7-9 *Definite Proportions*

In the preceding experiment you found that all of the copper oxide produced had at least one property in common. Every sample had the same relative masses of copper and oxygen in it, within experimental error. Further experiments would show that a similar result is found for all pure chemical compounds. As the result of such experiments, a law known as the **Law of Definite (or Constant) Proportions** has been established. This law states that **the masses of the elements in a pure compound always occur in a definite proportion.**

The discovery that a given compound always contains the same relative masses of its constituent elements was a very important one in establishing the science of chemistry. By 1812 chemists had agreed that experimental evidence showed this to be a law of chemistry. Along with the Law of Conservation of Mass, which you learned about in Chapter 2, the Law of Definite Proportions helped to suggest, most notably to John Dalton, that matter is made up of atoms. The Atomic Theory proposed by Dalton in 1808 was an attempt to explain why these laws should exist. (All disputes concerning the validity of the Law of Definite Proportions were not resolved until four years later.)

Dalton's atomic theory explained satisfactorily the experimental facts known in 1808, and has, with only minor modifications, proved to be consistent with the experimental facts discovered since. Today no chemist questions the existence of atoms, although no chemist has yet directly seen one.

Summary

There are three phases of matter: gas (or vapor), liquid, and solid. In general, substances can be transformed from one phase to another by changing the temperature (ice to water to steam and back, for example). The atoms and molecules that exist in the gas phase in general retain their identity when the gas condenses. Condensation occurs when the kinetic energy of the molecules becomes too low to overcome the small attractive forces that draw them together.

Pure substances are those of uniform composition that have constant and characteristic properties. Those pure substances (such as hydrogen, oxygen, or carbon) that cannot be changed or decomposed into two or more other substances consist of only one kind of atom and are called elements. Pure substances consisting of two or more kinds of atoms are called compounds. The determination of the composition of substances is called analysis or analytical chemistry.

Elements or simple molecules may be used to make compounds. This is called synthesis. It is found that the proportions by mass of the elements making up a compound are always the same. This is called the Law of Definite Proportions.

Self Check

1. How do you reason that solids and liquids are composed of atoms or molecules, as are gases?
2. How could you distinguish between a pure and an impure substance?
3. Distinguish between an element and a compound.
4. What do we mean by synthesis?
5. State in your own words the Law of Definite Proportions.

The Elements - Their Names and Symbols

7-10 The Elements

An element has been defined as a pure substance that contains only one kind of atom. There are over 100 elements known today. This means that about 100 different kinds of atoms have been recognized. Some of these elements occur in a pure form in nature and hence have been known for thousands of years. For example, the elements iron, silver, and gold were known to the ancients. During the late middle ages the alchemists referred to them by the Latin names *ferrum* for iron, *argentum* for silver, and *aurum* for gold. (Figure 7-12 shows some typical alchemists at work.)

Other elements occur free, that is, uncombined, in nature. The gases that comprise our atmosphere are primarily elements rather than compounds. But since these gases are mixed together, their separation and identification as pure elements came only in relatively recent times.

The majority of the elements do not occur uncombined in nature, to any great extent.

Figure 7-12 *When such alchemists as these were at work during the middle ages, Latin was the language of scholars. Some of the chemical symbols are derived from Latin names.*

John Dalton (1766–1844)

John Dalton, collecting marsh gas with one of his students.

JOHN DALTON is best remembered for his Atomic Theory of Chemistry, a collection of ideas that became the foundation of modern chemical science. Among the ideas that Dalton proposed were the following ones.

1. Elements are composed of atoms that cannot be further broken down. He said, "I have chosen the word 'atom' to signify these ultimate particles . . . The word 'atom' derives from the Greek *atomos* signifying uncut or indivisible."
2. Atoms of a given element are alike, but atoms of different elements are different. Dalton said, "All atoms of an element are perfectly alike in weight and figure."
3. In chemical reactions, atoms are neither created nor destroyed.
4. Chemical reactions take place between atoms or groups of atoms. Dalton called the smallest units of a compound "compound atoms", which today we call molecules.

Dalton soon came to realize that when atoms combine to form a compound, they do so in a simple ratio, but that for a given set of atoms, several simple ratios are possible. Therefore, a given set of atoms can be combined into several different compounds. This insight represented a major advance over earlier theories about the formation of chemical compounds.

Using the results of other men's analyses, Dalton formulated a table of relative atomic and molecular masses based on the reference mass of hydrogen as 1. He also suggested a new set of symbols to represent the known elements of the time. Both his table and his symbols were later superseded, but they were major steps in the advancement of chemistry.

When you consider the enormous influence that Dalton had on the progress of science, it is amazing to find that he was largely self-educated. His father was a poor weaver and young Dalton had to begin to support himself early in life. By the time that John Dalton had reached the age of *twelve,* he had progressed so well with his education that he was permitted by the village elders to start his own school in his home town. He continued to be a teacher and general tutor in the city of Manchester, England, for most of his life, but his heart was more in his scientific work than in his teaching. Dalton was elected to the Royal Society in 1822, but since his strict Quaker beliefs forbade his accepting any sort of honor for his work, the election was carried out without his knowledge.

Figure 7–13 *A large underground deposit of sodium chloride, NaCl, commonly known as salt, being mined in Louisiana.*

They are found chiefly as part of naturally-occurring compounds. (See Figure 7-13.) For this reason, the discovery of elements has depended mainly on the ability of chemists to decompose compounds into their elements. Before claiming them as newly discovered elements, chemists must purify them and show convincingly that they can in no way be further decomposed. From about 1700 to 1930, nearly all the elements that occur in nature, mainly in the form of their compounds, were discovered. Before that time, only about 13 elements were recognized. Since about the time of World War II, nearly a dozen new elements not found in nature have been made. These "synthetic" elements are produced by high energy nuclear processes occurring in cyclotrons and other particle accelerators, or in atomic reactors. (See Figure 7-14.)

Figure 7–14 *With the use of particle accelerators such as this one scientists can produce elements not found in nature.*

7-11 Symbols for the Elements

All of the elements have names, but they are also identified by means of *symbols*. Chemists have long used symbols for convenience in writing formulas and equations, since it is too cumbersome to write out the entire name each time an element, or an atom of the element, must be indicated. The names of the elements vary a little from one language to another, but symbols do not.

Most of the symbols are made up of the first letter of the English name together with one other letter from the name. The first letter of the symbol is capitalized but the second one is not. A few of the symbols consist of the capitalized first letter only. The symbols may be divided into four main groups.

1. A few symbols are derived not from the English name but from the old Latin (or other foreign) name. These present the greatest burden to the memory, but, fortunately, there are only eleven of them. They are listed, together with the common name and the ancient or foreign name of the element, in Table 7-1.

2. Some symbols are just the first letter of the English name of the element, as follows.

    B, boron   O, oxygen
    C, carbon   P, phosphorus
    F, fluorine   S, sulfur
    H, hydrogen   U, uranium
    I, iodine   V, vanadium
    N, nitrogen   Y, yttrium

3. Many symbols are the first two letters of the English name. The following list gives a few examples.

    Al, aluminum   Ne, neon
    Ca, calcium   Th, thorium

TABLE 7-1 Elements with Symbols Derived from an Ancient or Foreign Name

Common Name	Derivation of Symbol	Symbol
antimony	stibnum	Sb
copper	cuprum	Cu
gold	aurum	Au
iron	ferrum	Fe
lead	plumbum	Pb
mercury	hydrargyrum	Hg
potassium	kalium	K
silver	argentum	Ag
sodium	natrium	Na
tin	stannum	Sn
tungsten	wolfram	W

4. A few symbols are the first letter of the English name and one other letter which is not the second letter. Usually, this is because two or more elements have names with the same first two letters. For example, since calcium is Ca, cadmium cannot be. Instead, its symbol is Cd. Similarly, the symbol for thallium is Tl because Th is assigned to thorium. In some cases, there is no evident reason why the second letter is unused. For example, the only two elements that have names beginning with Z are zinc and zirconium. Neither, however, has the symbol Zi. Zinc is Zn, and zirconium is Zr.

Learning the symbols for the elements need not be a tedious, unpleasant task. Naturally, some effort must be made, but the trick is to take the right approach. It is foolish, except for students who have a very unique type of mind, to attempt to memorize the entire list, or any substantial part of it. Instead, simply learn and remember the symbols as they occur in the following pages of this book. A few, such as H and O, have already been introduced. Others will appear as needed. Only those of important, commonly-occurring elements will be used, and they will be used repeatedly, so as to reinforce their impression on your memory.

7-12 Molecular Formulas

The formula of a molecule simply shows what atoms are present and how many there are of each kind in the molecule. The molecular formulas for hydrogen, H_2, and oxygen, O_2, and water, H_2O, have already been used. Though they are simple, they are entirely typical, and thus they illustrate what simple things chemical formulas are. A summary of rules for writing formulas, given in somewhat more detail than previously, follows.

1. Each kind of atom present is represented by its symbol.
2. The number of atoms of a given kind, if there are more than one, is indicated by placing a subscript number immediately to the right of the symbol. The symbol alone without any following subscript represents one atom.
3. The sequence in which the various elements are listed is, unfortunately, rather arbitrary. It is not alphabetical. However, a certain sequence is often used as a matter of custom for many common molecules. Thus, water is normally written H_2O and not OH_2 (except for some special reason), although there is nothing wrong in principle with the second formula.

To illustrate further the writing of molecular formulas, formulas for some of the molecules that have already been mentioned in this book follow.

hydrogen chloride	HCl
ammonia	NH_3
nitric oxide	NO
nitrogen dioxide	NO_2
carbon dioxide	CO_2

Molecules that consist of only two atoms, either like or unlike, are called diatomic molecules, while all of the others are called polyatomic molecules. Molecules containing three atoms are often called triatomic molecules. In the above list, the diatomic molecules are HCl and NO. Hydrogen and oxygen molecules, H_2 and O_2, are also diatomic. NO_2 and CO_2 are triatomic, and also, of course, polyatomic.

The nitric oxide and nitrogen dioxide molecules, NO and NO_2, serve to illustrate the important point that the nature of a molecule depends not only on the kinds of atoms that it contains, but also on how many of each kind there are. NO and NO_2 are entirely different substances, even though both are composed only of nitrogen and oxygen atoms. In writing molecular formulas it is important to get both the subscripts and the atomic symbols correct.

Summary

More than 100 elements are known, about 90 of which occur in nature. There is an internationally accepted set of symbols used to represent these elements. Each symbol consists of one or two letters, the first letter being capitalized. Some of the commonest symbols, which you will immediately begin to use are: H, hydrogen; O, oxygen; N, nitrogen; C, carbon; Cl, chlorine; P, phosphorus.

The formula of a substance is a list of all the different kinds of atoms present in the pure substance with subscript numbers to tell how many there are of each. Each atom is represented by the symbol for the element. The correctness of the formula depends just as much on the numbers as it does on the symbols.

Self Check

1. How many atoms are there in a molecule with the formula H_3PO_4?
2. How much difference does it make if one of the subscript numbers in a formula is changed?
3. Write the formula for sulfur dioxide. Is this a diatomic, triatomic, or polyatomic molecule?

Representing Molecules and Chemical Reactions

7-13 Molecular Structures and Models

The molecular formula tells only what atoms are in the molecule. It does not give any information as to how the atoms are arranged relative to one another. Nevertheless, the formula is a very compact way to represent a molecule, or a substance, and for many purposes it is entirely adequate. There are times, however, when a **structural** formula, one that shows how the atoms are connected to each other, is useful.

A structural formula is written with each individual atom represented by its symbol and with lines connecting those atoms that are directly attached, or *bound*, to each other. For a diatomic molecule, of course, the structural formula is usually of no advantage. It is simply as a first illustration that the following structural formulas for a few common diatomic molecules are presented.

$$\underset{\text{hydrogen}}{\text{H---H}} \qquad \underset{\text{chlorine}}{\text{Cl---Cl}} \qquad \underset{\text{hydrogen chloride}}{\text{H---Cl}}$$

For polyatomic molecules, the structural formula gives a much more complete idea of what the molecule is like than does the ordinary molecular formula. Examples are given in Table 7-2. (Some atoms are shown joined by two lines in the table. The significance of these *double* bonds is discussed in Chapter 15. It is not necessary at this point to make any further distinction between double bonds and the more common single ones.)

The formulas of ethanol and dimethyl ether shown in Table 7-2 demonstrate the great value of structural formulas. Both of these molecules contain the same set of atoms, C_2H_6O. The two compounds are quite different substances, however, because the atoms are connected in different ways in the molecules, as shown by the structural formulas.

Molecules containing exactly the same set of atoms but differing in the way in which the atoms are arranged are called **isomers.** Thus, ethanol and dimethyl ether are isomers. Further examples of isomeric molecules will be discussed later.

While the structural formula is generally a much more informative way to represent the molecule than is the ordinary formula, it still suffers from a great disadvantage. It must be written in two dimensions, whereas the actual molecule is a three-dimensional structure. A molecule, then, is best represented by means of a three-dimensional model, or a picture of such a model.

TABLE 7-2 Formulas of Some Polyatomic Molecules

Molecule	Ordinary Formula	Structural Formula
ammonia	NH_3	H—N—H 　　| 　　H
hydrogen sulfide	H_2S	H—S—H
carbon dioxide	CO_2	O=C=O
methane	CH_4	H 　　| H—C—H 　　| 　　H
ethane	C_2H_6	H　H 　|　| H—C—C—H 　|　| 　H　H
ethanol	C_2H_6O	H　H 　|　| H—C—C—O—H 　|　| 　H　H
dimethyl ether	C_2H_6O	H　　　H 　|　　　| H—C—O—C—H 　|　　　| 　H　　　H

There are three common forms of molecular models, as shown in Figure 7-15. In each type, the atoms are represented by spheres. These can be made in different colors and sizes in order to distinguish between the different kinds of atoms. The crudest of the models is the one in which the spheres are connected by sticks (a ball-and-stick model). It is shown in Figure 7-15a. When springs are used in place of sticks, as in Figure 7-15b, a more realistic model is obtained (a ball-and-spring model). It shows that molecules are flexible, as they are known to be.

The third kind of model, shown in Figure 7-15c, is known as a space-filling model, since it shows not only how the atoms are attached to each other, but also represents the space actually occupied by each atom. This model is in many ways the most realistic one, since it shows how atoms that are not directly bound to one another may nevertheless be rather crowded together.

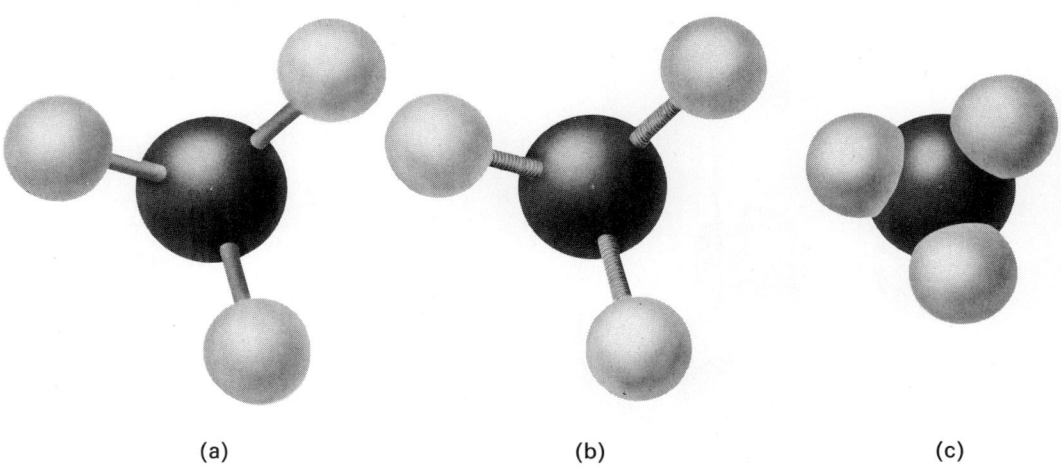

(a) (b) (c)

Figure 7-15 *Three types of molecular models. The ball-and-stick model (a) can serve for a crude representation. The ball-and-spring model (b) gives the correct idea of bond flexibility. The space-filling model (c) is in many ways the most realistic one.*

Several molecular models can be seen in Figures 7-16 and 7-17 and compared with the corresponding structural formulas given previously. Figure 7-16 shows ball-and-stick models of the isomeric C_2H_6O molecules. The space-filling models in Figure 7-17 give a realistic picture of the complete shape of the illustrated molecules. For example the models of H_2S and CO_2 clearly show that H_2S is a non-linear (bent) molecule while CO_2 is a linear (straight) one. While this difference can be indicated with the use of structural formulas, it is more clearly and unmistakably shown with the use of models.

7-14 Equations for Chemical Reactions

A **chemical reaction** is said to occur when one or more substances, called the **reactants,** are transformed into one or more different substances, called the **products.** A chemical equation is a way of representing what happens, using formulas for each of the substances involved in the reaction.

The combination of hydrogen and oxygen to form water is a chemical reaction. The reactants are hydrogen, H_2, and oxygen, O_2, and the product is water, H_2O. A chemical

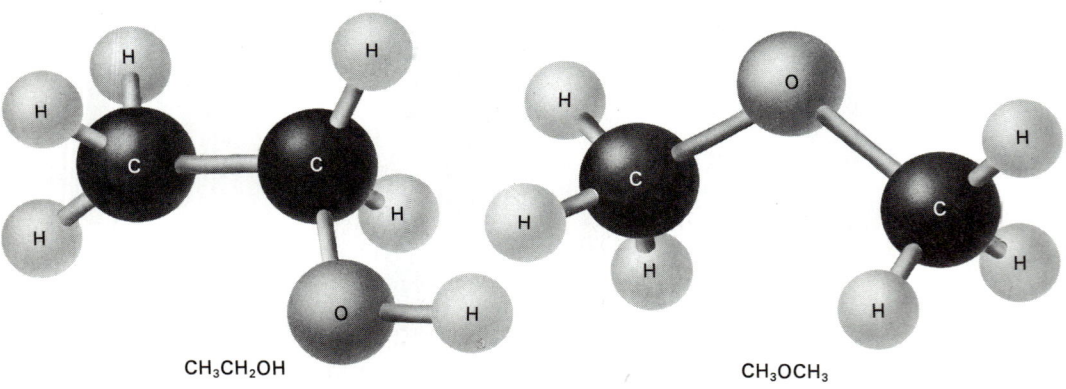

Figure 7–16 Ethanol (left) and dimethyl ether (right) contain exactly the same atoms, but the two compounds have quite different properties because of their different molecular structures. These structural differences are best shown by means of molecular models.

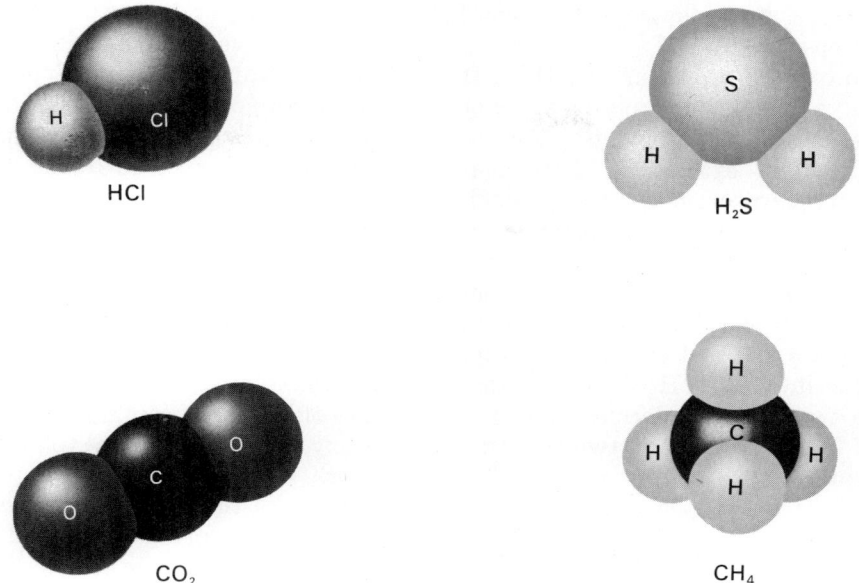

Figure 7–17 Space-filling models of some important molecules. They represent hydrogen chloride, HCl, hydrogen sulfide, H_2S, carbon dioxide, CO_2, and methane, CH_4.

equation was used in Section 5-14 to represent what happens in this reaction. The reaction will now be used as a specific example, in a general discussion of the rules for writing equations.

In order to write the equation for this reaction, the formulas of the reactants are written on the left, with a plus sign between them, and the formula of the product is written on the right. An arrow is drawn from left to right to show that reactants lead to products, giving:

$$H_2 + O_2 \longrightarrow H_2O.$$

This, however, is not a satisfactory chemical equation, because it is not "balanced." An equation that is balanced has the same number of atoms of each kind on each side. An equation that is not balanced is not correct.

A chemical equation must be balanced because atoms are conserved in chemical reactions. That is, when chemical changes or reactions occur, atoms are neither created nor destroyed. (As you learned in Section 2-12, there is no change in mass as the result of a chemical reaction.) The way in which the atoms are linked up to form molecules changes, but no atoms are obliterated, no atoms are created, and no atom is changed into any other kind. A **balanced equation** is simply one that conserves atoms.

The process of balancing the preceding equation (which really should not yet be called an equation, since it does not "equate" things), involves the following two common-sense steps.

1. Since two oxygen atoms are on the left side, it is necessary to have two water molecules on the right side in order to conserve oxygen atoms. To show that there are two water molecules, a coefficient of 2 is placed before the formula for the water molecule. That is, you write $2\,H_2O$. It is important to be clear about the fact that this coefficient multiplies the entire formula, not just the H that comes right after it. The "$2\,H_2O$" means that there are two entire molecules of water, a total of four H atoms and two O atoms. The "equation" now reads:

$$H_2 + O_2 \longrightarrow 2\,H_2O.$$

2. The equation is still incomplete because it shows four atoms of hydrogen on the product side and only two among the reactants. This, however, is easily corrected by placing a 2 in front of H_2:

$$2\,H_2 + O_2 \longrightarrow 2\,H_2O.$$

This is a balanced equation. It is the one given earlier, based on the results of gas volume measurements.

7-15 What a Chemical Equation Tells Us

There is a great deal more information packed into a chemical equation than may be evident at first sight. The most obvious thing that it tells is how many of each of the kinds of molecules are required in the reaction. The sample equation (Figure 7-18) shows that two molecules of hydrogen and one molecule of oxygen react to form two molecules of water. However, a mole of any substance contains the same number of molecules as a mole of any other substance. Thus the equation also tells us how many moles of each substance are involved in the reaction. The sample equation shows, then, that two moles of hydrogen combine with one mole of oxygen to form two moles of water. This is the result that was inferred earlier from gas volume measurements. Specifically, when the three substances were measured as gases at the same temperature and pressure, two volumes of hydrogen combined with one volume of oxygen to give two volumes of water vapor.

Once you think about the meaning of an equation in terms of moles, you can see that it also has meaning in terms of masses. The mass, in grams, of one mole of each substance can be calculated from its formula. Hence, the equation implicitly tells us what mass of each substance enters into or is produced by the reaction. This is one of the most valuable things that chemical equations tell us. In Chapter 19 this aspect will be discussed in greater detail.

7-16 Writing and Balancing Some Equations

In order that the concept of a chemical equation and what it implies may become firmly established in your mind, some additional examples will now be discussed.

A reaction that has been referred to earlier in this book is that of nitric oxide with oxygen to form nitrogen dioxide. This may serve as a second, still fairly simple, example for writing and balancing a chemical equation. First, the correct formulas for the molecules of all substances must be known. This reaction involves the following formulas.

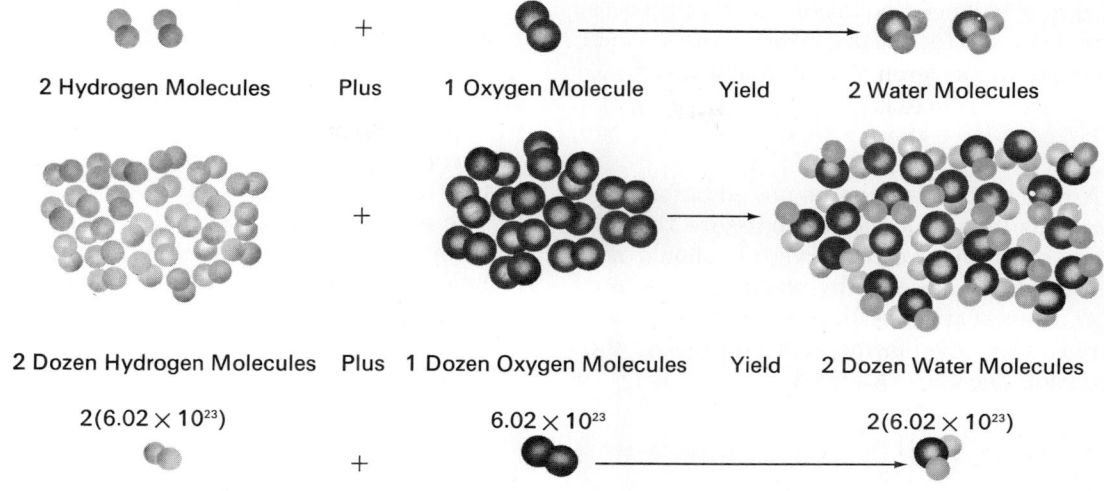

Figure 7–18 *A chemical equation shows us the proportional numbers of molecules taking part in a reaction. Here, the proportions range from 2 molecules : 1 molecule : 2 molecules to 2 moles : 1 mole : 2 moles.*

Reactants: nitric oxide, NO, and oxygen, O_2
Product: nitrogen dioxide, NO_2

The next step is to write the "equation" in unbalanced form:

$$NO + O_2 \longrightarrow NO_2.$$

Finally, coefficients must be assigned so as to conserve atoms. This produces a balanced equation, which is the only true equation. In this case, it can be seen that while N atoms are already conserved, O atoms are not. It may be noted that O atoms must come by twos on the right-hand side. Hence a first step toward a balanced equation would be to put a coefficient of 2 before NO, so as to give an even number, 4, of O atoms on the left-hand side. This gives:

$$2NO + O_2 \longrightarrow NO_2.$$

Now it is evident that if a coefficient 2 is also placed before NO_2, the equation becomes balanced:

$$2NO + O_2 \longrightarrow 2NO_2.$$

As a more complex example, let us consider the oxidation (burning) of methane, which is a principal constituent of the gas used for heating and cooking. Methane has the formula CH_4. Like most other substances consisting solely of carbon and hydrogen, it will burn in a liberal supply of oxygen to produce carbon dioxide and water. Thus, the first step toward a balanced equation for this process is to write the formulas for the reactants and products in the proper arrangement.

$$CH_4 + O_2 \longrightarrow CO_2 + H_2O$$

Now coefficients must be placed before the formulas in such a way as to assure that all kinds of atoms are conserved. It should not be difficult to see that by placing a 2 before H_2O, conservation of hydrogen atoms is assured. Then by placing a 2 before C_2, the equation becomes balanced.

$$CH_4 + 2O_2 \longrightarrow CO_2 + 2H_2O$$

This equation shows three basic things.
1. *The molecule relationships.* Two molecules of oxygen are needed to burn a molecule of methane completely. When this complete burning happens, one molecule of carbon dioxide and two molecules of water are formed.
2. *The mole relationships.* One mole of CH_4 requires two moles of O_2 in order to be burned completely, forming one mole of CO_2 and two moles of H_2O.
3. *The mass relationships.* The mass of one mole of each substance in the reaction can be calculated as follows.

$$\begin{aligned}CH_4&: 12 + 4(1) = 16 \text{ g}\\ O_2&: 2(16) = 32 \text{ g}\\ CO_2&: 12 + 2(16) = 44 \text{ g}\\ H_2O&: 2(1) + 16 = 18 \text{ g}\end{aligned}$$

Hence, 16 g of CH_4 require (2×32) g, or 64 g, of oxygen to burn completely. There are produced 44 g of CO_2 and (2×18) g, or 36 g, of water.

Summary

Molecules can be represented in several different ways, each of which has certain advantages and disadvantages. Molecular formulas are the most compact and useful in carrying out calculations, but they do not indicate how the different atoms are connected. Structural formulas are more cumbersome, but show the connections in a two-dimensional fashion. Molecular models show the entire three-dimensional structure of the molecule.

A chemical equation is a concise symbolic way to describe a chemical reaction. A chemical equation must be balanced in order to conserve atoms. That atoms are conserved is inferred from the experimental fact that mass is conserved in chemical processes. Balancing can be achieved by first writing formulas for all substances involved, in their proper places, and then systematically placing coefficients before the formulas so as to conserve all types of atoms. The complete equation shows the number of molecules, the number of moles, and the mass of each substance involved in the reaction.

Self Check

1. What are the advantages and disadvantages of a molecular formula as compared with a structural formula?
2. Is the following chemical equation balanced? If not, balance it.

$$Zn + HCl \longrightarrow ZnCl_2 + H_2$$

3. Why is it necessary to balance chemical equations?
4. On what evidence do scientists base the belief that all atoms are conserved in chemical processes?
5. What do molecular models show?

Chapter Summary

Matter exists in three phases (or states): gas (or vapor), liquid, and solid. In general, the phase in which a substance exists is determined by the kinetic energy of its molecules (or atoms) which, in turn, is controlled mainly by temperature. Therefore, substances can generally be transformed from one phase to another (ice to water to steam and back, for example) by changing the temperature. Such a change in phase does not usually cause any change in the identity of the atoms and molecules involved.

Pure substances are those of uniform composition that have constant and characteristic properties. Those pure substances that cannot be changed or decomposed into two or more other substances consist of only one kind of atom and are called elements. There are approximately 100 elements, about 90 of which are naturally-occurring.

Pure substances consisting of two or more kinds of atoms are called compounds. Analytical chemistry is concerned with the determination of the composition of substances. Synthetic chemistry is concerned with building up compounds from the elements or from other simpler compounds.

Experiments show that chemical compounds have definite constant proportions. Every pure sample of a given compound contains the same elements in the same proportions by mass.

There is an internationally accepted set of symbols used to represent the elements. In order to show what elements are present in a substance, the chemical symbols of each element are represented in a chemical formula to indicate all the kinds of atoms present in a molecule of that substance. Numerical subscripts, when needed, are placed after the symbol of an element to show how many atoms of that element are present in a particular molecule.

Molecules can be represented in several ways: written formulas, structural formulas, and molecular models. Chemical equations show that a chemical reaction takes place, and give the amounts of reactants needed and products produced. They also indicate conservation of mass. It is inferred that atoms are conserved in chemical reactions, since experimental evidence shows that mass is always conserved in chemical reactions.

Questions and Problems

1. Why do solid and liquid substances tend to vaporize as the temperature rises?
2. There are some gases that will not condense at pressures below a certain critical pressure, regardless of how far you lower the temperature. How might you account for this?
3. Is it possible for two substances to have

the same kinds of atoms in the same ratios and still have different properties? Explain.

4. What is an isomer? Examine the structural formulas below for isomers and, if you find any, write their molecular formulas.

(a)
$$H-\underset{\underset{H}{|}}{\overset{\overset{H}{|}}{C}}-\underset{\underset{H}{|}}{\overset{\overset{H}{|}}{C}}-H$$

(b)
$$H-\underset{\underset{H}{|}}{\overset{\overset{H}{|}}{C}}-O-\underset{\underset{H}{|}}{\overset{\overset{H}{|}}{C}}-H$$

(c)
$$H-\underset{\underset{H}{|}}{\overset{\overset{H}{|}}{C}}-O-H$$

(d)
$$H-\underset{\underset{H}{|}}{\overset{\overset{H}{|}}{C}}-\underset{\underset{H}{|}}{\overset{\overset{H}{|}}{C}}-O-\underset{\underset{H}{|}}{\overset{\overset{H}{|}}{C}}-H$$

(e)
$$H-\underset{\underset{H}{|}}{\overset{\overset{H}{|}}{C}}-\underset{\underset{H}{|}}{\overset{\overset{H}{|}}{C}}-O-H$$

5. Would ordinary beach sand be considered a pure substance? Explain.
6. What is in "pure" water?
7. Ammonia gas, NH_3, can be burned with oxygen gas, O_2, to give nitrogen gas, N_2, plus water, H_2O. A balanced equation for the reaction was arrived at by the following sequence of steps. Explain what was accomplished in each step.

$$NH_3 + O_2 \longrightarrow N_2 + H_2O$$
$$2NH_3 + O_2 \longrightarrow N_2 + H_2O$$
$$2NH_3 + O_2 \longrightarrow N_2 + 3H_2O$$
$$2NH_3 + 3/2\, O_2 \longrightarrow N_2 + 3H_2O$$

8. Propane gas, C_3H_8, is combined with oxygen gas, O_2, to yield the products, carbon dioxide, CO_2, and water, H_2O. In a stepwise fashion (as in Question 7), write and balance the equation.
9. Distinguish between the terms monatomic, diatomic, and polyatomic. Is a monatomic compound possible? If so, give an example; if not, explain.
10. What elements does SF_6 contain? How many atoms are in the molecule?
11. Discuss the importance of subscripts and coefficients in a chemical equation.
12. Consider the equation:
$$2C_2H_2 + 5O_2 \longrightarrow 4CO_2 + 2H_2O.$$
    (a) How many atoms are involved in the reaction?
    (b) How many moles of carbon dioxide are produced?
    (c) How many atoms of oxygen gas are consumed?
13. In your own words define or describe: (a) chemical symbol; (b) chemical formula (molecular); (c) chemical equation.
14. Suggest a reason why some compounds decompose before they can be brought to a temperature at which they are entirely converted to a gas.
15. How do Co and CO differ in meaning?
16. Suppose 10 hydrogen molecules and 10 oxygen molecules react. How many molecules of water are formed? What would be left over?
17. One million oxygen molecules combine with sufficient hydrogen molecules to form water molecules. How many water molecules are formed? How many hydrogen molecules are consumed?
18. Energy is produced in an internal combustion engine by burning hydrocarbons, such as C_7H_{16} and C_8H_{18}, with carbon dioxide and water being produced. Write balanced equations for the burning of these two hydrocarbons.

19. If yeast is added to a dilute solution of sucrose, $C_6H_{12}O_6$, at room temperature, a reaction occurs in which the sucrose is decomposed into carbon dioxide, CO_2, and ethanol, C_2H_6O. Write a balanced equation for this reaction.

20. One volume of hydrogen gas combines with one volume of chlorine gas to give two volumes of hydrogen chloride gas. The equation for this reaction is:

$$H_2 + Cl_2 \longrightarrow 2HCl.$$

  (a) According to this reaction, how many molecules of hydrogen chloride, HCl, can be formed from one molecule of hydrogen, H_2?
  (b) How many moles of hydrogen chloride can be formed from one mole of hydrogen?
  (c) Four molecules of chlorine, Cl_2, will produce how many molecules of HCl?
  (d) Eight moles of hydrogen chloride are formed from how many moles of chlorine?

21. The reaction between nitric oxide, NO, and oxygen, O_2, is written as follows.

$$2NO + O_2 \longrightarrow 2NO_2$$

  (a) Two molecules of nitric oxide give how many molecules of nitrogen dioxide, NO_2?
  (b) Three moles of NO can give how many moles of NO_2?
  (c) How many moles of oxygen atoms are there in two moles of NO?
  (d) How many moles of oxygen atoms are there in one mole of O_2?
  (e) How many moles of oxygen atoms are there in two moles of NO_2?
  (f) Make use of the answers to parts (c), (d), and (e) to verify that the reaction is written so as to conserve oxygen atoms.

22. (a) Write a balanced equation for the reaction in which nitrogen gas, N_2, and hydrogen gas, H_2, react to form ammonia gas, NH_3.
  (b) Verify that your equation conserves hydrogen atoms.
  (c) Verify that your equation conserves nitrogen atoms.

23. When ammonia is decomposed into nitrogen and hydrogen, the reaction absorbs heat. Written in terms of moles, the reaction is as follows.

$$2NH_3 + 22 \, \text{kcal} \longrightarrow N_2 + 3H_2$$

  (a) Two moles of ammonia produce how many moles of nitrogen?
  (b) The production of one mole of nitrogen absorbs how much heat?
  (c) The production of nine moles of hydrogen would absorb how much heat?

24. One step in the manufacture of sulfuric acid is to burn sulfur, S_8, in air to form a colorless gas with a choking odor. The name of the gas is sulfur dioxide, SO_2.
  (a) Write the balanced equation for this reaction.
  (b) Interpret the equation in terms of molecules.
  (c) Interpret the equation in terms of moles.
  (d) Two moles of sulfur, S_8, would produce how many moles of sulfur dioxide, SO_2?

25. When iron, Fe, rusts, it combines with the oxygen of the air to form iron oxide, Fe_2O_3. Which of the following are false?

(a) The equation is:
$4Fe + 3O_2 \longrightarrow 2Fe_2O_3$.
(b) The formula Fe_2O_3 represents five atoms.
(c) Oxygen gas is triatomic.
(d) The mass of the reactants equals the mass of the products.
(e) Atoms are conserved.

26. The copper oxide that you made (Section 7-8) has the formula CuO. Write a balanced equation for the reaction that you carried out.

27. Balance the equation for each of the following reactions. Begin on the basis of one mole of the substance underscored.
    (a) $\underline{Li} + Cl_2 \longrightarrow LiCl$
    (b) $Na + \underline{Cl_2} \longrightarrow NaCl$
    (c) $Na + \underline{F_2} \longrightarrow NaF$
    (d) $Ca + \underline{Br_2} \longrightarrow CaBr_2$
    (e) $\underline{O_2} + Cl_2 \longrightarrow Cl_2O$
    (f) $O_2 + \underline{Cl_2} \longrightarrow Cl_2O$

28. Balance the equation for each of the reactions involving oxygen. Begin on the basis of one mole of the substance underscored.
    (a) $Ni + \underline{O_2} \longrightarrow NiO$
    (b) $\underline{Ni} + O_2 \longrightarrow NiO$
    (c) $\underline{Li} + O_2 \longrightarrow Li_2O$
    (d) $\underline{N_2H_4} + O_2 \longrightarrow N_2 + H_2O$
    (e) $\underline{C_2H_2} + O_2 \longrightarrow CO_2 + H_2O$
    (f) $\underline{Cu_2S} + O_2 \longrightarrow Cu_2O + SO_2$
    (g) $\underline{FeS_2} + O_2 \longrightarrow Fe_2O_3 + SO_2$

29. Balance the following equations. If an equation is balanced as it appears below, simply write "Balanced."
    (a) $S + O_2 \longrightarrow SO_3$
    (b) $K + O_2 \longrightarrow K_2O_2$
    (c) $HgO \longrightarrow Hg + O_2$
    (d) $BaO_2 \longrightarrow BaO + O_2$
    (e) $CH_4 + Br_2 \longrightarrow CHBr_3 + HBr$

30. Repeat Question 29 for the following reactions.
    (a) $P + O_2 \longrightarrow P_4O_6$
    (b) $HgCl_2 + H_2S \longrightarrow HgS + HCl$
    (c) $Mg + HCl \longrightarrow MgCl_2 + H_2$
    (d) $MgCO_3 \longrightarrow MgO + CO_2$
    (e) $Na + Cl_2 \longrightarrow NaCl$
    (f) $Fe_3O_4 + H_2 \longrightarrow Fe + H_2O$

31. Determine the masses of one mole of each of the following to two decimal places.
    (a) CuS
    (b) Al_2O_3
    (c) $KMnO_4$
    (d) NH_4Cl
    (e) K_2HPO_4

32. Repeat the instructions to Question 31 for the following compounds.
    (a) $CdSO_4$
    (b) $HC_2H_3O_2$
    (c) SCl_2
    (d) C_7H_{16}
    (e) K_2PtCl_6

33. Write the full name of each element for which a symbol appears in Questions 31 and 32.

34. How many moles are contained in each of the following?
    (a) 98 g of H_2SO_4
    (b) 7 g of N_2
    (c) 7 g of CO
    (d) 0.051 g of NH_3

35. How many moles are contained in each of the following?
    (a) 2.45 g of H_2SO_4
    (b) 54 g of H_2O
    (c) 4.0 g of CH_4
    (d) 82 g of H_3PO_4

36. In the reaction below, how many moles of oxygen are required to burn 46 grams of sodium?

$$4Na + O_2 \longrightarrow 2Na_2O$$

37. In the reaction below, how many grams of CaO can be produced from 2 moles of oxygen?

$$2Ca + O_2 \longrightarrow 2CaO$$

38. Methane, CH_4, is the main constituent of natural gas. It burns in the presence of oxygen to form carbon dioxide and water.
    (a) Write and balance the equation.
    (b) One-eighth mole of methane produces how many moles of water vapor?
    (c) How many moles of water vapor would be produced by 4.0 grams of methane?

8 Patterns of Chemical Behavior

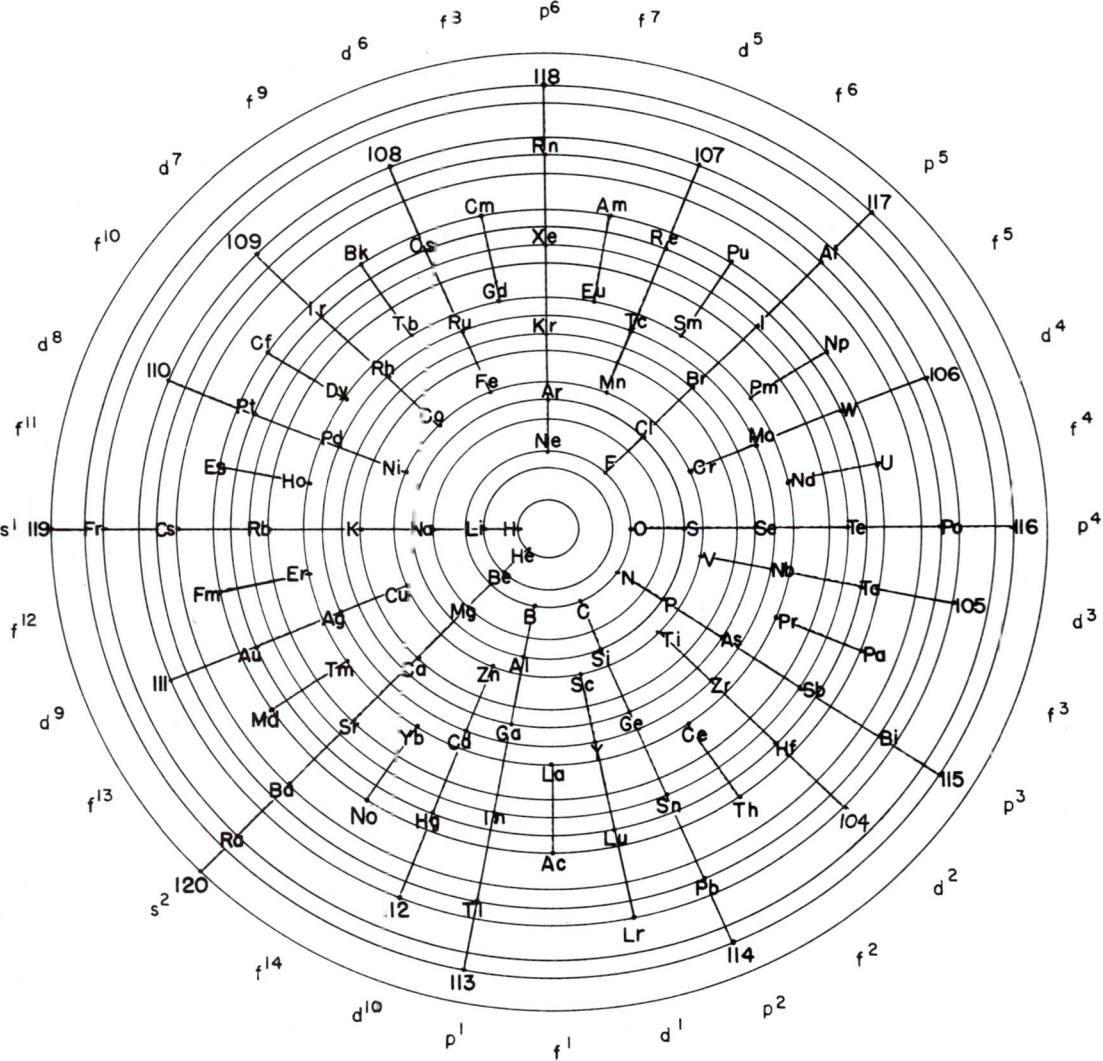

This arrangement of the elements is one of many that reflect observed patterns of chemical behavior.

So far you have been introduced to a few very simple chemical reactions and you have learned the basic rules for writing chemical equations. In the next few sections, the scope of your knowledge will be broadened as you examine a greater range of reactions. You will also be introduced to a greater range of compounds. In all of this, you should be looking for similarities and regularities that will help you to tie a variety of facts together in a sensible and systematic way. Such efforts will aid your understanding of one of the great generalizations of science: the periodic table of the elements.

Some Important Types of Reactions

8-1 Reactions with Oxygen

As you learned earlier, a chemical reaction is a change in which atoms that were originally linked up in certain ways become linked up in other ways. Before a chemical reaction begins, there are certain types of molecules, called the reactants, present. After the reaction has taken place, different molecules, called the products, are found. The number of atoms of each element does not change, but the way in which they associate to form molecules changes. Usually, heat or other forms of energy are produced or consumed as a chemical reaction occurs.

Of the few chemical reactions that you have already learned about, most involve the reaction of some element or compound with oxygen. Reactions with oxygen, especially the reactions of various elements with oxygen, are extremely important. It will be worthwhile to look at a few more. In this way you will also be introduced to some more elements.

The two following reactions of elements with oxygen have already been mentioned.

$$2H_2 + O_2 = 2H_2O$$
$$C + O_2 = CO_2$$

Practically all of the known elements will react with oxygen. Usually it is necessary to apply heat in order to get the reaction started. Some of the reactions are quite vigorous once they have started, as is the one shown in Figure 8-1. All of the elements

Figure 8-1 *Magnesium burns rapidly in air, giving off noticeable energy, chiefly in the form of light.*

Figure 8-2 *White phosphorus bursts into flame when exposed to the air. In other less reactive forms (red, black) of the element, the phosphorus atoms are bonded together in a different way. All three forms react with oxygen.*

known as metals will react with oxygen to form compounds called *oxides* Some examples follow.

Silver, Ag: $4Ag + O_2 = 2Ag_2O$
Lithium, Li: $4Li + O_2 = 2Li_2O$
Sodium, Na: $4Na + O_2 = 2Na_2O$
Potassium, K: $4K + O_2 = 2K_2O$

Beryllium, Be: $2Be + O_2 = 2BeO$
Magnesium, Mg: $2Mg + O_2 = 2MgO$
Calcium, Ca: $2Ca + O_2 = 2CaO$

Iron, Fe: $4Fe + 3O_2 = 2Fe_2O_3$
Aluminum, Al: $4Al + 3O_2 = 2Al_2O_3$

Tin, Sn: $Sn + O_2 = SnO_2$

The important thing to note is that the number of atoms of oxygen with which an atom of metal will combine is not the same for all metals. Clearly, the metals in the first group have a relatively small capacity to combine with oxygen. It takes two metal atoms to hold one oxygen atom. In the next group are listed three metals that have twice this capacity to combine with oxygen. Each metal atom reacts with and holds one oxygen atom. In the next two groups are metals with still greater combining capacities.

Not only metals can combine with oxygen. Elements that are not metals, such as carbon, can react with oxygen, too. Many chemical compounds can react with oxygen, as well. Most reactions with oxygen produce noticeable energy, usually in the form of heat and light, and so are known as **combustion** reactions.

Nonmetallic elements that react with oxygen include sulfur, S, and phosphorus, P.

$$S + O_2 = SO_2$$
$$4P + 5O_2 = P_4O_{10}$$

Phosphorus bursts into flame in air, giving off a dense white smoke, as shown in Figure 8-2. The P_4O_{10} molecule is the most complex oxide molecule given so far. It is because of this that the coefficients on the left side of the equation must be so large.

Of the many compounds that react with oxygen, the most important are those consisting of carbon and hydrogen, or of carbon, hydrogen, and oxygen. Compounds that are made up only of carbon and hydrogen are called **hydrocarbons**. The wax in a candle consists mainly of a mixture of hydrocarbons, one of which is pentadecane, $C_{15}H_{32}$. The following equation for the burning of pentadecane is typical of equations for the complete combustion of hydrocarbons.

$$C_{15}H_{32} + 23O_2 = 15CO_2 + 16H_2O$$

Some of the compounds that contain oxygen, in addition to carbon and hydrogen, are called **carbohydrates**. In these compounds the ratio of hydrogen to oxygen atoms is always two to one. The compound cellulose, which makes up forty to fifty percent of wood, is a typical carbohydrate. It consists of very large molecules, which are made up of small units with the formula $C_6H_{10}O_5$. The following equation shows what happens to one of these units on combustion.

$$C_6H_{10}O_5 + 6O_2 \longrightarrow 6CO_2 + 5H_2O$$

INVESTIGATION

8-2 Active Metals React with Water

Chemists have long known that some metals react with water. Since air and water are two of the most common substances around us, it is not surprising that early chemists tried to find out how elements and compounds react with these substances. You already know that water has some effect on iron and steel, since these substances rust more rapidly when they are wet. You will find out in this experiment how some of the more reactive (or active) metals react with water.

Most of the active metals have been known for some time. In 1807, Sir Humphry Davy discovered a new and very reactive element, which he named potassium. A few days later he isolated for the first time the element sodium, another reactive metal. Within one year Davy also discovered and isolated the elements magnesium, calcium, strontium, and barium, all of which were found to have properties somewhat similar to each other. Lithium was discovered by the Swedish chemist Johannes Arfvedson, in 1817. Robert Bunsen and Gustav Kirchhoff, in Germany, discovered cesium in 1860 and rubidium in 1861. They used a spectroscope to identify them. (See Chapter 11 for a description of a spectrograph, which is a type of spectroscope.) Friedrich Wöhler and Antoine-Alexandre-Brutus Bussy independently isolated metallic beryllium in 1828.

Chemistry and Our Environment
IRON AND ALUMINUM OXIDES

As you have seen in Section 8-1, both iron and aluminum react with oxygen to give oxides. These two oxides, Fe_2O_3 and Al_2O_3, have the same type of formula. However, iron oxide and aluminum oxide differ in one respect and this has a great bearing on the appearance of our environment.

Iron oxide is a soft, flaky, reddish substance. When formed in the presence of water—as it usually is—it is what we call rust. Actually, rust is not perfectly pure Fe_2O_3, but that need not concern us. Because of its nature, rust crumbles away as it is formed, and the metal goes right on rusting until it is all converted to rust. Iron articles that are to be exposed to both oxygen and moisture must be painted or otherwise coated, as a result.

Aluminum, on the other hand, does not need to be coated with paint. Why is this? Is it because it does not react with oxygen under ordinary conditions? No. It reacts even more rapidly than iron does, even if perfectly dry. The aluminum oxide, however, is hard and tough and clings to the surface. Once a coating of oxide, perhaps 20 layers of atoms thick, is built up, oxygen can no longer reach the rest of the metal underneath. Reaction with oxygen ceases, and the bulk of the metal is protected from oxygen and moisture. Moreover, the surface coating of Al_2O_3 is fairly attractive in appearance, looking much like the grey metal itself. This is why aluminum has become so very popular for so many purposes where the great strength of iron and steel are not required. All kinds of protective surfaces and structures that do not need to bear great weight, but which must withstand oxygen and moisture, are now made of aluminum. Aluminum is also light, that is, it has a relatively low density. The

The aluminum surface of a skyscraper. Its resistance to corrosion makes aluminum a popular building material.

combination of its lightness and its natural ability to protect itself against air and moisture make it ideal for constructing airplanes, and this is one of its most important uses. The chemical properties of aluminum and its oxide have been used to improve our environment.

Unfortunately, the same properties that are an advantage when wisely used can be a disadvantage when carelessly used. Because aluminum oxide gives a protective coating to the metal, aluminum objects must not be discarded carelessly. The aluminum tops for cans and bottles of beverages that thoughtless people leave in the country and along roadsides *will live longer than the people who leave them there*. Unless picked up, they will go on and on making our environment unsightly. Of course, no form of trash should ever be discarded carelessly by a responsible citizen. Articles made of aluminum, however, pose an aggravated problem because of their ability to survive.

While all of these elements have some similarities in the way they react, it is customary to group together lithium, sodium, potassium, rubidium, and cesium, since two atoms of each of these elements combine with one atom of oxygen.

PURPOSE

In this experiment you will compare the reactions of several metals with water.

MATERIALS

Safety shield	Litmus paper
Beaker, 250-ml	Ringstand, ring
Beaker, 400-ml	Wire gauze
Tongs or tweezers	Burner
Knife	Test tube
Paper towel	Utility clamp

Sodium, potassium, lithium, calcium, and magnesium
Stop watch or clock with sweep second hand
Sodium hydroxide, potassium hydroxide, and calcium hydroxide solutions

PROCEDURE

The first part of this experiment will be performed as a demonstration by your teacher, who will follow almost the same procedure that you will use in later parts of the experiment. Be prepared to record reaction times for the teacher demonstrations as well as for your own experiments.

Part A. (Teacher Demonstration)

Be sure to wear safety glasses for this experiment. Perform the demonstrations behind a safety shield or be sure that no observers are any closer than six or eight feet from the reaction.

1. Fill a 250-ml beaker about one-half full of water at room temperature.

2. Use tongs or tweezers to remove a piece of sodium from its container. With a sharp knife cut a cube-shaped piece of metal approximately 3 mm on an edge. Cut away any oxide coating, so that all surfaces have a metallic luster. Wipe off any oil with a piece of paper towel.

3. Drop the piece of sodium into the water in the beaker. Cover the beaker at once with a wire gauze. Use a stop watch or a clock with

a sweep second hand to measure the length of time needed for the reaction to go to completion.

4. When the reaction is finished, test the remaining solution with two pieces of litmus paper, one red and one blue. Note any color change.

5. Rinse out the beaker. Repeat the experiment with a piece of potassium about the same size as the piece of sodium used earlier.

Part B. (*Student Experiment*)

1. Fill a 250-ml beaker about one-half full of water at room temperature.

2. Use tongs or tweezers to remove a piece of lithium the size of a grain of rice from its container. As before, remove any oxide coating and wipe off any oil with a piece of paper towel.

3. Drop the piece of lithium into the water in the beaker. Use a stop watch or a clock with a sweep second hand to measure the length of time needed for the reaction to go to completion.

4. When the reaction is finished, test the resulting solution with both red and blue litmus paper. Note any color change.

5. Rinse out the beaker. Repeat steps 1-4 using a piece of calcium metal furnished by the teacher.

6. Rinse out the beaker. Repeat steps 1-4 using a piece of magnesium metal.

7. If any magnesium remains in the beaker after step 6, remove it from the water. Place the beaker of water on a wire gauze supported on a ring on a ringstand over a burner. Heat the water to boiling.

8. Remove the burner from below the beaker. Drop into the hot water the piece of magnesium from step 6. Note the rate of reaction, if any. Test the solution with red and blue litmus paper. Note any color change.

9. Test dilute solutions of sodium hydroxide, potassium hydroxide, and calcium hydroxide with red and blue litmus paper. Record any color change.

10. Fill a 400-ml beaker about two-thirds full of water at room temperature. Fill a test tube entirely full of water. Hold your thumb or finger over the open end of the test tube. Invert the test tube and lower it into the beaker of water. Remove your finger after the open end of the tube is below the surface of the water in the beaker. Clamp the test tube in position. Insert a piece of lithium or a piece

of calcium into the water at the open end of the test tube so that the bubbles of gas rise up into the test tube. (See Figure 8-3.) When the tube is about one-half full of gas, loosen the clamp and, using the clamp as a test tube holder, remove the tube from the water. Permit the remaining water to run out of the test tube. With the test tube still in an inverted position immediately hold its open end close to a burner flame. (Be sure to wear your safety glasses.) Note any reaction that may take place.

RELATED PROBLEMS

1. Look up the molar masses of lithium, sodium, and potassium. Do the same for magnesium and calcium.

2. For the first three elements, is there any relationship between the molar masses and the reaction rates for the reaction between these elements and water? (Remember that the shorter the time for the reaction, the faster was the reaction rate.) Describe any relationship that may exist.

3. Is there any relationship between the molar masses of magnesium and calcium and the reaction rates for the reaction between these elements and water? Describe any relationship that may exist.

4. In general, how did the reaction rates of magnesium and calcium compare with the rates for lithium, sodium, and potassium?

5. What two products were formed when each of the metals reacted with water? What was the evidence for your answer?

6. What general conclusion about reaction rates might you make after observing the behavior of magnesium in hot and cold water?

7. Based on its molar mass, how would you expect cesium to react with water? Would you expect beryllium to react more rapidly or more slowly than calcium?

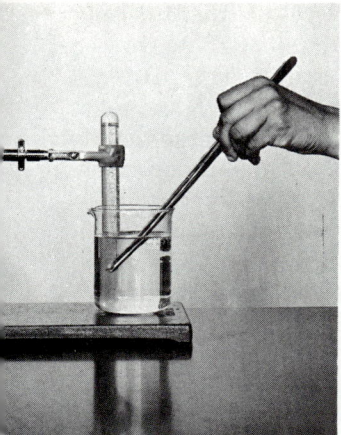

Figure 8-3

8-3 Comparing Reactions of Metals with Water

You have just seen that hydrogen gas is produced when an active metal reacts with water. The tests with litmus paper gave you a good clue to what the other product of the reaction was. However, you would have had to find out how much of each product was formed in order to write equations for the reactions.

When both the identities and the amounts of the products have been carefully determined, the following equations are found to be correct.

$$2Li + 2H_2O = 2LiOH \text{ (in solution)} + H_2$$
$$2Na + 2H_2O = 2NaOH \text{ (in solution)} + H_2$$
$$2K + 2H_2O = 2KOH \text{ (in solution)} + H_2$$
$$Ca + 2H_2O = Ca(OH)_2 \text{ (in solution)} + H_2$$
$$Mg + 2H_2O = Mg(OH)_2 \text{ (in solution)} + H_2$$

In each of the preceding reactions, the metal has liberated some of the hydrogen atoms that make up the water molecules. These atoms then form hydrogen molecules, H_2. For every water molecule that reacts, one hydrogen atom is liberated while the other remains attached to the oxygen atom. The resulting compounds, LiOH, NaOH, KOH, $Mg(OH)_2$, and $Ca(OH)_2$, are called *hydroxides*. The first three dissolve completely in the excess water. The $Ca(OH)_2$ and $Mg(OH)_2$ may not dissolve completely if there is only a small amount of water, since they are less soluble than the other hydroxides.

An important feature to note in the reactions just examined is that not all the metals react with water in the same proportions. For three of them, lithium, sodium, and potassium, it takes two moles of metal to form one mole of H_2. For calcium or magnesium, on the other hand, one mole of metal can liberate one mole of H_2.

Summary

Nearly all elements, both metals and nonmetals, react with oxygen to form oxides. Many compounds, especially hydrocarbons and carbohydrates, also react with oxygen. Quite often the reactions proceed with the emission of heat and light, in which case the process is called combustion. An important reaction of some metals, for example, lithium, sodium, potassium, magnesium, and calcium, is with water. Hydrogen gas is liberated and solutions of the metal hydroxides are formed. It is important to note that metals can be arranged into groups on the basis of how many atoms of oxygen they will hold in the oxide, or on the basis of how many moles of H_2 one mole of metal atoms can liberate from water.

Self Check

1. When an element combines with oxygen, what is the name of the compound that is formed?
2. What does a chemist mean by the term combustion?
3. Write a balanced equation for the combustion of the hydrocarbon C_8H_{18}, a component of automobile gasoline.

Some Elements Have Similar Properties

8-4 The Alkali Metals: A Chemical Family

You have seen two examples of how the elements lithium, sodium, and potassium resemble one another. They react with oxygen to form oxides that have formulas of the type M_2O. They also react with water to liberate hydrogen and form solutions of the hydroxides, MOH. (In both of the above general formulas, M represents any one of the metals that forms such a compound.) You may recall that silver also forms an oxide, Ag_2O, with the M_2O type of formula. However, silver does *not* react with water at all, not even slowly. You saw that the metal calcium reacts with water and oxygen, in much the same way as do lithium, sodium, and potassium. The formulas of its compounds, however, are different. A mole of calcium combines with more oxygen and liberates more hydrogen than does a mole of Li, Na, or K. Thus, the three elements, Li, Na, and K seem to have "family resemblances," not shared with other elements. Like brothers and sisters, they are not identical but are alike in many ways.

There are two more naturally-occurring elements, called rubidium, Rb, and cesium, Cs, which react very similarly to Li, Na, and K. Rubidium and cesium are a lot rarer than lithium, sodium, and potassium, and that is why you have not used them in the laboratory. All five elements have many similar physical properties in addition to their chemical similarities. They are all whitish, shiny, relatively soft metals. (See Figure 8-4.) They melt at fairly low temperatures. Cesium becomes a liquid at 28.7°C (about 85°F).

These elements are known as **alkali metals,** or sometimes just **alkalis.**

Figure 8-4 *Like the other alkali metals, potassium (above) can be readily cut, exposing a whitish, shiny surface.*

8-5 Another Chemical Family: The Halogens

Among the earliest elements to be discovered were three called chlorine, Cl, bromine, Br, and iodine, I. The physical appearance of these elements would not immediately give you the impression of a family relationship. Chlorine is a pale green gas, bromine is a deep red liquid, and iodine is a purple solid. However, as one examines these elements more carefully, as the chemists in the early part of the 19th century did, resemblances become very noticeable. Although bromine is a liquid and iodine is a solid, both of them are very volatile, that is, they evaporate to give a vapor very easily. Also, the molecular formulas of all these elements are similar, Cl_2, Br_2, and I_2. (How do you suppose chemists first discovered these molecular formulas?)

The chemical properties of these elements are very noticeably similar. Chlorine, bromine, and iodine are very reactive, and form compounds with nearly all of the other elements, especially with the metals. (See Figure 8-5.) The products generally have the

Figure 8-5 *Halogens react vigorously with metals. Here, chlorine is reacting with the metal antimony, Sb.*

same type of formula. For instance, sodium reacts with each one (let X represent Cl, Br, or I) as follows.

$$2Na + X_2 = 2NaX$$

The three resulting substances, NaCl, NaBr, and NaI, are white crystalline powders, which are soluble in water and have a salty taste. NaCl is, of course, common table salt.

A fourth element called fluorine, F, was discovered somewhat later than the three just mentioned. This element resembles chlorine in being a gas. It resembles all of the other three in forming a diatomic molecule, F_2. It reacts more vigorously than Cl_2, Br_2, or I_2, but forms many similar compounds. For example, it forms salts, called fluorides, with many metals. Salts that it forms with the alkali metals include NaF and KF.

The four elements F, Cl, Br, and I are called halogens. This name comes from the Greek root *halo*, meaning "salt," and *gen*, pertaining to "generation or formation." Thus, the word halogen means "salt-former," in keeping with the chemistry of these elements.

Summary

An important family of elements is the alkali metals (or alkalis). It includes lithium, Li; sodium, Na; potassium, K; rubidium, Rb; and cesium, Cs. These are soft white metals that react similarly with water, oxygen, and other substances.

Another family of elements is the halogens. It includes fluorine, F; chlorine, Cl; bromine, Br; and iodine, I. These are decidedly nonmetallic elements, being, respectively, a gas, another gas, a volatile liquid, and a volatile solid. They all react chemically in the same general way, with many other elements.

Self Check

1. What is the general equation for the reaction of any alkali metal, M, with O_2?
2. How would the halogens react with lithium? Write the equations.
3. Of the two families discussed, which are metals and which are nonmetals?

A Search for Regularities Leads to the Periodic Table

8-6 Other Elements Have Similar Properties

The years 1800–1860 saw the discovery of many new chemical elements, as you can see in Figure 8-6. Their reactions and the compounds they formed were studied. As information accumulated, chemists began to look for patterns that would help to organize and classify the growing body of facts. The most obvious things to look for were similarities in the types of reactions and in the formulas of compounds formed by different elements. In this way chemists began to recognize the existence of certain families of elements. In addition to the two families that have already been discussed, the alkalis and the halogens, several others were soon recognized. Two examples follow.

1. Calcium, strontium, and barium are soft, whitish metals, and they react similarly with many substances, such as oxygen, chlorine, bromine, and water. This is shown by the following equations, in which the symbol M represents any one of them.

$$2M + O_2 = 2MO$$
$$M + Cl_2 = MCl_2$$
$$M + Br_2 = MBr_2$$
$$M + 2H_2O = M(OH)_2 + H_2$$

2. Sulfur, selenium, and tellurium are all nonmetallic solids (although tellurium is slightly metallic) and they are only moderately reactive. When they do react, they give similar compounds. The following reactions, in which Z stands for S, Se, or Te, illustrate this.

$$Z + O_2 = ZO_2$$
$$Z + H_2 = H_2Z$$
$$Z + 2Na = Na_2Z$$

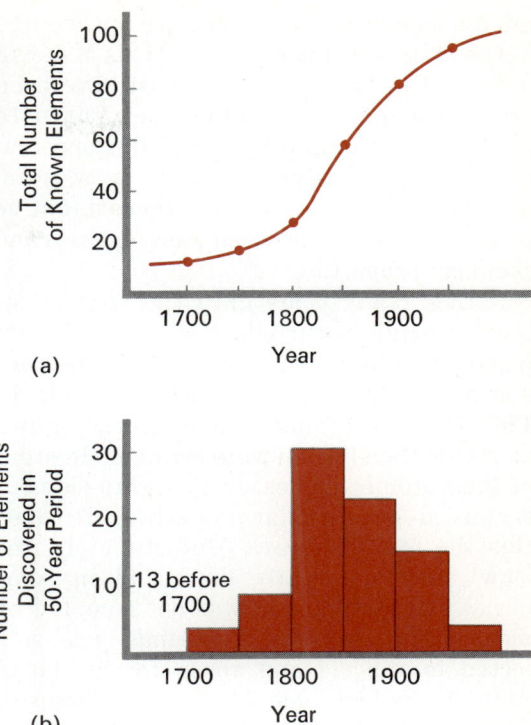

Figure 8-6 *Two graphs showing how the number of known elements has steadily increased (above) and the periods of greatest activity in discovering elements (below).*

The classification of elements into families was useful to chemists (and still is) for obvious and practical reasons. It helps to organize the data. It is often possible to make predictions, as a result. If a new reaction or a new type of compound is discovered for one element, the chemist can predict which other elements are most likely to behave similarly.

Some of the questions that began to stir the minds of chemists more and more in those early days were the following ones. Why do chemical families exist at all? What basic properties of atoms lead to the major family resemblances and the minor differences? Is there a guiding principle to account for the formulas that are found to exist?

The first person to spot a possible clue was a German chemist, Johann Döbereiner. Döbereiner noticed that families of elements frequently consisted of three members. He arranged the elements of each family in order of increasing molar mass. He called the groups triads, and brought them to the attention of fellow chemists in 1828. His original insight was that the atomic mass of the middle member was very close to the average of the masses of the other two. This is shown in Table 8-1 for three triads that have been previously discussed and that were known to Döbereiner. The importance of this observation was that it started a whole new trend of thought. It suggested that there might be some relationship between molar masses and chemical properties.

It took many years until the next major breakthrough occurred. Perhaps the first person to glimpse a more general pattern was an English chemist, John Newlands. In 1863, Newlands found that when the lighter elements then known were arranged in order of their atomic masses, every eighth element seemed to have similar properties. He called this the *law of octaves*. Unfortunately, the "law" was apparently disobeyed in some places and his proposal was not accepted by most other chemists. Newlands was subjected to ridicule and abuse for his ideas. Actually, he had come very close to discovering a principle of enormous importance in science, but the full accomplishment was to be the triumph of another chemist.

8-7 The First Periodic Table

In Russia, in 1868, a young professor in his early thirties, named Dmitri Mendeleev, had the task of writing a textbook for a new course in chemistry. The problem of how to organize the facts and classify the elements in a systematic way was thus forced upon him. His thinking ran along lines similar to Newlands', but with differences. He, like Newlands, felt that the properties of the elements must be a periodic function of their molar masses, since *periodic* means "occurring at regular intervals." However, he believed that the periods might vary in length instead of being always in octaves as Newlands had supposed. Mendeleev kept his thinking more flexible, and did not try to arrive at rigid conclusions too quickly.

He began by using two periods of seven elements but then he used three periods of seventeen! He did this in order to group together the elements with similar properties. In doing so, he was guided by the experimental facts.

Mendeleev had another bold idea. He recognized that in one sense the facts might be incomplete. There might be elements as yet undiscovered. In six places he found that there were not enough elements to make his periods work out. He had to leave blanks in order to group similar elements under one another.

Figure 8-7 shows Mendeleev's arrangement. We call this, or any similar table, a *periodic table*. This particular table, one of Mendeleev's later versions, has many interesting features. First of all, you can see that it places similar elements in columns. The

TABLE 8-1 Some of Döbereiner's Triads

Name	Molar Mass	Average of Largest and Smallest Mass
Calcium	40.1	88.7
Strontium	87.6	
Barium	137.3	
Sulfur	32.1	79.8
Selenium	79.0	
Tellurium	127.6	
Chlorine	35.5	81.2
Bromine	79.9	
Iodine	126.9	

Rows	I	II	III	IV	V	VI	VII	VIII
1	Hydrogen 1.0080							
2	Lithium 6.940	Beryllium 9.02	Boron 10.82	Carbon 12.01	Nitrogen 14.008	Oxygen 16.000	Fluorine 19.000 1886	
3	Sodium 22.997	Magnesium 24.32	Aluminum 26.98	Silicon 28.09	Phosphorus 30.975	Sulfur 32.066	Chlorine 35.457	
4	Potassium 39.1 / Copper 63.54	Calcium 40.08 / Zinc 65.38	Scandium 44.96 1879 / Gallium 69.72 1875	Titanium 47.90 / Germanium 72.60 1886	Vanadium 50.95 / Arsenic 74.91	Chromium 52.01 / Selenium 78.96	Manganese 54.93 / Bromine 79.916	Iron 55.85 Nickel 58.69 / Cobalt 58.94
5	Rubidium 85.48 / Silver 107.880	Strontium 87.63 / Cadmium 112.41	Yttrium 88.92 / Indium 114.76	Zirconium 91.22 / Tin 118.70	Niobium 92.91 / Antimony 121.76	Molybdenum 95.95 / Iodine 126.91	Technetium 99 1937 / Tellurium 127.61	Ruthenium 101.7 Rhodium 102.91 / Palladium 106.7
6	Cesium 132.91 / Gold 197.2	Barium 137.36 / Mercury 200.61	Lanthanum 138.92 / Thallium 204.39	Hafnium 178.6 1923 / Lead 207.21	Tantalum 180.88 / Bismuth 209.00	Wolfram 183.92 / Polonium 210 1898	Rhenium 186.31 1925 / Astatine 211 1940	Osmium 190.2 Iridium 193.1 / Platinum 195.23
7				Thorium 232.12		Uranium 238.07		

Figure 8-7 *Mendeleev's periodic table. Present-day molar masses are given. Elements unknown to Mendeleev, together with their dates of discovery, are given in the colored boxes, which were "gaps" in Mendeleev's table.*

families of elements, including Döbereiner's triads, fall in columns.

Second, you see that in two places the assumption that properties are a periodic function of molar mass appears to be violated. One case involves iodine and tellurium. Obviously, on the basis of chemical similarities iodine *has* to go under chlorine and bromine, and just as surely tellurium *has* to go under sulfur and selenium. But the molar masses appear to demand the reverse order.

Mendeleev had an answer for this. He proposed that one or both of the molar masses in question might be wrong. New measurements were quickly undertaken, but they confirmed the original values. Tellurium *is* heavier than iodine! The periodic table nevertheless gained acceptance, despite this (and one similar) anomaly. Mendeleev died in 1907, without ever having had the satisfaction of knowing that there is a logical explanation for the two "inversions." The basis of the explanation was not discovered until 1912. That is another story, which will be told in Chapter 13. You will see there that the correct order in which to place the elements is the order of their atomic numbers. This is very nearly identical with the ordering according to molar masses.

A third thing to notice about Mendeleev's periodic table is the blanks. As we mentioned before, Mendeleev took it as his guiding rule that similar elements had to fall into the same column. This required him to leave blanks. He assumed that there were undiscovered elements to fill the blanks. Having assumed this, he took a bold but logical next step. He used his table to predict what the missing elements would be like.

Not only do the vertical columns of the table contain similar elements, but as you read down a vertical column, you will see that the properties change smoothly. For example, among the alkalis reactivity toward water is slowest with lithium and fastest with cesium. The other elements, Na, K, and Rb, form an intermediate, graded series.

Mendeleev assumed that an analogous graded pattern of similarities could be expected in all columns, including those with gaps. A missing element was assumed to resemble the elements above and below it qualitatively, while being about halfway between the two with regard to quantitative properties. This, of course, harks back to Döbereiner's observations about the pattern of molar masses in his triads.

Mendeleev's success was fantastically good, as illustrated in Table 8-2. The element we now know as germanium was called by Mendeleev *eka*silicon, which means "beyond silicon." He chose this name for the element that should be the next member after silicon of the family: silicon, (*eka*silicon), tin, lead. He described this and several of the other missing elements (later named gallium, scandium, and polonium) with remarkable accuracy before they were actually isolated and studied.

8-8 The Modern Periodic Table

Mendeleev's arrangement of the periodic table, shown in Figure 8-7, is usually modified in present-day versions. In addition, contemporary tables are more complete because many more elements are now known. The known elements range from the lightest element (hydrogen) to a number of very heavy synthetic (that is, man-made) elements beyond uranium. A popular form of the periodic table as we know it today is

TABLE 8-2 The Properties of *Eka*silicon and the Element Germanium

Property	*Eka*silicon, as Predicted by Mendeleev	Germanium, as Actually Found, Later
Molar mass	72	72.6
Physical state	gray metal	gray metal
Density	5.5	5.36
Melting point	high	958°C
Formula of oxides	GeO_2	GeO_2
Density of oxide	4.7	4.70
Reaction with HCl	slight	none at 25°C

Dmitri Ivanovich Mendeleev (1834–1907)

AT THE TIME OF MENDELEEV'S birth, in Siberia, fifty-four elements were known to man. Thirty-two years later, when he became professor of chemistry at the University of St. Petersburg, only seven more had been discovered.

By the mid 1800's scientists had learned much about the chemical and physical properties of the known elements. The melting points and boiling points of many of them had been determined. The ability of the elements to react with hydrogen and oxygen had been measured. Many studies had been made on the electrochemistry of the elements. The ability of the elements to combine had been studied and recorded, and the molar masses of most of the elements had been determined.

With the publication of each new series of measurements, scientists tried to find some universal principle that could demonstrate a clear relationship between the elements and many of their properties. The efforts of many of these scientists had established that some groups of elements were very obviously related. These groups included the halogens (F, Cl, Br, I), the alkaline earth metals (Ca, Sr, Ba), the nitrogen group (N, P, As, Sb, Bi), the sulfur group (S, Se, Te), and elements related to platinum (Pt, Os, Ir). The problem of showing any relationship between the groups, however, had not been solved when Mendeleev became a professor of chemistry in 1866. He did know, however, that several scientists had grouped a number of elements into triads (groups of threes) based on their molar masses. In each case the middle element had a molar mass approximately one-half the sum of the other two elements.

When Mendeleev began to write a textbook entitled *Foundations of Chemistry,* he wanted to develop a relationship between the elements based on a mathematical principle rather than some other arbitrary scheme. Mendeleev arranged the elements in order of their molar masses, starting with the smallest. A definite periodicity in the properties and combining powers was apparent. This periodic table, published in 1868, showed how the properties and behavior of elements depended on their molar masses and made it possible to predict the properties and molar masses of elements not yet discovered. The accuracy of his predictions was proved when the missing elements with the predicted properties were discovered. Mendeleev's periodic table paved the way for future research that led to a much clearer understanding of the structure of matter.

shown in Figure 8-8. There are two main differences between this table and Mendeleev's.

The first difference is that the set of elements helium, He, neon, Ne, argon, Ar, krypton, Kr, xenon, Xe, and radon, Rn, appears in a column at the right. This family of elements, the noble gases, was entirely unknown when Mendeleev first devised a periodic table. They were only discovered in the years 1895-1900. Their molar masses are such that they could be placed either at the extreme left or the extreme right. In terms of present-day knowledge it seems more fitting to put them at the right, although they were at first put on the left.

The noble gases are a most unusual family of elements. Their common characteristic is not *how* they react with other elements, but rather, that, with few exceptions, they do *not* react. Until 1962, no one had succeeded in forming a single chemical compound from any one of these elements. It was generally believed that they were entirely incapable of entering into chemical reactions. In that year Neil Bartlett, a young British chemist working in Canada, obtained a compound containing xenon. Since then, several dozen more xenon compounds have been made, as well as a few compounds of krypton and radon.

The low reactivity of the noble gases is especially remarkable when we note that the families on each side, the alkali metals and the halogens, are both highly reactive. Each of the noble gases represents a very conspicuous and peculiar point in a list of the elements. They may, then, be considered as very natural and obvious anchor points for fitting a list of the elements, in order of increasing mass, into a logical framework.

The discovery of the noble gases serves as a good illustration of the importance of careful observation and diligent attention to details, however small. Early experiments on the composition of air by an English scientist, Henry Cavendish, showed that air is composed of approximately 79 percent nitrogen and 21 percent oxygen (by volume). Traces of carbon dioxide and water vapor were thought to account for any remaining gases present. In Lord Cavendish's experiment, however, after all the known gases had been removed or accounted for, there always remained a small bubble of gas from the original air sample. This bubble was so small (approximately $\frac{1}{120}$ of the original sample) that for many years it was thought to be some residual oxygen or nitrogen, or perhaps some unrecognized combination of them.

Nearly a hundred years later, in 1894, J.W.S. Rayleigh and William Ramsay repeated Cavendish's experiments. They obtained the same small bubble of residual "air," and decided to investigate this bothersome bubble further. Their refusal to accept the prevailing view concerning the bubble led them to the discovery of argon. By 1908 Ramsay and various coworkers had discovered the other noble gases, as well.

There is a second way in which the modern periodic table differs from Mendeleev's. It is wider, that is, more "stretched out." The reason for this is that the long periods, the ones running from potassium, K, to krypton, Kr, rubidium, Rb, to xenon, Xe, and cesium, Cs, to radon, Rn, have been spread out into extended single rows. In Mendeleev's table they were doubled up. Mendeleev and his contemporaries preferred the more compact arrangement for reasons that need not concern us now. The extended form makes more sense both in terms of the chemical similarities of the elements and with respect to the basic significance of the table. Its basic significance will become clear to you when the periodic table is discussed in further detail, in Chapter 13.

PERIODIC TABLE

1 H 1.00																	2 He 4.00
3 Li 6.94	4 Be 9.01											5 B 10.8	6 C 12.01	7 N 14.01	8 O 16.00	9 F 19.0	10 Ne 20.2
11 Na 23.0	12 Mg 24.3											13 Al 27.0	14 Si 28.1	15 P 31.0	16 S 32.1	17 Cl 35.5	18 Ar 39.9
19 K 39.1	20 Ca 40.1	21 Sc 45.0	22 Ti 47.9	23 V 50.9	24 Cr 52.0	25 Mn 54.9	26 Fe 55.8	27 Co 58.9	28 Ni 58.7	29 Cu 63.5	30 Zn 65.4	31 Ga 69.7	32 Ge 72.6	33 As 74.9	34 Se 79.0	35 Br 79.9	36 Kr 83.8
37 Rb 85.5	38 Sr 87.6	39 Y 88.9	40 Zr 91.2	41 Nb 92.9	42 Mo 95.9	43 Tc (99)	44 Ru 101.1	45 Rh 102.9	46 Pd 106.4	47 Ag 107.9	48 Cd 112.4	49 In 114.8	50 Sn 118.7	51 Sb 121.8	52 Te 127.6	53 I 126.9	54 Xe 131.3
55 Cs 132.9	56 Ba 137.3	57–71 See Below	72 Hf 178.5	73 Ta 180.9	74 W 183.9	75 Re 186.2	76 Os 190.2	77 Ir 192.2	78 Pt 195.1	79 Au 197.0	80 Hg 200.6	81 Tl 204.4	82 Pb 207.2	83 Bi 209.0	84 Po (209)	85 At (210)	86 Rn (222)
87 Fr (223)	88 Ra (226)	89–103 See Below	104 *	105 *													

57 La 138.9	58 Ce 140.1	59 Pr 140.9	60 Nd 144.2	61 Pm (147)	62 Sm 150.4	63 Eu 152.0	64 Gd 157.3	65 Tb 158.9	66 Dy 162.5	67 Ho 164.9	68 Er 167.3	69 Tm 168.9	70 Yb 173.0	71 Lu 175.0
89 Ac (227)	90 Th (232.0)	91 Pa (231)	92 U 238.0	93 Np (237)	94 Pu (242)	95 Am (243)	96 Cm (247)	97 Bk (249)	98 Cf (251)	99 Es (254)	100 Fm (253)	101 Md (256)	102 No (253)	103 Lr (257)

*There are no internationally-accepted names for these elements.

Figure 8–8 *The modern periodic table has the known elements arranged in order of increasing atomic numbers. This arrangement reveals important similarities and differences in the horizontal and vertical rows.*

8-9 *Properties Vary Smoothly within Groups*

You have seen that certain sets of elements, which would naturally be regarded as families because of their similar properties, form the vertical columns of the periodic table. These vertical sets of similar elements, which we have called families, are, more technically, called **groups**.

A very significant reason for the importance and usefulness of the periodic table is that it gives a list of the elements in each family or group in order of the trends in their properties. As we have already said, members of a family (people or chemical elements) are *similar*, but not *identical*. In dealing with the elements, we find that when those in each family or group are arranged in order of increasing mass, all of their other properties tend to be ranked in order as well. As we read down each group in the periodic table, we are reading off the group members in increasing order of molar mass. Let us see how the ranking or grading of other properties follows this order for several of the groups with which you have already become familiar.

TABLE 8-3 Some Properties of the Alkali Metals

Property	Lithium	Sodium	Potassium	Rubidium	Cesium
Atomic number	3	11	19	37	55
Molar mass	6.94	23.0	39.1	85.4	133
Boiling point (K)	1599	1162	1030	952	963
(°C)	1326	889	757	679	690
Melting point (K)	453	371	336.4	311.8	301.7
(°C)	180	98	63.4	38.8	28.7
Atomic volume, solid (ml/mole of atoms)	13.0	23.7	45.4	55.8	70.0
Density of solid at 20°C	0.535	0.971	0.862	1.53	1.90

TABLE 8-4 Some Properties of the Halogens

Property	Fluorine (F_2)	Chlorine (Cl_2)	Bromine (Br_2)	Iodine (I_2)
Atomic number	9	17	35	53
Molar mass	19.0	35.5	79.9	127
Boiling point (K)	85	238.9	331.8	457
(°C)	−188	−34.1	58.8	184
Melting point (K)	55	172	265.7	387
(°C)	−218	−101	−7.3	114
Atomic volume, solid (ml/mole of atoms)	14.6	18.7	23.5	25.7

TABLE 8-5 Some Properties of the Noble Gas Elements

Property	He	Ne	Ar	Kr	Xe	Rn
Atomic number	2	10	18	36	54	86
Molar mass	4.00	20.2	39.9	83.7	131	222
Boiling point (K)	4.2	27.2	87.3	120	165	211
Melting point (K)	—	24.6	83.9	116	161	202
Atomic volume, liquid (ml/mole of atoms)	31.8	16.8	28.5	32.2	42.9	50.5

In Table 8-3 are listed some of the physical properties of the alkali metals. In Tables 8-4 and 8-5 are similar data for the halogens and the noble gases. The melting and boiling points for all three groups are plotted in Figure 8-9. It can be seen that for each group, there is a steady, although not perfectly smooth, trend in each property with increasing molar mass. Note that the trends are sometimes up and sometimes down. The important point is not the direction, but the fact that consistent trends are present for each property in a given group. There are a few rare exceptions (note in Table 8-5 that the atomic volume of helium is out of line), but smooth trends are the rule for both chemical and physical properties among the elements of each group.

It is because of this rule that Mendeleev was able to anticipate fairly accurately the properties of missing elements. He knew that the element now called germanium would have to form an oxide of the formula GeO_2 because silicon and tin, which are above and below it, form SiO_2 and SnO_2. He estimated the density of GeO_2 by choosing a value between those of SiO_2 and SnO_2.

8-10 The Noble Gases Are Unreactive but Useful ★

Even though they are very unreactive, the noble gases are useful. Their inert character is precisely what makes them useful in certain of their applications. They do, however, have some interesting chemical reactivity.

All of the noble gases except helium are obtained by careful distillation of liquefied air. (See Section 7-2.) Although they occur in only very small percentages in the air, the total amounts available are large because there is so much air. Helium, the lightest of the noble gases, does not occur in significant quantities in the air. This is because the helium atom is so light that helium can escape

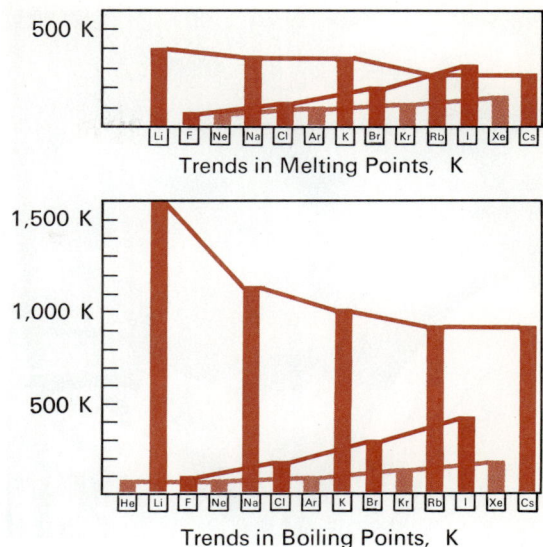

Figure 8-9 *Trends in melting and boiling points for three groups of elements: the alkalis, halogens, and noble gases.*

from the gravitational field of the earth. Helium occurs in underground pockets, however, often mixed with natural gas.

Probably you are already familiar with some of the uses for the noble gases. Helium is widely used in an "artificial air" mixture necessary in deep-sea diving. Otherwise, a painful experience, *caisson disease*, also known as the "bends," can occur when divers surface too rapidly. This disorder is due to gas bubbles that form in the blood when the high air pressure required at great depths is suddenly released. Nitrogen that had dissolved in the blood is then suddenly expelled and forms bubbles in the blood vessels. Helium, like nitrogen, will not react with the body chemically and, in addition, has the property of being considerably less soluble in the blood. Therefore when a helium mixture is used for breathing at deep-sea levels there is less opportunity for nitrogen to be present in the blood.

Ida Tacke Noddack (1896–)

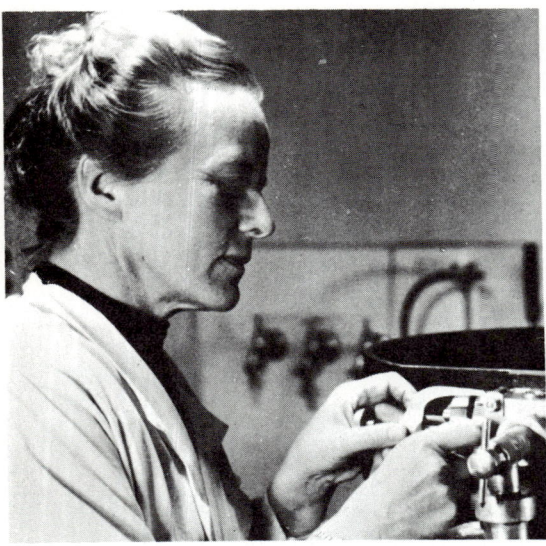

IDA NODDACK IS BEST KNOWN for the discovery of rhenium, element 75. She was a young chemistry graduate from the Technical University of Berlin when she decided to embark on this project, together with her husband-to-be, Walter Noddack.

Mendeleev's periodic table of 1868 left a number of gaps that Mendeleev predicted would be filled in with elements yet to be discovered. (See Figure 8–7.) In some instances Mendeleev had predicted the properties of unknown elements by their positions between elements of known properties, and the unknown elements were found in the following decades. Mendeleev was not able to predict the properties of elements 43 and 75, however, since manganese was the only known element in the same group.

Ida Tacke and Walter Noddack predicted the properties of these elements in 1922, after a long and careful study of the periodic table. They then decided to explore some ores of molybdenum and ruthenium (atomic numbers 42 and 44) and of tungsten and osmium (atomic numbers 74 and 76), other searchers for elements 43 and 75 having hunted unsuccessfully for them in ores of manganese. They worked with very many minerals, containing nearly all of the elements found in the middle of the periodic table, in the area surrounding the sought-after elements.

In 1925, they announced the discovery of elements 43 and 75, which they named masurium (Ma) and rhenium (Re). Masurium was named after Masuria, the native country of Walter Noddack's ancestors. Rhenium was named after the River Rhine (*Rhenus* in Latin), since Ida Tacke was born in the Rhineland. For decades other researchers were unable to confirm the existence of element 43 in nature, but there was no question concerning the existence of rhenium. (Indeed, despite its relative scarcity, rhenium has several important uses in current technology.)

By 1929, after working on nearly 660 kg of the ore molybdenite, Tacke and Noddack were able to purify one gram of rhenium. Even today, the total world reserve of rhenium is no more than some hundred tons. The rhenium remaining in the earth's crust is estimated to amount to only 0.001 parts per million of the crust. Element 43, later to be renamed technetium, was discovered in 1937 as an artificial element formed when molybdenum is bombarded with high energy particles in a cyclotron.

In 1934, Ida Noddack read of the recent work done by Enrico Fermi in bombarding uranium with neutrons and suggested that some of his observations might have been caused by the breaking apart, or fission, of uranium into smaller fragments. Her ideas were confirmed in 1939 when Hahn and Strassman demonstrated that uranium does undergo fission when exposed to neutrons.

Helium is also widely used in weather balloons and dirigibles. It replaces hydrogen gas in this use because, unlike hydrogen, it is nonflammable and hence not subject to ignition and subsequent explosion.

Neon is present in the atmosphere to the extent of approximately 0.002 percent. It is widely used in familiar electric signs known as neon signs. Small amounts of this gas under low pressure in glass tubes glow with a reddish-orange light when an electric current is passed through the gas. The colors given off by neon and other glowing gases are shown in Figure 9-9. By mixing gases in varying proportions, many different colors can be produced.

Argon is the most abundant of the noble gases in the atmosphere, comprising approximately one percent of dry air. Perhaps its most interesting modern-day use is as the gas that replaces air in the everyday incandescent light bulb. It greatly increases the life of the glowing filament of the bulb.

Few uses have been found for krypton, xenon, and radon. Radon is given off from the radioactive element radium and is itself radioactive. (Radioactivity is discussed in Section 10-9.) It is used in a form of treatment for cancer. The fluorides and oxides of xenon have some laboratory use as very strong oxidizing agents.

The noble gases are the least reactive of the elements. It was only in 1962 that the first compound of a noble gas was discovered. Prior to that it had been thought that no such compounds could exist. Whatever slight tendency there is for noble gases to form compounds should increase as the masses of the noble gases increase. This is an example of a chemical property that varies smoothly within a group. An explanation for this property will be given in Chapter 13.

The possibility of any noble gas atoms participating in chemical bonds should be greatest, then, for the noble gases with the greatest masses. Of the two heaviest noble gases, one (radon) is radioactive. This fact leads to special problems that you can better appreciate later. Thus, from a practical point of view, it is for xenon that compound formation would be most likely to be observed.

The first xenon compound was made using a rather obscure substance, platinum hexafluoride, PtF_6. This substance reacts with xenon to form various substances, such as $XePtF_6$. It was the discovery of this reaction by a young British chemist, Neil Bartlett, that immediately led other research workers to try the reaction of xenon with fluorine itself.

Xenon does react with fluorine, and, depending on conditions (pressure, temperature, ratio of F_2 to Xe) three different xenon fluorides can be made.

$$Xe + F_2 = XeF_2$$
$$Xe + 2F_2 = XeF_4$$
$$Xe + 3F_2 = XeF_6$$

These fluorides are colorless, volatile substances. (See Figure 8-10.) They are very reactive, and can be used to introduce fluorine into other substances.

It has since been shown that xenon forms a number of compounds with bonds to oxy-

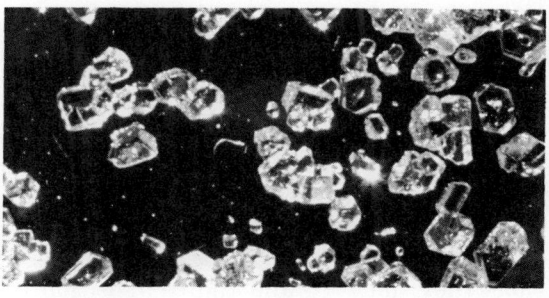

Figure 8-10 Crystals of the noble gas compound xenon tetrafluoride, XeF_4.

gen. An example is XeO_3, formed by reaction of XeF_6 with water.

$$XeF_6 + 3H_2O = XeO_3 + 6HF$$

There is no doubt that various fluorine and oxygen derivatives of radon can also be made. However, as anticipated, the radioactive decay of radon leads to great difficulties in studying these substances.

It is also known that krypton, though far less reactive than xenon, is capable of forming at least one compound, KrF_2.

Summary

Chemists began very early to notice that certain groups of 3 or 4 elements had marked similarities in their properties. It was later found that the occurrence of these groups was related to the molar masses of the elements. The Russian chemist Dmitri Mendeleev developed the first successful and widely accepted arrangement correlating molar masses with the grouping together of similar elements. His periodic table has been modified and refined over the past hundred years, but remains basically as he first proposed it. The modern periodic table lists the elements in order of increasing atomic number (which is also the order of increasing molar mass, with only a few exceptions), in a series of horizontal rows called periods. The lengths of the periods are such as to place sets of similar elements in vertical columns, called groups. In each group there is a uniform gradation of properties.

Self Check

1. Name, in order of increasing molar mass, the elements in each of the following groups: (a) the alkali metals; (b) the halogens; (c) the noble gases.
2. Why are the noble gases so named?
3. Give one reason why one might be tempted to classify calcium as an alkali metal and one reason why this classification could not be accepted.

Chapter Summary

There are certain especially important types of chemical reactions that metals and other elements undergo with oxygen, forming oxides. Metals can be grouped according to their capacity to combine with oxygen. It ranges from a small capacity for some (as in Na_2O) to a much greater capacity for others (as for SnO_2). Some metals react with water, liberating hydrogen gas and forming the metal hydroxides.

Careful comparison of the types of reactions they undergo and the composition of the products they form, enables us to sort the elements into families. Among the important families are: (1) The alkali metals: Li, Na, K, Rb, Cs; (2) The halogens: F, Cl, Br, I; (3) The noble gases: He, Ne, Ar, Kr, Xe, Rn. If all the elements are listed in order of their atomic numbers (which is the same as the order of their molar masses with only three exceptions), and this listing is arranged in horizontal rows of proper length, the periodic table is obtained. The lengths of the horizontal rows, called periods, are chosen so that families of elements form vertical columns. Each of these columns is called a group. The elements in each group are arranged, as a result, in order of increasing mass. This is nearly always the order in which their properties change uniformly.

Questions and Problems

1. What kind of evidence would you look for if you were trying to decide whether a chemical reaction does or does not take place when two substances are mixed with each other?
2. Which of the following changes would you consider to be chemical reactions?
   (a) Wheat is ground to make flour.
   (b) Water is changed to steam by boiling.
   (c) An electric current passing through water forms hydrogen and oxygen.
   (d) Food chars, producing a bad odor, when boiled dry on a stove.
   (e) Two colorless liquids produce an orange liquid when they are mixed together.
   (f) Candle wax solidifies when it is cooled.
3. Why have scientists found it convenient to use reactions of elements with oxygen and water to help in grouping elements together in families?
4. A number of elements in this chapter have been referred to as metals. If you were given an unknown substance, how would you tell whether it was a metal or not? What common properties of metals are you familiar with?
5. List two reasons why it is important for alkali metals to be stored under oil.
6. Complete and balance the following equations.
   (a) $Cs + O_2 \longrightarrow$
   (b) $Ba + O_2 \longrightarrow$
   (c) $Ga + O_2 \longrightarrow$
   (d) $Ge + O_2 \longrightarrow$
   (e) $Rb + F_2 \longrightarrow$
   (f) $Sr + Cl_2 \longrightarrow$
   (g) $Al + Br_2 \longrightarrow$
   (h) $Si + I_2 \longrightarrow$
7. Which of the following is *not* a correct formula for a substance at normal laboratory conditions? Explain the reasons for your choices. (a) H_2S; (b) $CaCl_2$; (c) He; (d) NaNe; (e) Al_2O_3; (f) $SiCl_3$.
8. A certain metal, M, reacts with bromine to form the compound MBr_3. Predict the formula of the oxide formed when M reacts with oxygen.
9. A certain nonmetal, Y, reacts with calcium to form the compound CaY. Predict the formula of the compound formed when Y reacts with lithium.
10. Why are gold and platinum called "noble" metals? Are the alkali metals also considered to be "noble" metals? Explain.
11. Why is the label "noble" gases more applicable than "inert" gases for the elements helium, neon, argon, krypton, xenon, and radon?
12. Describe the relationships between the melting points, boiling points, and atomic numbers for each of the following families of elements
    (a) The alkali metals.
    (b) The halogens.
    (c) The noble gases.
13. On the basis of your answers to Question 12, would you expect beryllium, Be, to have a higher or a lower melting point than barium, Ba? Explain.
14. Predict the relationship you would expect to find between the melting points of selenium, Se, and tellurium, Te. Explain.
15. If copper forms an oxide with the formula Cu_2O, similar to the M_2O oxides of the alkali metal family, why is copper not included in that family?
16. While the element francium with atomic number 87 may be found in nature, it is so rare that there are probably less than 25–30 grams of this element in the total crust of the earth. Even before enough of this element to work with had been prepared by nuclear bombardment techniques, chemists were confident they knew the chemical behavior this element would show. What predictions about this element would *you* make? Explain.

9 The Electrical Nature of the Atom

We see evidence of electrical charges and electricity around us all the time. There are, of course, all the man-made electrical machines and gadgets. Beside these, you experience electricity in other ways. For example, there is the "shock" you get on a dry day when you touch a door knob or shake hands with someone after crossing a carpeted room.

There is the flash of lightning. There is also the way your hair may stand on end if you brush it vigorously on a dry day.

In this chapter, the basic characteristics of electrical charges will first be explained. Then the ability of matter to become charged, and some of the properties of charged particles will be examined.

Matter Has Electrical Properties

9-1 Basic Characteristics of Electrical Charges

With a few simple pieces of equipment it is possible to discover some important things about electrical charges. Such equipment might include a rod made of hard rubber or plastic, a glass rod, two pith balls, a piece of fur, and a piece of silk. With these, it is possible to observe some of the basic evidence that has led to our present understanding of electrical charges.

If a plastic or hard rubber rod is rubbed vigorously with a piece of fur or flannel, it will readily attract lightweight objects such as small bits of paper. (See Figure 9-1.) The

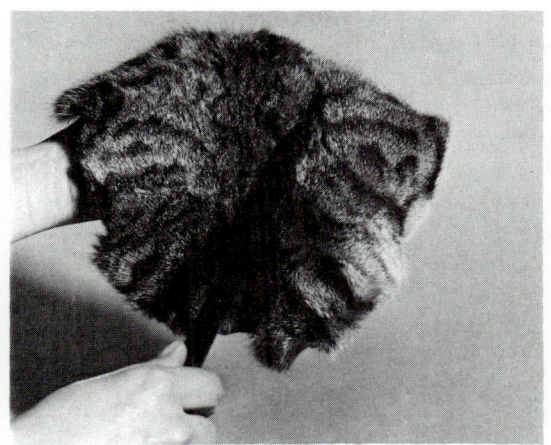

Figure 9-1 *Rubbing a hard rubber rod with a piece of fur (left) causes the rod to become electrically charged. As a result, it attracts small pieces of paper (right).*

 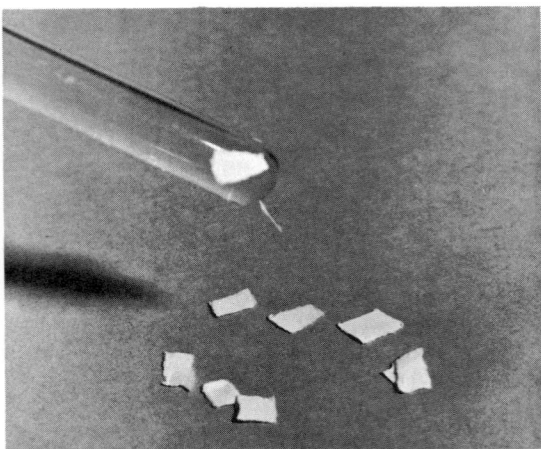

Figure 9-2 *A glass rod that has been rubbed with a silk cloth (left) attracts small pieces of paper (right) as strongly as does the charged rubber rod.*

same thing occurs after a glass rod has been rubbed with a silk cloth (Figure 9-2).

If two pith balls are suspended by pieces of thread, so that they can move freely, and are placed about an inch apart, some interesting observations can be made. Suppose you were to rub the rubber or plastic rod with fur and then touch the rod to each of the pith balls in turn. The pith balls would fly apart, as shown in Figure 9-3a. In other words, they would *repel* each other. Suppose you were to rub the plastic or rubber rod with fur and the glass one with silk, and then touch one ball with the rubber rod and the other with the glass rod. The pith balls would come together, as shown in Figure 9-3b. They would *attract* each other.

Some new property must be conferred upon the pith balls when they are touched by the rods. This new property causes them to attract or repel each other. This new condition can be described by saying that the balls possess electrical charges. Moreover, even from the simple experiments just discussed, it is clear that there must be at least two kinds of electrical charge.

Now let us assume that whenever hard rubber is rubbed with fur, the same type of charge is always produced. Let us also assume that whenever glass is rubbed with silk a charge of another type is consistently produced. Then it follows from the foregoing experiments that charges of the same type, *like* charges, repel each other, whereas charges of different types, *unlike* or *opposite* charges, attract each other.

Scientists have been able to detect only two kinds of electrical charge in all of their experiments. These two kinds of charge were given names, in order to differentiate between them. The names chosen were "positive" and "negative." Arbitrarily, the charge generated on the plastic or hard rubber rod by flannel or fur was called negative. The charge produced on the glass rod by rubbing it with silk was called positive.

The Electrical Nature of the Atom 221

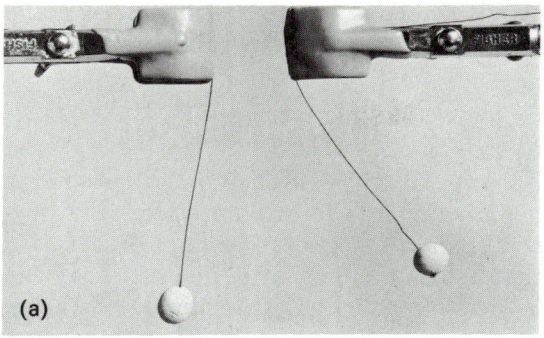 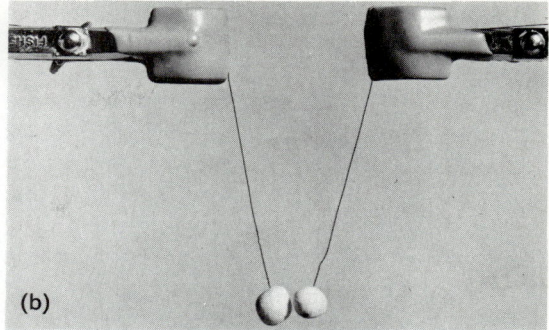

Figure 9-3 *When both pith balls are touched by the same charged rod, they fly apart (a). When each pith ball is touched by a different charged rod, they move together (b).*

9-2 Factors Determining the Amount of Attraction or Repulsion

It has already been noted that, in order to observe the attractive or repulsive force between them, the charged balls must be brought close together. Further investigation would show a definite regularity relating the distance between charged objects and the force that they exert on each other. The amount of force increases as the distance decreases, and decreases as the distance increases. Or, to say the same thing in another way, the force between electrically charged objects is inversely proportional to the distance between them, as shown in Figure 9-4.

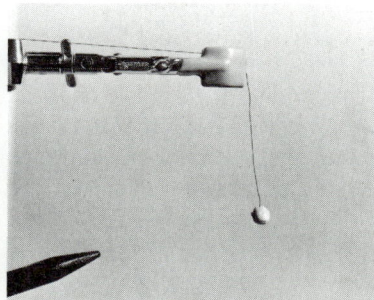 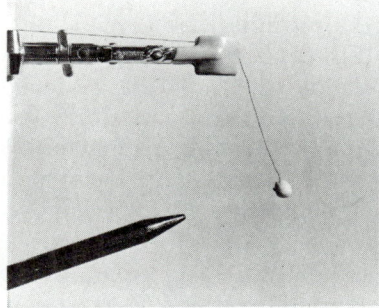 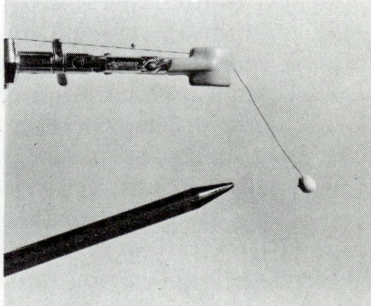

Figure 9-4 *The effect of distance on the force exerted between two charged objects can be seen in the experiment illustrated above, in which both objects have the same type of charge. As the distance between the objects is decreased (photos, from left to right), the repulsive force increases.*

Careful quantitative measurements have shown that this force, whether attractive or repulsive, is inversely proportional to the square of the distance, d, between the charged objects. That is, electrical force is proportional to $1/d^2$. If there is a certain force at a distance of x mm, at twice this distance, $2x$ mm, there is only one-fourth the force. At ten times this distance, the force is only $\frac{1}{100}$, or one percent, as great.

At a given fixed distance, experiments show that the force is directly proportional to the magnitude of each of the charges. If one charge is made n times as great, the force becomes n times as large. If each of the charges is increased, say one by a factor of 2 and the other by a factor of 3, the force is increased by both of these factors, that is, by $2 \times 3 = 6$ times.

When the influence of the magnitudes, q_1 and q_2, of each charge, and the distance, d, separating them are considered together, the following relationship results.

$$\text{Force is proportional to } \frac{q_1 \times q_2}{d^2}$$

This relationship is known as **Coulomb's law**. It was first stated by the 18th century French scientist Charles Augustin de Coulomb. (His name is generally pronounced by English-speaking people as coo-lom.) By now you may have noted the interesting similarity in form between Coulomb's law and the Law of Gravitation, given in Section 2-2.

Written in this form, Coulomb's law can also tell us whether the force is attractive or repulsive. If q_1 and q_2 are like charges, the product will be a positive number, because $(-\text{ times }-) = +$, and $(+\text{ times }+) = +$. If the charges are opposite, the product of q_1 and q_2 will be a negative number. Thus, a positive force is interpreted as repulsive and a negative force as attractive.

9-3 Neutralization of Charges

A body that has neither a positive nor a negative charge is termed electrically *neutral*. It is a matter of common observation that bodies are ordinarily neutral. They can be made to acquire charges in various ways, such as by the frictional effects that put charges on plastic and glass rods.

If two objects bearing opposite charges are brought into direct contact, the charges can mix. If the amount of negative charge on one object is exactly equal to the amount of positive charge on the other, there will be a complete cancellation of charge and the two objects will become neutral. If the two quantities of charge are not equal, then after the smaller one has neutralized its own equivalent of the larger one, there will remain an excess of the latter. The resultant charge will be spread uniformly over the two objects.

9-4 The Nature and Origin of Electrical Charges and Forces

The material discussed so far in this chapter naturally leads to questions about the nature of matter. As always, after a certain pattern of behavior has been observed and measured and some regularity has been discovered (Coulomb's law, in this case), it is natural and proper to wonder why the regularity exists. At this point, wondering why might take the form of two specific questions.
1. Why do electrical charges exert forces on one another?
2. Why is it possible to generate electrical charges on neutral objects?

The first of these questions is one to which there is no answer. That is, no model has been devised that can explain electrical forces in terms of a simpler or more familiar phenomenon. For this reason, the forces between electric charges are regarded as a fundamental property of matter. (Another fundamental property of matter is that it has mass.)

The second question is one that we can attempt to answer in terms of a model. In fact, an explanation can be worked out by developing in greater detail the particle model for matter, which has already proved so useful in explaining the properties of gases and the nature of chemical reactions.

The rather detailed atomic and molecular model of gases, as it was developed in Chapters 5 and 6, began with a basic qualitative idea. This idea was intended to account for some fairly simple qualitative facts. The idea of molecules flying around and colliding with the walls of their container was one with which we could explain everyday facts pertaining to the blowing up of balloons and tires. More detailed experiments (such as those first done by Boyle and Charles) then made it possible to refine the model, making it more detailed and quantitative.

Let us now do the same sort of thing in order to add a new dimension, an electrical one, to our atomic and molecular model. Let us first add a simple, qualitative idea that can provide an explanation of some simple, qualitative facts. Having done that, we can refine the idea (in Sections 9-5 to 9-9), making it more detailed and more useful.

The new idea is that atoms and molecules are electrically neutral but are made up of smaller particles. The smaller particles may be electrically-charged. Since molecules are made up of atoms, this idea reduces to the still simpler one that atoms contain charges, or **charged particles**, which are present in such numbers that the positive charges exactly equal or balance the negative charges.

The positively-charged particles will be called *protons* and the negatively-charged ones *electrons*. As you will see, later in this chapter, it is known that the amount of positive charge carried by one proton is precisely equal to the amount of negative charge carried by one electron. Therefore, the fact that an atom is neutral means that it contains equal numbers of protons and electrons.

Now we can explain how an object in its entirety can acquire an electrical charge. It is only necessary to postulate that the normal balance of electrons and protons is disturbed by some external force. In the case of the rods that became charged by rubbing, the explanation is simply that the frictional forces produced by rubbing dislodge and carry away some of the charged particles. Whether the negative particles (the electrons) or the positive particles (the protons) are transferred cannot be deduced directly and solely from the experiments with the rods. However, as you will learn later in this chapter, the atoms are so structured that, in all cases, it is the electrons that must be removed from one body and transferred to another. For example, when a plastic rod is rubbed with flannel, electrons are dislodged from the cloth and accumulate on the rod. As a result, the rod then has an excess of electrons, or a negative charge. The cloth is, of course, left with a positive charge. Conversely, when the glass rod is rubbed with a silk cloth, electrons are removed from the rod and acquired by the silk. Hence, the glass rod has a deficiency of electrons compared to the number needed to balance the charge of all the protons. The rod therefore has a positive charge. The silk, of course, has a negative charge because it has an excess of electrons. (The Van de Graaff generator shown

Figure 9-5 *A Van de Graaff generator builds up a high voltage by rubbing electrons off a surface and then storing them.*

in Figure 9-5 builds up a high charge by rubbing electrons off a surface, and then storing them.)

Neither the rods nor the cloths retain their charges very long because electrons leak off or onto them from their surroundings. In this way they regain electrical neutrality, which is the normal state for all matter. When humidity is high, leakage takes place very fast. The water vapor in humid air and the thin layer of moisture on objects facilitate the flow of electrons from one object to another, and between charged objects and their surroundings in general. Thus, on a warm, humid day it may not actually be possible to carry out the experiments with the rods. In cool, dry weather the experiments work very well.

The flow of electric currents through wires and other conductors of electricity can also be explained by assuming that matter consists of equal numbers of positive protons, which do not easily move, and negative electrons, which are rather mobile. When a conducting material is connected at one end to a source of electrons (the negative terminal of a battery or generator) and at the other to a positive terminal, which attracts electrons, a current of electrons flows. An example is given in Figure 9-6. Electrons are injected into the wire where it is connected to the negative terminal. Meanwhile other electrons are removed from the wire where it is connected to the positive terminal. As a result there is a stream of electrons flowing through the bulb, causing it to glow.

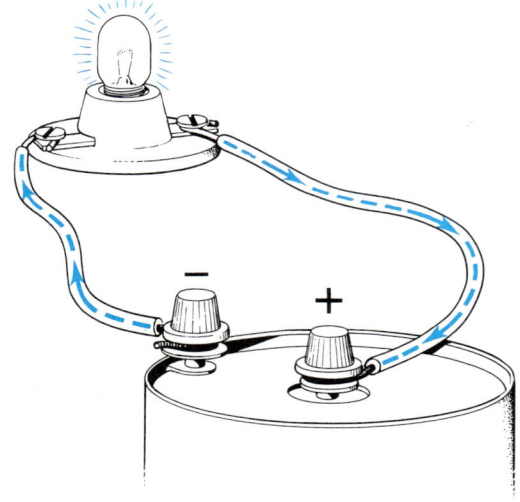

Figure 9-6 *The negative terminal of the dry cell provides electrons. They flow through the conducting wires to the positive terminal, lighting the bulb in transit.*

Summary

There are two kinds of electrical charges: positive and negative. Charges of the same kind (like charges) repel each other. Charges of different kinds (unlike or opposite charges) attract each other. The forces of attraction and repulsion are inversely proportional to the square of the distance between the charges and directly proportional to the magnitude of each charge. This is called Coulomb's law and can be stated, in algebraic form, as follows.

$$\text{Force varies as } \frac{q_1 \times q_2}{d^2}$$

When equal quantities of positive and negative charges are brought into direct contact, they neutralize each other, leaving the initially-charged objects electrically neutral.

Common facts about the charging of objects by frictional forces and the passage of electrical currents can be explained by means of a suitable model. According to the model, matter (atoms) is made up of equal numbers of positive particles (protons) and negative particles (electrons). The protons are fairly immobile (fixed), while the electrons can be readily dislodged and caused to move.

Self Check

1. How will the force between two electrically-charged objects be affected if the distance between them is increased to four times the original distance?
2. Bodies A and B, each bearing a negative charge of 3 units, are 3 cm apart. Bodies C and D have positive charges of 6 and 7 units, respectively, and are 2 cm apart. The force between C and D is how many times greater than that between A and B?
3. After a glass rod has been rubbed with a piece of silk, what is the charge on each object? Explain this in terms of the movement of electrons.
4. Which atmospheric conditions are more favorable for the buildup of electrical charges, cool and dry, or warm and moist? Explain.

Atoms Are Made of Positive and Negative Charges

9-5 More Detailed Evidence for Charged Particles

You have seen that normally-neutral bodies of matter can acquire positive and negative charges, and that certain kinds of matter, called metals, can carry an electric current. To account for these facts, you were offered the hypothesis that matter is made up of equal numbers of positively and negatively charged particles, called protons and electrons. The hypothesis was extended to include the idea that the electrons are more moveable than the protons and so bodies acquire charges by gaining or losing electrons.

Is there more evidence for these ideas? There are many different kinds of evidence that support and clarify the above hypotheses. Several of the most important ones will be considered here and in the next few sections.

You saw in Figure 9-6 that current flows through a light bulb connected to a battery. The connecting wires plus the bulb and battery form an electrical *circuit*. The *outer circuit* is the pathway by which electrons flow from the negative to the positive terminal of the battery. Of course, something also must be going on inside the battery to keep the circulation going. That process will be discussed in Chapter 24.

Electrons can flow through a circuit only when the resistance that it offers to that flow is relatively small. In the circuit of Figure 9-6, the resistance is low because both the connecting wires and the glowing part, or filament, of the light bulb are made of metal. Metals characteristically offer little resistance to current flow. They are good *conductors*. The filament, however, must have enough resistance to make it glow when current passes through it.

Suppose you were to disconnect a wire from one of the battery terminals. The light bulb would go out. No electron could flow through the circuit unless it had enough energy to jump across the gap between the disconnected wire and the battery terminal. Air offers far greater resistance to current flow than does metal. Unless the electron source were a very powerful one, the electrons would not have enough energy to jump across even a small air gap. Disconnecting the wire has created an *open* circuit.

Now consider the experimental set-up shown in Figure 9-7. Parts (a) and (b) both illustrate the same set-up, under different

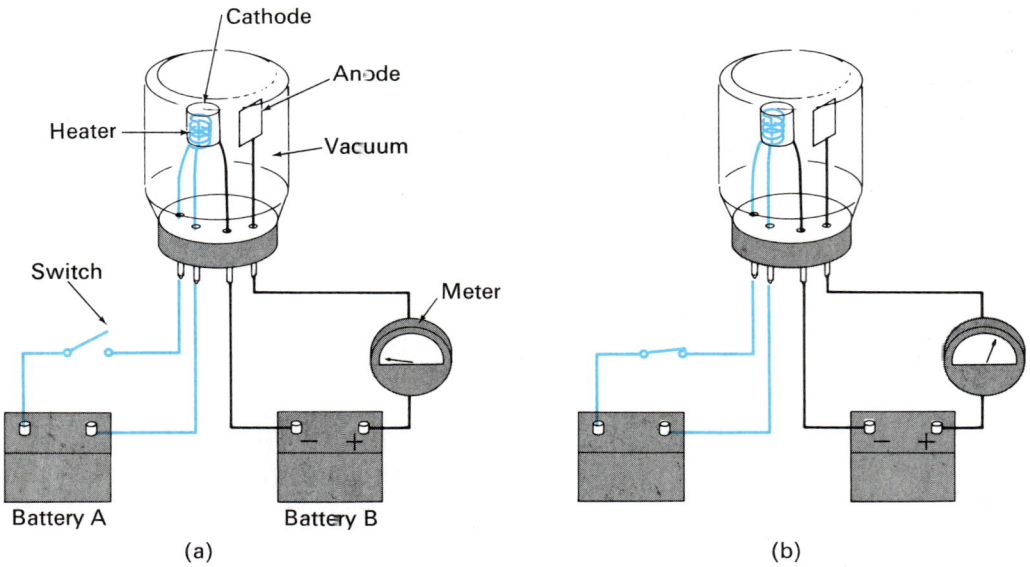

Figure 9-7 When the cathode heater circuit is closed, the cathode gets hot and gives off electrons, which are then attracted to the anode. Battery B makes the cathode negative and the anode positive. The current registers on the meter.

conditions. Each part of the figure shows two circuits, both of which run through the same vacuum tube. A vacuum tube is a glass envelope from which most of the air has been pumped out. Sealed into it are certain metallic parts that enable it to perform a useful function. You have probably seen vacuum tubes, such as the one shown in Figure 9-8, in a radio, television set, or some other electronic appliance. The *cathode* is connected to the negative terminal of the battery, and so has a negative charge when the tube is operating. The *anode*, being connected to the positive terminal, is positive.

One of the circuits connects Battery A with a coil of wire that gets hot, and glows, when current passes through it. In principle, this *heater* coil is like the ones found in electric toasters and electric irons. They all give off heat energy, primarily, when current passes through them. The switch in the circuit enables us to produce either an open circuit, as in Figure 9-7a, or a closed circuit, as in Figure 9-7b.

Let us call the circuit that includes the meter and Battery B the *main* circuit. The meter in Figure 9-7a shows that there is no current flowing in the circuit. This is not at all surprising, since the circuit is open between the cathode and the anode. Now, let us close the switch so that electricity can flow through the cathode-heating coil. The heat from this coil will cause the cathode to get hot. Now something will occur that may surprise you. When the cathode gets hot, current begins to flow in the main circuit, even though it still seems to be open. Why? Our hypothesis, or model, according to which atoms themselves consist of protons and electrons, can supply an answer. When the cathode is hot, electrons on its surface can acquire enough kinetic energy to jump out. Once they do that, they are then attracted by the positive electrode, the anode, and they flow across the gap to it. In this way the electric current is caused to flow across the gap. The circuit is closed by a stream of electrons flowing through the space between the heated cathode and the anode.

There is another, somewhat similar experiment that can give us additional evidence for, and information about, electrons. Consider the following experiment. A glass tube is fitted with electrodes so that a potential difference of 10,000 volts can be applied across a space filled with a desired gas at various pressures. Then a gas, such as neon, is placed in the tube. With the voltage applied, the gas will begin to conduct electricity when its pressure is reduced to about 0.01 atmosphere. The tube then glows with the familiar color of a neon sign. If a different

Figure 9-8 *In vacuum tubes, such as the one above, electrons can flow across nearly empty space.*

228 Chapter Nine

gas is used, the color is different, but otherwise, the behavior is about the same. (See Figure 9-9.)

If the pressure is reduced still further, to about 10^{-6} atmosphere, the glow from the gas disappears, but a fluorescent glow from the glass walls of the tube remains. Experiments show that the fluorescent glow is caused either by streams of particles or by light rays that travel from the negative electrode. This is most clearly seen in the Crookes tube, shown in Figure 9-10. The metal cross interrupts the stream and a shadow results on the large end of the tube. It is proof that such streams travel in straight lines from the cathode, the negative electrode. Do the streams consist of particles or light rays? The discharge tube shown in Figure 9-11 provides the answer. A fluorescent screen placed lengthwise in the tube makes the stream visible. In Figure 9-11a the stream is seen to be traveling in a straight line.

In Figure 9-11b, a magnet has been brought near to the tube. The poles of the magnet are positioned in a certain way, in

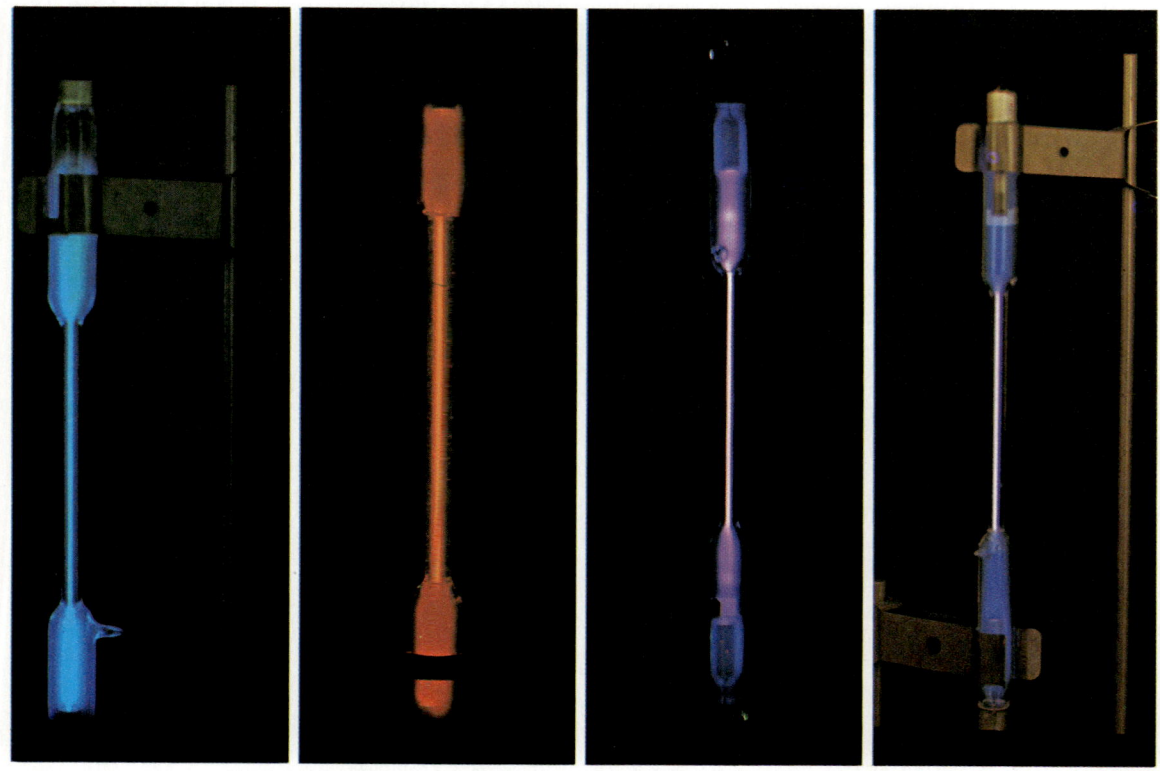

Figure 9-9 *Gases at low pressures (about 0.01 atm) glow when subjected to a high voltage. From left to right, the glowing gases shown are mercury, neon, argon, and helium.*

order that the stream must pass through the magnetic field. The **field** produced by a magnet is the region of space, concentrated chiefly around its poles, in which the magnet exerts its characteristic force. You have probably seen the effect produced by a magnetic field on any other magnet that enters it. More to the point of this discussion is the fact that a magnetic field will also exert a force on any moving *charged* particle that enters it. As you can see in Figure 9-11b, the stream is deflected by the magnet. It must, then, consist of charged particles. Light rays cannot be deflected in this way.

When an electric field is substituted for the magnetic field, the stream of particles is again deflected. An electric field, as you might guess from what was said about magnetic fields, exists in the space between electrically-charged plates. It too exerts a force on any charged particle that enters it. The direction in which the force acts depends on whether the charge of the entering particle

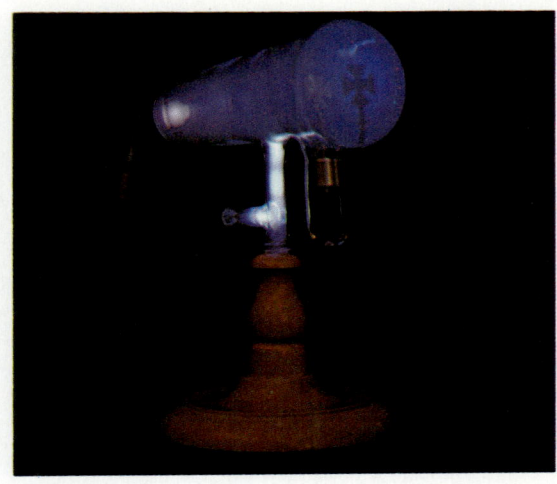

Figure 9-10 *The shadow due to the metal cross in the Crookes tube (above) shows that the stream producing the fluorescent glow must travel in a straight line from the cathode (at the left end of the tube) to the large end.*

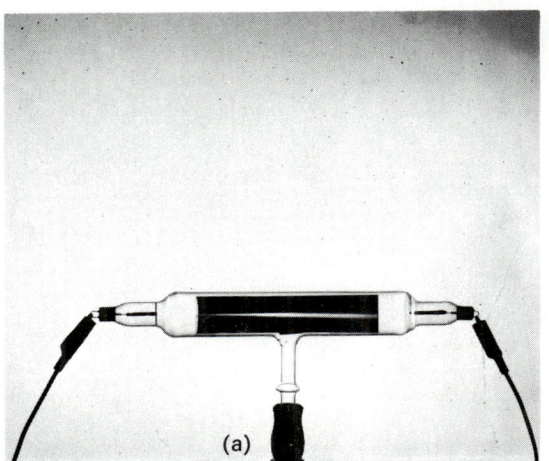
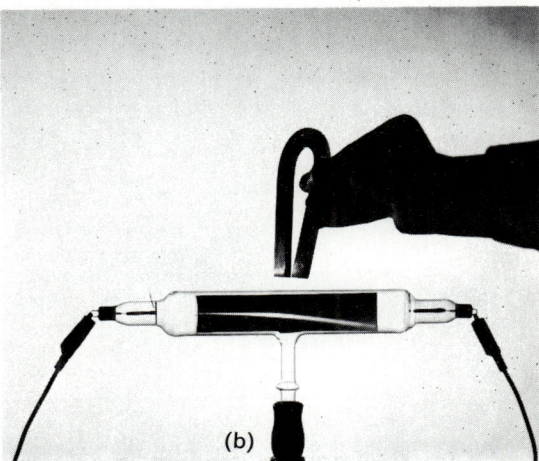

Figure 9-11 *The "cathode-ray" beam, made visible by the fluorescent screen within the tube, is seen to travel in a straight line (a). When a magnet is brought near the tube (b), the beam is deflected.*

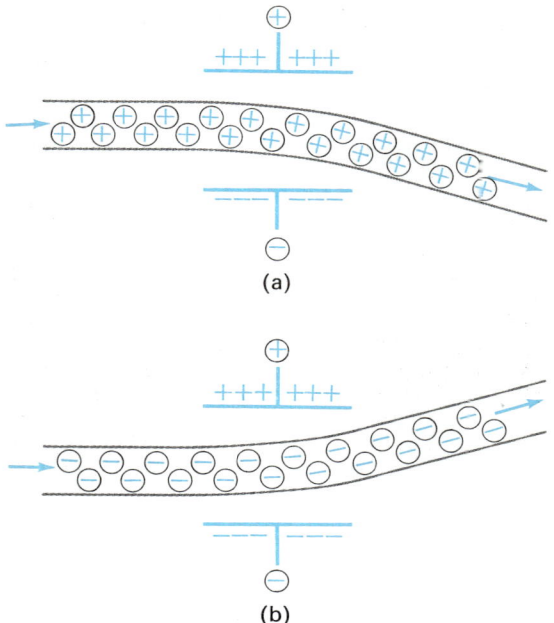

Figure 9-12 *Beams of positive and negative particles are deflected in opposite directions by an electric field.*

is positive or negative. (See Figure 9-12.) Experiments show that the stream of particles we have been describing is deflected towards the positively-charged plate. This proves that the stream is composed of negatively-charged particles. Evidently these particles are the electrons that we previously postulated.

Most of the things you have just learned about the behavior of electrons can be put to use in a device called a cathode-ray tube. A diagram of one is shown in Figure 9-13. Here again, a heated cathode is used to generate electrons. The anode now has a hole in it so that a narrow beam of electrons goes right on through it. Eventually, the beam strikes the fluorescent screen at the wide end of the tube, where it makes a bright spot. As the electron beam travels from the anode to the screen it is made to pass between two sets of metal plates, called deflection plates. By applying an electric field across these plates, the beam can be deflected both vertically and horizontally. Hence, the position of the spot on the screen can be moved.

Suppose an electric current that is varying in direction (that is, alternating current) or

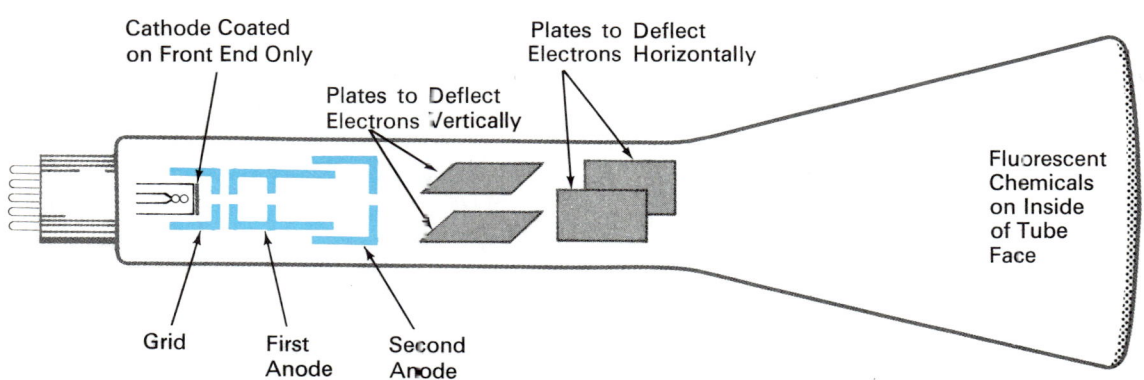

Figure 9-13 *Diagram of a simple cathode-ray tube.*

Robert Andrews Millikan (1868–1953)

ROBERT MILLIKAN received his Ph.D. degree from Columbia University in 1895, the same year that Roentgen discovered X rays. Working during a time when the scientific world was being rocked by major discoveries, he was particularly interested in the work done by J. J. Thomson in measuring the ratio of charge to mass for electrons. Both Thomson and another British scientist by the name of Townsend had made measurements of the charge on an electron late in the 1890's by studying the behavior of charged droplets of water. Their work was not very precise, and Millikan was sure that part of their difficulty lay in the evaporation of the water during their experiments.

By 1909 Millikan had developed his now-famous "oil drop experiment," which was based on the behavior of charged drops of oil in a strong electric field. He frequently spent many hours observing a single droplet of oil as it moved through his apparatus. When his experiments were complete, he was able to calculate the charge on an electron and also establish beyond any doubt that the charge on an electron was a fundamental unit of electricity.

Following his work on the charge on an electron, Millikan turned his attention to the photoelectric effect, in which light knocks electrons loose from the surface of metals. As a result of this work and his earlier work on the charge on an electron, he was awarded a Nobel prize in 1923.

Leaving the University of Chicago in 1921, Millikan went to the California Institute of Technology, where, as Chairman of the Executive Committee of the school, he helped the institution grow into one of the finest scientific schools in the world. For several decades he was extremely interested in cosmic rays. This work led to the discovery of other elementary particles.

An outgoing, warm-hearted, energetic man throughout his life Robert Millikan was an inspiration to all who knew him. His contributions to the field of atomic physics made it possible for us to come to our present understanding of the structure of matter.

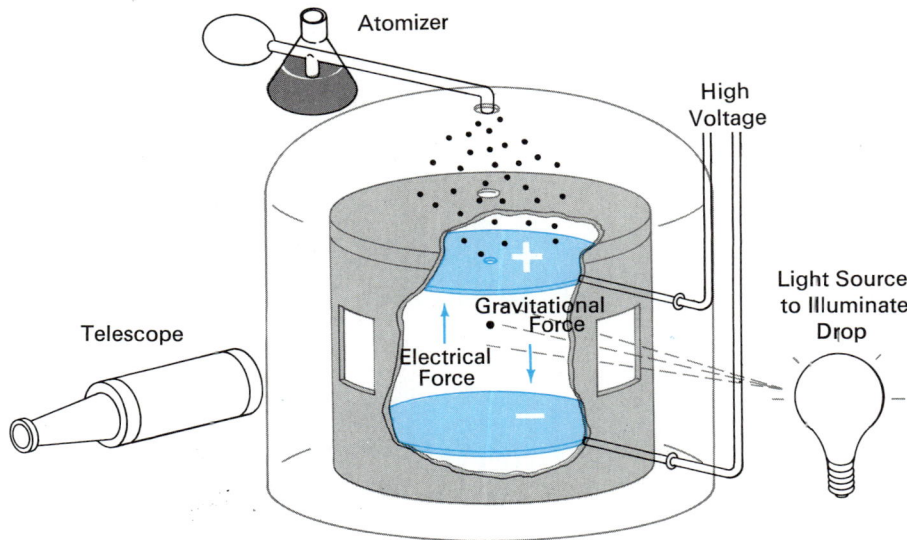

Figure 9-14 *Oil droplets fall through the hole in the positive electrode. Each droplet carries one or a few (n) excess electrons. By measuring the voltage required to suspend the drop against the gravitational force, the total charge (n times the charge of one electron) can be calculated.*

magnitude, or both, is supplied to the deflection plates. Then the pattern that the spot traces on the screen will show precisely how the current is varying. In this way a cathode-ray tube can be used to identify electrical "signals." In a slightly modified form, the cathode-ray tube is found in television sets, as the "picture tube."

9-6 The Charge and Mass of the Electron

You have learned that a stream of electrons is deflected when it passes through either an electric or magnetic field. It is possible to measure the curvature of the deflected stream very precisely. This measurement, together with other experimental data, such as the strength of the magnetic field, can be used to calculate an important number. That number is the ratio of the charge (e) of an electron to its mass (m). Only the ratio can be calculated from this experiment, since both e and m affect the amount of curvature, rather than either one alone, and neither can be eliminated in any mathematical treatment of the data. The value found for e/m is 1.759×10^8 coulombs per gram. (The coulomb is a unit of electric charge.)

The experiments that established the numerical value of e/m provided the first firm evidence for the existence of electrons. The importance of the experiments was soon recognized, winning honors, such as knighthood, for the scientist who performed them, Sir J. J. Thomson. With the numerical value of e/m established, it was evident that if either e or m could be measured separately, the individual values of both e and m would become known. In 1906, the same year that Thomson was awarded a Nobel prize for his work, an American physicist, Robert Millikan, devised a way of measuring the charge of the electron. The apparatus used for his experiment is shown schematically in Figure 9-14.

Tiny droplets of oil or some other liquid are sprayed into the upper part of the apparatus. A few droplets fall through a small hole into the lower chamber. During its production, an oil drop is very likely to become charged by friction, presumably by the gain or loss of one or more electrons. When an oil drop enters the lower chamber, a voltage is applied to the metal plates. If the oil drop is charged, its fall can be completely stopped by adjusting the voltage so that an upward electrical force, equal and opposite to the downward force of gravity, is exerted on the charged drop.

Millikan made thousands of determinations of the charge on drops of oil, glycerol, and mercury. The charge on the drop was sometimes positive and sometimes negative, but in every case its magnitude was some integral multiple of 1.602×10^{-19} coulomb. In no case was the charge any less than this. These experiments clearly demonstrate that the fundamental unit of electricity must be a charge of 1.602×10^{-19} coulomb. If the electron carries this fundamental unit of electricity (as scientists believe it does), the value of the charge on the electron must be 1.602×10^{-19} coulomb.

The charge on the electron can be used to calculate the mass of the electron. Using the result $e/m = 1.759 \times 10^8$ coulombs/gram, the mass of an electron is calculated as follows.

$$m = \frac{1.602 \times 10^{-19} \text{ coulomb/electron}}{1.759 \times 10^8 \text{ coulombs/g}}$$

$$m = 9.11 \times 10^{-28} \text{ g/electron}$$

9-7 How Thomson Measured the e/m Ratio ★

In this section you will learn in detail how the great British scientist J. J. Thomson measured the ratio of the charge, e, to the mass, m, of the electron. This is one of the landmark experiments in modern science. When Thomson's result for e/m was combined with the value for the charge, e, measured by the American physicist, Robert Andrews Millikan, the actual mass of the electron became known.

The basis of Thomson's experiment is the fact that if a charged particle moving along a straight path enters a uniform magnetic field it will then move in a circular path. (See Figure 9-15.) The radius, r, of the circle in

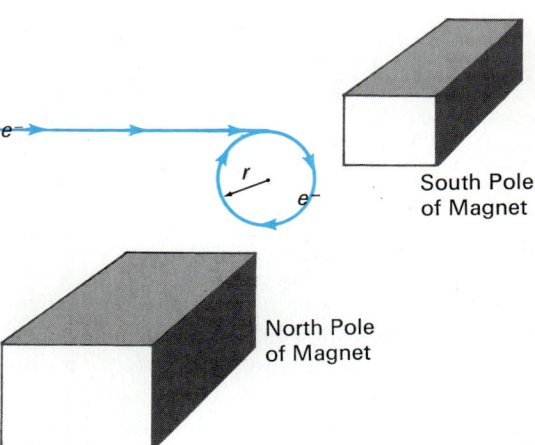

Figure 9-15 *A charged particle, such as an electron, entering a uniform magnetic field at right angles to the field direction will follow a circular path.*

which it moves is determined by the following four quantities.

1. The strength of the magnetic field, denoted B. The stronger the field, the more tightly the path will curve. As B increases, r decreases; r is proportional to $1/B$.
2. The charge of the particle. The greater its charge, e, the smaller the circle in which it moves. As e increases, r decreases; r is proportional to $1/e$.
3. The velocity, v, of the particle. The faster the particle is moving, the less a given magnetic field can deflect it. As v increases, r increases; r is proportional to v.
4. The mass, m, of the particle. The more massive the particle is, the less a given field can deflect it. As m increases, r increases; r is proportional to m.

When these four verbal statements are combined, the following equation, showing how the radius of the circular path depends on the four quantities, is obtained.

$$r = \frac{1}{B} \times \frac{1}{e} \times v \times m$$
$$= \frac{m \times v}{B \times e} = \frac{m}{e} \times \frac{v}{B} \quad (1)$$

Clearly, if we can measure r, v, and B, the ratio e/m can be calculated, since, by rearranging the above equation we get the following one.

$$\frac{e}{m} = \frac{1}{r} \times \frac{v}{B} \quad (2)$$

In the actual experiment, a stream of electrons generated by a hot filament and accelerated by a high voltage pass through a uniform magnetic field, as shown in Figure 9-16. In the absence of the field the electron beam would strike the fluorescent screen at A. When the field is applied, the beam is deflected a distance d, striking the screen at B. The path followed through the magnetic field by the electrons is an arc of a circle with

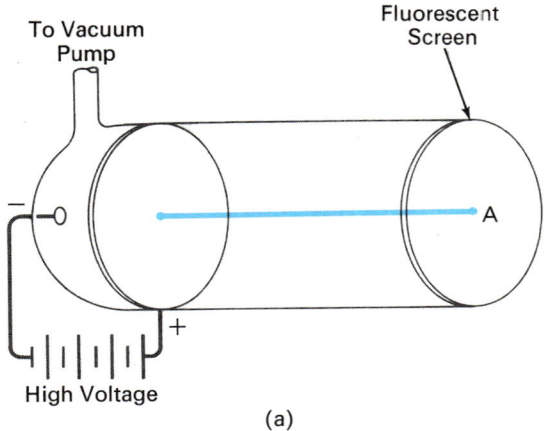

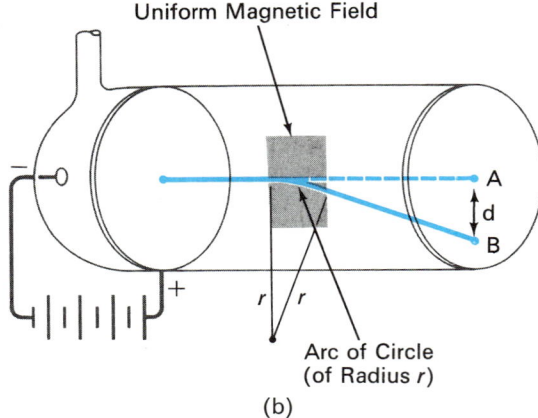

Figure 9-16 (a) An electron beam is directed in a straight path to the screen at point A. (b) When a magnetic field is applied, the path of the beam is deflected so that it strikes the screen at point B.

radius r. From the measured deflection, d, and the dimensions of the apparatus, this radius, r, can be calculated.

The magnetic field strength B can be measured independently.

Now all that is needed is the velocity, v. This is determined by the high voltage used to accelerate the electrons. The electrons acquire an amount of kinetic energy equal to their charge times the voltage, V, used. That is, their kinetic energy is equal to $e \times V$. The kinetic energy of a moving body is defined as $\frac{1}{2}mv^2$, where m and v are the mass and velocity of the body. Thus,

$$eV = \tfrac{1}{2}mv^2. \qquad (3)$$

This can be rearranged to

$$v^2 = \frac{e}{m} \times 2V. \qquad (4)$$

If we square both sides of equation (2), we get

$$\frac{e^2}{m^2} = \frac{1}{r^2} \times \frac{v^2}{B^2}.$$

If we now substitute the value of v^2 from equation (4) and then rearrange the equation, we obtain the following.

$$\frac{e^2}{m^2} = \frac{1}{r^2} \times \frac{\frac{e}{m} \times 2V}{B^2}$$

$$\frac{e^2}{m^2} \times \frac{m}{e} = \frac{1}{r^2} \times \frac{2V}{B^2}$$

$$\frac{e}{m} = \frac{2V}{r^2 B^2} \qquad (5)$$

Now e/m is expressed entirely in terms of quantities that can be measured in the experiment: V, the accelerating voltage; r, the curvature of the path; and B, the applied magnetic field.

9-8 Ions in Gases

We have been supposing that atoms are made up of protons and electrons in such a way that the total positive charge of all the protons is exactly equaled by the total negative charge of all the electrons. This makes the atom neutral. But we have also seen that electrons can be detached from atoms and experimented with by themselves. When electrons are removed, there should remain some incomplete atoms. Such atoms are missing an electron or even several electrons. These incomplete atoms will then be positively charged. They are called **ions**, or, more specifically, positive ions. It is possible to have negative ions too, as will be seen later. Experiments can be conducted in which positive ions are detected and their properties (charge and mass) measured. These experiments are similar to those already described for electrons. A gas-discharge tube, somewhat like those shown earlier, can be used, because measurements show that positive ions are present as well as electrons. Whereas electrons are attracted toward a positive electrode, the positively charged ions are attracted in the opposite direction, toward a negative electrode. These ions can be directed through an apparatus as a beam in the same way that the electron beam was produced and aimed along the fluorescent screen. Figure 9-17 shows a discharge tube

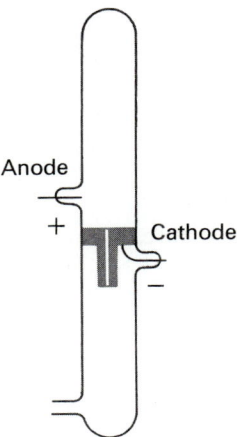

Figure 9-17 *A discharge tube of the Goldstein type. Positive ions pass downward through the channel in the cathode.*

that was designed by E. Goldstein, in 1886, for detecting positive ions. By deflecting beams of positive ions in electric and magnetic fields, the charges and masses of the positive ions can be measured.

The results of experiments of this type show two very important differences from measurements on electrons. (1) The charge/mass ratio for positive ions changes when the gas in the tube is changed. When the e/m measurement is made for electrons, the same value is obtained no matter what gas is used. (2) The charge/mass ratio for positive ions is very much smaller than e/m for electrons. These facts are interpreted to mean that the positive ions are ions formed from atoms of the gas in the tube. Since the electric charge is considered to arise from the removal of one or more electrons from an atom or a molecule, the value of the ratio (charge/mass) for positive ions depends upon the gas. Each type of atom or molecule has a distinctive mass.

Scientists have developed an instrument to measure the mass of positive ions, called the **mass spectrograph.** A diagram of one type is shown in Figure 9-18.

Positive ions can be produced and analyzed

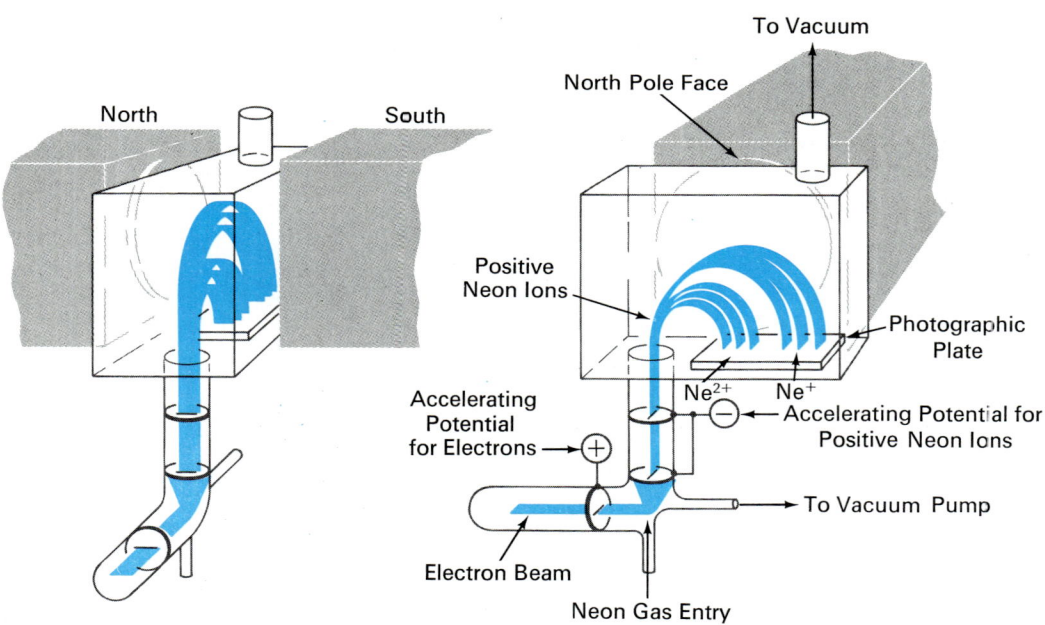

Figure 9–18 *A mass spectrograph enables scientists to find the masses of positive ions, in this case, neon ions, by measuring the curvatures of their paths in a magnetic field. Neon consists of three isotopes.*

with the use of a discharge tube, as you have seen. They are more commonly created, however, by bombarding gases with an electron beam. If the bombarding electrons have enough energy, they can knock one or more electrons out of any atom or molecule with which they collide. In this way positively charged ions are formed.

Figure 9-18 shows neon gas entering at the bottom of the apparatus. The gas passes through the electron beam, and some of the atoms collide with electrons, forming neon ions. Both Ne^+ and Ne^{2+} ions result and they are attracted to the slotted negative electrode. A beam of ions passes through this electrode into the magnetic field, at high speed. As the positive ions enter the magnetic field, they follow a circular path. (The deflecting force exerted on a charged particle moving within a uniform magnetic field is circular.) The amount of curvature of the path of an ion depends on both its mass and its charge for a given fixed magnetic field strength. The greater the mass of the ion, the greater the force that is necessary to deflect it. As a result, the amount of curvature is inversely proportional to the mass of the ion. The greater the charge on the ion, the greater the force that is exerted on it by the magnetic field. Hence the amount of curvature is directly proportional to the charge on the ion.

Each positive ion, then, follows a circular path fixed by its mass and charge. After they have circled through an arc of 180°, the ions are collected on a photographic plate. The impact of the ions with the photographic plate causes a reaction that leads to a darkening of the sensitized surface, just as exposure to light does. Such a record shows a spot or streak for each ion at a position fixed by the charge to mass ratio. If the instrument has been properly calibrated, the mass of each can be determined from its position on the plate. The record is called a mass spectrum.

When neon gas is put in the spectrograph, the resulting mass spectrum consists of two widely separated groups of three spots each. The three spots corresponding to the smaller amount of curvature, or deflection, are caused by neon ions with a single positive charge. The three spots corresponding to the greater amount of curvature are caused by doubly-charged ions. The three slightly separated spots for each ionic charge indicate that neon consists of atoms with three different masses. Atoms of the same element that have different masses are called *isotopes*. They will be discussed in detail in Chapter 10. The relative abundances of the isotopes can be determined by measuring the intensity of the spots caused by each of the ion beams.

9-9 *Ions in Solution*

The simplest way, in principle, to study ions is to observe them in a vacuum, as in the mass spectograph. However, the existence of ions was recognized long before such experiments were done. Although experiments with gaseous ions are simple and neat, in principle, very sophisticated apparatus is required to conduct them. On the other hand, evidence for ions in solution can be obtained by relatively simple means. Such evidence was obtained as early as 1830.

Many substances can be dissolved in water to give solutions that conduct electricity. Pure water itself does not conduct appreciably. Hence the dissolved substances must give rise to charged particles that can move when a voltage is applied. The substances that are dissolved are electrically neutral and the solutions as a whole are electrically neutral. It must, therefore, be assumed that the dissolved substances dissociate into charged particles in such a way that the total of all negative charges is equal to the total of all positive charges. A great deal of evidence shows that the charged particles in solution are not simply protons or electrons. Rather, they are positive and negative ions. The entire subject of substances that dissolve to form ions, and the properties of their solutions will be taken up in much more detail later in the book. The purpose here is simply to show that such solutions provide very direct evidence for the electrical nature of matter, and give additional details about it.

The mere fact that the solutions can conduct an electric current provides powerful qualitative evidence that matter is made up of electrical charges, just as does the ability of metals to conduct an electric current. However, there is much more interesting information to be obtained from the chemical changes that occur when an electric current passes through certain solutions. Experiments of the type you are about to do were first carried out by the great English chemist Michael Faraday in about 1830. Faraday drew essentially correct conclusions about his experiments, though he was not able to interpret them as fully as we can today.

INVESTIGATION

9-10 The Relation Between the Moles of Electrons and the Moles of Metals Used in Electrolysis

Near the beginning of the 1800's William Nicholson and Anthony Carlisle in England found that electricity could cause chemical changes to take place when passed through solutions. Using this information, Humphry Davy experimented with molten salts and found that, by passing electricity through them, he could decompose them. Such a process is known as *electrolysis*. Earlier you saw that water can be decomposed in a similar way.

Michael Faraday, who had been Davy's assistant at the Royal Institution, continued to improve on Davy's experiments for a number of years. By 1830 he had gathered enough data to be able to conclude that there was an exact relationship between the amounts of electricity used and the masses of chemicals reacting during electrochemical processes.

The electric circuit and the cells that you will use are shown in Figure 9-19. The source of electrons may either be a battery, as shown in the diagram, or a power supply that gives direct current. Direct current is necessary so that the electrons will move in only one direction in the circuit. The ammeter measures the rate of flow of the electrons in the circuit. The units shown on the face of the ammeter are amperes. One ampere is that amount of current flowing when 1.04×10^{-5} mole of electrons passes through any cross-section of the conductor in one second. This number of electrons has a total charge of one coulomb.

The source of direct current causes the electrons to be crowded onto one of the electrodes and to be drained away from the other. The circuit

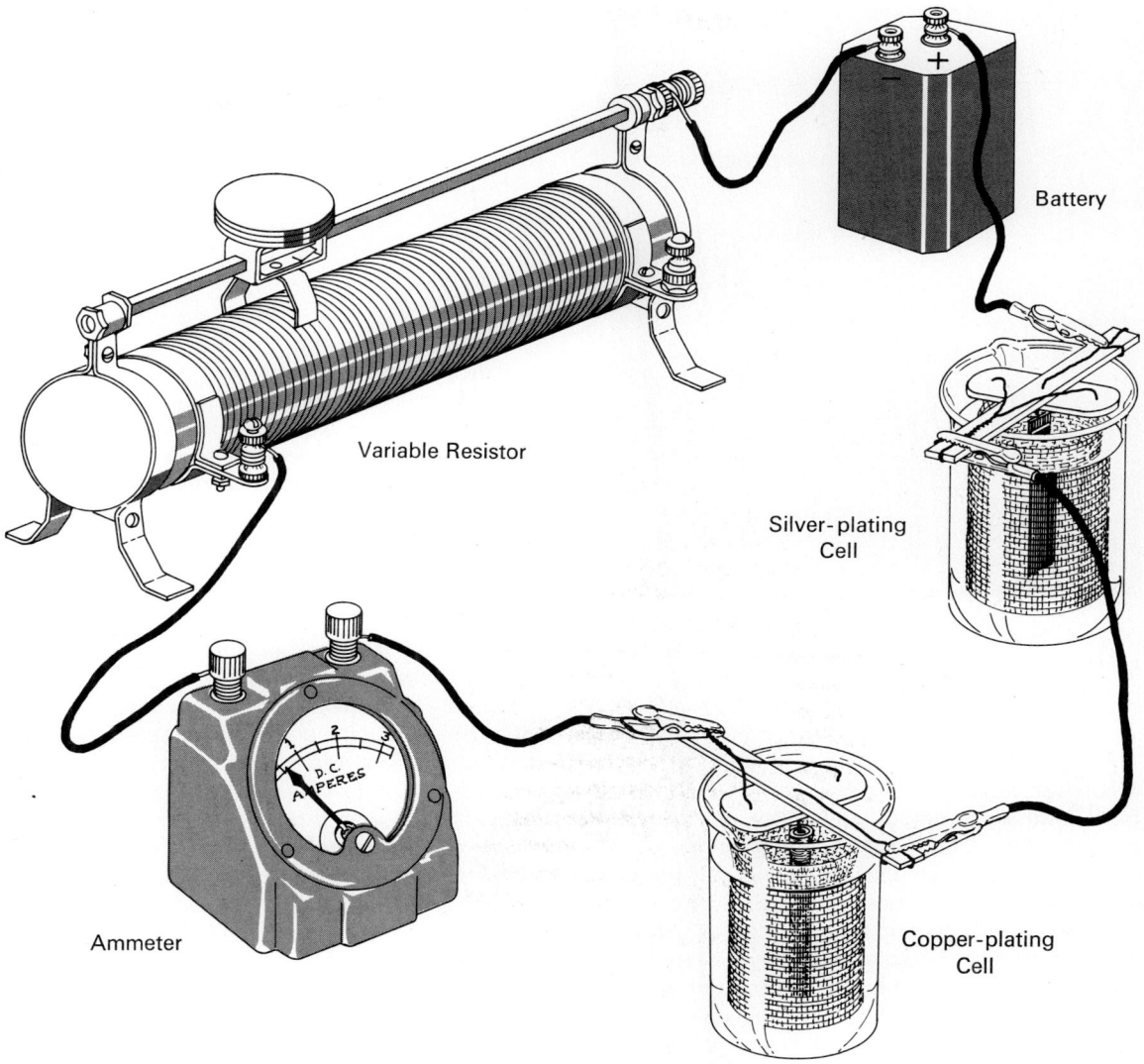

Figure 9-19 *Electric circuit and cells for this investigation.*

is complete when at one electrode an ion or molecule in the electrolyte solution accepts electrons, while at the other electrode an ion or molecule loses electrons. The electrode at which electrons are gained is the negative electrode, known as the *cathode*. Electrons are given up at the positive electrode, known as the *anode*.

Michael Faraday (1791–1867)

Michael Faraday, at work in the chemical laboratory at the Royal Institution.

MICHAEL FARADAY, whose father was a blacksmith in England, never had any formal education. At the age of 13 he became an errand boy for a bookseller and a year later was formally apprenticed for seven years to learn to be a bookbinder, stationer, and bookseller. Although he soon became an expert bookbinder, he also became interested in the contents of the books, particularly scientific ones, and began to perform some of the simpler experiments described in the books. After attending a number of public lectures and demonstrations on science, he applied to Sir Humphry Davy for a job in his laboratory, and was soon hired.

Working with Davy for a number of years, Faraday assisted his employer in the development of the Davy mine safety lamp, which was designed to prevent coal gas explosions. As time went on, Faraday began to carry out his own scientific explorations, which covered a wide field of knowledge. He first investigated compounds of carbon and chlorine, including the well-known compound carbon tetrachloride. He later succeeded in isolating the pure compound benzene, one of our most important hydrocarbons. For a time he experimented with steel alloys that could be used for making sharper and more durable cutting blades.

Faraday is best known for his work in electricity, magnetism, and electrochemistry. His discovery that a wire carrying an electric current would rotate when suspended above a fixed magnet was the basis for the invention of the electric motor. Later he found that a magnet would induce an electric current in a coil of wire when it was passed through the coil. This is the major principle behind the operation of all electric generators.

Following many outstanding experiments in which he passed electric currents through solutions of chemicals, Faraday proposed his fundamental law of electrochemistry which stated that "equal quantities of electricity discharge equivalent quantities of ions at the two electrodes, whatever these ions may be." He invented most of the terms used in electrochemistry today. Among these terms are: electrolyte, electrolysis, electrode, anode, cathode, ion, anion, and cation.

PURPOSE

The relationship between the number of moles of electrons flowing during the experiment, the number of moles of copper atoms that gain and lose electrons, and the number of moles of silver atoms that gain electrons is to be found.

MATERIALS

D.C. ammeter, 1 ampere
Variable resistor
4 alligator clips
Clock with sweep second hand
200 ml copper sulfate electrolyte solution
200 ml silver thiosulfate electrolyte solution
D.C. source, 6-volt battery or power supply
5 lengths copper wire, insulated, 20-30 cm long
2 pieces copper screen, 6 × 16 cm
Sheet lead strip, 2 × 7 cm, about $\frac{1}{16}''$ thick
Copper foil, 6 cm square, 0.002" thick, with 5 cm copper wire attached, or 80 cm #16 gauge copper wire

Centigram balance
2 electrode holders
2 beakers, 250-ml
200 ml acetone

PROCEDURE

1. Obtain a cylinder of copper screen to be used as the cathode and a sheet of copper wound into a spiral, or a heavy copper wire wound into a tight coil, to serve as the anode in the copper-plating cell. Handle the electrodes by the connecting wires only. Otherwise, your fingers may leave grease marks upon which the copper deposit will not adhere well.

2. Obtain a cylinder of copper screen that has a thin coating of silver to be used as the cathode and a strip of lead that will serve as the anode in the silver-plating cell. Again, do not touch the clean electrodes with your fingers.

3. Determine the mass of each of the clean, dry screen cylinders to the nearest 0.01 g.

4. Suspend the electrodes in separate 250-ml or 400-ml beakers by attaching them to a wooden holder or by clamping them with clothespins, as directed by your teacher. Be sure that the anodes are centered within the cylindrical cathodes and that the electrodes do not touch each other.

5. Make all of the connections in the circuit as shown in Figure 9-19,

except leave one of the wires to the variable resistor disconnected. Adjust the variable resistor so that its full resistance will be utilized. Ask your teacher to check your setup. (If a variable voltage power supply is used instead of the battery, the variable resistor will not be needed. Connect the negative terminal of the power supply directly to the ammeter. You may then adjust the current flowing through the circuit by varying the voltage output of the power supply.)

6. Add the electrolyte solution containing the copper ions to the beaker with the copper anode. Use enough solution to just cover the copper screen cathode.

7. In a similar manner add the solution containing silver ions to the other beaker.

8. Before proceeding, be sure that you have available a clock or timer with a second hand. Now make the last connection. Quickly adjust the current to one ampere by decreasing the resistance being used in the variable resistor or by increasing the voltage of the power supply. Record the time to the nearest second.

9. Let the current flow for 30 minutes. Watch the ammeter and keep the current as close to one ampere as possible by adjusting the variable resistor or the voltage control on the power supply.

10. Record the time to the nearest second as you disconnect one of the wires to break the circuit. Rinse the two cathodes by *gently* dipping each into a beaker containing cold water. Do not agitate the electrodes so vigorously that the metal deposits are dislodged.

11. Rinse the cathodes with acetone to remove the water droplets. Allow them to dry by evaporation for 2 or 3 minutes.

12. When the cathodes are thoroughly dry, determine their masses, to the nearest 0.01 g, on the same balance you used before.

DATA AND RESULTS

Tables, such as the following ones, in your notebook will help you to organize your data and results. Measure all masses in grams.

A. Copper-Plating Cathode
   Mass of screen with Cu coating
   Mass of screen at start
      Mass of Cu coating deposited
B. Silver-Plating Cathode
   Mass of screen with Ag coating

Mass of screen at start
　Mass of Ag coating deposited
C. Elapsed Time
　Time at end of experiment
　Time at start of experiment
　　Time elapsed for electron flow

Calculations

1. Calculate the masses, in grams, of copper, Cu, and silver, Ag, that were deposited. Include the uncertainties in your results.
2. Calculate the number of moles of each metal deposited.
3. Calculate the ratio: moles of Ag/moles of Cu.
4. Estimate the limits of error on the mole ratio.
5. Calculate the number of coulombs of electricity that were used. The rate of flow of electrons in a circuit is measured in amperes, the number of coulombs passing a given point in a second. Thus the total number of coulombs is found by multiplying the current, in amperes, by the elapsed time, in seconds. (Coulombs/sec × sec = total coulombs.)
6. Calculate the number of moles of electrons that were used. In the introduction to this experiment you found that 1.04×10^{-5} mole of electrons flows when a current of one ampere flows for one second. Therefore, moles of electrons equals 1.04×10^{-5} times the number of coulombs. (Moles of electrons/coulomb × coulombs = moles of electrons.)

RELATED PROBLEMS

1. How many moles of electrons are required to deposit one mole of Ag? to deposit one mole of Cu?
2. How many electrons does *each* silver ion gain when it is deposited on the cathode? How many electrons does *each* copper ion gain when it is deposited?

9-11 *Ions, Atoms, and Electrons*

The type of experiment that you have just done leads to some very important conclusions. In such experiments, it is found that the same amount of electricity always deposits the same number of moles of a given metal. This means that a certain definite electrical charge is required per atom of metal. Let us assume that an atom of metal gets deposited on the cathode when enough electrons are transferred to the positive ion to neutralize it.

The first thing that you can conclude from the experiment you just performed is that copper ions have twice as much positive charge on them as do silver ions. That is why a given number of coulombs can deposit only half as many moles of copper as of silver. The fact that the ratio of the charges on two different ions is a whole number (within experimental error) suggests that the charges come in well-defined "packages." Each ion, then, has a definite small number of such packages of charge. The ratios of charges on different ions will tend to be small whole numbers, or possibly half-numbers. For instance, there should be ratios such as $2:2 = 1$, $2:1 = 2$, $3:1 = 3$, $4:2 = 2$, or, maybe, $3:2 = 1\frac{1}{2}$. In short, the idea that the electrical charges of ions come in definite packages is the first conclusion suggested by electrolysis experiments.

But things get still more interesting! An experiment such as you have done, if performed with more accurate meters and timers, will show that it takes $96,500 \pm 100$ coulombs of electricity to deposit one mole of silver. Since there are 6.02×10^{23} (Avogadro's number) atoms in a mole, we can easily calculate how much negative charge had to be added to each silver ion to make it a neutral silver atom.

Charge per Ag ion
$$= \frac{(9.65 \pm 0.01) \times 10^4 \text{ coulombs/mole}}{6.02 \times 10^{23} \text{ atoms/mole}}$$
$$= (1.60 \pm 0.0017) \times 10^{-19}$$
$$= 1.60 \times 10^{-19} \frac{\text{coulomb}}{\text{atom}}$$

Does this number ring a bell? If not, refer back to Section 9-6. Within experimental uncertainty, it is the charge of one electron.

The following general conclusions can now be made.

1. Electricity is carried through wires and into solutions in "packages," that is, electrons.
2. Metal atoms form positive ions by losing one, two, or some other whole number of electrons.
3. The silver ion has a charge of $1+$ and the copper ion a charge of $2+$. Symbols for these ions are Ag^+ and Cu^{2+}.
4. If e^- is used to represent an electron, the following equations can be written for the reactions that occurred at the cathodes in your experiment.

$$Ag^+ + e^- = Ag$$
$$Cu^{2+} + 2e^- = Cu$$

Summary

Detailed evidence for the electrical nature of matter comes from many sources. Streams of electrons and positive ions can be shown to exist in vacuum tubes and gas-discharge tubes. By measuring the behavior of such streams in electrical and magnetic fields, the masses and charges of the ions and electrons can be determined. Studies of solutions that conduct electricity provide powerful evidence for the following: (1) the smallest unit of negative electrical charge is the electron, and (2) positive ions are formed by the loss of some integral number of electrons, usually one, two, or three, from an atom.

Self Check

1. What type of experiment shows that vacuum tubes and gas-discharge tubes contain streams of particles?

2. The masses of positive ions are determined with a mass spectrograph. How does this instrument work?

3. What is a positive ion?

4. What are the smallest packages of negative charge called?

Chapter Summary

Electrical charges are fundamental properties of matter. There are two kinds of charge, which we call positive and negative. A body possessing no charge is called neutral. The condition of neutrality results when positive and negative charges are present in equal amounts.

Like charges (that is both positive or both negative) repel each other, while opposite charges attract each other. Coulomb's law states that the force of attraction or repulsion increases in proportion to the product of the two charges, that is $q_1 \times q_2$. This law also states that the force decreases according to the square of the distance, d, separating the charges. Force is proportional to $1/d^2$. The complete algebraic statement of Coulomb's law is: Force = $(q_1 \times q_2)/d^2$. If the charges are alike, the force has a positive sign and is a repulsive force. If the charges are opposite, the force has a negative sign and is an attractive force.

The basic carriers of positive and negative charges are particles called protons and electrons, respectively. The positive charge on a proton is equal in magnitude and opposite in sign to the negative charge on the electron. Objects can acquire charges when the balance of protons and electrons is disturbed by some external force, such as rubbing.

An electric current is the movement of electrons through a body from a negative region toward a positive one. It is also possible to make a stream of electrons pass through a vacuum from a hot, negatively-charged electrode to a positively-charged electrode. The behavior of such a stream, which can be deflected by a magnetic field and blocked by a solid object, shows that electrons are charged particles. The charge and mass of the electron have been measured.

Atoms are neutral because they contain equal numbers of electrons and protons. One or more electrons can be detached from an atom. The remaining positively-charged particle is called an ion. Positive ions can be studied in gas-discharge tubes. Their masses can be determined by the paths they follow in a magnetic field. The instrument used for this purpose is called a mass spectrograph. In this way it is found that there may be atoms of the same element that have different masses. These are called isotopes.

Both positive and negative ions are also formed in certain solutions. When electrodes are placed in such a solution and a voltage applied, the positive ions, called cations, move toward the negative electrode, called the cathode. The negative ions, called anions, move toward the positive electrode, called the anode.

Questions and Problems

1. Describe several phenomena which support the belief that all matter is electrical in nature.
2. Since it is suggested that all matter contains positively and negatively charged particles, explain the fact that most matter in its ordinary state is electrically neutral.
3. When a plastic or hard rubber rod is rubbed with a piece of fur:
   (a) What kind of charge is acquired by the rod?
   (b) What kind of charge is acquired by the fur?
   (c) What was the original charge on each?
4. If you charged a comb electrically by running it through your hair, what would happen if the comb were then slowly brought close to your hair? Explain.
5. Briefly describe what must happen to an object in order to give it a positive electrical charge.
6. Is it possible to give an electrically neutral object a negative charge by removing one or more protons from its atoms? Explain your answer.
7. On the basis of Coulomb's law, what happens to the force between two charged bodies if the distance between the two bodies is cut in half? if the distance is tripled? if the distance between the two bodies remains the same but the charge on each body is doubled?
8. Why do scientists claim that there are only two types of electrical charge?
9. What is the difference between an atom and an ion?
10. Must all negative charges be removed from an object for it to become positively charged? Explain.
11. At one time scientists thought that electrical currents flowed from positive to negative. From what you have learned about electrons, show why this would be impossible.
12. Suppose that you have rubbed a hard rubber rod with a piece of fur and a glass rod with a piece of silk. Label each statement below true or false and briefly explain your reasoning.
    (a) The glass will attract the fur.
    (b) The silk will attract the fur.
    (c) The glass will attract the rubber.
13. In a mass spectrograph, which of the ions in the following pairs would be deflected by the magnets in an arc with the greater curvature: K^+ or K^{2+}; a light isotope of a Ca^{2+} ion or a heavier isotope of a Ca^{2+} ion? Explain.
14. Using the symbols Cu, Cu^{2+}, and e^- to represent a copper atom, a copper ion, and an electron, respectively, write an equation for the reaction that takes place at the copper anode in an electrolytic cell.
15. Today we know that two isotopes of the same element have the same numbers of protons and electrons in their atoms, but the atoms have different masses. Scientists in the 1920's assumed the existence of an uncharged particle, which they called a neutron. This particle was not detected experimentally until 1932. In their position, how might you have tried to explain the difference in masses of two isotopes of the same element?
16. Why was J. J. Thomson unable to calculate the mass of an electron until Millikan had determined the charge on an electron?
17. In the vacuum tube diode diagrammed in Figure 9-7, you found that an electric current flowed through the circuit when the cathode heater was turned on. What

would you expect to happen if the wires to the battery at the bottom were reversed so that the heated electrode became the anode and the cool electrode became the cathode. Explain. (After you have answered this question, you might like to look up the meaning of a vacuum tube rectifier, which provides direct currents in electronic circuits.)

18. Dust particles may be removed from air by passing the air through an electrical discharge and then between a pair of oppositely charged metal plates. Explain how this removes the dust. (Precipitators of this type are widely used on power plants and other industrial installations to remove fine particles from smoke in order that the air in the vicinity will not become polluted.)

19. When several oil drops enter the observation chamber of the Millikan apparatus, the voltage is turned on and adjusted. One drop may be made to remain stationary, but some may move up while others may fall. Explain these observations.

20. How many electrons would be required to weigh 1 g? What would be the mass of a "mole" of electrons?

21. If n coulombs will deposit 0.119 g of tin from a solution of $SnSO_4$, how many coulombs will you need to deposit 0.119 g of tin from a solution of $Sn(SO_4)_2$? (Note: Both compounds dissolve to form SO_4^{2-} ions.)

22. A circuit was connected in such a way that 0.6 ampere of current flowed continuously for 2.0 hours. In this time how many coulombs of electricity flowed? How many moles of electrons flowed? How many electrons flowed?

23. The gas in a neon sign tube begins to glow when the pressure in the tube is reduced to about 0.01, or 10^{-2}, atm. When the pressure is reduced even further, to about 10^{-6} atm, the glow disappears. If one mole of a gas occupies 24.0 l at 1 atm pressure and 20°C, find out how many molecules there would be in each cubic centimeter of the gas in the tube when the glow first appears. Also find out how many molecules would be left in each cubic centimeter of the gas when the glow disappears at the lower pressure. Assume the tube operates at a temperature of 20°C.

24. Two objects, each with an electrical charge of 3−, are suspended 2 mm apart. Two objects, each with an electrical charge of 4+, are suspended 3 mm apart.
    (a) Is the force between the first two objects repulsive or attractive?
    (b) Is the force between the second two objects repulsive or attractive?
    (c) Use Coulomb's law to show whether the magnitude of the electrical force between the negatively-charged objects is less than, equal to, or greater than the force between the positively-charged objects.

3

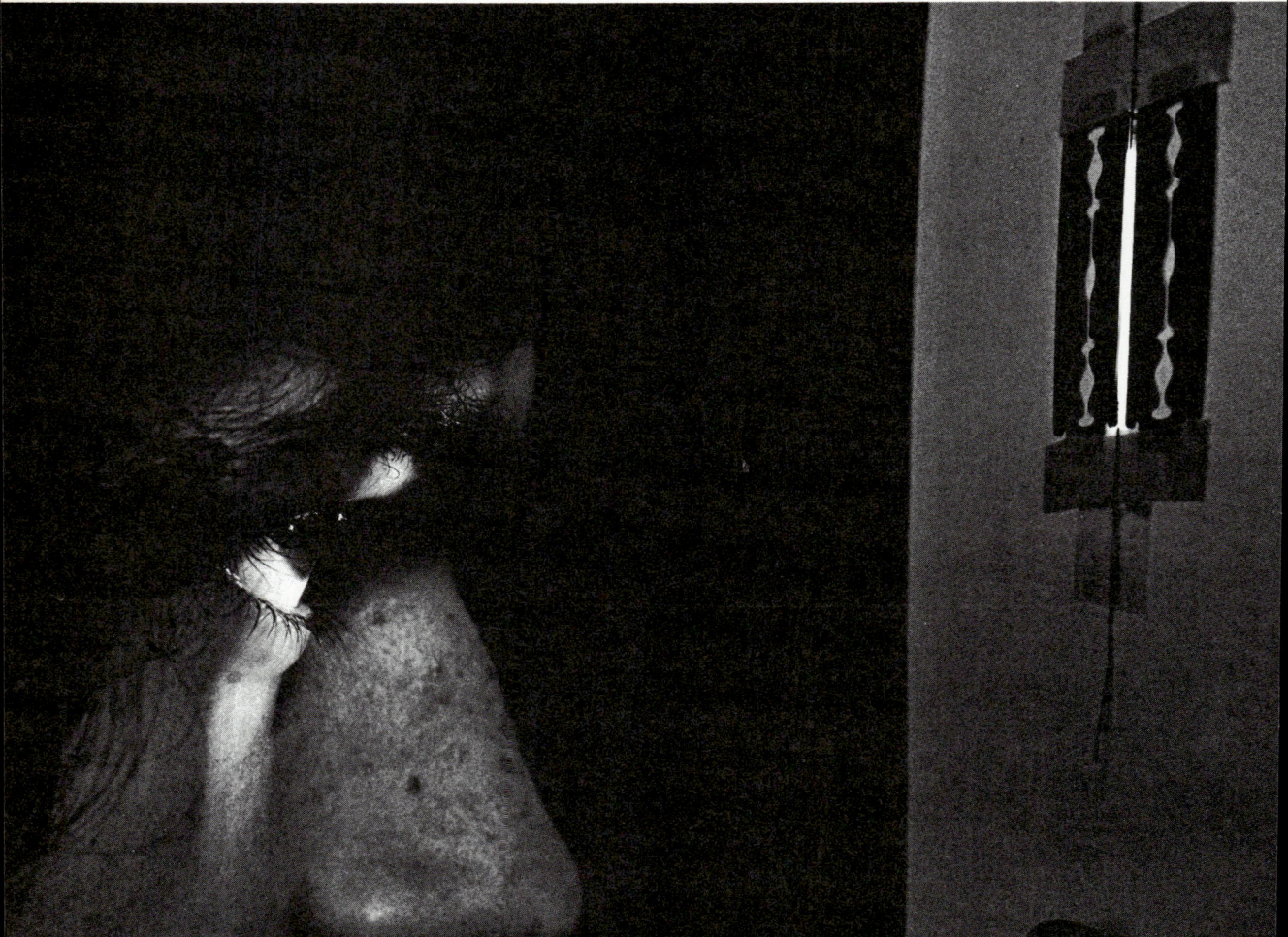

The Nature of the Atom

Chemistry is concerned with the reactions and properties of the various forms of matter. These forms of matter are all made up of atoms, in an almost infinite array of combinations. It will therefore be obvious that in order to understand chemistry we must understand atoms.

Understanding atoms is not easy. We cannot see individual atoms. Our knowledge is obtained indirectly, by many ingenious experiments and by logical deductions from these experiments. We also find that the inner workings of things as tiny as atoms cannot be described by the same physical principles that we use for large, everyday objects. A new set of principles, called quantum mechanics, had to be developed.

In this unit you will learn how atoms are built up of elementary particles, and how the energies and other important properties of atoms are governed by the principles of quantum mechanics. Finally, you will see how a knowledge of atomic structure enables you to understand why the periodic table provides such a useful correlation of the chemical properties of the elements.

10 The Structure of the Atom

Lord Ernest Rutherford, conducting a typical impromptu examination of his students, when he was director of the Cavendish Laboratory at Cambridge University.

The first questions we ask about atoms pertain to their basic design. How big are they? How are the positive and negative components arranged? Are there other particles besides negative and positive ones present in atoms? How can we account for the masses of atoms? We must know the answers to these and similar basic questions before we can hope to tackle even more subtle ones, such as: What energy states are possible for atoms? How do atoms combine to form molecules?

In this chapter you will see how scientists discovered, by experiment and careful reasoning, the basic design of atoms. You will see how an idea—a model—can be tested against experimental fact and either pass the test or be rejected. You will learn about the model we now accept, because it has passed all experimental tests.

Atoms Have a Definite Design

10-1 Rutherford's Scattering Experiments

Experiments discussed so far show that atoms are composed of charged particles called protons and electrons. You also know that electrons can be detached from the atoms to give positive ions. It is also possible for some atoms to acquire one or more extra electrons to become negative ions. The charges and masses of the various ions, and of the electron, can be measured. But a huge question remains! How are the positive and negative particles arranged in the atoms? What is the design or structure of the atom? You have not yet learned that, and neither had any scientist in the year 1898.

In 1898 J. J. Thomson, who had done many of the experiments that helped to characterize the electron, proposed the first detailed model of atomic structure. According to Thomson's model the atom was a sphere of positive electricity in which electrons were embedded like raisins in a cake. Of course, he "cooked up" this model in order to explain the experimental facts known to him at the time. Those facts included the ones that we have discussed so far, for example, the fact that electrons can be rather easily detached from atoms, whereas positive particles cannot. Although Thomson's model was consistent with the results of experiments, it was not useful in predicting or explaining the chemical properties of the atom. Finally, in 1911, a series of experiments performed by Ernest Rutherford in a laboratory at the University of Manchester in England showed that Thomson's picture of the atom had to be abandoned.

The experiment conducted by Rutherford and his co-workers involved bombarding gold foil with alpha particles. Alpha particles are small, positively-charged particles that are emitted by some radioactive elements. (The emission of alpha particles by radioactive elements is discussed in Section 10-9.) The apparatus used in Rutherford's experiment is shown in Figure 10-1. The alpha particles are produced by the radioactive decay of radium, and a narrow beam of these particles emerges from a deep hole in a block of lead. The beam of particles is directed at a thin metal foil, approximately 10,000 atoms thick. The alpha particles are detected by the light they produce when they collide with a scintillation screen. This screen is a plate covered with zinc sulfide, similar to the front of the picture tube in a television set. It is mounted on an arm in such a way that it can be moved around in a circle. The center of the circle is in line with the point at which the alpha particles strike the metal foil. A telescope is mounted behind the screen so that the very small flashes of light produced when individual alpha particles strike the scintillation screen can be seen and counted. Because the flashes of light are so faint, all observations must be made in a darkened room. The apparatus operates in a vacuum chamber in order to ensure that no deflections will be caused by the collision of alpha particles and gas molecules.

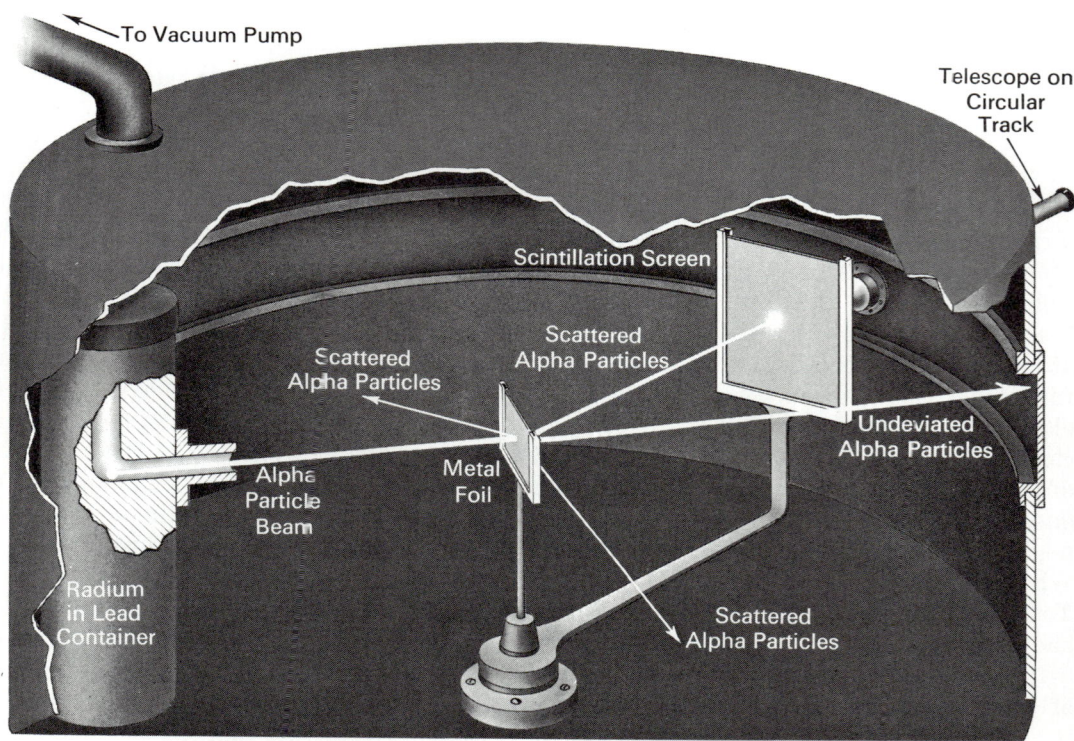

Figure 10-1 *Alpha-scattering apparatus. Alpha particles stream from the radium to the foil, within the evacuated container. Scattered particles are observed on the scintillation screen.*

The first observation made with this apparatus was that all the alpha particles apparently pass through the foil undeflected. Is such a result compatible with Thomson's model of the atom? It could be. Thomson's concept included the idea that the positive charge is distributed evenly throughout the entire volume of the atom, with the negative electrons embedded in it. Since electrons have very little mass, the positive part accounts for nearly all of the mass of the atom. Thus the Thomson model pictures the atom as a body of uniform density.

Imagine what the thin metal foil would be like if it were made up of Thomson atoms. The physical properties of most solids suggest that atoms lie very close together, so the metal foil would look something like that in Figure 10-2a. Of course, the real foil is 10,000 atoms thick. What would happen to the alpha particles if they were shot into a solid of such uniform density? At first you might think that they would be stopped, or deflected back, on colliding with the atoms. Since it was observed that the alpha particles went straight through the metal foil, we must consider the problem further.

If you were to shoot at a piece of cheese with a high powered rifle, the projectile would force its way through the cheese. The alpha particles produced by radium have very high kinetic energy and are very much like bullets from a high-powered rifle. Perhaps its very high kinetic energy allows an alpha particle to force its way right through the atoms of the metal foil. Since a rifle bullet fired into cheese would pass through undeflected, it seems reasonable to conclude that the alpha particle would also pass through the metal foil undeflected.

In summary, according to the Thomson model a metal foil should have essentially uniform density. If this were true, there would be no way for the bombarding alpha

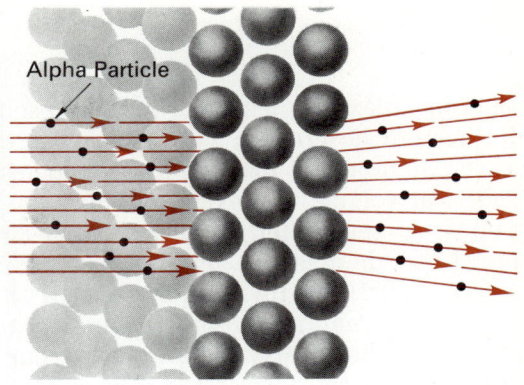

Figure 10-2a *Thomson's atomic model led Rutherford to expect the type of alpha-scattering shown above.*

particles to be deflected through large angles. At best the alpha particles might suffer slight deflections from many collisions with many atoms. The Thomson model predicts the scattering distribution shown in Figure 10-2a.

The first results of Rutherford's experiments seemed to be quite consistent with the Thomson picture of the atom. On more careful examination, however, an astounding discovery was made. By moving the screen along the circle surrounding the metal foil, Rutherford and his co-workers were able to observe that a very few scintillations occurred at many different angles. Some of these angles were nearly as large as 180 degrees. It was as if some of the alpha particles had rebounded from a head-on collision with an immovable object. In the words of Rutherford, "It is about as incredible as if you had fired a 15-inch shell at a piece of tissue paper and it came back and hit you." It was impossible to explain the simultaneous observation of large-angle and small-angle deflections by using the Thomson atom.

10-2 Rutherford's Atomic Model

In order to explain his experimental results, Rutherford designed a new picture of the atom. He proposed that the atom occupies a spherical volume approximately 10^{-8} cm in radius. At the center of each atom he proposed that there is a *nucleus* having a radius of about 10^{-12} cm. He further proposed that this nucleus contains most of the mass of the atom, and that it also has a positive charge that is some multiple of the charge on the electron. The region of space outside the nucleus must be occupied by the electrons. Figure 10-2b shows Rutherford's model. It requires that most of the volume of the atom be a region of very low density.

Using this kind of model of the atom, the deflection of alpha particles through both large and small angles can be accounted for. If alpha particles impinge upon a metal foil composed of atoms based on Rutherford's model, only a few of the particles would be appreciably deflected by the foil. The heavy, fast-moving alpha particles could brush past the lighter electrons without being deflected. Since most of the volume of the metal foil is relatively empty space, most of the alpha particles would pass through the metal undeflected. It is possible, however, for a few particles to be scattered through very large angles. Since both the alpha particle and the nucleus of the atom are positively charged, they exert a force of repulsion on each other. This force becomes large only when the alpha particle comes quite close to the nucleus. (Remember Coulomb's law.) Since the nucleus in question is much heavier than the alpha particle, it can deflect the alpha particle considerably, just as a steel post can deflect a rifle bullet. (Figure 10-3 shows a representation of alpha-scattering by a nucleus.)

Besides providing a qualitative picture of the atom, Rutherford's experiments provided

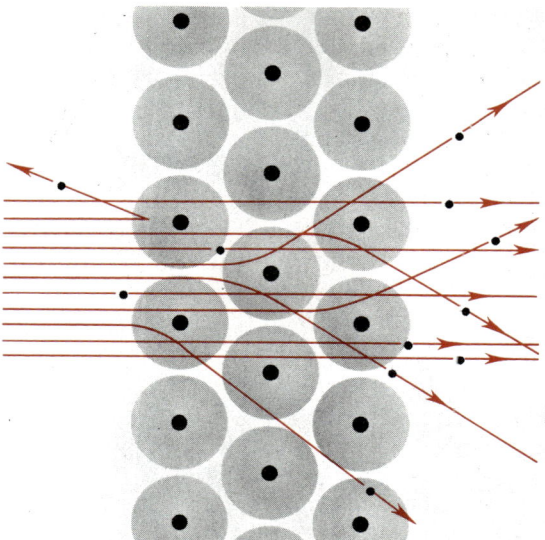

Figure 10–2b *The atomic model (above) proposed by Rutherford explained his experimental results.*

a way of measuring the charge of the nucleus. The force that a nucleus exerts on an alpha particle depends upon the magnitude of the charge on the nucleus, as well as the distance separating them. Rutherford showed how to relate the number of alpha particles scattered at any angle to the magnitude of the charge on the nucleus. The first measurements of the nuclear charge by this method were not very accurate, but they did show that different elements have different nuclear charges. By 1920 the alpha particle scattering experiments were so refined that they could be used to determine nuclear charge accurately.

10-3 The Nuclear Atom, A Successful Scientific Model

You have just seen how a great variety of evidence convinced scientists that atoms

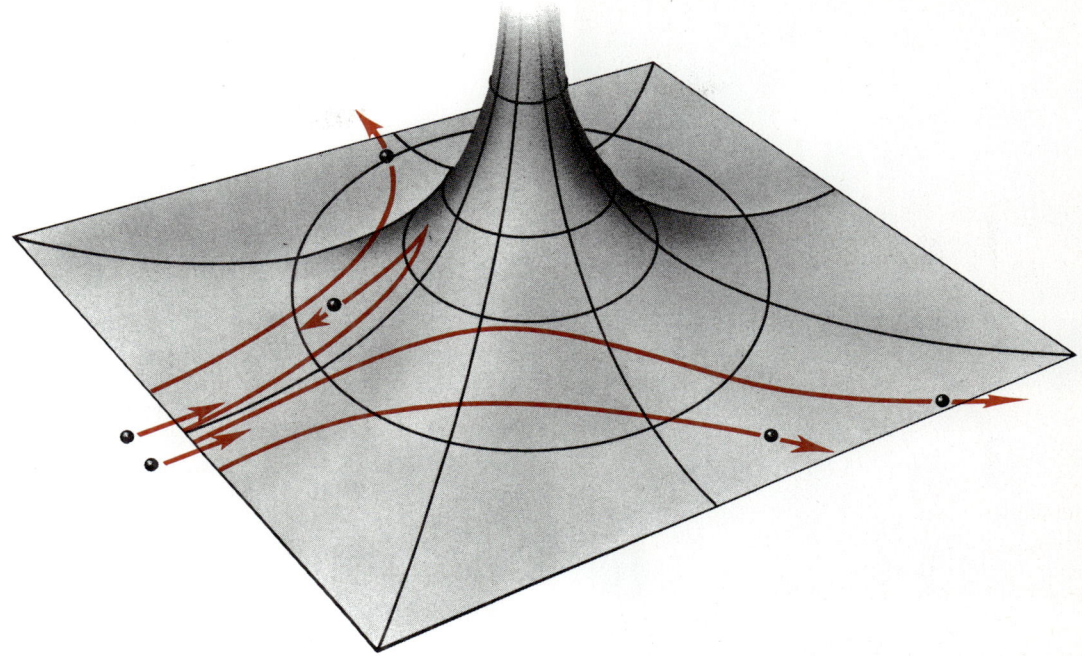

Figure 10–3 *A model that accurately represents alpha-scattering by a nucleus. The repelling energy encountered by an alpha particle becomes very great as it approaches the nucleus (the highest point in the model).*

must be made up of small charged particles. The negative particles are the **electrons.** The positive ones, about which you have learned very little as yet, are called **protons.** If the electrons and the protons have equal but opposite charges and are present in equal numbers in a normal atom, the fact that the normal atom is electrically neutral is accounted for. You have seen how Rutherford's scattering experiments showed that the positive charge present in an atom has to be concentrated in a tiny central nucleus. The electrons are then somehow spread out around this nucleus and take up most of the volume of the atom. Since the electron has very little mass in comparison to that of an entire atom, it also follows that most of the mass of an atom is to be found in the nucleus.

Rutherford's work established the essential features of the nuclear model of the atom. In the years since, an enormous amount of work, by both chemists and physicists, has helped to refine the model. Obviously, more understanding was needed on two points. (1) What is the nucleus like in detail? (2) How, in detail, are the electrons distributed about the nucleus? The second point is of such basic importance in chemistry that the next several chapters will be devoted to it. The rest of this chapter will summarize the main parts of the answer to the first question, and prepare the ground for a detailed examination of the behavior of the electrons.

Ernest Rutherford (1871–1937)

ERNEST RUTHERFORD probably had a greater influence in the field of atomic science than any other scientist during the past century. A man with a remarkable brain, Rutherford began his professional career during an exciting time in science. In the midst of all the new developments, Ernest Rutherford began research in a variety of fields that eventually led to a whole new concept of the structure of the atom.

Rutherford was born in New Zealand and received his undergraduate education there. After winning a scholarship to Cambridge University in England, he became an assistant to J. J. Thomson, the head of the Cavendish Laboratory. During this period he worked with Thomson on the experiments that led to the discovery of the electron. After three years he went to McGill University in Montreal, Canada, to direct the physics laboratory there. Within a year after the discovery of radioactivity he had discovered that the radiation from uranium was complex, consisting of "soft" rays, which were called alpha rays, and "hard" rays, which were called beta rays. Shortly after this, he and his assistant, Frederick Soddy, found that the alpha rays emitted by the radioactive element thorium were actually positively-charged nuclei of helium atoms. As a result of his continuing work in radioactivity, Rutherford received the Nobel Prize for Chemistry in 1908. His assistant, Soddy, became another outstanding scientist and received the Nobel Prize for Chemistry in 1921.

By 1908 Rutherford had returned to England where he became Professor of Physics at the University of Manchester. It was here that he carried out a series of alpha particle scattering experiments that resulted four years later in his formulation of a "planetary" model of atomic structure, perhaps his most significant work.

Later Rutherford became head of the Cavendish Laboratory at Cambridge and continued in this position until the end of his life. Rutherford's scientific career extended over a period of forty years. At its beginning radioactivity was a newly discovered phenomenon. By the time of his death, the atomic age had begun. In the field of modern science his work is classic and central to the concept of the nuclear atom.

Atomic Numbers. The positive charge of the nucleus is proportional to the number of protons that are present. In discussing electrical charge on the atomic scale, the positive charge of the proton is considered the unit of positive charge. Similarly, the negative charge of the electron is considered to be the unit of negative charge. These two units are of identical size but of opposite sign. Thus a nucleus that contains one proton has a charge of $+1$, and a nucleus that contains x protons has a charge of $+x$.

The number of protons in the nucleus is called the **atomic number** of the atom. Since an atom is electrically neutral, the atomic number is also equal to the number of electrons surrounding the nucleus. All atoms of a particular element have the same atomic number.

Hydrogen has the lightest atoms, with the smallest positive charge, $+1$. That is, there is only one proton in the nucleus of the hydrogen atom. Its atomic number is 1. If we say that a certain atom is a hydrogen atom, or if we say that the atom has an atomic number of 1, we are saying exactly the same thing in two different ways. The one statement implies the other. Similarly, the helium atom has a nuclear charge of $+2$. Its nucleus contains two protons (atomic number 2). Lithium atoms have atomic number 3 (nuclear charge $+3$), and so on. You will see that the reason different kinds of atoms have different chemical properties is, finally, because they have different nuclear charges and hence different numbers of electrons.

10-4 *Atomic Size*

As you learned in Section 10-2, Rutherford proposed that the nucleus has a diameter of about 10^{-12} cm, while an atom as a whole has a diameter of about 10^{-8} cm. The first number came out of his calculations based on the scattering results. Where does the second number come from? How *do* we know the sizes of atoms?

Since we cannot see atoms, the way to answer the question is not at all obvious. Actually, there is no way to answer it for an isolated atom. We can, however, determine how closely two identical atoms may approach one another. Then, if we assume the atoms to be spheres, the distance between the nuclei, when they have approached each other as closely as possible, tells us the diameter of this kind of atom. Even this way of answering the question is not without its difficulties, since the "distance of closest approach" depends on how much force or energy is used to push the atoms toward each other.

For practical purposes, the distance between adjacent nuclei (when an element is in its solid phase) can be considered to be approximately equal to the diameter of its atoms. (See Figure 10-4.) This is something that can be fairly easily determined if you know the atomic mass of the element and Avogadro's number (6.02×10^{23}), as well as the density of the solid substance.

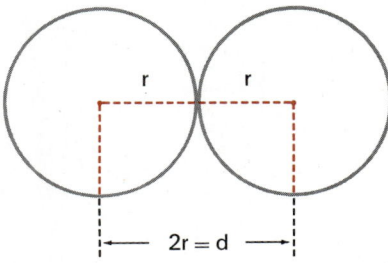

Figure 10-4 *The distance between adjacent nuclei, when an element is in its solid phase, can be considered to be approximately equal to the diameter of its atoms.*

As an example, consider the element silver. Its atomic mass is 108. Its density is 10.5 grams per cubic centimeter. Thus, one cubic centimeter of silver contains

$$\left(6.02 \times 10^{23} \frac{\text{atoms}}{\text{mole}}\right)\left(\frac{10.5 \text{ gm/cm}^3}{108 \text{ gm/mole}}\right)$$
$$= 5.85 \times 10^{22} \text{ atoms/cm}^3.$$

Suppose that n atoms are packed closely and uniformly together so that they fill a cube (Figure 10-5). Then there must be $\sqrt[3]{n}$ atoms along any one edge of the cube, as illustrated in Figure 10-6 (top). Thus, there are

$$\sqrt[3]{5.85 \times 10^{22}} = \sqrt[3]{0.0585 \times 10^{24}}$$
$$= \text{approx. } 0.40 \times 10^8 \text{ atoms/cm.}$$

Finally, if a string of about 0.40×10^8 atoms is one cm long, each atom must have a diameter (Figure 10-6, bottom) of about

$$\frac{1}{0.40 \times 10^8} = 2.5 \times 10^{-8} \text{ cm.}$$

Similar calculations for other atoms show that the sizes of atoms are in the range 1×10^{-8} to 5×10^{-8} cm. Because chemists and physicists so often discuss sizes of atoms and molecules, 10^{-8} cm is used as a unit of length. It is called an **ångström,** abbreviated **Å.**

The nucleus of the atom, however, has a diameter of only about 10^{-12} cm. That is, the diameter of the nucleus is only about 1/10,000 the diameter of the entire atom. If you imagine an atom enlarged to the size of the Houston Astrodome, or Yankee Stadium, then the nucleus would be about the size of an ant—and a small ant at that!

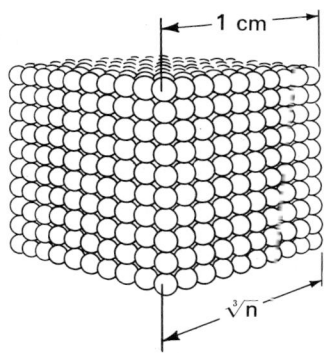

Figure 10-5 *A solidly packed cube of atoms. If its volume is one cm³ and there are n atoms in the cube, then each edge has $\sqrt[3]{n}$ atoms.*

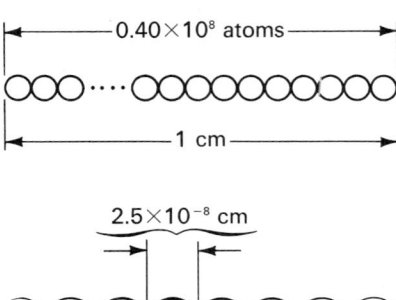

Figure 10-6 *A row of atoms (top), from the cube in Figure 10-5, contains, by calculation, 0.40×10^8 atoms. If the number of atoms in the row is known, the diameter of a single atom (bottom) can be calculated to be 2.5×10^{-8} cm.*

INVESTIGATION

10-5 Measurement of the Size of a Molecule

You have read that it is possible to estimate the size of an atom by measuring the density of an element and calculating the atomic diameter from the number of atoms in one cubic centimeter of the element. You can get a general idea of the size of a molecule by another indirect method.

If you have ever seen gasoline spilled on the surface of some water, you have probably noticed that the gasoline and water do not mix. Instead, the gasoline spreads out on the water in a very thin layer. It has been demonstrated that such a film spreads out on the surface of the water until it is just about one molecule thick.

Suppose you were to spread a known volume of an oil on the surface of some water and then waited for it to distribute itself as far as it would go, so that the layer would be just one molecule thick. You could then calculate the thickness of the layer that would be equal to the length or the diameter of the oil molecules. The calculation is easy since the volume of a solid figure equals the area of the base of the figure times its height or thickness. If the base of the figure is a circle, the formula becomes: Volume = $\pi \times$ radius^2 $\times$ thickness.

If you can assume that the molecules are tiny spheres, then the thickness of the oil layer will be equal to the diameter of the spherical molecules. Since the chances are very good, however, that the molecules are not spherical, the measurements will not give you an exact size for the diameter. But they will give you an indication of the order of magnitude of the dimensions of the molecules you happen to be examining.

In this experiment you will study a compound made of carbon, hydrogen, and oxygen, called oleic acid. This is an oily compound that does not mix with water, but forms a thin film on its surface. Because one drop of pure oleic acid would cover such an enormous area on the surface of the water, it is necessary to use a very dilute solution of oleic acid dissolved in alcohol. The dimensions of the oleic acid film on the water will be indicated by a fine dust of chalk or lycopodium powder spread on the surface of the water. When the oleic acid is put on the water, it will easily push the dust back until it stops spreading.

PURPOSE

You will measure the approximate size of an oleic acid molecule.

MATERIALS

Lycopodium powder or chalk dust   Meter stick
Shallow tray, cafeteria style, approximately 18 $\times$ 24 inches
Oleic acid-alcohol solution in dropper bottle
Graduated cylinder, 10-ml

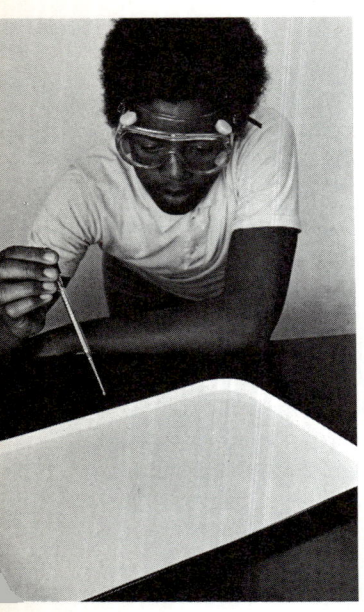

Figure 10-7

PROCEDURE

1. You will be given a small bottle equipped with a medicine dropper and filled with an alcohol solution of oleic acid. The concentration of the solution will be 0.005 cubic centimeter of oleic acid per cubic centimeter of alcohol solution.

2. Fill a large shallow tray with water to a depth of about one centimeter. Be sure the tray is clean before you start. If soap or detergent has been used to clean the tray, be sure that the tray has been completely rinsed before putting the water in it.

3. Sprinkle a tiny amount of lycopodium powder on the surface of the water. If lycopodium powder is not available, hold a chalkboard eraser over the tray of water and scratch or tap some of the chalk dust from the eraser onto the water.

4. In case you might wonder whether the alcohol could have some effect on the experiment, drop one or two drops of alcohol onto the surface of the water on the tray. Is the dust film disturbed? Why?

5. Use the medicine dropper to place a single drop of the oleic acid solution in alcohol in the center of the tray, as in Figure 10-7. The area covered by the spreading film will be more or less circular in shape, when produced with care. Measure the diameter of the film in several different directions. Find the average diameter and calculate the area of the film.

6. Add another drop to the center of the film in the tray. Measure the new diameter and calculate the new area. Repeat with a third drop. How are the three areas related? What can you conclude about the number of molecules lying on the water at any given point? From your results would you expect them to be piled on top of each other or spread out in a single layer? Why?

7. Fill your dropper with some of the alcohol solution of oleic acid. Count the number of drops needed to add exactly one cubic centimeter of the solution to a clean, dry 10-ml graduated cylinder. (Assume that 1 cubic centimeter = 1 milliliter.) Calculate the number of cubic centimeters in one drop of your solution.

DATA AND RESULTS

A table for recording the following data and results should be written in your notebook. Similar tables for the films produced by two drops and three drops of oleic acid should also appear there.

Film with one drop
Diameter 1 (cm)
Diameter 2 (cm)
Diameter 3 (cm)
  Total
Av. Diam. (cm)
Area (cm²)
Area per drop (cm²)

Calculations

1. From the results of steps 5 and 6, find the average area per drop.

2. From the number of drops of solution that you found to be in one cubic centimeter, calculate the number of cm³ of solution in one drop.

$$\frac{1 \text{ cm}^3}{\text{number of drops}} = \text{cm}^3/\text{drop}$$

3. Calculate the number of cm³ of pure oleic acid in one drop of alcohol solution. (cm³ of solution per drop × 0.005 cm³ oleic acid per cm³ of alcohol solution = cm³ of pure oleic acid per drop.)

4. Calculate the thickness of the oleic acid film. The volume of pure oleic acid used is equal to the volume of the film, and the volume of the film is equal to area of film times its thickness.

$$\text{Thickness of film} = \frac{\text{Volume of pure oleic acid}}{\text{Area of film}}$$

If the oleic acid molecule were spherical, the thickness of the film should be equal to the diameter of the molecule.

RELATED PROBLEMS

1. The formula for oleic acid is $C_{17}H_{33}COOH$. In what possible shapes could you imagine this molecule to exist?

2. In the introduction to this experiment you read that this experiment would give you an indication of the order of magnitude of the dimensions of the molecules you would be examining. What is meant by order of magnitude?

3. Based on the results of your experiment, what would be the area covered by one cubic centimeter of pure oleic acid if it were spread out in a film one molecule thick? If the film were circular, what would be its diameter in centimeters? in meters?

4. Suppose that the oleic acid molecule is not spherical, as you assumed for your calculations, but, instead, is built up of a long chain or string of 18 carbon atoms, with the hydrogen atoms distributed fairly evenly throughout the molecule in such a way that the molecule is about eight times as long as it is wide. On the basis of the measurements that you made in this experiment, what would be the dimensions of the molecule if you assumed that these long molecules "stand up" in the water so that their long dimension is perpendicular to the surface of the water?

10-6 Neutrons

Protons are in the nucleus and electrons surround it. Most of the mass of the atom is in the nucleus. These two statements imply that an electron has far less mass than a proton and this is, in fact, the case. Experiments can be performed in which the actual masses of individual protons and electrons are obtained. These experiments show that the proton has a mass about 1840 times that of the electron.

It is clear, then, why the nucleus furnishes practically all of the mass of an atom. However, except in the hydrogen atom, protons alone do not account for all of the mass of atomic nuclei. For example, a helium atom (atomic number 2) has two protons, while a hydrogen atom has one. Yet, experiments show that the helium atom has four times the mass, rather than twice the mass, of the hydrogen atom. Again, the fluorine atom has atomic number 9 and yet it has 19 times the mass of the hydrogen atom.

These and similar facts suggested that nuclei of atoms must contain some additional particle, or particles, having mass but no charge. In 1932, direct experimental evidence for such a particle was found. The particle is called a **neutron** and it has a mass almost identical to that of the proton.

It is now possible to explain in a simple and consistent way the masses of atoms. In the helium nucleus there are two neutrons as well as two protons, so its total mass is about four times the mass of one proton. In the fluorine nucleus there must be ten neutrons in addition to the nine protons, making a total mass about 19 times that of a single proton.

The total number of nuclear particles, neutrons plus protons, is called the **mass number** of an atom. For instance, the normal helium atom with two protons and two neutrons has a mass number of 4.

10-7 Basic Characteristics of the Nuclear Atom

The basic characteristics of the nuclear atom can now be summarized. Each atom is built up of a number of protons equal to its atomic number and the number of neutrons required to give, along with the protons, the atomic mass. The electrical neutrality and size of the atom are provided by the electrons, which are equal in number to the atomic number.

There is only one difficulty with this model. Why is it that two or more protons packed closely together in the nucleus do not exert such strong repulsive forces upon one another as to make the nucleus unstable? This is a very fundamental question, perhaps the most fundamental question that can be asked concerning the nature of matter. Nu-

TABLE 10-1 Some Fundamental Particles

Particle	Charge (electron units)	Approximate Mass (relative to proton mass)
Electron	−1	1/1840
Proton	+1	1
Neutron	0	1

clear physicists are grappling with it today. In a general way, the answer must be that the neutrons somehow serve to bind the nucleus together despite the repulsive forces between protons. There is no known nucleus that contains two or more protons that does not contain at least one neutron, and most stable nuclei contain more neutrons than protons. A detailed model for the binding role of the neutrons has yet to be devised.

The main properties of the fundamental particles composing atoms are summarized in Table 10-1.

10-8 Isotopes

All of the atoms of an element have the same nuclear charge. Do all of the atoms of an element have the same mass? Let us first consider a single element, hydrogen. Almost all hydrogen atoms do have the same mass: the sum of the mass of one proton and the mass of one electron. For these atoms the nucleus consists of a single proton. However, a small fraction of the hydrogen atoms (0.016 percent of them) have nuclei with masses approximately twice as great as the mass of the proton. To explain the mass of these hydrogen atoms, it is concluded that each of their nuclei consists of a neutron (charge zero, mass 1) and a proton (charge +1, mass 1). This kind of hydrogen atom is called hydrogen-2. It is also called deuterium. The two kinds of hydrogen atoms (having the same atomic number but different masses) are called isotopes. An isotope is designated by first specifying which element it is (usually by the symbol or name of the element) and then by the sum of the number of protons and the number of neutrons, that is, the mass number.

Isotopes are atoms of the same element that have different numbers of neutrons and thus different mass numbers. They are atoms of the same element because they have the same atomic number (number of protons). On the next page, Table 10-2 lists some common isotopes.

As implied by the percentage mentioned above, out of every 6250 atoms of hydrogen in nature only one is deuterium. This atom, although it is, chemically, a hydrogen atom, is often given the distinctive symbol D. In a sample of normal water there will be some molecules that contain a D atom and even a very few that contain two D atoms. About one molecule in 6,000 will be DHO instead of H_2O. Only about one molecule in 3.6×10^7 will be D_2O. It is not surprising that definite evidence for the isotope D and for the molecules D_2, DH, DHO, and D_2O was not obtained until 1932.

In that year an American chemist, Harold Urey, announced that he and his co-workers had proved the existence of D atoms. In the years since, a method of distilling water has

TABLE 10-2 Vital Statistics of Some Common Isotopes

Name	% Abundance in Nature	Atomic Number	Mass Number	Nucleus Composition*		Mass	Charge	Number of Electrons
Hydrogen-1	99.984	1	1	1p,		1	+1	1
Hydrogen-2	0.016	1	2	1p,	1n	2	+1	1
Helium-3	1.34×10^{-4}	2	3	2p,	1n	3	+2	2
Helium-4	100	2	4	2p,	2n	4	+2	2
Lithium-6	7.40	3	6	3p,	3n	6	+3	3
Lithium-7	92.60	3	7	3p,	4n	7	+3	3
Beryllium-9	100	4	9	4p,	5n	9	+4	4
Boron-10	18.83	5	10	5p,	5n	10	+5	5
Boron-11	81.17	5	11	5p,	6n	11	+5	5
Carbon-12	98.892	6	12	6p,	6n	12	+6	6
Carbon-13	1.108	6	13	6p,	7n	13	+6	6
Nitrogen-14	99.64	7	14	7p,	7n	14	+7	7
Nitrogen-15	0.36	7	15	7p,	8n	15	+7	7
Oxygen-16	99.76	8	16	8p,	8n	16	+8	8
Oxygen-17	0.04	8	17	8p,	9n	17	+8	8
Oxygen-18	0.20	8	18	8p,	10n	18	+8	8
Fluorine-19	100	9	19	9p,	10n	19	+9	9
Chlorine-35	75.4	17	35	17p,	18n	35	+17	17
Chlorine-37	24.6	17	37	17p,	20n	37	+17	17
Gold-197	100	79	197	79p,	118n	197	+79	79
Uranium-234	0.01	92	234	92p,	142n	234	+92	92
Uranium-235	0.71	92	235	92p,	143n	235	+92	92
Uranium-238	99.28	92	238	92p,	146n	238	+92	92

*p = proton, n = neutron

TABLE 10-3 Some Properties of Ordinary and Heavy Water

	H_2O (0.016% D)	D_2O
Density, g/cm³ (at 20°C)	0.9982	1.105
Freezing point, °C	0.00	3.82
Boiling point, °C	100.00	101.42

been engineered so carefully that it is possible to obtain, by the ton, so-called *heavy water*. This is essentially pure (99.9%) D_2O. Since the molar mass of D atoms is 2.0, the molar mass of D_2O is $(2 \times 2.0) + 16.0 = 20.0$. If we assume that the D_2O molecules crowd together just as do H_2O molecules, which have a molar mass of 18.0, the density of heavy water should be about 20/18 or 1.11 times that of ordinary water. As Table 10-3 shows, this is approximately correct. Table 10-3 also shows that heavy water and ordinary water differ slightly in other properties, such as freezing and boiling points.

Summary

According to the nuclear model, an atom consists of a nucleus surrounded by one or more electrons. The electrons carry all the negative charge and occupy essentially the entire volume, but furnish only a tiny fraction of the mass. At the center of this electron cloud is the nucleus, which consists of protons and neutrons. The protons carry all of the positive charge and together with the neutrons they furnish practically all of the mass of the atom. The masses of neutrons and protons are approximately equal, and the total number of protons and neutrons is called the mass number of the atom. The atomic number of the atom is the number of protons in the nucleus. It determines the chemical identity of the atom.

Atoms that have the same number of protons but different numbers of neutrons are called isotopes. They are atoms of the same element that differ by one or more mass units.

Self Check

1. What is the difference between mass number and atomic number?
2. Differentiate among protons, neutrons, and electrons. Which is the only atom that has no neutron?
3. What is the approximate order of magnitude of atomic diameters ($10^{-?}$ cm). What is the name given to this unit?

Atomic Nuclei May Decompose

10-9 *Radioactivity*

Not all atomic nuclei are stable. Unstable nuclei give off radiations of several types. These radiations were discovered toward the end of the 19th century. Their discovery initiated the entire study of atomic and nuclear structure that has led to an understanding of nuclear energy and the ability to use it.

When a substance contains nuclei that

Harold C. Urey
(1893–   )

HAROLD UREY HAS WORKED on a wide variety of problems in the fields of nuclear energy, the formation of planets, the geology of the surface of the moon, and the origin of life. He first gained prominence for his discovery of deuterium, a heavy isotope of hydrogen represented by the symbol $^2_1$H.

The concept of isotopes had first been proposed in 1913 by Frederick Soddy, a coworker of Ernest Rutherford. The isotopes of a number of elements had been discovered by 1930, but F. W. Aston, who had developed one of the first mass spectrographs, reported in 1927 that his measurements showed no trace of an isotope of hydrogen.

In 1931, scientists reviewing Aston's data felt that some of the measurements suggested the possibility of the existence of one part of $^2_1$H to 4500 parts of $^1_1$H. After considering this information, Urey calculated the theoretical vapor pressure of the suspected deuterium as compared to normal hydrogen. He concluded that there should be enough difference in the vapor pressures to make possible the separation of $^2_1$H from $^1_1$H by the fractional distillation of liquid hydrogen.

Between August and late November, 1931, Urey, along with George Murphy, one of his graduate students, and F. G. Brickwedde, carefully distilled and redistilled large quantities of hydrogen, saving the small amounts of the distillate that should contain the suspected isotope. When they placed this gas in an electric discharge tube and examined its spectrum, they found some very faint lines in the exact positions predicted for $^2_1$H, thus confirming the existence of deuterium. The amount of deuterium was found to be one atom of $^2_1$H to about every 6000 atoms of $^1_1$H. The most amazing part of the entire sequence of events is the fact that Aston's data, which had started Urey on the search for deuterium, was later found to be erroneous.

spontaneously decompose, giving off radiation, the substance is said to be **radioactive**.

Nearly every unstable nucleus gives off one or more of the following three kinds of radiation.
1. α (alpha) particles. These are identical to the nuclei of helium atoms. They consist of two protons and two neutrons.
2. β (beta) particles. These are electrons, ejected from the *nucleus* at varying speeds.
3. γ (gamma) rays. These are not particles, but a form of short-wave radiation similar to, but more energetic than, X rays. They usually accompany the α and β particles and are, in general, emitted whenever any significant change in the structure of a nucleus occurs. The emission of only γ rays provides a way for a nucleus to rid itself of excess energy without changing the number of neutrons or protons present.

We shall not consider the reasons why some nuclei are unstable and thus give rise to radioactivity. We shall examine only the results of radioactive decay, that is, how the identity of the atom changes when such decay occurs. Any change in the nucleus that changes the number of protons changes the identity of the atom. Both α-decay (the emission of α particles) and β-decay (the emission of electrons) cause a change in the atomic number.

When a nucleus emits an α particle, it loses two protons and two neutrons simultaneously. It is therefore converted into another element with an atomic number two units less and a mass number four units less.

This process can be expressed in symbols similar to those used to write chemical equations. For each element, you begin with the standard chemical symbol, expressing its mass number as a superscript and its atomic number as a subscript. Thus for the isotope of uranium (atomic number 92) of mass number 235, you write $^{235}_{92}U$. Similarly, you represent the α particle by $^{4}_{2}He$, since it is actually a helium nucleus. Strictly speaking you ought to write $^{4}_{2}He^{2+}$, since the helium nucleus alone is an ion of charge 2+. However, in equations describing nuclear changes it is customary to neglect the changes in outer electrons. Ultimately, any ions formed gain or lose electrons, as necessary, to form neutral atoms.

If $^{235}_{92}U$ undergoes an α-decay, it becomes an isotope with mass number 231 of the element with atomic number 90. The element with atomic number 90 is thorium, Th. Hence, the α-decay of uranium can be described by the following nuclear equation.

$$^{235}_{92}U \longrightarrow {}^{231}_{90}Th + {}^{4}_{2}He$$

Strictly speaking, the symbols for the elements are unnecessary, since the atomic numbers indicate what element is involved. However, the symbols are nearly always used as in the above equation.

In β-decay, the nucleus emits an electron. How can this be if there are no electrons in the nucleus? It is now known in detail how the electron is generated. You may, however, simply suppose that in some way a neutron breaks down into a proton, which remains in the nucleus, and an electron, which is ejected. Thus, when a nucleus undergoes β-decay, its atomic number increases by one unit but its mass number does not change (one neutron lost, but one proton gained). Cobalt-60, $^{60}_{27}Co$, is an isotope that decomposes by β-decay. The process leads to the isotope of mass number 60 of the element with an atomic number one unit greater, nickel. The nuclear reaction can be written as follows.

$$^{60}_{27}Co \longrightarrow {}^{60}_{28}Ni + {}^{0}_{-1}e$$

10-10 Nuclear Reactions and Nuclear Energy

The fact that nuclei remain intact during chemical reactions indicates that the nuclear binding forces and energies must be far greater than the forces and energies involved in chemical reactions. Experiments show that nuclear forces and energies are about a million times greater than chemical forces and energies. This is the reason for the great interest in nuclear reactions as a source of energy.

The energy of nuclear reactions has been popularly called "atomic energy." This is not a very accurate term, since it could equally well describe any kind of energy derived from atoms which, of course, includes the energy of chemical changes such as burning gasoline. The proper term for the enormous energies available in nuclear reactions is nuclear energy.

The two forms of radioactive decay already described, α-decay and β-decay, are two kinds of nuclear reactions. They occur spontaneously in certain nuclei that are inherently unstable. There are other nuclear reactions that are made to occur under certain conditions. An important example is nuclear fission. This is the kind of nuclear reaction responsible for the operation of atomic reactors, such as the one shown in Figure 10-8, and atomic bombs.

A nuclear reaction of the fission type can be represented by the following equation.

$$^{235}_{92}\text{U} + ^{1}_{0}\text{n} \longrightarrow ^{141}_{56}\text{Ba} + ^{92}_{36}\text{Kr} + 3^{1}_{0}\text{n} + \text{energy}$$

This means that when a neutron, represented by $^{1}_{0}\text{n}$ (mass number 1, charge 0), penetrates a $^{235}_{92}\text{U}$ nucleus, the resulting instability causes the nucleus to split into two main pieces, releasing several new neutrons plus energy. In one of the many possible

Figure 10–8 *Cutaway view of a research reactor in Idaho, showing the loading of fuel into its "south half." The two halves are brought together to start a self-sustaining fission reaction.*

fissions, the main pieces are atoms (or ions) of barium, Ba, and krypton, Kr, here the isotopes $^{141}_{56}\text{Ba}$ and $^{92}_{36}\text{Kr}$. The energy released when a mole of uranium-235 undergoes fission ("fishes," as nuclear scientists say in their slang) is about 10 million times as great as the energy released when a mole of carbon is burned!

10-11 Calculating the Energy of Nuclear Reactions ★

The quantities of energy inherent in atomic nuclei and released in certain nuclear reactions are so enormous that they have a measurable mass equivalent. It can be calculated with the use of Einstein's equation, $E = mc^2$. (Review Section 3-10, if necessary.)

By means of especially accurate mass spectrographs, similar to the one described in Section 9-8, it is possible to measure the masses of individual atoms very precisely. It

TABLE 10–4 Accurate Masses of Some Atoms and the Neutron

Atom or Particle	Mass (g/mole)
^3_1H (tritium)*	3.01605
^2_1H (deuterium)	2.01410
^4_2H (helium)	4.00260
^1_0n (the neutron)	1.00867

*This is the third known isotope of hydrogen. It is radioactive and occurs only in minute traces in nature. It can be made in atomic reactors.

is known, for example, that the masses of some atoms and the neutron are those shown in Table 10-4.

A possible nuclear reaction of the type that actually goes on in the sun and stars is the following *fusion* reaction.

$$^3_1H + ^2_1H \longrightarrow ^4_2He + ^1_0n$$

Total masses   5.03015   5.01127

It is seen that the total masses on the two sides of the equation are *not* identical. Evidently, some mass has been lost. It must appear as energy, in order that the Law of Conservation of Mass and Energy be obeyed. From Einstein's equation, we can determine exactly how much energy will be released when one mole of ^3_1H combines with one mole of ^2_1H to form a mole of ^4_2He.

The decrease in mass is:
$$5.03015 - 5.01127 = 0.01888 \text{ g/mole.}$$

$$\begin{aligned} E &= mc^2 \\ &= (1.888 \times 10^{-5} \text{ kg}) \times (3.00 \times 10^8 \text{ m/sec})^2 \\ &= 1.70 \times 10^{12} \text{ joules} \\ &= 4.06 \times 10^{11} \text{ calories} \end{aligned}$$

This is about 20,000 times the energy obtained by burning a ton of coal. There is a very practical reason why the above nuclear fusion reaction, or a similar one, is not actually used to generate energy. Scientists do not yet know how to get the energy out in a controlled way. (One of the techniques that scientists are currently investigating for this purpose is illustrated in Figure 10-9.) The hydrogen bomb is an example of an uncontrolled nuclear fusion reaction (very similar to the one above). The sun itself generates its energy by nuclear fusion reactions. One of the great challenges to engineers and scientists is the controlled generation of energy by nuclear fusion.

Nuclear fission can also generate energy, and it is known how to control this process. The trouble with nuclear fission is simply that it produces a lot of highly radioactive matter which can be dangerous, and difficult to dispose of safely.

Figure 10–9 A 5-meter arc segment of the Scyllac controlled thermonuclear research device in New Mexico. The completed device will be doughnut-shaped and should produce temperatures of $5 \times 10^7 K$.

Chemistry and Our Environment
NUCLEAR ENERGY

ALREADY IN THE WESTERN part of the world, and increasingly everywhere, the patterns of human life depend on the production of vast amounts of energy. A minor fraction of this energy can be derived directly from nature, using dams, waterfalls, tides, underground steam, and solar heat as well as the muscles of man and various domestic animals. But nearly all of our energy is generated by more elaborate processes. Until very recently, these processes have all been based on the type of chemical reaction called combustion. Most electric-generating plants burn coal, buildings are heated by burning coal, gas, or oil, and most vehicles are powered by burning gasoline. In other words, man now depends mainly on burning fossil fuels—coal, gas, and petroleum—to generate energy. The supplies of these substances must have some definite limit. One day, the supplies will be exhausted. It is difficult to say exactly when because we do not know exactly how rapidly the rate of consumption will increase in the future, nor do we know for certain the total amount of all the available reserves. Still, it seems safe to say that fossil fuels can only last for perhaps one more century, unless we revert to a pre-industrial way of life very soon.

Even if the supply of fossil fuels were limitless, there are serious pollution problems associated with their use. As you have already seen (*Smog*, Chapter 6) much of the fuel used in generating plants contains small amounts of sulfur or sulfur compounds. This enters the atmosphere as SO_2.

For reasons such as those just discussed, the generation of electric power from nuclear energy has appeared increasingly attractive to many people. There are basically two ways to do this: fusion and fission. Both processes depend on the fact that nuclei of intermediate atomic numbers are more stable than those of very high or very low numbers. Thus when large nuclei, such as uranium, break up into several nuclei of intermediate size, energy is released. This is the basis of obtaining energy by nuclear fission. When very light nuclei combine to form heavier ones, energy is also released. This is the basis for energy generation by nuclear fusion.

At the present time practical discussion of nuclear energy is concerned with fission. Although fusion processes are in principle very attractive, the engineering problems associated with making them practical have not been solved. In order to bring the light nuclei together enormous energy must first be supplied. After the nuclei fuse, all of this energy will be given back, plus a great deal more. However, the design of a reactor in which the fusion can be made to occur and the energy then released converted into a useful form of thermal energy is still to be developed. It is not entirely certain at this time whether such a development will be possible, although proponents of fusion power have predicted that they will have solved their problems in principle by the 1980's and that there will be operating power plants by the 1990's.

Our discussion here is concerned primarily with fission reactors. The engineering feasibility of these is already proven. There are some already in use for power generation in the United States, England, and elsewhere. Small reactors of this type power the so-called nuclear submarines.

As you have seen in Section 10-10, the fission of uranium-235 releases a great deal of energy. There are also other nuclei, such as plutonium-239, that work satisfactorily as nuclear fuel. Furthermore, it is technically possible to operate a ''breeder'' reactor. This is a device in which the fission of ^{235}U or ^{239}Pu will not only generate energy but cause ^{238}U present to be converted into ^{239}Pu, thus breeding new fuel. Technically there appears to be no doubt that energy generation by nuclear fission is practical.

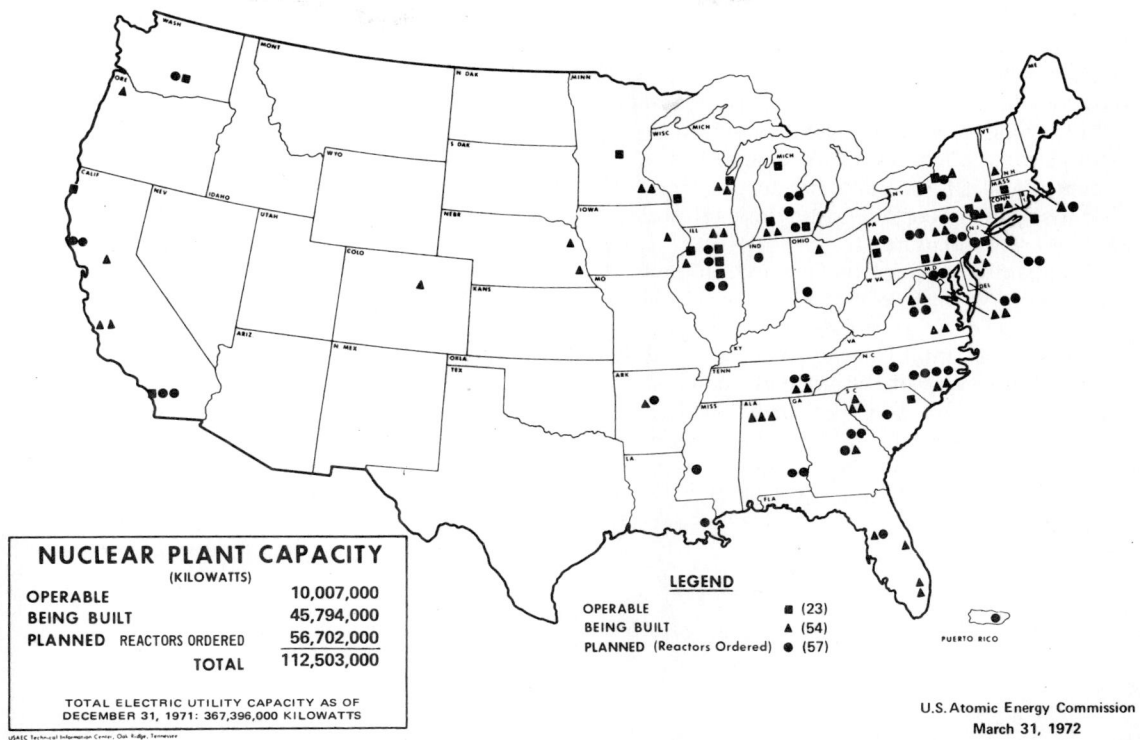

The locations of existing and planned nuclear power reactors in the United States in March, 1972.

The hard question that all citizens must consider has to do with the environmental impact of nuclear power generation, as compared to that of conventional power.

The following are advantages of nuclear power.
1. The procurement of fuel does not do any great violence to the environment, as compared, for example, to strip-mining of coal.
2. Atmospheric pollution is almost nil, as compared to the pollution resulting from burning of conventional fuel.
3. With the use of breeder reactors, it can be predicted that known fuel supplies will last for as much as 50,000 years.

The following are disadvantages of nuclear power generation.
1. Nuclear power plants cause more thermal pollution. That is, the amount of waste heat per unit of power generated is higher in a nuclear plant than in a conventional plant.
2. Nuclear fission produces highly radioactive "hot" wastes, which must be safely stored for thousands of years until they "cool off." It is not easy to do this in a manner that is certain to be safe for such long periods of time.

   Radioactive substances are extremely dangerous to all forms of life. The β and γ rays, particularly, destroy delicate tissues, such as bone marrow and red blood cells. High levels of radioactivity can cause death very quickly. Lower levels, which might arise from slow leakage or diffusion of stored wastes, could be the cause of cancer, anemia, and birth defects.
3. There is always some danger that a nuclear reactor will leak, overheat, or in some other way release radioactivity to the environment at a dangerous level.

As you can see, there are risks and disadvantages each way. The question of how far and how fast we should proceed with nuclear power plants, instead of increasing our generating capacity by conventional means, is not a simple one. To a high degree, the problem arises because we live in a very energy-demanding way. Population is still increasing rapidly and the per capita consumption of electricity is increasing even more rapidly. These two factors produce a steeply rising demand for more electricity.

There are those who respond to the problem *not* by deciding which direction we should take in future power generation. Instead they suggest that we try to bring both population growth and our demand for more and more power under control. This is a suggestion well worth consideration, too.

Summary

Naturally-unstable atomic nuclei undergo radioactive decay, emitting α or β particles and γ rays. α-decay consists of the emission of an α particle, or helium nucleus, and reduces the atomic mass by 4 and the atomic number by 2. In β-decay an electron is ejected from the nucleus. The result is an atom of the same mass but with an atomic number greater by 1. When γ rays are emitted from the nucleus, there is no change in atomic mass or number, but there is a release of energy.

Nuclear forces and energies are much greater than chemical forces and energies. For this reason nuclear fission, the deliberate splitting of an atomic nucleus, is a subject of great interest. It consists of injecting a neutron into the nucleus of an atom of high atomic mass and number to produce several atoms of lower atomic number, additional neutrons, and energy.

Self Check

1. What is given as the explanation for the emission of γ rays?
2. $^{214}_{84}$Po decays spontaneously by emitting α particles. Describe this by means of a nuclear equation.
3. $^{210}_{83}$Bi decays spontaneously by emitting β particles. Describe this by means of a nuclear equation.
4. List four different kinds of nuclear reactions.

The Golden Age of Atomic Discoveries (1895–1940)

Marie (1867–1934) and Pierre (1859–1906) Curie shared the Nobel Prize in Physics with Becquerel in 1903. In 1911, Marie was awarded the Nobel Prize in Chemistry.

IT IS HARD to know where to begin a discussion of the fantastic events that have taken place in the field of atomic and nuclear structure during the past seventy-five years. The moment you try to define the start of a "Golden Age" with the work of a certain scientist, you find at once that his work was based on the work of someone else who may have worked one year, or ten years, or fifty years earlier.

Perhaps the greatest explosion of knowledge in the field of atomic structure began with the discovery of X rays by William Roentgen (RENT-gen) in 1895. Roentgen was actually following up on the work begun as early as 1859 by Geissler (GYS-ler) and Plücker (PLIK-er) who had discovered cathode rays in gaseous discharge tubes. A number of other scientists had worked with these discharge tubes in the years between 1859 and 1895 and had undoubtedly produced X rays, but Roentgen was the first one to have noted the fluorescence produced by X rays on substances outside the tubes.

The discovery of radioactivity by Becquerel (*Bek*-uh-REL) in France one year later was a direct result of Roentgen's work. Becquerel suspected that the glowing of a piece of uranium ore in the dark after it had been exposed to sunlight might be caused by the formation of X rays by the action of the sunlight on the ore. His discovery that uranium ore produced radiations that exposed photographic plates even when the ore was not exposed to sunlight was the first time that anyone had observed radioactivity.

It did not take Becquerel long to discover that there apparently were several sources of radioactivity in his uranium ore. He assigned Pierre and Marie Curie, (KUR-ee), two of his graduate students, the task of identifying these sources. Within two years, they discovered two new elements, radium and polonium.

In the meantime, Rutherford had discovered alpha rays and beta rays coming from uranium. Becquerel and others discovered that these rays could be deflected by magnets. By 1900, Becquerel had proved that beta rays were identical to the particles in cathode rays, which were later shown to be electrons. Nine years later Rutherford showed that alpha particles were positively charged helium atoms.

The probing into atomic structure proceeded along several different lines. Rutherford and others used the radiations from radioactive sources to bombard a variety of materials. This culminated in Rutherford's alpha particle scattering experiments that led to the concept of the nuclear structure of the atom in 1911.

J. J. Thomson during this period made his measurements of the ratio of the charge to the mass of the electron, and R. A. Millikan successfully measured the charge on the electron with his "oil drop" experiments in 1910.

While all of this was going on, a number of other scientists were exploring the "Edison" effect, which described the emission of electrons from hot filaments. Still other scientists examined the photoelectric effect caused by the action of light on certain metals

In 1884, a Swiss school teacher named Balmer examined the lines of the hydrogen spectrum and found that there was an important mathematical relationship for the positions of the lines. (See Chapter 11.) This opened up a new field of research. For a number of years other scientists studied the spectra of many different elements under a variety of conditions. Their research revealed several other regular mathematical series describing the position of the lines on the spectrographic plates. These relationships were used later to help interpret the position of electrons around the nucleus of an atom.

M. S. Livingston (left) and Ernest O. Lawrence (right), standing beside the magnet for one of the earliest cyclotrons.

J. J. Thomson (1856–1940), at work in the Cavendish Laboratory.

By 1913, Niels Bohr of Denmark was able to propose the first successful theory of atomic structure, based on a densely packed nucleus, as described by Rutherford, and an electron circling the nucleus in a series of orbits. As more experimental data accumulated in following years, many modifications were made in Bohr's original ideas. Bohr's atomic model is discussed in Chapter 11.

In the years between 1907 and 1912, J. J. Thomson expanded his work on the measurement of the charge to mass ratio from measurements on electrons alone to measurements on all kinds of positive ions. This marked the start of the science of mass spectrometry and made possible the identification of isotopes of all the elements.

A very crude counter, first invented by Geiger (GY-ger) in 1908 and then perfected with Mueller in 1928, provided one way of detecting radiation. A British scientist, C. T. R. Wilson, invented the first expansion cloud chamber in 1911. This device, which is used for observing the tracks of fast-moving

Irène Curie and her husband, Frédéric Joliot. They shared the 1936 Nobel Prize in Chemistry.

charged particles, proved to be one of the most important instruments ever developed for use in nuclear physics.

Rutherford's bombardment of nitrogen with alpha particles, in 1919, resulted in the first man-made transmutation of an element. Soon other scientists began to search for more powerful tools to use in bombarding elements. This resulted in the first reliable, high voltage, electrostatic generator, developed by Robert J. Van de Graaff, in 1931, in the United States. In 1932 E. O. Lawrence and M. S. Livingston, also in the United States, constructed the first cyclotron. In the same year, Cockcroft and Walton in England produced a nuclear reaction for the first time with a high voltage accelerator.

The neutron, the existence of which Rutherford had predicted 12 years earlier, was discovered by Chadwick, in England, in 1932. The neutron provided scientists with an uncharged particle that did not need a high voltage accelerator for penetrating the nu-cleus of an atom. Within two years Irène Joliot-Curie and Frédéric Joliot (Zhaw-LEEOH) in France found that they could make elements radioactive by exposing them to neutrons. Further work by these same scientists, in addition to work by Enrico Fermi in Italy, Otto Hahn, Fritz Strassmann, Lise Meitner (MYT-ner), and Otto Frisch in Germany, resulted in the discovery of the fission of uranium in 1939. Glenn Seaborg and Edwin McMillan produced and isolated the synthetic elements neptunium and plutonium in 1940.

It would take much too long a time to describe all of the scientific discoveries in the field of atomic structure made during the period from 1895 to 1940. The main purpose of this discussion has been to show how scientists have used the work begun by others as a starting place for their own research and to illustrate how many different types of knowledge have been brought together to give us our present concept of atomic structure, which surely will be modified in the years to come.

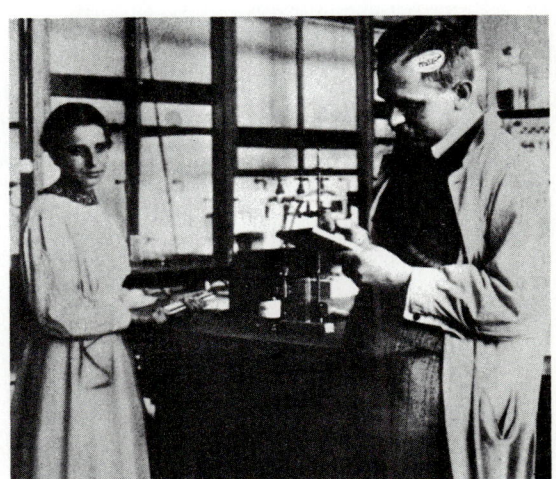

Lise Meitner and Otto Hahn, research collaborators for 30 years.

Chapter Summary

Atoms are all built on a basic plan represented by the nuclear model. They include a tiny nucleus made up of protons and neutrons. Neutrons are neutral particles with almost the same mass as protons. The only atom with no neutron in its nucleus is the most common hydrogen isotope. Its nucleus consists of a single proton. All other nuclei contain at least one neutron in addition to one or more protons. The volume of the atom is taken up mainly by the electrons that surround the nucleus. As many electrons surround the nucleus as there are protons within it, making the atom as a whole neutral.

The diameters of atoms are of the order of one ångström (Å) unit, which is 10^{-8} cm.

The number of protons (and hence, of electrons) determines what element the atom belongs to and is called the atomic number of the atom or element. The total number of protons and neutrons in the nucleus determines, approximately, the mass of the entire atom, since protons and neutrons weigh 1840 times as much as an electron. This total number of nuclear particles is called the mass number of the atom.

Atoms with more or less electrons than the number of protons in the nucleus have, respectively, a negative or positive charge and are called ions.

Atoms that have the same number of protons (and so, are atoms of the same element) but have different numbers of neutrons are called isotopes.

Processes in which nuclei are changed involve energies about a million times as great as those of chemical changes. One class of nuclear processes includes the spontaneous decay of unstable nuclei, in which α particles (helium nuclei), β particles (fast electrons), or γ rays may be emitted. Another type of nuclear reaction is fission, in which a neutron enters a nucleus and splits it into fragments, causing the release of neutrons and energy.

Questions and Problems

1. What is an alpha particle? How can alpha particles be produced for observation?
2. Why was it necessary for Rutherford's scattering experiments to be carried out in a vacuum?
3. What three basic conclusions can be drawn from the observations made in Rutherford's scattering experiments?
4. What observation during Rutherford's scattering experiments was consistent with Thomson's model of the atom? What was inconsistent with Thomson's model? Why?
5. Why did Rutherford conclude that the radius of the nucleus of an atom is only one ten-thousandth of the radius of the atom itself? What made him believe that the nucleus is so densely packed?
6. What do protons and neutrons have in common? In what ways are they different?
7. What is the difference between the atomic number and the mass number of an atom?
8. If an electrically neutral atom has an atomic number of 10, which of the following statements must be true about the atom? Briefly explain your reasoning.
   (a) It has 10 neutrons.
   (b) There are 10 electrons surrounding the nucleus.
   (c) There are 10 protons in the nucleus.
   (d) The atomic mass of the atom is 10.
   (e) The atomic mass of the atom is 20.
9. An atom of one isotope of uranium has an atomic number of 92 and a mass num-

ber of 238. How many electrons does this atom contain? Approximately what fraction of the total mass of this atom is made up of electrons?

10. How do isotopes of one element differ from each other? How are they the same?

11. Compare the properties of alpha radiation, beta radiation, and gamma radiation produced by radioactive substances.

12. In Figure 10-10 a radioactive source is placed in the bottom of a deep hole drilled in a lead block. The source produces alpha, beta, and gamma radiation, all of which come out of the hole in a straight line into a sealed chamber containing a vacuum. When these rays pass between two oppositely charged metal plates, it is found that the beta rays are deflected and follow a curved path as shown by the dotted line. Make a simple copy of the diagram. Sketch in the paths that you would expect to be followed by the alpha and the gamma radiation under the same conditions. Explain.

13. The element radium, Ra, ejects an alpha particle during its radioactive decay. An isotope of bismuth, Bi, ejects a beta particle. Complete the following two equations. First determine the atomic number and the mass number of the products formed. Identify the products by locating their respective atomic numbers on a periodic table.

    (a) $^{226}_{88}\text{Ra} \longrightarrow {}^{4}_{2}\text{He} + \underline{\qquad}$

    (b) $^{214}_{83}\text{Bi} \longrightarrow {}^{0}_{-1}e + \underline{\qquad}$

14. A fission reaction is started by a neutron entering the nucleus of a uranium atom. Why is it much easier for a neutron to enter an atomic nucleus than it would be for a proton to enter a nucleus?

15. List the following properties for an atom of each of the two naturally-occurring isotopes of carbon, $^{12}_{6}\text{C}$ and $^{13}_{6}\text{C}$: (a) number of protons; (b) number of neutrons;

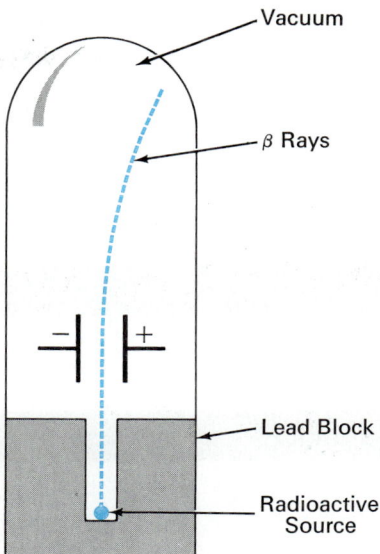

Figure 10-10

(c) number of electrons; (d) nuclear charge.

16. An average dimension for the diameter of a nucleus is 1×10^{-12} cm and for the diameter of an atom, 1×10^{-8} cm. Determine the ratio of atomic volume to nuclear volume.

17. Assume that the nucleus of a fluorine atom is a sphere with a radius of 5×10^{-13} cm. If fluorine has atomic number 9 and mass number 19, calculate the density (in grams per cubic centimeter) of matter in the fluorine nucleus.

18. The density of copper metal is approximately 8.9 grams per cubic centimeter. The molar mass of copper is 63.54 grams per mole.

    (a) How many atoms are in one cubic centimeter of copper?

    (b) How many atoms would be on one edge of a one centimeter cube of copper?

    (c) What is the approximate diameter of a copper atom?

11 Interactions Between Light and Atoms

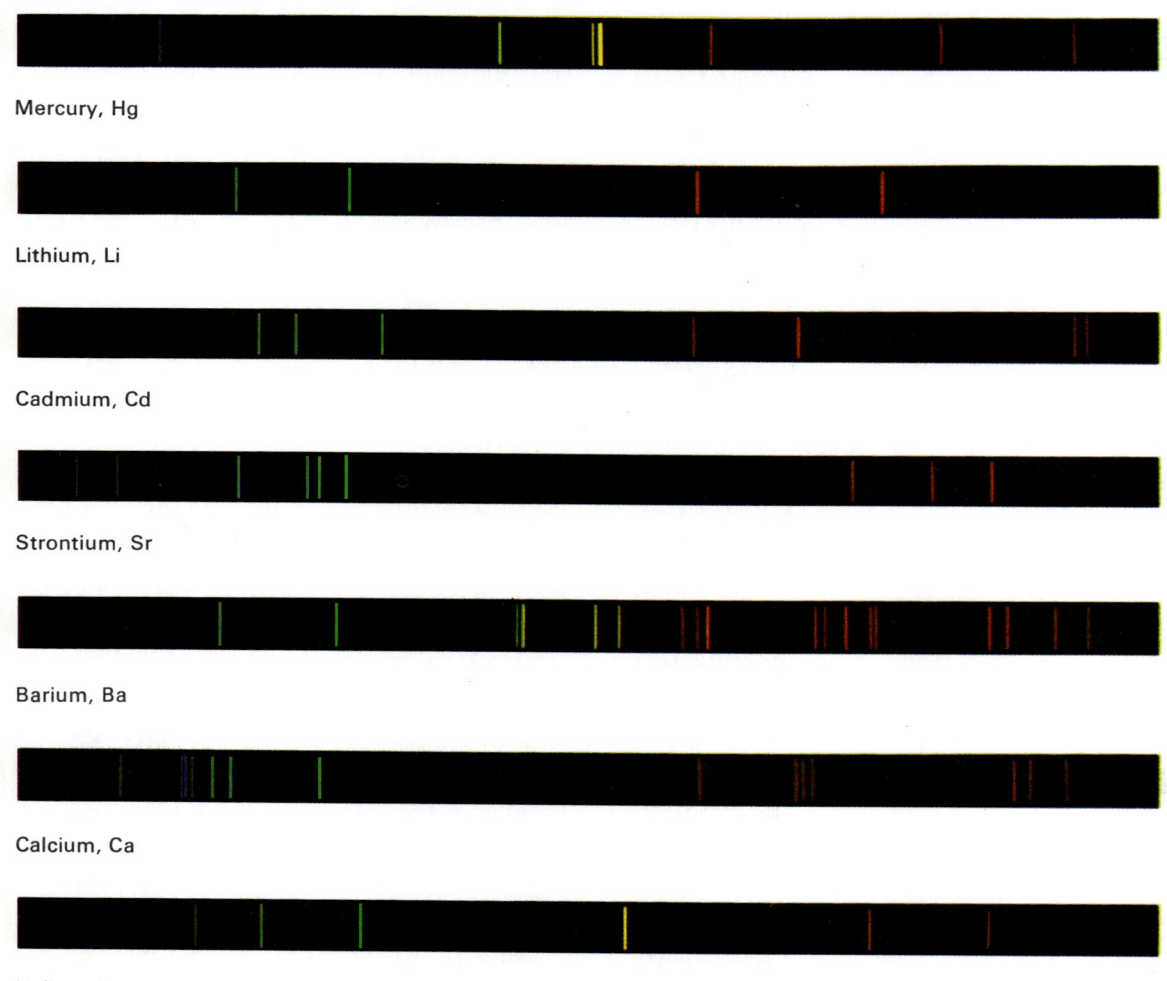

Mercury, Hg

Lithium, Li

Cadmium, Cd

Strontium, Sr

Barium, Ba

Calcium, Ca

Helium, He

Characteristic line spectra such as these are produced by highly energized gaseous elements. Line spectra have provided important clues to the understanding of atomic structure.

Light is one of the many forms that energy can take. All forms of matter are capable of giving off light under some circumstances and absorbing it under others. One of the most useful ways of gaining information about the structure of atoms and molecules is to study in detail how they absorb and emit light. Careful measurements of these processes have helped scientists to understand how electrons are arranged around the nuclei of atoms and how much energy the electrons have in each of various arrangements. In this chapter we shall find out about some of the most basic experiments of this type and how they are interpreted. Because light plays such a key role in these experiments, we shall begin by examining the nature of light itself.

The Nature of Light

11-1 Light as a Wave Motion

You have certainly seen rays, or beams, of light. What *is* a beam of light? Your general experience may have given you some ideas about the nature of light, such as the following ones.

1. Even a strong light does not seem to exert any force on you. Light is probably not a form of matter, since it does not appear to have mass.
2. When standing in strong sunlight, you feel warmth. Evidently the light carries energy and this energy becomes heat when it is absorbed by your skin. Perhaps light is just energy. If so, how does it travel?
3. Light comes in a whole range of colors. How can you explain the different colors of light?

To account for the above observations and answer the questions raised, scientists have developed a model for light. According to their model, light has the characteristics of *wave motion*. To understand what this statement means, consider a very familiar form of wave motion, water waves approaching and breaking on a beach. Figure 11-1 (next page) shows two measurements that might be made: the distance between crests and the time between waves.

The distance between crests is called the wavelength. The time between waves, t, is the period of time necessary for a complete wave to pass a given fixed point. Its reciprocal, $1/t$, the number of waves passing a given fixed point per second, is more commonly used. It is known as the **frequency** and its units are waves per second. The Greek letter *nu*, ν, is its symbol.

Light travels through space in a way similar to that in which waves travel over a water surface. Light is a species of electromagnetic radiation, that is, it is composed of electric and magnetic fields traveling together through space (or through some other medium, such as glass). The strengths of the electric and magnetic fields are constantly

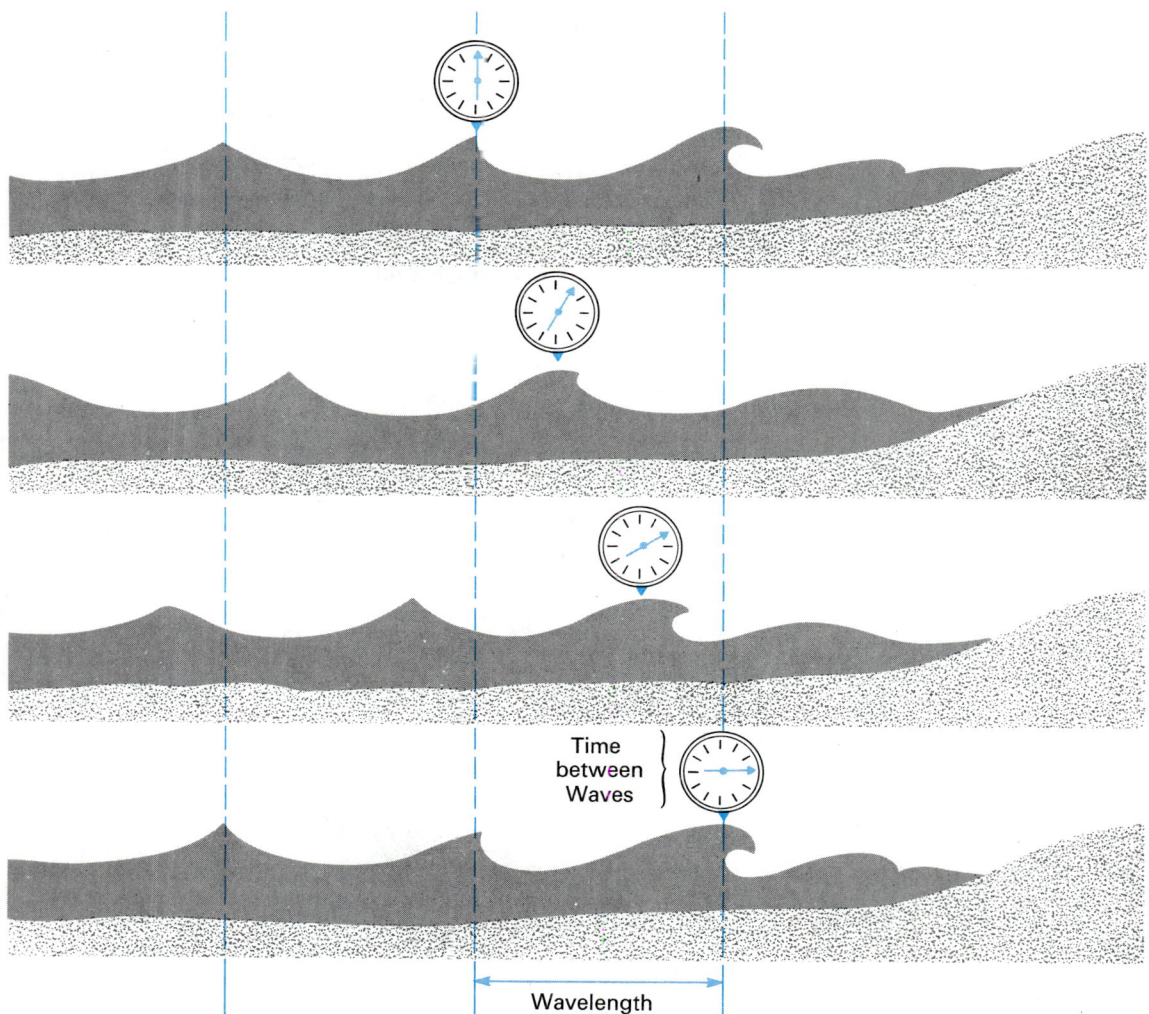

Figure 11-1 *Water waves illustrate two important properties of all waves: wavelength and frequency. Frequency is the number of waves passing a point in one second.*

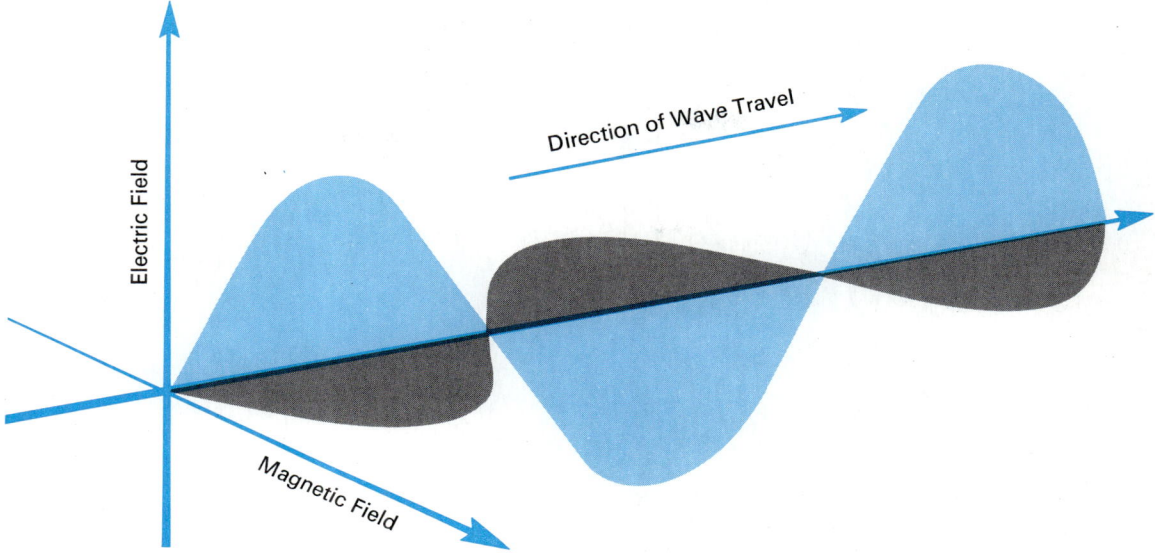

Figure 11-2 *In electromagnetic radiation energy travels in the form of simultaneous electric and magnetic fields that vary in wave-like fashion. As with water waves, there are crests and troughs, and both wavelength and frequency are characteristic properties.*

changing. These variations in field strength correspond to the water-level variations in the water wave analogy.

As you can see in Figure 11-2, the electric and magnetic fields are perpendicular to one another. Both fields are perpendicular to the direction in which the "light waves" are moving. Similarly, the rise and fall of water waves at a given point are perpendicular to the direction of wave motion.

The **wavelength** of light waves is the distance between successive maximum values of the electric field. In terms of the water wave analogy, it is the distance from crest to crest. The symbol for wavelength is another Greek letter, *lambda*, λ.

What sort of experiments suggested to scientists that light has a wave nature? Some of the most powerful, direct evidence for the wave nature of light is obtained from experiments on *diffraction* and *interference patterns*. Before discussing such experiments, let us think about how waves can interact.

Suppose you allow two beams of light to merge. What will the resulting combined beam be like? That depends on how the waves in one beam are related to those in the other. Figure 11-3 shows several possibilities. (The waves represent variations in electric field magnitudes, in a single plane.) In (a) the peaks of one beam coincide exactly with the peaks of the other. This results in maximum reinforcement and hence a strong combined beam is produced. In (b) the peaks of one beam coincide exactly with the troughs of the other. This causes complete cancellation. In (c) we see an intermediate case in which there is partial cancellation. All

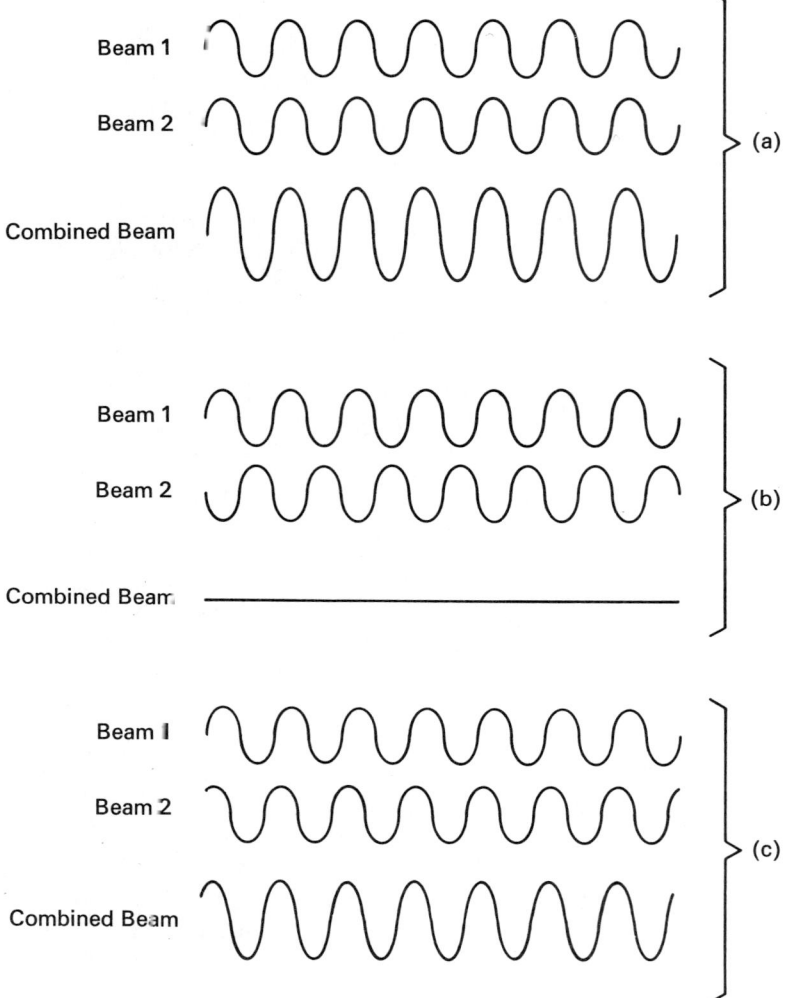

Figure 11-3 *Interference patterns result when (a) two waves are fully reinforcing each other, (b) two waves are fully canceling each other, (c) two waves are partially canceling each other.*

of these effects are called interference effects.

When a plane of waves approaches a barrier with only a small opening in it, some of the waves leak through the opening. Instead of a straight beam of parallel waves, what emerges is a rounded pattern that spreads out. This bending of the waves is called **diffraction.** Figure 11-4 shows this phenomenon, as exhibited by water waves. The analogous behavior of light is represented in Figure 11-5.

Interactions Between Light and Atoms 283

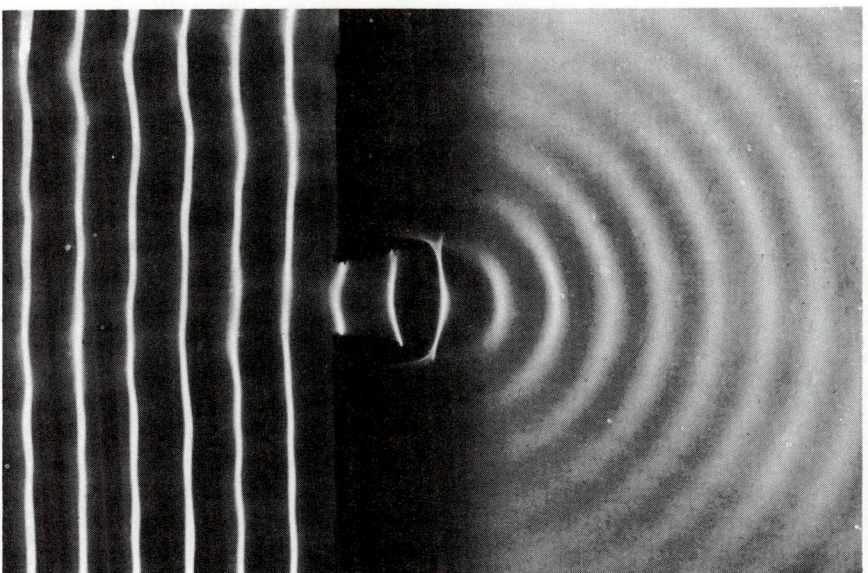

Figure 11-4 *This photo shows water waves passing from left to right through a narrow opening. Such a bending of waves is called diffraction.*

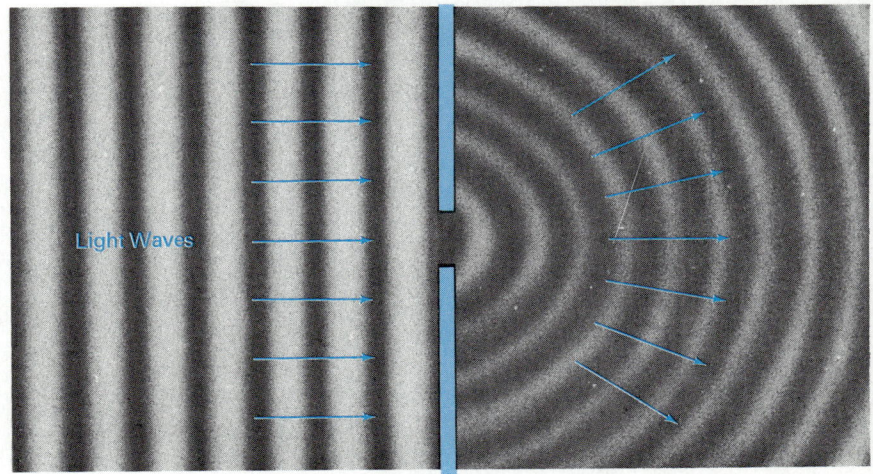

Figure 11-5 *Light waves passing through a small opening also undergo diffraction, as the above diagram indicates.*

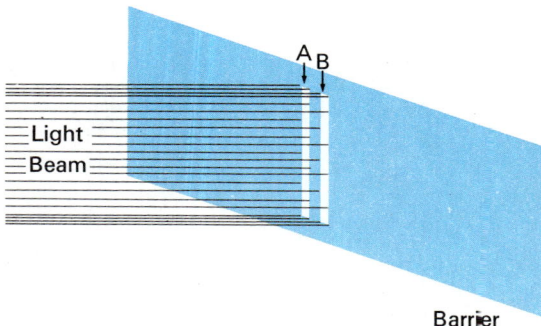

Figure 11–6 *A light beam approaching a pair of parallel slits. The slits are separated by a distance similar to the wavelength of the light.*

By combining diffraction and interference effects, it is possible to produce direct, visible interference patterns with light. Such patterns show very strikingly that light has wave properties. Suppose that a narrow beam of light passes through two parallel vertical slits, as shown in Figure 11-6. The beam that emerges from each slit would be expected to undergo diffraction, as you learned earlier. What happens as the two diffracted light beams flow into one another?

The photograph of water wave interference (Figure 11-7) shows the sort of effect that could be expected. Notice that in some places the waves seem to have been destroyed. If you could measure the crests in the places where the waves are not destroyed, you would find the crests higher than the crests for one wave alone.

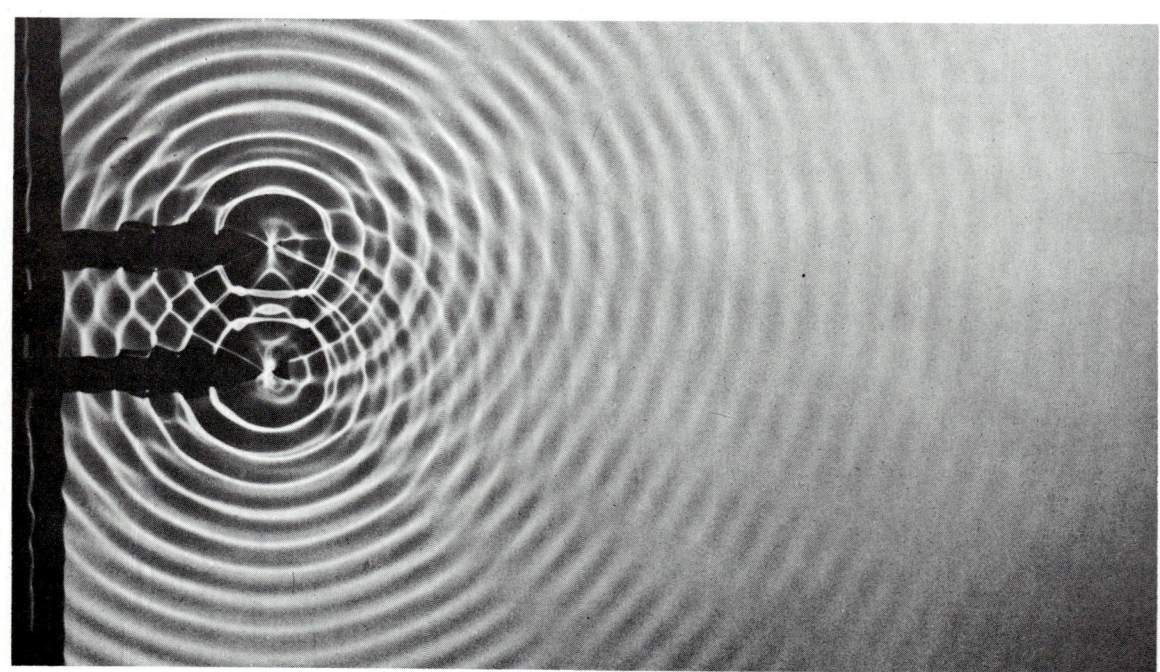

Figure 11–7 *Interference produced by two merging series of water waves.*

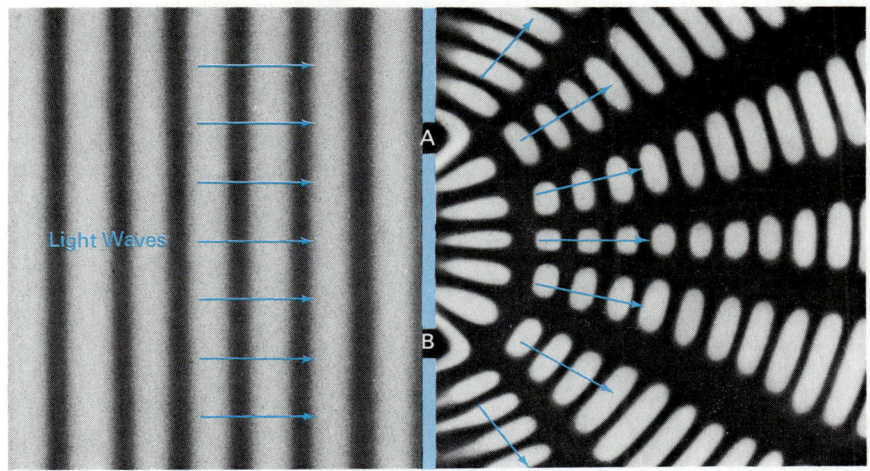

Figure 11-8 *Top view of the wave crests of diffracted light beams, showing interference. Arrows show lines of maximum reinforcement.*

A similar result should be obtained with the diffracted beams of light. At some points there should be strong reinforcement, and, at others, varying degrees of cancellation. The expected interactions are given in Figure 11-8, which represents an overhead view of the arrangement shown in Figure 11-6. By placing a screen, or your eye, at a certain distance beyond the two slits, you can see an interference pattern. A typical interference pattern produced by light is shown in Figure 11-9.

Figure 11-9 *An interference pattern that was produced with red light.*

INVESTIGATION

11-2 Does Light Behave Like a Series of Waves?

Several years ago Eugene Burdick wrote a book called *The Ninth Wave*. The title of the book was based on observations made by a number of surfers that at fairly regular intervals an unusually large wave would approach a beach. The surfers would eagerly wait for this wave. While there is no scientific evidence that the ninth wave is always a big one, a wave larger than the average does appear regularly. This is caused by the crests of two or more different sets of waves approaching the shore together. The combination of these smaller waves produces a larger one. There are also times when a wave nearing the beach is smaller than the average. This is caused by the crest of one wave approaching the shore together with the trough of another wave. The combination of a crest and a trough produces either a very small wave or no wave at all.

If light produces any effects similar to waves in water, perhaps it behaves at times as though it is made up of a series of waves. In order to decide whether light has wave-like characteristics, you will examine what happens when two beams of light interact with each other. The two beams of light will be produced by passing light through two very narrow, parallel slits. If light is made up of waves, what would you expect to observe at a point where two wave crests came together? at a point where a crest and a trough came together?

PURPOSE

You will study what happens when two beams of light come together and will determine whether there is any resulting pattern that suggests that light is made up of waves.

MATERIALS

Piece of cardboard, 10 × 10 cm
2 double-edge razor blades
Light source
Red filter
10 cm heavy thread or light opaque fishing line
10 cm masking tape or cellophane tape

Blue or green filter
Metric ruler
Compasses

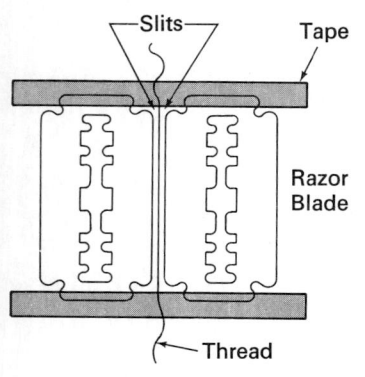

Figure 11-10

PROCEDURE

1. Cut a hole about 2 cm square in the center of a piece of cardboard about 10 cm square. The dimensions do not have to be measured carefully, nor does the hole need to be cut neatly. Across the hole tape two razor blades and a piece of heavy thread or a piece of lightweight opaque fishing line, as shown in Figure 11-10. Make the slits as small as you can. You should just barely be able to see light through them.

2. Set up a narrow vertical light source that is at least three meters away from you. This source can be a single straight filament light bulb. If this is not available, a vertical slit can be cut in a piece of

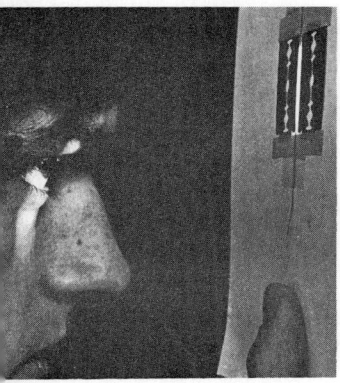

Figure 11-11

cardboard or poster paper and lit by means of an ordinary light bulb behind it.

3. Hold the piece of cardboard with the razor blades and the piece of thread very close to your eye and look through the slits at the vertical light source, as in Figure 11-11. Make a sketch in your notebook of what you see.

4. Place a red filter in front of the light source. Look through the slits again. Is what you observe the same as, or different from, your observations in step 3?

5. Repeat the experiment with a green or blue filter. If several light sources are available, look first through one filter and immediately look through another filter while the image is still fresh in your mind.

RELATED PROBLEMS

1. Explain what you saw through the slits. Did you see anything to suggest that light may behave like a series of waves?

2. Did the color of the light have any effect on what you saw through the slits? How might you explain this?

3. On a separate piece of paper make a drawing similar to Figure 11-12. To do so, place a compass point first in the middle of slit A and then in the middle of slit B. Using these two points as centers, draw two semicircles, each with a radius of 0.5 cm. Using the same two centers, draw a series of concentric semicircles, each with a radius 0.5 cm

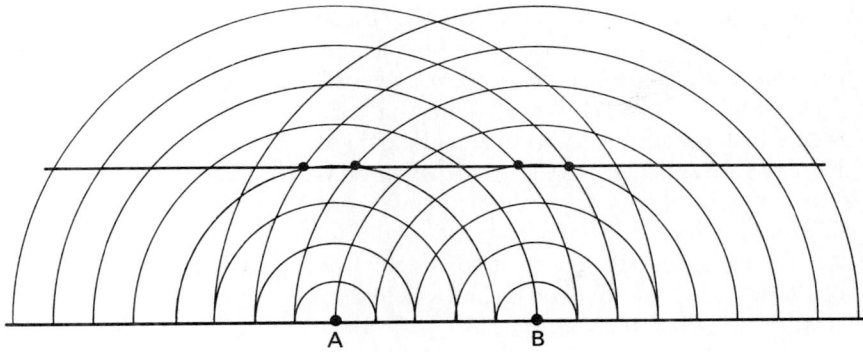

Figure 11-12

larger than the preceding one, until you have drawn circles with radii of 4.0 cm.

4. Make another drawing like the preceding one, with the exception that each series of concentric semicircles should have a radius 1.0 cm larger than the preceding one.

5. In each of your drawings draw a straight line 2.0 cm from the original line and parallel to it. These two straight lines should intersect all the concentric semicircles that you have drawn.

6. In each of your drawings mark with a heavy dot each point where two concentric semicircles intersect on or very close to the straight lines you have drawn.

7. Now consider the series of semicircles with a spacing of 1.0 cm between them to represent wave fronts of the red light and the series of semicircles with a spacing of 0.5 cm between them to represent wave fronts of the green or blue light. Do you find any relationship between the spacing of the points you have marked on your drawings and the observations you have made in the experiment?

11-3 Light Is Made Up of Different Frequencies

In the experiment you have just done, or watched your teacher demonstrate, you have seen interference patterns. Such patterns demonstrate very graphically that light has wave properties. But the experiment showed something else, as well. *The interference pattern changed when the color of light changed.* Since the details of the interference pattern must be determined by the wavelength of the light (assuming that the slit system is kept the same), this experiment shows that different colors of light correspond to different wavelengths. A closer analysis of the results would show that red light has a longer wavelength than blue light.

Experiments show that all colors of light travel at the same speed in a vacuum, and at virtually the same speed in air. The speed with which light travels is related to its wavelength and its frequency in a simple way. Remember, the frequency, ν, is the number of crests that pass a certain point every second and the wavelength, λ, is the distance between crests. The speed of light, c, must be equal to the number of crests arriving per second times the distance between crests, as follows.

$$\text{cm/sec} = \frac{\text{cm}}{\text{crest}} \times \frac{\text{crests}}{\text{sec}}, \quad \text{or} \quad c = \lambda\nu$$

The speed of light has been measured to be 3.00×10^{10} cm/sec in air. From this important relationship you can see that if red light has a greater wavelength than blue light, it must have a lower frequency.

We can learn more about the relationship between color and the frequency of light with the use of an instrument called a **spectrograph**. A simple one is shown in Figure 11-13. The spectograph is an instrument that

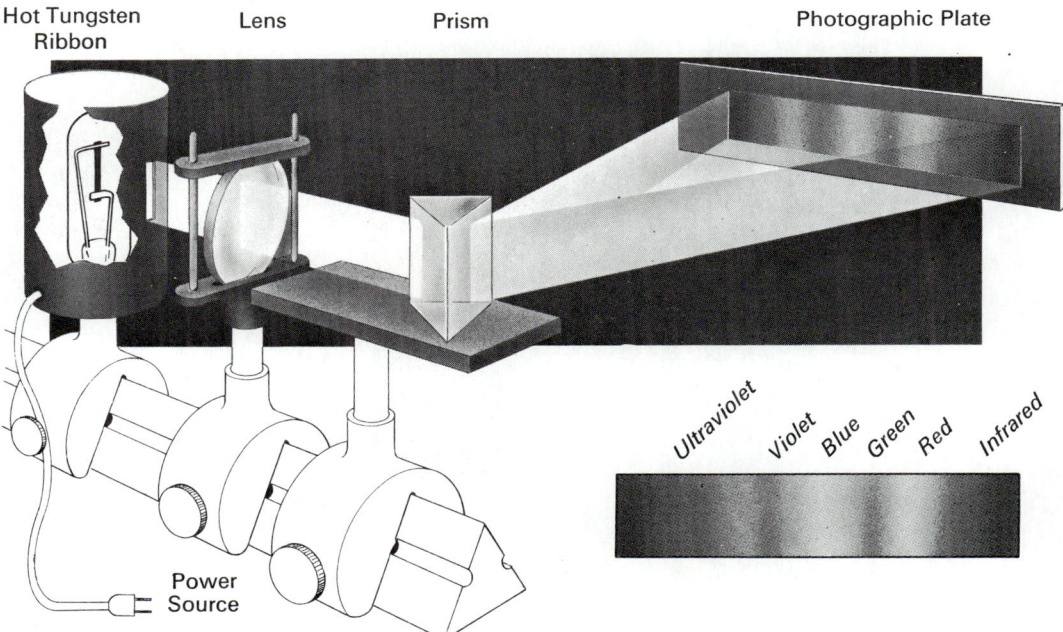

Figure 11–13 *An operating spectrograph (top) displays the visible spectrum by passing white light (focused into a beam of parallel rays) through a glass prism. Glass refracts (bends) different frequencies through different angles, separating the visible colors shown on the screen. The drawing (bottom) shows the essential parts of the spectrograph.*

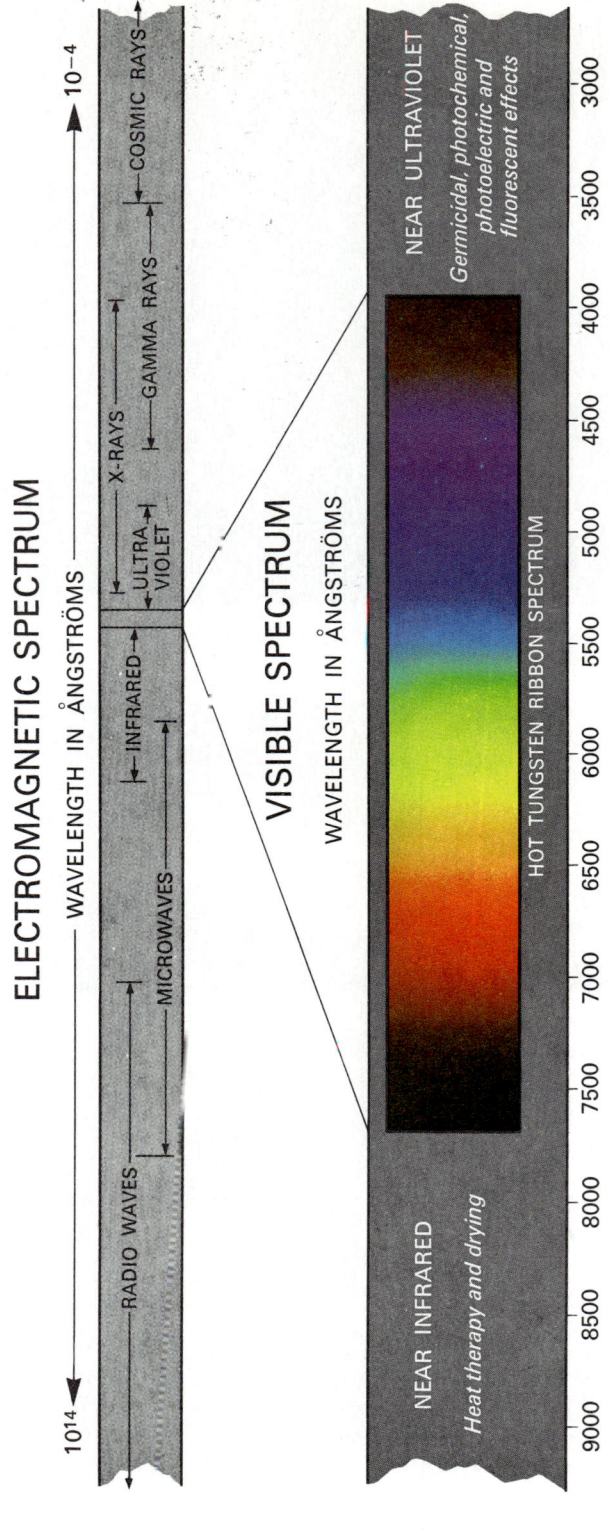

Figure 11-14 *The full electromagnetic spectrum (top) contains only a small visible portion (bottom).*

reveals the frequency composition of light. It separates from each other the various frequencies found in ordinary light, making it possible to measure each of those frequencies. In the set-up shown, the light emerging from a narrow slit is focused into a beam of parallel rays by the lens. This beam is passed through the prism. All of the light is **refracted,** that is, *bent,* by the angular prism. Different frequencies, however, are bent through different angles. The result is that the frequency composition of the light entering the slit can be learned from the pattern focused on a photographic film. In Figure 11-13 the light source is a tungsten ribbon heated to a temperature near 1000°C by an electric current.

The separation of light into its component frequencies produces a spectrum. This spectrum can be recorded on photographic film. The darkness of the film (on development) is determined by the light intensity, just as in an ordinary camera. From the spectrum you find that different colors correspond to different frequencies. Blue light, for example, has a frequency of about 7.5×10^{14} waves per second. Red light has a lower frequency, about 4.3×10^{14} waves per second. In recent years the frequency unit has become generally known as the **hertz (Hz).** This unit replaces not only waves per second, but such equivalent expressions as cycles per second and vibrations per second. It is equivalent to reciprocal seconds, that is, 1/sec, or sec^{-1}. The name honors Heinrich Hertz, who first produced radio waves, nearly 100 years ago.

The experiment shown in Figure 11-13 points up another extremely important fact. The photographic film is darkened at larger angles than those at which blue light appears and at smaller angles than those at which red light appears. This implies that the light emitted by the hot ribbon of tungsten includes frequencies that are not detected by

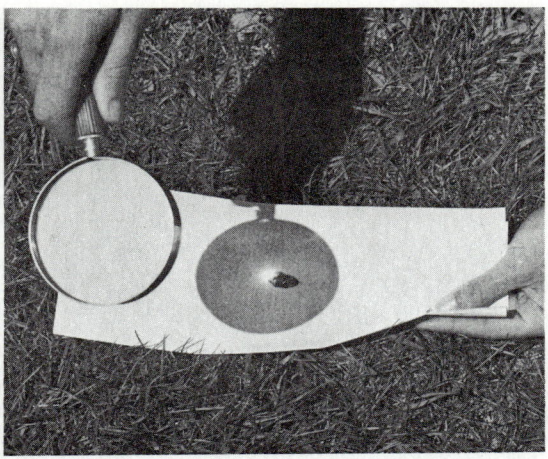

Figure 11–15 *Sunlight focused through a hand lens gives evidence that light is a form of energy. The light is transformed into heat, burning the paper.*

the human eye. The frequencies lower than the frequency of red light are called **infrared** (meaning "below the red") frequencies. The frequencies higher than the frequency of violet light are called **ultraviolet** (meaning "beyond the violet") frequencies.

Scientists now realize that electromagnetic radiation extends over an enormous range of frequencies. This range is much wider than the rather narrow region of the spectrum visible to the human eye. Figure 11-14 shows the range that is commonly studied and the familiar names given to various spectral regions.

11-4 *Light Is a Form of Energy*

Light is a form of energy. This statement is consistent with quite a bit of common experience. For example, you may have used a hand lens to focus light rays on paper, setting it afire as in Figure 11-15. This phe-

Max Planck (1858–1947)

HEAT IS SUCH AN IMPORTANT factor in our lives that scientists for years have been interested in studying heat and the many natural phenomena associated with it.

When you turn on an electric toaster, you can see the heating element begin to glow with a dull red color. As it gets hotter, the color changes to a brighter red. The filament of an electric light bulb becomes "white hot" at extremely high temperatures.

Scientists in the second half of the 19th century studied the color changes associated with heat very intensely. They examined the distribution of the wavelengths of energy emitted at different temperatures by many different substances. They got the same curves regardless of the material they used. When they tried to predict the shape of these radiant energy curves, for specific temperatures, using the methods of classical physics, they failed.

Max Planck, a scientist at the Physics Institute in Berlin, derived an equation in 1900 that made it possible to plot curves showing the relationship between temperature and the distribution of energy with great accuracy. He was not satisfied with this equation, however, because it only correlated the data but did not explain it.

Planck then proposed that energy is given off by tiny vibrating particles that can have only certain definite energies. Such oscillators radiate energy only when they change from one allowable energy value to a lower one. He assumed that energy is not continuous, but is given off in little bundles or quanta defined by the equation, $E = h\nu$, in which E is energy, h is a constant (known today as Planck's constant), and ν is the frequency of the oscillator.

Planck's proposal was so radical that few scientists paid much attention to it until Albert Einstein used this idea in explaining the photoelectric effect, in 1905. The Quantum Theory was the major factor in the development of Bohr's atomic model in 1913, and has withstood every test put to it up to the present day.

The 1918 Nobel Prize in Physics was awarded to Planck "in recognition of the services rendered by him to the development of physics by his discovery of the elementary quanta."

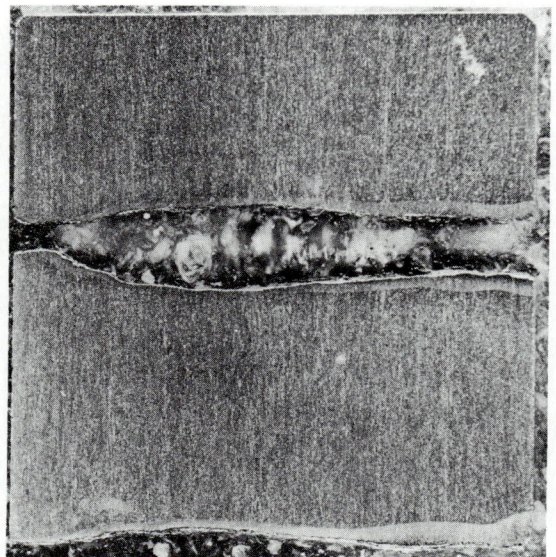

Figure 11-16 *This sample of the metal titanium was cut by laser light acting with a stream of oxygen gas.*

nomenon is put to work in huge solar furnaces. Such furnaces achieve temperatures of many thousands of degrees and can melt the most refractory (that is, difficult to melt) materials. The temperature rise in the paper or in the refractory material is caused by the absorption of light. A temperature rise means that energy has been absorbed. This energy must have come from the light. In some very intense laser beams, the energy is enormous. (See Figure 11-16.)

The qualitative idea that light is energy naturally leads to a consideration of the quantitative aspect. How much energy is contained in light? The answer is simple in form, but not in concept. Light energy comes in tiny packages. Each package, called a **photon,** contains an amount of energy, E, that depends on the frequency of the light.

This statement is contained in the following equation.

$$E = h\nu$$

The quantity h is called **Planck's constant.** It is merely a conversion factor that enables us to express frequency, ν, in energy units. If frequency is expressed in $\sec^{-1}$ and energy in joules, Planck's constant is equal to 6.6×10^{-34} joule-sec. A full account of the experimental evidence that led to the above equation is a fascinating story, one that you may hear in your physics class. Our present interest, however, is in the application of this equation in the interpretation of the hydrogen spectrum. This spectrum gives a record of the frequencies of light emitted by a hydrogen atom. Using the equation $E = h\nu$, scientists have obtained much information about the energies of hydrogen atoms, as you will discover in the next section.

11-5 Photons and the Photoelectric Effect

It may strike you that some of the foregoing statements seem inconsistent. First, by means of diffraction and interference effects, it was shown that light has wave-like properties. Then it was stated that light also behaves as though it comes in little packets, or particles. The two concepts, waves and particles, are in a certain sense opposites. Yet both are needed to account for all of the observed properties of light.

The dual nature of light has been a source of wonder and puzzlement to scientists for a long time, and it still is. Asking why light has both wave-like and particle-like properties is similar to asking why electrical charges attract and repel each other. There is really no answer to such basic questions. All that can be done is to describe these basic properties as precisely as possible.

Chapter Eleven

Chemistry and Our Environment
SOLAR ENERGY

If you look back over a long enough period of time, the source of nearly all our energy is or has been the sun. This comes about in several ways.

First of all, we eat two kinds of food: other animals, and plants. Most of the other animals we eat live entirely on plants; the cow is a good example. Even those animals, such as fish, that feed on smaller animals are eventually dependent on plants. Sooner or later as we trace back each *food chain* from large and complex animals to smaller and simpler ones we find a creature that has an entirely vegetable diet. Plants themselves do not eat anything. Their source of energy is light from the sun. We shall discuss how they absorb and use this solar energy later. The important point is that all of the energy that we use in our own bodies can be traced back to the sun's energy.

Of course, we use much larger quantities of energy than that which man or beasts of burden can supply. These vast amounts of energy are obtained by burning coal, oil, and gas. But how did coal, oil, and gas originate? By the decay of ancient vegetable matter! Again we trace our energy sources back to the sun, with plants as the important intermediaries.

How do plants perform their role of capturing the sun's energy? We do not know the full answer to that question. The processes involved still require much study to be understood fully, and this is an important field of chemical research. However, we do know the broad outlines of the process. The process is called *photosynthesis*, a word derived from *photo* (light) and *synthesis* (putting together). In short, photosynthesis is a process of putting together with light.

What is it that is put together in the process of photosynthesis? The answer: carbohydrates. In Chapter 16 you will learn in detail what carbohydrates are. For the present, it will suffice to know that they are complex molecules consisting of carbon, hydrogen, and oxygen, usually in the proportions $C_x(H_2O)_y$. Actual carbohydrates have formulas such as $C_6H_{12}O_6$ (the sugar glucose), $C_{12}H_{22}O_{11}$ (common table sugar, sucrose), and $(C_6H_{11}O_5)_x$, where x is a very large number, for starch and cellulose.

Plants make these carbohydrates by photosynthesis in a complex series of reactions that we can condense into the following simple equation.

$$x CO_2 + y H_2O + \text{energy} = \underset{\text{carbohydrate}}{C_x(H_2O)_y} + x O_2$$

This reaction is reversible. In the process of photosynthesis, plants take in carbon dioxide, CO_2, from the air, and water. They combine them, using solar energy, to make carbohydrates. Oxygen is released to the air. When carbohydrates (or other compounds containing carbon and hydrogen) are burned, we have the reverse of this reaction, the process of combustion. The products of combustion are CO_2, water, and energy in the form of heat.

It is important to note that when plants make carbohydrates by photosynthesis, they necessarily remove CO_2 from the air and return oxygen to it. This is as important as the making of carbohydrates. Man is constantly using up oxygen in the burning of coal and oil, and producing CO_2 and water. Nature maintains balance in the environment because plants remove CO_2 and water from their environment and put oxygen back into the air.

The way in which we currently use solar energy is thus an indirect, two-step process. First, plants use solar energy to make carbohydrates. Second, man uses the carbohydrates, or other substances derived from

Model of a solar power farm on a typical desert terrain. The rows of collectors noticeably darken the desert. This power station would use turbine waste heat to desalinate seawater. 1. Sea water 2. Fresh water 3. Desalting plant 4. Cooling tower 5,6. Thermal storage 7. Oil reserve 8. Maintenance 9. Turbine 10. Boiler

Artist's conception of how the collectors on a solar-power farm would look.

them as fuel to generate heat. The heat or thermal energy can then be converted into electrical or mechanical energy.

This indirect way of using solar energy has several disadvantages. (1) Combustion of coal and oil produces considerable air pollution. (2) We are burning up in a few centuries coal, oil, and gas that have accumulated over millions of years. The supplies of these substances are not being replaced, and will eventually be exhausted. (3) Plants are not producing usable carbohydrates at a rate comparable to our use of fuels. On average, plants only use about 1–2 percent of the solar energy that falls on them.

It is natural to wonder whether we cannot learn to use solar energy in a more direct way. While it does not seem likely that all of man's energy requirements (at least with his present mode of life) could ever be met by direct use of solar energy, it is possible that an important contribution could be made. Direct conversion of solar energy to thermal or electrical energy would generate no pollutants and involve no problem of radioactive wastes. But can it be done?

At present we do not know. Some things are clear. Small devices of the "rooftop" type for individual homes, though much discussed in the past, are uneconomic and require excessive maintenance. Moreover, most houses are not located where solar energy is sufficiently abundant. Practical direct utilization of solar energy can only succeed if it is done on a large scale in the sunniest parts of the earth. The most imaginative proposals so far describe "solar power farms." These could be located in places like the great western deserts of the United States, for example around the conjunction of California, Arizona, and Nevada, or in the Rio Grande valley in Texas and New Mexico. The possible appearance of such a solar power farm is shown in the illustrated model. Such a facility could be built on about 5000 square miles of land and might produce about one million megawatts of electrical power.

There are basic scientific problems to be solved before such an energy farm can become a reality. Devices capable of converting solar energy directly to electricity have been developed for spacecraft systems. However, in such applications the cost—several hundred dollars per watt of electricity—was unimportant. It does not seem likely at present that the cost of producing electricity by this method can ever be reduced enough for everyday large-scale use. New discoveries in chemistry and physics might alter that situation some day.

At present, however, the only attractive approach is to use solar energy to power high-temperature steam generators. These, in turn, produce electricity. There are two major problems to be solved, even in this approach. One is to develop the necessary solar-to-thermal converters. These are devices that can absorb sunlight and be raised to a fairly high temperature (about 500°C). The development of these, which requires the skills of chemists, physicists, and engineers, is underway. The other problem is how to store electrical energy from the time it is generated until the time it is used. Remember, the solar energy farm will produce electricity only during the day, but much of the use of it will occur at night. Most likely the energy will have to be stored by means of some chemical reaction that can be made to go in one direction as it absorbs energy and can be reversed later to release the energy.

What forces scientists to give a dual description of light? You have already seen some of the evidence suggesting that light is wave-like. What evidence exists for the particles, or photons? There are a number of phenomena in which the existence of photons is indicated. The most conclusive proof undoubtedly comes from the phenomenon known as the *photoelectric effect.*

Experiments to demonstrate and measure the photoelectric effect employ the type of apparatus shown in Figure 11-17. The main facts follow.

1. When light strikes the surface of certain metals it causes electrons to be ejected. Some of these electrons then travel through the vacuum to the other electrode, called the plate, and return to the emitter through the wire. A current is registered by the sensitive ammeter. This is shown in Figure 11-17a.
2. The brighter the light, the greater the current.
3. If a source of voltage is introduced as shown in Figure 11-17b, giving the plate a negative charge, less current flows. As the voltage is raised the current is reduced further until, finally, a voltage is reached that suppresses the current entirely.

The obvious explanation for these observations is that the energy of the light is somehow converted to kinetic energy of electrons in the metal. As a result, these electrons can overcome the force that normally holds them to the surface of the metal. It is perfectly natural to expect that more light per unit area would eject more electrons per unit area, as observed. It is also perfectly understandable that the opposing voltage can retard or even suppress entirely the flow of electrons from the emitter to the plate.

So far, there is nothing incompatible with light being merely a continuous series of waves. But consider the following further results.

4. Suppose the voltage is made just high enough to reduce the current to zero, for a given light intensity. If the intensity of the light is then increased, current still

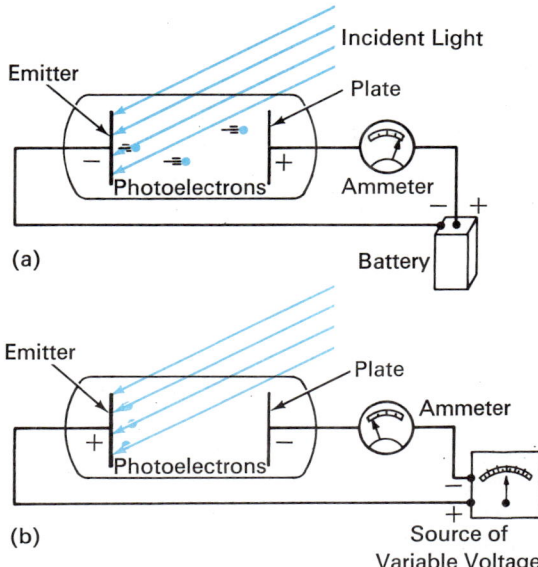

Figure 11-17 *In (a) electrons ejected by light reach the positively-charged plate. In (b) the ejected electrons are kept from reaching the plate by its high negative potential.*

does not flow. No matter how much brighter the light becomes no current flows!

5. With the voltage still set so as to just cut off all current, suppose that the frequency of the light is changed. If the frequency is lowered, still no current flows. But if the frequency is increased, current again begins to flow. If the voltage remains unchanged, at this higher light frequency the amount of current is again proportional to the intensity of the light.

The last two observations forced scientists to propose that, in spite of its wave nature, light also behaves like a stream of separate particles, or "photons." The basic assumptions can be summarized as follows.

1. Each photon in light of a certain frequency is a packet, or *quantum*, as it is sometimes called, of energy. As stated earlier, the amount of energy is given by Planck's equation, $E = h\nu$.
2. An electron can be ejected from the metal surface when it is struck by a photon, provided that the photon has at least enough energy to overcome the force binding the electron to the metal. The application of voltage in opposition to the direction in which the electrons must flow, as shown in Figure 11-17b, is equivalent to increasing the "binding energy" that must be overcome.

If there is a given binding energy, E_b, and if the photons striking the metal have energy, E, there are two possibilities. If $E \geqslant E_b$, electrons can be ejected, with kinetic energy equal to $E - E_b$. If $E < E_b$, the electrons will never be able to "lift off" from the surface of the metal. Assume that an electron acquires the full energy of a photon that strikes it. Assume, further, that the probability of two photons striking one electron at the same time is too slight to be concerned about.

It is easy to explain the experimental results in terms of the photon model. When the frequency of light shining on a metal surface is such that $E < E_b$, it makes no difference how intense the light becomes. Increasing the light intensity increases the number of photons striking the surface per unit time. But the energy per photon stays the same, and so no photon can "launch" an electron from the metal surface. It is rather like having a cannon with a maximum range of 3 miles, when you need to bombard a target 3.5 miles away. It doesn't do any good to fire the cannon over and over, or to get 50 more such cannons. The target still remains out of range.

You can easily see that so long as the energy of the photons is even a shade in excess of E_b, electrons can be ejected, and the number of electrons ejected must increase as the number of photons per second increases.

Albert Einstein, in 1905, proposed the photon model to explain the photoelectric effect. It was for this, rather than his more famous theory of relativity, that he received the Nobel Prize in Physics, in 1921. Einstein convinced himself and all other students of the problem that there was no way to account for the facts in terms of a pure wave description of light.

Einstein's explanation of the photoelectric effect was proven experimentally by R. A. Millikan in 1916. In 1923 Millikan was awarded a Nobel prize for this work and for his determination of the charge on an electron.

Summary

Light is a wave-like transfer of energy through space. Its wave properties are described by (1) its wavelength, the distance between one wave crest and the next, and (2) its frequency, the number of individual waves passing a given point during a given time (cycles per second, or hertz). Diffraction and interference effects demonstrate the wave character of light and show that light of different colors has different wavelengths and frequencies. Frequency, ν (sec^{-1}), and wavelength, λ (cm), are related to the speed of light, c (cm/sec), by the equation $c = \lambda\nu$. A spectrograph separates light into its different component frequencies and spreads them out, forming a spectrum.

The amount of energy associated with light is proportional to its frequency, as indicated by the equation $E = h\nu$, in which h is a basic constant of nature called Planck's constant, and ν is the frequency, in sec^{-1}.

Self Check

1. Give several examples supporting the statement that light is a form of energy.
2. What evidence indicates that the spectrum of visible light is only a small part of the electromagnetic spectrum?
3. What is the expression for the speed of light, c, in terms of wavelength, λ, and frequency, ν?

Atoms Emit Only Certain Frequencies of Light

11-6 Atomic Spectra

All atoms will emit light if they are energized by heating or by a high voltage electric discharge. You have already seen, in Figure 9-9, some examples of the light emitted by various atoms. In everyday life you have probably encountered the purplish light emitted by mercury atoms and the yellow light emitted by sodium atoms in the street lights called mercury lamps and sodium lamps, respectively.

The light emitted by a particular atom has a characteristic color. Obviously, this means that each atom emits light predominantly in one part of the visible region. The particular

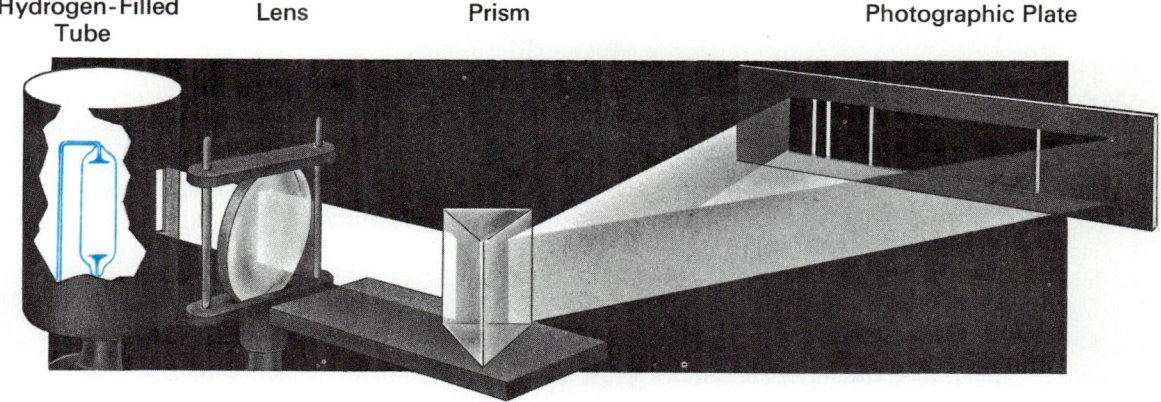

Figure 11-18 *When a high voltage passes through hydrogen gas, only certain frequencies of light are emitted. The spectrograph reveals the frequencies of the visible light as lines.*

pattern of frequencies of light emitted by an atom is called its spectrum. No two types of atom have the same spectrum. Atomic spectra are like fingerprints. They are characteristic of the type of atom that emits them.

Atomic spectra can be studied using a spectrograph, as shown in Figure 11-18. When this is done, a remarkable observation is made:

Atomic spectra consist of narrow lines.

In other words, each atom emits only a few sharply defined frequencies of light. Perhaps your teacher will be able to demonstrate some line spectra for you. In any case, several atomic spectra are shown on the first page of this chapter, and Figure 11-19 shows the hydrogen spectrum. You can see the line nature of these spectra.

Since light is a form of energy, an atom must be *losing* energy when it emits light. The fact that it emits only light of certain precisely defined frequencies means that it loses only certain precisely defined quantities of energy. With a knowledge of the frequencies, the energies can be calculated using Planck's equation, $E = h\nu$. Such a calculation is shown in the example on the next page.

Hydrogen, H

Figure 11-19 *The actual line spectrum produced in the visible region by hydrogen.*

Example

One of the purple lines in the hydrogen spectrum, shown in Figure 11-19, has a wavelength of 4103 ångströms. To what energy, in kilocalories per mole of hydrogen atoms, does this correspond? The calculation may be done in the following series of steps.
1. Convert ångströms to centimeters.
2. Convert the wavelength in centimeters to frequency.
3. Convert frequency to energy in joules.
4. Convert joules per atom to kilocalories per mole.

The actual calculations follow.
1. One ångström = 10^{-8} cm. Therefore, 4103 Å = 4.103×10^{-5} cm.
2. To convert λ (cm) to ν (sec^{-1}), use the equation $c = \lambda \nu$. First rearrange it to $\nu = c/\lambda$. ($c = 3.00 \times 10^{10}$ cm/sec)

$$\nu = \frac{3.00 \times 10^{10} \text{ cm/sec}}{4.103 \times 10^{-5} \text{ cm}}$$

$$= \frac{30.0 \times 10^9 \text{ sec}^{-1}}{4.103 \times 10^{-5}}$$

$$= 7.31 \times 10^{14} \text{ sec}^{-1}$$

3. Use Planck's relationship, $E = h\nu$. ($h = 6.63 \times 10^{-34}$ joule-sec)

$$E = (6.63 \times 10^{-34} \text{ joule-sec})$$
$$\times (7.31 \times 10^{14} \text{ sec}^{-1})$$
$$= 4.85 \times 10^{-19} \text{ joule}$$

4. First, multiply the number of joules per atom by Avogadro's number to get joules per mole. Then divide by 4.18 to get calories per mole. Finally, divide by 10^3 to get kilocalories per mole.

$$E = \left(4.85 \times 10^{-19} \frac{\text{joule}}{\text{atom}}\right)\left(6.02 \times 10^{23} \frac{\text{atoms}}{\text{mole}}\right)$$

$$= \frac{(4.85)(6.02)(10^4) \text{ joules/mole}}{4.18 \times 10^3 \text{ joules/kcal}}$$

$$= 69.8 \text{ kcal/mole}$$

Scientists discovered the line nature of atomic spectra many years ago. Of course, they wondered why the spectra should be of this character and why the lines for each atom are where they are. Since it was found that hydrogen atoms have the simplest line spectra of all, efforts to understand atomic spectra naturally focused on this particular case.

11-7 The Hydrogen Spectrum

Figure 11-20 shows some of the lines in the emission spectrum of hydrogen atoms. One set of lines is in the visible spectrum. The other set lies at higher frequencies and is in the ultraviolet region. In Figure 11-20 the positions of the lines are given in wavelength, in frequency, and also in the energy corresponding to each frequency. The energies are expressed in kilocalories per mole of atoms, since this is a very common unit of energy in chemistry.

You have seen that rather high energies are necessary to cause atoms to emit light. The gas discharge tubes shown in Figure 9-9 require about 10,000 volts to glow. What does this energy do to the atoms in producing the glow? Why does the glow produce only certain definite frequencies of light, as shown by the line spectra?

In order to answer these questions, scientists were forced to rely on a model. In devising a model to explain the hydrogen spectrum, they made two fundamental assumptions. One was that the electron is responsible for both the absorption and the emission of energy by the atom. The second was that the electron can have only certain "allowed" energies with the atom. The latter assumption sounds rather arbitrary but, as you will see, it works. Niels Bohr, who produced the first explanation for the origin of the atomic spectrum of hydrogen, made a very amusing statement in a similar context.

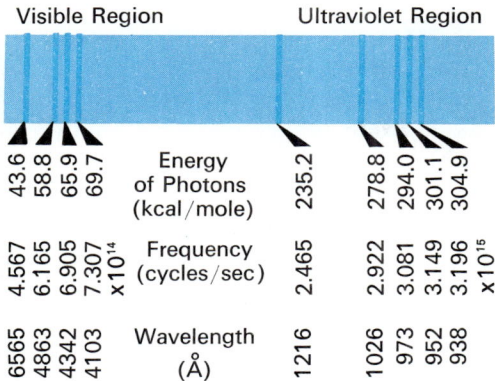

Figure 11–20 *The location of hydrogen lines in the visible and ultraviolet regions. Other lines exist, but they are too closely spaced to illustrate.*

"We are all agreed that the theory is crazy. The question that divides us is whether it is crazy enough to have a chance of being correct."

What sort of energy does an electron in an atom possess? As Rutherford showed, the negatively-charged electron is attracted to the positively-charged nucleus. Rutherford was able to give us the approximate volumes occupied by the electron(s) and the nucleus, but was not able to tell us what an electron is doing. Whatever an electron is doing, it must be attracted to the nucleus. The amount of attraction depends on the magnitude of charge on the nucleus, and that on the electron, and on the square of the distance between them. Since neither the electronic nor the nuclear charge is changing in magnitude, the distance between the electron and the nucleus must change when the energy of the electron changes. The electron, then, has potential energy that is a function of its distance from the nucleus.

The electron must also have some kinetic energy. Otherwise, it would be drawn to the nucleus. An analogy that may be suggested to you involves a satellite orbiting the earth. If it did not have kinetic energy, it would be drawn to earth immediately, due to the gravitational field of the earth. While this analogy suggests a comparable balance of two forces, one kinetic and one potential, it should not be inferred that the electron "orbits" the nucleus just as a satellite orbits the earth. As you will learn later, it is now thought to be impossible to tell precisely what an electron in an atom is doing. So a satellite-like orbit is not a suitable model, however attractive such a mental picture might be.

From what we know and have imagined so far, energy changes within the atom involve changes in the position of the electron relative to the nucleus. Now we need only construct a model that fits the above assumptions, and, most importantly, the experimental facts. In order to take the next big step we can use a staircase.

11-8 *A Model for Explaining the Hydrogen Spectrum*

A staircase is a special kind of potential energy system. You can increase your potential energy relative to the floor by ascending a staircase, just as you can increase your potential energy relative to sea level by climbing a hill, or by taking off in a rocket. But the staircase allows you to assume only certain definite heights relative to the floor. You cannot stop and rest between steps. Hence, only certain potential energies are "allowed" to you.

Some kind of staircase might help you to visualize the potential energies available to an electron in an atom, then. What macroscopic object might be useful for representing the electron? As you have seen, electrons act like little particles, in gas discharge tubes, at least. The simplest sort of shape to

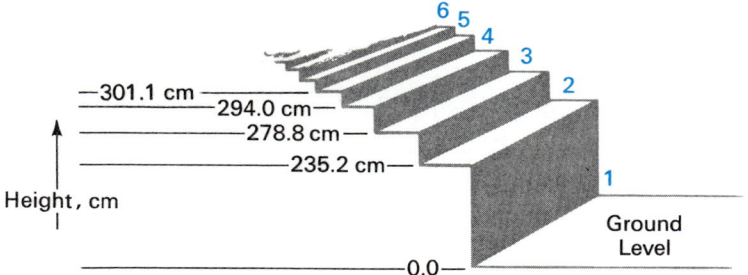

Figure 11-21 Our model of electron energy levels is an imaginary staircase with steps of progressively decreasing height. The distances (in cm) between the steps are numerically equal to the energies of the lines found in the ultraviolet spectrum of hydrogen.

imagine for little particles is a spherical one. In order that the little spheres be objects familiar to you in everyday life, think of them as tennis balls.

A special type of staircase and a tennis ball are the macroscopic objects that will be used to help you to understand the atomic processes that give rise to the hydrogen line spectrum. It is very difficult to devise a model, consisting of everyday macroscopic objects, that closely resembles the minute entities called atoms. Still, the following analogy clearly illustrates certain crucial features of the pattern of energies in the hydrogen atom.

Imagine a staircase of somewhat unorthodox design, as shown in Figure 11-21. Its most peculiar feature is that the space between the steps is not constant, as in ordinary stairs. Instead, it regularly decreases as the staircase rises. Secondly, each step slopes very slightly toward the front so that a round object such as a tennis ball could not remain very long on any step. Instead, it would soon roll off and fall to one of the lower steps. As before, it would not remain long on the lower step. It eventually would make its way from any upper step on which it might be located down to the ground. In dropping down, it might stop temporarily on one or more of the in-between steps. A final feature of the peculiar staircase is that it rises indefinitely, with an infinite number of steps that keep getting closer and closer together.

The various steps of the staircase are the heights, or levels, on which the tennis ball can remain for some short period of time. The only place on which the ball can remain indefinitely is the ground. If the ball is thrown upwards, it will land on some upper step, and, for a short time, have a perfectly well-defined level. However, it will soon roll off this higher step to a lower one. If it strikes a vertical part of the staircase hard enough, it will bounce back so vigorously that it will shoot out over all the intervening steps and fall straight back to the ground. In this case, it loses all its height in a single fall. The distance of fall depends on the height of the step to which it was thrown in the first place.

Let us suppose that the ball is thrown many times, in a fairly random way, so that it lands many times on each of the upper steps. Each time, the ball promptly falls. Sometimes it falls from the step it is on

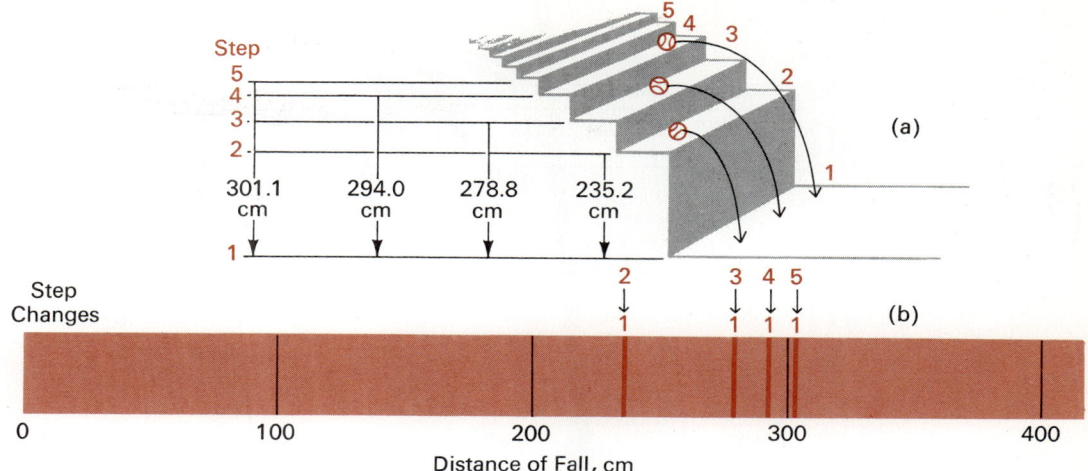

Figure 11-22 *Successive steps in our staircase model are at different distances from the ground, as shown in (a). A horizontal plot of these distances (b) looks very much like a part of the hydrogen spectrum.*

directly back down to the ground. Other times it stops temporarily on one or more of the intervening steps. Each fall, whatever its route, is from a particular, well-defined height to the ground.

You have seen that only a certain number of changes in position, from a higher level to ground level, are possible for a tennis ball on a staircase. The ball can fall from step 2 to the ground, or from step 3 to the ground, and so on. Each fall takes the tennis ball from a higher to a lower potential energy. Thus, there are only a certain number of possible energy changes for the tennis ball. The lines on the hydrogen spectrum, similarly, represent only a certain number of frequencies. Each of these frequencies corresponds to a definite energy, presumably the energy lost as an electron falls from a higher to a lower potential energy within the atom.

The staircase model is at least qualitatively consistent with the microscopic processes that we have been attempting to understand.

Can the model be made quantitative? Suppose the distance between each step and the ground is made numerically equal to one of the energies in the hydrogen spectrum. The energies represented by the lines found in the ultraviolet region, as given in Figure 11-20, will be used. The distances will be expressed in centimeters, since that is a familiar unit of length. (See Figure 11-21.)

Now consider possible energy transitions, using the quantitative staircase model. In Figure 11-22a, some of these transitions are shown. The numerical value of each of these transitions is plotted on a horizontal scale in Figure 11-22b. The result is that Figure 11-22b has the same appearance as part of the hydrogen spectrum. The series of lines

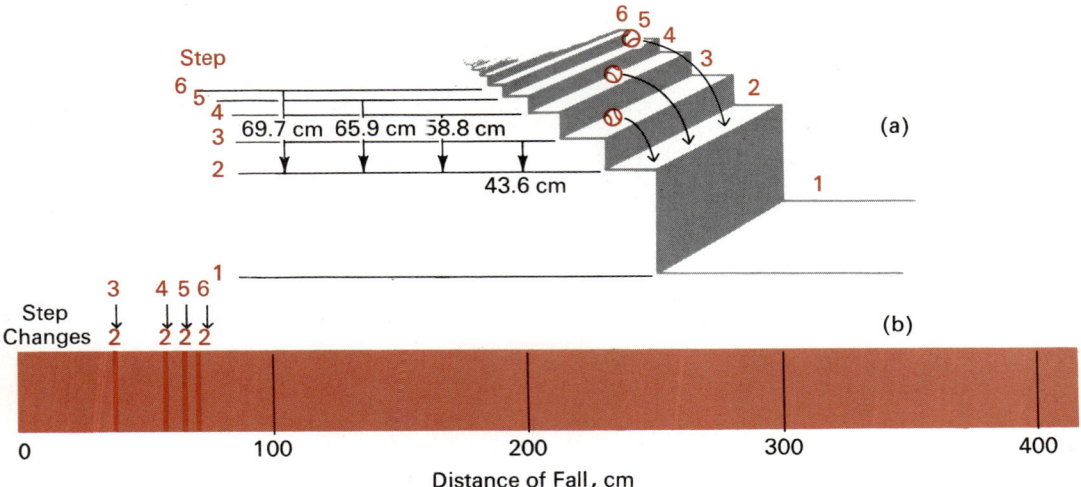

Figure 11-23 *The distances that a ball would fall to step 2 rather than to ground are shown in (a). When these distances are plotted on a horizontal scale (b), they look very much like another part of the hydrogen spectrum.*

is analogous to those found in the ultraviolet region of the hydrogen spectrum. The lines in this region have the highest frequencies of any of the lines in the hydrogen spectrum, and, hence, the highest energies as well.

This analogy is nothing to be very pleased or impressed with. The heights of the steps were deliberately adjusted so that they would correspond to the lines of the spectrum. The value of the model will depend on whether we can now get out of it something that we did not deliberately build into it. In this particular case, can the model account for the existence of the next series of lines, those in the visible region of the spectrum? Can it be used to predict their frequencies?

To find out, consider the times that the ball falls from the step where it initially lands, down to step 2, instead of all the way to step 1 (ground). The possible distances of fall will then be those shown in Figure 11-23a. If these distances are plotted on a horizontal scale, a series of lines analogous to the lines in the visible region of the hydrogen spectrum are obtained. (See Figure 11-23b.) Moreover, the relative positions of the lines are quantitatively the same in each case. Clearly, the line of lowest energy in the visible series corresponds to the distance of fall from step 3 to step 2. The energy change from the heights of the steps can be calculated as follows.

$$278.8 - 235.2 = 43.6$$

All the other lines of this series can be similarly calculated, as Table 11-1 shows.

A little thought will show that the model predicts many other converging series of lines. The next series consists of those changes resulting when the ball drops from

TABLE 11-1 Calculated Changes in Energy

Step Change	Energy Change (kcal)	
	Calculated	Observed
#3 → #2	278.8 − 235.2 = 43.6	43.6
#4 → #2	294.0 − 235.2 = 58.8	58.8
#5 → #2	301.1 − 235.2 = 65.9	66.0
#6 → #2	304.9 − 235.2 = 69.7	69.8

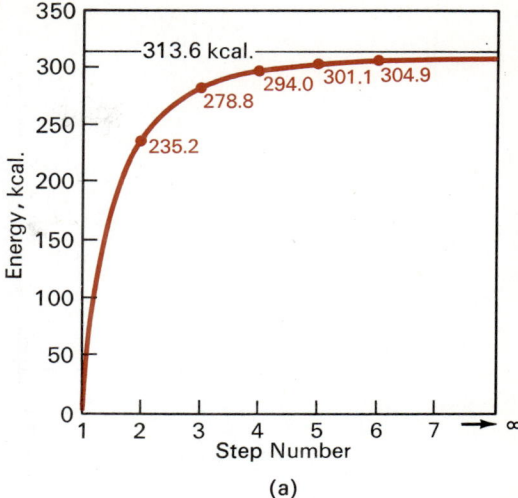

Figure 11-24a *The energies of the lines in the hydrogen spectrum have been plotted against the step number, with step 1 (ground) the zero energy level.*

the various upper steps, such as 4, 5, 6, and so on, down to step 3. Other possible series involve drops from higher steps down to step 4, and so on. Such series can actually be observed with spectrographic equipment suitable for making measurements in the infrared region of the spectrum. The positions of the lines are exactly what we would predict from the staircase model, thereby establishing the ability of the model to predict properties that can be observed in the laboratory.

11-9 The Energy Levels of a Hydrogen Atom

The staircase model adequately represents the following characteristics of the spectrum and energy states for hydrogen atoms.
1. It shows that the electron can be raised from its lowest or most stable level (the ground level) to only certain specific levels, represented by steps 2, 3, 4, and so on, which are above this ground level.
2. It shows that the electron cannot remain on any one of these upper levels very long but must fall back to lower ones and eventually to the ground level.
3. It shows that the process of falling back takes place in discrete steps, that is, from a higher step to a lower one, never to or from points between steps. The distances of fall on the staircase correspond to the amounts of energy emitted by the atom, in the form of light, as the electron drops back to a lower level.
4. The model shows just what series of emission lines are possible in the spectrum and how they are related, quantitatively, to each other.

Scientists actually deduced the scale of energies for the hydrogen atom in a way quite similar to our use of the staircase model. However, they used somewhat more sophisticated terms. For example, the staircase steps we have used, they called *energy levels*. We shall next transform our staircase model into the more conventional type of presentation, called an energy level diagram.

In Figure 11-24a the energies of the lines in the ultraviolet region of the spectrum (equal to distances of fall on the staircase,

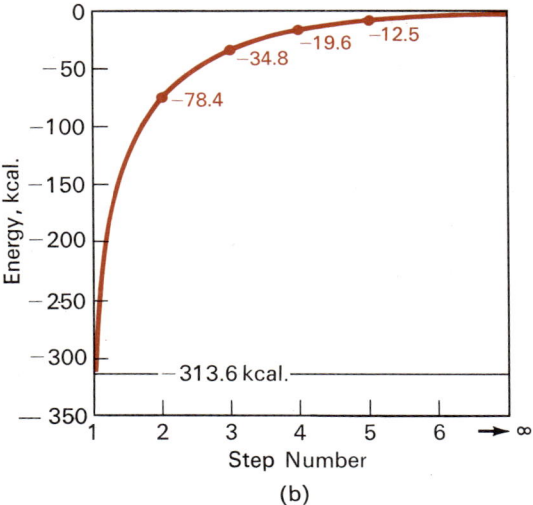

Figure 11-24b *The scale has been changed from the one used in Figure 11-24a such that the maximum energy level is now zero and the energies are expressed as negative numbers.*

in centimeters) have been plotted against the step numbers. A smooth curve was then drawn through these points. First, look at the ordinate (vertical) scale as it appears on the left side of the graph. It is arranged in the usual way, with zero at the bottom. An energy of zero has been assigned to step 1, an energy of 235.2 kcal to step 2, and so on, which exactly corresponds to Figure 11-22a.

The curve that passes through these points has a very definite and important property. It begins by rising very steeply, but then progressively flattens out. As the step numbers increase, the curve becomes more and more nearly flat. It can be shown, even at very high step numbers, that it will not rise above a value of 313.6 kcal. In fact, for a step number of ∞ (infinity) it would just reach this value of 313.6 kcal. A horizontal line has been drawn at this particular energy value.

It is possible to use a different ordinate scale for the same plot by making the maximum possible value of energy zero and running the scale downwards, using negative numbers. A graph using such an ordinate scale is shown in Figure 11-24b. In this graph, step 1 lies at an energy level of −313.6 kcal. This method of arranging the ordinate scale may seem arbitrary. But you will soon see that, from this point of view, the best interpretation of the arrangement of steps can be given. Such an interpretation is now accepted by scientists as part of the fundamental theory of the behavior of atoms and molecules.

Figure 11-25 shows a vertical plot of the energy levels of the hydrogen atom in which the method just described for assigning energy values is used. Note that the *differences* between the various levels are just the same, even though the energies now assigned to the levels are different. Of course, the energies represented by the observed spectral lines correspond to differences between energy levels, rather than the energies of individual levels. Thus, the scale on which zero is the maximum energy is a perfectly valid one for representing the experimental facts, even though it is not the first or simplest method that comes to mind.

In Figure 11-25, the numbers originally assigned to the steps are now used to identify energy levels. The lowest level is number 1, with the others following in order of increasing energy (2, 3, 4, etc.) and a possible upper limit at ∞, where the energy is zero.

Long after this energy level diagram for the hydrogen atom had been established, scientists still pondered its significance. Finally in the late 1920's, a mathematical scheme was developed that explained the facts. This scheme, called *quantum mechanics*, is discussed in the next chapter.

11-10 Bohr's Model of the Hydrogen Atom

The clues provided by the line spectra prompted scientists to attempt an explanation of how the electron in a hydrogen atom behaves. The first explanation that met with any success was given in 1913 by a great Danish physicist, Niels Bohr. The detailed model of the hydrogen atom that Bohr proposed had to be given up later. But it is worth describing because it was a crucial step on the way to the more accurate model that scientists accept today. Bohr's work also merits our attention because it shows how scientific theory forges ahead at certain critical times, as the result of bold and imaginative ideas that simultaneously build on past work and also take a new direction.

By the year 1912 it was known that the hydrogen atom is composed of a proton and an electron and that these particles are attracted to each other by reason of their opposite electric charges. Physicists felt that they should be able to calculate the properties of such a combination. After all, the laws of motion of macroscopic bodies had been studied for centuries. The behavior of electrically-charged bodies was also thoroughly understood on the macroscopic scale. Yet scientists could not explain why a hydrogen atom exists, let alone why it should have any particular value of energy. In fact, the laws that had been deduced for macroscopic bodies gave a convincing (though incorrect) prediction that the nuclear atom is unstable, and the electron should collapse into the nucleus.

At this point Niels Bohr made a fresh start. He faced the fact that an explanation is a search for likenesses between a system

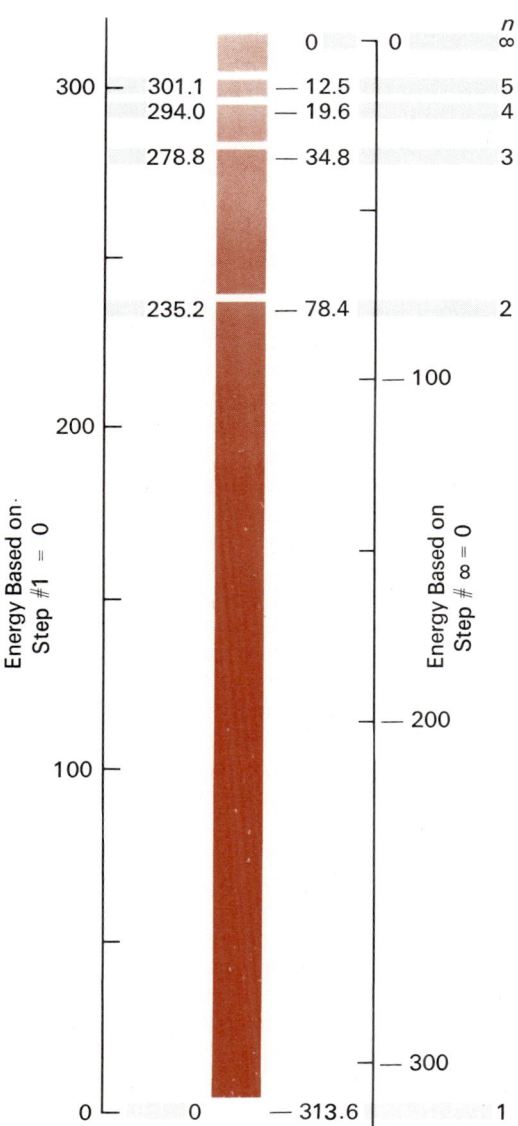

Figure 11–25 The data plotted in Figure 11–24 (a and b) are shown in a vertical energy level diagram. Energy levels are numbered (like the steps) from one to infinity. The scale on the left has one as the zero energy level. The scale on the right has infinity as the zero energy level, and so all other energies on it are negative quantities.

Niels Bohr
(1885–1962)

BOHR'S PROPOSAL FOR a model of the hydrogen atom is an outstanding illustration of the way in which many diverse investigations can be related to solve a complex problem.

Almost a century had passed from the time that John Dalton, in 1803, had proposed his Atomic Theory until Max Planck described his Quantum Theory, in 1900. For years scientists had been studying the spectra of many elements, but had not succeeded in correlating any of these spectra with the structure of an atom.

Bohr, a citizen of Denmark, was working with J. J. Thomson at the Cavendish Laboratory in Cambridge, England, when Ernest Rutherford announced his nuclear model of the atom in 1911. Shortly after this, Bohr joined Rutherford in his laboratory at Manchester and spent the next two years working out the details of his own model of an atom.

Although Rutherford had proposed an atom with a densely-packed nucleus and electrons revolving around the nucleus, the classical laws of electricity and motion could not show any relationship between the structure of the atom and the spectrum it produced when it was excited.

In a brilliant move, Bohr assumed that Planck's Quantum Theory could be applied directly to the behavior of the electrons in an atom. He calculated the energy needed to remove an electron from a fixed orbit around the nucleus, using Planck's constant. His result agreed very well with the experimentally measured ionization energy.

Since Planck had stated that energy could be emitted only in certain fixed quanta, Bohr assumed that the electrons moving around a nucleus could move only in a limited number of fixed orbits. As long as the electrons remained in these orbits, they emitted no energy. If they jumped from one orbit to another, they would gain or lose energy only in certain fixed amounts. When Bohr calculated the frequency of light that should be emitted by an excited hydrogen atom based on these assumptions, his results agreed excellently with experimental results. Although Bohr's model provided such close agreement only with the properties of the hydrogen atom, his work paved the way for later theories that explained the behavior of electrons in many other elements.

Bohr received the Nobel prize in 1922. In later years he participated actively in the development of atomic energy. Becoming increasingly concerned about the uses to which atomic energy might be put, he helped organize the 1955 Atoms for Peace Conference in Geneva and the European Center for Nuclear Research (CERN).

under study and a well-understood model system. An explanation is not good unless the likenesses are strong. Niels Bohr suggested that the mechanical and electrical behavior of macroscopic bodies ("classical" physics) does not lead to a completely suitable model for the hydrogen atom. He proceeded to seek a new model that did not contradict the known facts.

He began by supposing that the structure of an atom (the arrangement of the electrons around the nucleus) is determined by its energy. To agree with the facts, Bohr proposed that only special atomic structures can exist and called these special structures "stationary states." Each such state he characterized by a particular energy. If, then, the set of special atomic structures did exist, he predicted that a corresponding set of energies would be found. Here Bohr departed from the older atomic models that permitted structures corresponding to all energies.

The most stable state of an atom should be the one in which it has the lowest energy. Bohr took it to be a fundamental fact of nature that an atom can exist in its most stable state indefinitely. Even though this fact could not be explained (the earlier laws of physics having predicted that the atom should collapse), it had to be accepted because it was consistent with experimental results.

Bohr also proposed that although the lowest energy state of the atom is its most stable state, the atom can be excited to its higher "allowed" energy states by absorbing energy. The excited atom would not remain in this condition for long. It would soon lose its extra energy by emitting light. Since only certain levels of energy were assumed to be possible, only certain energy changes could occur. The energy change of the atom must be equal to the energy of the light emitted and must agree with the equation $E_2 - E_1 = h\nu$. Consequently, the frequency of the light emitted by an atom would be entirely determined by the values of the allowed energies of its electrons.

These ideas were so revolutionary that they would not have been accepted were it not for the fact that Bohr was able to propose a way to calculate exactly the energy levels for the hydrogen atom. The foregoing ideas are entirely correct and are fully accepted today. But, the specific model, which we are about to describe, had to be abandoned, even though it enabled Bohr to calculate correct energies for the hydrogen atom.

In order to visualize the hydrogen atom and to calculate its energy states, Bohr set up a series of circular "orbits" for the electron around the nucleus. These orbits were analogous to the orbits of the planets around the sun. He made two arbitrary assumptions, the first of which has already been mentioned.

1. In its most stable orbit the electron can keep going forever. It does not plunge into the nucleus, as classical physics would require.
2. In each orbit the angular momentum of the electron (its mass times its velocity times the radius, mvr) is restricted to the values $h/2\pi, 2h/2\pi, 3h/2\pi, \ldots$

Combining the second assumption with laws of physics that were fully accepted, he was able to calculate the energy of the electron in its most stable orbit. This most stable orbit is the smallest one, and Bohr assigned it an index number, n, of 1.

Bohr designated the energy of the electron in its most stable orbit $-R$. (Remember the energy scale that was discussed in Section 11-9, in which all the energies were assigned negative values.) Then he showed that the energies of the electrons in other orbits ($n = 2, 3, 4\ldots$) are equal to $-R/n^2$. Using this energy expression, he was able to account for the lines in the hydrogen spectrum.

Figure 11-26 shows how Bohr explained the two line series appearing in the visible and ultraviolet. The energy of each line is given by the difference between the energy in the lowest state (n_l) and that in the upper state (n_u), from which the electron starts. The expression is as follows.

$$\text{Energy Emitted} = \frac{-R}{(n_u)^2} - \frac{-R}{(n_l)^2}$$

$$= R\left(\frac{1}{(n_l)^2} - \frac{1}{(n_u)^2}\right)$$

For the series of lines in the ultraviolet, $n_l = 1$.

$$\text{Energy Emitted} = R\left(1 - \frac{1}{(n_u)^2}\right)$$

For the first few lines the following energies result.

$$E_1 = (1 - \tfrac{1}{4})R = \tfrac{3}{4}R$$
$$E_2 = (1 - \tfrac{1}{9})R = \tfrac{8}{9}R$$
$$E_3 = (1 - \tfrac{1}{16})R = \tfrac{15}{16}R$$

You can easily see that these energies form a series of values that get closer and closer together, gradually approaching a limit of R. This is exactly how the observed series of lines looks. By using any one line to calculate the value of R, Bohr was able to calculate precisely the positions of all the other lines in that series, and in several other series, as well.

From Figure 11-24a, it can be inferred that $R = 313.6$ kcal. Using this value for R, the energies of the first three lines of the ultraviolet series are calculated as follows.

$$E_1 = \tfrac{3}{4}(313.6) = 235.2 \text{ kcal}$$
$$E_2 = \tfrac{8}{9}(313.6) = 278.8 \text{ kcal}$$
$$E_3 = \tfrac{15}{16}(313.6) = 294.0 \text{ kcal}$$

They are in excellent agreement with experimental values (see Figure 11-20).

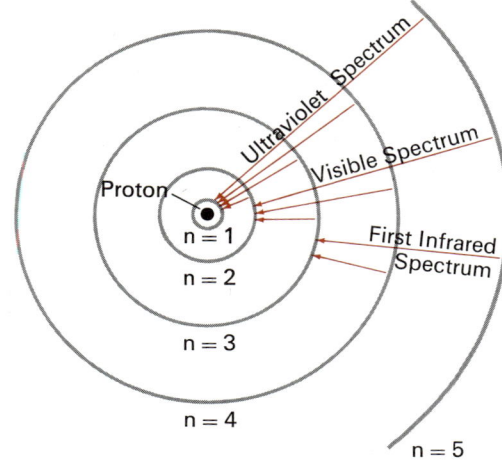

Figure 11-26 The electron orbits for the hydrogen atom, as proposed by Niels Bohr in 1912. The origins of three line spectra are shown.

11-11 Continuous Spectra ★

After all this discussion of the relationship between the lines discovered in the spectra of hydrogen and other elements, you may wonder why the spectrum of the hot tungsten filament, shown in Figure 11-13, shows a continuous blend of one color into the next with no prominent lines appearing. It is important to explain here that the bright line spectra appear only when the atoms of gases are excited. Such spectra are produced when a high voltage electrical discharge is passed through a tube containing some gas. They can also be observed when a sample of a substance is vaporized in a laboratory burner flame or in a high voltage arc or spark. In the gaseous state the atoms are separated far enough from one another so that they do not interfere with each other as the electrons drop back to their respective energy levels.

On the other hand, the atoms in solids are packed so closely together that the vibrations produced by heating are extremely complex. Evidently, electrons of one atom are influenced by the forces of adjacent atoms. The result is a continuous spectrum with no individual bright lines. As a matter of fact, the chemical composition of the solid has no effect on the nature of the spectrum it produces when it is heated. The temperature of the solid is the only factor that determines the nature of the spectrum. A piece of tungsten heated to 1000°C will produce the same colors as a piece of iron heated to the same temperature. Because of this property, it is possible to estimate the temperature of an object by observing its color. An object that is "white hot" is much hotter than one that is dull red.

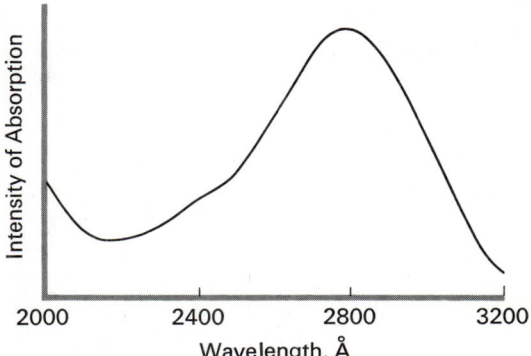

Figure 11–27 *The lowest energy absorption band of acetone, C_3H_6O.*

11-12 *Molecular Spectra* ★

You have seen how atomic spectra provide a way of learning about the energy levels in atoms. As you might expect, molecules also have spectra, and these can be used to find out about the energy levels in molecules. However, since molecules are more complicated than atoms, their spectra are more complicated. In fact, molecules have several different kinds of spectra.

Molecules, like atoms, have *electronic spectra*. These are the spectra caused by the absorption or emission of energy when electrons jump from one energy state to another in the molecule. The energy of an electron in a given energy state is determined by the attractive forces between the electron and the several nuclei in the molecule and by the repulsive forces between the electron in question and the other electrons in the molecule. The situation is analogous to that in an atom but it is more complicated because there are several positive centers attracting each electron instead of only one. It is thus much more difficult to calculate the electron energies. Still, electrons can be promoted from lower to higher energy levels when molecules absorb radiant energy. The processes are basically similar to those occurring in atoms.

The electronic spectra of molecules differ from the electronic spectra of atoms in an important way, however. Atomic spectra, as you have seen, consist of series of sharp lines. The molecular spectra consist of broad bands, as shown in Figure 11-27. (The graph shows some of the wavelengths of light *absorbed* by a sample of liquid, rather than those *emitted* by an excited gas, but the principle is the same. Only certain wavelengths can be emitted *or* absorbed, depending on the energy levels that can be occupied within the atom or molecule.)

The energies of the electronic states of molecules are not sharply defined because molecules have other forms of energy. These other forms are associated with vibrations and rotations. They are discussed later in this section. Despite the fact that molecular electronic spectra contain only broad bands,

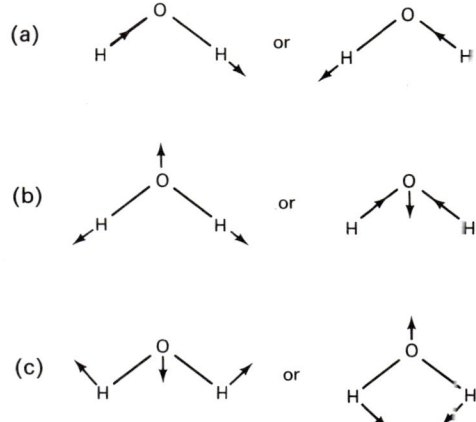

Figure 11-28 *Diagrams showing the three modes of vibration of a water molecule. In (a) the two O—H bonds stretch and contract "out of phase." In (b) the O—H bonds stretch and contract "in phase." In (c) the H—O—H angle opens and contracts.*

they still provide valuable information about the energies of electrons in molecules. Chemists study molecular spectra very carefully in order to understand better how electrons are arranged in molecules and how they serve to bind the atoms together.

The bonds in a molecule are not stiff and rigid. They are more like springs, and they can be stretched and compressed. The angles between bonds are not rigidly fixed either, but instead they can expand and contract. Because molecules are flexible in this way, they are constantly undergoing vibrations. A careful mathematical analysis shows that for every molecule these vibrations have a definite form. For the water molecule there are three basic *modes* of vibration. These are shown in Figure 11-28. As you can see, two of these modes consist principally of stretching and compressing the O—H bonds. The other consists mainly of bending the H—O—H angle.

Molecular vibrations can be observed because they give rise to spectra. Each mode of vibration has a certain frequency, f. Normally, the molecule executes the vibration with this frequency. However, the molecule has vibrational excited states, in which one, or more, of the modes takes place with a frequency of $2f, 3f, 4f$, and so on. If a molecule is exposed to radiation of frequency f, it may, in general, although not always, absorb energy. It gets excited from its normal state, in which a vibrational mode has a frequency f, to one in which that mode has a frequency $2f$. The frequencies of molecular vibrations range from about 50 to about 400 cm^{-1}. Radiation having frequencies in this range is called infrared radiation.

If a molecule such as H$_2$O, which has three modes of vibration, with frequencies f_1, f_2, and f_3, is exposed to radiation in the range of $f_1, f_2,$ and f_3, it will absorb radiation of just those frequencies. The experiment in which this absorption of energy at these frequencies is conducted uses a *spectrometer*. The basic design of a spectrometer is shown in Figure 11-29. By turning the grating (or a prism) the frequency of the light is varied. So long as the frequency of the light reaching the sample does not correspond to any of the vibrational frequencies of the molecule, it is not absorbed. It passes through. However, whenever the frequency of the light reaching the sample is equal to f_1, f_2, or f_3, some of it is absorbed. Molecules absorbing the light go into excited vibrational states. The absorption reduces the amount of light striking the detector and a dip, or *absorption band*, is produced on the recorder chart. The frequencies of these bands correspond to the differences $2f_1 - f_1 = f_1$, $2f_2 - f_2 = f_2$, and $2f_3 - f_3 = f_3$. In this way the values of the vibrational frequencies can be determined by experiment. The vibrational absorption spectrum of water molecules is shown in Figure 11-30. This type of

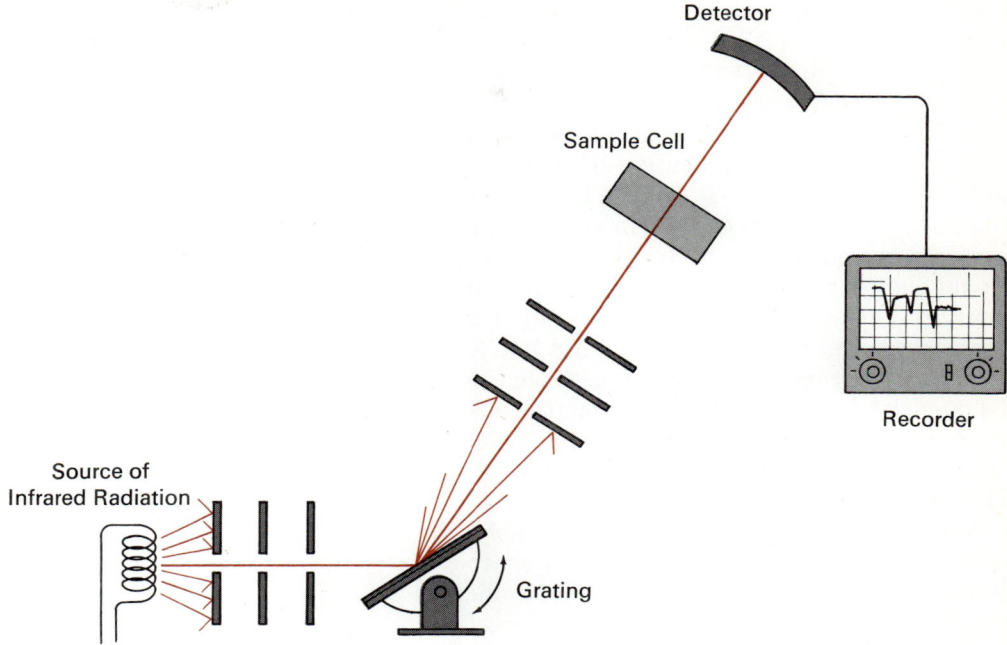

Figure 11-29 *The essential parts of an infrared spectrometer.*

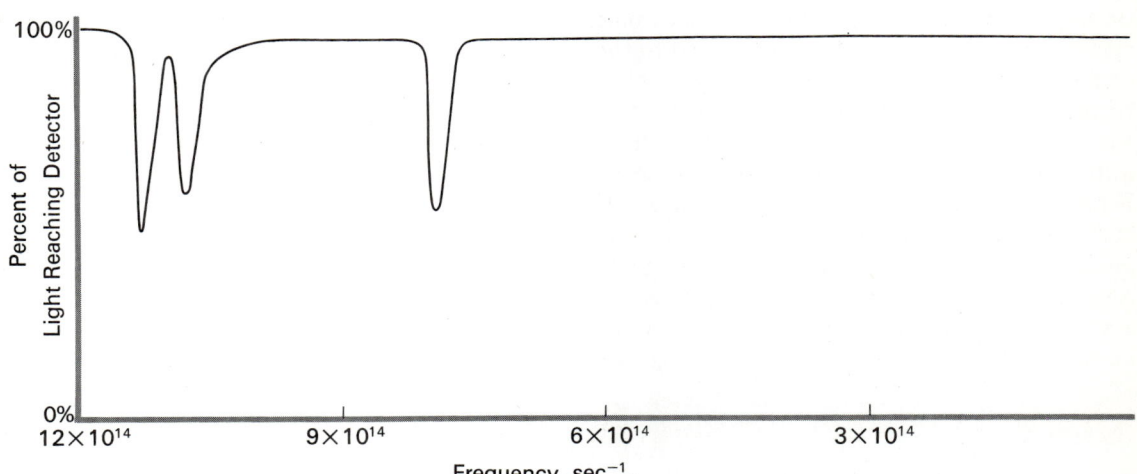

Figure 11-30 *A simplified representation of the vibrational (infrared) spectrum of gaseous H_2O.*

spectrum is often called an *infrared* spectrum of the molecule.

In addition to electronic and vibrational spectra, molecules also have rotational spectra. Molecules continually rotate about axes that pass through their centers of mass, as indicated in Figure 11-31 for the H_2O molecule. They can absorb energy in such a way as to be raised from a state of lower rotational energy to a state of higher rotational energy. The energy changes involved are quite small and correspond to quanta in the *microwave* region of the spectrum. Since the frequencies of rotation depend on the arrangement of the atomic masses in the molecule, microwave spectra are a source of information about molecular structures.

The reason the electronic spectra of molecules consist of broad bands instead of sharp lines can now be appreciated. Each electronic excitation is accompanied by a number of changes in vibrational and rotational energies. Since atoms have no vibrational and rotational energies, their electronic energy changes are simple and sharp and give rise to sharp spectral lines, as shown in Figure 11-32a. The separations between the vibrational energies accompanying each electronic state of a molecule are small compared to the energy difference between different electronic states, as shown in Figure 11-32b. Furthermore, the vibrational levels have different patterns for each electronic state. Finally, with each vibration there is a whole set of rotational energies. Therefore each electronic transition in a molecule is actually a whole collection of transitions from and to a whole series of vibrational and rotational energy states. Thus the electronic transition energy is spread out into a broad band, like the one in Figure 11-27.

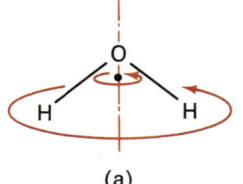

(a)

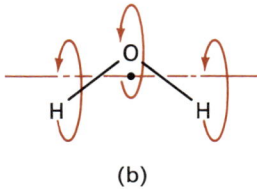

(b)

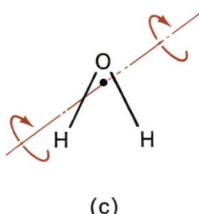

(c)

• Center of Mass of the H_2O Molecule

Figure 11-31 *The three rotational motions of water molecules. The three axes about which rotation occurs are mutually perpendicular and meet at the center of mass of the molecule.*

Summary

The emission spectrum of the hydrogen atom is the simplest of all atomic spectra. Efforts to understand atomic spectra in general were therefore first concentrated on the hydrogen atomic spectrum. It had been noted that the lines in this spectrum form regular patterns,

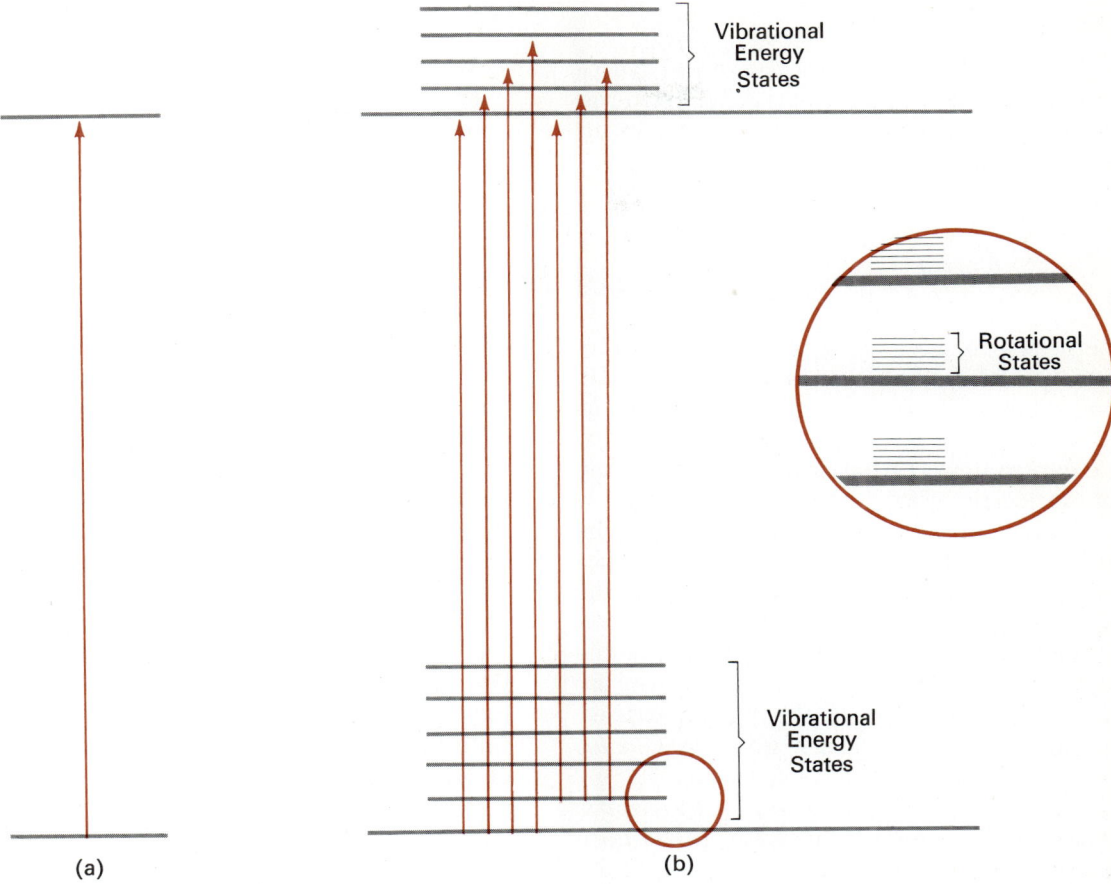

Figure 11-32 (a) *An electronic transition in an atom.* (b) *The complex set of transitions from one set of vibrational states in a molecule to another set.*

or series, and that there are numerical relationships between the frequencies (and hence energies) of the lines in different series. These relationships can be visualized in terms of a strange staircase with steps that get smaller as we ascend. A bold step toward understanding why the hydrogen spectrum is as it is was taken by Niels Bohr. He proposed a model of the atom in which the electron could be in one of many circular orbits around the proton. Only in the smallest of these orbits is the electron indefinitely stable. When it absorbs energy it goes to one of the outer orbits, but it soon drops back to a lower one, and eventually to the lowest one. Each drop to a lower energy level is accompanied by the emission of a photon. Bohr's model enabled him to calculate the energies of the electron in the various orbits. The differences in energy between orbits were found to be equal to the energies of the observed spectral lines.

Self Check

1. How do the emission spectra of the hydrogen atom and other atoms differ from the spectrum emitted by a glowing tungsten ribbon?
2. Describe the appearance of a series of lines in the hydrogen emission spectrum.
3. What is the relationship between the energy of an electron and the number of its orbit according to Bohr's model of the hydrogen atom?

Chapter Summary

Light is a form of energy. It has the characteristics of both waves and particles. Its wave nature is seen in the phenomena of diffraction and interference, while its particle nature shows up particularly clearly in the photoelectric effect.

Because of its wave nature, light may be specified by its wavelength, λ, the distance between peaks, and by its frequency, ν, the number of peaks passing a given point per second. The speed of light, c, which is the same for all frequencies, in air, is given by $c = \lambda\nu$. The energy of the light particles, called photons or quanta, is given by $E = h\nu$, where h is called Planck's constant.

When atoms are energized by an electric discharge they emit light. After it has been passed through a spectrograph, the light appears in the form of narrow lines at specific frequencies. Each type of atom (that is, each element) has its own characteristic line spectrum. The hydrogen line spectrum is relatively simple. It consists of several series of converging lines, one in the ultraviolet, one in the visible, and others at lower frequencies. To account for this, we must assume that the atom can exist in a whole series of energy states. The lowest one is the only one that can persist. Thus, when an atom is raised to one of the higher states, it soon falls back to a lower one, and, eventually, to the lowest one, emitting light corresponding to the various energy differences in the process.

Niels Bohr tried to account in detail for the nature of these various energy states, beginning with Rutherford's model of the atom. He proposed that the electron could stay indefinitely outside the nucleus without ever falling into it, and that there must be a set of well-defined energies that it can have. These proposals were a departure from the ideas accepted by physicists of his time, but are now considered to be correct. He went a step further and set up a model in which each energy state of the electron was described by a circular orbit about the proton. Although this model can correctly predict the energies of the states and the frequencies of lines in the hydrogen spectrum, it is no longer considered a correct description of the atom.

Questions and Problems

1. Draw a simple diagram representing a wave with several crests and troughs. Indicate clearly what is meant by (a) wavelength, (b) frequency, (c) speed of the wave.
2. Draw two more diagrams, similar to the diagram in Question 1, representing red light and blue light of the same intensity. Explain any differences in the two drawings.

3. Draw a diagram of a light wave produced by a very bright red light. Then show a wave produced by a faint red light. (The intensity of a light wave is analogous to the height of a water wave.) Explain.
4. Write an equation relating speed, wavelength, and frequency for a beam of light. What quantity in this equation is constant? Explain why long wavelengths have lower frequencies than short wavelengths.
5. Explain precisely the difference between diffraction and interference of light.
6. If ultraviolet and infrared light are not visible to the eye, what evidence do we have for their existence?
7. Use the equation $E = h\nu$ to explain why it is more dangerous to your eyesight to look at an ultraviolet light than it is to look at an ordinary incandescent light.
8. What is the photoelectric effect? What common piece of photographic equipment makes use of this effect?
9. A yellow light produced a current in the circuit shown in Figure 11-17 until an opposing voltage reduced the current to zero. In order to make a current flow again with the same opposing voltage, would you use a red light or a blue light? Explain your choice.
10. What is a photon? How do photons differ from each other? How are they the same?
11. If you closed your eyes and threw an empty tin can into the air, what evidence would there be that the can returned to the ground? (Assume it did not hit you!) What kind of waves would be produced? What attractive force had to be overcome to get the can into the air? Would the can have the most energy at the peak of its flight or on the ground?
12. When an electric spark or a high temperature flame forces an electron from its ground level to another energy level, what attractive force must be overcome? What evidence appears when the electron drops back to its starting place? What kind of waves are produced? At what point does the electron have the greatest amount of energy?
13. Using the energy level diagram of the hydrogen atom in Figure 11-25, explain in what part of the spectrum you would expect to find light produced by electrons dropping from levels 5 and 4 to level 3 and from level 5 to level 4.
14. Use the equation

$$E = R\left(\frac{1}{n_l^2} - \frac{1}{n_u^2}\right)$$

to calculate the energy emitted by electrons dropping from the third energy level to the second in the hydrogen atom. Now calculate the energy for the same electron drop by using Figure 11-25. Compare the results.
15. A helium atom has two protons and two electrons. How would you expect the first energy level of the two helium electrons to compare with the first energy level of a hydrogen electron? Explain.
16. The energy difference between two electron states is 46.12 kcal/mole. What will be the frequency of the light emitted when the electron drops from the higher to the lower energy state? Planck's constant = 9.52×10^{-14} (kcal-sec)/mole.
17. The frequency of a yellow line in an atomic spectrum is 5.2×10^{14} hertz. To what energy does this correspond (a) per mole of photons; (b) per photon?
18. Chemical bonds have energies of about 100 kcal/mole. Could they be dissociated by light of such energy as that found in Question 17?
19. Suppose there are two electron transitions, one involving 100 kcal/mole and the other 10 kcal/mole. What frequencies are involved? In what spectral region or regions do they occur?

12 The Atom According To Quantum Mechanics

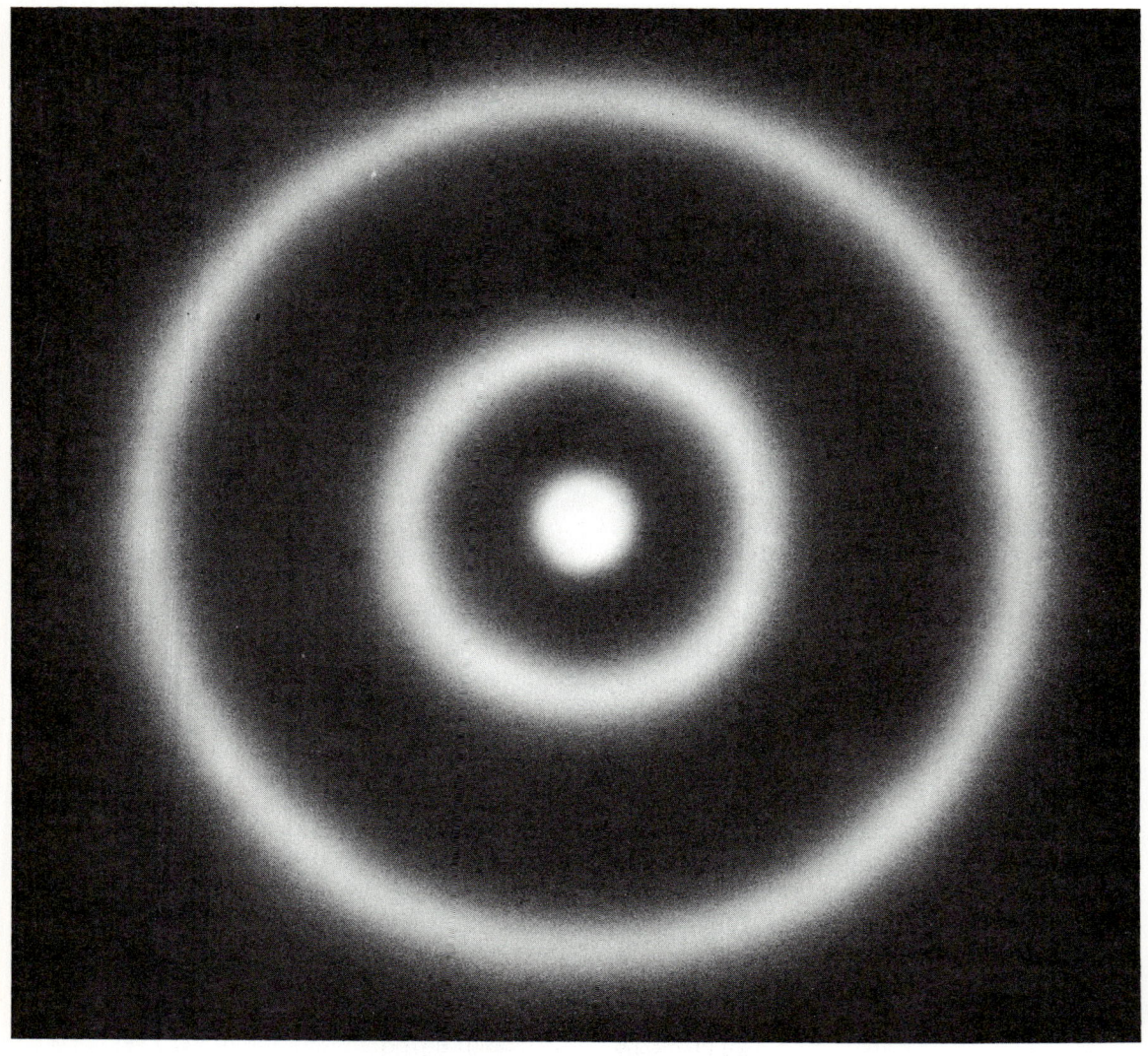

Wave intensities for the lowest energy electron in the third energy level of the hydrogen atom, as calculated from quantum mechanics.

In order to describe the atom and explain its behavior, scientists today use a theory called quantum mechanics. This theory was developed specifically to account for the behavior of very small bodies such as atoms and molecules, and their constituents, electrons and nuclei. Earlier "mechanics" had described successfully only the forces and energies associated with macroscopic objects.

Among the specific reasons for the development of quantum mechanics were the following ones. For one thing, the Bohr model of the atom was found to be unreliable except for the hydrogen atom. Furthermore, the experiments showed that electrons and other basic particles of matter have a wavelike character.

Using quantum mechanics, also called wave mechanics, a theory of atomic structure has been developed that applies to all atoms and accords with the wave properties of electrons. It forms the basis for understanding the periodic table and the nature of the chemical bonds in molecules.

From the Bohr Atom to the Atom of Quantum Mechanics

12-1 Bohr's Model Fails for Other Atoms

Although Bohr's model of the hydrogen atom was very successful in accounting for the spectrum of that atom, it was soon discovered that it could not explain the spectra of other atoms. It was not just a matter of getting incorrect values for the frequencies of spectral lines. There was the more serious problem that other atoms had much more complicated spectra, and the Bohr model could not say why this should be so. Naturally, after its brilliant success with the hydrogen atom, the model was not abandoned lightly. Great efforts were made to modify or extend it so that it would become generally useful. No such efforts were successful, however, and finally scientists had to admit, reluctantly, that this model could only contain a part of the truth about atomic structure. Something important was missing.

12-2 Matter Is Found To Have Wave Properties

We have noted that scientists were unable to make the Bohr model of the atom give an accurate and complete account of the spectra of atoms other than hydrogen. But they had other grounds for doubting this model, as well. In the Bohr model, the electron was thought of as a well-defined little particle traveling in one of a set of precisely-defined orbits about the nucleus. The assumption that the electron could be treated purely as a particle was shown to be wrong by the results of other experiments. Naturally then, a model of the atom based on such an assumption would have to have been doubted for that reason alone, even if it had worked better.

You will recall that in order to explain *all* the different ways in which light behaves, we are forced to think of it as having, somehow,

Figure 12-1 *A perspective view of a typical arrangement of atoms. This is the way the atoms are arranged in a number of metals, such as aluminum.*

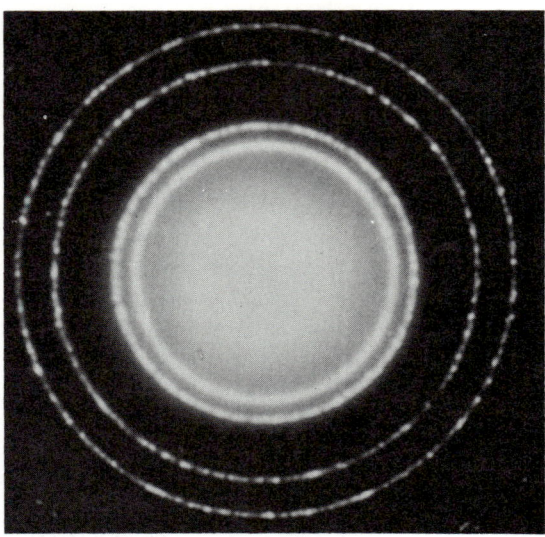

Figure 12-2 *This scattering pattern was produced when a beam of electrons was directed through aluminum foil.*

both particle-like and wave-like properties. In 1924, a French physicist, Count Louis de Broglie, wondered whether very small particles of matter might not also have wave characteristics when they moved rapidly. He derived an equation for the "wavelength" of the particle in terms of its mass and velocity. It turned out that for a beam of electrons accelerated by several kilovolts, the wavelength would be around 1 Å.

You have seen evidence for the wave nature of light. For example, light is diffracted when it passes through holes or slits of a size similar to its own wavelength. Visible light has wavelengths in the range of 0.0005 millimeter. Diffraction and interference effects can be observed using slits of the order of 0.1–0.01 mm, which are fairly easy to construct. But how can holes or slits as tiny as 1 Å, 10^{-8} cm, be constructed? The answer is in two parts: (a) they cannot be; (b) they do not need to be, because nature has provided them, ready-made.

The atoms in crystals are lined up in nice even rows (as in Figure 12-1) and these rows are a couple of ångströms apart. (The exact separation varies from one substance to another, and depends on the size of the atoms.) Therefore, if an electron beam does have wave character, and a wavelength of a few ångströms, one ought to be able to see diffraction and interference effects when such a beam passes through a thin slice of crystalline material. Such an experiment was done by Davisson and Germer, two American scientists at the Bell Telephone Laboratories, in 1926. They passed an electron beam through metal foil. A typical result, obtained with aluminum foil, is shown in Figure 12-2. Clearly, de Broglie's idea was right. Furthermore, the exact spacing of the rings in the diffraction pattern showed that his equation for calculating the wavelength of electrons was correct.

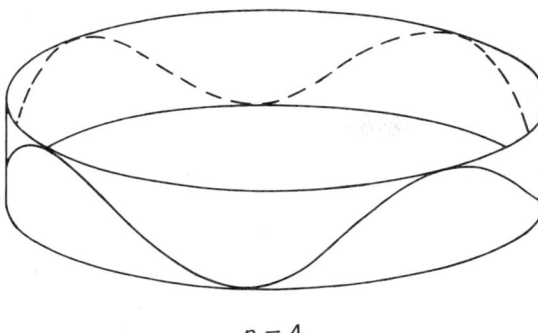

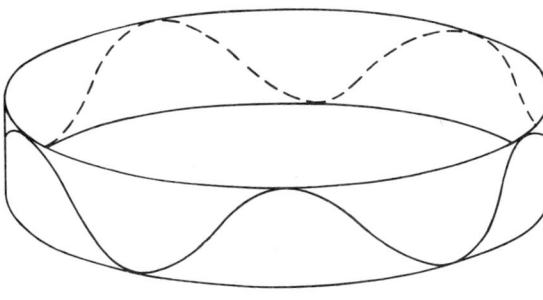

Figure 12-3 *Bohr orbits with electron waves. The orbit that Bohr labeled n has a wave of n wavelengths, according to de Broglie.*

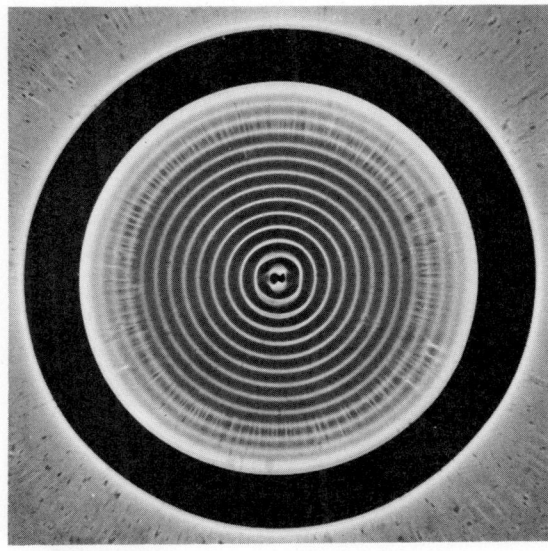

Figure 12-4 *A time exposure of standing waves in a circular ring. The entire container was vibrated, but the moving waves outside the ring are blurred out.*

De Broglie then applied his theory of the wave character of electrons to the hydrogen atom, and made a very interesting observation. He calculated that an electron traveling at the speed Bohr's model required in the lowest orbit has a wavelength exactly equal to the circumference of that orbit. For the second orbit, the circumference is equal to two wavelengths. In general, for the nth orbit, the circumference is found to be equal to n times the wavelength. This is illustrated in Figure 12-3. Thus, de Broglie's ideas led to a model of the hydrogen atom in which the orbits were replaced by wave patterns that exactly fitted the circumference. The wave patterns must therefore reinforce themselves, producing what are called "standing waves." (Standing waves can be observed, on the macroscopic level, in a number of situations. One example is shown in Figure 12-4.)

De Broglie's idea of standing waves was very interesting, and the tie-in of the wavelengths with the sizes of Bohr's orbits strongly suggested a connection between the wave and particle pictures. But there was still no complete theory of how the arrangement of the electron distribution was related to the energy states of atoms. Bohr's and de Broglie's ideas were like pieces of a jigsaw puzzle. They gave some hint of what the whole picture might look like, but not a complete view of it.

Louis Victor de Broglie (1892–)

SINCE THE MIDDLE 1600's, many members of the de Broglie family have been leaders in cultural, political, and scientific affairs in France. In 1929 Louis de Broglie was awarded the Nobel Prize in Physics for "his discovery of the wavelike nature of electrons."

For many years, scientists had observed that light under some conditions behaved as though it were made up of a series of waves. They measured the wavelength and frequency of light and were able to interpret such phenomena as refraction and diffraction by applying the principles of wave motion to their observations. In many instances, however, light behaved as though it were made up of tiny particles, often referred to as "corpuscles." Eventually the scientific community came to accept this dual nature of light.

In the 1890's J. J. Thomson measured the ratio of charge to mass of an electron. When Millikan in the early 1900's determined the charge on an electron, the mass of an electron was easily calculated. It seemed obvious that an electron was a particle with a fixed amount of mass.

As other scientists struggled to extend Bohr's model of the hydrogen atom to more complicated atoms, the concept of an electron possessing only properties of a particle became more and more difficult to accept. De Broglie studied Planck's quantum theory and Einstein's explanation of the photoelectric effect. In the middle 1920's he formulated a hypothesis stating that an electron with a given mass and speed should also possess a periodic behavior usually associated with waves. While he had no direct experimental evidence to confirm his theory, he worked out the mathematical details of the behavior that should be expected.

The genius of de Broglie's predictions was proved in 1927 when physicists Clinton J. Davisson and Lester H. Germer discovered that electrons could be diffracted by crystals. This could only take place if electrons actually did possess the properties of waves. This discovery led to the new science of quantum mechanics that became extremely helpful in examining and predicting the behavior of electrons in atoms.

Summary

The Bohr model of the hydrogen atom, in which the electron is viewed as a discrete little particle of negative charge revolving in one or another well-defined orbit about the nucleus, could not be extended to explain the spectra of other atoms. At the same time, it was found that electrons and other small "particles" also have wave properties. The Bohr orbits for the hydrogen atom turned out to correspond to standing wave patterns for the electron. The stage was set for a new comprehensive theory of atomic structure.

Self Check

1. What two main reasons can you give to explain why Bohr's model of the atom had to be given up?
2. Describe an experiment showing that a beam of electrons behaves like waves.

Quantum Mechanics Explains Atomic Structure

INVESTIGATION

12-3 *Probability*

The concept of probability affects each of us in many different ways. Each child that is born faces a certain probability of catching a contagious disease by the age of five years. In some societies this probability is greater than in others. The careless driver who fails to stop at a "Stop" sign gambles with the probability that there is no car coming in another direction. Life insurance rates are based on the probability that you will live to some predetermined age, based on the life histories of many thousands of other people. Bettors on horse races spend a lot of time studying the records of the horses in a particular race in an effort to increase the probability of their winning. Poker players attempt to determine roughly the probability that the other players are holding a stronger hand than they are. Probability is also important in the field of atomic structure, as you will learn in the next section. It has been determined that one can only estimate the probability of finding an electron at certain locations around a nucleus at any given instant. While it is impossible to duplicate the conditions inside an atom, in this experiment you will study the probability of certain objects coming to rest at given distances from a target. In this way, you will be better able to understand the probability of finding an electron.

PURPOSE

You will determine the most probable distance from a bull's-eye target that a penny will come to rest when it is pitched at the target from a fixed distance.

MATERIALS

Ruler   Graph paper
Chalkboard compasses or chalk and string
10 (or more) objects of uniform size (such as pennies or steel washers)

PROCEDURE

1. With a piece of chalk, draw a circle on the floor with a radius of about one inch. Now use a piece of chalk in a chalkboard compass or tied to a piece of string to draw on the floor a series of concentric circles with radii of 3, 5, 7, 9, 11, 13, and 15 inches, respectively. Inside the smallest circle write the number "1." Inside the spaces between the circles write in consecutive order the numbers "3," "5," "7," through "15."

2. In your laboratory notebook prepare a table with eight columns headed with the same numbers that are in the circles on the floor.

3. Now draw a line on the floor three feet from the center of the series of concentric circles. Stand behind this line and pitch pennies, one at a time, toward the center of the circle. Record the number of the space in which each penny lands. To be consistent, record the lower of the two numbers when a penny lands on a line. Do not count pennies that roll outside the target. Continue pitching pennies until you have recorded at least one hundred pennies that have landed within the outermost circle. Add up the numbers in each of the eight columns.

DATA AND RESULTS

Prepare a bar graph from your data. Under consecutive spaces on the horizontal axis write the numbers 1, 3, 5, 7, 9, 11, 13, and 15 to represent the numbers of the spaces on the floor. On the vertical axis prepare a scale to include the largest number of pennies landing in any one space. Plot your results on the graph.

Interpretation of Results

1. On your graph, what is the most probable distance from the center of the circle for a penny to come to rest? If you pitched one more penny, where would you expect it to land?

2. Compare your graph with those of other class members. What can you conclude from your comparison?

RELATED PROBLEMS

1. List several things you could do to increase the probability of the pennies landing in the center of the target.

2. If you were pitching small steel washers at a similar target with a magnet in the center, what differences would you expect to find when you plotted a graph of the results? In what ways would this be somewhat like an atom with its nucleus and surrounding electrons? How would it differ?

ADDITIONAL INVESTIGATION

Continue the experiment outside of class. Pitch five or six hundred more pennies and prepare a graph for each one hundred pennies landing on the target. Do you find any progressive change taking place from one graph to another? Does your skill seem to improve? Is there any point at which factors beyond your control seem to limit any improvement? What could some of these factors be?

12-4 Orbitals and the Principal Quantum Number

With the work of de Broglie and others showing the wave nature of the electron, the stage was set for the development of a new theory of matter. This theory is called *quantum mechanics*, or, sometimes, wave mechanics. It was developed in the years 1925-30 and was quickly proved capable of coping with the problem of how electrons are arranged in atoms and what their energy states can be.

With respect to electrons in atoms, the most important result obtained from quantum mechanics is a scheme of *quantum numbers*. The behavior of an electron in an atom can be specified by stating its quantum numbers. The most important quantum number is called the **principal quantum number, n.** This turns out to be identical to the number n in Bohr's model. The advantage of quantum mechanics is that it tells how to calculate the values of still other quantum numbers, all of which are needed to fully explain the structures and energies of atoms that are more complex than the hydrogen atom.

In quantum mechanics the electron is no longer thought of as a discrete particle in a well-defined orbit around the nucleus. The word orbit, as such, is not used. However, in order to have a name for the various states of the electron in an atom, the word "orbital" is used.

An **orbital** is a spatial region for the electron around the nucleus. Each orbital is identified by a set of quantum numbers, the most important of which is the principal quantum number, n.

There is a most important difference between Bohr's orbits and the orbitals of quantum mechanics. When we know everything that it is possible to know about an orbital, we still do not know the precise path the electron takes, nor precisely where the electron will be at a particular instant.

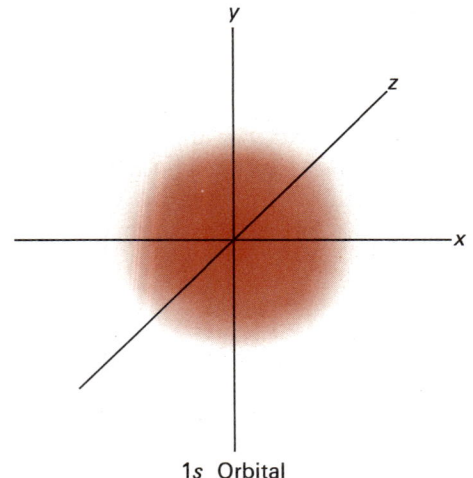

Figure 12-5 *The 1s orbital is represented by a cloud. The darker the cloud, the more likely it is that the electron will be found there.*

All that can be done is to predict the probability of finding an electron at a given point in space. This probability, considered over a period of time, gives an "averaged" picture of how an electron behaves. It is this description of the electron motion that is called an orbital.

Figure 12-5 shows how the most stable orbital for the electron in a hydrogen atom may be pictured, according to quantum mechanics. It does not show the actual location of the electron at any given time. Instead, it shows where the electron is most likely to be at any time. The denser the shading the more likely it is that the electron will be in that region. It can be seen that for this orbital, the electron is most likely to be near the nucleus. The probability of finding it farther from the nucleus decreases steadily, in all directions.

An orbital description of the motion of an electron contains the same information conveyed by the graph you obtained from the penny-pitching experiment, or by the holes made by darts in a dartboard. After a dartboard has been used in many games, the distribution of holes shows how successful earlier players have been in scoring. There are many holes near the bull's-eye and, moving away from it, there is a regular decrease in the number of holes per square centimeter of dartboard. The pattern of holes does *not* tell where the next throw is going to land. However, at any given distance from the bull's-eye, the "density" of dart holes (number per square centimeter) is a measure of the probability that the next throw will land there.

The holes in the dart board tell *only* the probability that a given throw will land at a particular distance from the bull's-eye. It does *not* tell the order in which the holes were made in the dartboard. In the case of the electron distribution, the orbital tells us the probability that an experiment designed to locate the electron will find it at a particular distance from the nucleus. It does *not* tell how the electron moves from point to point; that is, it does not describe its trajectory.

There is another way to interpret the meaning of an orbital. Because of its wavelike nature when in motion, the electron, as it moves about the nucleus, is considered to be spread out over a large region of space, rather than being a small, discrete particle. Therefore, a description of what it is doing must be a "smudged-out" picture, an "electron cloud."

The equations that quantum mechanics gives for orbitals all depend on the principal quantum number. There is, in general, more than one orbital for each principal quantum number. The rule is that there are n^2 orbitals of quantum number n. For $n = 1$, there is only one orbital. For $n = 2$, there are $n^2 = 2^2 = 4$ orbitals, and so on.

All orbitals of a given principal quantum number have the same energy in the hydrogen atom (though not in others). Just as in the Bohr model, this energy is given by the following equation.

$$E = -\frac{R}{n^2} = \frac{-313.6}{n^2} \text{ kcal/mole}$$

12-5 The Uncertainty Principle ★

As you have seen, the electronic structure of an atom, as described by quantum mechanics, is not a simple, sharp picture. It is not possible to say that, in the hydrogen atom, for example, the electron is at a certain point, at a certain time, and moving with a certain velocity. This is why Bohr's picture of the hydrogen atom had to be abandoned. He had assumed that such statements could be made about electrons in atoms.

The reason for the unavoidable diffuseness of our description of the electron in the hydrogen atom, or of the electrons in any other atom or molecule, is expressed in a fundamental law of physics. This law is called the *Uncertainty Principle*. It may be written or expressed mathematically in various ways. For the present discussion, the appropriate form follows.

$$(\Delta x)(\Delta v) = h/m$$

The letters x and v stand for the position and velocity, respectively, of a particle. The symbol Δ means "the uncertainty in" the quantity following it. Thus Δx is the uncertainty in our measurement of the position, and Δv means the uncertainty in our knowledge of the velocity. The letter h is Planck's constant, 6.63×10^{-27} g·cm²·sec⁻¹, and m is the mass of the particle. Since h is such a small number, it is only for particles of very small mass that the uncertainty is important. For instance, suppose you were to throw a stone with a mass of 50 grams. The product of the uncertainties in its position and velocity could be calculated as follows.

$$(\Delta x)(\Delta v) = \frac{6.63 \times 10^{-27} \text{ g} \cdot \text{cm}^2 \cdot \text{sec}^{-1}}{50 \text{ g}}$$

$$= 1.33 \times 10^{-28} \text{ (cm)}\left(\frac{\text{cm}}{\text{sec}}\right)$$

The product of the uncertainties in the position (in cm) and in the velocity (in cm/sec) is extremely small. According to the uncertainty principle, we can have information on the position and velocity of macroscopic particles, such as the 50-gram stone, to an accuracy far beyond what we are ever likely to want or need. The uncertainty principle applies to macroscopic objects, but the limitations it sets are of no practical importance.

For very tiny particles, such as electrons, however, the limitations on our knowledge are of key importance. An electron has a mass of only 9.10×10^{-28} g. Thus, the product of the uncertainties is as follows.

$$(\Delta x)(\Delta v) = \frac{6.63 \times 10^{-27}}{9.10 \times 10^{-28}} \text{ cm}^2/\text{sec}$$

$$= 7.3 \text{ cm}^2/\text{sec}$$

Consider now an electron within an atom. The diameter of an atom is about 10^{-8} cm. If we require the electron to be within a sphere of that diameter, we are saying that we know the position of the electron to within an uncertainty, Δx, of about 10^{-8} cm. How fast is the electron moving? If we put $\Delta x = 10^{-8}$ cm into the uncertainty relationship and solve for Δv we get the following.

$$\Delta v = (7 \text{ cm}^2 \text{ sec}^{-1})/(10^{-8} \text{ cm})$$

$$= \text{about } 10^9 \text{ cm/sec}$$

The uncertainty in the velocity of the electron is enormous, namely, about a billion cm sec^{-1}. Simply by requiring the electron to be located somewhere within a sphere that is approximately the size of the atom, we must accept an enormous uncertainty as to its velocity. If we tried to limit the uncertainty in the position of the electron still more by restricting it to a particular orbital pathway around the nucleus, the uncertainty in its velocity would become even greater. Indeed, the uncertainty in the velocity is so great that it makes no sense to persist with this type of picture.

Because of the uncertainty principle, it is necessary to give up any hope of describing the distribution of electrons in a particular atom or molecule in the same sense that you can describe the position and velocity of a stone that you throw. This may at first sight seem to deny us all hope of understanding the inner workings of atoms and molecules, but that is not actually the case. As you have seen, we can give a description of an atom in which the electron is treated as a cloud of negative charge with a certain distribution around the nucleus. This may seem, as indeed it is, much less precise than an exact specification of the position and velocity of the electron. Yet it is adequate for the calculation of all the properties of an atom that we can actually observe or measure. We can calculate its energy, its angular momentum, the charge on an ion, and other important quantities. It is important to realize that a theory which allows us to calculate all of the things we can actually measure is an entirely satisfactory theory. There is no need, nor any use, in asking for a theory which deals with things that cannot be observed. In a very real sense, things that cannot be observed do not exist. At least, they are of no concern.

The basis for the uncertainty principle is, in fact, a consideration of how measurements are made. You know, for example, that you can measure the position and velocity of a macroscopic object, such as a stone, by observing its path. We can see the stone because it reflects visible light. The stone has dimensions of the order of centimeters, while the wave length of light is of the order 10^{-15} cm. The light waves are very small compared to the size of the macroscopic object.

How would you observe an electron? If you used visible light, you would see nothing. The electron is much too small. So you try what seems to be the obvious thing. You use radiation with a wavelength as small as, or smaller than, the size of the electron. You might try using some X rays of very short wavelength. You are still in trouble. The photons in radiation of such short wavelengths have so much energy that each time even one photon strikes an electron it gives it a tremendous push. The very attempt to measure the position and velocity of the electron will change both the position and the velocity. (See Figure 12-6.) The attempt to make this measurement introduces great uncertainty into the result. The uncertainty principle is simply a statement of this fact.

12-6 The Shapes of s Orbitals

Consider the lowest energy level of a hydrogen atom, for which $n = 1$. There are n^2 levels with this energy. Since $n = 1$, there is only one level. It corresponds to an electron distribution that is spherically symmetrical around the nucleus, as shown in Figure 12-5, and is called the 1s orbital. In the symbol 1s, the number one tells you that $n = 1$. The letter s tells you that the orbital is spherically symmetrical. An electron moving in an s orbital is called an s electron.

Since the letter s has been used for a spherically symmetrical orbital, you might as

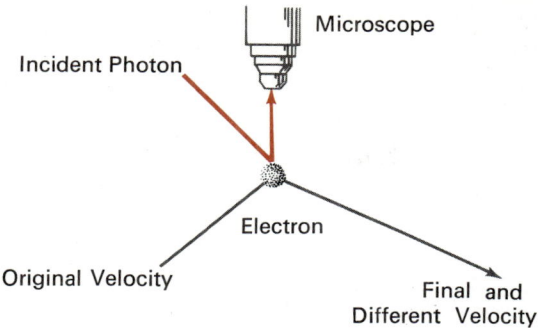

Figure 12-6 *The very attempt to observe an electron in a microscope would change its position and velocity.*

well think of s as an abbreviation for "spherical." Actually, the symbol was first used by early spectroscopists in describing a "sharp" series of lines in a particular spectrum. Other symbols that are used in designating orbitals originated in a similar way. The symbols p, d, and f, which you will soon encounter, originally came from the words "principal," "diffuse," and "fundamental," used by spectroscopists to designate other series of spectral lines.

The term "spherically symmetric," and the picture of an s orbital (Figure 12-5) must be clearly understood. Suppose you were to look for the electron somewhere on the surface of a sphere having a particular radius r_1, and the nucleus at its center. The probability of finding the electron at any one point on the r_1 sphere is the same as the probability of finding it at any other point on the r_1 sphere. The same would be true at a different radius, r_2, but the probability of finding the electron somewhere on the r_2 sphere would not be the same as that on the r_1 sphere. The chance of finding the electron depends upon the radius of the sphere. A $1s$ electron can be found anywhere from right at the nucleus to a great distance away, but it is most likely to be found approximately 0.5×10^{-8} cm (half an ångström) from the nucleus.

The next energy level corresponds to $n = 2$. According to the rule, there are $n^2 = 2^2 = 4$ different spatial arrangements with the same energy,

$$\frac{-313.6}{4} = -78.4 \text{ kcal/mole.}$$

One of them is again spherically symmetric and it is called the $2s$ orbital. Figure 12-7 shows the atomic $2s$ orbital. Here we find the reasonable result that the higher energy of the $2s$ electron permits the electron to spend more time far from the nucleus.

For every value of n, there is one spherically symmetric orbital.

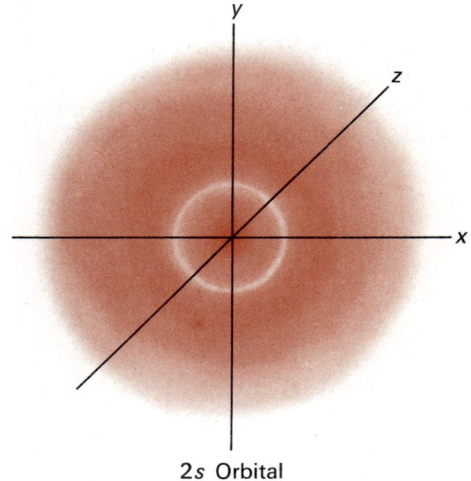

Figure 12-7 *The 2s orbital is represented by a spherical cloud having two dense regions, one close to and one farther from the nucleus.*

As n increases, the ns orbital places the electron, on the average, farther and farther from the nucleus.

12-7 p Orbitals

There is only one orbital corresponding to $n = 1$, the $1s$ orbital. For $n = 2$, there are four different spatial arrangements and one of them, the $2s$ orbital, has already been described. The other three are called $2p$ orbitals. An electron in a p orbital behaves in such a way that it is most likely found in either of two regions located on opposite sides of the nucleus. The motion of a p electron creates an electron distribution that is shaped somewhat like a dumbbell. The axis of the dumbbell can be placed along one of the three perpendicular Cartesian coordinate axes. Just as there are three distinct coordinate axes, there are three distinct p orbitals, each with its axis perpendicular to the other two. They are sometimes referred to as the p_x, p_y, p_z orbitals to emphasize their directional character. The p_x orbital is concentrated in the x direction. A p_x electron is more apt to be found near the x axis than anywhere else. The p_y orbital, on the other hand, is concentrated along the y axis. These directional characteristics are useful in explaining the geometrical properties of molecules. Figure 12-8 shows the electron distribution of the $2p$ orbitals. Since the three p orbitals can be considered to extend along three perpendicular axes (x, y, and z), p might be thought of as an abbreviation for "perpendicular."

Every energy level with n *greater than one has three* p *orbitals.*

As n increases, the np orbitals place the electron, on the average, farther and farther from the nucleus, but always with the axial directional properties shown in Figure 12-8.

12-8 d and f Orbitals

At this point you might recast the hydrogen atom energy level diagram to express what you know about orbitals. For a given value of n, there are n^2 total orbitals. For $n = 3$, there are $3^2 = 9$ orbitals, five more than accounted for by one $3s$ orbital and three $3p$ orbitals. These five new orbitals are called d orbitals and they have more complicated spatial distributions than do the p orbitals.

From the information that the numbers of s, p, and d orbitals are 1, 3, and 5, how many

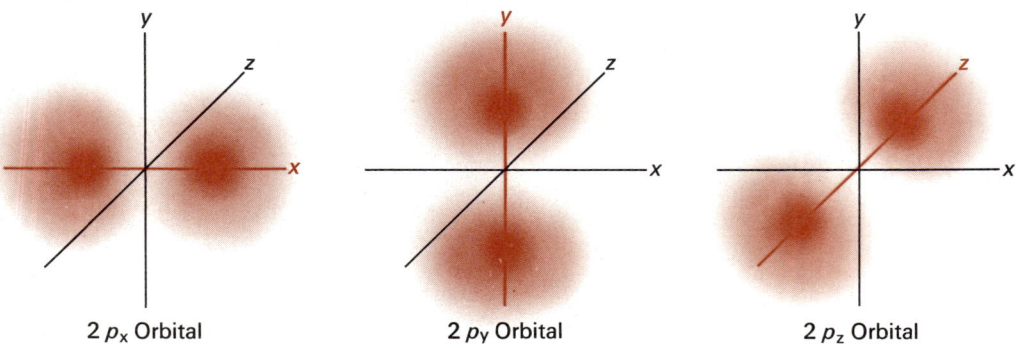

2 p_x Orbital   2 p_y Orbital   2 p_z Orbital

Figure 12-8 *The p orbitals are aligned with the familiar x, y, and z axes of the Cartesian coordinate system.*

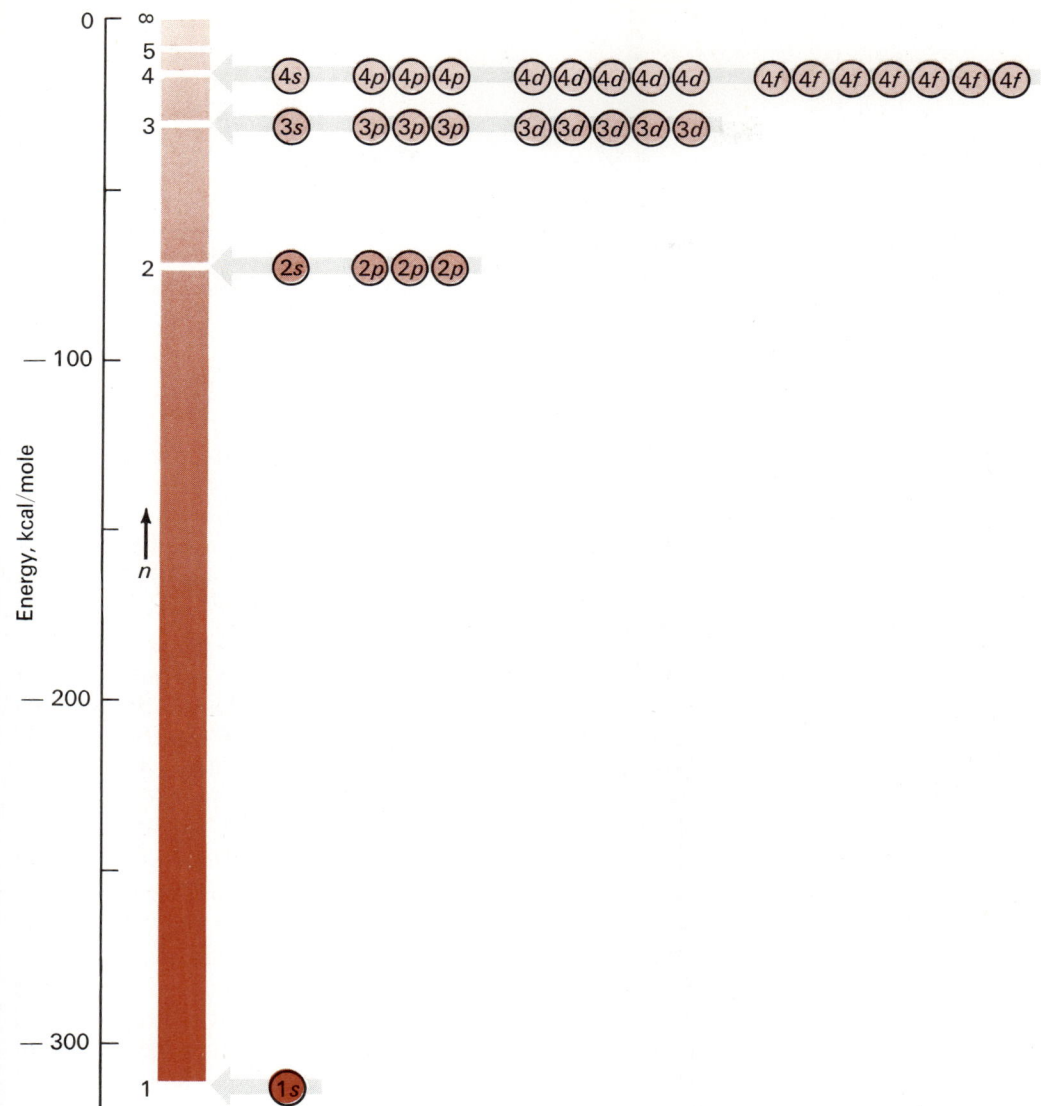

Figure 12-9 *This schematic diagram relates each possible orbital (shown as a circle) to the vertical energy-level scale at the left. Since for every given value of n there are n² orbitals, it is necessary to assign higher level orbitals, the d and f orbitals.*

of the next higher (f) orbitals would you expect? Verify your answer by calculating n^2 for $n = 4$ and comparing your answer to the total number of s, p, d, and f orbitals as shown in the schematic drawing, Figure 12-9.

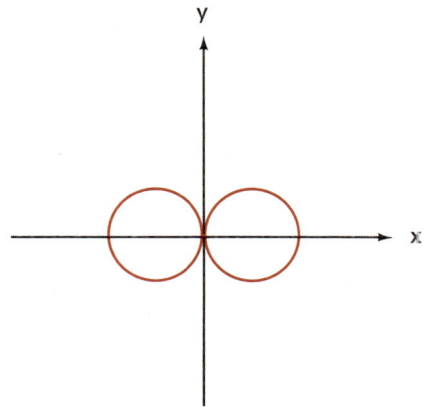

Figure 12-10 *A section, or cut, through the $2p_x$ orbital in the xy plane. The entire orbital can be generated by rotating this cut about the x axis.*

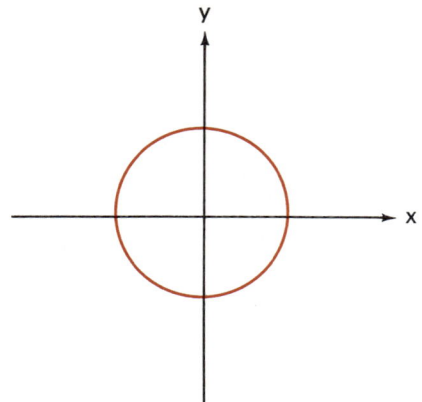

Figure 12-11 *A section, or cut, through the 1s orbital in the xy plane. The orbital can be generated by rotating the circle to produce a sphere.*

12-9 More About the Shapes of Orbitals

Figures 12-5, 12-7, and 12-8 are attempts to show where an electron in a particular orbital is most likely to be found with respect to the nucleus. This type of drawing is probably the best one for the purpose, but it is not an easy thing to draw. If you were trying to make sketches of orbitals on a piece of paper, you would find it difficult and tedious to make such shaded sketches. Chemists need to make sketches of orbitals quite often when discussing the bonds between atoms that hold molecules together. (This topic is discussed in Chapters 14 and 15.) As a result, a simpler type of drawing has been developed.

Suppose that instead of trying to show in great detail how the probability of finding the electron varies, we content ourselves with just showing the outline of the region within which the probability is fairly high. Suppose we say that we want to outline the region within which we expect to find the electron 90 percent of the time. This still poses a problem because the surface we want to represent exists in three-dimensional space, like the surface of an inflated balloon.

What can be done is illustrated in Figure 12-10, for the $2p_x$ orbital. The outline drawn is only a two-dimensional cut through he complete surface. In order to get the full shape of the orbital, you would have to rotate this outline around the x axis. That would generate a balloonlike surface, which is what you should really think of as representing the orbital. Similarly, the 1s orbital is represented by the sketch shown in Figure 12-11. Again the actual surface that you should think of would only be obtained if the circle were rotated so as to generate a sphere.

The advantage of these very simple line drawings is that they are convenient to draw on a piece of paper. You should always keep in mind their limitations, however. First of all, they simply hint at the full three-dimensional shape of the orbital. You must make up for this deficiency mentally. (The models

Figure 12-12 *Models showing the full three-dimensional shapes of s and p orbitals.*

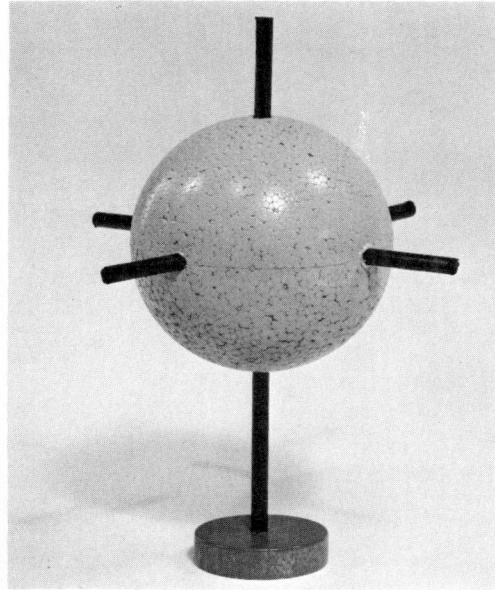

shown in Figure 12-12 give a better idea of the three-dimensional aspect.) Second, even in three dimensions, the surfaces are an over-simplification because there is some probability that the electron can be outside the surface, and also because the probability is not uniform everywhere within the solid.

There is another important point that should be made about orbitals. It is not necessary for you to be concerned about it for the time being, but it will be important later. The shapes of the orbitals, such as those you have just been looking at, are obtained by solving some fairly complex equations. Quantum mechanics is actually the theory of how to write down and solve the necessary equations. When the equations are solved they do not give the shapes of the orbitals *directly.* Instead, they give what are called *wave functions.* The actual probability of finding the electron at each point in space is given by the *square* of the wave function.

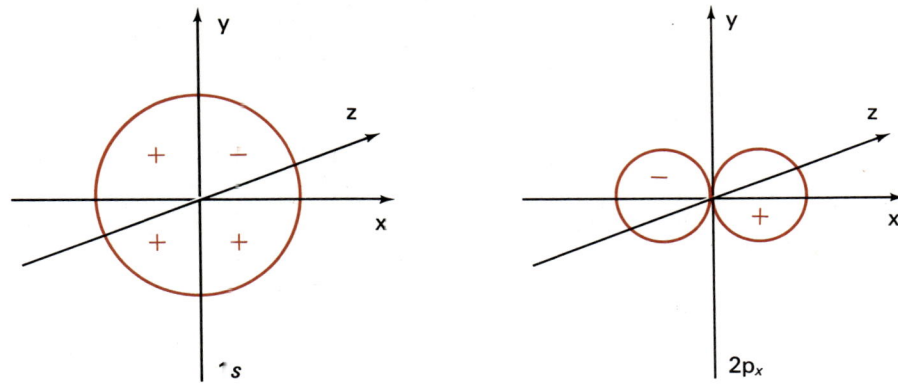
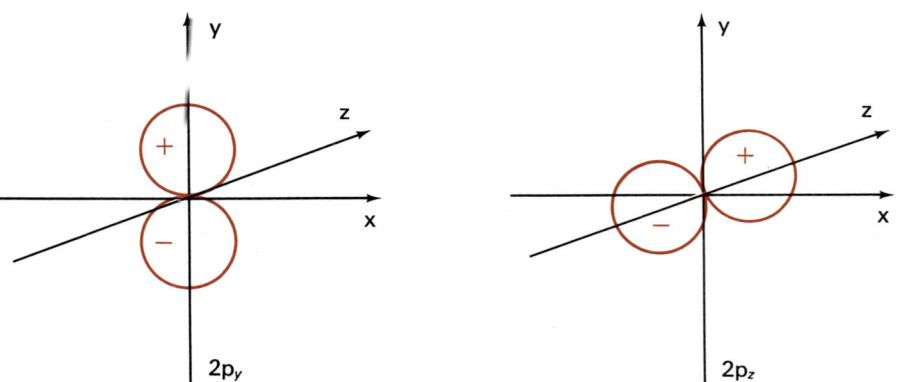

Figure 12-13 *Sketches of the actual wave functions for the 1s and 2p orbitals, showing the algebraic signs of the functions.*

The orbital pictures that have been given for the s and p orbitals are plots of the squares of the wave functions for the orbitals.

The wave functions themselves can have positive or negative signs at different places. Since the square of a number is always positive, even if the number is negative, the probability of finding an electron at a given point is always positive (or zero). This is as it must be, since the idea of a negative probability has no meaning. However, pictures of the wave functions themselves show some negative regions as well as positive ones. As you will see in Chapter 15, there are times when we need to know which regions of the wave functions are positive and which regions are negative.

Figure 12-13 shows the actual wave functions for the $1s$ and $2p$ orbitals. The s wave function is positive everywhere. On the other hand, each p orbital is one-half positive and one-half negative. The square, of course, is everywhere positive. The probability of the electron in a p orbital being found in either half is the same. The wave functions have opposite signs in the two halves and this will make a difference in certain connections, as you will see later.

Summary

According to quantum mechanics, the electron in a hydrogen atom must occupy one or another of many orbitals. An orbital differs from a Bohr orbit in not being a precise path along which the electron moves. Instead, it is a region in which the electron is most likely to be found. The various orbitals are identified by quantum numbers, of which the most important is the principal quantum number, n. The number of orbitals with quantum number n is n^2. Therefore, there are $1, 4, 9, \ldots$, orbitals of principal quantum numbers $1, 2, 3, \ldots$, respectively. For every value of n there is one spherically symmetrical orbital called an s orbital. For every principal quantum number from 2 on, there is also a set of three orbitals called p orbitals. These orbitals are shaped like dumbbells, with one lying along each of the Cartesian $(x, y, \text{ and } z)$ axes. Beginning with principal quantum number 3 there are sets of d orbitals, and beginning with $n = 4$ there are sets of f orbitals.

Self Check

1. How many orbitals are there with principal quantum number 2? What are the symbols for them?
2. How do the shapes of the p orbitals differ from the shapes of s orbitals?
3. In what ways do the three p orbitals of a given quantum number n differ from each other?

Chapter Summary

The Bohr picture of the hydrogen atom could not be extended to account for the properties, particularly the spectra, of more complex atoms. Scientists were forced to conclude that it lacked some essential features of a true picture. When experiments confirmed de Broglie's idea that moving electrons have the properties of waves and cannot be thought of simply as little particles of negative charge (as was assumed in Bohr's model) the stage was set for a more comprehensive theory of atomic structure. This was provided by the theory called quantum mechanics.

According to quantum mechanics, the whereabouts of an electron in an atom is described by an orbital. An orbital does not tell us precisely where an electron is at a given instant, nor precisely the path that it is following. Instead it tells us which are the regions around the nucleus where the probability of finding the electron is greatest. The orbitals can be classified, first, according to their principal quantum number, n. For each value of n there are n^2 orbitals. For every n there is an s orbital, which is spherical in shape. For $n = 2$ and higher, there is a set of three p orbitals. Each p orbital is shaped like a dumbbell, and lies along the x, y, or z axis.

Questions and Problems

1. In what ways did Bohr's atomic model fail?
2. What did Bohr assume about the nature of electrons for his model?
3. How did de Broglie's ideas about the electrons differ from Bohr's?
4. How did Davisson and Germer confirm de Broglie's theory about the nature of electrons? What properties of electrons are similar to the properties of a light beam?
5. What is a standing wave? What happens to the crests and troughs of a standing wave? How were standing waves used to modify Bohr's model of a hydrogen atom?
6. Why is the expression *quantum mechanics* applied to the study of the structure of atoms? Explain the term *quantum* as well as the term *mechanics*.
7. How is the principal quantum number, n, of any atom related to Bohr's model of the hydrogen atom?
8. Explain carefully the difference between the terms *orbit* and *orbital*.
9. How is probability related to the concept of an orbital?
10. Compare the probability of finding an electron at a distance of 1 Å from the nucleus of any hydrogen atom with that of finding it at a distance of 2 Å from the nucleus of the same atom.
11. If two spheres with different radii r_1 and r_2 were drawn with the nucleus of an atom as their center, how would the probability of finding an electron on the surface of the sphere with radius r_1 compare with the probability of finding another electron on the surface of the same sphere? How would the probability of finding an electron on the surface of the sphere with a radius r_1 compare with finding it on the surface of sphere with radius r_2?
12. What is the characteristic shape of all s orbitals? How many s orbitals are possible for each principal quantum number?
13. What is the maximum number of p orbitals possible for each principal quantum number? Draw three perpendicular axes x, y, and z. Sketch in the shapes of the p orbitals that are possible for a single principal quantum number.
14. What is the relationship between the principal quantum number and the maximum number of orbitals possible for that energy level? How many orbitals are theoretically possible when $n = 7$?

15. What is the maximum number of orbitals possible for a principal quantum number of 4? How many s, p, d, and f orbitals would there be?
16. Electrons have been located only in s, p, d, and f orbitals in atoms in their lowest energy states. If another set of orbitals exists, beyond the f orbitals, how many orbitals would you expect to find in that set?
17. How might you explain the fact that p orbitals are arranged symmetrically around the nucleus?

13 The Atomic Basis for Periodic Properties

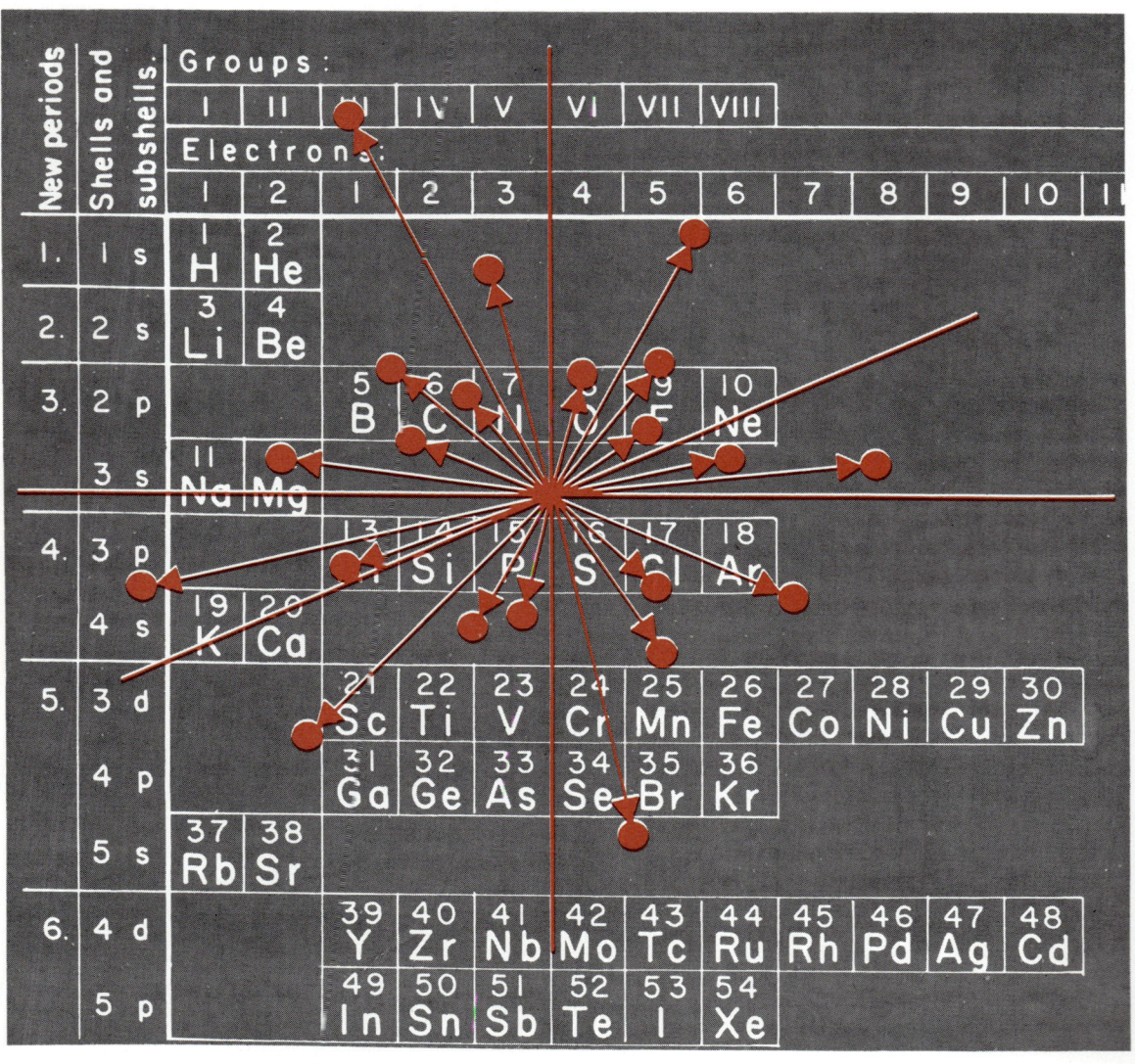

As you saw in Chapter 8, the periodic table was developed to correlate the observed chemical properties of the elements. It was based on experience and arranged to fit facts, but no one knew *why* this particular arrangement fit the facts. Since the chemical behavior of atoms must depend on how the atoms are structured, a complete and correct theory of atomic structure ought to enable us to understand the periodic table. Now that we have the quantum mechanical picture of the atom, including a knowledge of orbitals and their quantum numbers, we may return to the periodic table. We may now hope to see why this particular arrangement correlates the properties of the elements.

Building Up Complex Atoms

13-1 The Helium Atom, and Electron Configurations

What happens when there are two or more electrons around a nucleus? What arrangements do they adopt? Let us begin by considering the next most complex atom after hydrogen, the helium atom. The helium atom has two protons in its nucleus and hence two electrons that must occupy orbitals outside the nucleus.

Our approach to this problem will be based on the assumption that an electron in any atom will have available to it a set of orbitals like those in the hydrogen atom. Naturally, there must be some differences. When there is more positive charge on the nucleus, its greater attraction for the electron will cause all the orbitals to be at lower energies. Also, when two or more electrons are present, each one will be influenced by the repulsion of the other electron(s) as well as by the attraction to the nucleus. This may lead to some modification of the orbitals, or some changes in their relative energies. However, it is reasonable to expect a similar, though not identical, set of orbitals in all atoms. Let us adopt this assumption and see how it works out.

Assume, then, that for any atom the most stable available orbitals are arranged as shown in Figure 13-1 (next page).

In the following discussion, we shall be considering the addition of electrons to atoms, one at a time. Such a building-up process cannot be carried out experimentally. It is very useful, however, to carry out the process mentally. The electrons will be assigned to orbitals on the basis of what has been learned from other experiments.

Let us start with the nucleus of the helium atom. It is the same as an alpha particle, of course, with a charge of $+2$. Suppose that we introduce the first electron and allow it to find the most stable orbital. That, of course, is the $1s$ orbital, just as in the hydrogen atom. The fact that the nucleus of the helium atom has a charge of $+2$ instead of $+1$ makes the $1s$ orbital, the closest one to the nucleus, an even more stable place for that electron. We now have a helium ion, He^+.

Now let us add the second electron. You might expect that it too would go into the $1s$ orbital, and it does. For this second electron, the $1s$ orbital is a somewhat less attractive place than it was for the first one, because it is already occupied by an electron.

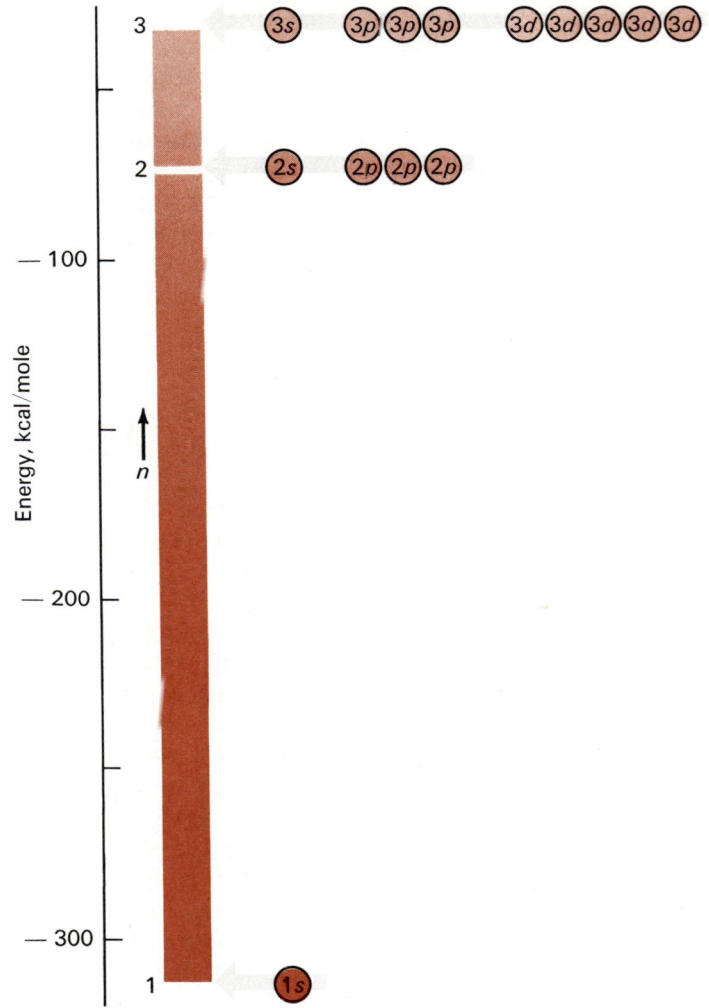

Figure 13-1 *The relative stabilities of atomic orbitals for the hydrogen atom can serve as a starting point in assigning electrons to orbitals for other elements.*

The two electrons, being of the same charge, repel each other. Nevertheless, the attraction of the +2 charge on the nucleus is the dominant force. Thus, both electrons occupy the 1s orbital in the most stable state, the **ground state,** of the helium atom.

The arrangement of electrons in any atom or ion is called the **electron configuration.** There is a simple shorthand way to state the electron configuration. Each occupied orbital is written down using the normal symbols, for example, 1s, 3p, 3d. If there is more than one electron present in that orbital or set of orbitals, the number present is written to the upper right. For instance, if there are four electrons in the 3d orbitals, we write $3d^4$.

Using this system, the ground state electron configuration of the helium atom is $1s^2$. The next most stable configuration of the helium atom would have one of the electrons occupying the next highest orbital, according to the pattern in Figure 13-1. This could be either the 2s or one of the 2p orbitals, judging from what is known about the hydrogen atom. Hence, the next most stable configurations of the helium atom would be 1s2s or 1s2p.

13-2 The Lithium Atom, and the Exclusion Principle

After helium, the next most complex atom is lithium. Lithium has an atomic number of 3, so the lithium nucleus has a charge of +3 and there are three electrons to occupy the orbitals. What is the most stable (ground state) configuration of these electrons?

You can safely assume that as electrons are added, one by one, to the bare nucleus, the first electron and then the second one will go into the 1s orbital, just as in the case of helium. With the very large positive charge, +3, on the nucleus, the 1s orbital of the lithium atom is a very stable place for a negative electron. Even with two electrons already there, you might think that this would still be the preferred location for the third one. It is a fact, however, that the third electron does not go into the 1s orbital. Instead, it occupies the 2s orbital. The ground state configuration of the lithium atom is $1s^2 2s$.

Even though it might have seemed that the third electron would have had lower energy in the 1s orbital, it did not go there. It went instead to the next lowest orbital. Apparently, the 1s orbital is *filled* when it contains two electrons. As we continue examining the electron configurations of atoms, we shall find that this same rule applies to all orbitals in all atoms. There are *no* exceptions. The general rule is:

No atomic orbital can ever contain more than two electrons.

The foregoing rule is often called **Pauli's exclusion principle,** in honor of the Austrian physicist Wolfgang Pauli. He first realized that this rule is one of the keys to understanding atomic structure.

The exclusion principle does not come naturally out of quantum mechanics. In fact, no one has yet explained why it *should* be so. But it is known, from experiment, that it *is* so. It is something like the fact that opposite charges attract each other. No one knows why, but it is known that they do.

It may perhaps have occurred to you to ask why the third electron was assigned to the 2s orbital rather than to one of the 2p orbitals. According to Figure 13-1, the 2s and 2p orbitals appear to have the same energy. In the hydrogen atom itself, that is true, but once there are electrons in the 1s orbital, it is no longer true. Because of their different shapes, the 2s and 2p orbitals are affected differently by the presence of electrons in the 1s orbital. The 2p orbitals become less stable than the 2s orbital. Similarly, the 3s and 3p orbitals, and also the $3d$ orbitals, cease to have identical energies when the lower orbitals are occupied. Consequently, it is necessary to draw a new diagram to replace Figure 13-1. In the new diagram (Figure 13-2) the relative energies of the orbitals are shown as they are for atoms having filled orbitals.

13-3 Atoms with Atomic Numbers Four Through Ten

You now have the basic ideas and information necessary to write the electron configurations for still more complex atoms. Consider next the element of atomic number 4, beryllium. As with He and Li, the first two electrons will go into the very stable $1s$ orbital. The next electron will go into the $2s$ orbital, and so will the fourth one. Thus, the configuration will be $1s^2 2s^2$.

Now that the $2s$ orbital is also filled, further electrons must go into the $2p$ orbitals. Therefore the element of atomic number 5, boron, will have the configuration $1s^2 2s^2 2p$. There are, of course, three different $2p$ orbitals, but they are all equally stable. The $2p$ electron may therefore be in any one of them, and at this point it makes no difference which one.

The element with atomic number 6 is carbon. It must have two electrons in its $2p$ orbitals. It is known from experiment that these two electrons will occupy different $2p$ orbitals, but it makes no difference which two. Thus, the configuration could be written as $2p_x 2p_y$, $2p_x 2p_z$, or $2p_y 2p_z$. We will use the first one arbitrarily, and write the entire ground state configuration as $1s^2 2s^2 2p_x 2p_y$. Writing it as $1s^2 2s^2 2p^2$ would not make clear the fact that the two $2p$ electrons are not in the same $2p$ orbital.

In the nitrogen atom, atomic number 7, there are three $2p$ electrons, and each of them occupies a different $2p$ orbital. The complete electron configuration of the nitrogen atom is $1s^2 2s^2 2p_x 2p_y 2p_z$.

Continuing in the same way you can obtain the following configurations for the next three elements.

Oxygen $\quad 1s^2 \quad 2s^2 2p_x^2 2p_y 2p_z$
Fluorine $\quad 1s^2 \quad 2s^2 2p_x^2 2p_y^2 2p_z$
Neon $\quad 1s^2 \quad 2s^2 2p_x^2 2p_y^2 2p_z^2$

In neon, all the orbitals with principal quantum numbers 1 and 2 are fully occupied. This type of configuration is called a **closed shell**. In this particular case, it is the "shell" of principal quantum number 2 that has just been closed. The helium atom also has a closed shell configuration, $1s^2$.

13-4 Atoms Beyond Neon

With the help of Figure 13-2, you can continue to determine the electron configurations of atoms. Consider those with atomic numbers 11-18. The element with atomic number 11 is sodium. Obviously, after the first 10 electrons have occupied the $1s$, $2s$, and $2p$ orbitals, as in neon, the 11th must go into the $3s$ orbital. Hence, for the sodium atom the configuration is $1s^2 2s^2 2p^6 3s$. From here on, the building-up procedure must obviously proceed in the same way as it did when the $2s$ and $2p$ orbitals were being filled. For the elements of atomic numbers 11-18, the configurations are as follows. ($[\text{Ne}] = 1s^2 2s^2 2p^6$)

Na $\quad [\text{Ne}]3s$
Mg $\quad [\text{Ne}]3s^2$
Al $\quad [\text{Ne}]3s^2 3p_x$
Si $\quad [\text{Ne}]3s^2 3p_x 3p_y$
P $\quad [\text{Ne}]3s^2 3p_x 3p_y 3p_z$
S $\quad [\text{Ne}]3s^2 3p_x^2 3p_y 3p_z$
Cl $\quad [\text{Ne}]3s^2 3p_x^2 3p_y^2 3p_z$
Ar $\quad [\text{Ne}]3s^2 3p_x^2 3p_y^2 3p_z^2$

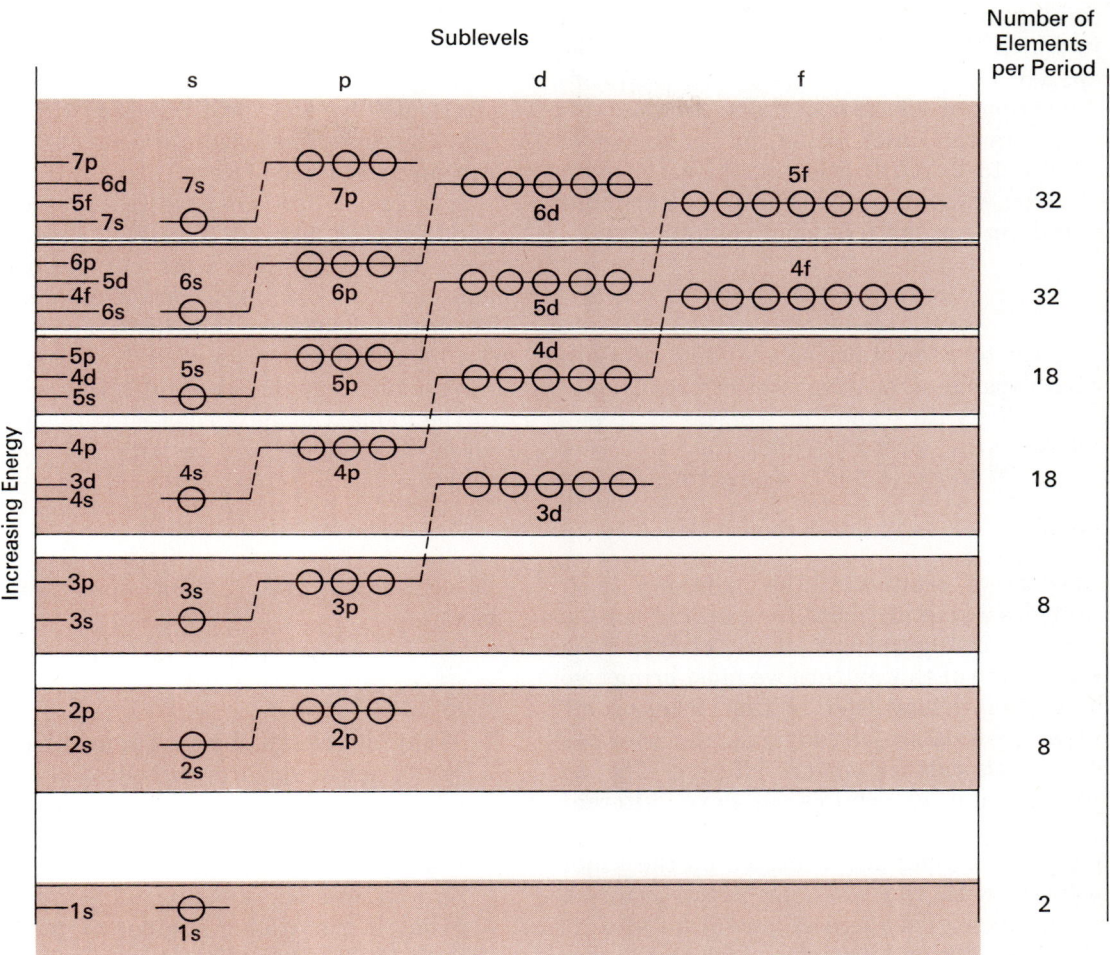

Figure 13-2 *The relative energies of atomic orbitals as they are in atoms with filled orbitals, as contrasted to the hydrogen atom. Colored bands show sets of orbitals with similar energies.*

Again a closed shell configuration has been reached. This particular one is the argon configuration. All atoms beyond argon have this configuration plus more electrons in still higher orbitals. You can see from Figure 13-2 that the next two elements after argon should have the following configurations.

Potassium [Ar]4s
Calcium [Ar]$4s^2$

But now you can see that something different comes next. Instead of proceeding to [Ar]$4s^2 4p_x$, in analogy to what has happened twice before, you now find the 3d orbitals coming into the picture. Because of the way in which the differently shaped orbitals, 4s, 4p, and 3d, are influenced by the [Ar] configuration of electrons, their energies increase in the order 4s, 3d, 4p, as shown in Figure 13-2. After calcium, then, the next electron will enter a 3d orbital. The element scandium has the following configuration.

$$\text{Scandium} \quad [Ar]4s^2 3d$$

A set of d orbitals includes five orbitals, each capable of holding 2 electrons. Hence, beginning with scandium, a series of 10 elements with electron configurations from $[Ar]4s^2 3d$ to $[Ar]4s^2 3d^{10}$ occurs. Once the 3d orbitals have been filled, the 4p orbitals are next in line, and through the next six elements the configurations proceed from $[Ar]4s^2 3d^{10}4p$, zinc, to $[Ar]4s^2 3d^{10}4p^6$, krypton. The closed shell configuration of krypton can be abbreviated [Kr].

It is easy to see that we can proceed beyond krypton, using Figure 13-2 as we have done so far, and thus work out the ground state electron configurations of more atoms. We shall not take the time to do this, because we have already proceeded far enough to have made clear the basic rules and patterns of the game. Our next concern is to see how all of this can be tied in with the periodic table.

Summary

The set of orbitals of the hydrogen atom provides a starting point for working out the electron configurations of all atoms. To a knowledge of this pattern we must bring two additional ideas. One is that when some inner orbitals are already full, the energies of orbitals farther out are affected. The ns, np, nd, and nf orbitals all have different energies. The ns orbital always has an energy lower than the energies of the other orbitals with the same principal quantum number, n. The second essential idea is Pauli's exclusion principle, which states that no orbital can ever contain more than two electrons.

Self Check

1. What is a closed shell configuration?
2. Write the first three closed shell configurations.
3. Using only Figure 13-2, write the configurations of the atoms with atomic numbers 28 and 34.
4. What is the atomic number of the atom with the configuration $[Ne]3s^2 3p$?

The Periodic Pattern of Electronic Configurations

13-5 The Recurrence of Closed Shells

In the previous sections you have seen how the process of building up atoms leads every so often to the attainment of a closed shell configuration. Specifically, you have seen that this happened in the following places.

Helium $1s^2 =$ He
Neon $1s^2 2s^2 2p^6 = [He]2s^2 2p^6 =$ Ne
Argon $1s^2 2s^2 2p^6 3s^2 3p^6 = [Ne]2s^2 2p^6 =$ Ar
Krypton $1s^2 2s^2 2p^6 3s^2 3p^6 4s^2 3d^{10} 4p^6$
           $= [Ar]4s^2 3d^{10} 4p^6 =$ Kr

When you look at Figure 13-2, you can see clearly that each of these closed shell config-

																H	He
																1	2
								Li	Be	B	C	N	O	F	Ne
								3	4	5	6	7	8	9	10
								Na	Mg	Al	Si	P	S	Cl	Ar
								11	12	13	14	15	16	17	18
K	Ca	Sc	Ti	V	Cr	Mn	Fe	Co	Ni	Cu	Zn	Ga	Ge	As	Se	Br	Kr
19	20	21	22	23	24	25	26	27	28	29	30	31	32	33	34	35	36
Rb	Sr	Y	Zr	Nb	Mo	Tc	Ru	Rh	Pd	Ag	Cd	In	Sn	Sb	Te	I	Xe
37	38	39	40	41	42	43	44	45	46	47	48	49	50	51	52	53	54
Cs	Ba	*	Hf	Ta	W	Re	Os	Ir	Pt	Au	Hg	Tl	Pb	Bi	Po	At	Rn
55	56		72	73	74	75	76	77	78	79	80	81	82	83	84	85	86

[*] = [La – Lu / 57 – 71]

Figure 13-3 *A table of the elements made by listing them in order of atomic number, with all closed shells in the right column. No gaps are left.*

urations is associated with the filling of a group of orbitals that lies within one of the colored bands. You can also see that two more closed shell configurations are to be expected within the range of this diagram. These will occur at the elements xenon (atomic number 54) and radon (atomic number 86).

Suppose we now take a list of all the elements up to radon in order of their atomic numbers and arrange it in horizontal rows. Suppose, further, that we arrange the rows in such a way that the last column contains all the elements with closed shell configurations. What we obtain is shown in Figure 13-3. Although its shape is somewhat irregular, this arrangement actually has many very close similarities to the modern or "long" form of the periodic table (Figure 8-8). It can easily be made exactly the same as the accepted form of the periodic table by just shifting a few elements to other columns.

This is done in accord with the rule that the columns are supposed to contain the elements that are most similar in properties. On this basis, lithium and sodium belong with potassium, K, rubidium, Rb, and cesium, Cs. Similarly, beryllium and magnesium resemble calcium, Ca, strontium, Sr, and barium, Ba, more than they do zinc, Zn, cadmium, Cd, and mercury, Hg. Because it is the lightest of the elements and is unique in having no electrons other than the one used in forming bonds to other atoms, hydrogen does not really resemble *any* other elements *closely*. However, in a number of ways it goes more naturally with the alkali metals (Li-Cs) than with the halogens (F-At) or any other family. It is therefore placed in the first column. Having made these moves, which do not change the order of the listing, we have the pattern in Figure 13-4. This is identical to the modern "long form" periodic table. It

H 1																	He 2
Li 3	Be 4											B 5	C 6	N 7	O 8	F 9	Ne 10
Na 11	Mg 12											Al 13	Si 14	P 15	S 16	Cl 17	Ar 18
K 19	Ca 20	Sc 21	Ti 22	V 23	Cr 24	Mn 25	Fe 26	Co 27	Ni 28	Cu 29	Zn 30	Ga 31	Ge 32	As 33	Se 34	Br 35	Kr 36
Rb 37	Sr 38	Y 39	Zr 40	Nb 41	Mo 42	Tc 43	Ru 44	Rh 45	Pd 46	Ag 47	Cd 48	In 49	Sn 50	Sb 51	Te 52	I 53	Xe 54
Cs 55	Ba 56	*	Hf 72	Ta 73	W 74	Re 75	Os 76	Ir 77	Pt 78	Au 79	Hg 80	Tl 81	Pb 82	Bi 83	Po 84	At 85	Rn 86

$[*] = \begin{bmatrix} La & - & Lu \\ 57 & & 71 \end{bmatrix}$

Figure 13-4 *A table similar to that in Figure 13-3, except that some elements are shifted to place all similar elements (that is, "families," such as the alkalis) in columns, thus leaving gaps.*

should now be evident to you that the periodic recurrence of closed shells of electrons can form a basis for obtaining the entire periodic table.

13-6 Properties Vary Periodically with the Atomic Number

You will recall from Chapter 8 that Mendeleev arrived at his original periodic table by listing the elements in order of increasing molar mass. Remember, at that time the idea of atomic number was not yet known. You may also recall that his listing in order of molar mass created some difficulties. Some pairs of elements, for example, iodine and tellurium, appeared in the opposite order from that dictated by their chemical properties. As you can see from Figure 13-3, a listing in order of atomic number leads to no such difficulties at all. Every element gets into the column in which it belongs on the basis of its chemical and physical properties. Evidently, it is the atomic number (not molar mass) and the electron configurations that form the fundamental basis for the chemical properties of an element and, hence, for its position in the periodic table. A young English scientist, Henry Moseley, discovered this shortly before the war of 1914-1918, in which he died.

Why does a list in order of molar masses give a *nearly* perfect periodic table? The reason can be found by considering the isotopes of the various elements. Most elements occur in nature as a mixture of isotopes. Hence, the molar mass that is found for them is an average of the masses of the individual isotopes. This is not very different from the situation you might find for the ages of students in a given high school class. There might, occasionally, be a Junior class in which everyone is 16 years of age, but more

Students in Class	Portion of Class
14 year olds	5%
15 year olds	5%
16 year olds	60%
17 year olds	15%
18 year olds	5%
19 year olds	10%

likely you would find a distribution of ages similar to the one given in the above table. The average age of students in this class will not likely come out to a round number. In this example, the average age will be

$$(0.05)14 + (0.05)15 + (0.60)16 + (0.15)17$$
$$+ (0.05)18 + (0.10)19 = 16.4.$$

Similarly, naturally-occurring tellurium consists of the percentages of isotopes given in Table 13-1.

When the actual masses of these isotopes are averaged in proportion to their percentages, just as was done in figuring the average age of the Junior class, 127.6 is the average molar mass.

You will note that the actual mass of each isotope is a little less than its mass number. Remember, the mass number is simply the total number of neutrons and protons in the nucleus. The actual masses of these particles are slightly less than one atomic mass unit. Therefore the actual mass of a given isotope is slightly less than its mass number.

The element iodine is one of the few that exists in nature entirely in the form of a single isotope ^{127}I. Hence, its average molar mass exactly equals the mass of this one isotope, which is 126.9.

From this example, you can see that the exact value of the average molar mass of any given element is not directly related to its atomic number or its electron configuration. The average molar mass is not therefore a proper and entirely reliable basis for periodicity in properties. It was fortunate for Mendeleev and his contemporaries that molar mass does generally run parallel with atomic number.

TABLE 13-1   The Isotopes in Naturally-Occurring Tellurium

Mass Number, M, of Isotope	Percent of Isotopes with Mass Number M	Molar Mass of Isotope
120	0.089	119.9
122	2.46	121.9
123	0.87	122.9
124	4.61	123.9
125	6.99	124.9
126	18.71	125.9
128	31.79	127.9
130	34.48	129.9

INVESTIGATION

13-7 The Solubility of Salts: A Search for Patterns

With the development of the periodic table and the organization of elements into families, scientists began to study the chemical and physical properties of many compounds to see if these properties varied in any related way. In some cases, properties of compounds made of elements in one family varied in a very predictable manner. In other cases, great differences appeared. An examination of these differences provided a deeper understanding of the structures of compounds and the effect that these structures would have on their behavior under a variety of conditions.

One property that can readily be compared for many substances is solubility in water. Solubility is the amount of a substance that will dissolve in a given amount of a liquid (the solvent) at a particular temperature.

PURPOSE

You will measure the solubility in water at room temperature of two series of compounds. The first series will include four salts formed by reacting sodium with each of the members of the halogen family—fluorine, chlorine, bromine, and iodine. The second series will consist of four salts formed by reacting potassium with the same members of the halogen family.

MATERIALS

Graduated cylinder, 10-ml   Ringstand, ring
Beaker, 100-ml   Wire gauze
Balance   Burner
Boiling stones   Graph paper
10 ml saturated sodium or potassium halide solution

PROCEDURE

The solubility of each of the following compounds will be measured: NaF, NaCl, NaBr, NaI, KF, KCl, KBr, KI. Saturated solutions of these compounds will be prepared by the teacher before the start of the laboratory period. You can assume them to be at room temperature at the time you use them. Each solution will be assigned to each of three or more students. When the experimental work has been completed, each student will report the solubility for his compound. Then an average figure for each compound will be calculated.

1. Place a boiling stone (or a piece of broken porcelain or glass) in a clean dry 100-ml beaker. Find the mass of the beaker and the boiling stone to the nearest 0.01 g.

2. Using a graduated cylinder, measure 8-10 ml of the solution assigned to you and then pour it into the beaker. Record the room temperature and the formula of your compound.

3. Find the combined mass of the beaker, boiling stone, and solution.

4. Place the beaker and its contents on a wire gauze on a ring above a burner, as in Figure 13-5. Heat very gently, to evaporate without spattering.

5. When the contents of the beaker appear dry, heat strongly for 2-3 minutes to be sure all moisture is gone.

6. Turn off the burner and allow the beaker to cool to room temperature. Find the new mass of the beaker and its contents to the nearest 0.01 g.

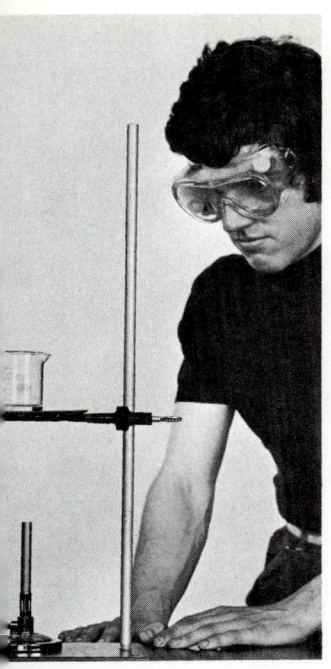

Figure 13-5

DATA AND RESULTS

You should prepare a table in your notebook for recording your results. (See the following list.) Also prepare a table for recording the class results. (See the following sample table.)

Room temperature, °C
Formula of compound used

Before Heating
Mass of beaker, stone, and solution (g)
Mass of empty beaker and stone (g)
Mass of original solution (g)

After Heating
Mass of beaker, stone, and dry solid (g)
Mass of empty beaker and stone (g)
Mass of dry solid (g)

Table for Recording the Class Results

Compound	Individual Solubilities (moles/100g H_2O)	Average Solubility (moles/100g H_2O)
(1) NaF		

Calculations

The solubility of each of the compounds will be determined in moles of compound dissolved per 100 grams of water.

1. To find the mass of water in your solution, subtract the mass of dry solid remaining after evaporation from the original mass of the solution.

2. Divide the mass of dry solid remaining by the molar mass of the compound you used. This will give you the moles of compound dissolved in the original solution.

3. Calculate moles of compound per 100 grams of water, from the moles of compound and the mass of water that were in your original sample.

4. List the solubilities (in your prepared table) found by all other class members for all compounds tested. If any one value in any of these groups appears to differ widely from other values within the same group, ask the teacher whether you should include this value in calculating the average solubility, or whether you should discard it as an erroneous result.

5. Prepare a graph showing moles of compound per 100 grams of water on the vertical axis (the ordinate) and the molar mass of the compounds on the horizontal axis (the abscissa). Label the axes clearly.

6. Plot the average solubility of each compound against the molar mass of the compound on your graph. Write the formula of each compound next to its point on the graph.

7. Join the four points for the sodium compounds consecutively with straight lines. Do the same thing for the potassium compounds.

RELATED PROBLEMS

1. What regularities appear on your graph?

2. Are there any irregularities in your results?

3. What properties of the compounds might affect the way they dissolve in water?

4. In what family of elements are sodium and potassium included?

5. What does this experiment suggest about your ability to predict properties of compounds formed by elements in a given family?

13-8 Electron Configurations Tie In with Properties

A knowledge of electron configurations enables us to comprehend why the periodic table exists and why it has periods of certain specific lengths. But it does more. It allows us to understand not only why the elements in a certain group are alike, but why the properties that they share occur in that group rather than in some other one.

The noble gases have as their outstanding property inertness, that is, a very limited tendency to form compounds with other elements. Their electron configurations, as you will see shortly, would lead you to expect this. On the other hand, the alkali metals and the halogens are extremely reactive. Again, their electron configurations would lead you to predict this.

To see why these properties are predictable from the electron configurations, we must first return to the idea of ions. Ions are atoms with too many or too few electrons to be neutral. A positive ion, or cation, is formed by removing one or more electrons from a neutral atom. A negative ion, or anion, is formed by adding one or more electrons to a neutral atom. Let us consider the process of positive ion formation. Sodium atoms, for example, readily form ions, as shown by the following equation.

$$Na \longrightarrow Na^+ + e^-$$

Energy must be expended to detach the electron from the sodium atom. The required energy is called the "ionization energy" and is usually given the symbol I.

If you reexamine Figure 11-25a, you will see that the curve formed by plotting the energy of each line of the hydrogen spectrum against its step number levels off and approaches the value of 313.6 kilocalories per mole as the step numbers get increasingly larger. Obviously, increasing the step number to infinity has the same effect as removing the electron from the atom. Since **ionization energy** is defined as the energy required to remove an electron from an atom, the ionization energy of hydrogen is 313.6 kilocalories per mole. Measurements of the lines of the spectrum of any element can be used to calculate the ionization energy of that element in the same way.

The ionization energies of all the atoms have been measured. If they are plotted against the atomic numbers of the first 38 elements, an important graph (Figure 13-6) results. While there are a lot of small jogs and spikes, the graph has an obvious main pattern. It has peaks (maxima) at all the noble gases and low points (minima) at all the alkali metals. These properties relate to the electron configuration patterns.

Refer to the electron energy diagram, Figure 13-2. Notice that each noble gas has a filled group of orbitals with similar energies. That is, all of the orbitals within the colored bands are filled up. An exactly filled group of orbitals of this type, which is called a closed shell, is particularly stable. As shown by the high ionization energies of the noble gas atoms, it is very difficult to break up such a group by removing an electron from it.

Following each noble gas, there is an atom that contains one electron more than the number required to give the very stable, filled group. As you can see in Figure 13-2, this additional electron must be placed way up in the next energy band. It is held there only loosely, and not much energy is required to remove it. An atom with this sort of arrangement of electrons has a low first ionization energy. The **first ionization energy** is the energy required to remove one electron.

When the first electron has been removed from one of the atoms following a noble gas,

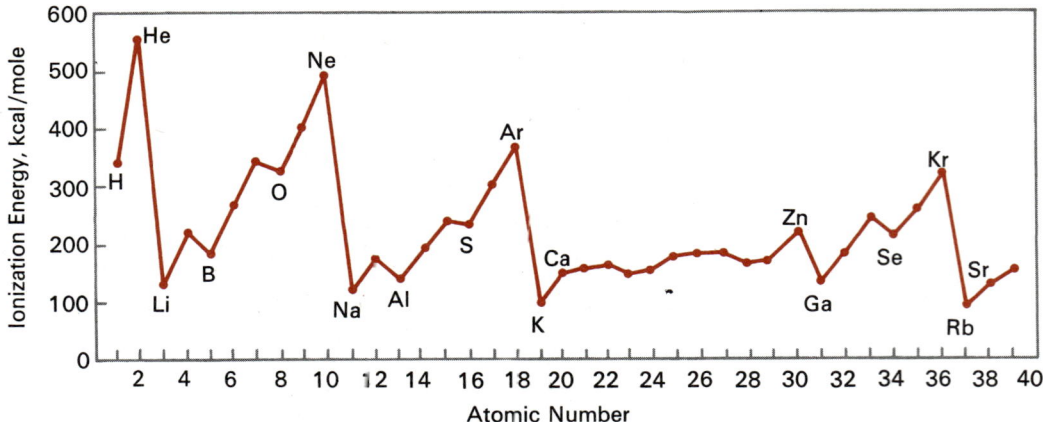

Figure 13-6 *The ionization energies of the atoms vary considerably, with high peaks at the noble gases and the lowest values for the alkali metals.*

the very stable closed shell characteristic of the noble gas atoms results. Besides the fact that a closed shell is in itself a very stable configuration and hard to break into, there is now an added difficulty for removing another electron. The removal of the first electron left a positive charge on the ion. This positive charge makes it even harder to move another negative electron away from the ion. The net result of all this is as follows. (1) For all of the alkali metal atoms, Li, Na, K, Rb, and Cs, it is easy to remove one electron and create the positive ions Li^+, Na^+, K^+, Rb^+, and Cs^+. (2) It is extremely difficult to remove any further electrons.

Two atomic numbers beyond each noble gas there is an atom that has two electrons more than the number needed to make up the stable group characteristic of the noble gas. Just as with the single extra electron in the alkali metal atoms, these two electrons occupy orbitals in the next band of energy. Not very much energy is required to remove one of them. Even the second one is not particularly hard to remove. It is harder to remove than the first, because of the positive charge already present due to removal of the first one. However, after the removal of these two electrons, the very stable noble gas group remains and this is extremely difficult to break into. It therefore turns out, quite logically, that for the elements that lie two places beyond each noble gas, the following process is easy and characteristic. It forms the basis for most of their chemical behavior.

$$M \longrightarrow M^{2+} + 2e^-$$

The elements in question are those in the second column of the periodic table: Be, Mg, Ca, Sr, Ba, and Ra. This group, or family, of elements is called the **alkaline earth metals**.

At this point, let us stop and compare what you have just learned about the properties, due to their electron configurations, of several groups of elements with the chemical behavior of these same groups. First, you can see that the inertness, the very limited tendency of the noble gas atoms to form compounds or ions, is understandable. They have the very stable closed shells of electrons that are difficult to break up. In addition, the fact that the main chemical properties of the alkali metals and the alkaline earth metals result from the formation of 1+ and 2+ ions, respectively, is in perfect accord with what can be predicted from their electron arrangements.

The particular stability of the electron groups found in the noble gases can also help us to understand the properties of the halogen atoms, and why it is that atoms with these particular properties should all be found in the column just to the left of the noble gases. Each halogen atom has one less electron than the noble gas atom that follows it. Because the closed shell of electrons that occurs in the noble gas atoms is especially stable, the halogen atoms have a distinct tendency to add one electron and thereby complete groups like those in the noble gases. Of course, the addition of one electron confers a negative charge and so an anion is formed. The following equation shows this, for chlorine.

$$Cl + e^- \longrightarrow Cl^-$$

The electron arrangements of the halogen atoms enable us to understand why it is characteristic of this family of elements to form compounds in which there are the anions F^-, Cl^-, Br^-, and I^-. We can also understand why a family of elements with this property must occur immediately to the left of the noble gases in the periodic table.

The preceding discussion of how ionization energies and electron configurations are related can help you to predict some chemical formulas. Specifically, you can now readily write formulas for many important compounds formed by the alkali and alkaline earth elements and the halogens. As you have seen, the alkali metal atoms easily lose one electron and the halogen atoms have a strong tendency to gain one electron. So you might expect that alkali metals and halogens would react to form compounds according to equations such as the following ones.

$$2Li + Cl_2 \longrightarrow 2LiCl$$
$$2K + Br_2 \longrightarrow 2KBr$$
$$2Na + F_2 \longrightarrow 2NaF$$

As you learned in Chapter 8, the preceding equations correctly describe experimental results. Other alkali-halogen reactions can also be predicted accurately. The products of the reactions all have formulas consisting of one alkali metal cation and one halogen anion (a *halide* ion). The formulas do not actually show that ions are present, but they are. In other words, the formula NaCl could be written Na^+Cl^-, but it is not customary to do so.

Since alkaline earth metal atoms easily lose *two* electrons to give *di*positive cations, it would be logical to expect that equations and formulas such as the following would be correct.

$$Ca + F_2 \longrightarrow CaF_2$$
$$Ba + Cl_2 \longrightarrow BaCl_2$$
$$Sr + Br_2 \longrightarrow SrBr_2$$

In fact, these equations, and similar ones, exactly describe the actual reactions and

Henry G. J. Moseley (1887–1915)

HENRY MOSELEY'S FAME was achieved in only two years of research ending in 1914, when he enlisted in the British army at the start of World War I. His work during those two brief years, however, produced results that contributed significantly to our understanding of the structure of the atom.

Shortly after Roentgen discovered X rays in 1895, Barkla, working at the University of London, found that the wavelength of X rays depends on the element used as the positively-charged target inside an X-ray tube. In 1911, Rutherford showed that the nucleus of an atom is a densely-packed, positively-charged particle. This led the Dutch scientist, van dan Broek, to suggest that the charge on the nucleus should be an integral multiple of the charge on the hydrogen nucleus. To test this theory, Rutherford asked Moseley, who was one of his students, to see if he could find any relationship between the charge on a nucleus and the wavelengths of the X rays produced by the element.

Moseley began his work by constructing twelve X-ray tubes, each having a different element as a target. Later he tested thirty more elements. The X rays produced by Moseley's samples were diffracted by a crystal and focused on photographic plates in an operation similar to that taking place in an optical spectrograph. Moseley found that the spectrum of each element was composed of two major lines. When he placed his exposed plates next to each other and compared them, Moseley discovered that the lines on the plates formed a series of regular steps. Each line was displaced from the position of the preceding one by about the same amount for each step. (See Figure 13–8.)

In interpreting his results, Moseley stated, "We have here a proof that there is in the atom a fundamental quantity, Z, which increases by regular steps as we pass from one element to the next. This quantity, Z, can only be the charge on the central positive nucleus, of the existence of which we already have definite proof." He went on to say that Z is the same as the number of the place occupied by the element in the periodic system. When he discovered gaps between the regularly spaced steps of his results, he predicted that these spaces belonged to undiscovered elements. It soon became evident that the arrangement of the periodic system according to Moseley's *atomic numbers* was more satisfactory than Mendeleev's original arrangement according to molar masses. Within a few years it was discovered that the atomic numbers corresponded to the number of protons in the nucleus of an element.

Moseley never knew how important his work had been. He was killed in a battle at Gallipoli in Turkey in August, 1915.

compounds formed. The alkaline earth halides all have formulas consisting of one alkaline earth cation and two halide ions.

13-9 How Moseley Determined Atomic Numbers ★

You have seen that the true basis of the periodic classification of the elements is their atomic numbers. This is simply the number of electrons in the atom, or the number of protons in the nucleus. The second definition is more fundamental since the identity of the element remains the same whatever its state of ionization or the type of chemical bond it forms. The number of protons in the nucleus is constant regardless of the chemical state of the element.

Since the charge of the nucleus is due entirely to the protons, the atomic number also gives the charge on the nucleus, in units of the charge on the proton.

Once the atomic number of an element is known, there can be no uncertainty about its correct position in the periodic table. But, how were atomic numbers first determined?

A way to do this was discovered and successfully used for a number of elements by the young English physicist H. G. J. Moseley, in 1913. His method depended on the measurement of the X-ray spectra of the elements.

Many solid forms of matter can be made to give off X rays if they are placed on the target (anode) of an X-ray tube. The design of such a tube is shown in Figure 13-7. A cathode and a target are within an evacuated tube. A high voltage is applied so that electrons emitted from the cathode acquire great kinetic energy before striking the target. The electrons striking the target have enough energy to eject electrons from the innermost (1s) orbital of some of the atoms. X rays are emitted when electrons from higher energy orbitals fall into the vacancy in the 1s orbital. They are sometimes called "K" X rays, since they involve electrons in the "K level," that is, the lowest, or innermost, energy level. These X rays are then made into a fine beam by means of some slits. This beam passes into an X-ray spectrometer, which is a device that can measure the wavelengths of X rays. It is similar in principle to the spectrometers used for visible light, except that it uses a crystal instead of a grating or a prism.

What Moseley discovered was that the wavelengths of the K X rays of the elements get shorter as the charge on the nucleus increases. Actually, there are two K X rays, called K_α and K_β, which are emitted together. The direct experimental results are

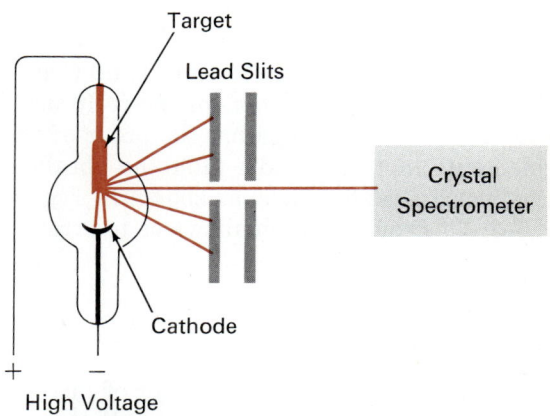

Figure 13-7 *X rays are generated, by the tube at the left, when high-energy electrons from the cathode strike the metal target (the anode). The resulting X rays are formed into a narrow beam by the slit system. Then the X-ray beam passes into the detector, a crystal spectrometer, where the wavelength of the X rays can be measured.*

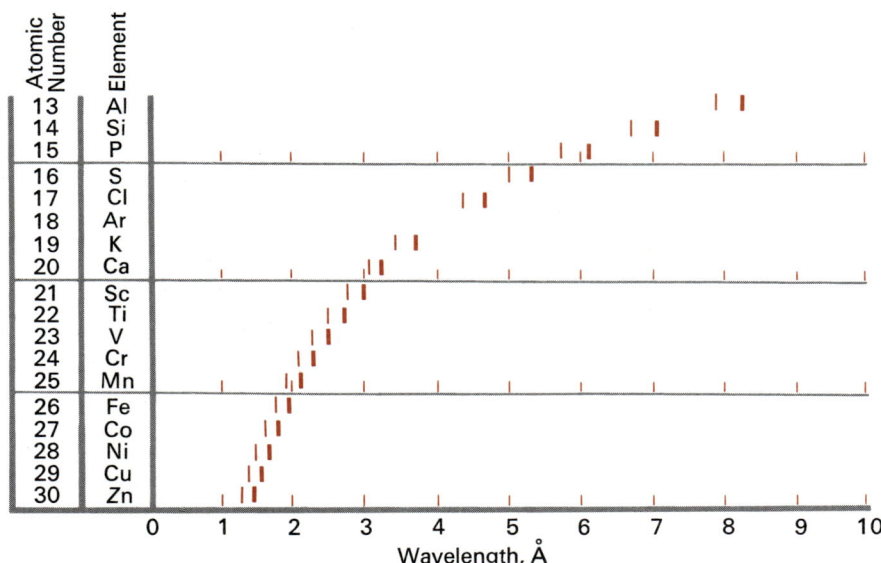

Figure 13-8 *There is a regular change of wavelength of X-ray emission lines for a series of elements, as Moseley first demonstrated.*

shown in Figure 13-8. Moseley found that if he plotted the square roots of the reciprocals of the wavelengths against the order of the elements in the periodic table, he obtained perfectly straight lines, as shown in Figure 13-9. This sort of graph allows us to arrange the elements in order of increasing atomic number. In each of the cases in which the molar masses of two elements are reversed, the atomic numbers, as found from the graph (the Moseley diagram) are in the correct sequence.

Summary

We can list the elements in order of increasing atomic number, then arrange them into horizontal rows of varying length. The row lengths are chosen in such a way that elements with similar electron configurations occur in columns. As a result, we get an arrangement that is identical to the periodic table. This shows that the basis for the periodic variation in properties and the existence of families of elements is the pattern whereby atomic orbitals are filled.

Mendeleev's original listing, in order of increasing molar masses, occasionally proves inconsistent with the need to put sets of similar elements into columns. These inconsistencies can be explained by the distribution of isotopes. The electron configurations also help us to understand why certain families of elements have their particular properties. Ionization energy is the amount of energy needed to detach the most loosely bound electron from a gaseous atom.

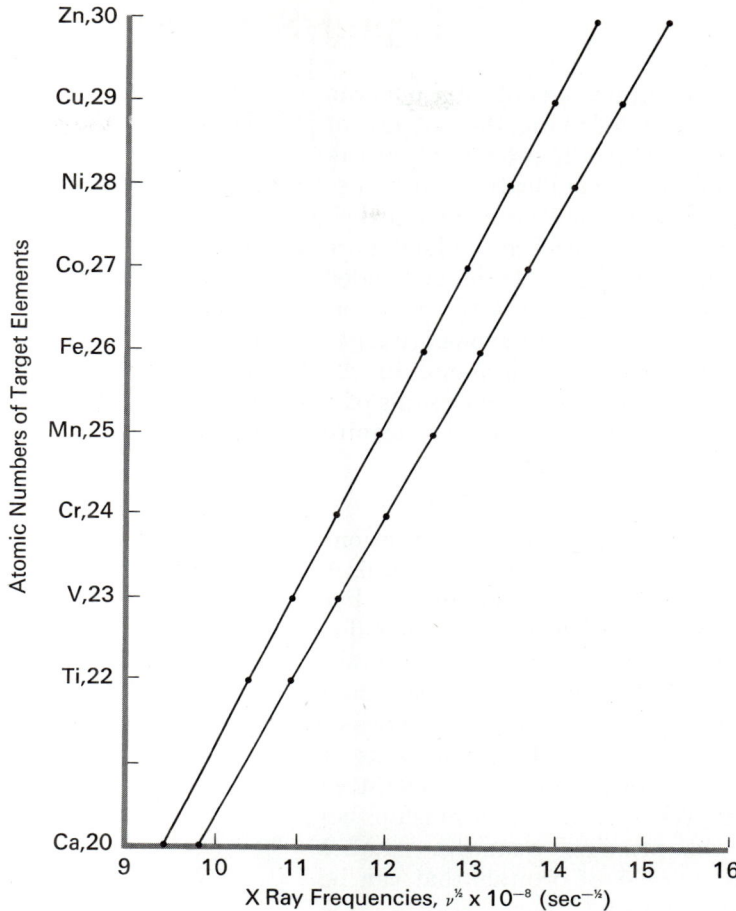

Figure 13-9 *This sort of graph, sometimes called a Moseley diagram, shows the order in which the atomic numbers of the elements increase.*

Self Check

1. Describe the set of electron configurations that we first put in a vertical column in order to get a periodic table based on electron configurations? What do we call such configurations?

2. What family of elements has these configurations?

3. Why are the ionization energies of alkali metal atoms much lower than those of noble gas atoms?

Chapter Summary

The electron configurations of all atoms can be worked out by following the pattern of orbitals found in the hydrogen atom, although with certain modifications. It is also necessary to respect Pauli's exclusion principle, which says that no orbital can hold more than two electrons. As electrons are added to inner orbitals, beginning with the 1s orbital, the pattern of the outer ones changes somewhat from the pattern found in the hydrogen atom. In general, the orbitals of a given principal quantum number acquire different energies such that their stabilities decrease in the order s, p, d, f. The $1s^2$ configuration and all those that end with ns^2np^6 are called closed shells. When a list of all the elements in order of increasing atomic number is arranged in rows so that all the atoms with closed shell configurations fall in a single vertical column, the periodic table is obtained. The fundamental reason for the periodic arrangement of the elements is to be found in the electronic configuration of the atoms.

Questions and Problems

1. Show complete electron configurations for atoms with atomic numbers 3, 6, 9, 12, 15, 18, 21, 25, 29, 33, 36, and 37. For the last five elements you need not distinguish among p_x, p_y, and p_z orbitals.
2. Make a table listing the principal quantum numbers through four, the types of orbitals, and the number of orbitals of each type possible for each quantum number. What is the relationship between the principal quantum number and the number of orbitals that can be associated with it?
3. What is Pauli's exclusion principle? Describe its importance to the understanding of the periodicity of the elements.
4. Consider the following electron configurations.

    $1s^2 \quad 2s^22p^6 \quad 3s$ (1)
    $1s^2 \quad 2s^22p^6 \quad 6s$ (2)

    Which of the following statements is false?
    (a) Energy is required to change (1) to (2).
    (b) (1) represents a sodium atom.
    (c) (1) and (2) represent different elements.
    (d) Less energy is required to remove one electron from (2) than from (1).
5. Consider a neutral atom with the following electron configuration.

    $1s^2 \quad 2s^22p^5 \quad 3s$

    Which of the following is false?
    (a) The atom has atomic number 10.
    (b) The atom is not in the most stable configuration.
    (c) The atom must gain energy to change to $1s^2 \quad 2s^22p^6$.
    (d) The 1s and 2s orbitals are filled.
6. Write the electron configuration for the highest energy level containing any electrons for each of the following, in its lowest energy state. (a) Potassium atom, K; (b) Potassium ion, K^+; (c) Fluorine atom, F; (d) Fluoride ion, F^-; (e) Helium atom, He; (f) Oxygen atom, O; (g) Magnesium ion, Mg^{2+}
7. What must be done to a 2s electron to

make it a $3s$ electron? What happens when a $3s$ electron becomes a $2s$ electron?

8. By examining Figure 13-6, determine which group of elements has the lowest ionization energies. Which group of elements has the highest ionization energies? Which group of elements consistently has the second highest ionization energies?

9. From Figure 13-6, explain why cesium would be preferred for use in a photoelectric cell over lithium or sodium.

10. In normal chemical reactions lithium forms an ion by losing an electron. On the other hand, fluorine forms ions in chemical reactions by gaining an electron. Explain why these two elements behave so differently in the formation of ions.

11. Which has a greater ionization energy, a sodium atom, Na, or a sodium ion, Na^+? Explain.

12. To form a sodium ion, Na^+, the ionization energy required is 118 kcal. To form a magnesium ion, Mg^+, the ionization energy required is 175 kcal. Explain why 1091 kcal are required to go from Na^+ to Na^{2+}, while only 345 kcal are needed to further ionize Mg^+ to Mg^{2+}.

13. The electron configuration for lithium is $1s^2 2s$, and for beryllium it is $1s^2 2s^2$. Refer to Figure 13-2 to estimate the relative ease of removing first one, then a second electron from each of these atoms. Explain your answer.

14. The first four ionization energies of boron atoms are as follows: $I_1 = 191$ kcal/mole; $I_2 = 578$ kcal/mole; $I_3 = 872$ kcal/mole; $I_4 = 5962$ kcal/mole. Write the complete electron configuration of a boron atom and show how these ionization energies are related to the electron configuration.

Chemical Bonds and Molecular Structure

For thousands of years men have pondered the question: What holds particles of matter together? As long ago as the fifth century B.C., the Greek scientist-philosopher Democritus (de-MOK-ri-tus) conceived of atoms as ultimate particles held together by hook-like arms. Only within the past fifty years have scientists really begun to understand the nature of the chemical bond—the forces holding atoms together.

Atoms bonded together in clusters that persist long enough to have characteristic properties of their own are called molecules. In this unit we shall study the way in which atoms cling together. We shall begin by examining the chemical bond, which is the arrangement of electrons (not hooks!) that can tie two atoms together. We shall then see how these bonds allow groups of atoms to form molecules, with very definite geometrical structures. We shall also consider how solids, liquids, and solutions are formed from atoms and molecules.

14 Chemical Bonds

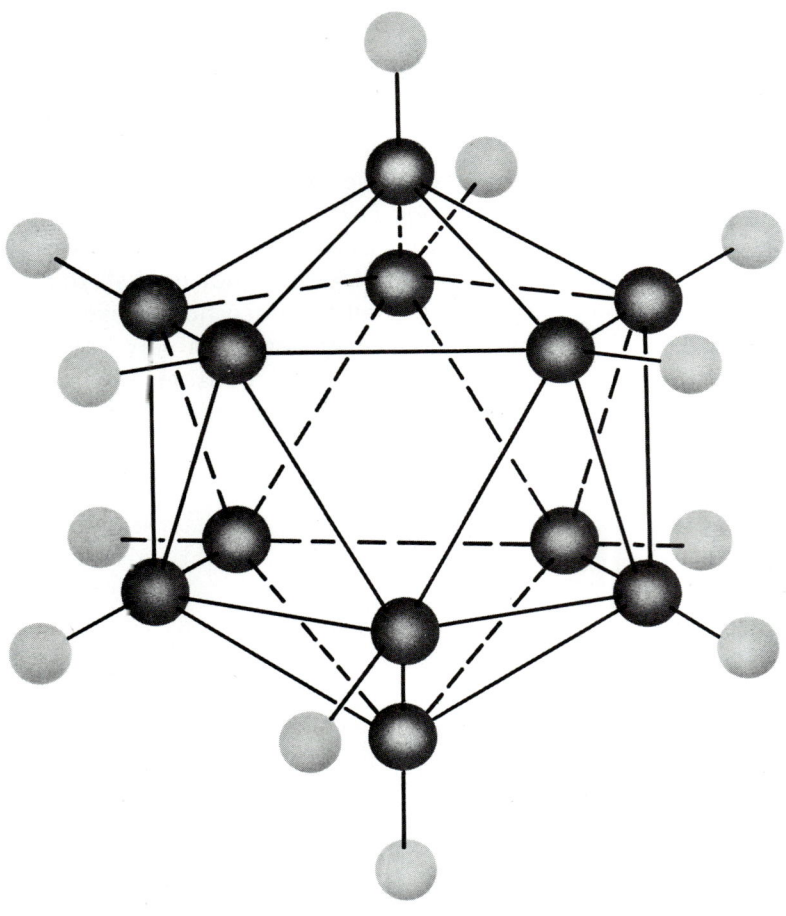

Molecular structures such as the one above can be explained with an understanding of chemical bonds.

You have already seen that a knowledge of the electron configurations of the atoms enables us to understand why the periodic table takes its particular form. The periodic table was originally worked out on the basis of experimental knowledge. That is, it was based on how the various elements react with each other and what kinds of compounds they form. Putting these two ideas together you might conclude that, starting with a knowledge of the electron configurations of atoms, it should be possible to explain the nature of the forces that hold atoms together in molecules. In other words, from electron configurations it should be possible to understand the nature of chemical bonds. Such an understanding will be developed in this chapter.

Electrical Forces Hold Atoms Together

14-1 The Hydrogen Molecule

Under normal conditions of temperature and pressure, hydrogen is a gas. At standard temperature and pressure, its measured density is 0.0898 gram per liter. Applying Avogadro's hypothesis, you can calculate from its density that one mole of hydrogen, which occupies 22.4 liters at STP, has a mass of 2.016 grams. Since the molar mass of hydrogen atoms is 1.008 grams per mole, you can conclude that under ordinary conditions hydrogen gas exists as diatomic molecules. Of course, we concluded earlier, on the basis of other evidence, that such is the case. Now, we shall consider why it is that hydrogen is more stable in the form of diatomic molecules than as atoms.

If the temperature is raised to several thousand degrees, hydrogen molecules decompose, as follows.

$$H_2(g) \longrightarrow H(g) + H(g)$$

Since energy must be expended in order to carry out the above reaction, the molecule H_2 must be more stable (has a lower energy) than two separated atoms.

This chemical bond (and every chemical bond) forms because the energy is lower when the atoms are near each other.

To see why the energy is lower when the atoms are near each other, let us consider the interactions among the electric charges of the atoms. Figure 14-1 (next page) represents molecule formation. Quantum mechanics tells us that the $1s$ orbital of each hydrogen atom has spherical symmetry before reaction. This is suggested by the shading in Figure 14-1. Yet, at any instant the electron can be pictured at some particular point, as shown by the negative charge of electron 1 located at a distance r_{1A} from nucleus A. The energy of hydrogen atom A can be explained in terms of the average attraction between electron 1 and nucleus A. This is fixed by the average of the distance between the two, r_{1A}. Likewise, electron 2 and nucleus B attract each other. Now consider the new electrical interactions present after the two atoms have moved close together. Electron 1 now is attracted to both protons. Electron 2 also is attracted to both protons. This is the "glue" that holds the two atoms together. The chemical bond in H_2 forms

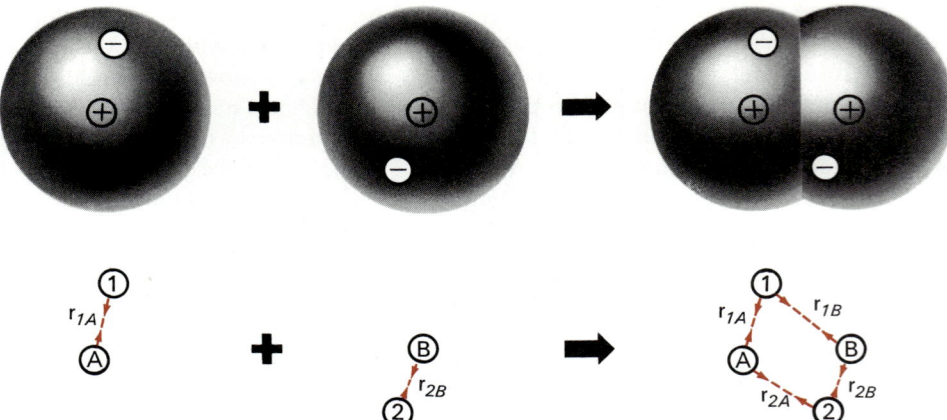

Figure 14-1 Hydrogen atoms form the hydrogen molecule because the moving electrons are simultaneously attracted to both nuclei. A lower energy state and, hence, greater stability results.

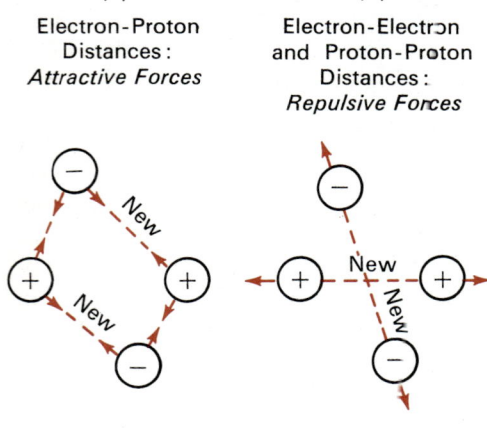

Figure 14-2 Analysis of the attractive forces between electrons and hydrogen nuclei (a) shows that they exceed the negative-negative and positive-positive repulsions (b).

because each of the two electrons is attracted to two protons simultaneously. This arrangement is energetically more stable than separated H atoms, in which each electron is attracted to only one proton.

Figure 14-2a shows a possible set of the electron-proton distances as they might be seen if it were possible to make an instantaneous photograph. Such distances fix the attractions that cause the chemical bond. But it is well to remember that there are also repulsions caused by the approach of the two atoms, as shown in Figure 14-2b. The two electrons repel each other and the two protons do the same. These repulsions tend to push the two atoms apart. Which are more important, the two new attraction terms or the two new repulsion terms? Experiment shows that the new attraction terms must dominate, since a stable chemical bond is formed. This is not to say that the repulsions do not exist. The proton-proton repulsions prevent the two hydrogen atoms from approaching even closer. The stable bond length in the hydrogen molecule is fixed by a balance between the forces of attraction, Figure 14-2a, and the forces of repulsion, Figure 14-2b.

The reason that attraction predominates over repulsion when two hydrogen atoms come together may be understood by recognizing that the electrons are mobile. They do not occupy fixed positions, but move about throughout the molecule. They occupy positions like those shown in Figure 14-1, in which each electron is closer to both nuclei than to the other electron. The two electrons preferentially move away from positions in which they would be near each other. They are said to "correlate" their motion so as to remain apart, reducing the electron-electron repulsion.

14-2 The Chemical Bond as an Overlap of Half-filled Orbitals

The discussion of chemical bonding in the hydrogen molecule can be simplified with the aid of Figure 14-3 (next page). Figure 14-3a shows the electron distribution in cross-section. The electron distribution extends outward, far from the nucleus, uniformly in all directions. But the distribution is concentrated near the nucleus, so attention ought to be focused on this region of the $1s$ orbital. Consequently, we represent the $1s$ orbital as a circle, with a radius large enough so that there is about a 90 percent probability of finding the electron within it, at any given instant.

As you learned in Chapter 13, an orbital can accommodate either one or two electrons, but no more. Figure 14-3b shows ways to represent an empty $1s$ orbital, a $1s$ orbital containing one electron, and a $1s$ orbital containing two electrons.

In Figure 14-3c is shown the interaction of two hydrogen atoms. Each atom has a single electron in a $1s$ orbital. As the two hydrogen atoms approach each other, the circles overlap. In this region of overlap the two electrons are shared by the two protons (as shown by the deeper tone in the area common to both circles).

As before, it is necessary for the electrons to be simultaneously and, on the average, equally attracted to both nuclei in order that attractive forces exceed repulsive forces. The sharing of electrons that occupy the region of overlap between the two orbitals tends to accomplish this. Therefore, a useful way to visualize a chemical bond is to think of the overlap of two orbitals. In the overlap region electrons are at least partially shared between two nuclei. However, the mere possibility of overlap of orbitals is not in itself an assurance that a chemical bond will result. It is also necessary that there be just the right number of electrons.

In the hydrogen molecule, there are just two electrons. Each atom enters into the process of bond formation by supplying one orbital, containing one electron. The bond then results from the sharing of two electrons between two atoms. The region in which the electrons are in the mutual possession of both atoms is the region of orbital overlap.

The most common way in which a chemical bond is formed is summarized as follows. Two atoms, each with a half-filled orbital, approach each other until their orbitals overlap. The two electrons are then shared between them in the region of overlap.

A bond formed by the equal sharing of electrons, such as the one described in the preceding paragraph, is known as a **covalent bond.** The prefix "co" in "covalent," of course, suggests sharing, as it does in many other words that should readily come to mind. The stem of the word, "-valent," comes from "valence," a noun long used by chemists in describing the combining capacity of an element.

(a) A Simplified Representation of a 1s Orbital:

(b) A Simplified Representation of Orbital Occupancy:

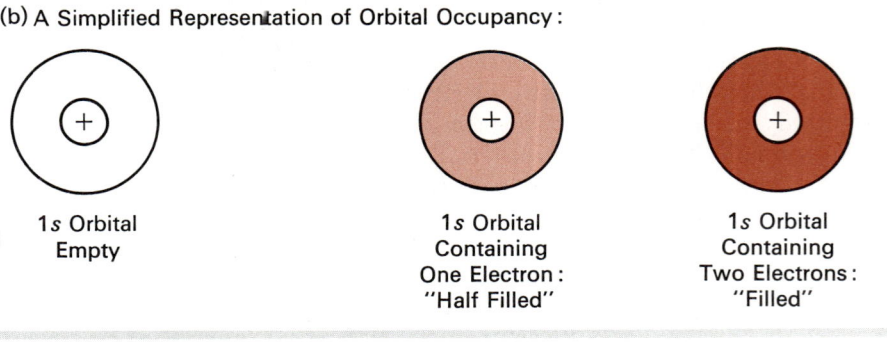

(c) Overlap and Bonding of the Hydrogen Molecule:

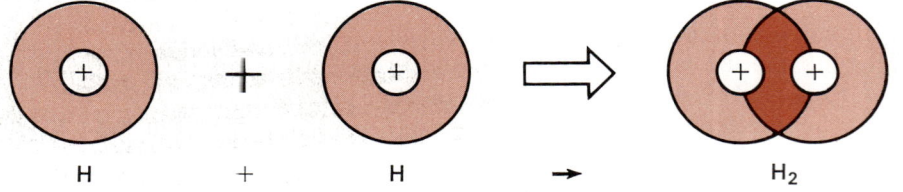

(d) Absence of Overlap for Two Helium Atoms:

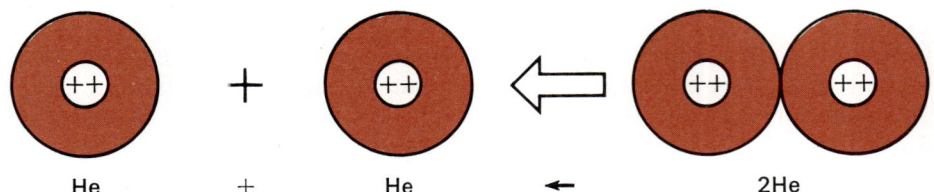

Figure 14-3 *Information concerning orbital occupancy and overlaps between atoms can be given in terms of diagrams like those above. They provide one way to represent the covalent bond of the hydrogen molecule (c) and the absence of bonding between helium atoms (d).*

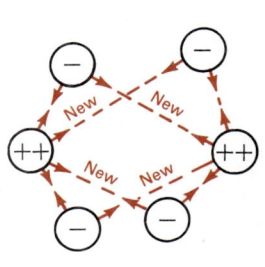 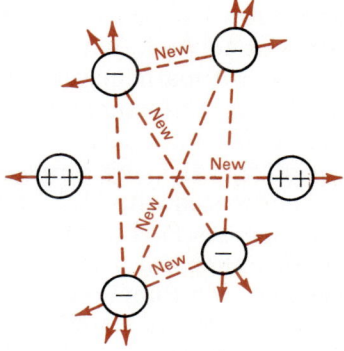

(a) Electron-Proton Distances: *Attractive Forces*

(b) Electron-Electron and Proton-Proton Distances: *Repulsive Forces*

Figure 14-4 *When two helium atoms approach one another, new forces of attraction result (a). There are also new forces of repulsion (b). The repulsive forces dominate and no helium molecule forms.*

A question that may occur to you is: What happens if orbitals overlap, but contain a total number of electrons other than two? When only one electron is present, a bond is still formed. Usually it is so weak, and the resulting molecule so reactive, that it is not of practical importance, at least at this stage in your study of chemistry. For example, when a hydrogen atom and a hydrogen ion are brought together, a hydrogen molecule ion, H_2^+, is formed. In the gas phase, at high temperature, H_2^+ is stable relative to $H + H^+$, but no compounds containing it can be isolated. It is not stable enough to resist reacting with other atoms and molecules. Similarly, bonds in which three electrons are shared occur. They also are not, in general, stable enough to be of practical importance.

When each atom involved has a filled orbital, so that *four* electrons would occupy any overlap region, a situation that deserves special attention results. This possibility is discussed in the following section.

14-3 Interaction between Helium Atoms

A measurement of its density shows that helium is a monatomic gas. Molecules of He_2 do not form. What difference between hydrogen atoms and helium atoms accounts for the absence of bonding in helium? The answer to this question must lie in the attractive and repulsive electrical interactions between two helium atoms when they approach each other. There would, of course, be four electrons, each one attracted to each nucleus. Figure 14-4a shows the attractive forces in one of our hypothetical instantaneous snapshots. Figure 14-4b shows the repulsive forces. Taking score, you find in Figure 14-4a eight attractive interactions, four more than in the separated atoms. In addition, there are seven repulsive interactions, five more than in the separated atoms. By experiment, chemists have learned that the four new forces of attraction are not sufficient to outweigh the five new repulsions. A chemical bond does not form.

The explanation, then, of the bonding in H_2 and the absence of bonding for two He atoms lies in the relative magnitudes of attractive and repulsive forces. Quantum mechanics can be put to work, with the aid of advanced and difficult mathematics, to calculate these quantities. Unfortunately, solving the mathematics presents such an obstacle that only a handful of the very simplest molecules have been treated with high accuracy. Nevertheless, chemists have long been able to decide whether chemical bonds can form, without appealing to a computer.

Figure 14-3d shows the simplified representation of the interaction of two helium atoms. Each helium atom is fully shaded before the two atoms approach. This is to indicate there are already two electrons in each of the 1s orbitals. As a result, the two overlapping orbitals would have their full complement of four electrons (two per orbital, in accordance with Pauli's principle). It is an experimental fact, however, that two full orbitals do not gain in stability on overlapping. No bond forms.

The lack of bonding between helium atoms illustrates the following general rule.

When two atoms have available only completely filled orbitals, the electron-electron repulsion prevents overlap and no bond can be formed between them.

14-4 Representations of Chemical Bonding

Chemists propose, then, that chemical bonds can form if electrons can be shared by two atoms, using partially filled orbitals. Forms of shorthand notation have been developed that aid in applying this rule. These provide representations of the bonding.

One way of representing covalent bonding involves the use of orbital-occupancy diagrams. First, available orbitals are shown as circles, just as in an energy level diagram, such as Figure 13-1. In order to save space, however, the orbitals are arrayed horizontally, with energy increasing from left to right, as in the following example.

$$\begin{array}{cccc} & 1s & 2s & 2p \\ H & \bigcirc & \bigcirc & \bigcirc\bigcirc\bigcirc \end{array}$$

The spaces between circles represent differences in energy.

The electron population of a given atom is indicated by using a diagonal line to represent an electron and by placing such lines in the proper orbitals. The hydrogen atom is represented as follows, in its most stable state.

$$\begin{array}{cccc} & 1s & 2s & 2p \\ H & \oslash & \bigcirc & \bigcirc\bigcirc\bigcirc \end{array}$$

Further examples, showing helium and lithium atoms, follow.

$$\begin{array}{cccc} & 1s & 2s & 2p \\ He & \otimes & \bigcirc & \bigcirc\bigcirc\bigcirc \\ Li & \otimes & \oslash & \bigcirc\bigcirc\bigcirc \end{array}$$

Orbital Representation of Chemical Bonding. The rule about covalent bond formation can be applied quite simply through an orbital representation.

$$\begin{array}{cccc} & 1s & 2s & 2p \\ H & \oslash & \bigcirc & \bigcirc\bigcirc\bigcirc \\ H & \oslash & \bigcirc & \bigcirc\bigcirc\bigcirc \end{array}$$
Bond Can Form

$$\begin{array}{cccc} & 1s & 2s & 2p \\ He & \otimes & \bigcirc & \bigcirc\bigcirc\bigcirc \\ He & \otimes & \bigcirc & \bigcirc\bigcirc\bigcirc \end{array}$$
Bond Cannot Form

In this representation there is no need to consider the next higher energy level cluster, the $2s$ and $2p$ orbitals. For hydrogen and helium these are much higher in energy, as the diagram indicates, and can give rise only to extremely weak attractions.

Electron Dot Representation of Chemical Bonding. The sharing of electrons can be shown by representing electrons as dots placed between the atoms. The formation of H_2 can be shown as follows.

$$H \cdot + H \cdot \longrightarrow H:H$$

Line Representation of Chemical Bonding. As you learned in Section 7-13, chemists often represent a bond between atoms with the use of a line (or lines). Chemists used this representation long before the role that electrons play in chemical bonding was understood. The line system fits in well with current bonding theories and is still widely used. A single line is equivalent to a shared electron pair.

All of the foregoing representations will be used in subsequent discussions of bonding.

14-5 The Bonding of Fluorine

An explanation of chemical bonding is of value only if it has wide applicability. Let us examine its usefulness in considering the compounds of the second row elements, beginning with fluorine.

Under normal conditions of temperature and pressure, fluorine is a gas. From gas density experiments it is deduced that a molecule of fluorine contains two atoms. There is a chemical bond between the two fluorine atoms. Do our bonding rules agree with these experimental facts?

Orbital Representation of the Bonding of Fluorine. A fluorine atom has nine electrons. Seven of these electrons occupy a partially filled cluster of energy levels, as you can see in the following diagram.

This cluster contains one electron less than its capacity permits. Fluorine, then, has the ability to share one electron with some other atom having similar ability. If, for example, another fluorine atom approaches, the two fluorine atoms might share a pair of electrons and form a covalent bond, as shown in the following diagram.

After the second atom has approached and is sharing its electron, the orbitals of each fluorine atom are "filled." No additional bonding capacity remains. Hence F_2 does not add a third or fourth atom to form F_3, F_4, and so on.

Now consider the possibility of the bonding that might occur when a fluorine atom encounters a hydrogen atom. Again fluorine has an opportunity to share electrons with a second atom having a partially filled orbital. The result is represented in the following diagram.

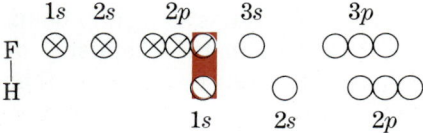

By sharing a pair of electrons with hydrogen, fluorine has again "filled" its orbitals. It has no further bonding capacity. The same is true of the hydrogen atom, though for this atom there is but one orbital to consider, the 1s orbital. Since no partially filled orbitals remain for either atom, there is no more bonding capacity and a stable compound is formed, HF.

In each of these molecules, F_2 and HF, the fluorine atom forms one bond. In F_2 the bond is to a second fluorine atom, and in HF it is to a hydrogen atom. This single-bonding capacity is described by saying that fluorine is *univalent*.

You have just seen that the fluorine atom forms bonds using an orbital in the second energy cluster. Its 1s orbital is of no use in bonding because it is already filled. It cannot, therefore, overlap, or share an electron, with another atom. Other orbitals outside of the $2s2p$ set are entirely empty and cannot make any contribution to forming bonds either. Thus, in considering how a fluorine atom can form bonds, we concentrate our attention entirely on the energy cluster containing partially filled orbitals. We shall refer to the orbitals within such a cluster as **valence orbitals.** "Valence" means bonding ability. It is the valence orbitals that are responsible for bonding. We shall refer to the electrons that occupy the valence orbitals as **valence electrons.**

Electron Dot Representation of the Bonding of Fluorine. In using the electron dot representation of chemical bonds, it is necessary to show only the valence electrons. In fluorine there are seven, since the pair of electrons in the 1s orbital is so tightly bound that it plays little role in the chemistry of fluorine. Using this representation, then, the reaction between two fluorine atoms is shown as follows.

$$:\!\ddot{F}\!\cdot\; +\; \cdot\ddot{F}\!: \longrightarrow\; :\!\ddot{F}\!:\!\ddot{F}\!:$$

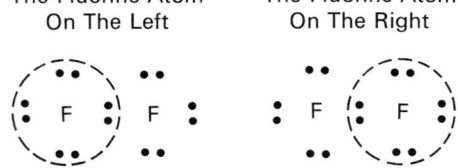

Figure 14–5 *In the fluorine molecule each atom achieves the stable noble gas configuration.*

From this you can conclude that a covalent bond can form between two fluorine atoms. A census of the electrons owned or shared by either of the fluorine atoms shows that the valence orbitals are filled. For the fluorine atom on the left, for example, there are electrical attractions involving eight valence electrons. Six of these electrons belong solely to itself and two of them are shared with the fluorine atom on the right, as shown in Figure 14-5. Since eight electrons is just the capacity of the 2s and 2p valence orbitals, each fluorine atom has approximated the electron configuration of neon. You have seen that the electron configuration of any noble gas must be an energetically stable one, since those gases are so unreactive.

The formation of hydrogen fluoride can be represented by the following electron dot picture.

$$:\!\ddot{F}\!\cdot\; +\; \cdot H \longrightarrow\; :\!\ddot{F}\!:\!H$$

Again, a census of electrons near each of the atoms (Figure 14-6) shows that this is a stable arrangement. True, the hydrogen atom has, close at hand, only two electrons, whereas fluorine has eight. This is energetically desirable, however, because hydrogen has only one valence orbital, the 1s orbital. Two electrons just fill this orbital.

The bonding of a fluorine atom to another fluorine atom, or to a hydrogen atom, can be explained in terms of electron-sharing that fills partially filled valence orbitals. This sharing makes the molecule, F₂ or HF, energetically more stable than the separated atoms would be. *The energy stability results from the shared electrons being attracted simultaneously to both positive nuclei.* A chemical bond results.

Eight Valence Electrons Are Involved with The Fluorine Atom

Two Valence Electrons Are Involved with The Hydrogen Atom

Figure 14-6 *In the hydrogen fluoride molecule both hydrogen and fluorine reach noble gas configurations.*

Summary

A molecule is a cluster of atoms that persists long enough to have characteristic properties. Atoms of the noble gases have very stable electron configurations and, thus, do not readily form molecules.

Elements other than the noble gases have atoms with electron configurations that produce less stable atoms. They are thus, in varying degrees, more chemically reactive. In an attempt to achieve the stable electron configuration of the noble gas atoms, many atoms will "cooperate" with other atoms, either of the same or of different elements. The most common way in which this cooperative operation is carried out is for two or more atoms to share electrons. Sharing creates a common bond that tends to hold the atoms together, forming a molecule. A bond created by this method of sharing is called a covalent bond.

Self Check

1. Why is energy lower when atoms are near each other than it is when they are farther apart?
2. Why do the repulsive forces between the electrons and the protons of two different atoms not prevent molecules from forming?

How Bonding Occurs in Second Row Elements

14-6 *The Bonding Capacity of Oxygen Atoms*

The second row of the periodic table consists of the following elements.

Li   Be   B   C   N   O   F   Ne

You have already seen how fluorine is able to form just one bond, and why neon can form no such bond. The abilities of the other elements in this row to form bonds vary greatly, but certain trends are apparent. With the ideas about chemical bonds that have been developed in this chapter, you have a way of explaining these trends.

The neutral oxygen atom has eight electrons. Six of these partially fill the $2s$ and $2p$ orbitals. Two are in the filled $1s$ orbital. Therefore, oxygen has six valence electrons. The $2s$ and $2p$ orbitals are the valence orbitals. They can accommodate the valence electrons in two ways, as follows.

In order to understand which of the two configurations is more likely to be correct, you should recall the spatial arrangement of p orbitals. Each orbital protrudes along one of the three Cartesian axes (as shown in Chapter 12). If the electrons had the orbital occupancy of the first diagram, two electrons would occupy the p orbital protruding along the x axis (p_x), two electrons would occupy the p orbital protruding along the y axis (p_y), and the p_z orbital would be empty. In the second scheme, there would be two electrons in p_x, one in p_y, and one in p_z. Since electrons repel each other, you might expect that the second configuration, which keeps the electrons farther apart, would be lower in energy. Experiment shows that it is, and so the discussion of the bonding of oxygen will be based on the second orbital-occupancy arrangement.

Suppose a hydrogen atom approaches an oxygen atom in its most stable state. Each atom has partially filled valence orbitals. Electron sharing can occur, placing electrons close to two nuclei simultaneously. Hence a stable bond can result. This is shown in the following representations.

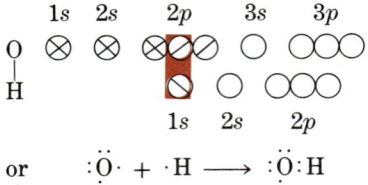

In either representation, there is residual bonding capacity remaining in the species OH. As you can see in the upper diagram, the third $2p$ orbital has a capacity for one more electron. This means that more bonding can occur. In the dot diagram, there are only seven electrons near the oxygen atom. This atom would be more stable if it could add one more electron. Either representation shows that OH should be able to react with another hydrogen atom, as follows.

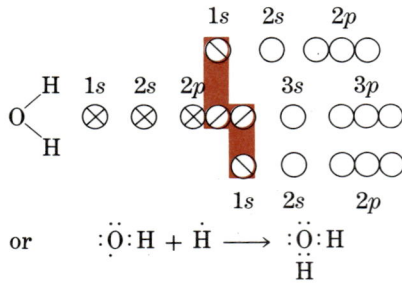

This produces the compound H_2O. Either representation shows that the bonding capacity of oxygen is expended when two bonds are formed. Oxygen is thus said to be *divalent*. The compound H_2O is extremely stable. Each of the atoms in H_2O has filled its valence orbitals by electron sharing.

Reaction Between Two OH Molecules. Though OH is reactive, it is a cluster of atoms with sufficient stability to be identified as a molecule. It is present in a number of high-temperature flames, for example. Its chemistry might be expected to be like that of fluorine atoms. Compare their electron dot formulas.

$$:\!\ddot{\underset{..}{F}}\!: \quad :\!\ddot{\underset{..}{O}}\!:\!H$$

Since two fluorine atoms react, forming a covalent bond, two OH molecules can be expected to do the same sort of thing.

$$:\!\ddot{\underset{..}{O}}\!\cdot \;+\; \cdot\!\ddot{\underset{..}{O}}\!: \;\longrightarrow\; :\!\ddot{\underset{..}{O}}\!:\!\ddot{\underset{..}{O}}\!:$$
$$\;\;\;\;\text{H} \quad\quad\; \text{H} \quad\quad\quad\;\; \text{H} \;\; \text{H}$$

The preceding reaction yields the compound H_2O_2. This is the formula of the common substance hydrogen peroxide. The structure of H_2O_2 can more conveniently be shown as follows.

$$\begin{array}{c} \text{H} \\ | \\ \text{O}\!-\!\text{O} \\ | \\ \text{H} \end{array}$$

Oxygen-Fluorine Compounds. It is a simple matter to predict that oxygen will form a stable compound with two fluorine atoms, OF_2. The orbital representation is as follows. (In this representation, and in subsequent ones, empty orbitals that are much higher than the valence orbitals are omitted, for convenience.)

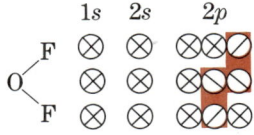

The electron dot representation is

$$:\!\ddot{\underset{..}{O}}\!:\!\ddot{\underset{..}{F}}\!:$$
$$:\!\ddot{\underset{..}{F}}\!:$$

Again note that oxygen is divalent.

14-7 The Bonding Capacity of Nitrogen Atoms

The unbonded nitrogen atom is most stable when it has the maximum possible number of partially filled valence orbitals. This keeps the electrons as far apart as possible, as did the most stable arrangement for the oxygen atom. The most stable state of the nitrogen atom is represented as follows.

$$\;\;\;\; 1s \;\;\; 2s \;\;\; 2p$$
$$_7\text{N} \;\; \otimes \;\; \otimes \;\; \oslash\oslash\oslash$$

The diagrams should lead you to predict that nitrogen is *trivalent*. There is ample evidence to support such a prediction. For example, nitrogen reacts with hydrogen to form the gas ammonia. Ammonia is a stable compound with the formula NH_3. Nitrogen also reacts with fluorine, producing another gas, NF_3. The electron dot formulas follow.

$$\begin{array}{cc} \text{H} & :\!\ddot{\underset{..}{F}}\!: \\ :\!\ddot{\text{N}}\!:\!\text{H} & :\!\ddot{\text{N}}\!:\!\ddot{\underset{..}{F}}\!: \\ \text{H} & :\!\ddot{\underset{..}{F}}\!: \end{array}$$

ammonia   nitrogen trifluoride

14-8 The Bonding Capacity of Carbon Atoms

There are a number of orbital occupancies that might be considered for the carbon atom. Two possibilities follow.

$$\;\;\;\; 1s \;\;\; 2s \;\;\; 2p$$
$$_6\text{C} \;\; \otimes \;\; \otimes \;\; \otimes\oslash\oslash$$
$$_6\text{C} \;\; \otimes \;\; \otimes \;\; \oslash\oslash\oslash$$

By the conventional argument that electrons repel each other, the second configuration should be more stable than the first. The second places one electron in each of the p_x and p_y orbitals, whereas the first places the two electrons in the same orbital, p_x. It is an experimental fact that the second configuration is more stable than the first.

The chemistry of the carbon atom in this state can now be predicted. It should be divalent, forming compounds CH_2 and CF_2. Consider one of these, CH_2.

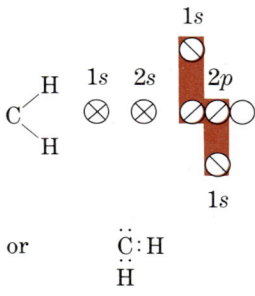

Here is a situation that you have not met before. After using the two available partially filled orbitals to form covalent bonds with hydrogen atoms, the carbon atom still has a vacant valence orbital. In the electron dot formula you can see that the carbon atom finds itself near only six electrons in CH_2. But the valence orbitals will accommodate eight electrons. *Because one valence orbital is completely vacant, you can expect CH_2 to be reactive.*

The CH_2 molecule is an extremely reactive one. It is considered to be stable because it does not spontaneously break up into smaller fragments. Like the previously discussed H_2^+ molecule-ion, however, it reacts very readily with other chemical species, forming new chemical species with much lower energies than that of CH_2.

Both CH_2 and the analogous compound CF_2 exist, but both are too reactive to be prepared in pure form, that is, "isolated." Their great reactivity suggests that carbon might produce more stable compounds if all four of its valence orbitals were used in bonding. This leads us to consider a third orbital occupancy.

With this orbital occupancy, a carbon atom has four half-filled valence orbitals. It is somewhat less stable than that in which the valence electrons have the configuration $2s^2\,2p_x 2p_y$. That is because it takes energy to raise, or "promote," a $2s$ electron to the somewhat higher energy $2p_z$ orbital. On the other hand, the promotional energy is not very large, and in return for it the carbon atom acquires the capacity to form four covalent bonds. Each covalent bond increases stability, more than compensating for the energy investment in promoting one of the $2s$ electrons. With this orbital occupancy, carbon can share pairs of electrons with, for example, four hydrogen atoms or four fluorine atoms. In doing so, it forms the stable compounds methane and carbon tetrafluoride. In these, as in most of its stable compounds, carbon is *tetravalent*.

methane   carbon tetrafluoride

14-9 The Bonding Capacity of Boron Atoms

The boron atom, shown in the two following diagrams, presents the same sort of option in orbital occupancy as does carbon.

```
           1s  2s   2p
   (a) ₅B  ⊗   ⊗   ○○○
   (b) ₅B  ⊗   ⊘   ⊘○○
```

The electron configuration (b) is somewhat higher in energy than (a). It is necessary to promote a $2s$ electron to the $2p$ state to obtain (b). In return, however, the boron atom gains bonding capacity. Whereas a boron atom can form only one covalent bond in configuration (a), it can form three in configuration (b). Since each bond lowers the energy, electron configuration (b) is favored in the chemistry of boron.

You would expect the boron atom to be trivalent, then. You might predict that there should be molecules such as BH_3 and BF_3.

```
      H              :F:
      ..             ..
      B:H            B:F:
      ..             ..
      H              :F:
                     ..
```

Notice, however, that in each compound there remains a completely vacant valence orbital. The orbital representation for BH_3, for example, would be shown as follows.

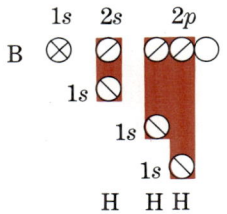

The vacant $2p$ orbital is reminiscent of the one found in CH_2. Since CH_2 is very reactive, presumably BH_3 should be the same. Such is the case. There is only indirect evidence establishing the existence of BH_3. Instead, boron forms a series of unusual compounds with hydrogen, the simplest of which is called *diborane*, B_2H_6. The structures of B_2H_6 and the other more elaborate boron-hydrogen compounds are quite complex, and the bonding in these molecules involves considerations outside the scope of this chapter. These molecules are still being actively investigated by research chemists and are truly on the frontier of chemical knowledge.

Although a vacant valence orbital makes BH_3 a very reactive compound, the analogous compound BF_3 is stable. Boron trifluoride is a gas that is readily prepared in pure form. The explanation of how the fluorine atoms are able to satisfy in part the bonding capacity of the vacant $2p$ orbital (though hydrogen atoms cannot), like a study of boron-hydrogen compounds in general, cannot be given at this point. It is sufficient to observe here that BF_3 is the most stable boron-fluorine compound and that it demonstrates the trivalent bonding capacity of boron.

14-10 The Bonding Capacity of Beryllium Atoms

The beryllium atom, like boron and carbon, can promote an electron in order to form more chemical bonds. The following figure shows the resultant configuration.

```
           1s  2s   2p
   ₄Be     ⊗   ⊘   ⊘○○
```

You should expect to find molecules in the gaseous state such as BeH_2 and BeF_2. These molecules have been detected. On the other hand, beryllium has the trouble boron has, only in a double dose. It has *two* vacant valence orbitals. As a result, BeH_2 and BeF_2 molecules, as such, are obtained only at extremely high temperatures (above 1000 K). At lower temperatures the presence of vacant valence orbitals causes the formation of

TABLE 14-1 The Fluorides of the Elements in the Second Row of the Periodic Table

Second Row Elements	Fluoride Formula	Melting Point (K)	Boiling Point (K)
Li	LiF	1115	1949
Be	BeF_2	1073*	1073*
B	BF_3	146	172
C	CF_4	89	145
N	NF_3	66	144
O	OF_2	49	128
F	F_2	50	85

*Changes directly from solid to vapor at 1073 K and atmospheric pressure.

structurally more complicated solids. In these the beryllium atom is able to make use of all its valence orbitals.

14-11 The Bonding Capacity of Lithium Atoms

There is little new to be said about the bonding capacity of a lithium atom. With just one valence electron, it should form gaseous molecules LiH and LiF. Because of the vacant valence orbitals, these substances should be expected only at extremely high temperatures. You may recall from earlier chapters that the lithium atom can readily lose its valence electron completely to form the ion Li^+. This ion plays a significant role in lithium chemistry. Simple covalent molecules like LiF and LiH are known only in the gas phase, and even in these molecules ionic forces play a major role, as explained in Section 14-13.

Table 14-1 summarizes the formulas and the melting and boiling points of the stable fluorides of the second row elements. In each case, the formula given in the table is the actual molecular formula of the species found in the gas phase.

14-12 A Word of Caution about the Word "Valence"

Very often the word "valence" is used in discussing chemical bonding. Unfortunately this word has been used as a noun to mean a number of different things. Valence has been used to mean such things as the charge on an ion and the total number of atoms to which a particular atom will form bonds. Perhaps the most widely accepted definition of the word is that it is the number of univalent atoms (such as H or F) with which an atom can combine in a chemical reaction. It is clear that a word with so many meanings might confuse a discussion of chemical bonding. For this reason the word valence has not been used as a noun thus far. Once the basic ideas of chemical bonding are clearly in mind, however, it is helpful in everyday practical problems, in which formulas and equations must be written, to speak of the valence (or, at least, the usual valence) of an element. In Chapter 19, in which formula writing and equation balancing will be taken up in detail, the word valence will be useful. The basic ideas about chemical bonds, as they are being studied here, should never be lost sight of, however.

In this chapter, the *valence orbitals* have been defined as those comprising an energy cluster that contains partially filled orbitals. *Valence electrons* have been defined as those electrons found in valence orbitals. In both cases "valence" is used as an adjective.

Summary

The ability of each of the second row atoms to form electron-pair bonds can be related to the electron configuration of the atom. The atoms of Li, N, O, and F form as many bonds as there are singly occupied orbitals in the most stable configuration of their atoms. For Be, B, and C, more than this number of bonds are formed because some of the electrons that are normally placed two in an orbital can easily be promoted into empty valence orbitals. Thus Be, which has $2s^2$ as its most stable configuration, can change to a $2s2p$ configuration. It then has two singly occupied orbitals and can form two bonds. Similarly, boron can go from $2s^22p$ to $2s2p_x2p_y$ and form three bonds, and carbon can go from $2s^22p_x2p_y$ to $2s2p_x2p_y2p_z$ and form four bonds.

Self Check

1. The molecule NH_2 has residual, unused bonding capacity and is extremely reactive. The molecule N_2H_4 (hydrazine) is much more stable. Draw an electron dot representation of the bonding of hydrazine. Draw its structural formula.
2. Draw orbital and electron dot representations for each of the following molecules: OF, O_2F_2, HOF, HFO_2. Which of these should be the most reactive?
3. Draw electron dot formulas for the molecules CH_3, CF_3, CHF_3, CH_2F_2, CH_3F. Which will be extremely reactive?
4. Draw an electron dot formula and a structural formula for the molecule, C_2H_6 (ethane), which forms if two CH_3 molecules are brought together. Explain why C_2H_6 is much less reactive than CH_3.

Some Bonds Have Ionic Character

14-13 The Bonding in Gaseous Lithium Fluoride

The chemical bond in the hydrogen molecule, H_2, has been termed covalent. This indicates that electrons are shared so that they are simultaneously and, on the average, equally near two nuclei. This makes the two hydrogen atoms more stable and that is why a chemical bond results.

All chemical bonds occur because electrons can be placed simultaneously near two nuclei. Yet it is often true that the shared electrons are not shared equally. Sometimes the electrons, though close to both nuclei, tend to be nearer to one nucleus than to the other. You can see why by contrasting the bonding in gaseous fluorine, F_2, and in gaseous lithium fluoride, LiF.

The bonding in the F_2 molecule has already been discussed. Since neither of the fluorine atoms can pull an electron entirely away from the other, they compromise by sharing a pair of electrons equally. How does the chemical bonding in the lithium fluoride molecule compare with this?

As mentioned earlier, lithium has one valence electron. It can share a pair of electrons with one fluorine atom, as shown in the following representations.

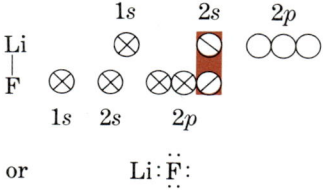

A stable molecular species, LiF, can be expected. The term "stable" again means that energy is required to disrupt the molecule. The energy of the molecule is lower than that of the reacting atoms because the bonding electron pair feels, simultaneously, the attractions of both the lithium nucleus and the fluorine nucleus. This is not to say, however, that the electrons are shared equally. After all, the lithium and fluorine atoms attract the electrons to a different extent. This is shown by the ionization energies (I) of these two atoms (g means "gas").

$$F(g) \longrightarrow F^+(g) + e^-(g)$$
$$I = 401.5 \text{ kcal/mole}$$

$$Li(g) \longrightarrow Li^+(g) + e^-(g)$$
$$I = 124.3 \text{ kcal/mole}$$

Clearly the fluorine atom holds its electron much more strongly than does the lithium atom. As a result, the electron pair in the lithium fluoride bond is more strongly attracted to the fluorine atom than to the lithium atom. The energy is lower if the electrons are closer to the fluorine atom. When the bonding electrons move closer to one of the two atoms, a bond is said to have ionic character.

In the most extreme situation, the bonding electrons move so near one of the atoms that this atom has virtually the electron distribution of the negative ion. This is the case in gaseous LiF, which can be shown as follows.

$$\overset{+}{Li} \, \overset{..}{\underset{..}{F}}\, \overset{-}{} \qquad \text{or} \qquad Li^+ F^-$$

When such formulas provide a useful basis for discussing the properties of a molecule, the bond in that molecule is said to be an ionic bond.

Contrasting Covalent and Ionic Bonds. The fluorine molecule is held together by the energy gained from placing a bonding pair of electrons near two fluorine nuclei simultaneously. The electrons move about in the molecule in such a way that, on the average, they are distributed symmetrically between and around the identical fluorine nuclei. This symmetrical distribution is to be expected, since the two fluorine nuclei attract the bonding electrons equally. The lithium fluoride molecule also is held together by the energy gain that results from placing a bonding pair of electrons near the lithium and fluorine atoms simultaneously. In this case, however, the electrons move in such a way as to remain closer to the fluorine than to the lithium atom. Fluorine attracts the bonding electrons more strongly than does lithium.

It is now evident that there is but one principle underlying the formation of a chemical bond between two atoms. *Bonds form because electrons are placed simultaneously near two positive nuclei.* The term *covalent bond* indicates that the most stable distribution of the electrons (as far as energy is concerned) is symmetrical between the two atoms. When the bonding electrons are somewhat nearer to one of the atoms than the other, the bond is said to have ionic character. The term *ionic bond* applies when the electrons are displaced so much toward one atom that it is a good approximation to represent the bonded atoms as a pair of ions near each other. Figure 14-7 shows schematically how the electron distributions are pictured in covalent, partially ionic, and ionic bonds. The figure shows how the electrons might look in an instantaneous snapshot. In each type of bond, the electron-nucleus attractions account for the energy stability of the molecule.

The Electric Dipole of the Ionic Bond. The drifting of negative electric charge toward one of the atoms in an ionic bond causes a charge separation. This can be represented crudely as in the last drawing in Figure 14-7. The molecule is electrically positive at the lithium end and electrically negative at the fluorine end.

As a result, it is said to possess an electric dipole. Any such molecule, possessing two separate centers (or poles) of electric charge is known as a polar molecule. Each polar molecule is electrically neutral, since its charge centers, or poles, are equal in magnitude, though opposite in sign.

Polar molecules have some important properties that are different from those

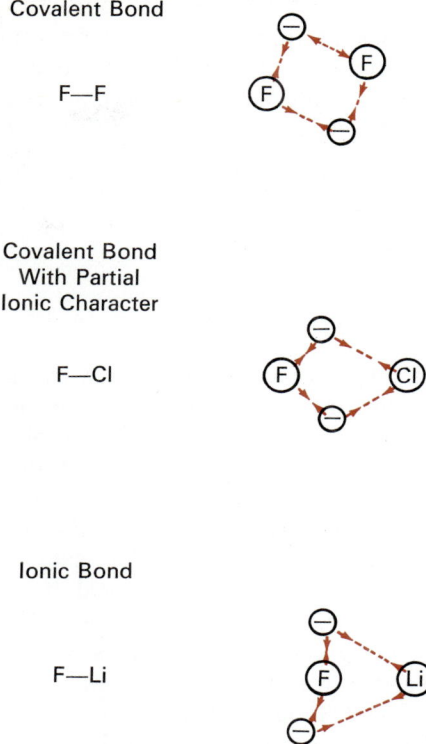

Figure 14-7 *The covalent bond (top) presents a neutral mass of charges to approaching particles. The center molecule has partial ionic character, and the bottom molecule is ionic.*

shared by nonpolar molecules. For example, the forces that polar molecules exert on one another are much stronger than those operating among nonpolar molecules. You can probably guess why this should be the case. If not, you will find out in Chapter 17.

There are several simple ways of representing a polar molecule. Some of them are shown in Figure 14-8. The last one, the arrow representation, is quite commonly used. Conventionally, the head of the arrow represents the negative end of the molecule. The directional property of the arrow implies that the force this molecule exerts on another molecule depends on the direction of approach of the second molecule.

14-14 Ionic Character in Bonds to Fluorine

The effects just discussed will be at work in the bonds that fluorine forms with other elements. The ionization energies of the elements give a rough clue to the magnitudes of electron-nuclear attraction. Table 14-2 compares the ionization energies of fluorine and other second row elements. The last column describes the type of chemical bond formed in each case. The trend from covalent to ionic bonding (and vice versa) is due to the decreasing (or increasing) ionization energies of the elements bonded to F. The trend in bond type is an important influence on the trend in chemical properties of these compounds.

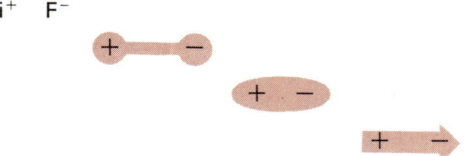

Figure 14-8 *The ionic bond of lithium fluoride creates a dipole with positive charge at the lithium nucleus and negative charge at that of fluorine. The arrow is a frequently used symbol for a dipole.*

14-15 Ionic Character in Bonds to Hydrogen

The element hydrogen can be characterized as a family by itself. Often its chemistry distinguishes it from the rest of the periodic table. This is the case when attempts are made to predict the ionic character of bonds to hydrogen.

The ionization energy of the hydrogen atom, 313.6 kcal/mole, is quite close to that of fluorine. A covalent bond between these

TABLE 14-2 Bond Types in Some Fluorine Compounds

Compound	Bond	Ionization Energies (kcal/mole)		Bond Type	
		Element Bonded to F	Fluorine		
FF	F—F	F 401.5	401.5	Covalent	
OF_2	O—F	O 313.8	401.5	slightly ionic	↑
NF_3	N—F	N 335	401.5	increasing ionic character	increasing covalent character
CF_4	C—F	C 259.5	401.5		
BF_3	B—F	B 191.2	401.5		slightly covalent
BeF_2	Be—F	Be 214.9	401.5		
LiF	Li—F	Li 124.3	401.5	Ionic	

Linus C. Pauling (1901–)

THE INFLUENCE OF Linus Pauling on the modern world has extended far beyond the walls of his laboratory. Pauling and Madame Marie Curie are the only two people to have been awarded two Nobel prizes. He received the 1954 Nobel Prize in Chemistry "for his research into the nature of the chemical bond" and the 1962 Nobel Peace Prize "for opposition to all warfare as a means of solving international conflicts."

Early in his career at the California Institute of Technology, where he has done most of his scientific work, he used the methods of X-ray diffraction (described in Chapter 15) to measure the distance between atoms in molecules and crystals. This led him into an intensive study of the structure of these substances and the forces holding them together. In 1939 he published *The Nature of the Chemical Bond,* which exerted an enormous influence on research in all aspects of chemistry. In this book he described the nature of covalent and ionic bonds, examined the part that hybrid orbitals play in bonding, and proposed the theory of resonance to explain some of the peculiarities in the behavior of benzene and other molecules. (Hybrid orbitals and resonance are discussed in Chapter 15.)

During his years at the California Institute of Technology he helped make this school one of the great biochemical research centers of the world. He applied the techniques of physical chemistry to the solution of many medical problems. He studied the structure of proteins, including red blood cells, and proposed the term "molecular disease" for certain types of mental illness.

While he was working with problems associated with heredity, he became increasingly concerned about the problems of nuclear warfare and the threat of serious genetic damage caused by radioactive fallout from the testing of nuclear weapons in the atmosphere. In early 1958 he presented a petition to the United Nations with the signatures of 9235 scientists calling on the U.N. to ban any further atmospheric testing of these weapons. His efforts led to an almost immediate moratorium on atmospheric testing and eventually resulted in the signing of a test ban treaty by more than 100 members of the United Nations.

two atoms in HF would thus be expected. Actually, the properties of HF show that the molecule has a significant electric dipole, indicating ionic character in the bond. The same is true in the O—H bonds of water and, to a lesser extent, in the N—H bonds of ammonia. *The ionic character of bonds to hydrogen cannot be predicted from its measured ionization energy.*

The properties of a number of compounds containing hydrogen have been examined. On the basis of these properties, the ionic character of bonds to hydrogen has been estimated. It is roughly what would be expected for bonds to an element with an ionization energy of about 200 kcal/mole. Thus, the hydrogen-fluorine bond in HF is ionic. The electrons are skewed toward the fluorine atom, leaving the hydrogen atom with a partial positive charge. Hydrogen acts like an element with lower ionization energy than fluorine. The same is true, but to a lesser extent, for hydrogen when it is bonded to oxygen and to nitrogen. The carbon-hydrogen bond has only slight ionic character. At the other end of the periodic table, gaseous lithium hydride is known to have a significant electric dipole, but with the electric dipole turned around. In LiH the electrons are skewed toward the hydrogen atom, leaving the lithium atom with a partial positive charge. This is in accord with the low ionization energy of lithium, 124.3 kcal/mole, well below the value of 200 kcal/mole that has been assigned to hydrogen. For your purposes, it suffices to discuss the bonding of hydrogen in terms of an *apparent ionization energy near 200 kcal/mole.*

INVESTIGATION

14-16 *Covalent and Ionic Compounds*

In an earlier experiment (Section 9-10) you passed an electric current through a solution that contained ions. If there had been no ions present, no current would have flowed in the circuit.

In this chapter you have found that covalent bonds are formed when two atoms share electrons between them. They may be shared in such a way that the electrons, on the average, are equally near the two nuclei of the atoms. You have also seen that bonds become ionic in character when the bonding electrons are located much closer to one atom than to another.

It would seem reasonable to wonder whether electrons in ionic bonds could get so close to one nucleus and so far from the other that ions would be present in the melted substance. If this turned out to be the case, the liquid should conduct an electric current and you could be quite sure that the bonds joining the atoms were ionic in character. Conversely, if the liquid did not conduct an electric current, you could conclude that there were no ions present and that the bonds must be mainly covalent.

Once you have determined the nature of the bonds in a substance, it would be interesting to see if there is any relationship between the character of the bonds and the way the substance behaves when it is dissolved in water.

PURPOSE

In this experiment your teacher will measure the conductivities of two groups of related materials. The first group will include sodium chloride, NaCl, carbon tetrachloride, CCl_4, and hydrogen chloride, HCl. The second group will include sodium hydroxide, NaOH, and methanol, CH_3OH.

When the members of the first group have been added to water, the solutions will be tested for the presence of chloride ions by the addition of a dilute solution of silver nitrate. Silver nitrate solution contains silver ions, which combine with chloride ions to form a white precipitate. Hydroxide ions, in the second group, will be tested for by the addition of a dilute solution of ferric chloride, which contains ferric ions. Ferric ions produce a characteristic precipitate in the presence of hydroxide ions.

MATERIALS

Ringstand and ring
Burner
Pipestem triangle
2 porcelain crucibles
5 test tubes, 18- × 150-mm
1 g sodium chloride crystals
1 g sodium hydroxide pellets
5 ml carbon tetrachloride
5 ml methanol
2 pieces #16 copper wire, 10 cm long
3 insulated wires, with alligator clips
Battery, 6-volt, or direct current power supply
Ammeter, or 6-volt flashlight bulb with suitable mount

5 ml solution of HCl in toluene
1 ml silver nitrate solution
1 ml ferric chloride solution
2 rubber stoppers
Distilled water
Electrical insulating tape
Utility clamp
Wire gauze, asbestos center

Figure 14–9

PROCEDURE

1. Attach two pieces of #16 copper wire, each about 10 cm long, to opposite sides of a #4 or #5 rubber stopper. Attach them by means of a piece of electrical insulating tape wrapped several times around the stopper. Leave about 1 cm of wire sticking above the stopper on each side for making connections, as in Figure 14-9. Use a utility clamp to attach the stopper with its electrodes to the top of a ringstand.

2. Connect one copper electrode to a terminal of an ammeter or a 6-volt flashlight bulb, using insulated wire equipped with alligator clips. Connect the other terminal of the ammeter or bulb to a 6-volt battery or power supply. Connect the remaining terminal of the

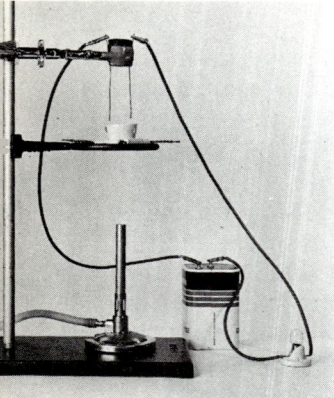

Figure 14–10

electrical source to the other copper electrode. (See Figure 14-10.) Test the apparatus by gently pushing the lower ends of the copper electrodes together. If the bulb lights or the ammeter needle is deflected, the apparatus is working properly.

3. Use an iron ring, attached to the ringstand, to support a porcelain crucible on a pipestem triangle over a burner, as shown in Figure 14-10.

4. Fill the crucible to a depth of about 0.5 cm with crystalline sodium chloride, NaCl.
   Be sure to wear your safety glasses for steps 5–8.

5. Heat the crucible and its contents strongly until the sodium chloride melts.

6. Remove the burner and lower the electrodes briefly into the molten salt. Note whether the substance conducts an electric current.

7. When the crucible has cooled enough for the salt to solidify, use crucible tongs to lift the hot crucible from the triangle. Set it aside on an asbestos-centered wire gauze to cool. Wash the copper electrodes with water and dry them.

8. Place another crucible on the pipestem triangle. Add 4 or 5 sodium hydroxide, NaOH, pellets. Melt them carefully with the burner. Test the conductivity of the molten substance as before. Set the crucible aside to cool for use in step 15.

9. Add about 5 ml of distilled water to the crucible containing the cooled sodium chloride. Heat gently to dissolve the NaCl.

10. Pour the salt water into a clean test tube. Add 2-3 drops of dilute silver nitrate solution, to test for the presence of chloride ions. Observe the results.

11. Carefully clean all salt from the crucible. Dry the crucible and add about 5 ml of carbon tetrachloride, CCl_4. Test the conductivity of the liquid.

12. Pour the CCl_4 into a clean test tube. Add 4-5 ml of distilled water and 2-3 drops of silver nitrate solution. Stopper with a clean rubber stopper and shake to mix. Record your results. Pour the mixture into the drain and rinse well with water.

13. Clean and dry the crucible. Add 5 ml of a solution of hydrogen chloride, HCl, dissolved in toluene. Test its conductivity.

14. Pour the mixture into a clean test tube. Add 4-5 ml of distilled water and 2-3 drops of silver nitrate solution. Mix thoroughly. Record the results. Rinse the solution down the drain.

15. Add 5 ml of distilled water to the crucible of NaOH from step 8. Warm gently to dissolve some of the NaOH. Pour the solution into a clean test tube. Add several drops of ferric chloride, $FeCl_3$, solution, to test for the presence of hydroxide ions. Mix, and observe the results.

16. Place 5 ml of methanol, CH_3OH, in a clean crucible. Test its conductivity.

17. Pour the methanol into a clean test tube. Add several drops of ferric chloride solution and observe the results.

RELATED PROBLEMS

1. Which of the compounds appeared to be joined by covalent bonds and which by ionic bonds?

2. Consider the two pairs of compounds, $NaCl$-CCl_4 and $NaOH$-CH_3OH. Does there seem to be any relationship between the nature of the bonds holding these compounds together and the distance between the locations of the bonded elements in the periodic table? (In the second pair consider the bonds between Na and O and between C and O.)

3. Consider the relative positions of H and Cl in the periodic table. Does the bonding that you found in HCl seem to be consistent with the bonding found in the other two Cl compounds?

4. Compare the electronic structures of the Na and H atoms. How might the sizes of these two atoms account for the difference found in the bonds they form with Cl?

5. Which of the compounds you tested appeared to form ions when mixed with water?

6. On the basis of the very limited data that you have collected, what tentative conclusions might you draw concerning the behavior of ionic and covalent compounds when they are mixed with water?

A Question to Wonder About
How can a compound joined by covalent bonds form ions when it dissolves in water?

ADDITIONAL INVESTIGATIONS

If time permits, determine the conductivity of a variety of other available materials. Before running your tests, try to predict in advance what your results will be. Base your predictions primarily on the positions of the bonding elements in the periodic table.

Chemistry and Our Environment

DISPOSAL AND RECYCLING OF WASTES

Along with all the desired goods and services available to us in our present-day industrialized society, there are undesired wastes, dangerous by-products, and residues. Until recently the general attitude toward the wastes, the by-products, and the residues has been to dump them, to ignore them, to assume that in some magic way nature could and would absorb them harmlessly. There is no such magic. If we keep pouring trash and poison into our environment, we shall soon have a trashy, poisoned environment.

In this article some of the main aspects of "the disposal problem" will be examined. In subsequent articles several of the more prominent specific problems will be discussed in detail.

Degradability. A very important concept in understanding how various materials influence our environment is degradability. A degradable substance is one that the normal agencies of nature, such as bacteria, molds, moisture, heat, cold, sunlight, and the oxygen of the air, can decompose into simpler substances, which are absorbed safely by the environment.

The most important means of decomposing natural wastes are biological processes, the action of microorganisms and molds. Materials susceptible to this form of decomposition are said to be *biodegradable.* Materials produced by natural processes, or directly from natural sources, are almost always biodegradable. Many of the substances that are produced by elaborate chemistry are not, or, at least, they are degraded only very slowly.

To illustrate degradability and nondegradability, consider wood and plastics. Wood consists largely of cellulose. When wood falls to the ground it almost immediately begins to decay, by a series of complex biological processes. Not only is the wood decomposed, thus disposing of waste, but positive good results. The degradation products of wood, leaves, and other cellulose-based plant materials become humus, the nonmineral part of the soil. Humus is essential to good soil, and enriches it.

When a piece of plastic—and that includes synthetic fibers as well—is discarded, there is no such natural disposal of the waste, and no beneficial effect. All plastics and synthetic fibers are composed of "unnatural" molecules. Thus the biological environment cannot degrade them, and cannot use them. They are, in the purest sense of the word, junk. They can, of course, be burned, and tons are, every day. That, however, only contributes to air pollution, since many toxic gases are formed when plastics are incinerated.

Solid Wastes. Discarded solid articles constitute an ever-increasing form of pollution. In 1920 the U.S. per capita production of solid wastes was about 2.7 pounds per day. In 1970 it reached 5 pounds per day. Multiply 5 pounds per day by the U.S. population (215 million) and then by 365 and you get about 200 million tons a year. This is counting only consumer waste, such as paper, cans, bottles, tires, junked cars, and plastic articles. Industrial solid wastes reach nearly the same tonnage.

There are two reasons why we cannot continue to just throw all this away. One is that we are constantly using up irreplaceable natural resources, such as iron ore and petroleum, in order to make new articles to replace those we have junked. The second is that we are literally running out of acceptable places to dump our junk.

What are our practical alternatives? Basically, two. (1) Produce less trash. (2) Reclaim

and new ways to reclaim and recycle much of the unavoidable trash production of an industrialized society.

Recycling. This means recovering the material, such as metal, rubber, and paper, in worn-out and discarded articles and treating it so it can be used again. There are many reasons why recycling is desirable. It helps preserve natural resources. It reduces pollution. It helps deal with the solid waste disposal problem. As yet, recycling processes are still in an early stage of development and are not widely used. Recycling of paper wastes has probably progressed further than for other materials. Recycling of paper is discussed in more detail in the environmental article in Chapter 18.

In all recycling processes, separating the different types of waste is a major problem. The average load of trash contains paper, plastics, rubber, glass, aluminum, iron, copper, and perhaps other materials, all mixed together. Many ingenious ideas have been suggested to carry out a separation. In one comprehensive scheme, the bulk waste is first shredded, converting everything into small particles. Then in a chamber of turbulent air, called an "air classifier," the denser materials, mainly metals and glass, are separated from the less dense ones, mainly paper and plastics. The denser materials can then be further separated, since metals are far denser than glass. The iron particles can then be separated from the other metals magnetically.

Recycling of metals does not require any fundamentally new technology, but there is certainly need for improvements in the detailed processing. One extremely interesting possibility, which is currently under study by engineers, is the use of combustible waste to generate power. This is different from merely burning it. The idea is to use it as fuel in place of coal, oil, or gas.

The hardboard held by the scientist at the left was made from shredded newspapers (like those in the bottle held by the scientist at the right) and plastic scrap. The cabinet in the foreground, including its partially-opened door, was made with the hardboard of the type shown.

and recycle everything we can. It seems certain that before long economic considerations will force us to do both of these things, although the present economic philosophy seems to encourage our current junk-producing, junk-discarding way of life. Right now would seem to be the time to start the scientific and engineering research needed to develop new ways of reducing trash production

Water Pollution and Chemical Poisons. Man has traditionally regarded the nearest body of water, such as a lake, a river, or the ocean, as a suitable place to dispose of his sewage and a variety of liquid wastes. In small and industrially primitive societies this is an acceptable practice. In our present society it can lead to disaster.

A special problem that has arisen in recent times is the persistence of certain chemical poisons deliberately introduced into the environment by man. These include a variety of substances used in agriculture, and insecticides are an especially important example. In a later article (Chapter 16) this problem will be discussed in some detail.

Another special problem has resulted from the industrial dumping of mercury into various bodies of water, where it is converted into toxic compounds by bacteria.

14-17 *Electron Density Maps* ★

In the preceding discussions of chemical bonds it has been stressed that bonding is basically due to the presence of electrons between the nuclei that are bonded. These electrons are simultaneously attracted to two nuclei. This lowers the energy of the group of atoms as compared to the energy of the separate atoms, in which each electron is attracted to only one nucleus. In this section you will see how the presence of electrons between the nuclei can be represented accurately with pictures. These pictures are produced by computers from very complex quantum mechanical equations. As you know, we cannot pinpoint the positions of individual electrons in a molecule Therefore, the pictures will be *maps of electron density*. They will show where electrons are most likely to be found.

Before turning to the electron density maps themselves, let us consider the principle that will be involved in such maps in a more familiar context. An excellent analogy to an electron density map is a contour map of some area of the earth's surface. A contour map shows how the height of the land varies from place to place.

As an example, consider the hill shown in Figure 14-11a. This hill is represented on the map, Figure 14-11b, by the *contour lines*. To see what these contour lines mean, imagine that you could cut horizontal slices through the hill, beginning at the bottom and taking a slice every 20 feet. Figure 14-11c shows how these slices would look. You can see that the lines on the contour map are lines connecting all points that have the same height. Note that where contour lines are close together on the map the ground is rising steeply. Note also that the smallest contour loop is nearest the top of the hill. As the surrounding loops get larger they correspond to lower and lower ground.

The same contouring principle can be used to represent accurately the distribution of electron density in molecules. A simple illustration is provided by the hydrogen atom. As you have already seen, the electron density in a $1s$ orbital is highest at the nucleus and decreases steadily with increasing distance from the nucleus. This is represented by the contour map in Figure 14-12. The contours here are drawn a little differently from those on the map we have just been discussing. Instead of each contour representing an equal decrease in the magnitude of electron

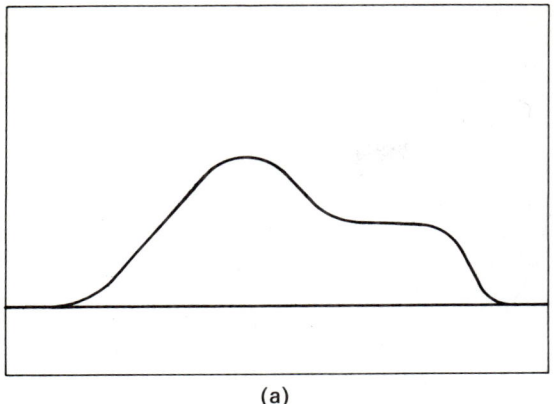

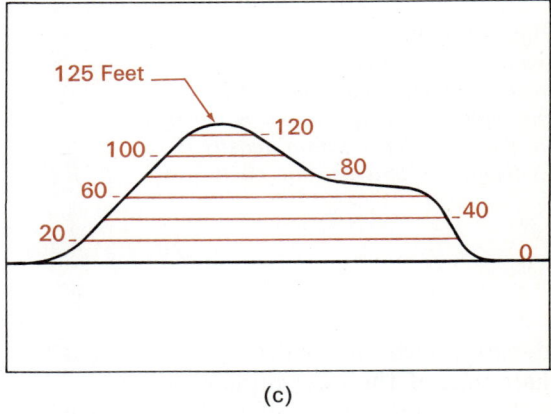

Figure 14–11 *A hill can be represented on a map by contour lines. A profile of a hill is shown in (a), and its contour-line representation in (b). These contour lines "slice" the hill into horizontal slabs, as shown in (c).*

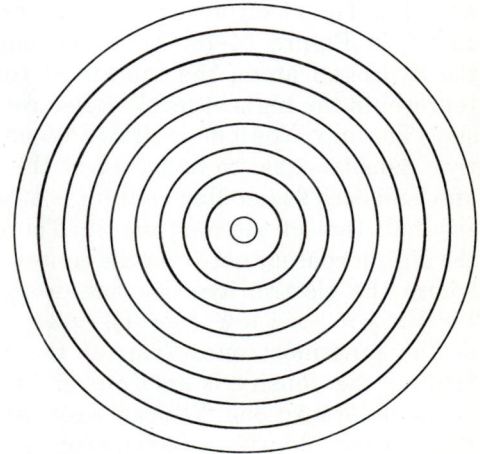

Figure 14–12 *A set of contour lines showing the electron density in the 1s orbital. Each line represents a density equal to half that of the next one in.*

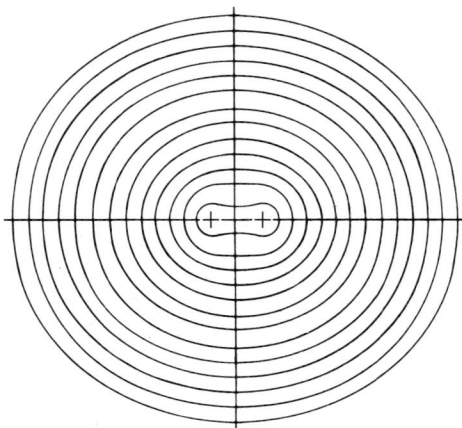

Figure 14–13 *A set of contours representing the electron density in the hydrogen molecule. Note that the highest density contour is pinched in, showing that electron density is concentrated between the nuclei.*

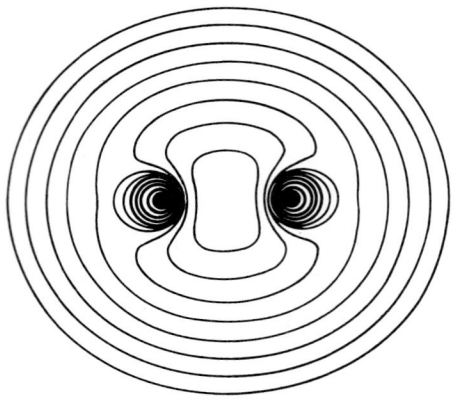

Figure 14–14 *Electron density contours for the lithium molecule, Li_2. The very dark areas about the nuclei represent 1s electrons, which stay around their own nuclei. The 2s electrons are shared and form the bond.*

density, each one is drawn at a magnitude half that of the next higher one.

The simplest molecular illustration of an electron density map is the hydrogen molecule, H_2. The electron density contours are shown in Figure 14–13. As in the map for the hydrogen atom, the innermost contour represents the highest level of electron density. The next contour is drawn at an electron density equal to half that of the inner one, and each succeeding contour is at a density half that of the next one in. Therefore, the eighth contour out from the center shows where the electron density has dropped to $\frac{1}{2} \times \frac{1}{2} \times \frac{1}{2} \times \frac{1}{2} \times \frac{1}{2} \times \frac{1}{2} \times \frac{1}{2} = (\frac{1}{2})^7 = \frac{1}{128}$ that of the innermost one. Even at the fourth contour the density is only one-eighth that of the innermost one. You can see, therefore, that electron density is heavily concentrated near the nuclei. It is particularly high between the nuclei as shown by the pinching in of the innermost contour.

A complete, three-dimensional idea of the electron density distribution in this molecule is obtained by rotating the set of contours about a line passing through the two nuclei. In other words, the "bond" is symmetrical about its axis.

As another illustration of the use of electron density contour maps, let us look at the lithium molecule, Li_2. In this molecule the bond is formed mainly by overlap of the 2s orbitals. The pair of 2s electrons is shared between the two atoms. The pairs of 1s electrons on each lithium atom play practically no role in the bonding. Figure 14–14 shows the distribution of the two electrons that were in the 2s orbitals of the individual lithium atoms. You can see that several of the contours of highest density lie largely within the region between the nuclei. This shows that electron density is concentrated in this region, thus forming a bond.

Finally, it is interesting to look at what happens as two helium atoms approach each other. As you know, a bond is not formed, since the 1s orbital of each helium atom is already filled. As Figure 14-15 shows, when the helium nuclei are the same distance apart as the hydrogen nuclei in the H_2 molecule, the contours of highest electron density still surround the individual nuclei. This means that electron density tends to remain on the individual atoms rather than concentrating in the region between them. This accords with the fact that the two helium atoms do not form a bond.

Figure 14-15 *Electron density contours for two helium atoms that are close together. Note that the highest contours still close around individual nuclei. This shows a lack of electron-sharing and, hence, a lack of bonding.*

Summary

When two or more atoms group together to form a relatively stable molecule, the bonds that hold these atoms in place result from a sharing of electrons among the atoms. This sharing helps create electron configurations similar to those of the very stable noble gases.

When the ionization energy of one atom is considerably higher than that of another, the atom with the higher energy exerts a very strong force of attraction on the shared electrons. The electrons are, then, captured by the atom with the higher ionization energy, forming two oppositely-charged ions. The opposite charge of the two ions creates a strong attraction between them. The resultant molecule is said to have an ionic bond.

There are many intermediate cases in which the bonding electrons are shared unequally, but not to the extreme degree that constitutes an ionic bond. In such cases, the covalent bond is said to have ionic character. All bonds that are ionic or that have ionic character are polar. They have electric dipoles in which there is negative charge in excess at one end of the bond and positive charge in excess at the other.

Self Check

1. How do covalent bonds and ionic bonds differ? In what ways are they similar?
2. Why are diatomic molecules formed from different atoms generally polar?

Chapter Summary

Chemical bonds form between atoms because electrons are shared. This sharing, in which two electrons are simultaneously attracted to two atoms, makes the energy lower when the two atoms are close than when they are apart. In general, the overlap of two half-filled orbitals, which results in a pair of electrons being shared by two atoms, gives rise to a bond. Atoms such as helium or neon, in which all orbitals are either completely filled or completely empty, cannot normally form bonds. The capacity to form bonds by the second row elements, Li through Ne, can be understood by examining the populations of the valence orbitals, namely, the 2s and 2p orbitals. If the electrons in a bond are shared equally between the two atoms, as in Li_2 or F_2, the bond is purely covalent. If the electron pair goes entirely over to one atom, as in Li^+F^-, the bonding is called ionic. When the electron pair is shared unequally, the bond is said to be covalent with ionic character. Bonds between unlike atoms nearly always have some ionic character, since the two atoms will seldom have exactly equal attraction for the shared electrons.

Questions and Problems

1. How would the nature of matter be affected if the repulsive forces between all atoms were greater than the attractive forces?
2. If you had a piece of steel stuck to a strong permanent magnet, what would you have to do to remove the steel from the magnet? Which arrangement would contain more energy—the piece of steel stuck to the magnet or the piece of steel six inches away from the magnet? Which arrangement is more stable? Explain.
3. Compare a hydrogen molecule, H_2, to the piece of steel stuck to a magnet. Explain why two hydrogen atoms have more energy than a hydrogen molecule.
4. Which one of the statements below is *false* as applied to the following equation?

$$H_2(g) + energy = 2H(g)$$

   (a) Hydrogen atoms release energy when they combine to form hydrogen molecules.
   (b) Two grams of $H(g)$ contain more energy than two grams of $H_2(g)$.
   (c) Per unit mass, $H(g)$ would be better fuel than $H_2(g)$.
   (d) The bright line spectrum of $H_2(g)$ is the same as the bright line spectrum of $H(g)$.
5. Which would be more stable—a flask full of hydrogen atoms or a flask full of helium atoms, under the same conditions of temperature and pressure? Explain.
6. Why does hydrogen form H_2 molecules, while helium atoms do not combine with each other? Explain in terms of attractive and repulsive forces and also in terms of incomplete and complete orbitals.
7. What energy condition must exist if a chemical bond is to form between two approaching atoms?
8. What orbital conditions and electron conditions must exist if a chemical bond is to form between two approaching atoms?
9. What are the molecular species present in gaseous neon, argon, krypton, and xenon?
10. Determine the number of attractive forces and repulsive forces in LiH.
11. With the use of electron dot symbols, represent the first 20 elements in the periodic table.
12. Show electron dot formulas for the three compounds that could be formed by combining Li with each one of the following elements: Cl, S, and P. Do the same for Ca and for Al.
13. Give the orbital and electron dot representations for the bonding in the molecules Cl_2, HCl, Cl_2O.

14. Using the electron dot representation, show a neutral, a negatively charged, and a positively charged OH group.
15. Draw the electron dot representation of the molecule N_2H_4, hydrazine.
16. Draw the orbital representations of sodium fluoride, NaF, and beryllium fluoride, BeF_2.
17. In general, what conditions cause two atoms to combine to form (a) a bond that is mainly covalent, (b) a bond that is mainly ionic, (c) a polar molecule?
18. Draw orbital representations for the highest energy level electrons of each of the alkali metals. Show electron dot symbols for each of these elements. Using the above information, explain why these elements are so chemically reactive.
19. Repeat Question 18 for the halogens.
20. On the basis of your answer to Question 18, explain which member of the alkali family should have the lowest ionization energy. Which of these elements should be the most reactive?
21. Which element in the halogen family would you expect to be the most reactive? Why?
22. Which compound would you expect to be more ionic in character—LiF or NaF? Explain.
23. Knowing the orbitals carbon uses for bonding, use the periodic table to predict the formula for the chloride of silicon. What orbitals does silicon use for bonding? Draw an electron dot formula for silicon chloride.
24. What type of bonding would you expect to find in MgO? Explain.
25. From Table 14-2, explain what general trend in ionization energy seems to occur with increasing atomic number of elements in the second horizontal row (the second period) of the periodic table. In general, would you expect compounds formed between elements close to each other in the periodic table to be more ionic or less ionic than compounds formed between elements widely separated from each other? Why would this be true?

15 The Shapes of Molecules

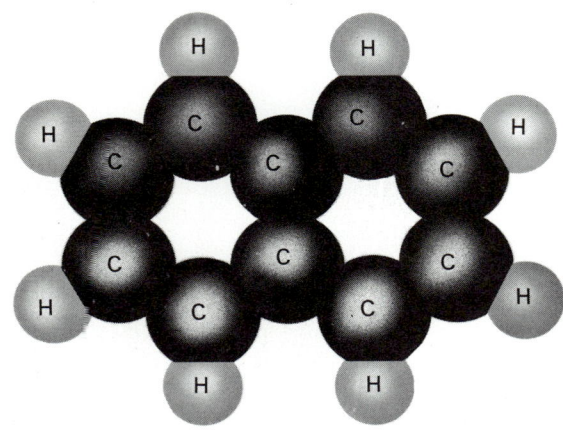

An electron-density map of naphthalene, obtained from X-ray diffraction studies, and the corresponding molecular model.

In Chapter 7 we briefly examined models of a few important molecules, such as O_2, N_2, CO_2, NH_3, and H_2O. These models brought out a point of basic importance that we did not pursue at the time, having other ideas to develop. Molecules have characteristic shapes. The shape of a molecule often helps us to understand its properties. The theory of bonds that we have developed in the preceding chapter can be extended to help us understand the shapes or structures of molecules. It can even help us to predict what shape or structure a molecule might be expected to have. In this chapter some of the basic relationships between bonding and molecular structure will be developed.

Electron Pairs Form Bonds in Definite Directions

15-1 *Experiments Tell Us about Molecular Architecture*

The **structure** of a molecule is the way in which its atoms are linked up, and the shape of the resulting arrangement. Before considering how the architecture depends on the bonding, we should consider the interesting question of how scientists learn about the architecture of molecules. Molecules are, of course, too small to see. How, then, do chemists know whether H_2O is straight or bent (it is bent) or whether CO_2 is straight or bent (it is straight)? Furthermore, even if chemists know that a molecule is bent, how do they find out the exact value of the angle?

The experimental methods by which such information is obtained are among the most ingenious accomplishments of modern science. Most of them involve some fairly sophisticated physics to be fully understood. Therefore, only one of the most important ones will be described here, briefly.

X-ray Diffraction by Crystals. The determination of crystal structure by means of X-ray diffraction is the most important of the techniques. It is based on the fact that nearly every substance will form crystals—if not at room temperature, then at lower temperatures. For instance, water is a liquid at room temperature, but forms a crystalline solid, ice, at 0°C. Carbon dioxide is a gas at 25°C, but at −70°C it condenses to a solid, "dry ice." The important thing about crystals is that they contain molecules lined up, one after the other, in a definite, orderly way. This orderliness is often evident on a macroscopic level. Compare the sodium chloride crystal and the corresponding model in Figures 17-2 and 17-3.

When a beam of X rays strikes a crystal, the X rays are scattered in a definite, orderly pattern. Figure 15-1 illustrates the process.

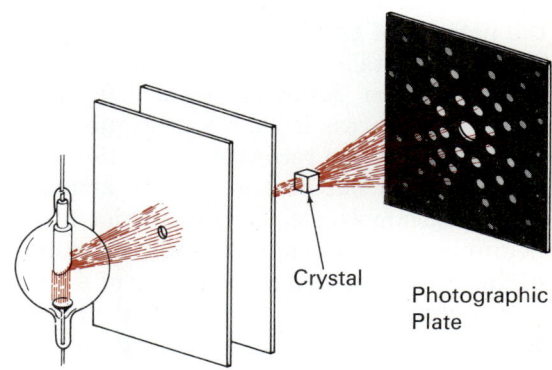

Figure 15-1 *A thin beam of X rays striking the crystal is diffracted in specific directions by the arrangement of atoms within the crystal.*

396 Chapter Fifteen

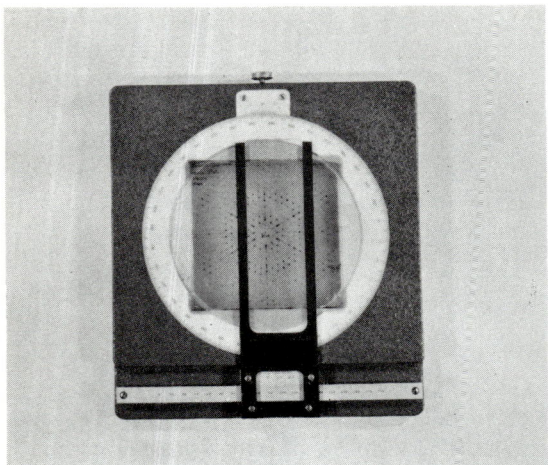

Figure 15-2 *A typical X-ray diffraction photo taken with a precession camera. The photo is mounted on a special film-measuring instrument.*

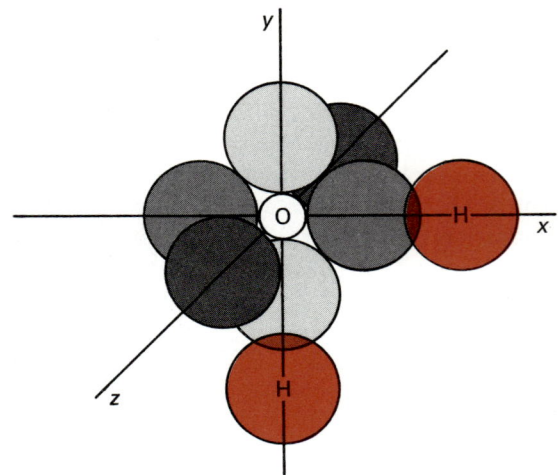

Figure 15-3 *The spatial arrangement of p orbitals on the oxygen atom, with hydrogen atoms covalently bonded to two of them, accounts for the shape of a water molecule.*

It can be seen that this pattern has two characteristics. (1) The spots are arranged in a definite pattern. (2) The spots have different intensities. From the spacings of the rows and columns of spots one can figure out what the pattern of the molecules, atoms, or ions is in the crystal.

The pattern resulting from one X-ray diffraction process is shown in Figure 15-2. From the way the recorded intensities vary, one can work out how atoms are arranged within each molecule. The illustration at the beginning of this chapter shows an example of the way in which a contour map of a molecule, obtained from its scattering pattern, makes it possible to build a model of the structure of a molecule.

Obviously, once it is known where all the atoms in a molecule are located relative to one another, all the bond lengths and all the angles between bonds can be evaluated.

15-2 The Shapes of Some Molecules Formed by Oxygen and Nitrogen

The orbital representations of the bonding in H_2O and in OF_2 suggest that two p orbitals of oxygen are involved in the bonding. Figure 15-3 shows the spatial arrangement assigned to the p orbitals (assuming that they are like hydrogen atom orbitals). If this spatial arrangement persisted after the bonds were formed, the molecular shape would be fixed, as shown. The molecule would be bent, with the angle near 90°. The same would be true for OF_2. The bond angles found experimentally are as follows.

$$\underset{H \quad\quad H}{O} \quad \angle H—O—H = 104.5°$$

$$\underset{F \quad\quad F}{O} \quad \angle F—O—F = 102°$$

Thus, the idea that p orbitals are used to form the bonds correctly predicts that the H_2O and OF_2 molecules are bent, rather than straight (linear). It does not, however, predict the bond angles very accurately.

The same type of argument can be extended to some molecules formed by nitrogen, such as NH_3 and NF_3. In Section 14-7 it was suggested that nitrogen would use one of its p orbitals to form each of the three bonds. From Figure 15-4 you can see that this would lead to a pyramidal shape for NH_3 and NF_3 (rather than planar). The HNH and FNF angles should be close to 90°. In actual fact, NH_3 and NF_3 are pyramidal. The measured bond angles follow.

$\angle$ H—N—H = 107°

$\angle$ F—N—F = 103°

Again our approach gives the correct shape, although the prediction of bonding angles is less accurate than we might wish.

The approach we have been using works moderately well for compounds of divalent oxygen and trivalent nitrogen. It works even better for the heavier elements in the oxygen and nitrogen groups of the periodic table. For example, sulfur, which comes below oxygen, has the following configuration.

1s 2s 2p 3s 3p

This leads us to expect H_2S to have the electron dot formula

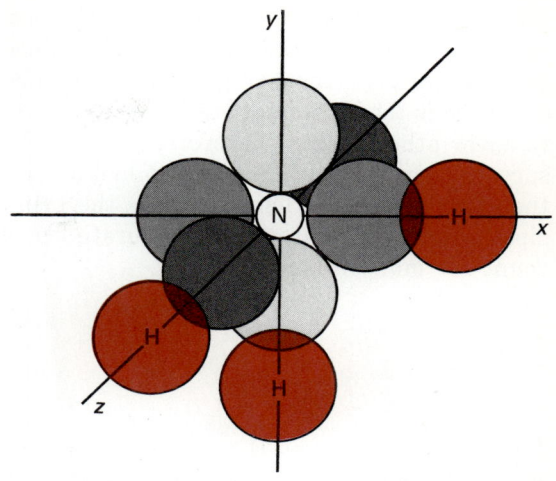

Figure 15-4 *The spatial arrangement of p orbitals on the nitrogen atom, with hydrogen atoms covalently bonded to all three of them, accounts for the shape of an ammonia molecule.*

and a predicted bond angle, again, of about 90°. For H_2S the actual angle is very close to this value, 93°.

Phosphorus has the following configuration.

1s 2s 2p 3s 3p

Hence, we should expect the following electron dot formula.

The observed bond angle is 93°, quite close to the expected value of 90°.

You have just seen that for the elements in the oxygen and nitrogen groups the shapes of molecules, such as H_2O, H_2S, NH_3, NF_3, and PH_3, can be predicted by assuming that the central atoms use pure p orbitals to form bonds. As was noted, this approach does not always predict the actual bond angles very accurately. Another problem with this approach is that it cannot easily be extended to cover other cases. Therefore, it is necessary to introduce other ways of looking at the same molecules. This will have the principal advantage that we can accurately account for many other cases.

15-3 The Tendency for Electrons To Form Pairs

Let us look again at the electron dot formulas for some of the molecules already discussed.

$$\overset{..}{\underset{H}{\text{O}}}\text{:H} \qquad \overset{\overset{..}{:}\overset{..}{\text{F}}:}{\underset{:\overset{..}{\text{F}}:}{\text{:N:F:}}} \qquad \overset{H}{\underset{H}{\text{:P:H}}}$$

In all cases there are four electron pairs around the central atom. This is also true for compounds of carbon and fluorine, such as the following.

$$\overset{H}{\underset{H}{\text{H:C:H}}} \qquad :\overset{..}{\text{F}}\text{:H}$$

In the few simple molecules that we have looked at so far, we see that orbitals are generally occupied by pairs of electrons. Of course, no more than two electrons *can* occupy one orbital, according to Pauli's exclusion principle. There is, however, a marked tendency for each orbital to contain two electrons rather than one. Part of the reason for this is that two electrons in the same orbital interact with each other. They can do this because they have a property called *spin*.

Since we cannot see electrons, we obviously cannot see them spinning. Scientists know in an indirect way that electrons have the property they call spin. It is possible to show directly, by experiment, that electrons behave like tiny magnets. This can be understood by using the idea of spin.

You know that the electron consists of a certain definite quantity of negative charge. Charge by itself is not associated with magnetism, but *moving* charge is. Figure 15-5a shows the magnetic field created about a straight wire, when the wire is conducting a current. A somewhat more complicated pattern of magnetic force results when the wire is in the form of a loop, as in Figure 15-5b. The pattern or field has an easily predictable effect on substances that are affected by magnetic forces, however. That effect is just the same as the one that would be produced by a tiny bar magnet, centered in the middle of the loop. In order to illustrate this fact an imaginary bar magnet is shown in a wire loop in Figure 15-6a.

Suppose, now, that you were to move a real (not imaginary) bar magnet close to a conducting wire loop. It would be easy to predict the effect of the field on such a magnet. You would only have to consider what the effect of the imaginary magnet would be. Since like magnetic poles repel and unlike magnetic poles attract, the north pole of the real magnet would be repelled in the direction indicated by the arrow in Figure 15-6b.

Since the arrow in the diagram shows how the field will affect a nearby north pole, we could simply use an arrow (as in Figure 15-6c) to denote the direction of the force exerted by the field. That would be far less difficult than drawing the lines of force shown by the iron filings in Figure 15-5b, and would eliminate the need for sketching in an imaginary bar magnet, as well. Scientists commonly do use such an arrow when they want to indicate simply the direction,

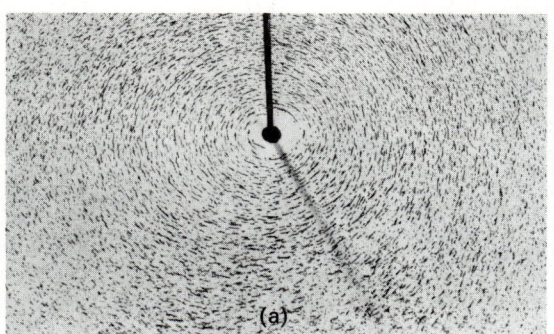 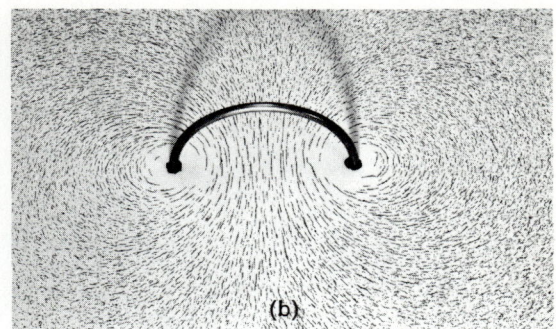

Figure 15-5 *The iron filings in the above photographs align with the magnetic field, revealing the field pattern. The magnetic fields about a current-carrying straight wire (a) and a current-carrying wire loop (b) are shown.*

and often the magnitude, of a force. Such a symbol has a name that is probably familiar to you. It is known as a *vector*.

You can see that a magnetic field should exist about an electron, if you assume that the cloud of negative charge is rotating about an axis that passes through it. In this view the outer parts of the charge cloud are circulating in a loop, just as the current in the loop of wire did. A magnetic field is thereby generated. The magnetic field of the electron can also be represented by a vector.

There are two possibilities for electron spin, clockwise or counterclockwise. The resulting fields can be represented by vectors that point either up or down, as shown in

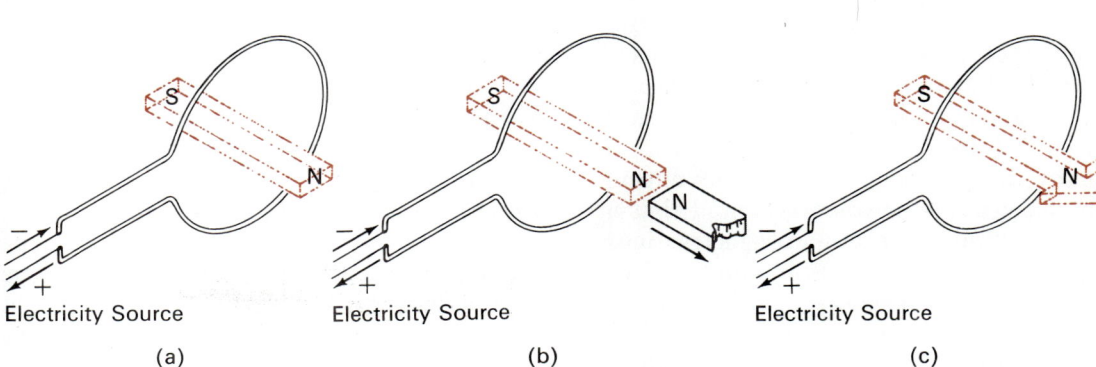

Figure 15-6 *An electric current flowing in a loop of wire sets up a magnetic field like that due to a bar magnet (a). This field will repel a real bar magnet (b) and can be represented by an arrow called a vector (c).*

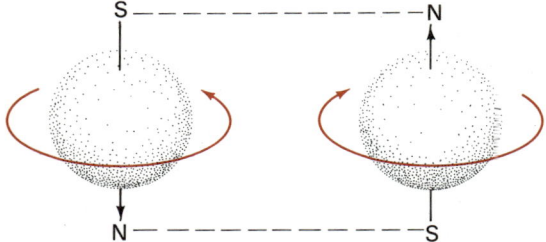

Figure 15-7 *Electrons with opposite spins have their magnetic poles lined up so as to attract.*

Figure 15-7. Another way of expressing Pauli's exclusion principle is to say that when two electrons occupy the same orbital, their spins must point in opposite directions. When this happens, their magnetic fields are lined up as shown in Figure 15-7 with opposite poles adjacent. This leads to an attractive force between them that helps to make the formation of electron pairs favorable.

Before going on to the next section, let us first note briefly some of the experimental evidence that led to the concept of electron spin. Scientists first began to suspect some very complex motion of particles inside an atom after Pieter Zeeman (in the Netherlands, in 1896) had examined the spectrum of atoms that were excited in a strong magnetic field. Instead of finding single lines in the spectrum, Zeeman found some lines that were split into pairs. Johannes Stark, in Germany, noticed somewhat similar effects seventeen years later when he found that a strong electrical field also splits spectral lines into pairs. In 1921, Otto Stern and Walter Gerlach, also in Germany, shot vaporized atoms through a strong magnetic field. They found that atoms of elements such as hydrogen, lithium, silver, copper, and gold, all of which have single electrons in their highest energy levels, divided into two separate beams when they passed through the magnetic field. Atoms of zinc, cadmium, and mercury, with two electrons in their highest energy levels, did not form split beams. Stern and Gerlach concluded that the single electrons must have been spinning in one direction in some atoms and in the opposite direction in other atoms. This would cause them to move in opposite directions in a magnetic field. If the two highest energy electrons of the other atoms were spinning in opposite directions in the same atoms, the effects of the opposite spins would cancel out in the magnetic field and the beams would not split.

15-4 The Electron-Pair Repulsion Theory of Molecular Shapes

Once a set of electron pairs has been formed in the orbitals of a given atom, each pair tends to behave as a unit, remaining closely coupled and occupying the same region of space. Other electron pairs are excluded from that region because of the electrostatic repulsion between the -2 charges carried by each electron pair. Electron pairs therefore tend to distribute themselves so as to keep the average distance between pairs as great as possible. At the same time, they cannot get too far from the nucleus because they are attracted by its positive charge.

Suppose that four electron pairs are trying to get as far as possible from each other, while each staying the same distance from the nucleus. What will be their arrangement in space? As Figure 15-8 shows, if all four pairs were to lie in one plane, a square would be the best arrangement. However, with three-dimensional space available they can get even farther apart than that. If we move those in one set of opposite pairs up and those in the other set down, as is also shown in Figure 15-8, the average distances be-

The Shapes of Molecules   401

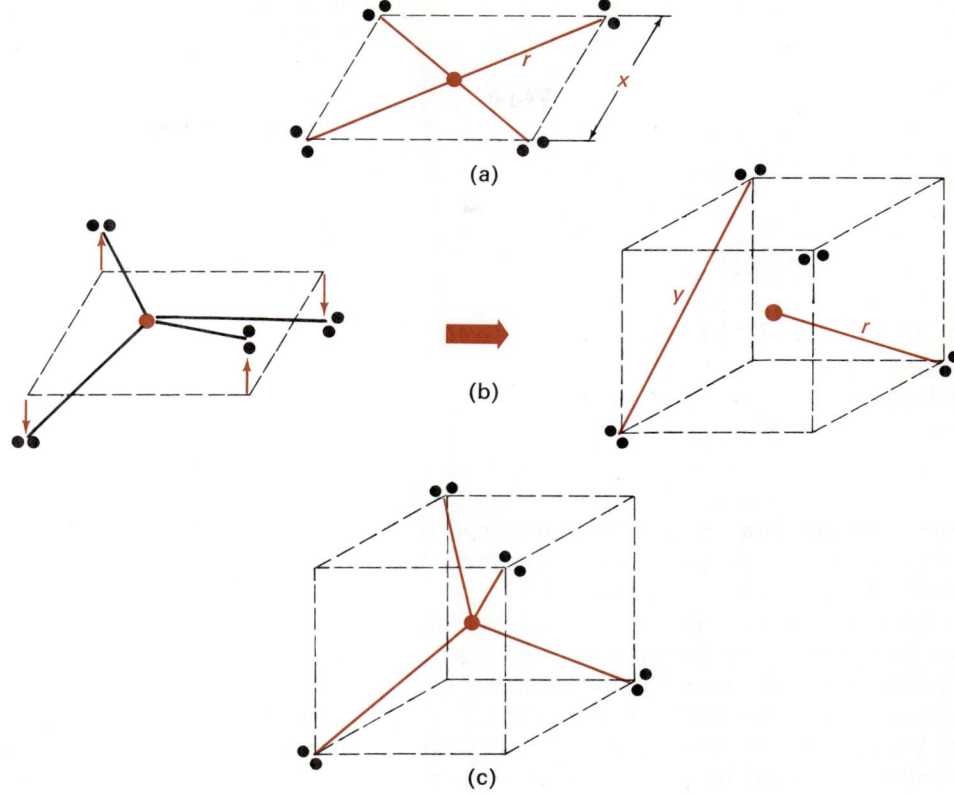

Figure 15-8   *Four electron pairs surrounding a nucleus in a square (a) are separated by a distance x. If they move as shown in (b) to a tetrahedral arrangement (c), they can keep the same distance, r, from the nucleus, but get farther, y, from each other.*

tween pairs will increase. The best possible arrangement will be reached when this process is continued just until all of the six distances between pairs become equal. If at this point we draw connecting lines between all the electron pairs, a solid figure called a tetrahedron (Figure 15-9, next page) is created. This arrangement of the four electron pairs is called a **tetrahedral** arrangement. This arrangement minimizes the total repulsion between the electron pairs.

The simple assumption that is being introduced at this point to explain the shapes of molecules follows.

The electron pairs in the valence orbitals of an atom will tend to arrange themselves so as to minimize interelectronic repulsions.

When there are four such electron pairs, the tetrahedral arrangement is the one they will adopt. In the majority of molecules there

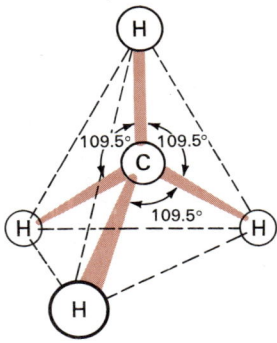

Figure 15-9 *In a perfectly regular tetrahedron, the angle made by two vertices and the center is 109°28′.*

are four electron pairs, so the tetrahedral array is by far the most important one. A few other cases will be examined later.

In a perfectly regular tetrahedron, the angle made by two vertices and the center is 109°28′, as shown in Figure 15-9. For our purposes we can round this off to 109°.

We are now in a position to predict the shapes of molecules in which the central atom has four electron pairs. Three such predictions are shown in Figure 15-10. Just as before, it is predicted that molecules such as H_2O and NH_3 will be bent and pyramidal, respectively. We can now predict, at least as a starting point, that the angles will be 109°. For H_2O and NH_3 this is fairly close to the actual values of 105° and 107°, respectively.

We can also predict the shape of the methane molecule, CH_4, and other such molecules formed by carbon. These molecules should have the shape of a tetrahedron. Experiment shows that they do. In this case, the prediction of the bond angles is not merely close. It is exactly correct.

The electron-pair repulsion model that has just been developed is flexible enough to account for some of the finer features of molecular structures. For example, it can account for the fact that, in all of the bent and pyramidal molecules formed by elements of the oxygen and nitrogen groups, the angles are less than the tetrahedral value of 109°. In order to understand how it does so, you should first recall that electrons in atoms are not like tiny dots of negative charge but act

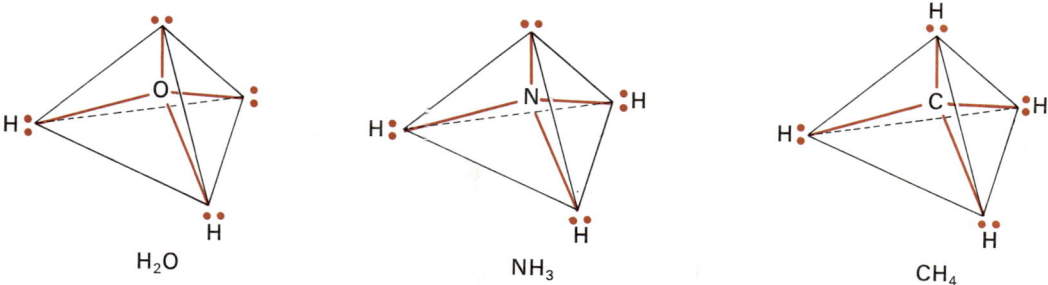

Figure 15-10 *The predicted shapes for H_2O, NH_3, and CH_4 molecules, if the four pairs of electrons are in each case placed at the vertices of a tetrahedron.*

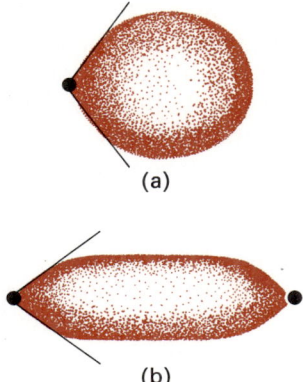

Figure 15-11 *The electron cloud of an unshared electron pair (a) takes up a larger part of the volume around an atom than does a shared pair (b).*

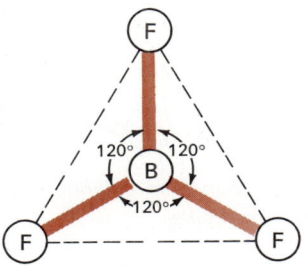

Figure 15-12 *The molecule BF_3 forms an equilateral triangle that is shown lying flat in the plane of this page.*

more like small clouds of negative electricity. Naturally, electron pairs are larger negative clouds. The size of these clouds will vary somewhat, however. When an electron pair is shared between two nuclei in a bond, the corresponding electron cloud is less diffuse than that corresponding to an unbonded electron pair. This is because the two nuclei sharing an electron pair draw the negative charge fairly strongly into the internuclear area, while an unshared pair is attracted by the central atom alone. This difference is illustrated in Figure 15-11. A bigger cloud will naturally tend to take up a greater share of the space around the central atom. As a rule:

Unshared electron pairs take up more room than shared, or bonding, pairs of electrons.

It naturally follows from this rule that the angles formed at the nucleus by unshared pairs will be greater than 109°. Those formed by shared pairs, by the same reasoning, will be less than the ideal tetrahedral angle of 109°. Thus you can predict that the bond angles on all molecules such as H_2S, H_2O, NH_3, and PH_3 should be less than 109°, as indeed they are.

You can also expect that in a molecule formed by carbon, in which the four surrounding atoms are not identical, the six angles will not all be exactly the same. For instance, in the CH_3F molecule the size of the electron clouds due to the electrons in the C—H bonds should be different from that of the electron pair in the C—F bond. Hence, the H—C—H angles should be somewhat different, and experiment shows that they are.

The electron-pair repulsion model also enables us to understand the structures of molecules in which the central atom has only three, or two, electron pairs. You will recall that, even when a $2s$ electron is promoted to a $2p$ orbital, the boron atom has only three unpaired valence electrons available to form bonds. When three bonds have been formed the boron atom is surrounded by three electron pairs. What arrangement of these pairs will maximize the average distance between them? It is the arrangement in which they lie at the vertices of an equilateral triangle. Thus, as shown in Figure 15-12, a molecule

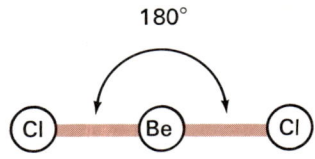

Figure 15-13 *Beryllium chloride, $BeCl_2$, forms a linear molecule that is perfectly symmetrical.*

such as BF_3 should have the shape of a planar equilateral triangle. This shape is the one found for it, by experiment.

One would also expect BH_3 to have this shape, but here things get a bit more complicated. The BH_3 molecule is so unstable that it lives only a fraction of a second. It is therefore very difficult to study its shape. There is no reason to doubt that it is a planar equilateral triangle. What happens to BH_3 molecules is that they join together to form B_2H_6 molecules.

The element beryllium has only two valence electrons and therefore only two pairs of electrons are present when two bonds have been formed. It should be rather obvious that the way for these two electron pairs to get farthest apart is by their being on opposite sides of the Be atom. Therefore, you would expect beryllium to form *linear* molecules, as illustrated in Figure 15-13 for $BeCl_2$.

15-5 Hybrid Orbitals

There is still another way of predicting and understanding the geometry of molecular structure. Again we can use the idea of definite orbitals on the central atom holding the electron pairs, but we must go beyond the simple notion that the orbitals are necessarily pure *s* or *p* orbitals. Instead, we must use certain new combinations of these orbitals. These combinations, which mix the *s* and *p* orbitals together, are called **hybrid** orbitals. Perhaps the easiest way to illustrate the formation of hybrid orbitals is to reconsider the formation of linear molecules by beryllium.

The ground state configuration of the beryllium atom is $1s^2$. In order for the atom to form bonds at all, one of these electrons must be promoted to a *p* orbital. Let us assume that it is the p_x orbital. It would be possible for two univalent atoms, let us say hydrogen atoms, to approach the beryllium atom and form bonds by overlap of their orbitals with the $2s$ and $2p_x$ orbitals of the beryllium atom. But this does not actually happen. The reason is that greater overlap of orbitals can occur if the $2s$ and $2p_x$ orbitals are first mixed into two similar but oppositely directed hybrid orbitals, as shown in Figure 15-14. Greater orbital overlap allows fuller sharing of electrons and so results in stronger bonds.

In order to understand the mixing of the pure orbitals and the shapes of the resulting hybrid orbitals, you should remember that the two lobes of the *p* orbital have opposite algebraic signs. Therefore, when we simply add the *s* orbital to the *p* orbital we get a hybrid orbital that is very small and weak along the negative *x* axis, but very strong and concentrated along the positive *x* axis. The buildup along one direction at the expense of the other means that in this one direction a much stronger bond can be formed than a pure *s* or *p* orbital alone could form, in any direction.

If now we reverse the sign of the *p* orbital and then add it to the *s* orbital, we get a hybrid that is strongly concentrated in the direction of the negative *x* axis. As a result, we have an orbital that is exceptionally well-suited to form a bond in the direction opposite to the previous one. What has hy-

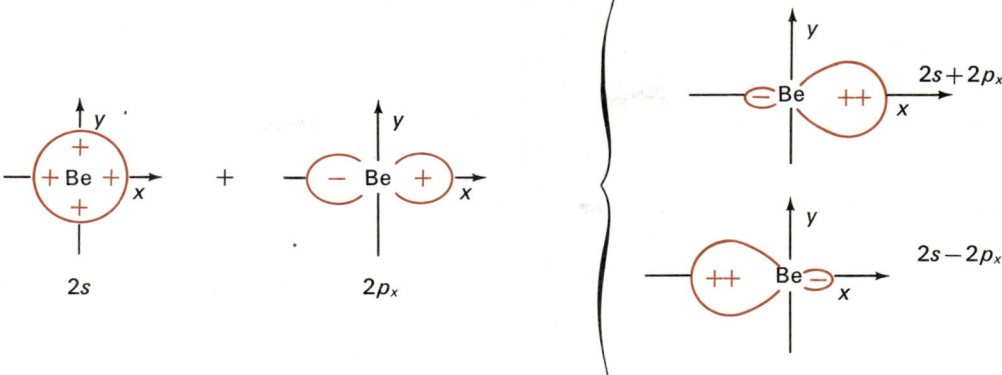

Figure 15-14 *A 2s and a $2p_x$ orbital can combine to form two oppositely-directed sp hybrid orbitals, as illustrated here for the beryllium atom.*

bridization accomplished? The total bond-forming power of the separate atomic orbitals, which is proportional to their ability to overlap with the orbitals of other atoms, is now concentrated in particular directions instead of being spread out over a large region of space. That is why the formation of hybrid orbitals, a process called **hybridization,** is favored. Stronger bonds can be formed in certain particular directions by hybrid orbitals than by the individual atomic orbitals from which they are formed.

The pair of hybrid orbitals formed by beryllium from one s and one p orbital, are known as sp hybrids. When you look at the directions in which these two hybrid orbitals are concentrated, you can see that they make an angle of 180° with each other. These particular hybrids enable beryllium to form two strong bonds, and the resulting molecule is linear. This is the same geometry predicted by the electron-pair repulsion model.

The hybridization model can also be applied to the case in which there are three valence electrons. Boron has three valence electrons, in its second energy level, with the distribution s^2p. One of the s electrons can readily be promoted to a vacant $2p$ orbital, giving the configuration sp_xp_y. But the pure valence orbitals do not provide for effective bonding, in definite directions. In order to do so, they become hybridized. Figure 15-15, on the next page, shows how this works out. You can see that the three hybrid orbitals, called sp^2 hybrids, are directed toward the vertices of an equilateral triangle lying in the xy plane. Thus, a molecule such as BF_3 would be expected to be planar. This, again, is the same prediction as that given by the electron-pair repulsion model, and it is in agreement with experimental fact.

Finally, we come to the cases in which there are four valence electrons, in compounds of carbon such as methane, CH_4. The configuration in which these four electrons are all in separate orbitals must be $sp_xp_yp_z$. Thus we should be able to find a set of four hybrid orbitals, sp^3 hybrids, which gives the greatest concentrations in particular directions. The algebra necessary for doing this,

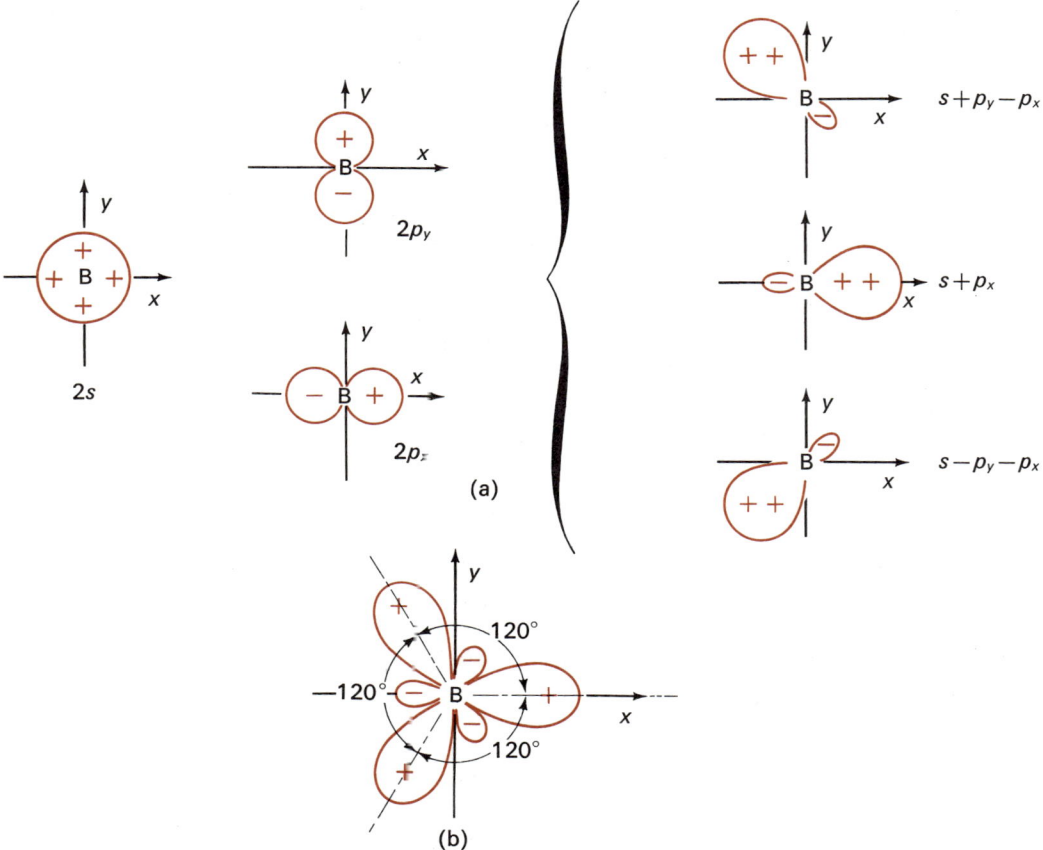

Figure 15-15 When an s orbital and two p orbitals are combined into equivalent hybrid orbitals (sp² hybrids), three orbitals making angles of 120° with each other result.

because it involves working in three dimensions, gets somewhat complex. The result, however, was worked out many years ago and takes a very simple form. It turns out that the four hybrid orbitals, as shown in Figure 15-16, are directed toward the vertices of a tetrahedron. Again, this model leads to an answer in accord with that obtained from the electron-pair repulsion model. Methane, CH_4, and other molecules, such as CF_4, should have the shape of a tetrahedron.

In molecules such as ammonia, NH_3, and water, H_2O, the valence orbitals of the central atom end up with four pairs of electrons. This result is like that found for methane, except that one or two of the electron pairs are not used in bonding. We may assume that sp^3 hybrids are formed, but that only two or three of them are actually used to form bonds. The use of even a few hybrids to form bonds gives a stable molecule because the hybrid orbitals are concentrated in particular directions. They can form stronger bonds in these directions than could any pure, unhybridized atomic orbital. We can now see why our very first attempt to explain the shapes of the H₂O and NH₃ molecules, in which we assumed that pure p orbitals were used, did not work out in a fully satisfactory way. Rather than using pure p orbitals, the central atom, O or N, mixes in the $2s$ orbital to give hybrid orbitals, and these form stronger bonds. The bond angles in H₂O and NH₃ do not have exactly the tetrahedral value of 109°, as we have seen. This can be attributed to the fact that, under these particular circumstances, all four sp^3 hybrid orbitals are not actually needed to form bonds. The best results can be obtained for the two or three bonds that *are* formed by using slightly different proportions of s and p orbitals in hybridizing. This in turn means the directions of the bonds are not exactly the same as the tetrahedral directions.

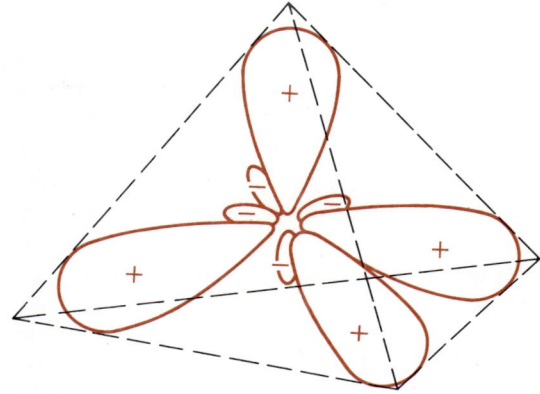

Figure 15–16 *The sp^3 hybrid orbitals.*

Summary

There are various kinds of experiments, especially the diffraction of X rays by crystals, which show us that molecules have definite structures. We can understand why most simple molecules have the shapes they do by considering the properties of electrons and orbitals. For molecules such as H₂O and NH₃, we expect the bonds to be formed by p orbitals of the O and N atoms. This naturally leads us to expect these molecules to be bent and pyramidal, respectively. However, a more general way to predict molecular shapes is to recognize that electron pairs will tend to arrange themselves so as to be as far apart, on the average, as they can. One of the most important cases is that of four electron pairs, which tend to lie at the corners of a tetrahedron. Still another model of the electron distribution in molecules is based on the idea that individual s and p orbitals will tend to mix, or hybridize, so as to give new orbitals. These new orbitals are strongly concentrated in the directions in which bonds are to be formed. This model also leads to the prediction of tetrahedral structures where four bonds are to be formed, and in general gives predictions very similar to those from the electron-pair repulsion model.

Self Check

1. How does the electron-pair repulsion model explain the fact that BF₃ is planar, while NF₃ is pyramidal?
2. How would you describe the directions in which hybrid orbitals of the sp, sp^2, and sp^3 types are concentrated?
3. How many faces and how many edges does a tetrahedron have? What is the shape of each face?

Double and Triple Bonds

15-6 Double Bonds

In all the cases we have considered so far, each pair of atoms has been held together by sharing between them *one* pair of electrons. Such a bond is called a single bond. It is also possible for *two* electron pairs to be shared between a given pair of atoms. This forms a stronger type of bond called a *double bond*.

To see how this can come about, let us suppose that each of two carbon atoms has formed bonds to two hydrogen atoms. This would give two CH_2 groups, as shown in Figure 15-17a. Each of the carbon atoms still has the capacity to form two more electron pair bonds. Suppose that these two CH_2 groups approach each other. One electron from each carbon atom may be used to form a single bond between these two atoms, as shown in Figure 15-17b. However, each carbon atom still has one electron that has not been used in bonding. These remaining two electrons can also be shared between the two carbon atoms. The resulting arrangement can be represented by the electron dot formula shown in Figure 15-17c.

Now that we have the electron dot formula, the next step is to find out what the shape of the molecule should be. This particular molecule, with the formula C_2H_4, is called *ethylene*. It is not very hard to predict its shape. Just a slight extension of the ideas we have already developed for molecules having only single bonds is required.

The basic idea we shall use is this. *Whenever there are four electron pairs in the valence orbitals of an atom, we expect them to take up, or try to take up, a tetrahedral arrangement around that atom.*

If each carbon atom in ethylene is to have four electron pairs around it in a tetrahedral arrangement and if, at the same time, two of the electron pairs are to belong to both carbon atoms simultaneously, the complete arrangement must be that shown in Figure 15-18. The resulting molecule has all six atoms in one plane. The HCH angles are about 109° and, as you can prove with a little geometry, the CCH angles are about 125°. This turns out to be a fairly accurate prediction. Experiments show that the ethylene molecule is indeed perfectly planar. As often happens with our simple models, the predicted angles are a little off, though not greatly so. The actual CCH and HCH angles in ethylene are both close to 120°.

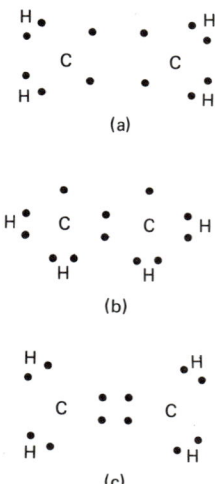

Figure 15-17 (a) Two —CH_2 groups, each with two electrons not used in bonding. (b) The two —CH_2 groups have combined to form a single bond between the carbon atoms. (c) The remaining two electrons now have also become shared by the carbon atoms, thus forming a double bond.

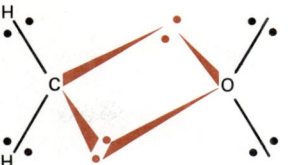

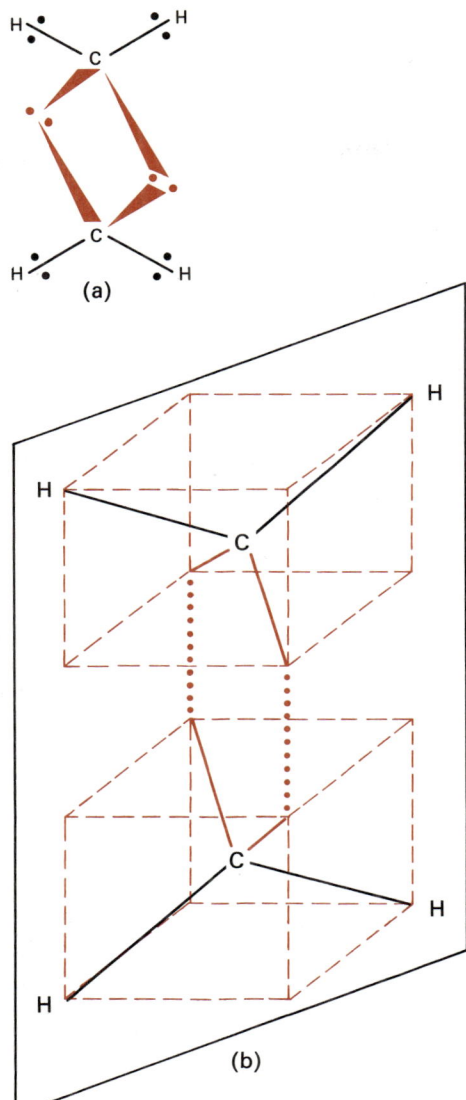

Figure 15-19 *The spatial distribution of atoms and electrons in the formaldehyde molecule.*

Figure 15-18 *When two tetrahedral sets of carbon orbitals combine, as in the formation of ethylene, all atoms (2 C atoms, 4 H atoms) lie in a plane, and a double bond is formed. A conventional shorthand representation of ethylene is given in (a). Its actual geometry is shown in (b).*

As another example of double bonding, let us consider what can happen if a CH_2 group and an oxygen atom are to be combined. The electron dot formula that results if both C and O use all of their valence electrons in bonds follows.

$$\begin{array}{c} H \\ \ddot{\,}\,\ddot{\,} \\ :C::\ddot{O}: \\ H \end{array}$$

If we now use our idea of having each of these atoms surrounded by a tetrahedral set of four electron pairs, we must put the molecule together as shown in Figure 15-19. Again, we get a molecule that is planar. Experiments show that *formaldehyde*, CH_2O, does indeed consist of planar molecules. Our angle predictions are not exactly right, but they are within 5 to 10 degrees of the true values.

Finally, let us look at the familiar carbon dioxide molecule, CO_2. The carbon atom can share two of its electrons with each of the two oxygen atoms. This gives a molecule with two double bonds, represented by the following electron dot formula.

$$:\ddot{O}::C::\ddot{O}:$$

The shape of this molecule can be predicted if we again assume that the four electron pairs surrounding each atom will tend

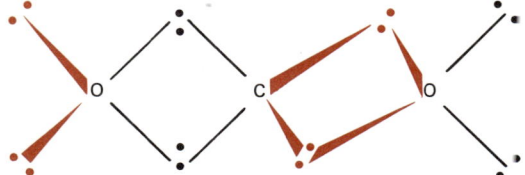

Figure 15-20 *The spatial distribution of atoms and electrons in carbon dioxide.*

to lie, as nearly as possible, at the vertices of a tetrahedron. We get the arrangement shown in Figure 15-20 and you can see that it is linear. The CO_2 molecule is known by experiment to be linear.

15-7 Triple Bonds

There are molecules in which *three* electron pairs are shared between two atoms. Such an arrangement is called a *triple bond*. One of the simplest and most important examples is afforded by the nitrogen molecule, N_2. You have already seen that the bonding capacity of nitrogen is 3, because its electron configuration is $1s^2 2s^2 2p_x 2p_y 2p_z$. Thus, one nitrogen atom combines with three other univalent atoms, such as H or F, to form molecules, such as NH_3 or NF_3. When two nitrogen atoms combine with each other, the formation of the bond may be expressed in either of the following ways.

$$\begin{array}{cccc} & 1s & 2s & 2p \\ N & \otimes & \otimes & \boxed{\otimes\otimes\otimes} \\ \parallel\!\!\parallel & & & \\ N & \otimes & \otimes & \boxed{\otimes\otimes\otimes} \end{array} \quad \text{or} \quad :N:::N:$$

One of the reasons that the element nitrogen, which makes up about four-fifths of our atmosphere, is so unreactive is the presence of this triple bond.

For a given pair of atoms, a double bond is stronger than a single bond and a triple bond is stronger yet, as illustrated by the following bond energies for nitrogen to nitrogen bonds. The **bond energy** is the amount of energy required to break a mole of bonds.

$$\begin{array}{ll} N\!:\!N & 38 \text{ kcal/mole } N_2 \\ N\!:\!:\!N & 100 \text{ kcal/mole } N_2 \\ N\!:\!:\!:\!N & 226 \text{ kcal/mole } N_2 \end{array}$$

Because single, double, and triple bonds are progressively stronger, it is not surprising that they are also progressively shorter. Two pairs of electrons can hold a given pair of atoms closer together than one pair of electrons can. Three pairs of electrons can pull them even closer together. This is illustrated by the following N to N distances for the single, double, and triple bonds between nitrogen atoms.

$$N\!:\!N \quad 1.40 \text{ Å} \qquad N\!:\!:\!N \quad 1.20 \text{ Å} \qquad N\!:\!:\!:\!N \quad 1.09 \text{ Å}$$

In order to see what effect a triple bond has on the shape of a molecule, let us consider a molecule of the compound acetylene. Its electron dot formula follows.

$$H\!:\!C\!:\!:\!:\!C\!:\!H$$

If the four electron pairs around each carbon atom are to lie at the vertices of a tetrahedron, the molecule must have the shape shown in Figure 15-21. From this it is seen that the acetylene molecule should be linear. Experiment shows that it is.

Because double and triple bonds are very strong, you should not assume that they must be unreactive. The multiple bonds in

$$H\!:\!\ddot{C}\!:\!:\!\ddot{C}\!:\!H \quad \text{and} \quad H\!:\!C\!:\!:\!:\!C\!:\!H, \text{ for example,}$$

are quite reactive. The following equations show reactions that they undergo, rather rapidly, when exposed to bromine, Br_2.

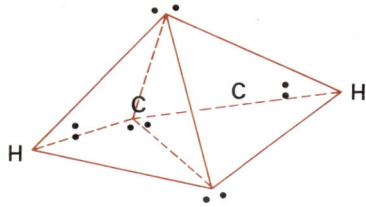

Figure 15-21 *Two tetrahedra sharing a face represent a triple bond. This gives rise to a linear molecule, as shown here for acetylene, HC≡CH.*

H₂C::CH₂ + Br₂ ⟶ BrH₂C:CH₂Br
  ethylene   dibromoethylene

HC:::CH + 2Br₂ ⟶ Br₂HC:CHBr₂
 acetylene   tetrabromoethylene

As you learned in Section 14-4, chemists often represent an electron-pair bond as a single line, for convenience. Similarly, double and triple bonds are commonly shown as two lines, or three lines, respectively. Table 15-1 illustrates this usage.

15-8 Resonance

There is still one important problem concerning molecular shape to be considered. We can illustrate it with an important ion, the *nitrate* ion, NO_3^-. First, let us see how to write an electron dot formula for the nitrate ion. Because the entire ion has a charge of 1− there must be one extra electron present, in addition to the electrons that the neutral atoms have brought with them. Suppose we begin by placing this additional electron on the nitrogen atom. We can then write a formula for nitrogen showing three electron pairs. Each of these electron pairs may be shared with an oxygen atom.

$$:\!N\!:\ \xrightarrow{e^-}\ :\!\ddot{N}\!:^-\ \xrightarrow{3\ddot{O}:}\ :\!\ddot{O}\!:\!\ddot{N}\!:\!\ddot{O}\!:$$

This electron dot formula takes proper account of the charge, and gives each oxygen atom a full complement of valence electrons. But it is unsatisfactory in one important respect, since it leaves the nitrogen atom

TABLE 15-1 Ways of Representing Some Compounds Having Multiple Bonds

Name of Compound	Molecular Formula	Electron Dot Formula	Structural Formula
ethylene	C₂H₄	H:C::C:H (with H's above and below)	H₂C=CH₂
acetylene	C₂H₂	H:C:::C:H	H—C≡C—H
formaldehyde	HCHO	H₂C::Ö:	H₂C=O
carbon disulfide	CS₂	:S̈::C::S̈:	S=C=S

with only six electrons. There is an easy solution to that problem, however. We have only to move an electron pair from one of the oxygen atoms into the space between the N atom and an O atom. Then the nitrogen will have a complete set of valence electrons, and so will all the oxygen atoms.

$$:\!\ddot{\text{O}}:$$
$$\phantom{:\ddot{\text{O}}:}\text{N}::\!\ddot{\text{O}}^{-}$$
$$:\!\ddot{\text{O}}:$$

From this electron dot formula, we could predict that the nitrate ion should be planar, just as we predicted that the formaldehyde molecule would be planar. Our prediction would be correct in this case, too. The nitrate ion *is* planar. However, because the electron dot formula shows one double bond and two single bonds it leads us to expect that the nitrate ion should have one N to O distance shorter than the other two. But experiments have shown that, without a doubt, the nitrate ion has the shape of an equilateral triangle. All three N to O bonds are of equal length.

Chemists have developed two ways of dealing with this problem and others like it. One way is discussed in this section, and the other one is treated in Section 15-9. In both, chemists begin by observing that there is really no reason why there should be a double bond in one place instead of the other two. The oxygen atoms all start out alike. Why should one of them become different from the other two? The actual structure shows that there is no difference.

One way to explain this is to write the three following formulas for the nitrate ion.

$$:\!\ddot{\text{O}}:^{-}:\!\ddot{\text{O}}:^{-}:\!\ddot{\text{O}}:^{-}$$
$$\text{N}\longleftrightarrow\text{N}\longleftrightarrow\text{N}$$
$$:\!\ddot{\text{O}}:\ddot{\text{O}}::\!\ddot{\text{O}}:\ddot{\text{O}}::\!\ddot{\text{O}}:\ddot{\text{O}}$$

Since no one of these is in any way preferable to the other two, chemists conclude that the way in which the electrons are actually distributed in the molecule is an average of all three. Since there is no one simple way to represent this average arrangement, they write the three separate ones and connect them with two-headed arrows. It is very important to interpret this representation correctly. You should *not* conclude that the double bond moves from one place to another. You should *not* infer that at a certain instant it is one oxygen atom, and a moment later another one, that is double-bonded to the nitrogen atom.

What *is* the correct interpretation of the preceding formulas? Chemists mean simply that all of the oxygen atoms are *always* partially double-bonded to the nitrogen atom. The actual electron distribution in the ion is said to be a **resonance hybrid** of the three individual electron dot formulas. The three individual representations of the bonding are said to be in **resonance**.

A very good way to keep the correct interpretation of the resonance concept straight is to think about the fact that a mule is a hybrid. Mules are not bred from mule parents. Instead they are bred from a donkey and a horse. A mule has some of the characteristics of a donkey and some of the characteristics of a horse, but it is at all times a mule. It does not change back and forth between being a horse and a donkey, but constantly shows characteristics of both.

15-9 *Molecular Orbitals* ★

You have just seen that the idea of resonance enables us to describe how electrons are distributed within molecules for which no single electron dot formula, with localized electron pairs, will do the job. There is another way to tackle that problem. Instead of starting with the idea that orbitals will always overlap in the space between a pair of atoms,

Robert Sanderson Mulliken (1896–)

ROBERT MULLIKEN demonstrated his early interest in science when he read his high school graduation speech entitled, "The Electron, What It Is and What It Does."

After graduating from the Massachusetts Institute of Technology in 1917 and earning his doctorate, he spent the early 1920's working on a variety of problems in chemistry. For a while he focused his attention on the separation of isotopes, but soon became interested in the relationship between band spectra (Section 11-12) and the structure of molecules.

When atoms of elements are excited by heat or high voltage electricity, they produce sharply defined bright-line spectra. The spectra exhibited by excited molecules are broader and more diffuse, appearing in the form of bands on a spectrographic plate. When compounds absorb radiation, bands appear in the regions where radiation of certain wavelengths is absorbed. We know today that the complexity of such spectra is caused by electronic, vibrational, and rotational motions in a molecule, but this was unknown territory in the 1920's. Mulliken's associates warned him to avoid this subject unless he planned to spend the rest of his life struggling with the problems involved.

Mulliken accepted the challenge and began to examine the properties of "isoelectronic" molecules, that is, different types of molecules having the same numbers of electrons in them. As he broadened his work to include other types of molecules, he began to see that the concept of a single pair of electrons bound tightly between two adjacent atoms was not as simple as the American chemist G. N. Lewis had suggested in 1916. He began to realize that many of these electrons were behaving like waves and spreading over considerable portions of the molecules. To account for the spaces occupied by these electrons in atoms and molecules he coined the terms "atomic orbital" and "molecular orbital" in 1932.

The basic concept of orbital theory as proposed by Mulliken is that there exist in atoms and in molecules certain discrete electronic energy levels or orbitals that are built up in a regular order as the complexity of the system increases. The molecular orbital theory has been particularly helpful in explaining the structures of benzene, the nitrate ion, and other species for which several equivalent electron structures can be drawn.

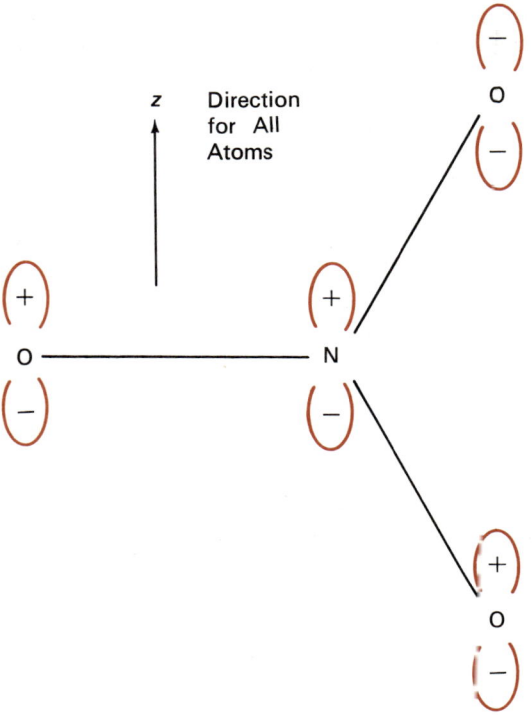

Figure 15-22 Parallel p_z orbitals in the nitrate ion.

An excellent example with which to illustrate this idea is the nitrate ion, which we have already considered from the resonance point of view. Let us begin, as we did before, by adding an electron to the nitrogen atom and then forming one electron-pair bond to each oxygen atom. From this point on, however, our approach will differ.

The three electron pairs around nitrogen that are being used to form bonds to the oxygen atoms can be considered to occupy a set of three sp^2 hybrid orbitals. As you have seen earlier, sp^2 hybrids are directed toward the corners of an equilateral triangle. If we assume that the nitrate ion lies in the xy plane of a Cartesian coordinate system, then it must be the p_x and p_y orbitals that are used to form these hybrids. The p_z orbital lies perpendicular to the plane of the three bonds, and hence perpendicular to the plane of the entire nitrate ion.

Let us now look at the distribution of electron pairs about each oxygen atom. One of the four pairs is located along the line between the oxygen atom and the nitrogen atom. This direction can be taken as the direction of one axis, let us say the x axis, of a coordinate system centered on the oxygen atom. The oxygen atom may then be considered to use its p_x orbital to form the electron pair bond to the nitrogen atom. The remaining three electron pairs occupy the s, p_y, and p_z orbitals. If we define the rest of the coordinate system on the oxygen atom so that the y axis lies in the plane of the nitrate ion, then the p_z orbital of each oxygen atom must lie perpendicular to the plane. Thus the p_z orbitals on all four of the atoms of the nitrate ion will be parallel. You can see this in Figure 15-22. (For clarity, only the p_z orbitals are shown.)

Figure 15-22 indicates that the p_z orbital of the nitrogen atom can overlap equally well with the p_z orbitals of each one of the oxygen

which means that each electron pair must be localized between some particular pair of atoms, we can take a more flexible point of view. In some cases, an orbital in a given atom is geometrically capable of reaching orbitals in two or even three more atoms equally well. When this is the case, we can assume that there will be equal overlap with all of the orbitals that can be reached equally well. This will lead to the formation of a giant orbital extending over more than two atoms. Sometimes such orbitals will extend over the entire molecule. That is why they are usually called molecular orbitals, even when they do not actually extend over the entire molecule.

atoms. There would be no justification for assuming that it overlaps with any one of them in preference to the others. We are thus led to picture a molecular orbital, spread uniformly and symmetrically over the entire nitrate ion, as shown in Figure 15-23. The electrons of the p_z orbitals of the oxygen atoms occupy this molecular orbital, with an equal contribution from each oxygen atom. On the average, each oxygen atom contributes one third of an electron pair. A molecular orbital, like an atomic orbital, can be occupied by no more than one electron pair. In this case, in which there are three equally well qualified pairs available to contribute to the occupancy of one molecular orbital, equal and partial contributions must be taken from each. As you have seen, there is an electron-pair bond between the nitrogen atom and each oxygen atom, in the xy plane. Considering this bond together with the p_z bonding, you can see that each nitrogen to oxygen bond is made up by an average of $1\frac{1}{3}$ electron pairs. This is the same conclusion we reached using the resonance picture.

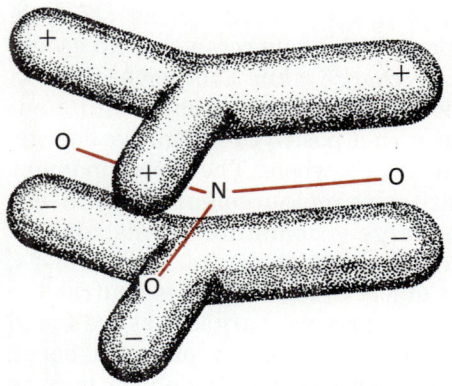

Figure 15-23 *A molecular orbital in the nitrate ion, formed by overlapping of the p_z orbitals shown in Figure 15-22.*

Summary

When one electron pair is shared between two atoms, a single bond is said to exist. It is possible to have more than one electron pair shared between two atoms. When there are two shared pairs, we have a double bond, and when there are three, we have a triple bond. Double and triple bonds are stronger than single bonds, and the triple bond is strongest of all for any given pair of atoms. Stronger bonds involve shorter interatomic separations. For some molecules it is not possible to write a single electron dot formula that adequately describes the shape of the molecule and the electron distribution within it. We must then resort to one of two more elaborate models. We obtain one model by first writing all of the different, but equivalent, electron dot formulas. We consider the true electron distribution to be given by an average or mixture, called a resonance hybrid, of all of them. Alternatively, some portion of the electrons may be assigned to molecular orbitals. Molecular orbitals spread over three or more atoms, and at times over all atoms in the molecule, rather than being restricted to the region between one particular pair of atoms.

Self Check

1. Draw an electron dot formula for the hydrogen cyanide molecule, HCN. What kinds of bonds does it contain?
2. The SO_2 molecule is bent and the sulfur to oxygen bonds are equal in length. Use the resonance model to account for this.
3. Write an electron dot formula for the molecule called ethane, C_2H_6.

Molecular Shape and Dipole Moments

15-10 Definition of a Dipole Moment

A **dipole moment** exists whenever the distribution of electric charge on an object is not uniform, so that one region has an excess of negative charge while another has an excess of positive charge. The object in its entirety has no net positive or negative charge. It is neutral as a whole. The dipole moment is a quantitative measurement of the extent to which charge is nonuniformly distributed. Consider the object shown in Figure 15-24. Since the object in its entirety is neutral, the positive and negative charges that are found at each end of the object must be equal. Though it is not charged, the object may be said to have polarity, or to be polarized. It has a positive pole and a negative pole. The degree of polarization will obviously depend on two things. (1) It depends on the amount of the charge that resides at each pole. (2) It depends on the distance between the two poles. These two factors are combined mathematically into the definition of a dipole moment.

The **dipole moment** is defined as the product of the charge, q, and the separation of the poles, d. Dipole moment equals qd.

The meaning of the word "dipole" should certainly be clear to you by now. The word "moment", however, may strike you as a rather arbitrary term to use. It is a word commonly used in science, particularly in physics. **Moment** may be defined as a tendency to produce motion about an axis of rotation. A molecule with a dipole moment will tend to turn on its axis when it is exposed to an electrostatic force. If you hang a short rod from a thread tied to its middle, nothing will happen if you bring a charged body near it. If you could charge the rod in such a way that one of its ends had a positive charge on it and the other end had a negative charge, the rod would rotate on its axis when another charged body was brought near it.

Another important property of dipoles is that, in addition to their magnitude, given by the dipole moment, they have a definite direction. As you have seen, in Figure 15-24, the distance d is a straight line joining the two charge centers. A dipole, then, can be represented by a vector. Scientists arbitrarily assign the arrowhead to the negative end of a dipole. The length of the vector is proportional to the magnitude of the dipole moment. Representation of dipoles by vectors is very important because it enables us to figure out what the net dipole will be when several dipoles are assembled in a particular geometrical relationship to each other.

Several examples of combining dipoles into a resultant (net) dipole, using vector addition of the individual dipoles, are shown in Figure 15-25. Such combinations will always have a magnitude different from either of the two contributing individual dipoles. The combinations may also have a different direction.

15-11 How Molecules Can Have Dipole Moments

You have already seen, in Section 14-13, that a bond between two unlike atoms will,

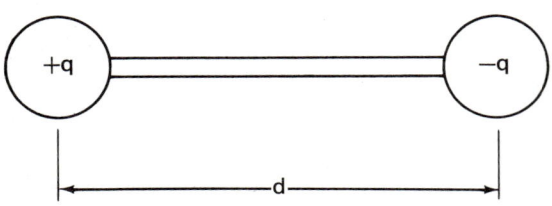

Figure 15-24 *A dipole representation.*

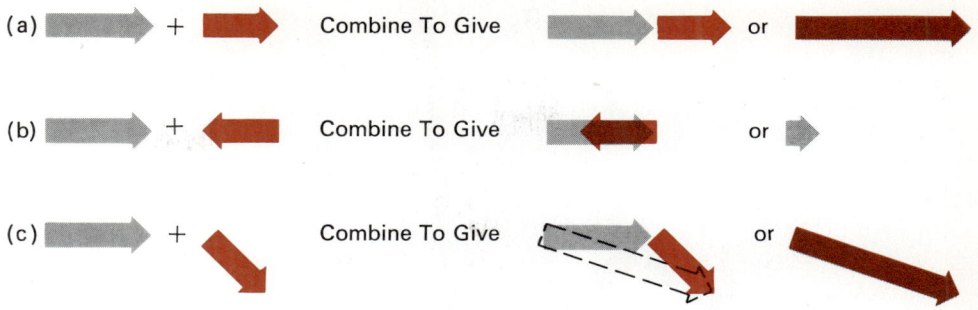

Figure 15–25 *Dipole symbols can be added as shown here. In (a) two dipoles pointing in the same direction produce their arithmetic sum. When oppositely directed (b), the net difference is as shown. When the dipoles form angles, the arrows are added geometrically to give the net effect.*

in general, have some ionic character. The two atoms will not share the electron pair (or pairs) equally. Instead, the negative charge of the shared electrons will lie closer to one of the atoms. A bond with some ionic character, that is, with some unequal sharing of electrons, must be polar. One end will be negative and the other positive. There will thus be a dipole moment lying along the direction of the interatomic axis. All diatomic molecules consisting of unlike atoms have dipole moments. In the upper part of Table 15-2 the dipole moments of a few such molecules are listed.

For more complex molecules containing polar bonds, the presence or absence of a dipole moment in the molecule depends on the shape of the molecule. For this reason, the measurement of dipole moments can often tell us something about the shape of a molecule. For example, all triatomic molecules of the type XYX, such as BeF_2, H_2O, CO_2, and OF_2, to name ones that you have already seen, must fall into one or the other of two classes. Either they are bent, or they

TABLE 15–2 Dipole Moments of Some Molecules

Molecule	Dipole Moment (D)*
Diatomic	
HCl	1.08
HF	1.91
CO	0.10
Triatomic	
BeF_2	0
CO_2	0
OF_2	0.30
H_2O	1.85
Tetra-atomic	
BF_3	0
NF_3	0.25
NH_3	1.47

*The debye, D, is the dipole moment that results when + and − charges, each equal to the charge of an electron, are separated by a distance of one ångström.

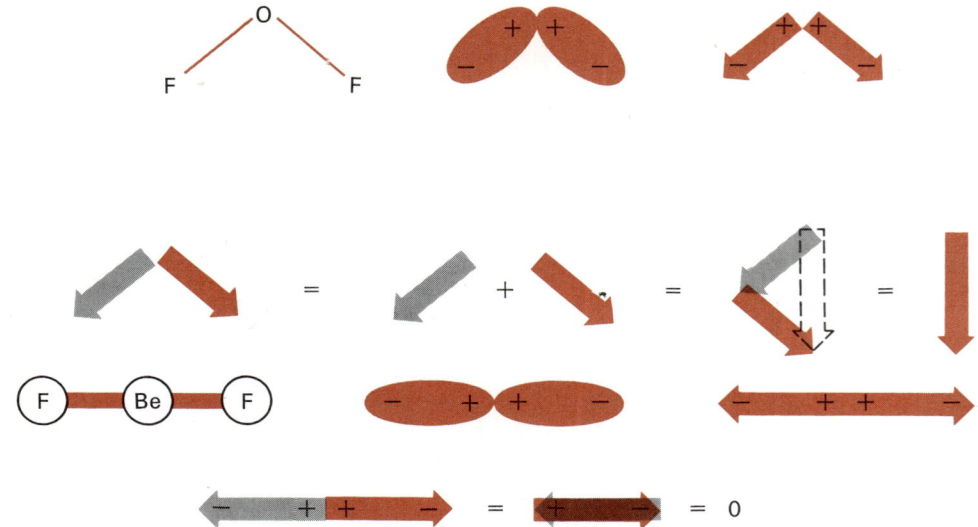

Figure 15-26 *The OF$_2$ molecule (top) has a net dipole moment, as indicated by the addition of the two bond dipole vectors. The net dipole in the BeF$_2$ molecule (bottom) is zero, since the two oppositely directly bond dipoles cancel each other exactly.*

are linear. As Figure 15-26 shows, when such a triatomic molecule is bent, the vector sum of its bond dipoles adds up to a net (resultant) dipole moment for the molecule as a whole. On the other hand, when an XYX molecule is linear, the X-Y and Y-X bond dipoles exactly cancel each other, and the molecule as a whole has no dipole moment. Thus, measuring the dipole moments of XYX-type triatomic molecules indicates whether they are linear or bent. For the molecules mentioned above, the results are shown in Table 15-2.

For tetra-atomic molecules of the type XY$_3$, there are two main possibilities for shape. (1) The molecule may lie entirely in one plane, having the shape of an equilateral triangle with the X atom at the center. (2) The molecule may have the shape of a triangular pyramid, with the Y atoms forming the base and the X atom at the apex. Some XY$_3$-type molecules that you have already seen are BF$_3$, NF$_3$, and NH$_3$. Figure 15-27 shows how the individual bond vectors add up for the two possible structures. You can see that the planar molecules will have zero dipole moment, regardless of how polar the individual bonds may be. The pyramidal molecules should always have at least a small dipole moment. Again, Table 15-2 shows the experimental results. From these results you can conclude that BF$_3$ is planar, and that NF$_3$ and NH$_3$ are pyramidal.

15-12 Structural Isomers

One of the things that makes chemistry a fascinating and often very complex game is that, in all but the simplest cases, any given collection of atoms can be connected up in

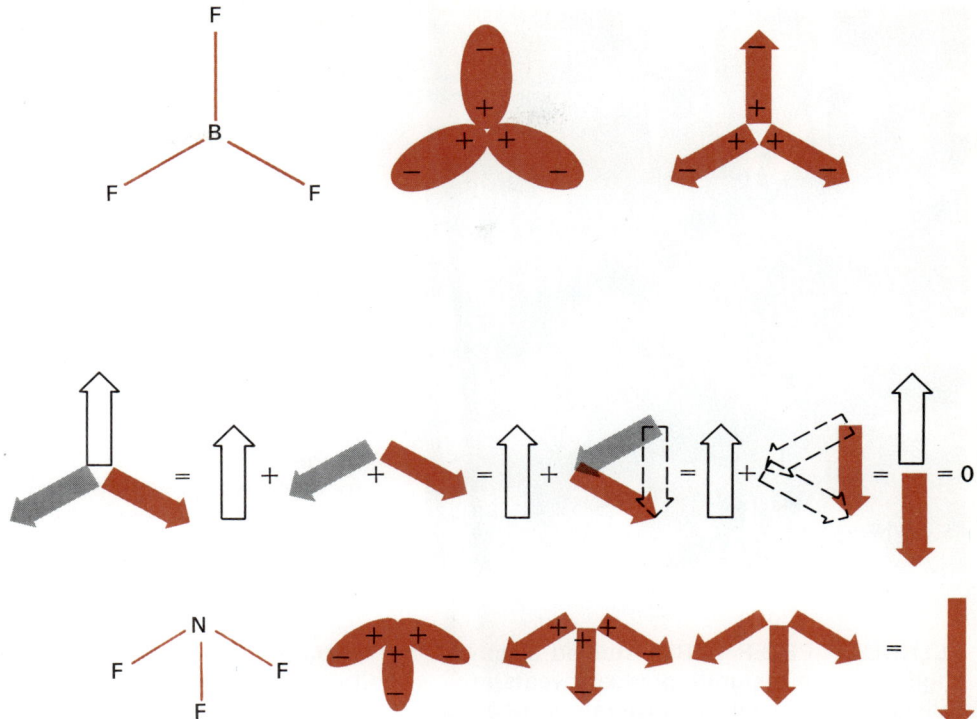

Figure 15–27 *The three equal and equally-spaced dipoles forming the BF₃ molecule (top) produce a net dipole of zero. The dipole vectors of the individual N—F bonds give a resultant molecular dipole in pyramidal NF₃ (bottom).*

more than one way. In other words, there can be two (or more) different molecules that contain exactly the same number of each kind of atom. In the next chapter you will see this illustrated again and again. Right now we merely want to introduce the idea and give a few examples.

The active component of all intoxicating beverages is a substance commonly called alcohol. For reasons that you will learn in the next chapter, this is not a satisfactory scientific name. Chemists call it *ethanol*. When ethanol is analyzed it is found to consist only of the elements carbon, hydrogen, and oxygen. Further, if the exact proportions by mass of the three elements are determined, it is found that they are in the ratio: 24 parts of carbon to 6 parts of hydrogen to 16 parts of oxygen. By measuring the mass of a known volume of ethanol vapor, and using

Peter Joseph Wilhelm Debye (1884–1966)

ALTHOUGH PETER DEBYE studied electrical engineering in his undergraduate years in college, he received his doctorate in physics at the University of Munich in 1908. Within three years he was recognized as such an outstanding scientist that he was appointed to the position at the University of Zurich that Albert Einstein had just left.

Very early in his career Debye became interested in the effect of electrical forces on molecules. He found that if the distribution of the electrical charge on a molecule is not symmetrically arranged, the molecule tends to undergo a rotating motion within the electrical field. The molecule then aligns itself in certain directions consistent with the strength of the field and the charges on the ends of the molecule. By measuring the changes in the electrical field when molecules were introduced, Debye was able to determine what he called the *dipole moment* of the molecules. By measuring the distribution of charges on many molecules, Debye and others have been able to learn much about the structure of these molecules and to predict how these molecules will behave under a variety of conditions. Today we use the *debye* as the primary unit for measuring dipole moments.

Prior to 1916, one of the major disadvantages to the X-ray diffraction analysis of crystalline substances was the necessity to grow large, perfect crystals. Debye was one of the first to discover that X-ray diffraction methods would work perfectly well on powdered crystals prepared by conventional methods. The elimination of the time-consuming step of growing crystals made X-ray diffraction a far more important analytical tool.

During his work with X-ray diffraction, Debye discovered that ions actually existed in salt crystals before they were dissolved in water. The Swedish chemist Arrhenius had studied solutions of salts in the 1880's. Arrhenius concluded that, since their solutions did not conduct electricity as well as expected, those salts did not dissociate completely into ions in solution. Debye, as a result of his own studies of polar molecules, proposed that salts did dissociate completely in solution, but that their ions became surrounded by clouds of polar solvent molecules that hindered their movement in an electrical field.

Debye won the 1936 Nobel Prize in Chemistry for his work on dipole moments. Although he served as Director of the Kaiser Wilhelm Institute in Berlin for several years, he found it increasingly difficult to work there under the Hitler regime. In 1940 he left Germany and accepted a professorship at Cornell University.

TABLE 15-3 Some Physical Properties of Ethanol

Property	Description
Melting point	−117.3 °C
Boiling point	78.5 °C
Density (at 20°C)	0.789 g/ml
Color	Colorless
Solubility	Soluble in water in all proportions

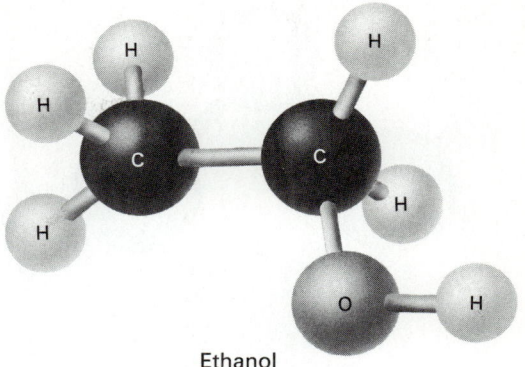

Figure 15-28 A "ball-and-stick" model of the ethanol molecule.

the gas laws that you studied earlier, you could find that the molar mass of ethanol is 46.

You can find the formula of ethanol, using the preceding experimental results. One mole of ethanol has a mass of 46 grams. Of this mass, 24 grams must be carbon atoms, 6 grams must be hydrogen atoms, and 16 grams must be oxygen atoms. The molar mass of carbon (that is, the mass of a mole of carbon atoms) is 12.0 grams. Hence, a mole of ethanol must contain two moles (24 grams) of carbon atoms. Similarly, there must be six moles of hydrogen atoms (molar mass × 1.0) and one mole of oxygen atoms (molar mass = 16.0). Each individual molecule of ethanol, then, must contain 2 carbon atoms, 6 hydrogen atoms, and 1 oxygen atom. Its formula must be C_2H_6O.

Aside from its commonly recognized physiological properties, ethanol has other characteristic physical properties. As with all pure substances, these are constant for all samples of the substance. A material possessing all of them may reasonably be assumed to be this particular substance, ethanol. Table 15-3 lists some of the characteristic properties of ethanol.

The final thing we might wish to know about ethanol is the structure of the molecule. From a variety of measurements, chemists have shown conclusively that the bonding and shape of the ethanol molecule are as pictured in Figure 15-28. It can also be represented as

$$\begin{array}{c} \text{H} \quad \text{H} \\ | \quad | \\ \text{H}-\text{C}-\text{C}-\text{OH}. \\ | \quad | \\ \text{H} \quad \text{H} \end{array}$$

The story so far is a rather straightforward one. From the composition and molar mass, the formula is deduced for a substance that can always be identified by its own characteristic properties, such as color, odor, boiling point, and density. But now we come to an interesting complication. There is another substance that has exactly the same formula, C_2H_6O, but different properties. While ethanol is a water-like liquid at room temperature, the other compound is a gas at room temperature, with a boiling point of −27°C. Obviously these are two different substances. They must therefore consist of different molecules. Yet these different molecules are put together using exactly the same set of atoms. The second

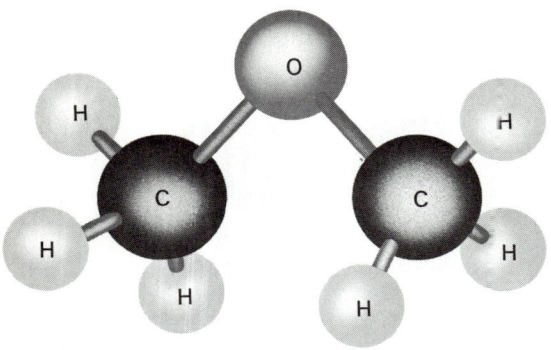

Dimethyl Ether

Figure 15-29 *A "ball-and-stick" model of dimethyl ether.*

substance, also of molecular formula C_2H_6O, is called dimethyl ether.

The structure of the dimethyl ether molecule has been determined by various kinds of measurements and is shown in Figure 15-29. It can also be represented as

$$H-\underset{\underset{H}{|}}{\overset{\overset{H}{|}}{C}}-O-\underset{\underset{H}{|}}{\overset{\overset{H}{|}}{C}}-H.$$

The molecules of ethanol and dimethyl ether are isomers. As you learned in Section 7-13, isomers are molecules consisting of the same set of atoms, arranged differently in each case. To be more precise, isomers of this particular kind are called **structural** isomers. This term signifies that the isomers differ in having different sets of bonds. The bonds in ethanol are the following.

5 C—H bonds
1 C—C bond
1 C—O bond
1 O—H bond

The same bonds are not present in dimethyl ether, however. It has the following ones.

6 C—H bonds
2 C—O bonds

Structural isomers also exist even when the sets of bonds are the same, but occur in a different order. Consider the following pair of molecules.

$$H-\underset{\underset{H}{|}}{\overset{\overset{H}{|}}{C}}-\underset{\underset{H}{|}}{\overset{\overset{H}{|}}{C}}-\underset{\underset{H}{|}}{\overset{\overset{H}{|}}{C}}-O-\underset{\underset{}{}}{\overset{\overset{H}{|}}{C}}-H$$

10 C-H bonds
2 C-C bonds
2 C-O bonds

$$H-\underset{\underset{H}{|}}{\overset{\overset{H}{|}}{C}}-\underset{\underset{H}{|}}{\overset{\overset{H}{|}}{C}}-O-\underset{\underset{H}{|}}{\overset{\overset{H}{|}}{C}}-\underset{\underset{H}{|}}{\overset{\overset{H}{|}}{C}}-H$$

10 C-H bonds
2 C-C bonds
2 C-O bonds

The two molecules are certainly different even though they contain the same set of bonds. They differ in the sequence of the bonds. In the first case, going along the "backbone" of the molecule you find C—C—C—O—C, while in the second case you find C—C—O—C—C.

15-13 *Geometric Isomers*

For an example of a different set of isomers, let us go back to the ethylene molecule. As

Figure 15–30 Each C—F bond in tetrafluoroethylene has its own dipole moment, as shown. However, their arrangement in the molecule is such that they cancel each other out.

you may recall from Section 15-6, this molecule is planar. There are many molecules that are related to ethylene in having the pair of carbon atoms connected by a double bond, but differ from it in having something other than four hydrogen atoms bonded to the carbon atoms. For instance, instead of four hydrogen atoms there may be four fluorine atoms or four chlorine atoms. All such molecules are planar and, as a little thought will show, they have no molecular dipole moment, even though the individual bonds are polar. The four C—H, C—F, or C—Cl bond moments are so directed that they cancel one another out, as shown for the C_2F_4 molecule in Figure 15-30.

Suppose, however, that only two of the hydrogen atoms are replaced, let us say by fluorine atoms. The formula will now be $C_2H_2F_2$. However, you can see that there are three different ways in which to choose the two hydrogen atoms that are to be replaced. They are shown in Figure 15-31. It obviously makes a great deal of difference whether you replace both hydrogen atoms on the same carbon atom or one on each. The fact that isomers arise from this cause is not difficult to accept. This is an illustration of structural isomerism.

Maybe it is not quite so obvious that there are two ways to replace one hydrogen atom on each carbon atom. It should, however, be

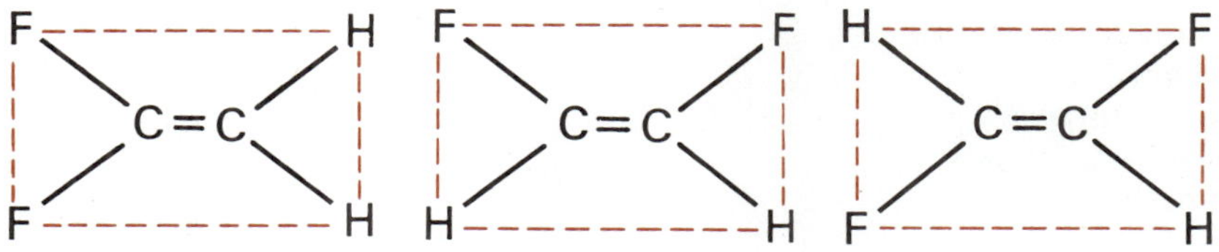

cis *trans*

Figure 15–31 Three isomers have the formula $C_2H_2F_2$.

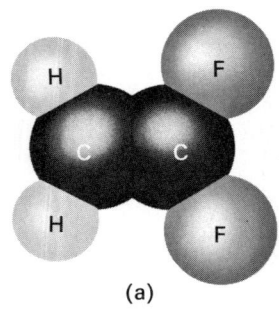

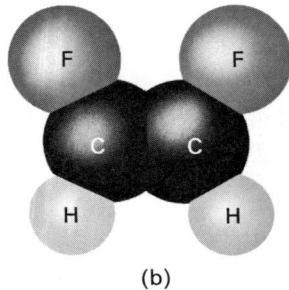

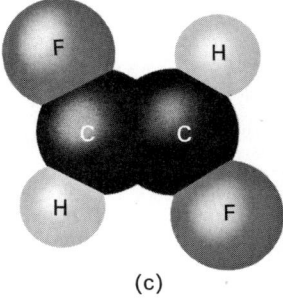

Figure 15-32 *The three isomers of difluoroethylene. (a) is 1,1-difluoroethylene. (b) and (c) are the cis and trans isomers of 1,2-difluoroethylene.*

clear from Figures 15-31 and 15-32 that the two molecules are different and should have different properties. In fact, it would be quite easy to tell the difference between these two

isomers, and to tell which one is which, by measuring their dipole moments. The isomer designated *cis*, which means "on the same side," has a dipole moment. The isomer designated *trans*, which means "across," has no dipole moment. In the *trans* arrangement, the two C—F bond moments cancel each other and so do the two C—H bond moments.

Isomers of this type are called **geometric** isomers, to distinguish them from structural isomers. Two (or sometimes more) geometric isomers have the same types of bonds, in the same order. But, the relationship of certain atoms in space differs. Therefore, the geometry differs and so they are called geometric isomers.

It is important to realize that *cis* and *trans* isomers can only occur when there is a *double bond* between the carbon atoms. It is not possible to convert the *cis* isomer to the *trans* isomer without breaking the double bond, or at least one half of it. The four electron pairs around each carbon atom must remain in a tetrahedral arrangement. This makes it impossible to twist the molecule. The models that you will build later (Section 15-14) should make this clear. Only by removing at least one of the toothpicks in the double bond, which corresponds to breaking the bond, can one end of the molecule be rotated. Since breaking bonds costs a great deal of energy, it does not happen by itself. The *cis* and *trans* isomers therefore do not interconvert. They remain as separate, distinct, and different molecules.

When carbon atoms are connected by a single bond only, geometric isomers are not possible. Look at the two structures shown in Figure 15-33. They look like geometric isomers, but they are not. Why? Because, without breaking any bonds, they can change into each other. Only rotation about the C—C bond is needed and this can occur easily. Again, your ball-and-toothpick models will show this clearly.

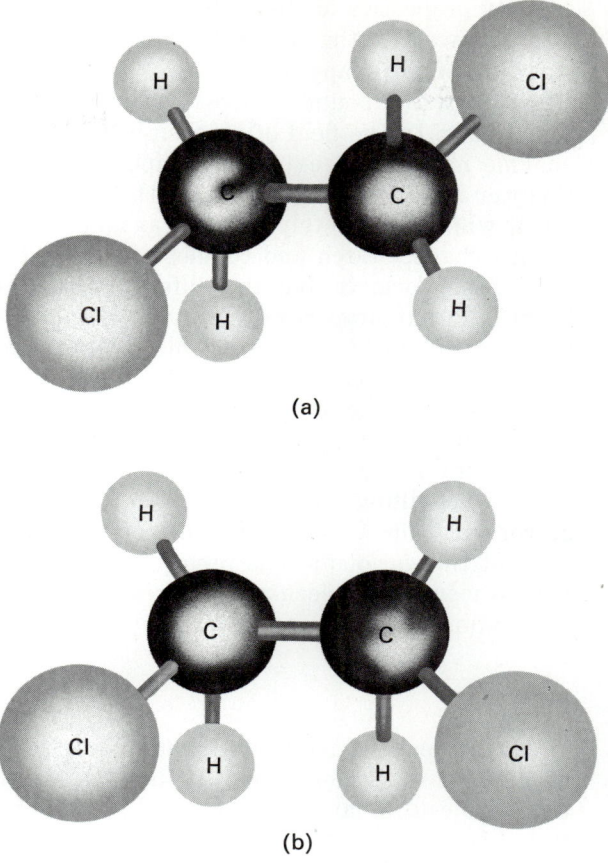

Figure 15-33 *Two arrangements of the atoms in 1,2-dichloroethane. These are easily interconverted by rotation about the C—C bond and are not considered to be isomers.*

INVESTIGATION

15-14 Molecules and Their Shapes

When two atoms join to form a diatomic molecule, the size of the atoms and the distance between their nuclei may vary. The basic shape, however, is the same for all such molecules. It is a three-dimensional figure formed by two intersecting spheres. (See Figure 15-34, next page.)

Molecules with more than two atoms may assume a great variety of shapes. These shapes are determined by the number and size of the atoms, the number of electrons, the types of orbitals available for bonding in each atom, and the number of bonds joining the atoms.

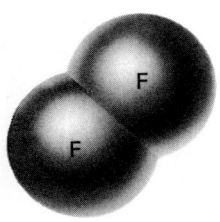

Figure 15-34 *A space-filling model of the fluorine, F_2, molecule.*

Many properties of compounds depend on the shapes of their molecules, as well as the forces holding them together. Thus, an understanding of the shapes of molecules is important in predicting properties of new compounds and explaining the behavior of many known compounds. In the field of biochemistry, for example, the shape of an enzyme molecule is extremely important. (Enzymes are discussed in Chapter 21.) A molecule may have the right number and type of atoms, but it will be completely inactive as an enzyme if it does not have the correct shape. Starch and cellulose (discussed in Chapter 18) have the same basic formula, but the different shapes of their molecules give them different properties. By modifying the shapes of the molecules involved, chemists have been able to strengthen enormously some of our common plastics.

In this experiment you will construct a number of styrofoam models of molecules. As you can see from one such model, Figure 15-35, these models are similar to the ball-and-stick models, rather than the more exact space-filling models. Also, due to the limited number of sizes of styrofoam balls available, it will not be possible to show the relative sizes of the various atoms precisely. Your models will, however, help you to visualize molecular shapes and to understand various aspects of bonding.

PURPOSE

You will construct models of a number of molecules based on the arrangement of the electrons in their atoms. The models will include those molecules having single, double, and triple bonds. You will also make models of isomers. From your models you will be able to predict how differences in bonding orbitals and sizes of atoms affect the shapes of molecules. Moreover, you will be able to predict which molecules have dipole moments.

MATERIALS

16 styrofoam balls, $\frac{3}{4}''$ diameter
2 styrofoam balls, $1''$ diameter
7 styrofoam balls, $1\frac{1}{8}''$ diameter
1 styrofoam ball, $1\frac{1}{2}''$ diameter
1 styrofoam ball, $2''$ diameter
24 round toothpicks, sharp on each end

Protractor
Metric ruler
1 gram absorbent cotton

PROCEDURE

1. Fill in the chart described in the "Data and Results" section, down through the row "Orbitals used in bonding" (before coming to lab).

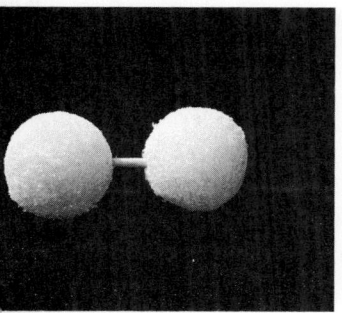

Figure 15-35

2. Construct models of each of the seven hydrogen compounds listed in your table. Assume that the orbitals used will be spread as far apart as possible around the central atom. It is known that the atoms in the second period of the periodic table have different sizes in covalent compounds, decreasing from a radius of 1.23 Å for Li to 0.72 Å for F. Nevertheless, use $1\frac{1}{8}''$ balls to represent each of these atoms and $\frac{3}{4}''$ balls to represent hydrogen atoms. A toothpick stuck into adjacent balls will represent a pair of electrons in a single bond between two atoms, as in Figure 15-35. (You will need seven $1\frac{1}{8}''$ styrofoam balls, sixteen $\frac{3}{4}''$ balls, a wad of absorbent cotton, and about 24 toothpicks.)

  a. Before completing the CH_4 model, first insert four toothpicks to a depth of 1 centimeter in the central atom in such a way that they all lie in one plane with their outer ends forming the vertices of a square. Measure the distance between the ends of all four pairs of adjacent toothpicks and find the average distance. (This procedure should correct for slight irregularities in the construction of your model.) Now remove the four toothpicks and do your best to insert them again into the central atom in such a way that all four toothpicks are spaced symmetrically around the atom and form nearly equal angles with each other. (This is a little tricky and may take several trials before you have them spaced properly.) Be sure each toothpick is inserted to a depth of 1 centimeter in the central atom. Again, find the average distance between adjacent ends of the toothpicks.

  b. In constructing models of NH_3, H_2O, and HF, use the same orbitals as you did with CH_4. Use a ball of cotton on the end of a toothpick to represent each unshared pair of electrons.

3. Make a model of an H_2S molecule, using a $1\frac{1}{2}''$ ball for sulfur and $\frac{3}{4}''$ balls for hydrogen. Compare this model with your H_2O model, made with a $1\frac{1}{8}''$ ball.

4. Use a $\frac{3}{4}''$ ball to represent F, a $1''$ ball to represent Cl, a $1\frac{1}{2}''$ ball for Br, and a $2''$ ball for I. Make models of HF, HCl, HBr, and HI. Use short pieces of toothpicks and push the balls together until they touch. (These models will not be exactly proportional to the sizes of the true molecules, but will serve our purposes.)

5. Dismantle the models you have made. Now construct a series of carbon, hydrogen, and oxygen compounds using $1\frac{1}{8}''$ balls for carbon, $\frac{3}{4}''$ balls for hydrogen, and $1''$ balls for oxygen. Show only the bonding electrons. Do not represent unshared electrons. Use two toothpicks for double bonds and three toothpicks for triple bonds, as shown in Figure 15-36. When you complete a model, write a structural formula for the compound formed.

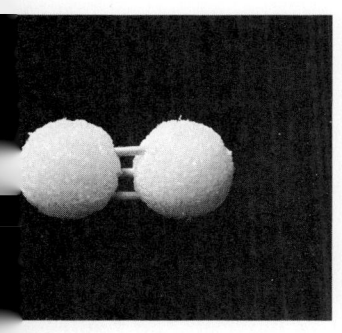

Figure 15-36

Construct molecular models having the following atoms.

  a. 2 C's, 6 H's
  b. 2 C's, 4 H's
  c. 2 C's, 2 H's
  d. 1 C, 2 H's, 1 O
  e. 1 C, 4 H's, 1 O
  f. 2 C's, 6 H's, 1 O (Two different forms.)

6. Label two of the 1″ balls with small pieces of masking tape with Cl written on them to represent chlorine atoms. Use these to construct three different isomers of dichloroethylene ($C_2H_2Cl_2$). Write structural formulas for each isomer. Show which are *cis* and *trans* forms.

DATA AND RESULTS

Before constructing any models prepare a chart like the following one on a sheet of full-size notebook paper. Turn the paper so that the long side is at the top, and list the symbols of the elements that appear in the second row of the periodic table (Li–F) at the top.

Element   Li, Be, etc.

Atomic Number
Electron configuration
Electron dot symbol
 of element
Electron dot representation of hydrogen compound
Orbitals used in bonding
Drawings of models
Approximate bond angles
Does molecule have dipole moment?
Shape of molecule—linear, bent, planar, pyramidal, tetrahedral.

RELATED PROBLEMS

1. Which of the two forms of CH_4 that you made is the more stable one? Why?

2. Explain the effect that the unbonded electrons have on the bond angles in NH_3 and H_2O. Compare these bond angles with those in

CH$_4$. Which of these three compounds should have the largest bond angles?

3. Suggest a possible reason why the actual H-S-H angle of an H$_2$S molecule is about 93°, while the H-O-H angle of an H$_2$O molecule is about 104.5°.

4. Recall that the force between two charges is inversely proportional to the square of the distance between them. Then, consider the compounds HF and HI, for which you made molecular models. Would you expect fluorine or iodine to have the greater attraction for the bonding electrons?

5. Consider the four compounds, HF, HCl, HBr, and HI. Which would you expect to be held together by the strongest bonds? Which bond should be the most ionic in character? Which molecule should have the greatest dipole moment?

6. Which of the isomers of dichloroethylene should have a dipole moment?

15-15 Another Way of Describing Double and Triple Bonds ★

You have seen that a double bond can be represented by the sharing of an edge between two tetrahedra. This very simple description is useful and accounts for many of the basic features of a double bond. It helps us to understand why rotation about a double bond is highly restricted, thus leading to *cis* and *trans* isomers. However, this simple picture does not provide an entirely adequate description of double bonds. There are some features of molecules with double bonds, particularly their spectra, that chemists can understand more easily in terms of a different description. This alternative way of looking at a double bond (and, also, a triple bond) will now be explained. It is called the sigma (σ) and pi (π) description.

Let us first use the ethylene molecule, H$_2$C=CH$_2$, as an illustration. Experiment shows that the ethylene molecule is planar.

The plane of the molecule can be taken as the xy plane in a Cartesian coordinate system, as shown in Figure 15-37. A carbon atom in its lowest energy state has an electron configuration $1s^2 2s^2 2p^2$. First, an electron in the carbon atom must be promoted

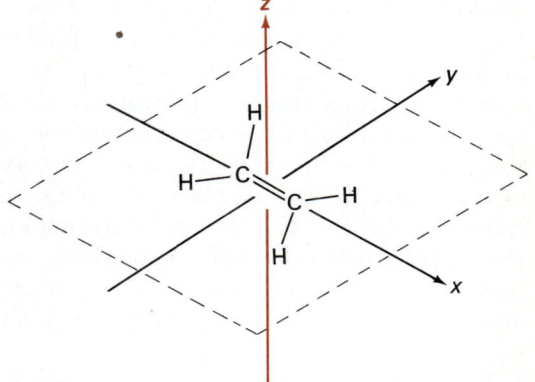

Figure 15-37 *The ethylene molecule in the xy plane of a cartesian coordinate system.*

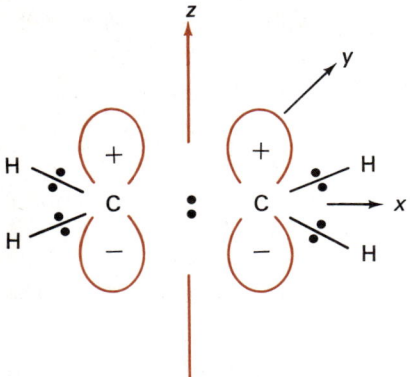

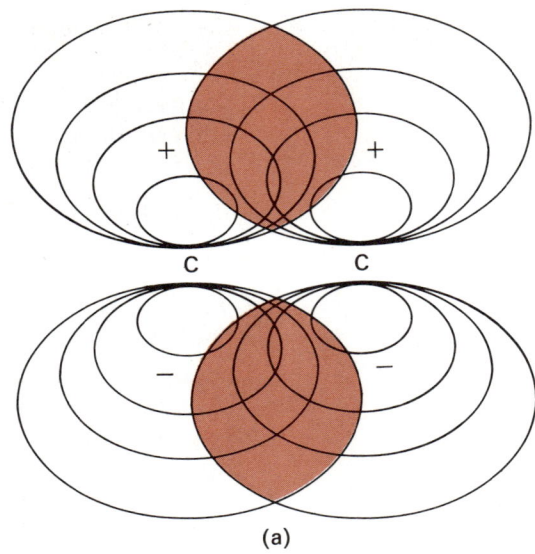

(a)

Figure 15-38 *Each carbon atom has formed a set of three single bonds using electron pairs in the plane of the molecule. The p_z orbital on each carbon atom is also shown.*

to obtain the configuration $1s^2 2s 2p_x 2p_y 2p_z$. The electrons in the $2s$, $2p_x$, $2p_y$, and $2p_z$ orbitals can now all be used to form bonds. Henceforth we shall omit mention of the $1s$ electrons, since these will play no role in forming bonds.

Now, instead of assuming that the four orbitals of the second shell are hybridized to form a set of tetrahedral orbitals, we proceed as follows. We assume that only the $2s$, $2p_x$ and $2p_y$ orbitals become hybridized. These sp^2 hybrid orbitals on each carbon atom can then be used to form single bonds to three other atoms in the xy plane. In ethylene, these will be the two hydrogen atoms and the other carbon atom. Each carbon atom still has its p_z orbital, which contains an unused electron. The situation up to this point is depicted in Figure 15-38.

Figure 15-38 does not really represent the situation accurately. At the actual C—C distance the $2p_z$ orbitals of the carbon atoms

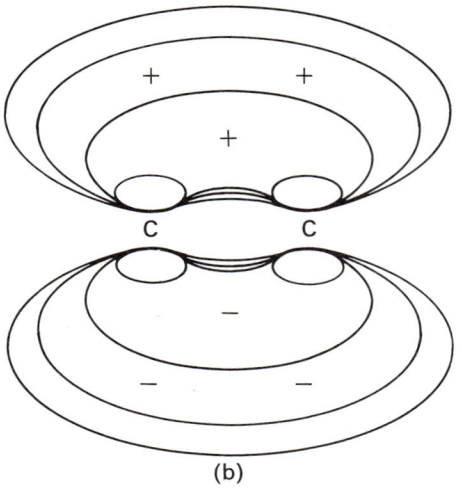

(b)

Figure 15-39 *Part (a) shows how the two p_z orbitals actually overlap. Part (b) shows the overall contours of the overlapping orbitals, the π orbitals of the molecule.*

overlap each other considerably. This is shown in Figure 15-39a. It is possible to represent the distribution of electron density in the regions of both $2p_z$ orbitals jointly by a single set of contours, as shown in Figure 15-39b. This set of contours represents what we call a π (Greek letter pi) bond. It is called a π bond because it is comparable in shape to a p orbital. The Greek letter π is equivalent to the English letter p. The two electrons that were initially present, one in each of the $2p_z$ orbitals, now become paired in this π orbital.

The double bond between the carbon atoms is thus depicted as having two different components. One is the bond formed by overlap of lobes of the sp^2 hybrid orbitals in the xy plane. This bond is called the σ (Greek letter sigma, corresponding to the letter s in our alphabet) bond. The designation σ is used because this type of bond can be formed by the overlap of two s orbitals. The bond in the hydrogen molecule is a σ bond. It is symmetrical around the bond axis, as shown in Figure 15-40a. A σ bond can also be formed by overlap of two p orbitals when these orbitals are of the type that point toward each other, as shown in Figure 15-40b. Sigma bonds can also be formed by the overlap of an s and a p orbital, or of any two sp, sp^2, or sp^3 hybrid orbitals, or by any combination of one of the hybrid orbitals and an s or p orbital.

As already shown in Figure 15-39, π bonds are formed by the overlap of p orbitals that have their axes parallel. The basic difference in shape between σ and π orbitals is best seen in Figure 15-40. For a π orbital the electron density is concentrated in two distinct regions on opposite sides of the line between the bonded atoms.

The σ-π description of a double bond accounts quite naturally for restricted rotation. The maximum overlap between the two p_z orbitals occurs when their z axes are exactly parallel. Any rotation of one end of the molecule relative to the other decreases the overlap and hence requires an increase in energy.

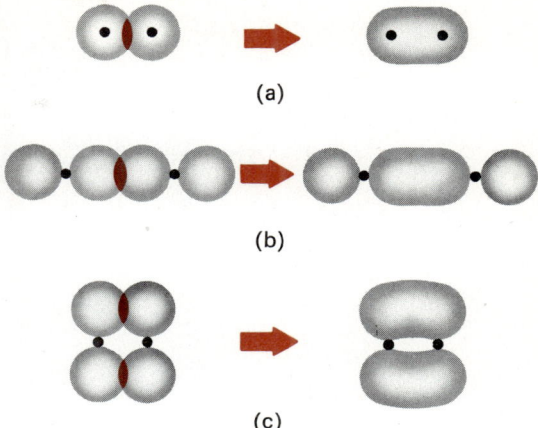

Figure 15-40 (a) Formation of a σ bond by overlap of two s orbitals. (b) Formation of a σ bond by the overlap of two p orbitals. (c) Formation of a π bond by the overlap of two p orbitals.

Triple bonds can also be described in the σ-π fashion. In the case of the nitrogen molecule, for example, we begin with two atoms each having a $2s^2 2p_x 2p_y 2p_z$ electron configuration. If these two atoms approach each other along the x axis, a σ bond is formed by overlap of their p_x orbitals, like the one shown in Figure 15-40b. The two p_x electrons become paired. If the orientation of the y and z axes of the two atoms is as shown in Figure 15-41a (next page), two π bonds may be formed. The two p_y orbitals overlap and so also do the two p_z orbitals. Thus, a triple bond is considered to consist of a σ bond and two π bonds. This is shown for nitrogen in Figure 15-41b (next page).

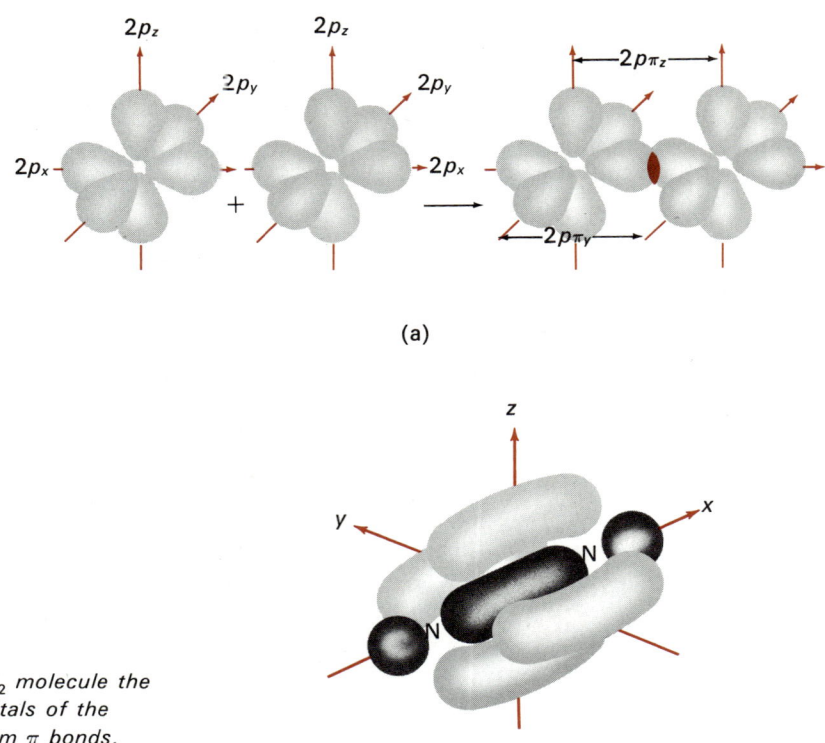

Figure 15–41 In the N_2 molecule the pairs of $2p_y$ and $2p_z$ orbitals of the two atoms overlap to form π bonds, while the $2p_x$ orbitals form a σ bond.

Summary

A dipole results when the charges on a body that is neutral as a whole are not uniformly spread over it. Such a body has more positive than negative charge on one side or end, and an equal excess of negative over positive charge on the other side or end. A quantitative measure of polarity is the dipole moment, obtained by multiplying the magnitude of the unbalanced charges by the distance between them. All bonds between unlike atoms have a dipole moment, though it may, by chance, be small in a few cases. When a molecule contains two or more polar bonds, the molecule as a whole may have a dipole moment or it may not. It depends on whether the individual bond moments add to give a resultant dipole vector or whether they are so directed as to cancel one another out. Because of this, molecular shapes can often be ascertained by measuring molecular dipole moments.

Two or more molecules that contain the same set of atoms connected in different ways are called isomers. Isomers are different substances, with quite different properties.

Self Check

1. Name three diatomic molecules that have dipole moments and three that do not.
2. Should the molecule CH_3F have a dipole moment? What about CH_2F_2?
3. Suppose one C—H bond of ethanol is to be replaced with a C—F bond. How many isomers will there be? Answer the same question for dimethyl ether.

Chapter Summary

It is known from many types of experimental evidence, especially by X-ray diffraction studies of crystalline materials, that molecules have definite shapes. These shapes are determined by the directions in which the bonds formed by atoms tend to point. It is possible to understand, and to a considerable extent to predict, these preferred bond directions, and hence the shapes of molecules. This is done either by considering the shapes of the atomic orbitals used by each atom when forming bonds or by recognizing that electron pairs repel one another and will arrange themselves around each atom accordingly. Still another model involves the hybridization of atomic orbitals. With hybridization, orbitals are produced that are better suited than the pure s or p orbitals for overlapping with orbitals on other atoms. All of these ways of analyzing the problem lead to the same shapes for most molecules, although the various approaches often predict slightly different values for the bond angles.

There may be two, or even three, pairs of electrons shared by two atoms. In this way double and triple bonds are formed. There are some molecules in which no single way of arranging the electrons is better than all others. In such cases the actual electron distribution in the molecule must be thought of as a mixture, or resonance hybrid, of all possible individual distributions. (This situation can also be handled by constructing molecular orbitals. These are orbitals that extend over three or more atoms and often over the entire molecule.)

Bonds between unlike atoms are, in principle, polar. When several of these are present in a molecule, the molecule as a whole may or may not have a dipole moment. It depends on whether the shape of the molecule allows the individual bond dipoles to give a net resultant dipole, or whether they exactly cancel one another out. Structures can often be inferred from whether or not there is a molecular dipole moment.

Two or more molecules consisting of the same set of atoms connected in different ways are called isomers. Sometimes isomers differ in the kinds of bonds they contain. For example, ethanol contains a C—C bond and an O—H bond as well as five C—H bonds and a C—O bond. Dimethyl ether, which is an isomer of ethanol, contains neither C—C nor O—H bonds. Instead, it contains more C—O and C—H bonds than does ethanol. In other cases the same number of each kind of bond is present, but the way in which the bonds are arranged, relative to each other, is different. This is illustrated by the *cis* and *trans* isomers of $C_2H_2F_2$.

Questions and Problems

1. When you predict the H—O—H angle in a molecule of water on the basis of the p orbitals only, what angle do you expect? Why?
2. Why might you predict the H—N—H angle in NH_3 to be the same as the H—O—H angle in water?
3. Explain why you would expect an H_2S molecule to have nearly the same shape as an H_2O molecule.
4. On the basis of the electron-pair repulsion theory, how might you account for the fact that the H—S—H angle in H_2S is about 93°, while the H—O—H angle in H_2O is about 105°?
5. Why does a beam of atoms of copper (atomic number 29) split into two beams in a magnetic field, while a beam of atoms of zinc (atomic number 30) does not split into two beams?
6. The magnetic field produced by a spinning electron can be represented by a vector. Explain the terms "magnetic field" and "vector." Use vectors to represent the motion of two airplanes, one flying north at 200 mph and the other flying southeast at 300 mph.
7. How do opposite poles of two magnets affect each other? How do like poles of two magnets affect each other? Explain why two electrons spinning in opposite directions favor the formation of an electron pair.
8. Why do electron pairs try to distribute themselves as far apart as possible in space, when we have just seen that electrons spinning in opposite directions tend to attract each other?
9. Refer to Figure 15-8. Explain why the distance y between the vertices of a tetrahedron is greater than the distance x between the vertices of a square, when the distance r from the central atom to the vertices is the same in both geometric forms.
10. Why do pairs of unshared electrons take up more room than shared, or bonding, electrons? Use this concept and Figure 15-11 to explain why the H—N—H angle of NH_3 is less than the H—C—H angle of CH_4.
11. If you label the unshared pair of electrons in the NH_3 molecule in Figure 15-10 as X, would you expect the X—N—H angle to be greater than or less than one of the H—N—H angles? Explain.
12. Why does a molecule such as CH_2F_2 *not* form a tetrahedron having equal angles between the central atom and the vertices?
13. Predict the shapes of $AlCl_3$ and $MgBr_2$ molecules. Use the electron repulsion theory as well as the hybrid orbital theory to defend your answers.
14. Using the concept of hybrid orbitals, predict the shape of a PCl_3 molecule.
15. Show electron dot formulas for the following compounds: CH_4, C_3H_6, C_3H_4, CH_3OH, $(CH_3)_2CO$, C_2H_5CN.
16. Benzene has the formula C_6H_6. It has been shown that the six carbons form a ring joined by covalent bonds with one H on each carbon atom. Write two different electron dot formulas for benzene and explain how they can be in resonance with each other.
17. Draw an electron dot representation for the NH_4^+ ion. What shape do you predict this ion will have?
18. Predict the type of bonding in the ion BF_4^-, and the shape of the ion.

19. In which of the following compounds is the bond between the carbon atoms the strongest: CH_3CH_3, CH_2CH_2, or CHCH? Explain.
20. Considering the comparable oxygen compound, predict the shape of the H_2S_2 molecule. What bonding orbitals are used?
21. Predict the structure of N_2F_2 from the electron dot representations of the atoms and the molecule.
22. Consider the two compounds CH_3CH_3 (ethane) and CH_3NH_2 (methylamine). Why does CH_3NH_2 have a dipole moment while CH_3CH_3 does not?
23. Consider the following series: CH_4, CH_3Cl, CH_2Cl_2, $CHCl_3$, CCl_4. In which cases will the molecules have dipole moments? Support your answer by considering the bonding orbitals of carbon, the shape of the molecules, and the symmetry of the molecules.
24. Write structural formulas for all isomers with the formula C_6H_{14}. Use single lines instead of two dots to represent single bonds between atoms. Note that writing a different shape on paper does not represent another isomer if the atoms are arranged in the same order.

16 Molecules Formed by Carbon

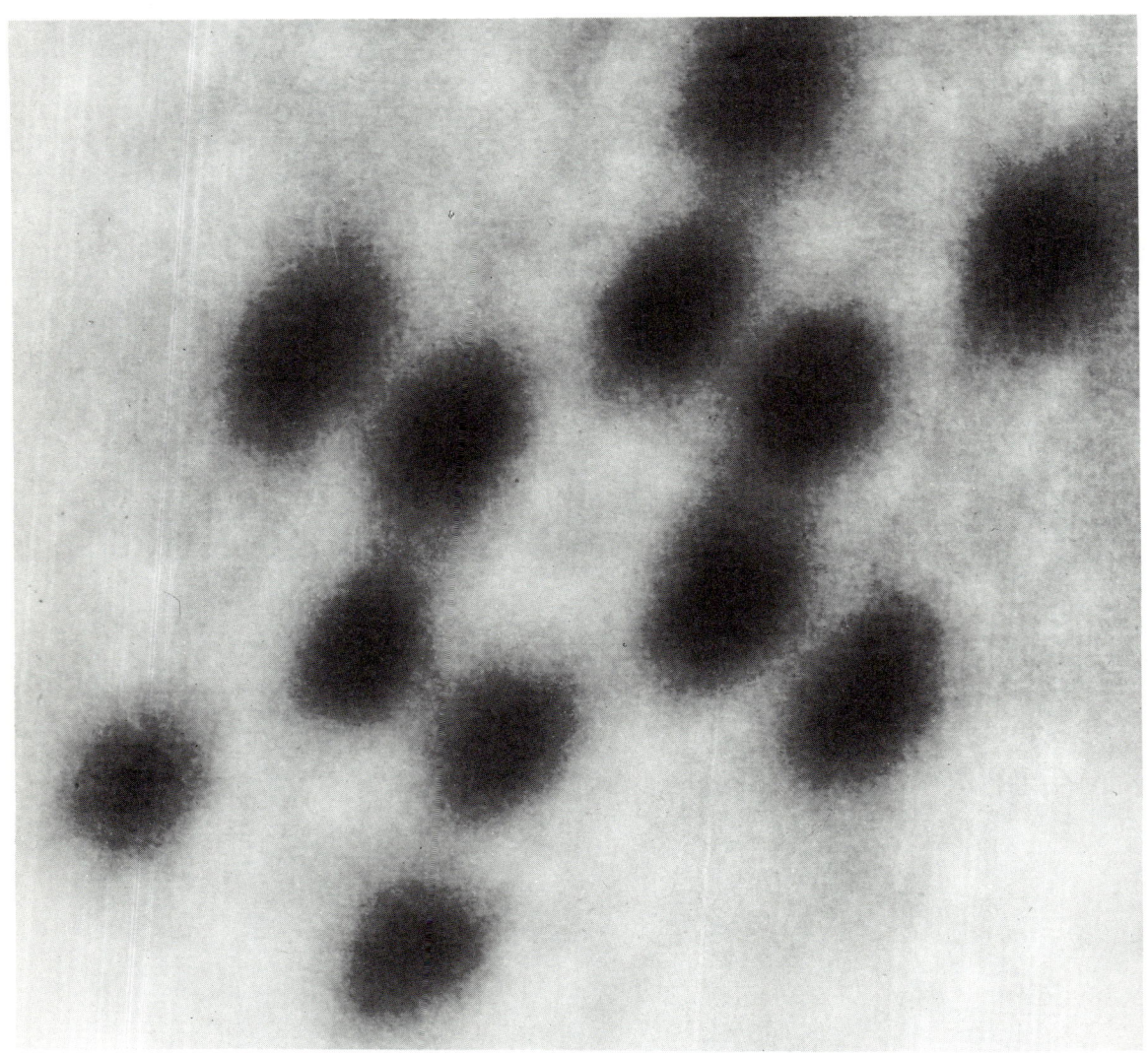

This "photograph" of a hexamethylbenzene molecule, a typical molecule formed by carbon, was obtained by converting X-ray diffraction data into a visual signal. The magnification factor is 10^8.

The chemistry of the element carbon has special importance. Nearly all of the chemicals that make up living things consist mainly of carbon atoms, along with some atoms of hydrogen, oxygen, and nitrogen. While many other elements are essential to life, the starring role is played by carbon. Besides its important place in the chemistry of living things, carbon also features prominently in other familiar and important parts of our environment. Coal, oil, gas, most plastics, paint, paper, wood, and many other materials are rich in carbon and its compounds. Of course, for many of these their carbon content is due to their ultimate origin in living things. In order for us to understand many of the most important chemical processes in nature, we must become familiar with the basic chemistry of the element carbon.

Organic Compounds Are Made of Carbon Atoms

16-1 Sources of Carbon Compounds

Compounds that contain carbon constitute by far the largest single known group of chemical compounds. Many are produced in the "chemical factories" represented by living things, both plant and animal. Living organisms synthesize thousands of complex substances, including, among others, sugars, starches, proteins, hormones, antibiotics, waxes, oils, and dyes. Because of their origin in living organisms, such compounds are called organic compounds.

Chemists have learned to synthesize a large number of carbon-containing substances, too. By analogy with those compounds produced by living organisms as part of their life processes, those carbon-containing compounds created in the laboratory are also known as organic compounds. Thus, **organic** compounds are those that contain carbon.

The naturally-occurring sources of carbon important to the organic chemist are coal, natural gas, and petroleum. From these the chemist can isolate the substances necessary for the study and synthesis of the hundreds of thousands of new compounds that each year are added to the roster of known organic substances.

Coal. This form of carbon is found in varying quantities throughout the world. It is a black substance that apparently has been produced by decay of plant materials for countless numbers of years, dating back to prehistoric times. (See Figure 16-1, next page.) Built up gradually, layer by layer, these materials have been subjected to considerable extremes of temperature and pressure over the centuries, resulting finally in the deposits now identified as coal.

These deposits are not of uniform composition. They range from a very hard variety known as anthracite coal, which can contain as much as 95 percent pure carbon, to bituminous coal. The latter is a much softer variety containing considerably less pure carbon and characterized by the presence of larger quantities of volatile impurities. Peat represents a relatively early stage of coal formation. It contains an abundance of varying kinds of impurities, including much unchanged plant material. Frequently it contains as little as 50 percent carbon.

Figure 16-1 *Primeval forests similar to this one provided most of the material found in present coal and oil deposits.*

When substances are heated to high temperatures in the presence of oxygen (as in air), they frequently combine with the oxygen. Coal and other substances containing combustible organic materials burn in the presence of air. Carbon dioxide and water are produced when coal undergoes complete combustion.

A process known as *destructive distillation* involves the heating of coal, or other substances, to high temperatures in the absence of air. Coal decomposes under these conditions, and a series of more volatile components distill away, leaving a solid residue, coke. Coke is essentially pure carbon and contains the nonvolatile mineral substances present in the original mixture. It is widely used in the extraction of certain metals from their ores. The most important of these processes is the conversion of iron oxides to iron in the manufacture of steel.

The volatile components distilled off in the destructive distillation of coal consist primarily of two mixtures: coal gas and coal tar. These mixtures contain a variety of organic compounds that serve as important raw materials for the production of the thousands of organic compounds vital to modern chemical industries.

The destructive distillation of wood produces a relatively pure form of carbon called charcoal and a series of volatile components different from those produced from coal.

Natural Gas and Petroleum. Other important sources of organic raw materials for the chemical industries include underground deposits of natural gas and of petroleum

Figure 16-2 *A fractional distillation tower used for separating components of crude oil.*

(sometimes known as crude oil). Natural gas is a mixture of compounds and is found in certain sandstone or other porous rock formations. As much as 90 percent of natural gas may be methane (marsh gas). The gas used for cooking and heating consists mainly of this colorless, odorless methane. The distinctive odor of the gas used in homes is due to a sulfur-containing compound added as an impurity before the gas is distributed for commercial consumption. This addition of an impurity with a distinctive odor is required by law and was brought about because of accidents resulting from the undetected escape of the odorless methane gas.

Petroleum is a mixture of organic compounds ranging from relatively light volatile liquids to heavier, viscous, tarlike substances. Crude oil serves as the major source of the gasoline used in internal combustion engines. Modern gasolines are mixtures of compounds resulting from the fractional distillation of crude oil. *Fractional distillation* is a process in which a mixture of liquids is boiled and the most volatile components are collected first. As the temperature rises, the less volatile components are boiled off and collected. Figure 16-2 shows a fractional distillation tower used for separating components of crude oil.

Many additives have been developed by industrial chemists to improve the smooth-burning (antiknock) qualities of gasoline. One of the best known is tetraethyl lead, $Pb(C_2H_5)_4$. Gasoline can also be made to burn more smoothly by changing the shapes and sizes of the molecules in gasoline produced by fractional distillation.

Many industrial processes have been developed to increase the yield of gasoline from petroleum. These include "polymerization" (in which the very smallest molecules are combined to make larger ones), and "cracking" (in which the larger molecules are broken down into the medium sizes desired in gasoline).

As you have learned, the chemistry of carbon is generally referred to as organic chemistry. The chemistry of all of the remaining elements is called **inorganic** chemistry. However, there is no sharp dividing line between organic and inorganic chemistry. A few of the compounds containing carbon are usually considered to belong to the field of inorganic chemistry rather than to that of organic chemistry. These include carbon dioxide, CO_2, carbon monoxide, CO, and the carbonates and cyanides, examples of which are sodium carbonate, Na_2CO_3, and sodium cyanide, $NaCN$. However, these inorganic substances represent an almost insignificant fraction when compared to the vast number of carbon-containing compounds of organic chemistry.

16-2 Composition and Structure of Organic Compounds

The unique ability of carbon to form great numbers of compounds is primarily due to two factors. (1) One factor is the structure of the carbon atom itself. (2) The other factor is the ability of carbon atoms to form long chains containing bonds between carbon atoms.

As you have already learned in the immediately preceding chapters, the carbon atom has four electrons available for forming bonds. When each of these electrons becomes paired with an electron from another atom, four single bonds are formed. The carbon atom is then surrounded by four electron pairs. These electron pairs arrange themselves in space so as to maximize the average distance between them. As a result the four bonds are directed toward the vertices of a tetrahedron. A carbon atom can also form double and triple bonds to other atoms, including other carbon atoms. This ability creates still more possibilities for structural variations.

The ability of carbon atoms to form strong bonds to each other is a unique property. For all other elements, the formation of bonds to different kinds of atoms is at least as favorable as the formation of bonds to other atoms of its own kind. Only carbon has the characteristic property of forming molecules containing long chains, rings, and complex three-dimensional arrays made up of the same type of atom.

Almost all organic compounds contain hydrogen atoms in addition to carbon atoms. Atoms of oxygen, sulfur, and nitrogen also appear in the chemical formulas of many important organic substances.

With all of the preceding possibilities in mind, you can readily see that in identifying and understanding an organic compound, the chemist is faced with a complex problem. He must determine not only what kinds of atoms are present, but also the total number of each kind of atom present in each molecule, and, finally, the structural arrangement of the atoms within that molecule. These three problems are otherwise described as determining the *empirical formula*, the *molecular formula*, and the *structural formula*.

16-3 Determining Empirical and Molecular Formulas

Chemists determine the empirical formula by experiments that reveal the relative number of atoms of each element present in the molecule. For example, analysis of an organic compound called ethane reveals that it is made up entirely of carbon and hydrogen. Further experiments show that the molecule contains one carbon atom for every three hydrogen atoms. Hence the simplest possible formula for a molecule of ethane is CH_3. This is its empirical formula.

A type of experiment that can be used to determine the empirical formula of ethane involves burning a sample of ethane in oxygen. The following results are obtained when all of a 30-gram sample of ethane is burned.
1. Eighty-eight grams of CO_2 are formed. Since the molar mass of CO_2 is 44, the 88 grams represent 88/44 or 2 moles of CO_2. Hence, there must have been two moles of carbon in the ethane sample.
2. Fifty-four grams of H_2O are formed. Since the molar mass of H_2O is 18, the 54 grams represent 54/18 or 3 moles of H_2O. Hence, there must have been six moles of hydrogen atoms in the sample of ethane.

There were two moles of carbon atoms and six moles of hydrogen atoms in the ethane sample. It follows that in the compound ethane there must be one carbon atom for every three hydrogen atoms. The empirical formula must be CH_3.

We have seen that ethane contains carbon and hydrogen, and that they are present in a ratio of three hydrogen atoms to one carbon atom. But we have not learned how many carbon and hydrogen atoms there are *per molecule*. The empirical formula only tells us that *if* there is one carbon atom per molecule, then there *must* be three hydrogen atoms per molecule. Similarly, if there are two carbon atoms per molecule, there must be six hydrogen atoms per molecule, and so on. Thus, the molecular formula could be CH_3, C_2H_6, C_3H_9, or any other multiple of the empirical formula. Even $C_{100}H_{300}$ is a possibility, though admittedly a rather remote one.

In order to find the molecular formula, we must also know the *molar mass*. In the case of ethane we are dealing with a substance that is a gas at room temperature. Hence it is possible to determine the molar mass by measuring the volume of a known mass of the gas at a fixed pressure and temperature. Such an experiment, and the calculation of results from it, is similar to experiments that you actually carried out earlier in this course.

Ethane has a molar mass of 30.0. Since the empirical formula, CH_3, would give a mass of only 15.0, it is clear that this is not the actual formula of the molecule. Instead, it must be multiplied by a factor of 30.0/15.0 = 2. Thus the molecular formula of ethane is C_2H_6. Figure 16-3 shows a molecular model and the structural formula of ethane.

Consider the following example, which shows a calculation of the molar mass of ethane. It was found by experiment that a 12.2-gram sample of ethane occupies a volume of 10.0 liters at 27°C and 1.0 atm.

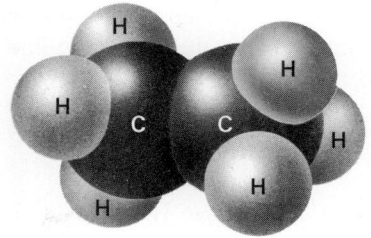
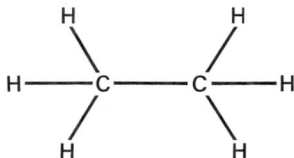

Figure 16-3 A space-filling model of ethane, C_2H_6, and its structural formula. The molecular formula of a compound must be known before its structural formula can be determined.

Example

$m = 12.2$ g $\qquad T = 27°C$ (300 K)
$P = 1.00$ atm $\qquad V = 10.0$ liters
$R = 0.0820 \dfrac{\text{liter-atm}}{\text{mole-K}}$

$PV = NRT$

$N = \dfrac{PV}{RT} = \dfrac{(1.00)(10.0)}{(0.0820)(300)}$

$N = 0.406$ mole ethane

Since 0.406 mole of ethane has a mass of 12.2 grams, then one mole of ethane must have a mass of 12.2/0.406 = 30.0 g.

In order to understand the chemistry of organic compounds, it is absolutely essential to know and understand the *structures* of the molecules as well. However, any consideration of molecular structure must begin with the knowledge of how many atoms of each kind make up the molecule. In other words, we must know the molecular formulas of the substances we deal with. That is why we have stressed the difference between the empirical and molecular formulas and have indicated the importance of knowing the molar mass as well as the empirical composition of organic compounds.

Summary

The element carbon forms an enormous number of compounds that are essential to organic (life) processes. As a result, the chemistry of most of the compounds formed by the element carbon is called organic chemistry.

In addition to its occurrence in the molecules of living things, carbon is found in other places in nature, most notably in coal, which is practically pure carbon, and in petroleum. Both of these substances have formed beneath the surface of the earth over thousands of years by the operation of high temperatures and pressures on what was formerly living matter. Both the large number and the complexity of organic compounds are due chiefly to two facts. Carbon can form a tetrahedral set of four single bonds as well as double or triple bonds. Also, it has a unique capacity to form bonds to other carbon atoms. Thus, long chains, rings, and other kinds of arrays of carbon atoms are formed.

Organic compounds practically always contain hydrogen and usually also contain oxygen or nitrogen, as well as some other elements.

The molecular formula of an organic compound is very important. It is not usually the same as the empirical formula, which simply gives the relative numbers of the various atoms present. The molecular formula can be found if the empirical formula and the molar mass are known.

Self Check

1. What is the distinction between organic and inorganic chemistry?
2. A compound is found to have the empirical formula CH_2 and a molar mass of 70.0. What is the molecular formula of the compound?
3. What two features of carbon account for its ability to form the large number of organic compounds?

Hydrocarbons Provide the Framework of Organic Compounds

16-4 Nomenclature

The number of known organic compounds already runs well over a million, and new compounds are being synthesized every day. If such a large body of information is to be studied with any ease or efficiency, some system for distinguishing among existing compounds and for naming new ones is essential. The problem of establishing a naming system (nomenclature) for organic compounds is further complicated by the fact that many of the earliest compounds discovered were given names that are still in common use. These names are entirely arbitrary. That is, they do not fit into any systematic scheme of classification. The serious student of organic chemistry is left with no choice but to become acquainted with two sets of names. There are names derived from the presently accepted logical systems of organic nomenclature in addition to the older, sometimes more familiar, names that have been in use over the years. For example, the organic compound heretofore called ethanol is the familiar ethyl alcohol, or grain alcohol, present in alcoholic beverages, food flavoring, and many other substances.

What are some other organic compounds, and how were they named? Briefly, all organic compounds fall within two broad divisions. One includes the various **hydrocarbons,** compounds composed only of carbon and hydrogen atoms. The other division consists of "derivatives" of the hydrocarbons, compounds in which one or more hydrogen atoms have been replaced by other atoms or groups of atoms. In the sections that follow, you will be introduced to some of the members of the vast family of organic compounds and to some aspects of their interesting and characteristic chemical behavior.

16-5 *Saturated Hydrocarbons*

Compounds consisting solely of carbon and hydrogen atoms bonded together by *single bonds only* are called **saturated** hydrocarbons. Such compounds are also known as the **alkane** series or the **paraffin** series. As so much of the nomenclature for organic compounds is derived from the names of the members of this series, it would be desirable for you to become very familiar with the names and formulas (next column) of the first ten alkanes. Figures 16-4 and 16-5 (next page) show space-filling models and structural formulas for some of these alkanes.

methane	CH_4	hexane	C_6H_{14}
ethane	C_2H_6	heptane	C_7H_{16}
propane	C_3H_8	octane	C_8H_{18}
butane	C_4H_{10}	nonane	C_9H_{20}
pentane	C_5H_{12}	decane	$C_{10}H_{22}$

Figure 16-4 shows the structural formulas and models of two saturated hydrocarbons. Beginning with butane, these hydrocarbons exist as isomers. The possibilities for isomerism arise because a carbon atom can form 1, 2, 3, or even all 4 of its single bonds to other carbon atoms rather than to hydrogen atoms. Obviously, the nomenclature required to distinguish between a large number of isomers is fairly complex. We shall not try to deal with it here, except for two simple rules that will be useful later. The isomers containing the completely "straight" chains of carbon atoms are called the normal (*n*) isomers, such as *n*-pentane in Figure 16-5a. Those isomers that contain only one branch at the end of the chain are called the *iso* isomers. An example of such an isomer is isopentane, which is shown in Figure 16-5b.

It can be seen that the saturated hydrocarbons we have just considered all have formulas in which there are twice as many hydrogen atoms as carbon atoms, plus two. It is easy to see why this is so by looking at the normal isomers in each case. Each carbon atom that is not on the end has two hydrogen atoms. Each of the end carbon atoms must have not only two hydrogen atoms but a third one to "end" the chain.

Suppose, however, that we have a chain of CH_2 groups, and instead of finishing off each end with one hydrogen atom we curl the chain into a loop and allow the two end carbon atoms to form a bond between them. This satisfies the combining capacities of all the atoms and should make a stable molecule. In fact, it does. Such molecules meet all the

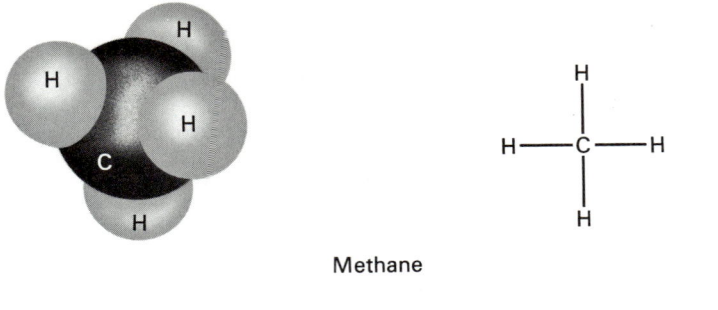

Methane

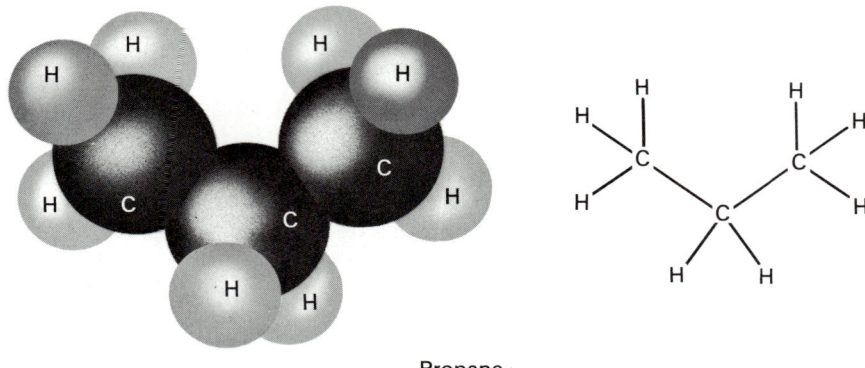

Propane

Figure 16-4 *Space-filling models and structural formulas for methane, CH_4, and propane, C_3H_8.*

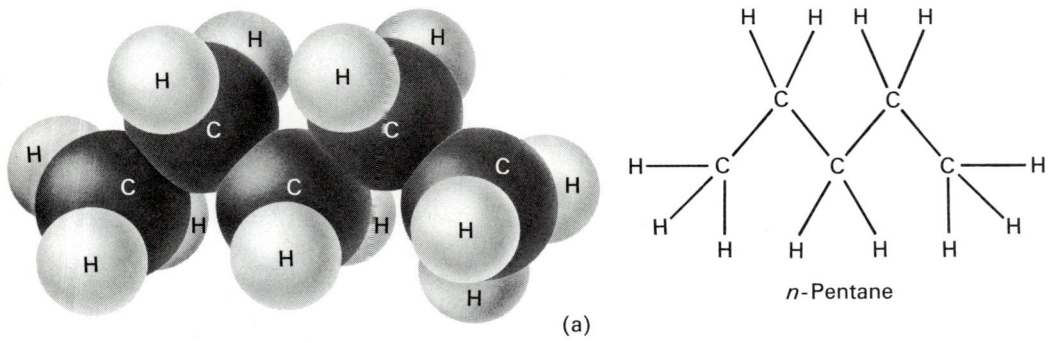

n-Pentane

(a)

Figure 16-5a *Space-filling model and structural formula for n-pentane, C_5H_{12}.*

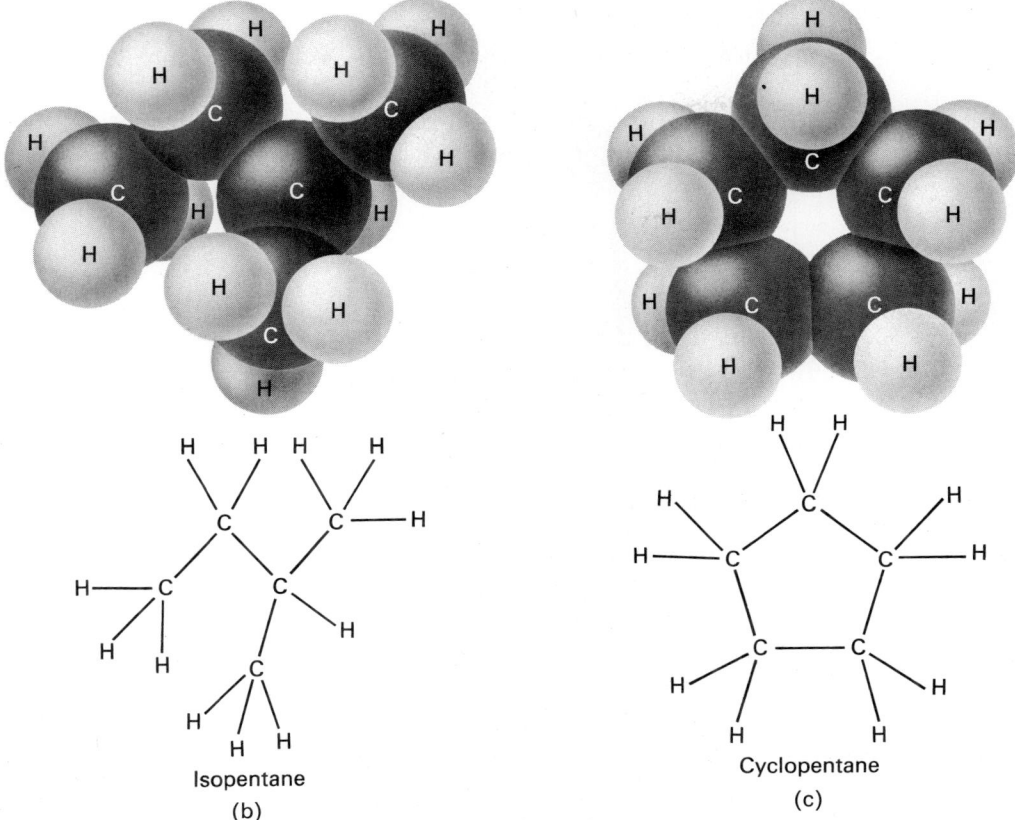

Figure 16–5b,c *Space-filling models and structural formulas for isopentane, C_5H_{12}, and cyclopentane, C_5H_{10}.*

requirements for being considered saturated hydrocarbons: they contain only C and H atoms and have only single bonds between C atoms. To distinguish them from the straight- and branched-chain compounds, which have formulas of the type C_nH_{2n+2}, we use the prefix *cyclo*. Thus the cyclic compound shown in Figure 16-5c is called cyclopentane. Note carefully that its formula is C_5H_{10}, whereas the formulas for the noncyclic pentanes are C_5H_{12}. Cyclopentane is *not* an isomer of the others. The cyclic saturated hydrocarbons are different from the noncyclic ones. They have a different type of general formula: C_nH_{2n}.

The preferred general name for the entire class of saturated hydrocarbons is *alkanes*. (For the cyclic ones the term is *cycloalkanes*.) This is a simpler and more systematic name than "saturated hydrocarbon," and it will fit into a system of naming that we shall develop as we go along.

The alkanes are the principal compounds present in natural gas and in petroleum.

TABLE 16-1 Some Properties of Saturated Hydrocarbons (Alkanes)

Saturated Hydrocarbon	Molecular Formula	Melting Point (°C)	Boiling Point (°C)
methane	CH_4	−182.5	−161.5
ethane	CH_3CH_3	−182.8	−88.6
propane	$CH_3CH_2CH_3$	−189.9	−44.5
n-butane	$CH_3CH_2CH_2CH_3$	−138.3	−0.5
isobutane	$CH_3\text{—}CH\text{—}CH_3$ $\quad\quad\;\; |$ $\quad\quad\; CH_3$	−159.6	−11.7
n-hexane	C_6H_{14}	−95.3	68.7
cyclohexane	C_6H_{12}	+6.6	80.7
n-octane	C_8H_{18}	−56.5	125.7
n-octadecane	$C_{18}H_{38}$	+28.2	306.0

Under normal conditions, the alkanes with low molar masses are gases. Their boiling points are given in Table 16-1. Gasoline is composed mainly of highly branched alkanes containing from six to ten carbon atoms. Paraffin waxes are usually alkanes having from 20 to 35 carbon atoms.

The saturated hydrocarbons are relatively inert, except at high temperatures. Combustion is almost the only important simple reaction of the alkanes. This reaction, however, makes the hydrocarbons one of our most important energy sources.

The absence of reactivity of saturated hydrocarbons is an important aspect of their chemistry. This inertness accounts for the fact that the chemistry of organic compounds is mainly concerned with so-called *functional* groups. These groups of atoms are usually much more reactive than the carbon "skeleton" to which they are attached. Thus it can be assumed that the skeleton will remain intact and unchanged throughout the reaction. The more important functional groups and some of their chemistry will be discussed later in this chapter.

INVESTIGATION

16-6 *Some Properties and Reactions of Hydrocarbons*

In the preceding section you learned that compounds composed of only carbon and hydrogen are called hydrocarbons. If all the carbon atoms in a hydrocarbon molecule are connected by single bonds, the compound is called a saturated hydrocarbon. The more carbon atoms there are in a hydrocarbon molecule, the greater the number of isomers. Hydrocarbons containing at least one double or triple bond are called unsaturated hydrocarbons. They will be discussed in more detail in Section 16-7.

PURPOSE

In this experiment you will examine the properties and reactions of some saturated hydrocarbons and their isomers. You will also compare the properties and reactions of unsaturated hydrocarbons, hydrocarbons with a linear arrangement of carbon atoms, and hydrocarbons with carbon atoms arranged in a ring.

MATERIALS

Thermometer, $-10°C$ to $150°C$
Ringstand and ring
Wire gauze
Burner
Beaker, 100-ml
Test tube, 18- $\times$ 150-mm
Rubber stoppers for small test tubes
2 utility clamps, one with three prongs

6 test tubes, 13- $\times$ 100-mm
6 ml 0.1 M Br_2 in CCl_4
Test tube rack
Wax pencil or labels
6 boiling stones
50 ml vegetable oil

5 ml of one of the following, as directed by your teacher
   n-heptane   2,2-dimethylpentane
   n-octane   2,2,4-trimethylpentane
   2-methylhexane n-hexane

1 ml of each of the following
   n-hexane   cyclohexene
   1-hexene   benzene
   cyclohexane   toluene

3 ml of each of the following solutions
   0.01 M $KMnO_4$ 6 M NaOH

(M indicates the concentration, in moles per liter. Concentration is the amount of a specified substance in a unit amount of another substance. This concentration unit is discussed in Chapter 17.)

Special Precautions. Hydrocarbons are very flammable. Keep the dispensing bottles away from your laboratory desk and burners. Bring to your work area only what you need to work with for one test at a time. There is little chance for your boiling point liquid to ignite if you extinguish your burner as soon as your liquid boils. If the liquid does ignite, merely blow it out.

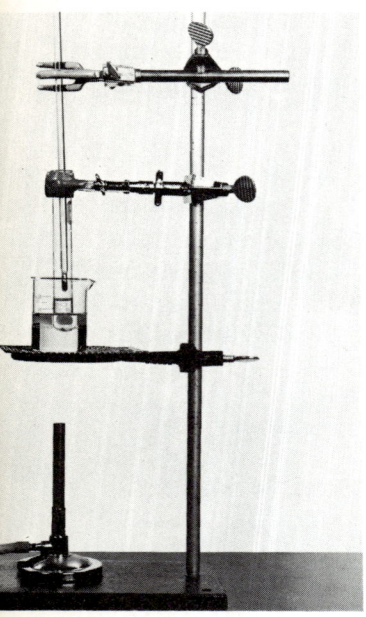

Figure 16-6

PROCEDURE

Part A.

1. Place about 50 ml of vegetable oil in a 100-ml beaker. Set the beaker on a wire gauze on a ring over a burner.

2. Place about 5 ml of a hydrocarbon assigned to you in an 18- × 150-mm test tube. Drop a single boiling stone into the liquid in the test tube.

3. Clamp the test tube to the ringstand with a utility clamp. The lower end of the test tube should be immersed about 1-2 cm in the vegetable oil.

4. Use a three-prong clamp to suspend a thermometer ($-10°$C to $150°$C) inside the test tube. Be sure that the lower part of the bulb is about 1 cm above the surface of the liquid in the tube. Figure 16-6 shows the assembled apparatus.

5. Heat the beaker containing the vegetable oil until the liquid in the test tube just starts to boil. Turn off the burner at once. Read the thermometer while the liquid is boiling vigorously. Record the result.

6. Remove the thermometer from the test tube. Wipe the oil from the outside of the test tube. Then run water over the outside of the test tube to cool its contents. Pour the cool liquid into the waste receptacle designated by your teacher.

7. Report your result to the rest of the class. Be sure to record the boiling points determined by other class members. You should have at least one boiling point for each of the following compounds: 2,2-dimethylpentane, 2-methylhexane, *n*-heptane, 2,2,4-trimethylpentane, *n*-hexane, and *n*-octane.

Part B.

1. Label a clean, dry 13- × 100-mm test tube for each of the hydrocarbons to be tested: *n*-hexane, 1-hexene, cyclohexane, cyclohexene, benzene, and toluene. Add about 10 drops of the appropriate hydrocarbon to each test tube.

2. Prepare about 6 ml of a 0.005 M alkaline potassium permanganate solution by adding 3 ml of 0.01 M $KMnO_4$ to 3 ml of 6 M NaOH.

3. Add 20 drops (about 1 ml) of this solution to each of the test tubes containing the different hydrocarbons. Place a stopper in each test tube and shake the contents gently to obtain greater contact between the two phases. Note any changes in the color of the contents after about one minute. To help you see a color change, place a sheet

of white paper behind the test tube. To discard the solutions, pour them into the center of the drain and flush with large amounts of running water.

4. Place about 10 drops of each of the same six hydrocarbons into six properly-labeled clean test tubes.

5. Add about 20 drops (about 1 ml) of 0.1 M Br_2 in carbon tetrachloride, drop by drop, to each test tube. Place a stopper in each test tube and shake the contents occasionally as you add the bromine solution. Note any changes in color. Continue adding the bromine to those hydrocarbons showing a color change until the bromine color persists.

RELATED PROBLEMS

1. Write structural formulas for the six compounds used in the boiling point determinations. Recall that the n prefix in front of heptane and others stands for "normal." It means that these are "linear," unbranched molecules. The numerical prefixes denote the position of the carbon atom on which a group is placed. The compound 2,2-dimethylpentane has two methyl groups (CH_3—) attached to the second carbon atom of a pentane molecule.

2. What relationship exists between the size of the linear (n) molecules and their boiling points?

3. Which of the six compounds are isomers of each other? How do the boiling points of branched molecules compare with the boiling point of their linear isomer?

4. Write structural formulas for the hydrocarbons tested with $KMnO_4$ and Br_2. The molecular formulas for cyclohexane, cyclohexene, and benzene are C_6H_{12}, C_6H_{10}, and C_6H_6, respectively. Each molecule has six carbons connected in a ring. Place the hydrogen atoms and bonds in such a way that each carbon atom has four bonds attached to it. Toluene is very similar to benzene, but has one CH_3— group attached to a carbon atom in place of an H atom. The compound 1-hexene, with the molecular formula C_6H_{12}, has two less hydrogen atoms than n-hexane, C_6H_{14}. The numerical prefix denotes the first carbon in the chain having a double bond attached to it.

5. Which of the hydrocarbons reacted readily with the alkaline solution of $KMnO_4$? Which reacted with the bromine solution?

6. From your results, what can you conclude about the reactivity of saturated hydrocarbons? about the reactivity of unsaturated hydrocarbons?

Questions To Wonder About

1. What factors might account for the differences between the boiling points of the linear (n) isomers and the branched-chain isomers of compounds having the same molecular formulas?

2. How might you explain the fact that all unsaturated hydrocarbons with a ring structure do not react similarly with $KMnO_4$ and Br_2?

16-7 Unsaturated Hydrocarbons

As you learned in Section 16-6, there are unsaturated hydrocarbons, as well as saturated ones. **Unsaturated** hydrocarbons contain double or triple bonds, or both. The meaning of "unsaturated" will be explained later in this section.

The naming of compounds with double and triple bonds is systematic, just as in the case of saturated hydrocarbons. The compounds containing a double bond are called **alkenes**. The *-ene* ending is an indication that they are unsaturated compounds. (The names of unsaturated compounds that contain triple bonds end in *-yne*.) Table 16-2 shows the first few alkenes and their systematic names. Note that these follow very closely the systematic names for the alkanes. The first of the alkenes, shown in Figure 16-7, has the systematic name ethene. Its traditional name, however, is ethylene. The traditional name is so commonly used by chemists that it should be remembered. Generally, the simplest compounds of a given class have names that were first used long before there was any systematic set of names. Their traditional names are still commonly used.

Hydrocarbons containing a triple bond are called **alkynes**. Table 16-3 lists the simplest

TABLE 16–2 Some Alkenes and Their Properties

Name	Molecular Formula	Melting Point (°C)	Boiling Point (°C)
ethene (ethylene)	$H_2C=CH_2$	−169.2	−104.0
propene	$H_2C=CHCH_3$	−185.2	−47.8
1-butene	$H_2C=CHCH_2CH_3$	−185.4	−6.3
cis-2-butene	H_3CCH_3 $C=C$ HH	−138.9	3.7
trans-2-butene	H_3CH $C=C$ HCH_3	−105.6	0.9
isobutene	$H_3C-C=CH_2$ $\|$ CH_3	—	−6.6
1-dodecene	$H_2C=CH(CH_2)_9CH_3$	−35.2	213.4

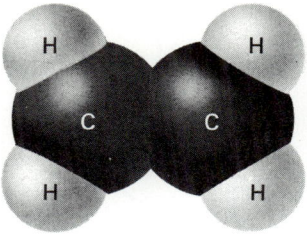

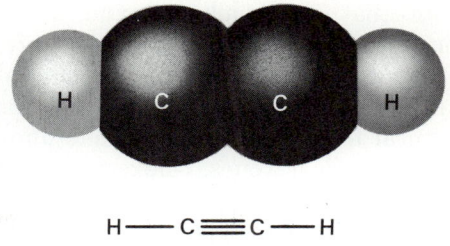

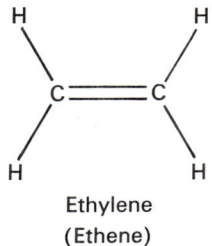

Ethylene
(Ethene)

Figure 16-7 *Space-filling model and structural formula for ethylene (or ethene), C_2H_4.*

Acetylene

Figure 16-8 *Space-filling model and structural formula for acetylene, C_2H_2.*

alkynes. Once again, the first member of the series has a name that does not fit into the system, but is the one that nearly everyone uses. This compound, HC≡CH, is called acetylene. Its structure is shown in Figure 16-8. One of its common uses is in welding torches. Such torches are often called oxyacetylene torches, because they burn acetylene in a stream of pure oxygen. This produces an extremely hot flame capable of melting all forms of iron and steel.

The possibilities for isomerism among the alkenes are even greater than among the alkanes. Consider, for example, the four isomers of butene, C_4H_8, given in Table 16-2. You can see first of all that the position of the double bond in the chain can vary. One possibility is 1-butene, the number 1 indicating that the double bond is between an end carbon atom (1) and its neighbor (2). When the double bond is between carbon atom number 2 and its other neighbor (3), the molecule is called 2-butene. You can see further, that there are *two* possible 2-butene's due to *cis-trans* isomerism. Finally, there is an isomer called isobutene in which the carbon chain is branched rather than straight.

TABLE 16-3 Some Alkynes and Their Properties

Name	Molecular Formula	Melting Point (°C)	Boiling Point (°C)
ethyne (acetylene)	HC≡CH	−81.8	−83.6*
propyne	HC≡CCH₃	−102.7	−23.2
2-pentyne	CH₃C≡CCH₂CH₃	−101.0	55.5

*Goes directly from solid to vapor (*sublimes*) at a pressure of 1 atm.

Note that cyclic alkanes and noncyclic alkenes are isomers. For example, C_6H_{12} is the formula for both cyclohexane and all of the isomeric hexenes. Figure 16-9 shows the structural formulas of cyclohexane and two of the isomeric hexenes, 1-hexene and 2-hexene.

Alkynes display a type of isomerism similar to that of the alkenes, due to the placement of the triple bond in the carbon chain. The compound 2-pentyne is given in Table 16-3. One of its isomers is 1-pentyne. Alkynes have no *cis* and *trans* isomers, however. As shown in the 2-pentyne formula, the two bonds formed by a —C≡C— group are both in line with the C≡C bond. The reason for this linearity of alkynes was explained in Section 15-7.

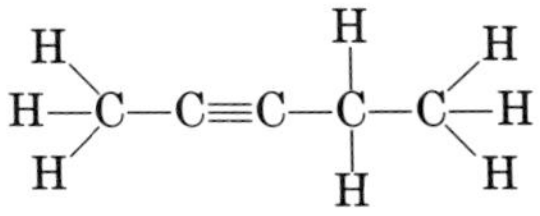

2-pentyne

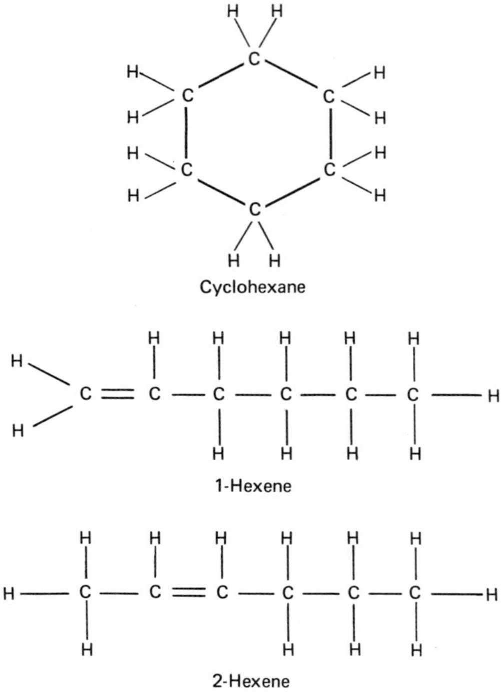

Figure 16-9 *Structural formulas of some isomers having the molecular formula C_6H_{12}.*

Why are alkenes and alkynes called unsaturated hydrocarbons? In contrast to the relatively inert alkanes, unsaturated hydrocarbons are quite reactive. Their most characteristic type of reaction is *addition* of other atoms to the double or triple bonds. It is as if one of the two components of the double bond and either one or two components of the triple bond can "open up." The carbon atoms can then form bonds to additional atoms. The following equations are examples of this type of reaction for ethylene and butyne. Because alkenes and alkynes have the ability to combine with still more atoms, they are called unsaturated. Thus, *un*saturated hydrocarbons have room for additional atoms, while saturated hydrocarbons do not.

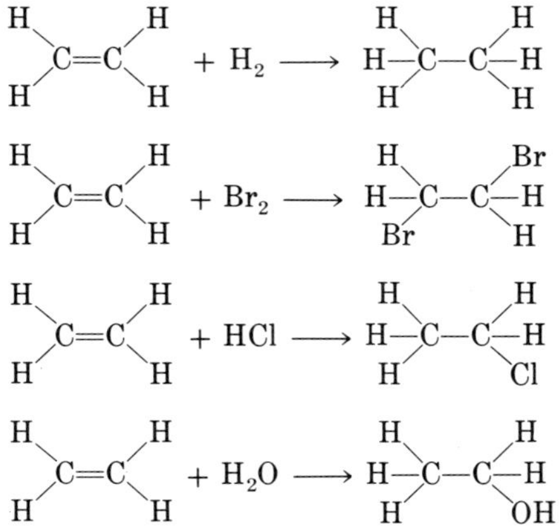

$$H_3C-C\equiv C-CH_3 + H_2 \longrightarrow \underset{H}{\overset{H_3C}{>}}C=C\underset{CH_3}{\overset{H}{<}} + \underset{H}{\overset{H_3C}{>}}C=C\underset{H}{\overset{CH_3}{<}}$$

$$H_3C-C\equiv C-CH_3 + 2H_2 \longrightarrow H_3C-\underset{H}{\overset{H}{C}}-\underset{H}{\overset{H}{C}}-CH_3$$

$$H_3C-C\equiv C-CH_3 + HCl \longrightarrow \underset{Cl}{\overset{H_3C}{>}}C=C\underset{H}{\overset{CH_3}{<}} + \underset{Cl}{\overset{H_3C}{>}}C=C\underset{CH_3}{\overset{H}{<}}$$

16-8 Systematic Names for Noncyclic Hydrocarbons

Noncyclic hydrocarbons are named in a systematic way, as Table 16-4 indicates. Except for the few nonsystematic names such as ethylene and acetylene, there is a consistent pattern of stems used; only the endings vary. The stem tells you how many carbon atoms are present, and the ending tells you whether the compound is an alkane, alkene, or alkyne. The same stems are used in naming compounds that have atoms of other elements attached to a hydrocarbon "skeleton." Knowing the stems will aid you in naming and in recognizing by name both hydrocarbon derivatives and the hydrocarbons themselves.

Note that, beginning with five carbon atoms, the stems are all systematically derived from the Greek or Latin words for the

TABLE 16-4 Summary of Names for Hydrocarbons

Number of Carbon Atoms	Stem	Alk*ane*	Alk*ene*	Alk*yne*
1	meth-	methane	—	—
2	eth-	ethane	ethene (ethylene)	ethyne (acetylene)
3	prop-	propane	propene	propyne
4	but-	butane	butene	butyne
5	pent-	pentane	pentene	pentyne
6	hex-	hexane	hexene	hexyne
7	hept-	heptane	heptene	heptyne
8	oct-	octane	octene	octyne
9	non-	nonane	nonene	nonyne
10	dec-	decane	decene	decyne

Friedrich A. Kekulé von Stradowitz (1829–1896)

AT AN EARLY AGE Friedrich Kekulé had decided to study architecture. However, during his student days he attended several lectures by the famous German chemist Liebig, and was so impressed that he decided to become a chemist himself. His initial training in architecture was not wasted. It had generated in Kekulé a habit of always considering the form and structure of materials with which he worked.

After a short stay at the University of Heidelberg, Kekulé moved to the University of Ghent in Belgium where he became a professor of chemistry in 1858. He was dissatisfied with the practice of representing compounds by their molecular formulas only, because these formulas told nothing about the structure of the molecules. He proposed that the tetravalent carbon atom could connect with other carbon atoms to form chains. He also suggested that the bonds between atoms could be represented by short dashes. Although a Scottish chemist by the name of Couper made a similar proposal at about the same time, the formulas came to be called Kekulé formulas.

Kekulé's formulas made it possible to represent complicated chemical compounds simply. Before the advent of these structural formulas, there had been no way to explain isomers. Now isomers could be distinguished clearly. Despite many recent ideas about chemical bonding, these structural formulas have been a valuable tool for chemists and are still being used in the same form more than one hundred years later.

While the molecular formula for benzene, C_6H_6, had been known for some time, Kekulé was unable to use his formulas to illustrate the structure of this common substance. One day while dozing in front of his fireplace in his room in Ghent, he dreamed of snakelike chains of atoms dancing and whirling in front of his eyes. He reported later, "Suddenly one of the serpents caught its own tail and the ring thus formed whirled exasperatingly before my eyes. I woke as by lightning, and spent the rest of the night working out the logical consequences of the hypothesis." The result of this dream was the now familiar ring-shaped molecule of benzene.

Kekulé's fame made it possible for him to organize the First International Chemical Congress in Karlsruhe in 1861. He was one of the first chemists to define organic chemistry as the chemistry of carbon compounds, a concept that broadened the boundaries of that science beyond merely the study of compounds produced by living organisms.

numbers, many of which should already be familiar to you. For example, a *hex*agon is a *six*-sided polygon, an *oct*et is a composition for *eight* musical instruments (or is a group of *eight* instruments that play the music), and a *dec*ade is a period of *ten* years.

16-9 Aromatic Hydrocarbons: Benzene and Its Derivatives

Another important group of compounds, quite different from the ones just described, are those unsaturated compounds called aromatic hydrocarbons. The simplest example is the compound **benzene,** a cyclic compound with six carbon atoms in the ring, having the formula C_6H_6. The benzene ring is planar and has the shape of a regular hexagon as indicated in the following structural formula.

A difficulty arises when you attempt to represent the bonding in benzene and its derivatives. You might do it in either of the following ways.

Both of these structures satisfy the formal valence rules for carbon, but there is a serious fault. Each structure shows three of the carbon-carbon bonds as double bonds, and three as single bonds. There is a wealth of experimental evidence to indicate that this is not true. Instead, it is found that any one of the six carbon-carbon bonds in benzene is the same as any other.

This situation should sound somewhat familiar. Exactly the same kind of problem was encountered with the nitrate ion, in Chapter 15. Here again, no single arrangement of bonds is capable of properly representing the true structure and electron distribution in the molecule.

There are two ways to handle such a situation. One way is to use the method of resonance. Thus, you can say that benzene is to be represented by neither one of the above formulas alone, but rather by both of them at once. The true distribution of electrons is a resonance hybrid of both.

Another way to show the electron distribution is by using "molecular orbitals," rather than atomic orbitals. In the atomic orbital model of benzene, each of the three electron pairs forming the second bond in the double bonds are located between one particular pair of carbon atoms. In the molecular orbital model, the electrons are spread uniformly over the entire hexagon of carbon atoms in two rings lying above and below the plane of the hexagon, as shown in Figure 16-10 (Molecular orbitals are discussed in Section 15-9.)

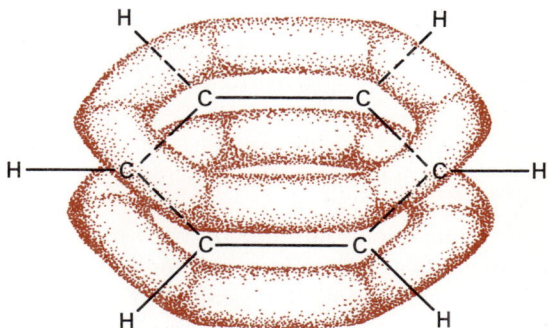

Figure 16-10 *The shaded areas show how electrons are spread, in donut-shaped molecular orbitals, above and below the molecular plane of benzene.*

Whichever model, resonance or molecular orbital, you choose to account for the uniform spread of electrons around the entire benzene ring, you need a convenient way to represent the electron distribution when writing formulas. Chemists usually draw a hexagon to stand for the six carbon atoms and the attached hydrogens, and draw a circle in the center to indicate the distribution of electrons over and under the ring, as follows.

The Substitution Reactions of Benzene. Benzene shows neither the typical reactivity nor the usual addition reaction of ethylene. Benzene does not react with bromine to give a simple addition across a double bond. Instead, the reaction is as follows.

$$\text{C}_6\text{H}_6 + \text{Br}_2 \longrightarrow \text{C}_6\text{H}_5\text{Br} + \text{HBr}$$

bromobenzene

In this reaction, called *bromination*, one of the hydrogen atoms has been replaced by a bromine atom. It is important to notice that the characteristic electron distribution of the ring, as represented by the circular symbol, is not affected. Although this distribution of electrons is a kind of unsaturation, it is much less liable to react that way than the simple double and triple bonds of alkenes and alkynes.

The characteristic reaction of aromatic molecules such as benzene is *substitution* of the hydrogen atoms and not addition to the carbon atoms. Another example of the substitution reactions characteristic of benzene occurs with nitric acid and is called *nitration*.

$$\text{C}_6\text{H}_6 + \text{HONO}_2 \longrightarrow \text{C}_6\text{H}_5\text{NO}_2 + \text{H}_2\text{O}$$

The substitution reaction provides the main path by which a multitude of compounds are prepared by the organic chemist. It enables him to add functional groups to the benzene molecule. These groups can then, by further reaction, be made into still other groups, or be combined with some other element or compound. By means of such reactions a number of very important chemical compounds can be produced. Many are in daily use throughout the world. Several important benzene derivatives are identified in Table 16-5.

Summary

The tremendous number of known organic compounds (well over a million) obviously need to be named and classified in some systematic way. Since many organic compounds still retain the familiar (though arbitrary) names originally given them, it is often necessary to learn both the old and the presently accepted nomenclature.

One important category of organic compounds is the hydrocarbons, compounds containing only hydrogen and carbon atoms. This group can be subdivided into saturated and unsaturated hydrocarbons, according to the type of bonding holding the atoms together. The saturated hydrocarbons are relatively inert except at high temperatures. In

TABLE 16-5
Some Benzene Derivatives

Structure	Name	Use
(benzene ring with NH_2)	aniline	basic raw material for production of many dyes and drugs
(benzene ring with OH)	phenol	germicides and disinfectants
(benzene ring with $O-CO-CH_3$ and COOH)	aspirin	pain reliever
(benzene ring with OH, OCH_3, and CHO)	vanillin	flavoring
(benzene ring with CH_3CH_2O and $NH-CO-CH_3$)	phenacetin	pain reliever (in headache remedies)
(benzene ring with OH, OH para)	hydroquinone	photographic developer
(benzene ring with NH_2 and $CO-OCH_2CH_2N(C_2H_5)_2$)	"Novocaine" (procaine)	local anesthetic
(benzene ring with $HC=CH_2$)	styrene	monomer for preparation of polystyrene plastics

Percy Lavon Julian
(1899–1975)

been able to determine its structure or to prepare it synthetically. Julian learned that soybeans contain physostigmine, and decided to tackle the problem of synthesizing the drug.

Julian began to work on physostigmine as soon as he returned to the United States. Four years later, he and Joseph Pikl, who had accompanied him from Vienna, finally announced the structure and the means of synthesizing this complicated compound.

physostigmine

AFTER GRADUATING from the State Normal School for Negroes in Montgomery, Alabama, Percy Julian, grandson of a former slave, continued his studies at DePauw University in Indiana, from which he graduated in 1920 at the top of his class.

Despite his outstanding record at DePauw, no university was willing to award a graduate fellowship to a black student. Frustrated by this rejection, Julian decided to teach chemistry at Fisk University, in Nashville, Tennessee. Within two years he won a fellowship in chemistry at Harvard University, where again he achieved a brilliant record. At this point he applied for a position as teaching assistant, so that he could work toward his doctorate. Prejudice again stood in his way. Six years later, after several other teaching positions in colleges for black students, he was awarded a fellowship to study in Vienna and received his Ph.D. in 1931.

While in Vienna, Percy Julian became interested in physostigmine, a drug used in treating glaucoma. Glaucoma is a disease of the eye that produces enormous pressure within the eyeball. Physostigmine had been isolated and identified in 1865, but nobody had

Following this success, Percy Julian was appointed to the position of chief chemist and director of research for the Glidden Company, a large manufacturer of paints and varnishes. This appointment paved the way for the employment of other blacks.

Julian continued to work with soybeans and produced many useful substances from them, including hormones. Naturally-occurring hormones are produced by living organisms. These very important substances control many body functions, such as growth, reproduction, and metabolism. One of the hormones that Julian prepared was cortisone, which is used in the treatment of rheumatoid arthritis. The cortisone that Julian produced from cortexelone, a soybean compound, cost only a few cents per gram. Previously cortisone had been available only from natural sources, at a cost of several hundred dollars per gram.

contrast, the unsaturated hydrocarbons are quite reactive. They readily add atoms from molecules such as H_2, HCl, and Br_2.

A third subdivision is another unsaturated group, the aromatic hydrocarbons (benzene and its derivatives). These compounds have structures characterized by cyclic or ring formation of their atoms. The rings are conveniently represented by a hexagon with a circle inside. The circle implies that there is equal (in benzene) bonding between all adjacent pairs of carbon atoms. Aromatic hydrocarbons undergo chiefly substitution reactions, in which a hydrogen atom is replaced by another atom or group of atoms.

Self Check

1. Describe the bonding in a saturated hydrocarbon. How do you account for the relative inertness of this group of organic compounds?
2. Describe the bonding in an unsaturated hydrocarbon. How do you account for the relative reactivity exhibited by such molecules?
3. Draw the two formulas for benzene that show localized double bonds and state why neither of them is a good way to represent the bonding. Draw another symbol to represent the benzene molecule.

Functional Groups Determine the Behavior of Organic Molecules

16-10 General Characteristics of Functional Groups

Nearly all of the reactions and physical properties of ethanol, C_2H_5OH,

$$\begin{array}{c} H\ \ H \\ |\ \ \ | \\ H-C-C-OH, \\ |\ \ \ | \\ H\ \ H \end{array}$$

can be explained by the behavior of the —OH part of the molecule—its functional group. The reactions occur at the —OH group, while the other part of the molecule, C_2H_5, often remains intact. These reactions suggest that the ethanol molecule consists of two important parts. The part that changes during a chemical reaction is the **hydroxyl** group, —OH. The part that does not change is the **ethyl** group, the formula for which follows.

$$\begin{array}{c} H\ \ H \\ |\ \ \ | \\ H-C-C- \\ |\ \ \ | \\ H\ \ H \end{array}$$

This concept of the structural integrity of the hydrocarbon group is an important one in organic chemistry. It focuses attention on the groups that change, the **functional groups.** If the chemistry of a particular functional group is understood for one compound, it is a good assumption that very similar chemistry, in its main features, will be found for another compound containing this same functional group. For this reason, compounds with the hydroxyl group are given a family name, **alcohols.** The rest of the molecule (the carbon skeleton) has relatively little effect and remains intact throughout the reactions involving the functional group.

The following equations illustrate one way in which the hydroxyl group found in alcohols reacts.

$$CH_3CH_2OH + HBr \longrightarrow CH_3CH_2Br + H_2O$$
$$CH_3CH_2OH + HCl \longrightarrow CH_3CH_2Cl + H_2O$$
$$CH_3CH_2OH + HI \longrightarrow CH_3CH_2I + H_2O$$

Chemists say that the —OH group has been displaced and that the halogen atom has been substituted for it. You can see that the group CH_3CH_2— has remained intact in all of these reactions. You may recall an earlier reference to ethane, C_2H_6. Looking at the structural formula of ethane, you see that it is simply the CH_3CH_2— group attached to a hydrogen atom.

$$\underset{\text{ethyl group}}{H-\overset{\overset{H}{|}}{\underset{\underset{H}{|}}{C}}-\overset{\overset{H}{|}}{\underset{\underset{H}{|}}{C}}-} + -H \longrightarrow \underset{\text{ethane}}{H-\overset{\overset{H}{|}}{\underset{\underset{H}{|}}{C}}-\overset{\overset{H}{|}}{\underset{\underset{H}{|}}{C}}-H}$$

Compounds containing the ethyl group can be named accordingly. For example, C_2H_5Br is called ethyl bromide, and C_2H_5OH is called ethyl alcohol. Systematic names for the same compounds are bromoethane and ethanol, respectively. Ethyl bromide, ethyl alcohol, and the other ethyl compounds can be thought of as having been derived from ethane. This involves simply substituting for one of its hydrogens —Br, —OH, or some other functional group. Thus, these compounds are called **derivatives** of ethane. Ethane is considered the parent hydrocarbon for a whole series of related compounds.

The name "ethyl" is derived from the name of the parent hydrocarbon, ethane. In the same way, the name of the **methyl** group (CH_3—) is derived from methane, CH_4

(Figure 16-4). Of course, as the size of the hydrocarbon group increases, isomers occur. Propane can give rise to either of the following two groups.

$$\underset{n\text{-propyl group}}{CH_3CH_2CH_2-} \qquad \underset{\text{isopropyl group}}{\overset{H_3C}{\underset{H_3C}{>}}CH-}$$

It is important to recognize that these "skeleton" groups are not substances that can be isolated and bottled. They must always form *parts* of molecules. Their own composition and structure remain intact during reactions involving other parts of the molecules. This way of classifying organic groups is a useful and convenient one, but it must be kept in mind that in the reactions described, the ethyl group is not actually formed as a distinct substance.

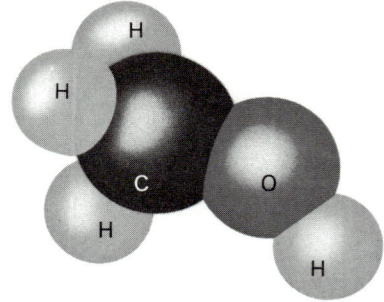

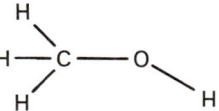

Methanol

Figure 16-11 *Space-filling model and structural formula for methanol, CH_3OH, the simplest alcohol.*

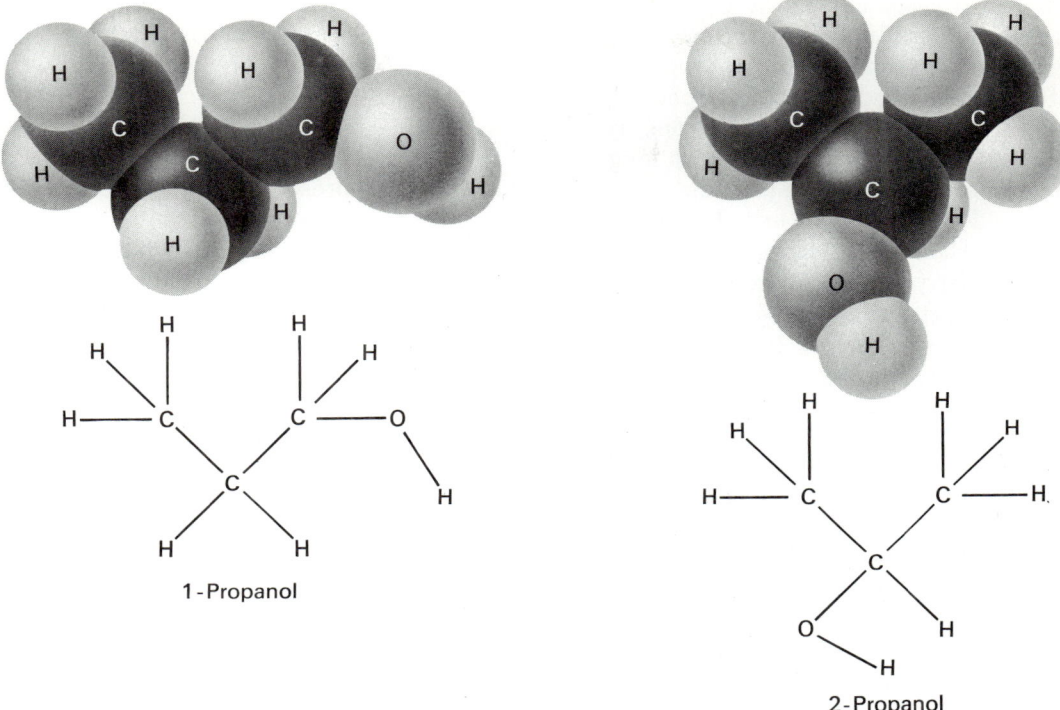

Figure 16–12 *Space-filling models and structural formulas for the isomers 1-propanol and 2-propanol.*

16-11 Alcohols

Shown in Figure 16-11 are the structural formula and a molecular model for the simplest alcohol, methanol, or methyl alcohol. Compare its formula with that of ethanol, or ethyl alcohol, which follows.

$$H-\underset{\underset{H}{|}}{\overset{\overset{H}{|}}{C}}-\underset{\underset{H}{|}}{\overset{\overset{H}{|}}{C}}-OH$$

You can see that these two alcohols are derived from methane and ethane, respectively, by replacing one hydrogen atom of the hydrocarbon with the hydroxyl group. A whole series of alcohols are derived by replacing one hydrogen of a hydrocarbon with an —OH group. The —OH group is the functional group that is characteristic of all alcohols. For example, from propane you get propanol, also called propyl alcohol, $CH_3CH_2CH_2OH$.

If you think carefully about deriving an alcohol from propane, you will realize that it can be done in two ways. There is the way that was just indicated, that is, by substituting —OH for —H on one of the end carbon atoms. However, you can also substitute —OH for one of the H atoms on the middle carbon atom. The two molecules, shown in Figure 16-12, are different, but they have the same molecular formula,

C_3H_8O. They are isomers. In order to give them names that distinguish between them, they are called 1-propanol and 2-propanol. The latter is also called isopropyl alcohol.

It is possible to have more than one H atom replaced by a hydroxyl group, so long as the hydrogen atoms are not on the same carbon atom. For example, if one H atom on each of the carbon atoms of propane is replaced by —OH, you get the common substance glycerol (glycerin).

$$\begin{array}{c} \text{H} \quad \text{H} \quad \text{H} \\ | \quad | \quad | \\ \text{H}-\text{C}-\text{C}-\text{C}-\text{H} \\ | \quad | \quad | \\ \text{OH} \; \text{OH} \; \text{OH} \end{array}$$

One important group of reactions of alcohols involves substances that "oxidize" them. Such reactions are called "oxidations." When an alcohol is oxidized, it loses hydrogen atoms. These hydrogen atoms combine with oxygen atoms to form water molecules. In addition, oxygen atoms may also be added to the molecule, forming new bonds to the carbon atoms. The maximum degree of oxidation of an alcohol, or of the original hydrocarbon itself, is the process of combustion, with which you are already familiar. Complete combustion converts all C and H atoms to carbon dioxide and water, as shown in the following equation for the burning of ethanol.

$$C_2H_5OH + 3O_2 \longrightarrow 2CO_2 + 3H_2O$$

This reaction represents what happens in an alcohol lamp, or in one of those little cans of a substance commercially known as Sterno, or "canned heat."

Of more immediate interest is what happens when an alcohol is less severely oxidized. Less severe oxidation can be carried out in a chemical reaction in which the alcohols are treated with certain inorganic chemicals called *oxidizing agents*. Among these are the compounds potassium dichromate, $K_2Cr_2O_7$, and potassium permanganate, $KMnO_4$. You will learn more about these substances, and others like them, later. For the moment, you need only observe that they contain lots of oxygen. They are capable of giving up these oxygen atoms to substances such as alcohols.

Depending on just how the reactions are carried out, oxidation of alcohols can be made to yield one of several types of compounds. These new compounds contain new types of functional groups. You will consider each of the possibilities in succeeding sections.

INVESTIGATION

16-12 Some Properties and Reactions of Alcohols

You have seen that alcohols, as well as hydrocarbons, can exist in a great variety of sizes and isomeric forms. You have also learned that reactions of hydrocarbons depend on the size and structure of their molecules. Could different sizes and shapes of alcohol molecules affect their behavior in an analogous way?

PURPOSE

This experiment will allow you to explore the effect of size and structure on some of the properties of alcohols, including their ability to react with a variety of reagents.

MATERIALS

35 ml 12 M hydrochloric acid
20-25 cm copper wire, 22-gauge
2.0 grams potassium dichromate
10 ml 3 M sulfuric acid
6-8 ml concentrated acetic acid
50 ml vegetable oil
1.0 gram salicylic acid
Blue litmus paper
3 ml 1-pentanol
8 test tubes, 13- × 100-mm
10 ml of each of the following alcohols
    methanol   1-butanol
    ethanol   2-butanol
    1-propanol 2-methyl-2-propanol
    2-propanol
40 ml 0.015 M potassium dichromate in 3 M sulfuric acid
15-20 drops concentrated sulfuric acid
7 rubber stoppers, solid, size 0, to fit 13- × 100-mm test tubes
1-2 grams sodium metal (for teacher use only)
Three-prong utility clamp or other suitable clamp for thermometer

Test tube, 18- × 150-mm
Test tube rack
Ringstand, ring
Burner
Wire gauze
Boiling stones
Beaker, 100-ml
Beaker, 250-ml
Graduated cylinder, 100-ml

PROCEDURE

Part A.

Determine the boiling point of the alcohol assigned to you, in the same way that you determined the boiling point of a hydrocarbon in Section 16-6. Class members will measure the boiling points of methanol, ethanol, 1-propanol, 2-propanol, 1-butanol, 2-butanol, and 2-methyl-2-propanol. Record in your notebook all of the boiling points measured.

Part B.

1. Add 20 drops of ethanol to a clean, dry 13- × 100-mm test tube. Add 5.0 ml of 0.015 M potassium dichromate in 3 M sulfuric acid. Shake once to mix the contents. Observe the changes that take place during the next several minutes. Keep this test tube and its contents as a color reference for following experiments.

2. Repeat step 1. This time measure the amount of time required for the solution to turn the same color as the contents of the reference test tube. To make the color comparison easier, hold the two test tubes next to each other against a piece of white paper. Since the color change is gradual, you will have to estimate when the two shades are the same.

464   Chapter Sixteen

3. Repeat step 2 with 1-propanol, 2-propanol, 1-butanol, 2-butanol, methanol, and 2-methyl-2-propanol.

Part C.

1. Rinse out the test tubes. Add about 1 ml (20 drops) of each of the alcohols to the labeled test tubes. Add about 5 ml of 12 M HCl to each test tube. Stopper the test tubes and shake for a minute. Look for the presence of any cloudiness that would indicate the presence of a slightly soluble product.

2. (To be performed by the teacher.) Add about 1 ml of each of the above alcohols to seven clean, *dry* test tubes. Drop a thin slice of freshly cut metallic sodium into each test tube. Note any immediate reaction and any changes several minutes later.

Part D.

1. Place about 2 ml of methanol in a small test tube. Make a small spiral of bare copper wire by winding several turns of 22-gauge wire around the point of your pencil. Leave about 15 cm of striaght wire to serve as a handle. Hold the coil of the wire in the burner flame until it is red hot (Figure 16-13a). Thrust the hot wire into the test tube until it is just above the level of the alcohol (Figure 16-13b). Note the change in the appearance of the copper coil. Cautiously, smell the contents in the test tube by wafting some of the vapors to your nose with your hand. Compare the vapors with those of methanol. The new substance formed is formaldehyde, HCHO, which you may recognize as the liquid used to preserve specimens in the biology laboratory.
(*Caution:* In steps 2-5, point the test tube away from yourself and from others. Wear safety glasses.)

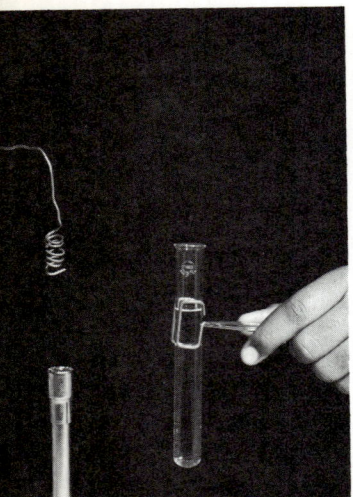

Figure 16-13a

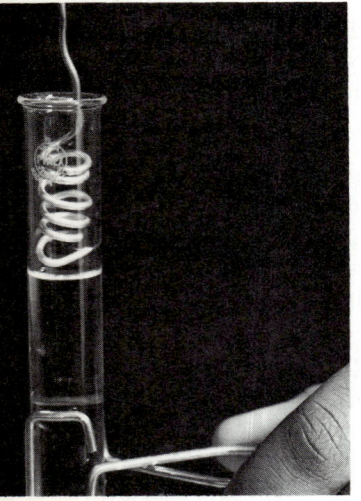

Figure 16-13b

2. Transfer about 2.0 grams of potassium dichromate into an 18- × 150-mm test tube. Add about 10 ml of 3 M sulfuric acid. Attach the test tube to a ringstand with a utility clamp. Add about 125 ml of water to a 250-ml beaker. Using a ring attached to the same ringstand, place the beaker of water on wire gauze. Place the burner under the beaker. Lower the test tube with its potassium dichromate solution into the water. Place a thermometer in the test tube and clamp it in place. (See Figure 16-14.) Heat the beaker of water until the temperature of the dichromate solution is about 60-65°C. Very carefully add *one* drop of ethanol to the dichromate solution. Observe the results. Carefully add more alcohol, one drop at a time, until you have added a total of 8 drops. Remove the thermometer from the test tube. Using the utility clamp as a test tube holder, remove

the test tube from the water bath (the beaker and the water). Smell its contents carefully by wafting. Compare the odor with that of ethanol. Heat the test tube and its contents directly over the burner until it boils gently. Remove the burner and again carefully smell the vapors. Alternate boiling the solution and smelling the vapors until a definite change of odor takes place. Several times during the boiling, lower a piece of moist blue litmus paper into the fumes at the top of the test tube. Note whether the litmus changes color. Keep your water bath set up for the next steps.

3. Add 2-3 ml of concentrated acetic acid to an 18- × 150-mm test tube. Add an equal volume of ethanol. Add 4-5 drops of concentrated sulfuric acid to the mixture. Shake gently to mix. Use a utility clamp to suspend the test tube from the ringstand in such a way that the lower part of the test tube is immersed in the water bath. Bring the water in the water bath to a boil and heat the test tube and its contents for several minutes. Remove the burner. Using the utility clamp as a holder, raise the test tube out of the hot water and carefully smell its contents. If there is no apparent change, heat the test tube *cautiously* over the burner for 1-2 minutes. Again carefully smell its contents.

4. Add just enough crystals of salicylic acid to fill the curved bottom of a clean, dry 18- × 150-mm test tube. Add about 3 ml of methanol and 4-5 drops of concentrated sulfuric acid. Shake gently at once to mix the contents. Heat the test tube and its contents in a water bath in exactly the same way as you did in step 3. Continue heating until you get a definite change in odor.

5. Repeat step 3, using 1-pentanol (*n*-amyl alcohol) instead of ethanol. Heat until you get a definite change in odor.

RELATED PROBLEMS

1. Write structural formulas for the seven alcohols you used in Parts A, B, and C. The numerical prefixes denote the number of the carbon atom in the parent hydrocarbon chain to which the hydroxyl group or other substituted group is attached.

2. Examine the formulas you have written. Which compounds have the greatest similarities in their structures? Which compounds are isomers of each other?

3. What relationships exist between the boiling points and the structures of the alcohols? What relationships exist between the reactions of the alcohols and their structures, as determined in Parts A, B, and C?

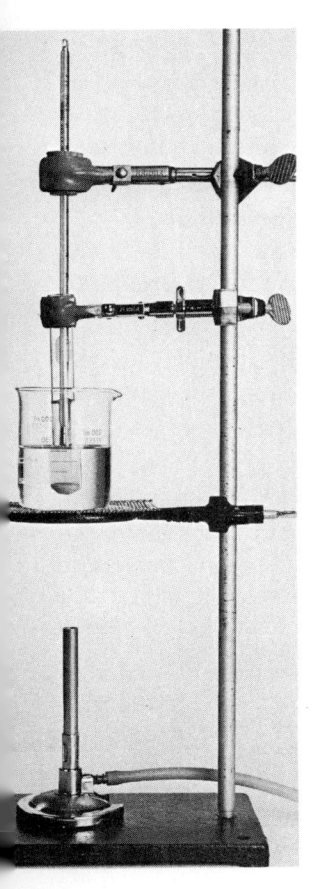

Figure 16-14

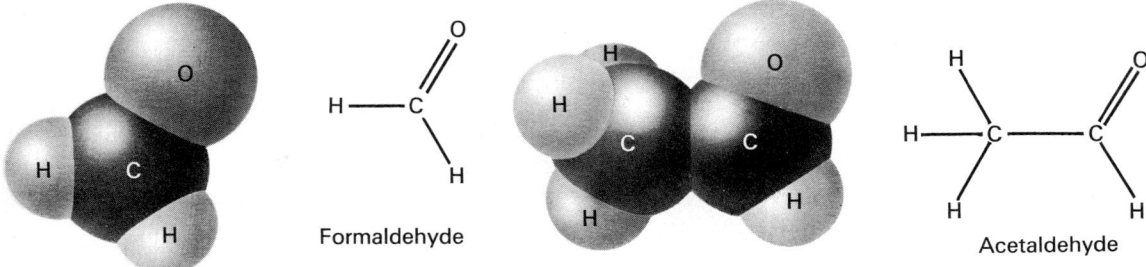

Figure 16-15 *Space-filling models and structural formulas for formaldehyde and acetaldehyde.*

4. When an alcohol reacted with concentrated hydrochloric acid, an alkyl chloride having the general formula RCl was formed. Write an equation to show this reaction. What other product must also have been formed?

5. By comparing the differences of reactivity of the various alcohols with concentrated hydrochloric acid, what can you conclude about the relative strengths of the bonds holding the hydroxyl groups to the carbon atoms in each molecule?

6. What gas was evolved when sodium reacted with several of the alcohols? If you are not sure, how would you test the gas to establish its identity? What other products were also formed?

7. Why did the copper wire become discolored when it was heated by the burner? Compare the formulas of methanol and formaldehyde. (See Figure 16-15 for the formaldehyde formula.) How could the conversion of methanol to formaldehyde be related to the change of the darkened copper back to its original lighter color? What else must have been formed in this reaction?

8. In step 2 of Part D, was there any similarity between the odor of the first product formed by the reaction between ethanol and potassium dichromate and the odor of formaldehyde? What might you conclude from this? What did the final odor produced in this reaction remind you of? How can you account for the change in color of the litmus paper?

9. The pleasant-smelling compounds formed in steps 3, 4, and 5 of Part D are known as *esters*. They are formed when an alcohol reacts with

an acid. Assume that water is also formed in this reaction, and write a general equation to represent the reaction, letting ROH stand for an alcohol and R'COOH stand for an acid.

16-13 Aldehydes

The least severe oxidation of an alcohol with its —OH group on an end carbon atom gives rise to a new compound called an **aldehyde**. This type of reaction is illustrated by the oxidation of ethanol with potassium dichromate.

$$CH_3CH_2OH + [O] \xrightarrow{K_2Cr_2O_7} CH_3CHO + H_2O$$

This equation shows that oxygen from $K_2Cr_2O_7$ reacts with the alcohol to form an aldehyde and water. It does not show how $K_2Cr_2O_7$ changes, but such changes do not concern us at the moment.

What has happened is that the group

$$-\underset{H}{\overset{H}{C}}-OH$$

has been oxidized to the aldehyde group

$$-C\overset{O}{\underset{H}{\diagdown}}$$

as the following reaction shows.

$$-\underset{H}{\overset{H}{C}}-O-H + [O] \longrightarrow -C\overset{O}{\underset{H}{\diagdown}} + H_2O$$

You can see that two hydrogen atoms were removed from the alcohol. The name "aldehyde" actually was derived from *dehydro*genated *al*cohol.

The older names for some of the aldehydes are so well-established that they are universally used in preference to newer, systematic names. Thus, the one-carbon aldehyde is known by its common name, formaldehyde, and the two-carbon aldehyde by its common name, acetaldehyde. Beyond this, the names are derived from the names of the hydrocarbons with the same number of carbon atoms as the corresponding aldehyde. Thus, CH_3CH_2CHO is called propanal (from propane), and $CH_3CH_2CH_2CHO$ is called *n*-butanal, and so on.

Structural formulas and molecular models for formaldehyde and acetaldehyde are shown in Figure 16-15.

16-14 Carboxylic Acids

It is possible to oxidize an organic alcohol even more vigorously and produce still another family of organic compounds called **carboxylic acids** (organic acids). These acids are characterized by a functional group called the **carboxyl** group, which has the following structure.

$$-C\overset{O}{\underset{OH}{\diagdown}}$$

To illustrate the oxidation of an alcohol to a carboxylic acid, we can write the following equation for the oxidation of methanol by potassium permanganate.

$$CH_3OH + 2[O] \xrightarrow{KMnO_4} H-C\underset{OH}{\overset{O}{\lessgtr}} + H_2O$$

This reaction actually proceeds in two steps. In the first step an aldehyde is formed.

$$H-\underset{H}{\overset{H}{\underset{|}{C}}}-O-H + [O] \longrightarrow H-C\underset{H}{\overset{O}{\lessgtr}} + H_2O$$

In the second step, another oxygen atom reacts with the aldehyde to form an acid.

$$H-C\underset{H}{\overset{O}{\lessgtr}} + [O] \longrightarrow H-C\underset{OH}{\overset{O}{\lessgtr}}$$

Of course, it is possible to start with an aldehyde and oxidize it to an acid.

The acid obtained in the preceding reaction is the simplest possible carboxylic acid. It is called formic acid.

The alcohols that have more than one carbon atom can also be oxidized to the corresponding carboxylic acids. Ethanol, for example, can be oxidized to acetic acid. The structural formula and a molecular model for acetic acid are given in Figure 16-16.

You may not be aware of it, but you are undoubtedly familiar with some important carboxylic acids. Vinegar is an approximately five percent solution of acetic acid in water. If you have ever smelled rancid butter, you probably still remember the unpleasant odor. That odor is due to butyric acid, $CH_3CH_2CH_2COOH$.

One of the most important properties of carboxylic acids is that one or more atoms of the carboxyl group can be replaced by

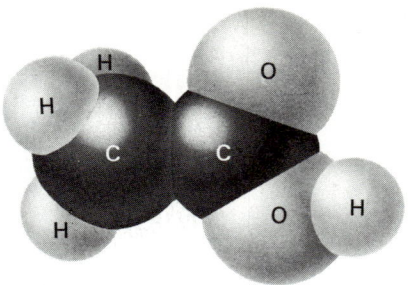

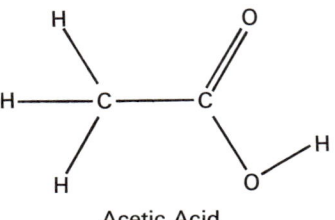

Acetic Acid

Figure 16-16 *Space-filling model and structural formula for acetic acid.*

other atoms or groups. In several later sections we shall discuss a few of the important types of compounds that can be obtained in this way.

16-15 Ketones

In the discussion of alcohols, you saw that when the carbon chain contains three or more atoms, isomers exist, since the —OH group may be either on an end carbon atom or on one of the middle ones. You saw, in Figure 16-12, that both 1-propanol (n-propyl alcohol) and 2-propanol (isopropanol or isopropyl alcohol) are derivatives of propane. You also learned that when an alcohol with an —OH group on an end carbon is mildly oxidized, an aldehyde results.

What happens when an alcohol such as 2-propanol is mildly oxidized? To answer this question, first write the structural formula for the alcohol in the following form.

Molecules Formed by Carbon 469

$$CH_3-\underset{CH_3}{\underset{|}{\overset{H}{\overset{|}{C}}}}-OH$$

You can then see that its oxidation, which follows, is similar to the one shown in Section 16-13.

$$CH_3-\underset{CH_3}{\underset{|}{\overset{H}{\overset{|}{C}}}}-O{-}H + [O] \longrightarrow \underset{H_3C}{\overset{H_3C}{>}}C=O + H_2O$$

The organic product in the equation is a substance called a **ketone**. A ketone differs from an aldehyde in that the ketone does not have a hydrogen atom attached to the carbon atom that makes the double bond to the oxygen atom, as does the aldehyde.

Both ketones and aldehydes contain

$$>C=O$$

which is called a **carbonyl** group. Organic acids contain this group as well. The thing that differentiates these three types of compounds is the nature of the groups attached to the carbonyl group. In each case there is a hydrocarbon group, conventionally designated R, on one side. Table 16-6 shows how the class of compound depends on the nature of the other group, X.

16-16 Esters

As you have learned, functional groups are important because they are the parts of an organic molecule that usually react most easily. One very important reaction between two of the types of functional groups that we have studied involves a carboxyl group and the hydroxyl group of an alcohol. The result of such a reaction, as you can see from the following equation, is an **ester**.

TABLE 16-6 Some Compounds Containing the Carbonyl Group

X in $R-\overset{\overset{O}{\|}}{C}-X$	Type of Compound	Formula
Another hydrocarbon group, R'	ketone	$R-\overset{\overset{O}{\|}}{C}-R'$
H	aldehyde	$R-\overset{\overset{O}{\|}}{C}-H$
OH	acid	$R-\overset{\overset{O}{\|}}{C}-OH$

$$CH_3\overset{\overset{O}{\|}}{C}{-}OH + HOCH_2CH_3 \longrightarrow$$
acetic acid   ethanol

$$CH_3\overset{\overset{O}{\|}}{C}{-}OCH_2CH_3 + H_2O$$
ethyl acetate

Esters of the tri-alcohol glycerol constitute an important group of compounds called **glycerates**. Practically all animal fats are made up of glycerates. The acid groups, that is, the portions of the glycerates derived from carboxylic acids, vary from fat to fat. If $RCOOH$ is the general formula for a carboxylic acid, then the following general formula represents a glycerate.

$$\begin{array}{l}R-\overset{\overset{O}{\|}}{C}-O-CH_2\\ R'-\overset{\overset{O}{\|}}{C}-O-CH\\ R''-\overset{\overset{O}{\|}}{C}-O-CH_2\end{array}$$

Robert Burns Woodward (1917–   )

ROBERT WOODWARD finished his formal education in record time. He graduated from Massachesetts Institute of Technology at the age of nineteen, earned his doctorate at the age of twenty, and became an instructor of chemistry at Harvard University at the age of twenty-one.

The effects of many naturally-occurring substances on living organisms have been known for a long time. While it is possible to extract many such compounds from their natural sources, the quality and yield may vary widely. It has been a goal of scientists to develop simple, economical, and dependable ways to produce such compounds synthetically, starting with simple molecules.

Quinine has been used for over three hundred years in the treatment of malaria. Extracted from the bark of the South American cinchona tree, it has a complicated structure involving a number of interconnected rings of carbon atoms. The English chemist William Perkin was trying to synthesize this compound in the 1870's when he discovered how to make the world's first coal tar dyes instead. The synthesis of quinine proved to be very difficult, but it was finally achieved by Robert Woodward with Doering, a fellow chemist, in 1944.

In 1951, Woodward succeeded in synthesizing, from simple compounds, cholesterol, a compound that appears to be partly responsible for hardening of the arteries. He followed this with the synthesis of a somewhat similar compound, cortisone, that has had spectacular success in the treatment of rheumatoid arthritis and other diseases. He has also produced strychnine, which is a complicated compound that contains seven ring structures, chlorophyll, the green pigment in plants, and lysergic acid and reserpine, both of which have unusual and important effects on the human brain.

In addition to their practical applications and enhancement of scientific understanding, Woodward's dramatic successes have paved the way for the synthesis of even more complicated compounds in the years to come.

One of the fats in butter contains butyric acid groups. Butter turns rancid when certain microorganisms cause the ester groups to be broken down according to the following reaction.

$$\begin{array}{l} H_2C-O-\overset{O}{\underset{\|}{C}}-CH_2CH_2CH_3 \\ HC-O-\overset{O}{\underset{\|}{C}}-CH_2CH_2CH_3 + 3\,H_2O \longrightarrow \\ H_2C-O-\overset{O}{\underset{\|}{C}}-CH_2CH_2CH_3 \end{array}$$

$$\begin{array}{cc} H_2C-OH & \\ HC-OH & + \;\; 3\,CH_3CH_2CH_2COOH \\ H_2C-OH & \\ \text{glycerol (glycerin)} & \text{butyric acid} \end{array}$$

Butyric acid is thus produced, and the characteristic odor of rancid butter results.

The glycerates present in most animal fats have acid groups with very long chains. These chains are mostly 15 to 17 carbon atoms long.

16-17 Soap and Detergents

All esters can be completely broken down into their alcohol and acid components when they are treated with a hot solution of sodium hydroxide. The alcohol is obtained as such, but the acid is obtained as its sodium salt. The following equation shows a typical reaction.

$$\underset{\text{ethyl acetate}}{CH_3CH_2O-\underset{\underset{O}{\|}}{C}-CH_3} \;+\; \underset{\text{sodium hydroxide}}{NaOH} \longrightarrow$$

$$\underset{\text{ethanol}}{CH_3CH_2OH} \;+\; \underset{\text{sodium acetate}}{CH_3COONa}$$

The sodium salt differs from a carboxylic acid by having a sodium atom instead of a H atom as part of its carboxyl group. There is a difference between the bond to Na and the bond to H. The O—H bond is mainly covalent, although it has some ionic character. The O—Na bond, on the other hand, is almost entirely ionic in character. The sodium salt is best represented as a combination of two ions.

$$\left[H_3C-C\underset{\ddot{\underset{..}{O}}:}{\overset{\ddot{O}:}{\diagup}} \right]^{-} Na^+$$

As stated in the preceding section, animal fats are glycerates containing very-long-chain acid groups. When they are broken down by treatment with hot sodium hydroxide solution, the products are glycerol and the sodium salts of the long-chain acids.

$$\begin{array}{l} H_2C-O-\overset{O}{\underset{\|}{C}}-C_{15}H_{31} \\ HC-O-\overset{O}{\underset{\|}{C}}-C_{15}H_{31} + 3\,NaOH \longrightarrow \\ H_2C-O-\overset{O}{\underset{\|}{C}}-C_{15}H_{31} \\ \text{glyceryl tripalmitate} \\ \text{(a fat)} \end{array}$$

$$\begin{array}{cc} H_2C-OH & \\ HC-OH & + \;\; 3\,C_{15}H_{31}COO^-Na^+ \\ H_2C-OH & \\ \text{glycerol} & \text{sodium palmitate} \\ \text{(glycerin)} & \text{(a soap)} \end{array}$$

Soaps, then, are sodium (and sometimes potassium) salts of long-chain acids. What is there about such molecules that makes them good cleaning agents?

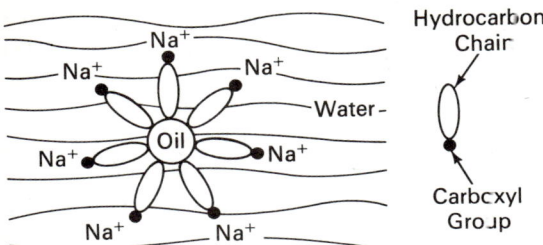

Figure 16-17 *A diagram representing soap action. Hydrocarbon chains are attracted to the oil and the ionic $CO_2^-Na^+$ groups prefer the water.*

The cleansing ability of soap depends on the fact that one end of a soap molecule is very much like a fat molecule, while the other end is ionic. Fats and oils do not mix with, or dissolve in, water, but different fats and oils will usually mix with each other. Most ionic compounds containing sodium ions will dissolve in water, but not in fats or oils. For example, sodium chloride dissolves very well in water, but not at all in an oil, such as olive oil. A soap molecule is effective because it can interact with fats, oils, and water. Its long chain of carbon atoms can associate with the molecules making up fats and oils, while its ionic end can associate with the water. Therefore, small particles or droplets of fatty and oily materials, instead of simply avoiding water, are drawn into it by the soap molecules, as shown in Figure 16-17.

It may surprise you to learn that chemists began to understand how soap molecules work only about 40 years ago. Once they gained this understanding, they realized that other molecules, some of which had not then even been made, should do the same job. What was required was simply a molecule with a long hydrocarbon tail and an ionic group at the end. There was no reason why the ionic group had to be a $-COO^-\,Na^+$ group. Any other group likely to be soluble in water would do as well. What spurred chemists to try and synthesize other materials was the hope of finding one that would work even better than soap itself. There was also the hope that a material might be found that would work at least as well as soap and cost less to make.

After a great deal of research it was found that the best combination of good properties and acceptable cost was achieved by using molecules that contain sulfur instead of carbon, $R\text{-}OSO_3^-\,Na^+$. Substances of this type are called **detergents.** Soap itself is a detergent, but the word is usually used for the newer materials, to distinguish them from soap.

A much greater amount of detergents than of soap is used in the United States. Detergents are somewhat cheaper and, for certain applications, better than soap. More important, the supply of soap is limited, since it all comes from the fat of animals. There is a limit to the supply of animal fats, even in a meat-eating culture. On the other hand, most detergents are derived from petroleum products, which are more generally available.

16-18 Functional Groups Containing Nitrogen: Amines and Amides

A very important functional group is the **amino** group, $-NH_2$. When the amino group is combined with an alkyl group, a compound known as an **amine** results. For example, an amino group together with a methyl group gives the simplest amine, CH_3NH_2, known as methylamine. Similarly, the compound $CH_3CH_2NH_2$ is an amine, ethylamine. Note that these two compounds are named as derivatives of ammonia (called *amine* in Latin).

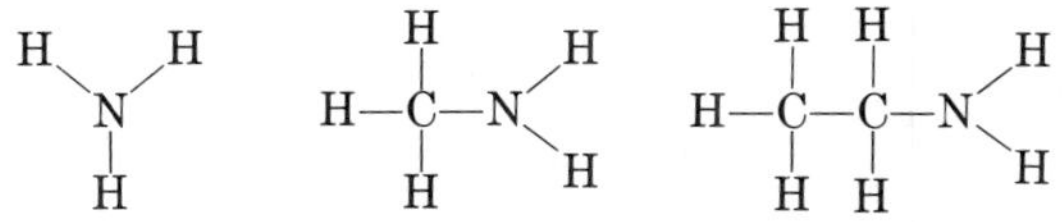

ammonia (amine)

methylamine (aminomethane)

ethylamine (aminoethane)

The foregoing compounds are also given names suggesting that they are derivatives of hydrocarbons. These systematic names, aminomethane and aminoethane, are seldom used for the simplest amines, but systematic names are generally used for more complex ones.

For naming, or for any other purpose, it does not matter whether you think of amines as derivatives of a hydrocarbon or of ammonia. Still another viewpoint can be helpful when you consider a second important class of compounds that contain the $-NH_2$ group. You can think of amines as containing the $-NH_2$ group where the corresponding alcohol would have its $-OH$ group.

When an $-OH$ group that forms part of a carboxyl group is replaced by $-NH_2$, an **amide** is formed. Replacing the $-OH$ group in formic acid, for example, with $-NH_2$ gives the simplest amide, formamide.

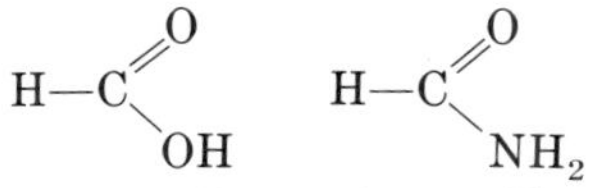

formic acid

formamide

One way of making an amide is by reacting an ester with ammonia, as shown by the following equation.

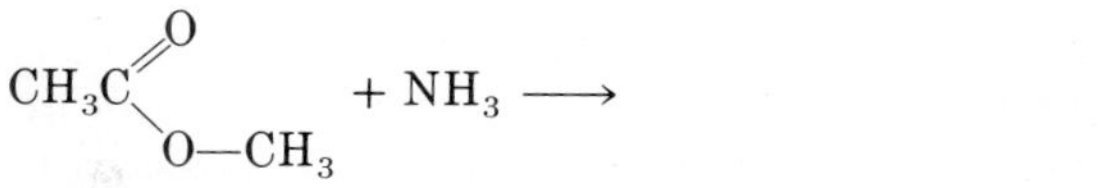

$CH_3C(=O)O-CH_3$ + NH_3 ⟶

methyl acetate

$CH_3C(=O)NH_2$ + CH_3OH

acetamide

It is also possible to carry out a very similar reaction using an amine instead of ammonia.

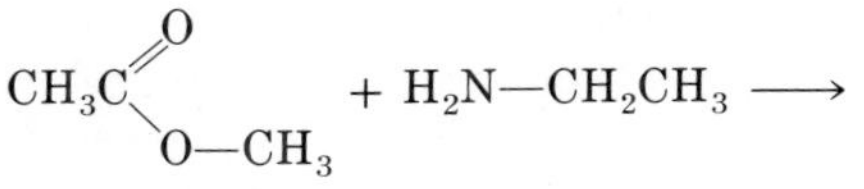

$CH_3C(=O)O-CH_3$ + $H_2N-CH_2CH_3$ ⟶

methyl acetate

$CH_3C(=O)N(H)-CH_2CH_3$ + CH_3OH

N-ethylacetamide

This type of product is called a nitrogen-substituted amide. The particular one just shown is called N-ethylacetamide. ("N" stands for nitrogen.) Amides are of great importance because the grouping

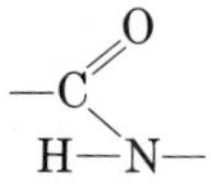

$$-C(=O)-N(H)-$$

is the basic structural element in the large molecules called proteins, which we shall discuss in Chapter 18.

Chemistry and Our Environment

PERSISTENT INSECTICIDES

Insecticides are chemicals that are fatal to insects while doing as little harm as possible to other forms of life. (The word insecticide means "insect killer.") Some insects are carriers of disease. For example, mosquitos carry malaria and tsetse flies carry sleeping sickness. Many other insects do great damage to food crops as well as to other plant life. It has, therefore, always been considered beneficial to human welfare for chemists to discover chemical compounds that are effective and safe insecticides.

Until 1939, few insecticides were known, and these were not practical for large-scale use. They consisted, generally, of arsenic-containing compounds, sodium fluoride, and a few organic compounds. They had to be eaten by the insects and thus were useful only on a small scale to control insects that eat leafy plants.

The great change in the nature and use of insecticides dates from 1939, when a Swiss chemist named Paul Hermann Muller discovered DDT. DDT (*D*ichloro-*D*iphenyl*T*richloro-ethane) is named as a derivative of ethane. The accompanying structural formula should enable you to see why. (The benzene rings are known as "phenyl" groups.)

DDT was the first *broad-spectrum contact* insecticide. This means that it is effective against a great variety of insects (nearly all, in fact) and that it is effective merely on contact. It does not have to be eaten by the insect, but rather, it penetrates the outer coating of the body. The toxic action of DDT and other contact pesticides is not fully understood. Most likely they affect the nerves.

In view of the fact that DDT contains many carbon-to-chlorine bonds, numerous similar chlorinated hydrocarbons have been tested. Some of them have been found to be effective insecticides and have been nearly as widely used, but not so well known, as DDT. For example, there are chlordane and heptachlor, the formulas of which are also shown.

In addition to being broad-spectrum and contact insecticides, these chlorinated substances are also of the type called *persistent*. This means that they are only slowly biodegraded, if at all. Once applied, they stay around for a long time. Even though they are gradually washed away by rain, the molecules either remain intact or are only partly degraded to substances that are still toxic to at least some forms of life.

After some 30 years of use, DDT and similar insecticides are now the subject of earnest controversy. As with so many of our environmental problems, it is a question of whether good effects outweigh the bad ones, and also whether there might not be some completely different approach that is superior. Let us look at some of the principal pluses and minuses involved in the insecticide problem.

Before DDT and similar insecticides were available, starvation, due to insect-caused crop losses, and insect-carried diseases killed hundreds of thousands of people each year. Accurate figures are lacking, but those who favor continued large-scale use of such insecticides contend that without their continued use starvation and disease would again drastically increase in many parts of the world.

Protection of human life is the obvious, and real, argument that can be made for the continued use of DDT and similar insecticides. However, there is another side to the picture. Many of the facts that make up this other side were first brought to public attention by Rachel Carson in her famous book *Silent Spring*.

DDT and similar chemicals are unquestionably seriously toxic to other forms of life besides insects. In this sense, they fail the test

Structural formulas for three widely-used insecticides.

down (metabolized) in the digestive systems of animals. Instead, it accumulates. Thus, the concentration of DDT in the food at the bottom of the food chain may be low, but at each stage it becomes more concentrated. Larger fish, birds, and small mammals can therefore take in quite large amounts.

Chlorinated hydrocarbons, such as DDT, are well known to be very poisonous to fish, birds, and small mammals. There is now plenty of evidence to show that DDT and its relatives have had serious effects on fish and birds all over the world. It is true that there is not yet any direct evidence that chlorinated insecticides have actually harmed human beings. But it is well to remember that we are at the top of the food chain, and that concentration of DDT from the soil and ground water up the food chain is still increasing.

Until a few years ago there was no evidence that DDT harmed birds. However, evidence that DDT is profoundly dangerous to birds has since rapidly accumulated. It gets into their livers, and at a certain level, which has been reached or exceeded in many, many places in the world, it interferes with the formation of the egg shells. The shells become thin and, at worst, they scarcely form at all. Such eggs generally cannot survive to produce young. Thus the birds cannot reproduce, and whole generations of some species in some places have been lost. Birds, of course, do more to control insects than do insecticides, so that a serious reduction in bird populations could do more to increase insect populations than the insecticides have done to control them!

A further serious problem is that insects gradually develop immunity to insecticides. In the long run, large-scale use of insecticides probably will not solve many of our insect control problems. It may even make them worse. Other safer methods of insect control will probably have to be developed.

of the ideal insecticide. However, their effects on other animals have taken time to show up. The time that was required was the time necessary for the residues of these persistent insecticides to become concentrated and passed up the *food chain*. Primitive forms of life living in the soil, in lakes and streams, and in decaying vegetable matter absorb the DDT. These primitive organisms are then eaten by larger ones. DDT is not broken

TABLE 16-7 Regularities in Names of Alkanes, Alcohols, and Amines

Number of Carbon Atoms	Alkanes	Alcohols	Amines
1	CH_4 methane	CH_3OH methanol methyl alcohol	CH_3NH_2 methylamine
2	CH_3CH_3 ethane	CH_3CH_2OH ethanol ethyl alcohol	$CH_3CH_2NH_2$ ethylamine
3	$CH_3CH_2CH_3$ propane	$CH_3CH_2CH_2OH$ 1-propanol propyl alcohol	$CH_3CH_2CH_2NH_2$ 1-propylamine
4	$CH_3CH_2CH_2CH_3$ butane	$CH_3CH_2CH_2CH_2OH$ 1-butanol butyl alcohol	$CH_3CH_2CH_2CH_2NH_2$ 1-butylamine
8	$CH_3(CH_2)_6CH_3$ octane	$CH_3(CH_2)_6CH_2OH$ 1-octanol octyl alcohol	$CH_3(CH_2)_6CH_2NH_2$ 1-octylamine

TABLE 16-8 Regularities in Names of Acids, Amides, and Esters

Number of Carbon Atoms	Acids	Amides	Esters
1	HCOOH formic acid	$HCONH_2$ formamide	$HCOOCH_3$ methyl formate
2	CH_3COOH acetic acid	CH_3CONH_2 acetamide	CH_3COOCH_3 methyl acetate
3	CH_3CH_2COOH propionic acid	$CH_3CH_2CONH_2$ propionamide	$CH_3CH_2COOCH_3$ methyl propionate
4	$CH_3CH_2CH_2COOH$ butyric acid	$CH_3CH_2CH_2CONH_2$ butyramide	$CH_3CH_2CH_2COOCH_3$ methyl butyrate
8	$CH_3(CH_2)_6COOH$ octanoic acid caprylic acid	$CH_3(CH_2)_6CONH_2$ octanamide caprylamide	$CH_3(CH_2)_6COOCH_3$ methyl octanoate methyl caprylate

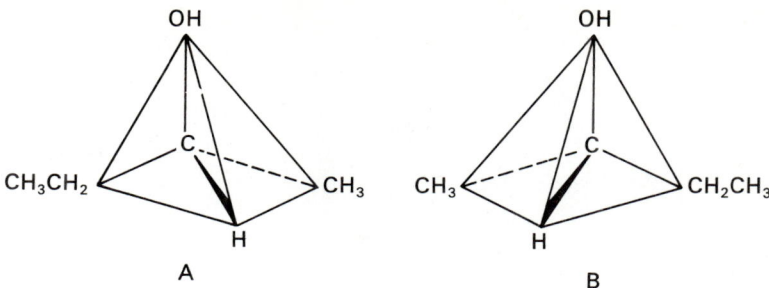

Figure 16–18 *The three-dimensional structures of the two molecules, A and B, discussed in the text.*

Tables 16-7 and 16-8 tie together the names for the various classes of organic compounds that we have been studying. Included in the tables are some of the more important members of each class, grouped according to the number of carbon atoms they contain.

16-19 Right- and Left-handed Molecules ★

You have already encountered two forms of isomerism, called structural isomerism and geometrical isomerism. One more form of isomerism is of great importance. Consider the two substances, A and B, with the following molecular structures. (Figure 16-18 shows clearly the structures represented by the following formulas.)

$$\begin{array}{cc} \text{CH}_3\text{CH}_2 \diagdown \quad \diagup \text{OH} & \text{HO} \diagdown \quad \diagup \text{CH}_2\text{CH}_3 \\ \text{C} & \text{C} \\ \text{H} \diagup \quad \diagdown \text{CH}_3 & \text{H} \diagup \quad \diagdown \text{CH}_3 \\ \text{A} & \text{B} \end{array}$$

Are they isomers, or are they identical?

They are not isomers in either of the two senses with which you are already familiar. They are not structural isomers, since they have their atoms arranged in the same way. Each contains the same subgroups, connected in the same way. They are also not geometrical isomers, since there is no possibility of *cis* or *trans* character in this molecule.

It is also true that samples of pure A and B are identical in all common physical properties such as color, melting point, boiling point, dipole moment, density, and solubility in water. A and B are also identical in all common chemical properties. For example, they can be oxidized to the identical ketone, $\text{CH}_3\text{CH}_2\underset{\underset{\text{O}}{\|}}{\text{C}}\text{CH}_3$. In all their common reactions they react in exactly the same way and at exactly the same rate. These properties are truly identical for A and B. The most accurate measurements that have been made show no differences.

Perhaps you are now about to conclude that A and B are identical substances. After all, you can see that they are not isomeric in either of the two ways we have discussed. Secondly, they are indistinguishable in all their common physical and chemical properties. What else could you ask in order to prove that two substances are identical?

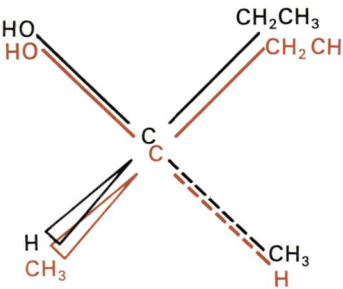

Figure 16-19 *If the* —OH *and* —CH$_2$CH$_3$ *groups of the two* C(CH$_3$)(C$_2$H$_5$)(H)(OH) *molecules are superimposed, the* —H *and* —CH$_3$ *groups will not match.*

There is one more thing you could ask. If two substances are truly identical, a molecule of one should be perfectly superimposable upon a molecule of the other. Try that test with molecules A and B. Turn molecule A around so that its central carbon atom, its hydroxyl group, and its ethyl group are superimposed on the corresponding groups of molecule B. The result is shown in Figure 16-19. As you can see, when these three parts of the two molecules are superimposed, the other two parts, the hydrogen atoms and the methyl groups, do not match. They are on opposite sides. A little thought should convince you that you can superimpose the central carbon atom and any two of the four groups around it for molecules A and B. The other two groups will then always be arranged oppositely.

Your comparison of the two molecules forces you to the conclusion that they are isomers. They have identical molecular formulas but their structures are not exactly the same. That is the basic definition of isomers. However, what you see here is a new type of isomerism, more subtle than the types with which you are already familiar. Isomers of the type we are discussing are called *optical* isomers. The reason for that name will be explained later.

If you look closely at the two isomeric molecules, A and B, as they were originally drawn, you will see that one is related to the other as are an object and its image in a mirror. In most cases a molecule and its mirror image are superimposible and therefore identical. In this case they are not. Perhaps the objects most familiar to you for which this is also true are your own hands. They are not really identical. They are mirror images of each other that cannot be superimposed. Molecules that have this same property are said to have "handedness," or to be *chiral* molecules. "Chiral" is derived from the Greek word for "hand."

The most important class of chiral molecules is the type you have just examined, molecules in which four different groups are arranged tetrahedrally around a central atom. In organic molecules the atom having four different substituents is usually a carbon atom, which is called an *asymmetric* carbon atom.

Why are such isomers called optical isomers? This is because one way to tell them apart is by means of an optical effect, that is, an effect having to do with light. When light is passed through certain devices called *polarizers* it emerges with its oscillating electric field (see Figure 11-2) confined to a single plane. The light is *plane-polarized*. When plane-polarized light passes through most forms of matter, the plane of polarization enters and emerges with the same orientation, as indicated in Figure 16-20. However, a sample consisting of chiral molecules (all of the same handedness) causes the plane to rotate. Right-handed molecules rotate it one way, and left-handed molecules rotate it the other way.

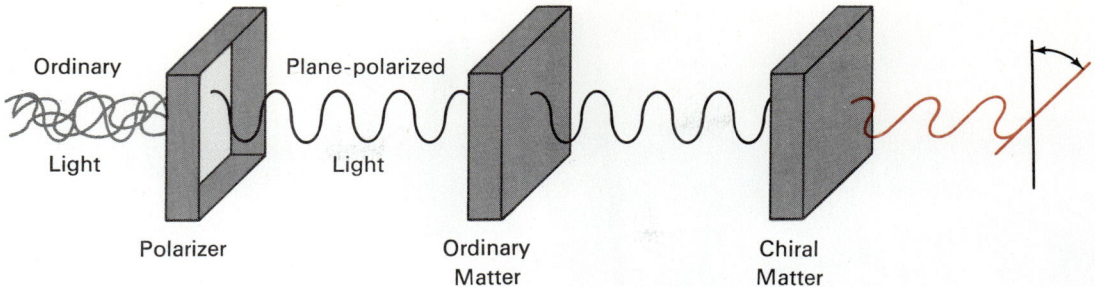

Figure 16-20 *Ordinary light (left) passes through a polarizer. It emerges plane-polarized. On passing through ordinary matter, the plane of polarization remains unchanged. On passing through chiral matter, the plane of polarization is rotated.*

The greatest importance of optical isomerism is in biological chemistry. In Section 16-18, the basic structural element in proteins was introduced. These linkages result from the joining together of "amino acids." Amino acids contain both a carboxyl (acid) group and an amine group. Glycine, the simplest amino acid, has the following formula.

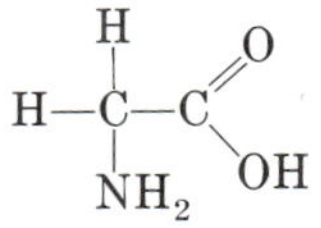

$$\begin{array}{c} H \\ | \\ H-C-C \\ | \quad\;\; \diagdown \\ NH_2 \quad OH \end{array}\;\;\;\diagup\!\!\!O$$

Other amino acids have some group more complicated than a hydrogen atom bonded to them. If any one of these groups is designated R, the following general formula for an amino acid results.

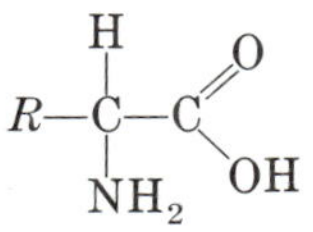

$$\begin{array}{c} H \\ | \\ R-C-C \\ | \quad\;\; \diagdown \\ NH_2 \quad OH \end{array}\;\;\;\diagup\!\!\!O$$

All amino acids, with the exception of glycine, can exist as optical isomers, as shown in Figure 16-21. The designations L and D stand for "Levo" and "Dextro." Originally, they were based on the direction of rotation of the plane of polarized light at a particular wavelength. However, they are now used in a more fundamental sense to specify the true (so-called "absolute") configuration of the groups around the asymmetric carbon atom. The drawings in Figure 16-21 can be considered as defining L and D.

All naturally-occurring amino acids have the L configuration. All of the biochemical reactions in which amino acids are made or

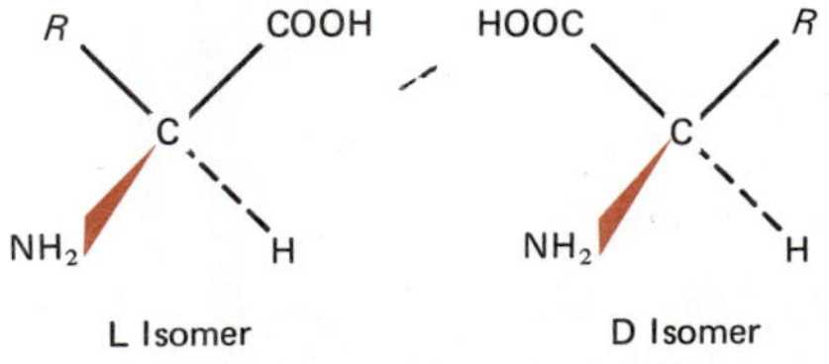

Figure 16-21 *The two optical isomers of an amino acid.*

Figure 16-22 *Glucose is found in many foods. Cereal grains such as rice (shown being harvested), corn, wheat, barley, and oats are a major source of glucose.*

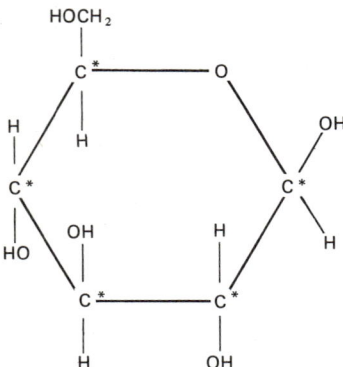

Figure 16-23 *The glucose molecule contains five asymmetric carbon atoms, as indicated by the asterisks.*

otherwise participate are specific for the L configurations. It would be impossible for you to live on a diet of D amino acids, except to the extent that your body could break them down and use the component atoms to rebuild L acids.

Many other molecules in life processes contain asymmetric carbon atoms. Some of them contain two or more. An example is the sugar glucose (Figure 16-22). Its structural formula, Figure 16-23, shows its five asymmetric carbon atoms. In order for the molecule to be the "real" glucose, the handedness at every one of these must be correct. At each one there are two possibilities and thus there are 32 possibilities for the molecule as a whole. Only *one* is correct.

Summary

The enormous amount of factual material that constitutes organic chemistry can be organized and made easier to understand and remember by thinking of molecules as composed of two parts. These parts are: (1) the reactive functional group, and (2) the relatively unreactive hydrocarbon group. The functional groups have characteristic ways of reacting. As a result, all compounds containing a particular functional group have similar (but not identical) chemical properties, regardless of the particular hydrocarbon group to which the functional group is attached. The hydrocarbon groups contain one less hydrogen atom than their respective parent hydrocarbons, thus permitting the attachment of functional groups. Compounds all containing a particular functional group constitute a family. Alcohols, for example, all contain the —OH group, directly attached to a hydrocarbon group.

Self Check

1. What is meant by preserving the "structural integrity" of a group in a reaction?
2. Name the three hydrocarbon groups CH_3—, CH_3CH_2—, and $CH_3CH_2CH_2$—.
3. Name some important functional groups you have just studied.

Chapter Summary

The chemistry of the compounds formed by the element carbon is called organic chemistry. Vast deposits of the element carbon, in a more or less pure condition, are found in nature as coal. Petroleum is a complex mixture of carbon compounds, mostly hydrocarbons. The molecules that make up living organisms are organic compounds, for the most part.

In order to organize the complicated chemistry of carbon compounds along manageable lines, it is necessary to have a system for classifying and naming the compounds. In order to classify and name organic molecules, we think of them as consisting of two parts: (1) a hydrocarbon group, which is relatively unreactive, and (2) one or more functional groups. The most common functional groups contain oxygen atoms, nitrogen atoms, or both, and determine the main characteristics of the compounds in which they occur. We can use R to represent a hydrocarbon group, such as an alkyl group. Thus the main types of organic molecules are the following: ROH, alcohols; RCHO, aldehydes; RCOR', ketones; RCOOH, carboxylic acids; RCOOR', esters; RNH$_2$, amines; RCONH$_2$ or RCONHR', amides.

Questions and Problems

1. List three important naturally-occurring sources of carbon.
2. How do petroleum chemists separate the molecules of many sizes found in crude oil?
3. The distribution of compounds consisting of small-, medium-, and large-size molecules in a typical batch of petroleum may not correspond to the market demands for these compounds. What can be done

to modify the compounds to meet the needs?

4. In what different ways do chemists process gasoline to decrease its tendency to "knock" in an automobile engine? How may one of these methods be environmentally harmful?

5. How does inorganic chemistry differ from organic chemistry?

6. What information is contained in an empirical formula? a molecular formula? a structural formula? Write these three formulas for ethane.

7. A sample of an unknown hydrocarbon with a mass of 0.29 gram is completely burned in oxygen to produce 0.88 gram of carbon dioxide and 0.45 gram of water. Find the empirical formula of the compound.

8. When a 1.00-gram sample of the hydrocarbon used in Question 7 is vaporized at a temperature of 100°C and a pressure of 1.00 atmosphere, its volume is found to be 0.527 liter. Find the mass of one mole of the compound and its molecular formula.

9. A molecule of a noncyclic saturated hydrocarbon and a molecule of an unsaturated hydrocarbon each have the same number of carbon atoms. Compare the number of hydrogen atoms in each. Why must unsaturated compounds have multiple bonds?

10. How can bromine, Br_2, be used to distinguish between a saturated and an unsaturated hydrocarbon? Write a structural formula for propene, C_3H_6, and show what product is formed when Br_2 reacts with it.

11. Why is the failure of a hydrocarbon to react readily with bromine not conclusive proof that the hydrocarbon is saturated?

12. Write structural formulas for three isomers of pentane, C_5H_{12}. (Note that bending the carbon chain does not produce a new form.)

13. Write structural formulas for all isomers of pentene, C_5H_{10}. Why are there more isomers for this compound than for pentane?

14. Although benzene, C_6H_6, is an unsaturated compound, what conclusions can you draw about its structure from its inability to react with bromine in the same way that ethylene does?

15. Explain why ⬡ is used to represent a benzene molecule?

16. Write an equation, using structural formulas, to show the substitution reaction between benzene and chlorine, Cl_2.

17. Draw structural formulas for all isomers of dichlorobenzene, $C_6H_4Cl_2$.

18. Give simple structural formulas of (a) an alcohol, (b) an aldehyde, and (c) an acid, each derived from methane. Repeat, using ethane, butane, and octane as the hydrocarbons from which the other compounds are derived.

19. Write structural formulas for all alcohols having the formula, C_4H_9OH.

20. Use structural formulas to show what happens when one mole of oxygen atoms from the oxidizing agent potassium dichromate reacts with 1-propanol.

21. Write structural formulas for 2-propanol and the compound formed by its gentle oxidation. How many moles of oxygen atoms are required to oxidize one mole of 2-propanol?

22. Write an equation to represent the vigorous oxidation of one mole of 1-propanol by oxygen atoms from a suitable oxidizing agent.

23. Given the structural formula for a simple ester, write the formula of the acid and

the alcohol from which it could most directly be made.

24. Suppose that you could put a tag on the oxygen atom marked with an asterisk in a molecule of ethanol, $C_2H_5O^*H$, so that you could trace its progress in a reaction. Show how you could determine experimentally if the entire —OH group or only the H atom is removed from the ethanol molecule when it reacts with acetic acid, CH_3COOH, to form an ester.

25. Using structural formulas, show an equation for the reaction between one mole of glycerol and three moles of acetic acid.

26. When 0.601 gram of a sample having the empirical formula CH_2O was vaporized at 200°C and 1.00 atmosphere pressure, the volume occupied was 388 ml. The same volume was occupied by 0.301 gram of ethane under the same conditions. What is the molecular formula of CH_2O?

27. One mole of the sample in Question 26 reacted with one mole of methanol to form an ester. Write the structural formula for the sample.

28. What types of organic compounds can be obtained by reacting an ester with ammonia?

29. Draw the structural formulas for all the compounds with the molecular formula $C_2H_3Cl_3$.

30. How is a typical soap molecule related to the fat molecule from which it is produced?

31. Compare the structures of amines, amides, and N-substituted amides.

32. Analysis of a pure compound frequently used as automobile antifreeze shows that it contains only carbon, hydrogen, and oxygen. A sample of this compound with a mass of 15.5 mg is burned, and the following masses of CO_2 and H_2O result.

Mass of sample burned	15.5 mg
Mass of CO_2 formed	22.0 mg
Mass of H_2O formed	13.5 mg

What is the empirical formula of this compound?

17 Solids, Liquids, and Solutions

Salvador Dali's the Persistence of Memory gains attention by contradicting the onlooker's sense of what is solid and what is not.

Is a window pane a liquid? Is a hard lump of paraffin a solid? Such questions might seem to have easy answers, but in fact they do not. A search for an answer must be sought in the forces acting between atoms and molecules. In this way, the scientific meanings of the words *solid* and *liquid* can be understood.

Up to now you have been concerned almost entirely with individual atoms or molecules and with the gases. In the gas phase, the atoms and molecules are almost completely independent of one another. They are too far apart and contain too much energy for the relatively weak forces of attraction to be effective. Most types of matter are not gases, but are solids, liquids, or mixtures of two or more kinds of atoms and molecules. These types of matter are more dense than gases. For this reason chemists call them *condensed* types of matter.

In this chapter you will learn about the interactions of atoms and molecules in condensed phases.

Solids, Liquids, and Solutions Have Special Properties

17-1 The Attractive Forces Causing Condensation of Gases

Any pure gas, when cooled sufficiently, will condense to a liquid. (Liquid nitrogen is shown in Figure 17-1.) At a lower temperature, it will become a solid. The temperatures at which such condensations occur vary greatly. For example, lithium fluoride gas at one atmosphere pressure condenses to a liquid when cooled below 1949 K. When the temperature is lowered to 1143 K, the liquid forms a clear crystal. In contrast, lithium gas at the same pressure must be cooled to 1609 K before it forms a liquid. This liquid does not solidify until the temperature is lowered to 459 K. The solid is a white, soft metal, not resembling crystalline lithium fluoride at all. Fluorine gas is equally distinctive. At one atmosphere pressure it must be cooled far below room temperature before condensation occurs, at 85 K. The liquid solidifies at 50 K. Why do these three materials behave so differently? A consideration of energy leads to the answers to this question.

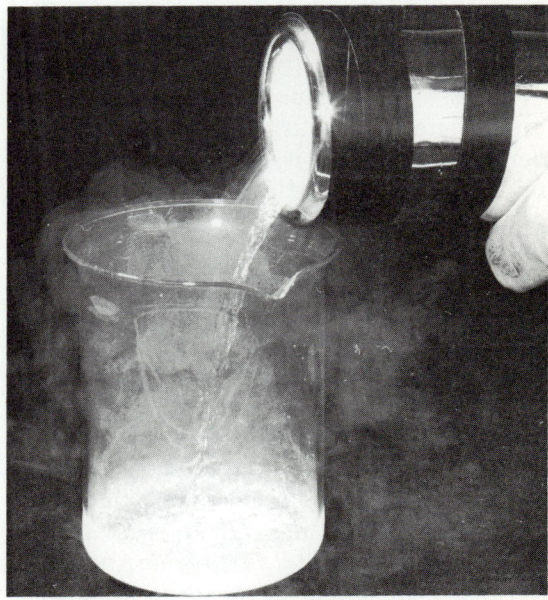

Figure 17-1 *Liquid nitrogen bubbles vigorously at room temperature and causes the smoke-like condensation of water vapor in the air. (Nitrogen boils at 77K.)*

As you know, two or more atoms remain near each other because their energies favor that arrangement. The attractive forces between them hold them in a position at which the overall energy is at a minimum. This is true whether the cluster of atoms is strongly or weakly bound, whether it contains a few atoms or 10^{23} atoms, whether the arrangement is regular (as in a crystal) or irregular (as in a liquid). The cluster of atoms is stable if and only if the total energy is lower when the atoms are together than when they are apart.

Their energies are lowered because their electrons are located between positive nuclei and are held there by forces of electrical attraction. The magnitude of the attractive forces varies greatly, depending on how close to these positive nuclei the electrons are able to approach. This approach distance is fixed by the way electrons occupy valence orbitals.

The electron occupancy of the valence orbitals determines largely the type of condensed matter that will result. It also determines such properties of substances as hardness, ease of melting, and electrical conductivity.

17-2 The Nature of Crystalline Solids

Nearly all pure solid substances normally occur (or can be made to occur) in crystalline form. Some crystals can be seen with the unaided eye. Others are so small that a microscope is needed in order to see them.

In essence, *a crystal is a regular, repeated arrangement of the particles (atoms, molecules, or ions) of which it is made*. More specifically, a crystal consists of a side-by-side repetition of one unit of atoms, molecules, or ions. The larger the crystal, the larger the number of repetitions of this unit.

Consider, for example, a crystal of sodium chloride (common table salt). Each of these

Figure 17-2 *A sodium chloride, NaCl, crystal, showing its cubic geometry.*

crystals is a minute cube about 0.5 mm on an edge, like the one shown in Figure 17-2. The cubic arrangement results from alternating Na^+ and Cl^- ions, as shown in the model in Figure 17-3.

This illustration involves a solid made up of ions. However, the fundamental idea of building up a crystal by orderly repetition of a small group of particles is entirely general. This is the way in which all crystalline solids are constructed. You will consider a number of other examples (molecular solids, network solids, and metals) later in this chapter.

A number of substances are ordinarily considered solids that do not fit the pattern of crystalline solids. Paraffin and ordinary glass are good examples. They do not have a regular extended pattern of atoms or molecules nor a fixed melting point. With this in mind, you would not consider them as "true" or crystalline solids. Instead, they are better characterized as very viscous liquids. (A "viscous" liquid flows very slowly, like cold molasses.)

Figure 17-3 *The arrangement of Na⁺ and Cl⁻ ions in the model at the left clearly shows the cubic geometry of NaCl. The closeup of a crystal face (right) reveals that each Na⁺ ion (small sphere) is surounded by chloride ions and each Cl⁻ ion is similarly surrounded by sodium ions.*

17-3 The Nature of Pure Liquids

A pure liquid is the liquid phase of a single pure chemical compound such as water or benzene. It is not as regular nor as rigid in its structure as is a solid. When a solid melts and forms a liquid, the molecules (or ions, or atoms) remain close together, but acquire enough energy to move around, slipping over one another. Over very short distances they may often arrange themselves in about the same way as they are arranged in the crystalline solid. Over *long* distances the ordering is not so fixed or precise. Because of the movement of the molecules, the liquid is not rigid but flows and pours.

In this connection recall Figure 7-2. From it you can see quite clearly a representation of the molecular motion in the liquid phase. It is important that you remember that the molecules of liquids, and those of solids and gases as well, are constantly in motion. This fact is important because the degree of molecular motion is the distinguishing characteristic of each phase of matter. The motion of the molecules of a liquid is intermediate between that of a solid and that of a gas.

Between crystalline solids and liquids there is a kind of material called a *glassy* substance. The most obvious example is glass itself. Glass does not have the extensive regular pattern of a solid. While you have never seen glass "flow" at room temperature it will, over many years, move slightly. For example, window panes in very old houses become thicker at the bottom and thinner at the top as the glass slowly flows in response to gravity.

17-4 The Nature of Solutions

A **solution** consists of two or more pure substances mixed together *at the molecular level*. If small particles of two substances, say powdered sugar and finely crushed salt, are mixed together, the result is not a solution. It is merely a physical mixture.

A solution has the same appearance and composition throughout. However, its properties depend on the relative amounts of the pure substances in it. For example, a solution made by dissolving only a small amount of sugar in water will flow freely (have low viscosity). If the proportion of sugar is high, the solution will be viscous and sticky.

Many people tend to think that solutions are liquids that are made by dissolving a solid in a liquid. However, the concept is much more general. Indeed, solutions that are themselves liquids can be obtained by dissolving gases or other liquids in a given liquid. For instance, ammonia or alcohol can be added to water to form solutions. There are also solid solutions. Perhaps the commonest kind of solid solution consists of two or more kinds of metal atoms and is called an **alloy**.

INVESTIGATION

17-5 The Packing of Atoms or Ions in Crystals

The solid state of matter usually consists of a regular arrangement of atoms, molecules, or ions. If you represent each building block as a point, then the crystal structure can be represented by a repeating pattern called a *space lattice*. In 1848, a scientist by the name of Bravais used arguments based entirely on geometry to demonstrate that only fourteen types of simple space lattices are possible.

PURPOSE

Using styrofoam spheres as building blocks, you will study some of the ways in which typical metallic crystals can be formed. Three types of packing will be investigated: hexagonal closest packing, face-centered cubic packing (also called cubic closest packing), and body-centered cubic packing. You will observe the number of nearest neighbors (the coordination number) of the particles in each of these structures.

In addition, you will investigate some of the possibilities of packing spheres of different radii into lattices that represent ionic crystals. You will build the rock salt lattice, $Na^+Cl^-(s)$, and the Wurtzite lattice, $Zn^{2+}S^{2-}(s)$. In doing so, you will observe how the ratio of the radii of the cations (the positively-charged ions) and the anions (the negatively-charged ions) determines the coordination number.

MATERIALS

13 1-inch spheres   13 $\frac{3}{4}$-inch spheres
60 1- to 2-inch lengths of pipestem cleaners
36 2-inch styrofoam spheres (one of these should be colored)

Solids, Liquids, and Solutions   489

PROCEDURE

Part A. Some General Considerations on Packing of Spheres

1. Determine how many two-inch spheres you can pack around a marked single sphere in the same plane. (Do you think this number is dependent on the size of the sphere? Check your prediction.)

2. Place additional loose spheres above and below the marked sphere (still surrounded by other spheres) so that they all touch it. How many nearest neighbors does it have? In other words, what is the coordination number for this type of closest packing?

3. Note that the three spheres in the top layer can occupy a position directly above the comparable bottom layer or they may be twisted through an angle of 60°. Either arrangement gives the same coordination number.

Part B. Model I: Hexagonal Closest Packing

1. Connect the groups of spheres you used in Part A with short lengths of pipestem cleaners to obtain the layers shown in Figure 17-4.

Figure 17-4

2. Place the layer of three spheres on the desk with the apex of the triangle facing you.

3. Now place the layer of seven spheres over the three spheres so that the center sphere fits closely into the depression of the first layer.

4. For the top layer, place another layer of three spheres over the center sphere of the second layer so that they are oriented directly over those in the first layer. Suppose a pattern such as this were expanded into space until billions of atoms were involved. The result would be a model of a very small crystal of metallic magnesium, zinc, or one of many other metals. Note the coordination number. Retain this model for use in Part D.

Part C. Model II: Cubic Closest Packing, or Face-centered Cubic Packing

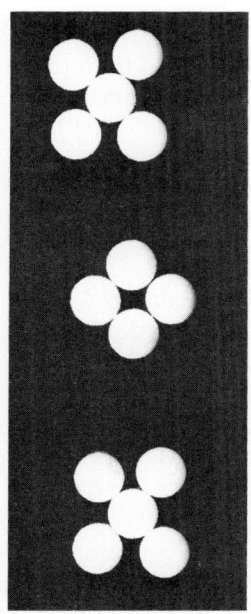

Figure 17-5

1. Construct the layers shown in Figure 17-5, using two-inch spheres and pipestem cleaners as before.

2. Place the first layer flat on the desk. Place the second layer on it with the spheres resting in the spaces between the corner spheres of the first layer. Now add the third layer so its spheres are directly over those in the first layer. Study this model carefully. Why is it called face-centered cubic? This packing is found in copper, silver, aluminum, and other metals.

Figure 17-6

Part D. Comparison of Hexagonal Closest Packing with Cubic Closest Packing

1. Return to Model I, hexagonal closest packing. Arrange it in such a way that the top layer is not directly over the first layer, but is rotated 60° with respect to it.

2. Rotate this model slightly and look for four spheres forming a square facing you. Now take the top layer off of Model II and place it against the four spheres you located on Model I. The spheres of this top layer should be lined up with those in the layer beneath the four spheres that you located in Model I. Note that you now have a face-centered cubic structure just like Model II, except that it is tilted from the vertical.

3. In light of the above comparison, is there a difference in coordination number between the two types of closest packing? Is there a difference in density when spheres of comparable size and mass are involved in each of these types of packing? Most metals crystallize in only one, not both, of these forms. What does this indicate about the directional nature of bonds between atoms of these metals?

Part E. Model III: Body-centered Cubic Packing

1. Construct the layers shown in Figure 17-6, using two-inch spheres. Be sure to leave about a one-fourth inch space between the spheres, as indicated.

2. Place the single sphere in the center of the first layer and then place the third layer so that its spheres are directly over the first layer. Study the symmetry of this model and justify its name. This type of packing is typical of the alkali metals, such as sodium and potassium. In terms of numbers of valence electrons, can you suggest any reason why Na and K crystallize in this form while most of the other metals crystallize in a close-packed form?

3. Below 906°C metallic iron crystallizes in the body-centered cubic form called alpha ferrite. Above this temperature the stable form is beta ferrite, a face-centered cubic arrangement. At 1401°C the crystal form changes back to a body-centered cubic solid called gamma ferrite. What is the coordination number of iron in each of these forms? Can you suggest any reason for these transitions, in terms of numbers of available bonding electrons?

Part F. The Sodium Chloride Lattice

1. Ionic crystals are formed by packing positive and negative ions alternately into a lattice. Sodium ions have a diameter of 1.90 Å

while that of chloride ions is 3.62 Å. Use one-inch spheres for Na^+ ions and two-inch spheres for Cl^- ions to approximate the relative sizes in this model.

2. Use Model II, with its two-inch spheres, for the face-centered cubic arrangement of the chloride ions. Insert the thirteen one-inch spheres representing the sodium ions into the holes between the chloride ions in each layer. Note that the Na^+Cl^- lattice is an interpenetrating set of face-centered cubes. One set is made up of Na^+ ions and the other set of Cl^- ions.

3. What type of ion surrounds each Na^+? each Cl^-? What is the coordination number of the spheres representing Na^+? Cl^-?

4. Note that to achieve this type of lattice there must be a favorable relation between the relative radii of the two spheres that will permit a given sphere to fit into a given hole in the lattice. What is the radius ratio for Na^+/Cl^- ions? Can you account for the stability of this type of packing in terms of interionic forces?

Part G. The Zinc Sulfide (Wurtzite) Lattice

1. The zinc ion has a diameter of 1.5 Å and the diameter of the sulfide ion is 3.7 Å. Use three-fourths inch spheres for the Zn^{2+} ion and two-inch spheres for the S^{2-} ion to approximate the relative sizes.

2. Use Model I, with its hexagonal closest packing orientation, to represent the lattice of the larger, sulfide (S^{2-}), ions. Attach one of the smaller (three-fourths inch) spheres representing the zinc ions (Zn^{2+}) directly above each of the larger spheres in each of the three layers of Model I. (Use short lengths of pipestem cleaners, or toothpicks.)

Place the large layer on the table top with the small spheres pointing down. Place one of the small layers on the large layer in such a way that the smaller spheres fit into alternate depressions. Invert the two layers. Then place the third layer, with its small spheres pointing up, atop the large layer. Place this third layer in such a way that each of its spheres is directly above a sphere in the bottom layer.

3. What is the coordination number of the spheres representing Zn^{2+} ions?

4. What is the radius ratio of zinc ions to sulfide ions?

RELATED PROBLEMS

1. Write a brief description of each type of metallic crystal packing studied.

2. Answer all questions raised in the Procedure sections. Label each answer by Part and step, for example, Q. A-1.

3. In one type of cubic packing, the spheres occupy about two-thirds of the available space and in the other they fill about three-fourths. Identify each type. Which is more dense? Which has the larger number of bonds?

4. From your consideration of the models constructed in Parts F and G, what can you deduce about the relationship between the radius ratio of ions and the coordination number in crystals? In which case is the number of interionic attractions around any given ion the greater?

5. Assume that you have a crystal XY with the sodium chloride packing arrangement. Assume, further, that the ions are the same size as Na^+ and Cl^-, respectively, but doubly charged. Would XY have a higher or lower melting point than NaCl? Suggest a real pair of crystals that meet the above criteria, that is, that differ only in ionic charge. Look up their melting points to check your prediction.

6. Suppose that you have a crystal AB with the sodium chloride packing arrangement. Suppose, further, that each of the ions has the same charge (A^+ and B^-) as Na^+ and Cl^-, but the radii of A and B are proportionately larger. Would AB have a higher or lower melting point than NaCl? Again suggest a real pair of crystals that meets the above criteria, and then look up their melting points.

Summary

The condensed forms of matter (solids, liquids, and solutions) are produced when atoms, molecules, or ions come very close together. In this way, short-range attractive forces can operate between them. These forces vary greatly in magnitude, thus accounting for the great variation in melting and boiling points of different substances.

Crystalline solids are built up by the orderly and regular repetition of a basic arrangement of a few atoms, molecules, or ions. In liquids, there is much less regularity at any instant of time. The particles move around more than they do in solids. Solutions are mixtures of uniform composition at the molecular level. They are made up of pure liquids, solids, or gases. Solutions themselves can be liquids or solids. The properties of a solution vary with the proportions in which its constituents are combined.

Self Check

1. What happens when any pure gas condenses?
2. What clue would you use to predict whether a substance has a high or low melting point?
3. State the characteristic features of the crystalline state.
4. On the molecular level, distinguish between a crystalline solid and a liquid.
5. What is the difference between a solution and a physical mixture?

Forces Between Atoms Vary in Strength

17-6 Interatomic Forces

You are already familiar with the way in which two bar magnets, magnetic dipoles, attract each other. This is illustrated in Figure 17-7a. In the same way, electric dipoles are attracted to each other, as shown in Figure 17-7b, opposite charges being drawn together. This type of force is called a dipole-dipole attraction.

For atoms, which are spherically symmetrical and have no permanent electric dipoles, and for nonpolar molecules dipole-dipole forces, as such, are not important. However, they do come into play in an indirect way.

There are forces between the atoms of the noble gases, such as helium, neon, and krypton. There are also forces between the molecules of elements such as fluorine, F_2, chlorine, Cl_2, and phosphorus, P_4, even though these atoms and molecules have no permanent dipole moments. The fact that all molecules of all these elements will condense to form liquids and solids shows that *some* forces exist.

The forces that cause the noble gases or F_2, for example, to condense are called *van der Waals* forces, after the Dutch scientist who studied them. They are very weak forces, so that when they are the only forces available, quite low temperatures are required in order for condensation to occur. Van der Waals forces are always present, but when other, stronger forces also operate, the van der Waals forces can generally be ignored.

The cause of van der Waals forces is *fluctuating* dipoles. As the electrons move about in an atom or molecule, at any instant the distribution may be distorted, that is, nonsymmetrical. At that instant, a small dipole exists. This momentary dipole will affect the electron distribution in an atom or molecule next to it. It will induce, that is, bring about, a dipole in its neighbor. However, it will induce an oppositely-oriented dipole and the two dipoles will attract each other, so long as they exist. These attractions are weak because the dipoles are small and of short duration. They also extend over only a very short range because a dipole in one molecule cannot induce a dipole in another molecule unless the two molecules are very close together.

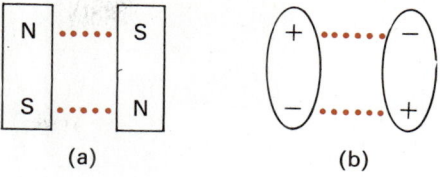

Figure 17-7 *Two types of dipole attraction:* (a) *magnetic dipoles,* (b) *electric dipoles.*

You now have a simple rule for predicting when a weakly-bound molecular liquid and a low-melting crystal will be formed by a given element. *If the element forms a molecule that gives each atom the orbital occupancy of a noble gas, then only van der Waals interactions among such molecules remain.* The low-melting crystals formed in this way are known as **molecular crystals,** since the smallest units in the space lattice are molecules.

Later in this chapter the factors that determine the magnitude of van der Waals forces will be discussed. For the moment, it is sufficient to observe that the elements forming van der Waals liquids and solids are concentrated in the upper right-hand corner

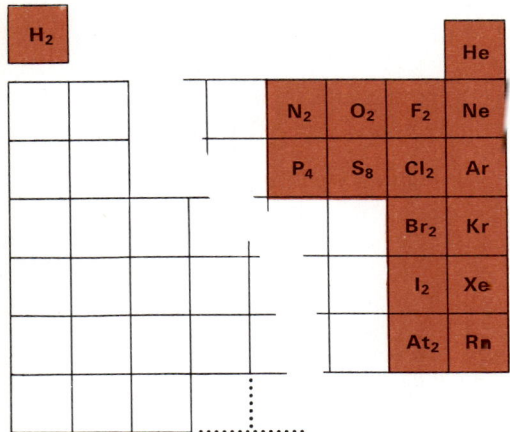

Figure 17–8 *Elements that tend to form liquids and solids through van der Waals interactions are concentrated in the upper right-hand section of the periodic table.*

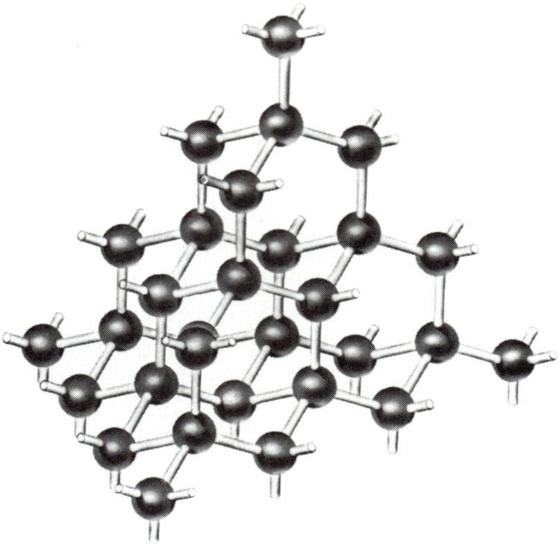

Figure 17–9 *Diamond is a pure form of carbon. Its structural network consists of a compact tetrahedral arrangement of atoms.*

of the periodic table, as shown in Figure 17-8. These elements are able to form stable molecules that completely satisfy the bonding capacity of each atom.

17-7 Covalent Bonds and Network Solids

Fluorine, F_2, oxygen, O_2, and nitrogen, N_2, all form molecular crystals. However, the next member of this row of the periodic table, carbon, presents another situation. There is no small molecule of pure carbon that satisfies completely the bonding capacity of each atom. As a result, carbon is bound in its crystal by a network of interlocking chemical bonds.

With one $2s$ and three $2p$ orbitals available for bonding, carbon can be expected to form a lattice in which each atom forms four bonds. As you saw in Figure 15-10, sp^3 bonding produces tetrahedral bond angles. This type of bonding is consistent with the experimentally-determined structure of diamond, shown in Figure 17-9.

Diamond is a naturally-occurring form of pure, crystalline carbon. Each carbon atom is surrounded by four others, arranged tetrahedrally. The result is a compact structural **network** (interconnected system) bound by normal chemical bonds. This arrangement explains the extreme hardness and the great stability of carbon in this form.

Graphite is another solid form of carbon. In contrast to the three-dimensional lattice structure of diamond, graphite has a layered

Johannes Diderick van der Waals (1837–1923)

COMBINING BOYLE'S LAW, Charles' law, and Avogadro's hypothesis leads to a general law for gases as you have seen. Johannes van der Waals, however, was not satisfied with this law, expressed as *PV = NRT,* because it did not provide accurate results for gases at high pressures.

After studying the problem for some time, van der Waals concluded that the failure of this law to work under all conditions was due to two factors. The first of these was the size of the gas molecules themselves. At low pressures the molecules are so far apart that their size is negligible compared to the total volume occupied by the gas. At high pressures, however, the total volume is far smaller, and the space occupied by the molecules amounts to a measurable fraction of the total volume. This means that the space remaining for them to move around in is much less. The second factor proposed by van der Waals was the possibility that molecules must have some slight attraction for each other when they are close together. This mutual attraction would have the same effect on the total volume as would a further increase in pressure.

To correct the original gas law for the factors of size and mutual attraction, van der Waals developed an equation that was able to predict with a high degree of accuracy the behavior of gases at high pressures. For one mole of a gas, his equation is written in the form:

$$\left(P + \frac{a}{V^2}\right)(V - b) = RT.$$

In this equation "*a*" is a constant related to the attractive force between the molecules, and "*b*" is a constant that depends on the size of the molecules of the gas being studied.

Once the validity of van der Waals' equation had been established, the concept of weak attractive forces between molecules, known today as *van der Waals' forces,* was extended to help explain certain types of behavior in liquids and solids. Van der Waals was awarded the 1910 Nobel Prize in Physics for his work in developing his famous equation.

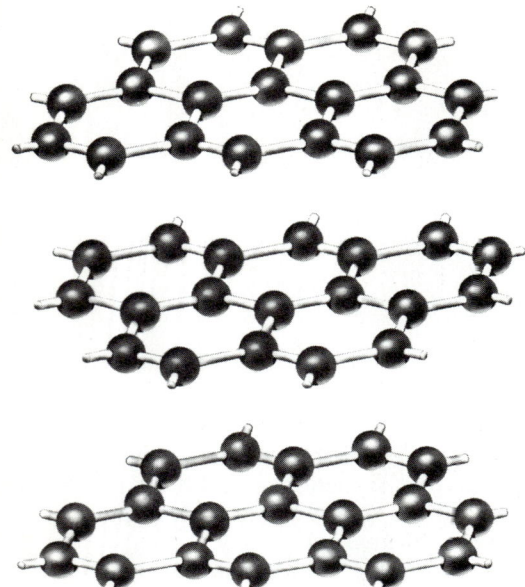

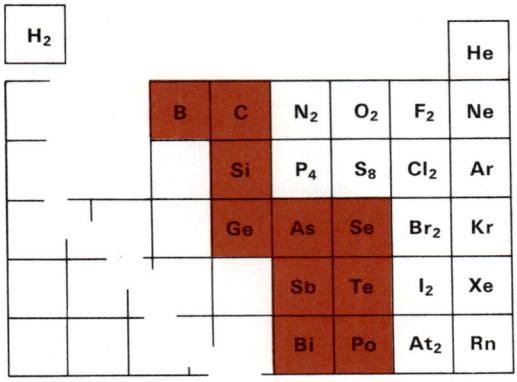

Figure 17-11 *Elements that form network solids involving covalent bonding are located on the right side of the periodic table between metallic and nonmetallic elements.*

Figure 17-10 *Graphite is a pure form of carbon that is different from diamond. It differs from diamond in that its structural network is a two-dimensional layered arrangement of carbon atoms.*

structure, as shown in Figure 17-10. Although each layer is strongly bound, only weak forces exist between adjacent layers. These weak forces make the graphite crystal easy to cleave and explain its softness and lubricating qualities.

The elements that form network solids lie on the right side of the periodic table. They border the elements that form molecular crystals on one side and those that form metals on the other. Thus their properties are often intermediate between those of the metals and of the nonmetals. In this borderline region classifications are sometimes difficult to make. Whereas one property may suggest one classification, another property may lead to a different conclusion. Figure 17-11 shows some elements that form solids that are neither wholly metallic nor wholly molecular crystals.

17-8 Metallic Bonding

You have considered solid forms of the elements fluorine, oxygen, nitrogen, and carbon. In each case, a solid is formed in which the bonding capacity is completely satisfied. Two elements of the second row (beryllium and lithium) are metallic. They do not have enough electrons to permit the complete use of the valence orbitals in covalent bonding. Furthermore, the ionization energies of these elements are quite low. *Two conditions are necessary for metallic bonding: vacant valence orbitals and low ionization energies.*

Characteristic Properties of Metals. Perhaps the most obvious metallic property is reflec-

Figure 17-12 *A silversmith forces solid metal to "flow" under his hammer, shaping it to his design. This property is called malleability.*

tivity or luster. With few exceptions (gold, copper, bismuth, manganese) all metals have a silvery-white color. This results from their reflecting all frequencies of light. As previously stated, in Chapter 11, the electron configuration of a substance determines the way in which it interacts with light. Thus, the characteristic reflectivity of metals indicates that all metals have a common type of electron configuration.

A second characteristic property of metals is high electrical conductivity. Again you find a metallic behavior that suggests a special electron configuration.

Metals also possess unusually high thermal conductivity, as anyone who has drunk hot coffee from a tin cup can testify. Among metals, the best electrical conductors are also the best thermal conductors. This is a clue that these two properties are somehow related. Once again, the electron configuration proves to be responsible.

Though the mechanical properties of the various metals differ, all metals can be drawn into wires and hammered into sheets without shattering. Metals, then, are said to be malleable, or workable. This fourth characteristic property of metals is illustrated in Figure 17-12.

Location of Metals in the Periodic Table. The location of the metals in the periodic table is shown in Figure 17-13, on the next page. The metals are located on the left side of the table, while the nonmetals are exclusively in the upper right corner. Furthermore, the elements on the left side of the table have relatively low ionization energies. You will see that the low ionization energies help to explain many of the features of metallic behavior.

Electron Behavior in Metals. What is the nature of the metallic bond? This bond, like all others, forms because the electrons can move simultaneously near two or more positive nuclei. How do the electrons in metals do this?

Consider a crystal of metallic lithium. In its crystal lattice each lithium atom has eight nearest neighbors around it. Because this atom has only one valence electron, it cannot form ordinary electron-pair bonds with all of these nearby atoms. However, it does have four valence orbitals available. Its electron and the valence electrons of its neighbors thus can approach quite near to its nucleus. In short, each lithium atom has an abundance of valence orbitals but a shortage of bonding electrons.

Consider the dilemma of the valence electron of a particular lithium atom. It finds

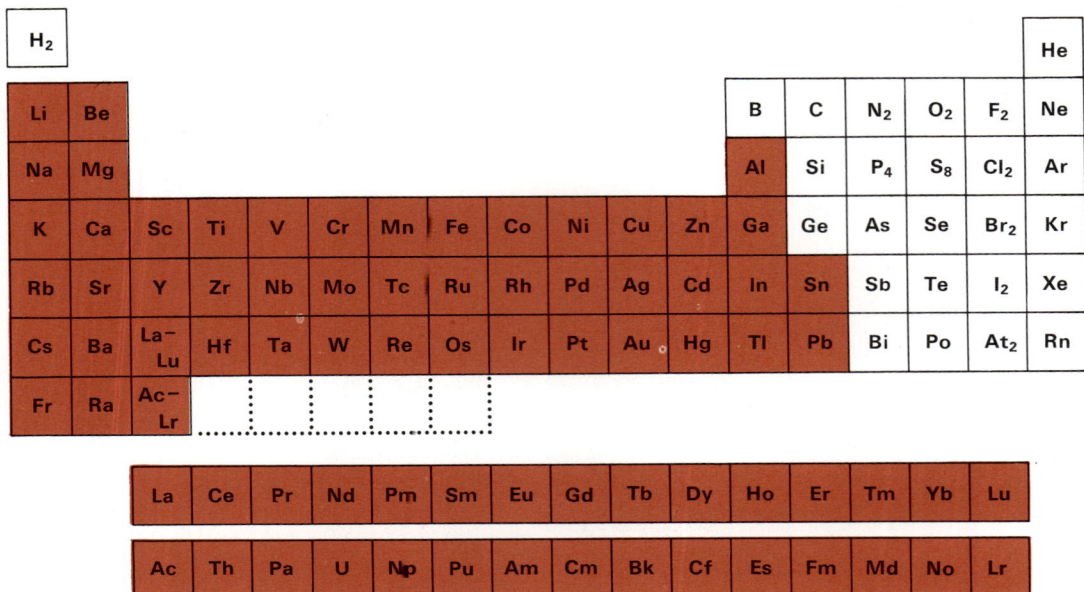

Figure 17-13 *Metallic elements are by far the most abundant. They extend from the extreme left side of the periodic table all the way over to the elements that form network solids.*

eight neighbor nuclei nearby, and it has complete freedom of movement in the empty valence orbitals around its parent nucleus. Everywhere the electron moves it finds itself between two positive nuclei. All of the space around a central atom is a region of almost uniformly low potential energy. Under these circumstances an electron can move easily from place to place. Each valence electron is virtually free to make its way throughout the crystal.

This argument leads us to picture a metal as an array of positive ions located at the crystal lattice sites and immersed in a "sea" of mobile electrons. The idea of a more or less uniform electron sea emphasizes an important difference between metallic bonding and ordinary covalent bonding. In molecular covalent bonds the electrons are localized in a way that fixes the positions of the atoms quite rigidly. The bonds are said to have *directional* character. The electrons tend to remain concentrated in certain regions of space. In contrast, the valence electrons in a metal are spread almost uniformly throughout the crystal. For this reason, the metallic bond does not exert the directional influence of the ordinary covalent bond.

How effectively does the electron sea bind the atoms together? You can answer this question by comparing the energy necessary to vaporize one mole of the metal to the free atoms with the energy required to break one mole of ordinary covalent bonds. The vaporization energy of a mole of one of the alkali metals is only one-fourth to one-third of the energy needed to break a mole of ordinary covalent bonds. This is not too surprising. Remember that the ionization energy of a free alkali metal atom is small. Hence, the valence electron in the free atom does not experience a strong attraction to the nucleus. Since the electron is not strongly attracted by one alkali metal atom, it is not strongly attracted by two or three such atoms in the metallic crystal. Because of this, the binding energy between electrons and nuclei in the alkali metal crystals is rather small. This results in relatively weak metallic bonds. You might expect that the metallic bond would become stronger in those elements that have a greater number of valence electrons and a greater nuclear charge. In these cases there are more electrons in the sea, and each electron is more strongly bound, owing to the increased nuclear charge. This argument is in accord with the experimental heats of vaporization shown in Table 17-1.

To choose a specific case, compare the heats of vaporization of magnesium and aluminum. The higher value for aluminum shows that the metallic bond is stronger when the number of valence electrons and the charge on the nucleus increase. Thus the strength of the metallic bond tends to increase as you go from left to right along a row of the periodic table.

Explanation of the Properties of Metals. The nonlocalized or mobile electrons account for the many unique features of metals. Since metallic bonds do not have strong directional character, many metals can be deformed easily without shattering their crystal structure. Under the influence of a stress, one plane of atoms may slip by another (see Figure 17-14). As they do so, the electrons are able to maintain some degree of bonding between the two planes. Metals can be hardened by alloying them with elements that have the

TABLE 17–1 Heats of Vaporization of Metals

Placement in Periodic Table	Metal	Heat of Vaporization (kcal/mole)
Second row	Li	32.2
	Be	53.5
	B	129
Third row	Na	23.1
	Mg	31.5
	Al	67.9
Fourth row	K	18.9
	Ca	36.6
	Sc	73
Fifth row	Rb	18.1
	Sr	33.6
	Y	94
Sixth row	Cs	16.3
	Ba	35.7
	La	96

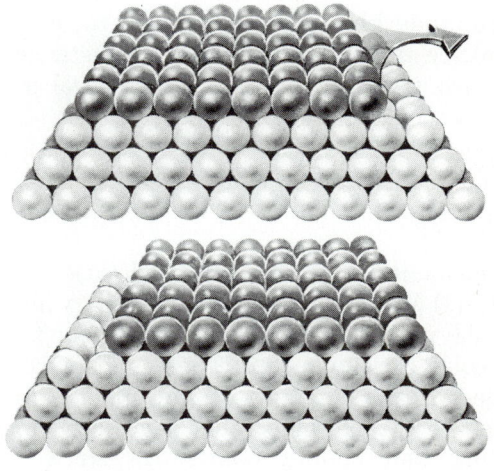

Figure 17–14 *The slippage of planes of metal atoms helps to explain some of the unique properties exhibited by the metallic elements.*

property of forming directed covalent bonds. Often just a trace of carbon, phosphorus, or sulfur will turn a relatively soft and workable metal into a very brittle solid.

Metals conduct electricity because some valence electrons are free to move throughout the solid. At the same time, these mobile electrons are effective in holding the crystal together. This is true because wherever they move, they are simultaneously close to two or more nuclei. In covalently bonded solids the electrons are strongly localized in the space between a particular pair of atoms. In order for these substances to conduct electricity, a great deal of energy must be supplied to remove the electrons from this region between the atoms. This amount of energy is not available in normal electric fields, so covalent substances do not normally conduct electricity.

The excellent heat conductivity of metals is also due to the mobile electrons. Electrons in regions of high temperature acquire large amounts of kinetic energy. These electrons move through the metal very rapidly and give up their kinetic energy to heat the crystal lattice in the cooler regions. Substances in which the electrons are highly localized conduct heat as small amounts of energy, transferred from one atom to its immediate neighbor. This is a slower process than is the conduction of energy by means of electron motion.

Finally, it is necessary to explain why metallic properties eventually disappear as one proceeds from left to right along a row in the periodic table. You have seen that there are two main reasons for the mobility of electrons in metals. (1) Electrons are readily removed from the atom (the ionization energy is low). (2) Electrons can be close to two or more positive nuclei just about anywhere in the crystal (there are numerous vacant valence orbitals). As the nuclear charge on atoms increases, the vacant orbitals become filled. The regions immediately between two nuclei thus become relatively more attractive to the electron, compared with all other regions. Electrons tend to be more and more localized in these regions. This results in the formation of normal covalent bonds with directional character.

In other words, the **metallic bond** is a sort of nondirectional covalent bond. It occurs when atoms have vacant valence orbitals and low ionization energies.

Summary

The elements form three general classes of solids. (1) Molecular solids are those in which only the weak van der Waals forces keep the molecules close together. These substances, formed by the elements in the upper right corner of the periodic table have very low melting and boiling points. (2) Network solids are formed by elements a little further to the left in the periodic table. These elements cannot form small molecules that satisfy all their normal valence requirements. Instead, they form the infinite networks responsible for such physical properties as great strength, hardness, and high melting points. (3) All the remaining elements are more or less metallic. The main characteristics of metals are shiny surfaces, high electrical conductivity, high heat conductivity, and deformability or workability. These properties can all be explained in terms of an orderly array of positive ions bathed in a sea of mobile electrons.

Self Check

1. Why is diamond very hard, whereas graphite is soft and slippery?
2. What are the two conditions necessary for metallic bonding?
3. How does the "sea of electrons" model explain why a metal conducts heat and electricity well?

Complex Forces Hold Compounds Together

17-9 Additional Forces in Compounds

You have seen that the pure elements can solidify in the form of molecular solids, network solids, or metallic solids. Compounds can also condense to these three types of solids. However, there is one main difference between elemental and compound solids. In a pure element the ionization energies of all atoms are identical, and electrons are shared equally. In compounds, the most stable electron distribution need not involve equal sharing. Because of this, electric dipoles may result. Since two bonded atoms can have different ionization energies, the electrons may spend more time near one of the positive nuclei than near the other. This charge separation can give rise to strong intermolecular forces of a type not found in molecules composed of pure elements.

17-10 Van der Waals Forces and Molecular Compounds

Though charge separations are possible in compounds, there are many molecular compounds that do not have appreciable electric dipoles. On cooling, these molecules behave much like the molecules of pure elements. If the bonding capacity of each atom is completely satisfied, then only the weak van der Waals forces remain between molecules. These weak interactions result in low-melting solids and low-boiling liquids that retain many of the properties of the gaseous molecules.

Three factors seem to be particularly important in determining the magnitudes of van der Waals forces: (1) the number of electrons, (2) the molecular size, and (3) the molecular shape. These factors operate in both elements and compounds. Their effect, however, is greater in compounds.

Van der Waals Forces and Number of Electrons. As you saw in Table 8-5, the melting and boiling points of the noble gases increase as the number of electrons increases. Elements and compounds with covalent bonding behave in the same way. Figure 17-15 (next page) shows this graphically. Figure 17-15a shows the melting and boiling point trends among the noble gases and among the halogens. The horizontal axis gives the row number, which is an index to the total number of electrons of the respective elements. Figure 17-15b refers to compounds with formulas CX_4. Here, the horizontal axis shows the row number of the outermost atoms in the molecule. These are the atoms that "rub shoulders" with neighboring molecules. Regarding van der Waals forces, it is quite important that CBr_4 has atoms from the fourth row of the periodic table on the "surface" of

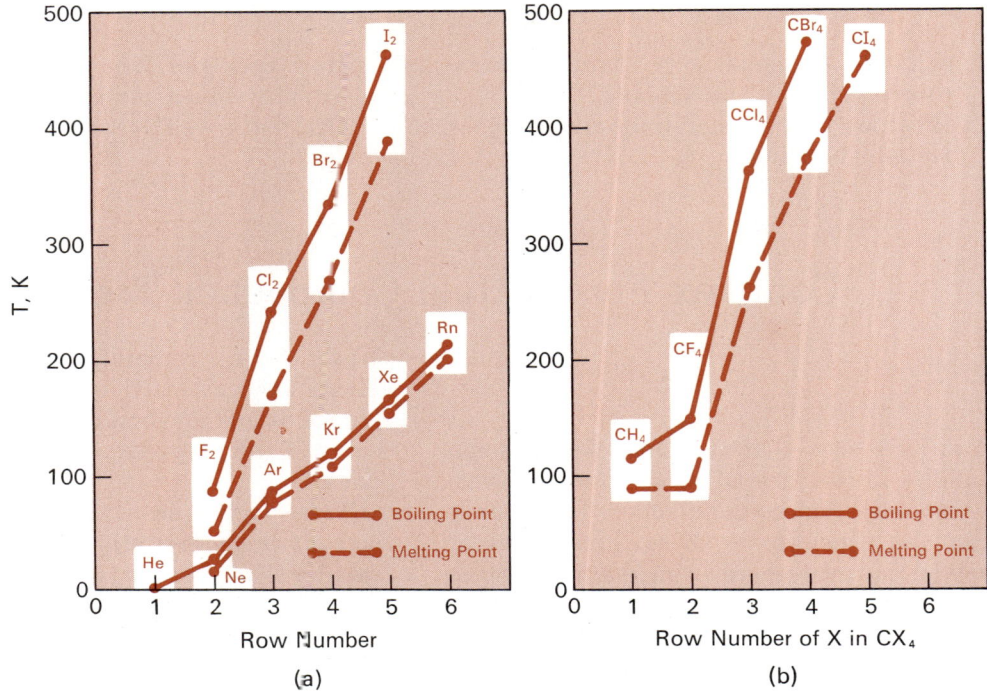

Figure 17-15 *These graphs show the relationships between row numbers in the periodic table and the melting and boiling points of (a) the noble gases and the halogens and (b) carbon compounds of formula CX_4.*

the molecule. It is somewhat less important that the central atom, carbon, is from the second row. This is true because the outermost atoms are most influential in fixing intermolecular forces. Thus, in molecular compounds, the greater the number of electrons the greater the van der Waals forces.

Van der Waals Forces and Molecular Size. Scientists have found that, among similar molecules, the larger the molecule, the higher is its melting point. For example, compare methane CH_4, and ethane, C_2H_6. Their exterior atoms are the same, hydrogen atoms. Still, the boiling point of ethane, 185 K, is higher than that of methane, 112 K. The same effect is found for C_2F_6 (boiling point, 195 K) and CF_4 (boiling point, 145 K). It is also found for C_2Br_6 (this substance decomposes at 483 K before it reaches its boiling point) and CBr_4 (boiling point, 463 K).

These differences in boiling points can be explained. As you learned in Section 17-6, van der Waals forces are weak. These weak interactions are effective only between the sections of different molecules that are in

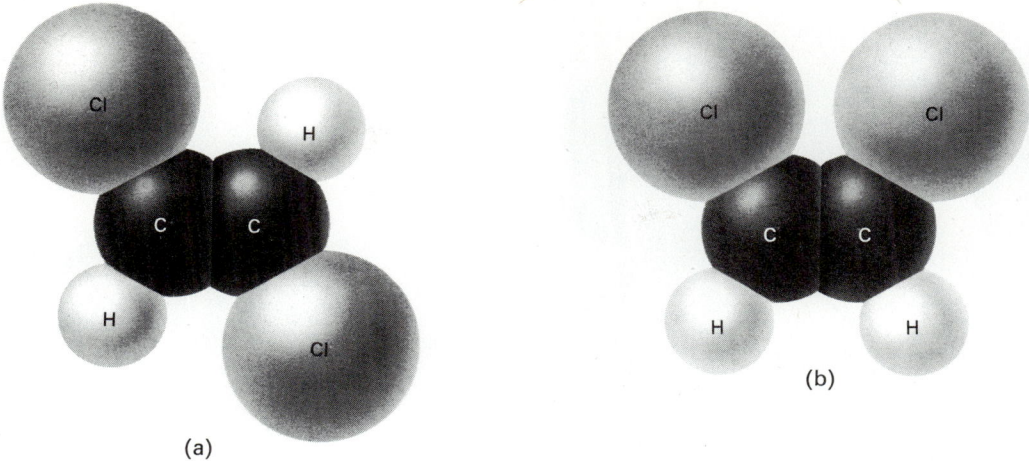

Figure 17-16 *Geometric isomers, of 1,2-dichloroethylene. The trans isomer (a) is more symmetrical than the "one-sided" cis isomer (b).*

close contact. In other words, van der Waals forces operate mainly between the surfaces of molecules. Among similar molecules, then, the larger the molecule—and, therefore, the larger its surface area—the stronger the van der Waals forces.

Van der Waals Forces and Molecular Shape. Substances with structures that have a high degree of symmetry generally have higher melting points than closely related compounds that lack this symmetry. There are striking examples of this among the double-bonded compounds. The *cis* and *trans* isomers of many such compounds have melting point differences due to differences in molecular shape. For example, *trans* 1,2-dichloroethylene has a higher melting point than has its isomer *cis* 1,2-dichloroethylene. The explanation for this is easy to understand. The symmetrical *trans* form can pack into an orderly crystal lattice in a neater and more compact fashion than the "one-sided" *cis* form molecules. (See Figure 17-16.) This and two other examples are given in Table 17-2, below.

TABLE 17-2 The Effect of Molecular Shape on Melting Point: *cis* and *trans* Isomers

Name	Formula	m.p. *cis*-(°C)	m.p. *trans*-(°C)
1,2-dichloroethylene	ClCH=CHCl	−80	−50
butenoic acid	$CH_3CH=CHCOOH$	15	72
fumaric, maleic acids	HOOCCH=CHCOOH	130	290

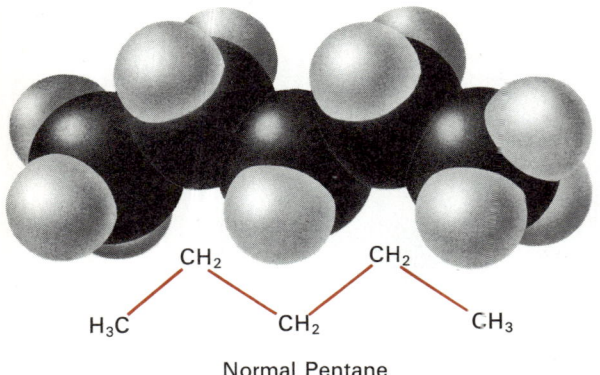

Figure 17-17 *Two structural isomers of pentane, C_5H_{12}. The different structures of their molecules are responsible for their very different sets of properties.*

Another example of the influence of molecular symmetry on physical properties is found in two structural isomers of the formula C_5H_{12}. These compounds, called *normal pentane* and *neopentane*, have molecular shapes that differ drastically, as shown in Figure 17-17.

The extended molecule, n-pentane, has a zig-zag shape. The van der Waals forces act between the external hydrogen atoms of one molecule and those of adjacent molecules. The large surface contact produces relatively large van der Waals forces. This results in a relatively high boiling point. On the other hand, this snakelike molecule does not pack readily in a regular lattice. Therefore, its crystal has a low melting point.

Contrast the highly compact symmetrical neopentane. This molecule is like a ball and readily packs into an orderly crystal lattice. Because of the resulting stability, it has a rather high melting point. Once melted, however, neopentane forms a liquid that boils at a temperature below the boiling point of n-pentane. Neopentane has less surface contact with its neighbors and hence is more volatile.

It is well to note that most of the compounds of carbon condense to molecular liquids and solids. Their melting points are generally low (below about 573 K). In fact, many carbon compounds boil below 373 K. The similar chemistry of the liquid and solid phases shows the retention of the molecular identities.

Some molecular solids, such as HCl and SO_2, have dipole moments. In these, there are forces between the molecular dipoles in addition to the van der Waals forces. Additional forces such as these usually increase the stability of the solid, but it is not easy to make comparisons.

17-11 Covalent Bonds and Network Solid Compounds

Compounds can form network solids. Since two or more different atoms are involved, there is much greater variety among the network solid compounds than among the network solid elements. Silica, with empirical formula SiO_2, is a network solid. Silica and other silicon-oxygen compounds make up about 87 percent of the earth's crust. Almost all common minerals contain substantial amounts of silicates, the general term for silicon-oxygen solids. (Actually, silicates contain very small amounts of other elements, which you need not consider here.) These are network solids that have interesting and important variations. Figures 17-18, 17-19, and 17-20 show in a schematic way the three types of network solids formed by silicon. The silicon is always tetravalent. However, in some of its compounds it forms very long silicon-oxygen-silicon chains. In others it forms large interlinked sheets. In still others it forms a very large three-dimensional network solid.

Many properties of silicates can be understood in terms of the type of network lattice that is formed. The "one-dimensional" network appears in Figure 17-18. Here the atoms within a given chain are strongly linked by covalent bonds. The chains, nevertheless, interact with each other through much weaker forces. This is consistent with the threadlike properties of many of these silicates. The asbestos minerals are of this type.

In a similar way, the sheets of the "two-dimensional" network silicates (shown in Figure 17-19, next page) are held together weakly. Hence these minerals cleave readily into thin, strong sheets. The micas and clays have this type of structure.

The three-dimensional network shown in Figure 17-20, on the next page, is silica (quartz). Like diamond, it is hard and has a high melting point. The various minerals that make up granite are of this type.

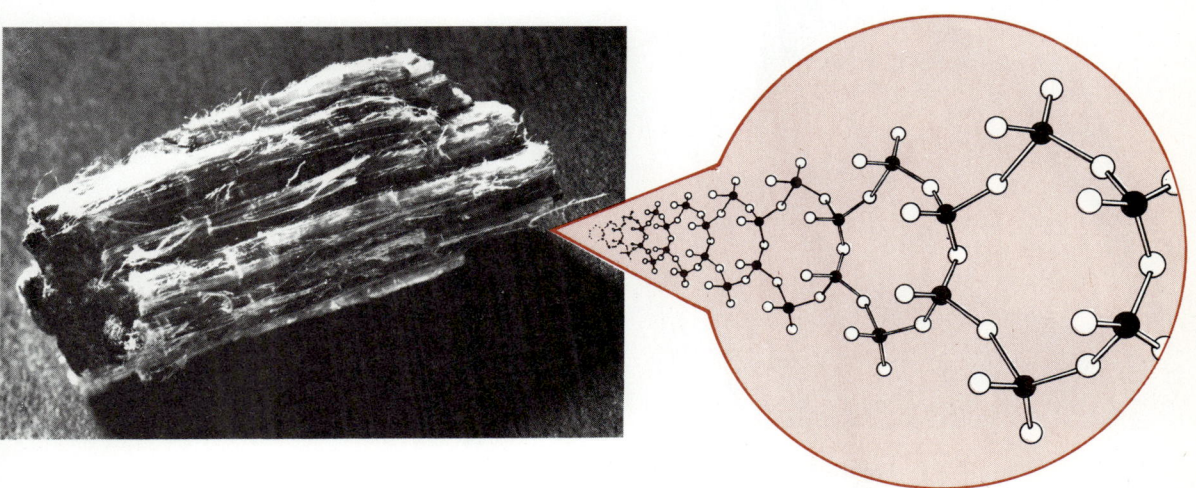

Figure 17–18 *Asbestos, a mineral, has a "one-dimensional" network of atoms.*

506   Chapter Seventeen

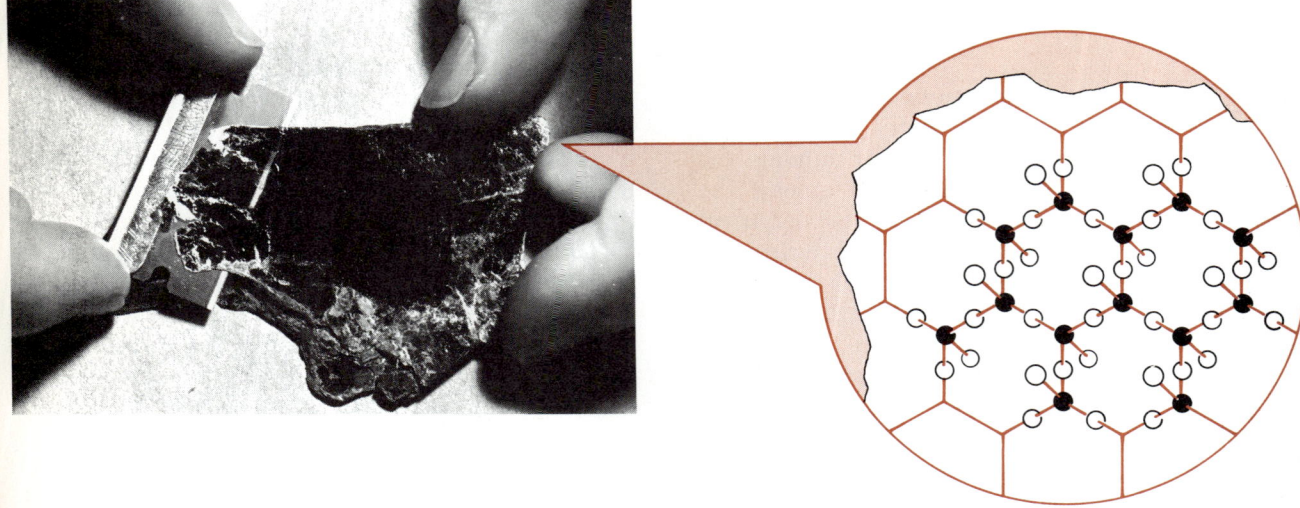

Figure 17-19 *Mica displays a "two-dimensional" network in which the sheets of atoms are held together weakly.*

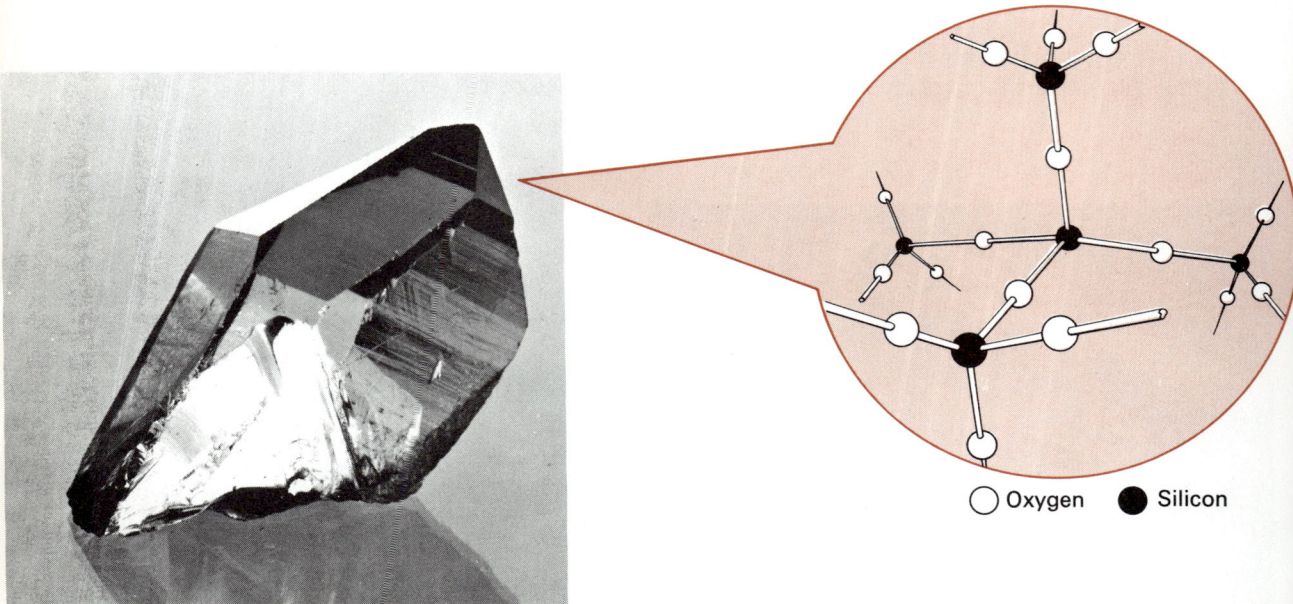

○ Oxygen   ● Silicon

Figure 17-20 *Quartz consists of a three-dimensional network of atoms, similar to that of diamond.*

Summary

Compounds, like elements, form molecular solids, network solids, or metallic solids, depending on the electron configurations of the atoms involved. Some important properties of compounds that form molecular solids depend upon the strength of van der Waals forces. Attractive forces operating between molecular dipoles may be present in addition to the van der Waals forces. The strength of van der Waals forces tends to increase with increasing numbers of electrons and with molecular size. More symmetrical molecules tend to form higher-melting, but lower-boiling, solids. Examples of network solid compounds are asbestos, mica, and quartz.

Self Check

1. How do the forces operating in molecules of pure elements differ from those operating in compounds?
2. What is the relationship between van der Waals forces and molecular size? and number of electrons?
3. Compare network solid compounds with network solid elements.

Ionic Solids Are Simply Explained

17-12 Factors Favoring Ion Formation

Solids built up of ions are formed by many compounds. They are never formed by a single element. Ionic substances have often been mentioned earlier in this book. The structure of a typical one, NaCl, was used in Section 17-2 to illustrate the regular structure of a crystalline solid.

Perhaps the first questions to be asked concerning ionic solids are: What compounds will form ionic solids, and why? The answers are actually rather simple. They have been given, at least in part, in the discussion of the properties of halogens and alkali metals presented in Chapter 8. The halides of the alkali metals are all excellent examples of ionic compounds. Thus, LiF is built up of Li^+ and F^- ions, CsI is built up of Cs^+ and I^- ions, and so on.

An ionic compound forms when a metal that can easily lose one or more electrons combines with a nonmetal that tends strongly to acquire one or two electrons. In this way, a noble gas electron configuration results.

The nonmetals that most readily acquire electrons to form negative ions of noble gas electron configuration are those in the extreme upper right corner of the periodic table. They are the following ones.

$$F, \text{ which gives } F^-$$
$$Cl, \text{ which gives } Cl^-$$
$$Br, \text{ which gives } Br^-$$
$$O, \text{ which gives } O^{2-}$$

All of the fluorides, chlorides, bromides, and oxides formed with the more metallic elements are ionic compounds. For instance, magnesium oxide, MgO, consists of Mg^{2+} and

O^{2-} ions, while scandium chloride, $ScCl_3$, consists of Sc^{3+} and Cl^- ions. On the other hand, such compounds as phosphorus trichloride, PCl_3, and carbon monoxide, CO, are not ionic. Ions do not form, because the elements P and C do not have any significant tendency to lose electrons. They are not metallic elements.

Another general class of ionic compounds is one that contains polyatomic ions. These are ions that have a very stable group of atoms and, as a group, bear a charge. Many of the common polyatomic ions are negative ions. Important examples follow.

CO_3^{2-}	carbonate ion
NO_3^-	nitrate ion
SO_4^{2-}	sulfate ion

The most common polyatomic positive ion is NH_4^+, the ammonium ion.

There are ionic compounds that contain either polyatomic negative ions, polyatomic positive ions, or both. The following examples show this.

Compound	Ions
Na_2SO_4	$2Na^+ + SO_4^{2-}$
$Fe(NO_3)_3$	$Fe^{3+} + 3NO_3^-$
NH_4Cl	$NH_4^+ + Cl^-$
$(NH_4)_2SO_4$	$2NH_4^+ + SO_4^{2-}$

The polyatomic ions are stable for the same basic reason that monatomic ions are stable. When the charge is included in these polyatomic ions, each atom has achieved a noble gas electron population. For example,

consider the nitrate ion. We have already discussed this ion in some detail, in Section 15-8, because it illustrates the ideas of resonance. One of the three equivalent electron dot formulas that enter into the resonance hybrid description of its electron distribution follows.

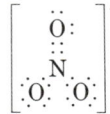

You can see that altogether there are 24 electrons present. Each of the four atoms is surrounded by a set of eight valence electrons (a neonlike configuration). This is possible only because of the presence of an extra electron. This electron causes the entire nitrate ion to have a charge of 1−. Without this there would be only 23 valence electrons (6 from each O atom plus 5 from the N atom). The added electron serves to complete a set of neonlike electron structures around all atoms in the unit. For this reason it is very strongly held, and the NO_3^- ion is very stable.

17-13 Structures of Ionic Compounds

Ionic compounds are built up of ions packed together into a compact, orderly array. Precisely how they will be packed in any given compound depends on several factors. These factors include their relative sizes and charges. However, the underlying principle is that the structure must bring oppositely-charged ions close together, while keeping ions of like charge separated.

The sodium chloride structure illustrates this principle. The ball-and-stick model in

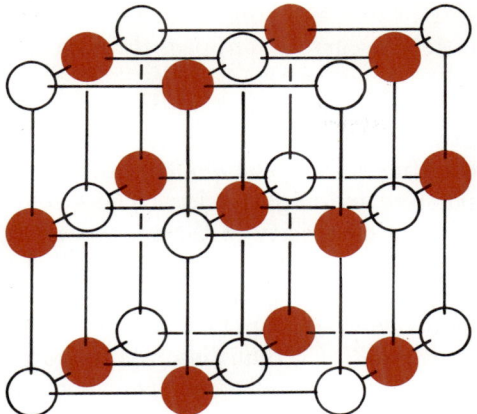

Figure 17–21 *The ball-and-stick model above shows the structure of sodium chloride, NaCl. The open circles represent one type of ion and the filled circles represent the other type. It is a three-dimensional checkerboard.*

Figure 17-21 clearly shows the spatial relationships. A close examination of this structure reveals that each Na$^+$ ion is surrounded by a set of six Cl$^-$ ions. Only beyond this set of oppositely-charged ions are there again to be found some Na$^+$ ions. The structure around each Cl$^-$ ion is the same system in reverse. Each Cl$^-$ ion is entirely surrounded by a set of six Na$^+$ ions. Only beyond these are Cl$^-$ ions found again.

Magnesium oxide has exactly the same structure, in which Mg^{2+} ions replace Na$^+$ ions, and O^{2-} ions replace Cl$^-$ ions.

Another example of an ionic structure is provided by calcium carbonate, CaCO$_3$. A common form of this substance is known as limestone. Here the Ca^{2+} ions alternate with the CO$_3^{2-}$ ions. No two ions of the same kind ever come directly into contact.

Another example of an ionic structure is the wurtzite, ZnS, that you made earlier (Section 17-5).

17-14 *Properties of Ionic Compounds*

As you know, ionic compounds are built up of charged particles. Because of this, they have characteristic properties that distinguish them from other kinds of solids (metallic solids, network solids, and molecular solids). These properties can be understood in terms of the ionic nature of the material.

Ionic solids are generally hard, though not tough, and have high melting points. These properties result from the electrical forces of attraction between oppositely-charged ions. These forces of attraction are quite strong because the ions are very close together in an ionic crystal. (Review Coulomb's law, in Section 9-2.) The structure of an ionic crystal results from maximizing these attractive forces and minimizing the repulsive ones between ions of like charge.

Although many ionic compounds are as hard as metals, they lack the toughness of metals. This brittle character of ionic substances (remember how easily table salt is crushed) can also be explained in terms of the forces involved. When stress is applied to a piece of metal, one layer of atoms can slide past another. Throughout this process, the mobile sea of electrons also flows in such a way as to maintain cohesive force between the two layers. However, the process is different when two layers in an ionic crystal slide by each other. The layers ultimately must reach a stage where the positive ions of one layer are all directly opposite positive ions in the other. At the same time, the negative ions of each layer are brought up against each other. At this point there are

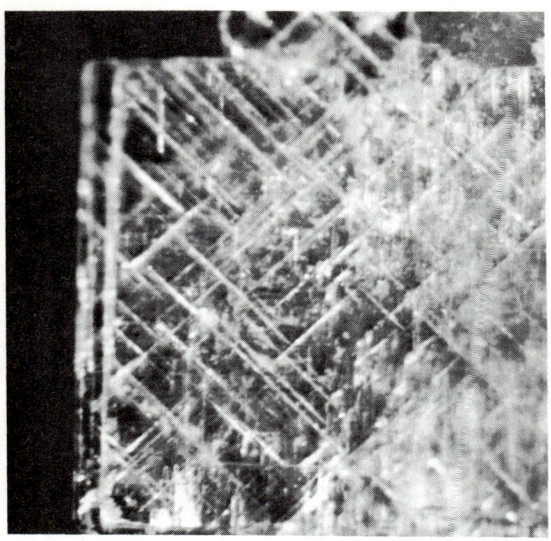

Figure 17-22 *As illustrated above, ionic crystals tend to crack along definite lines. This results from layers of ions of the same charge being forced together.*

tremendous forces of repulsion. This results in the separation of the two parts of the crystal along the plane between the sliding layers. The result is that the crystal cracks, as shown in Figure 17-22.

While ionic *solids* do not conduct electricity, they become good conductors in the *molten* (melted) state. In a solid ionic compound, each ion is strongly held by the forces on it from all the other ions. Therefore, no movement of charges is possible, even though a high voltage may be applied. Within each ion all the electrons are also strongly bound. For this reason, they cannot produce an electric current either. However, an important change occurs when the ionic substance becomes a liquid. Then the positive and negative ions become mobile and can be made to move by applying a voltage. Therefore an electric current *will* pass through the material.

The last main property of ionic compounds is that they dissolve in appropriate solvents. The resulting *ionic solutions* contain separated positive and negative ions. These solutions conduct electricity in the same way as do the molten ionic substances. Again, there is movement of the positive and negative ions when a voltage is applied. An ionic compound, such as NaCl, that dissolves in water to form conducting solutions is called an **electrolyte**. The nature of electrolytes and ionic solutions will be considered in more detail later in this chapter.

17-15 Some Quantitative Ideas about Ionic Crystals ★

You have already encountered the basic qualitative idea about the structures of ionic solids. Positive and negative ions are packed together so as to maximize the attractive forces between oppositely-charged ions and minimize the repulsive forces between ions of the same charge.

In an ionic solid each positive ion has as its nearest neighbors a set of negative ions, which surround it in a uniform, symmetrical way. Each negative ion is surrounded by a set of positive ions. The sodium chloride structure that you have studied shows this nicely. Each Na^+ ion is surrounded by six Cl^- ions. The arrangement of the six Cl^- ions, as you may recall, is octahedral. That is, if they were connected by lines, the lines would form the edges of an octahedron, with the Na^+ ion in the middle of it. The six Na^+ ions that surround each Cl^- ion are also arranged octahedrally.

Another alkali metal chloride is cesium chloride, CsCl. This has a different structure

Solids, Liquids, and Solutions 511

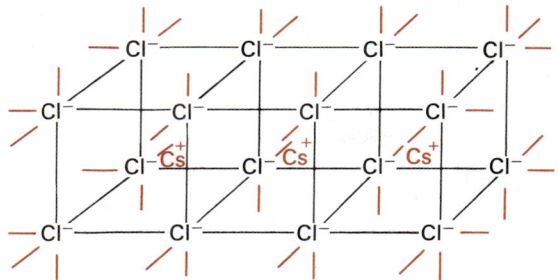

Figure 17-23 *In the cesium chloride structure, the Cl⁻ ions form an array of adjoining cubes and there is a Cs⁺ ion in the center of each.*

from NaCl, as shown in Figure 17-23. The same qualitative idea is evident. Each ion has only those of opposite charge as its closest neighbors. But the detailed arrangement is different. In the CsCl structure each ion is surrounded by *eight* ions of the opposite charge in a cubic arrangement. Why do these two compounds have different structures?

The Cs^+ ion is much larger than the Na^+ ion. You would expect this to be so because the element cesium is near the bottom of the alkali metal family while the element sodium is near the top. The radii of the Na^+ and Cs^+ ions, as well as the radius of the Cl^- ion, are presented in Figure 17-24.

The greater the number of negative ions that can be placed next to each positive ion, the more electrostatically favorable the ionic crystal will tend to be. However, the number of anions that can make contact with each cation is limited by the size of the anion. As soon as the anions begin to overlap each other, large anion-anion repulsive forces will arise. The ideal situation is to have the largest possible number of anions around each cation, without having the anions overlap.

The reason we focus our attention on the question of how many anions will fit around a cation, instead of the reverse, is because the cation is nearly always the smaller ion. For NaCl and CsCl this is so. The packing arrangement is restricted by how many of the bigger ions can be placed around the smaller one.

Let us now look at the geometrical relationships involved in the NaCl and CsCl structures. The key question is this: What is the largest radius that the anion can have, relative to the cation, in order that the structure can be formed without overlap of the

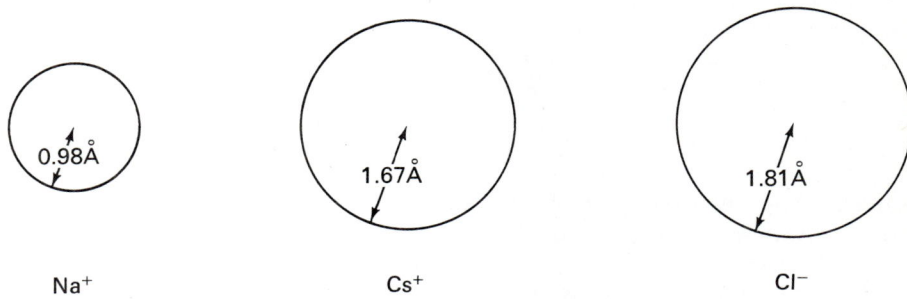

Figure 17-24 *The radii of the Na^+, Cs^+, and Cl^- ions.*

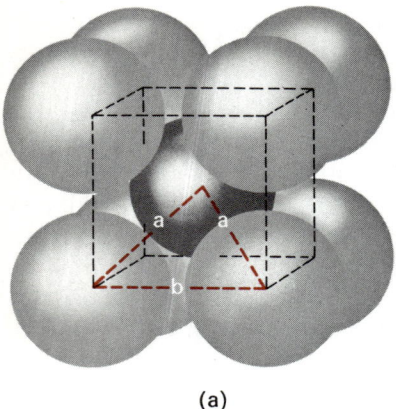

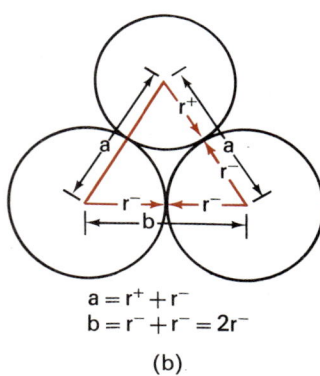

Figure 17–25 Geometry of the cesium chloride, CsCl, structure. The Cl⁻ ions shown in (a) should actually be just touching one another, as indicated in (b). They are separated slightly in (a) in order to show the geometry of the structure more clearly.

anions? For the CsCl structure, the geometric problem is presented in Figure 17-25.

The distance a is half the length of a body diagonal, d, of a cube. The distance b is the length of the edge of a cube. Using the Pythagorean theorem (twice), you can readily show that

$$d = \sqrt{3b^2} = (\sqrt{3})b.$$

Since $d = 2a$, there is the following relationship between a and b.

$$a = \frac{\sqrt{3}}{2} b$$

Now substitute for a and b the expressions in Figure 17-25b, which contain the radii r^+ and r^-, of the cation and anion, respectively. Then, carrying out some algebraic rearrangements, you get the following result.

$$a = \frac{\sqrt{3}}{2} b$$

$$r^+ + r^- = \frac{\sqrt{3}}{2} (2\, r^-)$$

$$r^+ = \sqrt{3}\, r^- - r^-$$

$$r^+ = r^- (\sqrt{3} - 1)$$

$$\frac{r^-}{r^+} = \frac{1}{\sqrt{3} - 1} = \frac{1}{1.732 - 1.000}$$

$$= \frac{1}{0.732} = 1.37$$

This means that when the radius ratio, r^-/r^+, is 1.37 the anions just touch each other and each anion just touches the cation in the CsCl structure. If the anion gets any larger, relative to the cation, the CsCl structure cannot be built without either (1) having the anions overlap if the anions all make contact with the cation, or (2) having the anion-to-cation distance become greater than $r^+ + r^-$ in order to avoid anion-anion overlap. In either case the structure would be less stable.

The radii given in Figure 17-24 lead to the following results.

CsCl $r^-/r^+ = 1.08$ (<1.37)
NaCl $r^-/r^+ = 1.85$ ($\gg 1.37$)

Since 1.85 is much greater than 1.37, NaCl cannot adopt the CsCl type of structure. Eight Cl$^-$ ions cannot fit around the Na$^+$ ion. However, with CsCl, the radius ratio is less than 1.37, and the structure is possible.

For the NaCl type of structure, the geometric problem is presented in Figure 17-26. Proceeding as before, you can get the following result.

$$b = (\sqrt{2})\,a$$
$$2r^- = \sqrt{2}\,(r^+ + r^-)$$
$$2r^- - \sqrt{2}\,r^- = \sqrt{2}\,r^+$$
$$(2 - \sqrt{2})\,r^- = \sqrt{2}\,r^+$$
$$\frac{r^-}{r^+} = \frac{\sqrt{2}}{2 - \sqrt{2}} = \frac{1.414}{0.586}$$
$$= 2.41$$

Since $r^-/r^+ = 1.85$ for NaCl, the type of structure that NaCl adopts is geometrically satisfactory. Six Cl$^-$ ions can fit around Na$^+$ in the NaCl structure. Eight Cl$^-$ ions cannot fit around Na$^+$.

It is true that, geometrically, CsCl could form the NaCl type of structure. This structure would not be stable relative to its actual structure, however, since it gives fewer contacts between oppositely-charged ions.

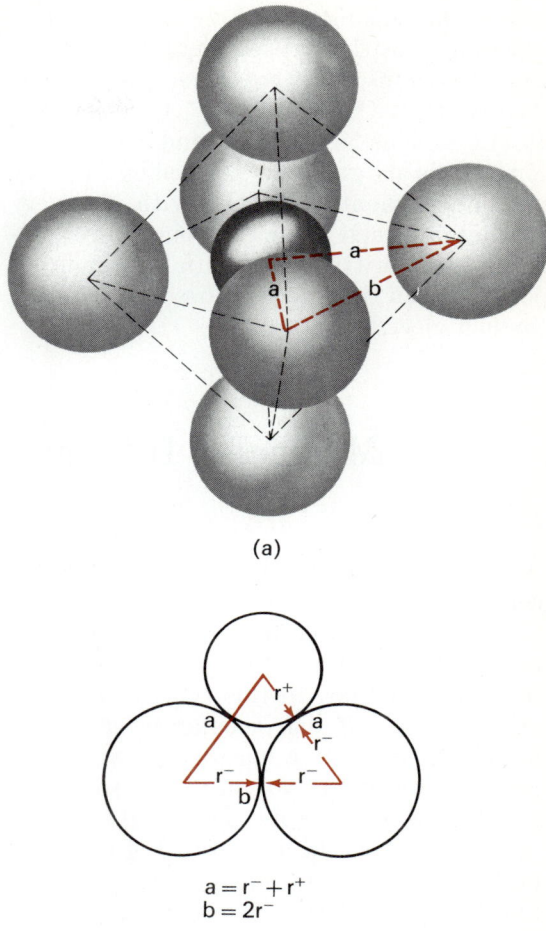

Figure 17-26 *The geometry of NaCl. If the ions shown in (a) are assumed to be just touching each other, the distances a and b can be expressed in terms of the ionic radii (b).*

Summary

Ionic compounds are built of positive ions and negative ions packed together to form a crystalline solid. In this state, the attractive forces between opposite charges are maximized while the repulsive forces between like charges are minimized. This means that, along any direction in such a substance, positive ions alternate with negative ions. Ions may be either monatomic, such as Na^+, Ca^{2+}, Al^{3+}, Cl^-, O^{2-}, or polyatomic, such as NH_4^+, NO_3^-, and SO_4^{2-}. Ionic compounds are generally formed by the very metallic elements together with the very nonmetallic ones. Ionic compounds are generally hard, brittle solids with high melting points. When they melt or dissolve, they can conduct electricity.

Self Check

1. What are the properties of ionic compounds?
2. Which elements can form ionic compounds? Where is this group of elements located in the periodic table?
3. Describe the underlying principle of ionic structure.
4. Write an electron dot formula for the sulfate ion and show why its charge is 2—.

Molecules Intermingle To Form Solutions

17-16 *Solvent Properties*

As noted earlier, in Section 17-4, a solution consists of two or more substances mixed together at the molecular level. There is one particular type of solution that is most common and familiar. In this type, a relatively small volume of one substance, called the **solute,** dissolves in a much larger volume of another, called the **solvent.** The solvent is almost always a liquid. The solute is most often a solid, as illustrated in Figure 17-27, but it may well be a gas or a liquid. There may be two or more solutes in the same solution. An important question to wonder about is: Why does a particular solvent dissolve some solutes but not others? A consideration of some characteristics of liquids will help answer this question about their *solvent properties*.

You read how molecules can have molecular dipoles in Section 15-11. Chloroform, $CHCl_3$, is one such *polar* molecule. It has about the same bond angles as methane, CH_4, and carbon tetrachloride, CCl_4. As you have seen, carbon with sp^3 bonding forms four tetrahedrally-oriented bonds. Cancellation of the four bond dipoles occurs when all

Figure 17-27 *The most common type of solution results when a solid (the solute) dissolves in a liquid (the solvent). Here, potassium permanganate crystals are dissolving in water.*

of the atoms attached to the central carbon atom are of the same element. A nonpolar molecule results. Suppose that one of the chlorine atoms in CCl_4 is replaced by a hydrogen atom. The polar molecule $CHCl_3$ results, since the four bond dipoles do not now cancel.

17-17 Molecular Dipoles

The forces between molecules are strongly affected by the presence of molecular dipoles. As noted earlier, two molecules that possess molecular dipoles tend to attract each other more strongly than do molecules without dipoles. One of the most important results of this is found in solvent properties.

Table 17-3 shows some solubility data for various solutes in two solvents, carbon tetrachloride and acetone, CH_3COCH_3. These two solvents differ in their polar properties. As noted in the preceding section, the central carbon atom in CCl_4 is surrounded by four bonds that form a regular tetrahedron. With this molecular shape, CCl_4 has a zero molecular dipole. In contrast, acetone has a bent structure, as you can see in Figure 17-28. As

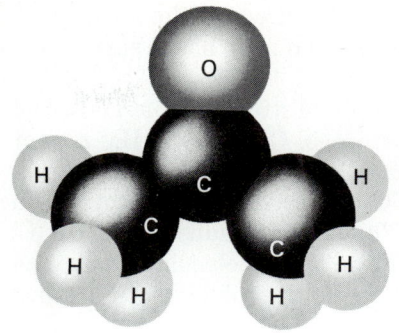

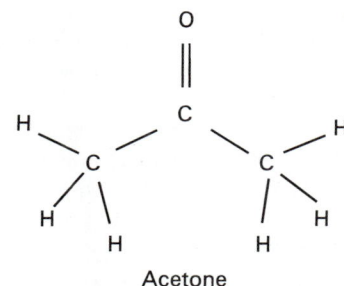

Acetone

Figure 17-28 *A space-filling molecular model and structural formula for acetone, CH_3COCH_3. Its structure helps to explain why the acetone molecule is polar.*

TABLE 17-3 Solubilities in Carbon Tetrachloride, CCl_4, and in Acetone, CH_3COCH_3 at 25°C (moles/liter)

Solute	Solute Polarity	Solubility in Solvent	
		CCl_4 (nonpolar)	CH_3COCH_3 (polar)
CH_4, methane	nonpolar	0.029	0.025
C_2H_6, ethane	nonpolar	0.22	0.13
CH_3Cl, chloromethane	polar	1.7	2.8
CH_3OCH_3, methyl ether	polar	1.9	2.2

a result of this structure and the polarity of the carbon-oxygen bond it has a significant electric dipole.

Contrast the solubilities in Table 17-3. The first two substances, CH_4 and C_2H_6, do not have molecular dipoles. In each case, the solubility in CCl_4 exceeds the solubility in CH_3COCH_3. The next two substances, CH_3Cl and CH_3OCH_3, do have molecular dipoles. In each of these cases, the solubility in acetone is the larger.

There is a reasonable explanation for the data in Table 17-3. When a solute dissolves, the solute molecules must be separated from each other and then become surrounded by solvent molecules. Furthermore, the solvent molecules must be pushed apart to make room for the solute molecules. Dipoles interact strongly with each other. For this reason, a polar molecule such as CH_3Cl is energetically more stable when surrounded by solvent molecules that are also polar. Hence, CH_3Cl has a higher solubility in the polar solvent acetone than in the nonpolar carbon tetrachloride. On the other hand, a nonpolar molecule such as CH_4 does not dissolve as well in a polar solvent. It finds it more difficult to wedge in between the strongly interacting molecules of a polar solvent. Thus, CH_4 has higher solubility in the nonpolar solvent, CCl_4.

The old (and true) saying that oil and water do not mix provides a familiar example of the dependence of solvent properties on molecular polarity. The molecules in oil are practically nonpolar, whereas water molecules are extremely polar. The nonpolar liquid does not dissolve any significant amount of the highly polar one, and vice versa.

17-18 Solubility of Electrolytes in Water

The dissolving of electrolytes in water is one of the most important solvent effects caused by electric dipoles. Crystalline sodium chloride is quite stable, as shown by its high melting point. Yet it dissolves readily in water. To break up the stable crystal arrangement, there must be a strong interaction between water molecules and the ions. This interaction can be explained in terms of the dipolar properties of water molecules.

When an electric dipole is brought near an ion, the energy decreases when the dipole orients itself so as to place unlike charges in proximity. Hence water molecules tend to orient around ions. The positive (hydrogen) end of the water dipole points inward if the ion carries negative charge. On the other hand, the negative (oxygen) end points inward if the ion carries positive charge. Figure 17-29 schematically shows this process, which is called **hydration**.

17-19 Boiling and Freezing Points of Solutions

Two properties of a solution that differ from those of the pure solvent are the temperatures at which freezing and boiling occur. As a general rule, when a solid is dissolved in a liquid the freezing point is lowered and the boiling point is raised. In Figure 17-30, on page 518, the rise in boiling point is illustrated.

These temperature effects are particularly useful to chemists. In solutions that contain a relatively small quantity of solute, temperature effects are proportional to the number of dissolved particles (molecules or ions). However, they are independent of what the dissolved particles are.

To illustrate this, consider how sugar depresses the freezing point of water. Sugar consists of molecules that neither dissociate nor associate in dilute solution. Experiments have shown that a solution resulting from one mole of sugar dissolved in 1000 g (1 kg) of water has a freezing point of $-1.86°C$. In

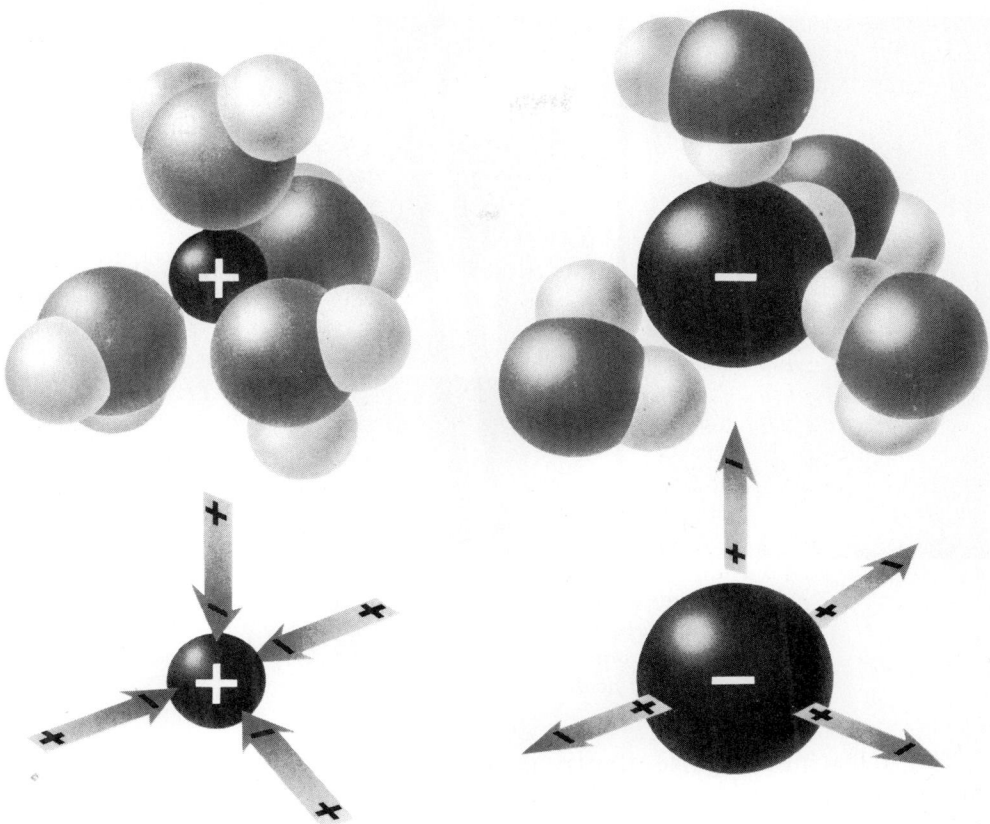

Figure 17-29 *In a water solution of an electrolyte, water molecules, which are electric dipoles, cluster around the electrically charged ions. The negative (oxygen) pole of the water molecule is attracted to the positive ion (left) and the positive (hydrogen) pole is attracted to the negative ion (right).*

general, this can be expressed in the following equation.

$$\Delta t = 1.86 \, m$$

Here, Δt stands for the decrease (in °C) in freezing point, and m for the number of moles of solute dissolved in each 1000 grams of water. A solution containing only one third of a mole of sugar per kilogram of water would have a freezing point of

$$\frac{1.86}{3} = -0.62°C.$$

This freezing-point depression effect has two important uses: (1) in the determination of molar masses and (2) in the determination of the number of ions formed by an ionic compound when it dissolves.

Figure 17-30 *Boiling water (left) has its boiling point raised when a solute is added (center). This results in a brief halt in boiling (right) until the temperature of the solution reaches the higher boiling point.*

The following example illustrates how molar mass can be determined. Consider the substance urea, which has the empirical formula CH_4N_2O. The mass of this unit is 60.0. But, is this the *molar* mass? Perhaps the molecule has the formula $C_2H_8N_4O_2$, that is, twice the empirical formula. In that case, its mass would be 120.0. In order to resolve this problem, you can prepare a solution containing 60.0 g of urea per 1000 g of water (or 6.00 g per 100 g of water). After it has been prepared, measure its freezing point. If 60.0 g represents the true molar mass, the freezing point will be $-1.86°C$. If the true molar mass is 120.0, this solution will contain only half a mole of solute per liter of solvent. Its freezing point will be lowered only $1.86/2 = 0.93°C$ below $0°C$. Experiment shows that urea actually has the molecular formula CH_4N_2O.

Now consider a solution of a typical ionic solute, NaCl, in water. As you know, this substance consists of an array of Na^+ and Cl^- ions, which become separated from each other when the substance dissolves in water. In this way, the electrical conductivity of the solution can be explained. Furthermore, the freezing-point depression of this solution gives evidence for the same description. The molar mass of NaCl is $23.0 + 35.5 = 58.5$. Therefore, a solution containing 5.85 g dissolved in 100 g of water would have a freezing point of $-1.86°C$ if the sodium chloride were present as NaCl units. Actually, the freezing point of such a solution is about $-3.70°C$. In other words, the freezing point depression, Δt, is about twice as great. This means that there are twice as many moles of solute particles as there are moles of the NaCl unit. Clearly, each NaCl unit produces two particles. That is consistent with your knowledge about solid NaCl and its solutions in water. Thus you can assume that these particles are Na^+ and Cl^- ions.

As you might have guessed, there is also a mathematical relationship for boiling point elevation. That relationship, however, will not be treated in this book.

17-20 Units for Expressing Concentration: Molarity

Concentration is the quantity of a substance within a specified quantity of another substance. The more solute that is dissolved in a constant amount of solvent, the more *concentrated* the solution becomes. On the other hand, the more solvent that is added to a constant amount of solute, the more *dilute* the solution becomes. When you concentrate a solution, then, the ratio of solute particles to solvent particles increases. When you dilute a solution, this ratio decreases.

Because the properties of a solution depend on its concentration, it is important to have clearly defined and convenient units for expressing concentrations. Actually, there are a number of useful ways of doing this. Each one is advantageous for a particular purpose. For example, in the preceding section the number of moles of solute per 1000 grams of solvent served as a measure of concentration.

The most generally useful way of expressing concentrations for most chemical problems is in terms of **molarity**. A one molar solution contains one mole of solute in one liter of solution. The abbreviation for molarity is M. Thus, if a liter of solution contains x moles of solute, we say it is an x molar or x M solution.

This unit is very useful. With it we can measure out a desired mass, or a desired number of moles, of solute by measuring out a certain volume of solution. This is an operation that can often be done more rapidly and conveniently than can the actual weighing of the pure solute.

Suppose, for example, that a chemist frequently needs to use about 0.15 mole of sodium hydroxide in about 100 ml of water. The molar mass of sodium hydroxide, NaOH, is 40.0 g. He could, each and every time, weigh out $0.150 \times 40.0 = 6.00$ g of solid sodium hydroxide, dissolve it in 100 ml of water, and add it to his reaction.

It is much simpler, however, for him to weigh out $1.50 \times 40.0 = 60.0$ g and dissolve it in enough water to make precisely one liter. Then he has one liter of a 1.50 M solution. Each time he needs 0.15 mole, he need only measure out exactly 100 ml of this solution. Since one liter contains 1.50 moles of NaOH, one tenth of a liter (0.100 l = 100 ml) contains one tenth of this number of moles. In other words, $0.100 \times 1.500 = 0.15$ mole.

Summary

The solvent properties of liquids depend very much on the polarities of their molecules. Liquids with polar molecules tend to be good solvents for other polar substances and poorer solvents for nonpolar substances. Conversely, nonpolar liquids dissolve nonpolar solutes best, and polar ones not so well.

In an extreme case, such as oil and water, solubility of the polar and nonpolar substance in each other may be very poor indeed. Highly polar solvents such as water are needed to dissolve ionic compounds.

Concentration is an important property of a solution. For our purposes, concentrations are expressed in molarities. A one molar (1 M) solution contains one mole of solute in one liter of solution.

Solutes raise the boiling point and lower the freezing point of a solvent in direct proportion to the quantity present per given mass of solvent. This proportionality is useful for measuring molar mass and for determining the number of ions into which an ionic substance dissociates in solution.

Self Check

1. Distinguish between polar and nonpolar substances.
2. Predict the order of increasing solubility of the following molecules in water: acetone, CH_3COCH_3; chloromethane, CH_3Cl; carbon tetrachloride, CCl_4.
3. How would you go about making up one liter of a 0.50 M NaCl solution?

Hydrogen Has Unique Bonding Properties

17-21 *Hydrogen Bonds*

In Figure 17-15 you saw that the boiling points of symmetrical molecules increase regularly as you go down the periodic table. Figure 17-31 shows the corresponding plot for some molecules possessing electric dipoles.

Consider first the boiling points of HI, HBr, HCl, and HF. The last, hydrogen fluoride, is far out of line, boiling at 19.9°C instead of below −95°C, as would be predicted by extrapolation from the other three. An even larger discrepancy exists between the boiling point of H_2O and the value you would predict from the trend suggested by H_2Te, H_2Se, and H_2S.

Could the extremely high boiling points of HF and H_2O be due to the fact that these are the smallest molecules of their respective series? No, this does not appear to be the explanation because corresponding discrepancies are not present in Figure 17-15. There must be some other explanation for these exceptional boiling points. There must be a new kind of force between the molecules of H_2O and of HF.

This same force is recognized in solid compounds. The most familiar example is solid H_2O, or ice. Ice has a crystal structure in

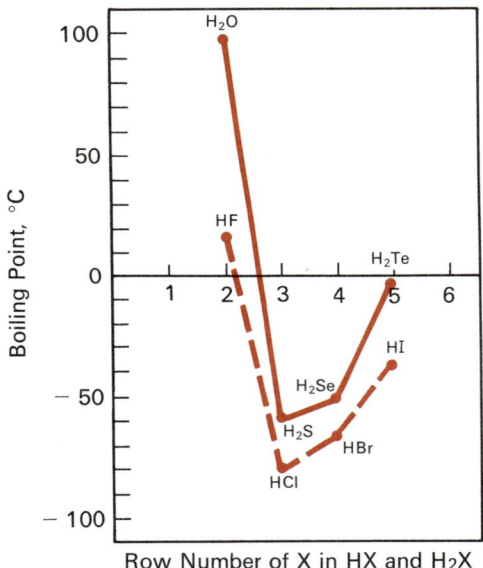

Figure 17–31 *This graph shows the trend in boiling points for the hydrogen halides (HX) and the hydrides of the group six elements (H_2X).*

which the oxygen and hydrogen atoms are distributed in a regular hexagonal crystalline lattice. This arrangement somewhat resembles the diamond lattice. A model of the ice structure is shown in Figure 17-32. Each oxygen atom is surrounded by four other oxygen atoms in a tetrahedral arrangement. The hydrogen atoms are found on the lines extending between the oxygen atoms in the following representation.

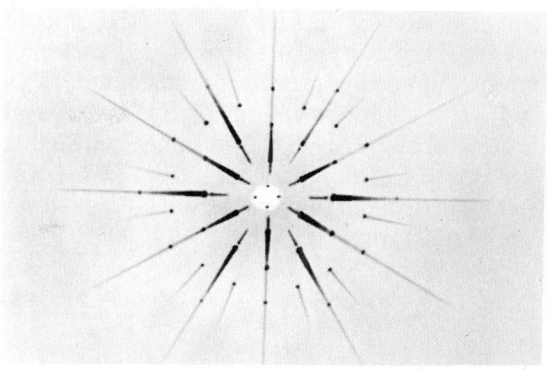

$$\diagup\!\!\!\!\text{O}\!-\!\text{H}\cdots\text{O}\diagdown$$

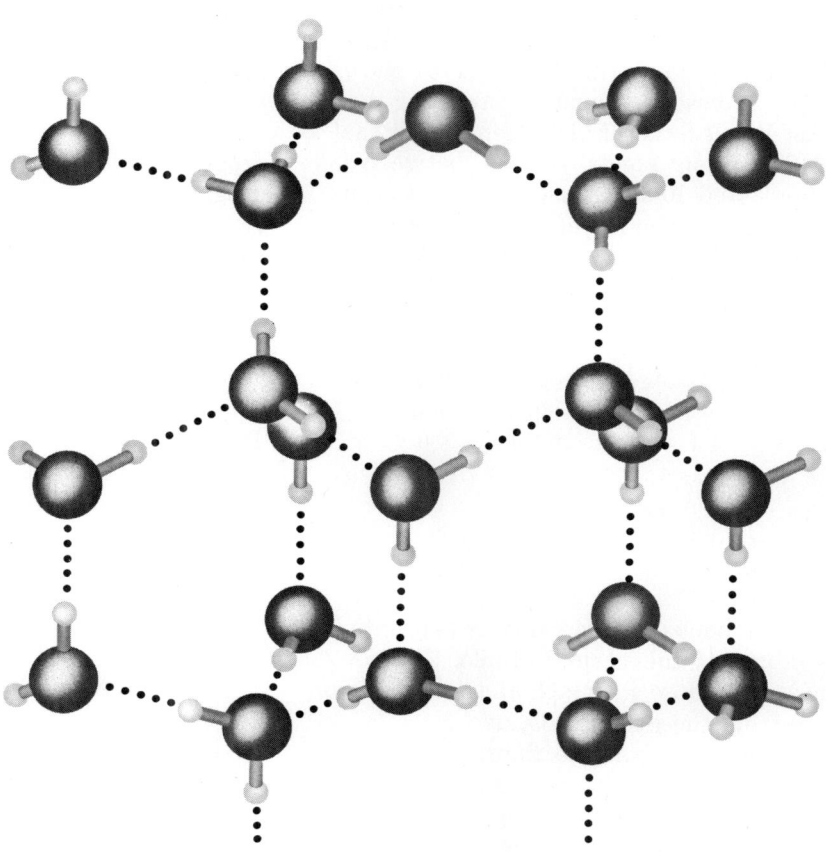

Figure 17-32 *The top photo shows the X-ray diffraction pattern produced by a single ice crystal. The model of the ice structure below it, in which the dotted lines represent hydrogen bonds, is consistent with the experimental results.*

The attractive force between —OH and O must be an example of the new kind of force, or bond, mentioned above. This bond joins the water molecules together in the crystal lattice of ice. This type of bond is called a **hydrogen bond**. It is usually represented in formulas by a dotted line, as in the preceding representation.

17-22 Energy of Hydrogen Bonds

The hydrogen bond is usually shown as a dotted line to indicate that it is much weaker than a normal covalent bond. Consideration of the boiling points in Figure 17-31 shows that the interaction must be stronger than van der Waals forces. Experiments show that most hydrogen bonds release between 3 kcal/mole and 10 kcal/mole upon formation. The energy of this bond places it between van der Waals and covalent bonds. Roughly speaking, the energies are in the following ratio.

van der Waals attractions	hydrogen bonds	covalent bonds
1	10	100

17-23 Where Hydrogen Bonds Are Found

Hydrogen bonds are found between only a few atoms of the periodic table. The commonest are those in which H connects two atoms from the group F, O, and N.

The hydrogen bond to fluorine is clearly evident in most of the properties of hydrogen fluoride. The high boiling point of HF, compared to those of the other halides, is due to hydrogen bonds. It is one of several pieces of data that indicate that HF does not exist in the liquid phase solely as separate HF molecules. Instead, there are aggregates of molecules, described in general terms as $(HF)_x$. Gaseous hydrogen fluoride contains the molecular species H_2F_2, H_3F_3, and so on, up to H_6F_6. Some single HF molecules are also present.

These species can be represented in a descriptive formula such as the following.

$$H\text{—}F\cdots H\text{—}F\cdots H\text{—}F\cdots H\text{—}F$$

One of the most important cases of hydrogen bonding occurs in water. In both liquid and solid water every molecule is hydrogen-bonded to its neighboring molecules, by as many as four hydrogen bonds. This is shown in Figure 17-33. Each water molecule can supply the hydrogen atoms to form two hydrogen bonds. It can also interact with hydrogen atoms of other molecules to complete two more hydrogen bonds. It is because of this extensive hydrogen bonding that water has many of its unique properties.

17-24 The Nature of the Hydrogen Bond

In the hydrogen bond the hydrogen atom is attached to two other atoms. Yet bonding rules state that the hydrogen atom, with only the $1s$ orbital for bond formation, cannot form two covalent bonds. What is the explanation for this second bond?

The best explanation for the hydrogen bond is based upon the polar nature of F—H, O—H, and N—H bonds. In each of these three bonds the electron pair is held closer to the F, O, or N atom than it is to the H atom. The bond therefore has a dipole, with the F, O, or N atom negative, and the H atom positive. It is then possible for the positive H atom of one molecule to closely approach the unshared electrons of a negatively charged atom in another molecule. The attraction between the opposite charges causes the formation of a hydrogen bond, as illustrated in Figure 17-34 for HF.

17-25 The Significance of the Hydrogen Bond

Hydrogen bonds play an important part in determining such properties as solubility, melting points, and boiling points. In addition, they affect the form and stability of crystal structures and play a crucial role in biological systems. An example of this role is the way water influences the chemical behavior of many biological molecules. Water molecules can attach themselves to other molecules by forming hydrogen bonds in either of two ways. The water molecule may supply the positively-charged hydrogen atom, as in the following example.

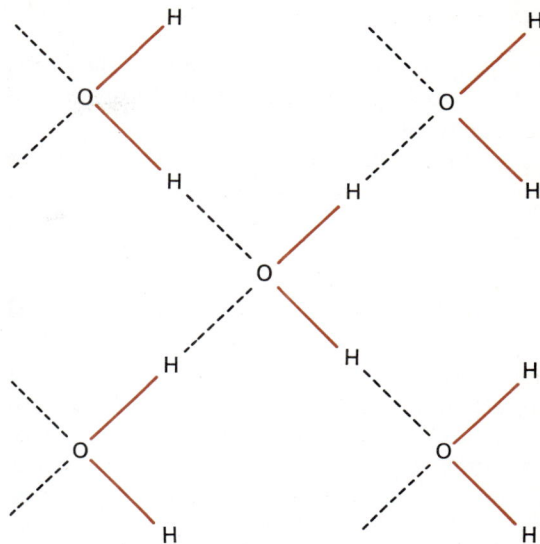

Figure 17-33 Hydrogen-bonding among water molecules.

It may, alternatively, supply the negatively-charged atom of oxygen to interact with the positively-charged hydrogen atom from the other molecule, as follows.

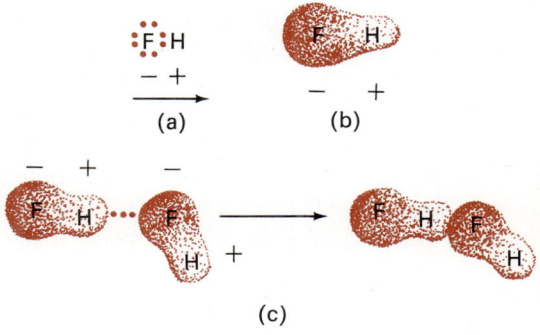

Figure 17-34 Hydrogen-bonding between HF molecules.

Summary

Substances containing O—H, N—H, and F—H bonds often form hydrogen bonds. These are weak linkages (about 5 kcal/mole). Hydrogen bonds form because the O—H, N—H, and F—H bonds are highly polar.

The H atom has considerable positive charge while the other atom has considerable negative charge. Hence, there is an additional attraction between the H of one molecule and the O, N, or F of another.

Self Check

1. Compare the boiling points of HI, HBr, HCl, and HF. Explain any apparent discrepancies.
2. Describe the relative energies of van der Waals attractions, hydrogen bonds, and covalent bonds.
3. How would you represent the bonding in a crystal of ice?

Chapter Summary

The forces that hold atoms and molecules together in liquids, solids, and solutions are of several main types. In any particular case the type depends on the properties of the atoms or molecules making up the substance. Crystalline solids are built up by the orderly and regular repetition of a basic arrangement of a few atoms, molecules, or ions.

Some solids contain infinite networks or sheets of atoms held together by covalent bonds. Examples of these solids are the element carbon and the compound silicon dioxide. Other solids and liquids are built up of discrete molecules (like F_2, P_4, CH_4, SO_2) within which each atom makes use of all its capacity to form covalent bonds. The molecules attract one another through the weak van der Waals forces of about 1 kcal per mole. Such molecular substances tend to have low melting and boiling points. With polar molecules, both van der Waals forces and forces between molecular dipoles cause molecular attraction.

A large number of elements with low ionization energies form metals. Metals consist of orderly arrays of atoms bound together by a mobile sea of electrons. These electrons come from the outermost orbitals of the atoms. Metals are characteristically shiny and malleable, as well as being good conductors of heat and electricity.

Ionic solids consist of orderly arrays of ions packed together in an alternating fashion. The ions may be monatomic (such as Ca^{2+}, Cl^-) or polyatomic (such as NH_4^+, SO_4^{2-}). Ionic solids are hard, high-melting, and brittle. When they melt or dissolve, they can conduct electricity by movement of the ions.

A solution consists of two or more pure substances mixed on the molecular level. The commonest type of solution consists of a smaller amount of a liquid or solid (the solute) dissolved in a larger amount of a liquid (the solvent). The capacity of a liquid to dissolve any given solute depends very much, although not entirely, on the polar nature of the two substances.

The extent to which the melting point of a solvent is depressed by a solute can be used to determine molar masses and the extent of ionization of solutes.

Hydrogen bonding is important in some solids and solutions containing —O—H, —N—H, and —F—H groups.

Questions and Problems

1. Explain why a liquid retains its fluid properties in the temperature range between its boiling point and its freezing point, but becomes a gas at higher temperatures and a solid at lower temperatures.

2. What is the major difference between the arrangement of particles in liquids and in crystalline solids? How do glassy substances differ from these two forms?
3. If very fine talcum powder is thoroughly mixed with a pure hydrocarbon such as n-pentane, how would you determine whether a true solution or a physical mixture is formed?
4. What is a space lattice? What is meant by the coordination number of a particle in a space lattice?
5. Use simple drawings to illustrate the difference between body-centered and face-centered cubic packing in a crystalline solid.
6. What causes van der Waals forces to exist between molecules? What major condition must be satisfied for molecules to be held together by van der Waals forces only?
7. Contrast the bonds between atoms in metals, in molecular solids, and in network solids with regard to (a) bond strength, (b) orientation in space, and (c) the number of orbitals available for bonding.
8. Aluminum, silicon, and sulfur are close together in the same row of the periodic table, yet their electrical conductivities are widely different. Aluminum is a metal. Silicon has much lower conductivity and is called a semiconductor. Sulfur has such a low conductivity it is called an insulator. Explain these differences in terms of valence-orbital occupancy.
9. Sulfur exists in a number of forms depending on the temperature and, sometimes, on the past history of the sample. Descriptions of three of these forms, designated A, B, and C, follow. A is the room temperature form. A changes to B above its melting point, 113°C. B changes to C on heating above 160°C.

(A)
(room temperature)
Crystalline solid
Yellow color, no metallic luster
m.p. = 113°C
Dissolves in CS_2, not in water
Electrical insulator

(B)
(113°C to 160°C)
Liquid
Clear, straw color
Viscosity (fluidity) about the same as water
Electrical insulator

(C)
(around 200°C)
Liquid
Dark color
Very viscous (syrupy)
Electrical insulator

Which of the following structures would most likely account for the observed properties of each of the three forms described above?
(a) A metallic crystal of sulfur atoms
(b) A network solid of sulfur atoms
(c) An ionic solid of S^+ and S^- ions
(d) A molecular crystal of S_8 molecules
(e) A metallic liquid such as mercury
(f) A molecular liquid of S_8 molecules
(g) A molecular liquid of S_n chains, with n a very large number
(h) An ionic liquid consisting of S^+ and S^- ions

10. Solid sulfur is made up of S_8 molecules. Each molecule has a cyclic structure. Phosphorus contains P_4 molecules and each molecule has a tetrahedral structure. On the basis of valence-orbital occupancy explain how these molecular forms can exist while fluorine can only exist as F_2. By comparing the molecular size and

shape of the sulfur and phosphorus molecules, which would you expect to have the higher melting point? What type of attraction exists between these molecules?

11. Discuss the ways in which heat and electricity are conducted by copper (a metal) and by glass (a network solid) in terms of valence-orbital occupancy and electron mobility.

12. The elements carbon and silicon form oxides with similar empirical formulas: CO_2 and SiO_2. The former sublimes (changes from solid to gas) at $-78.5°C$. The latter melts at about $1700°C$ and boils at about $2200°C$. From this large difference, propose the types of solids involved. Draw an electron dot or orbital representation of the bonding in CO_2 and SiO_2 consistent with your answer.

13. Assume you are given a sample of a white solid. Describe some simple experiments you could perform to determine whether or not the bonding involves primarily covalent bonds, ionic bonds, or van der Waals forces.

14. Assume elements A, D, E, and J have atomic numbers 6, 9, 10, and 11, respectively. Write the formula for a substance you would expect to form between each of the following. (a) D and J, (b) A and D, (c) D and D, (d) E and E, (e) J and J. In each case describe the forces involved between the building blocks in the solid state.

15. How would you account for the following properties in terms of the structures of the solids?
    (a) Graphite and diamond both contain carbon. Both are high melting, yet diamond is very hard, while graphite is a soft, greasy solid.
    (b) When sodium chloride crystals are shattered, plane surfaces are produced on the fragments.
    (c) Silicon carbide (carborundum) is a very high-melting, hard substance, used as an abrasive.

16. Consider each of the following in the solid state: sodium, germanium, methane, neon, potassium chloride, water. Which of the following statements describes each of these substances?
    (a) A solid held together by van der Waals forces that melts below room temperature.
    (b) A solid with a high degree of electrical conductivity that melts near $200°C$.
    (c) A high-melting, network solid involving covalently-bonded atoms.
    (d) A nonconducting solid that becomes a good conductor upon melting.
    (e) A substance in which hydrogen bonding is pronounced.

17. Predict the order of increasing melting points of the following substances, each of which contains chlorine: HCl, Cl_2, $NaCl$, CCl_4. Explain the basis of your prediction.

18. Identify *all* types of bonds that you would expect to find in crystals of each of the following substances. (a) Ar, (b) H_2O, (c) CH_4, (d) CO_2, (e) Si, (f) Al, (g) $CaCl_2$, (h) $KClO_3$, (i) NaCl, (j) HCN.

19. A solution containing 2.40 g of NaH_2PO_4 in 100 g of water has a freezing point of $-0.74°C$. What are the solute particles most likely to be?

20. A sugar named glucose has the empirical formula CH_2O. A solution of 6.00 g of glucose dissolved in 100 g of water lowers the freezing point to $0.59 \pm 0.05°C$ according to an actual measurement. Knowing that glucose does not dissociate into any ions when it is dissolved in water, calculate the molecular formula for glucose that best fits the experimental data.

21. Express the concentration of each of the solutions that are listed below in terms of molarity.

Solute	Mass of Solute (g)	Volume of Solution (ml)
$AgNO_3$	17.0	1000
$AgNO_3$	34.0	500
$AgNO_3$	340.0	3000
NaOH	50.0	600
NaOH	0.200	10

22. If enough water is added to 300 ml of a 0.200 M solution of NaCl to increase the volume to one liter, what is the molarity of the new solution? How many grams of NaCl are contained in the original 300 ml? How many grams of NaCl are contained in 300 ml of the final solution?

23. If 500 ml of a 1.00 M solution of KOH and 1500 ml of a 2.00 M solution of KOH are mixed and the volume brought up to precisely 2500 ml by adding more water, what is the molarity of the final solution? How many grams of KOH do 250 ml of this final solution contain?

24. Three bottles on the chemical shelf are marked A, B, and C. Each contains a colorless liquid. The labels that have fallen off the bottles read as follows.

Label No. 1
n-butanol
$CH_3CH_2CH_2CH_2OH$

Label No. 2
n-pentane
$CH_3CH_2CH_2CH_2CH_3$

Label No. 3
diethyl ether
$CH_3CH_2OCH_2CH_3$

A series of measurements made on the three liquids provided the data given in the table below.
Which liquid should be given Label No. 1? Label No. 2? Label No. 3? Explain how each type of measurement influenced your choices.

Measurements	Liquid		
	A	B	C
m.p. (°C)	−131.5	−116	−89.2
b.p. (°C)	36.2	34.6	117.7
Density (g/ml)	0.63	0.71	0.81
Heat of vaporization (cal/g)	85	89.3	141
Solubility in water (g/100 ml)	0.036	7.5	7.9

18 Giant Molecules

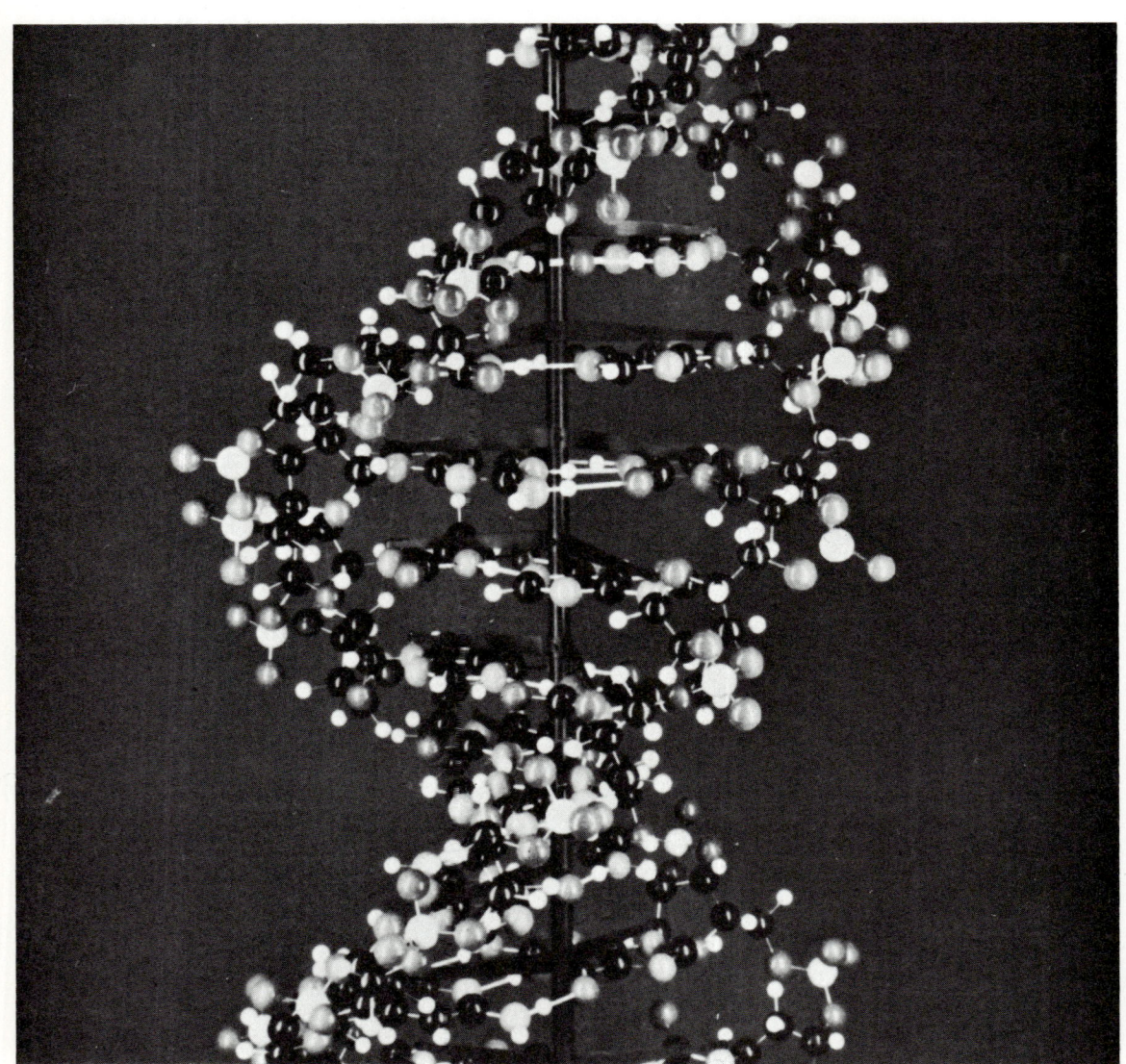

One of the giant molecules discussed in this chapter is the DNA (deoxyribonucleic acid) molecule, a portion of which is represented above.

The molecules you have considered in previous chapters have been small ones. Most molecules are relatively small, with molar masses of less than 1000. However, there are some extremely important substances that contain molecules with molar masses of 1,000,000 or more. Among these are the proteins and starches, which are vital constituents of living organisms, the plastics and fibers, which are so important in shaping the way we live, and even some minerals, such as asbestos. To understand chemistry, it is as important to understand big molecules as it is to understand small ones. In this chapter you will see how some of the main types of giant molecules are constructed and you will also learn why they have their useful properties.

Giant Molecules Are Built of Smaller Units

18-1 Giant Molecules Are Usually Polymers

Giant molecules surround us. They are found in nature and are made by man. These very big molecules are built in a systematic and organized way. One type of small unit or a very few types of small units combine, forming a large aggregate of the required size. Buildings are made in a similar manner. Many bricks, many stones, and many boards are put together in a regular way to build up walls, floors, and ceilings. By choosing a certain type of brick or board or stone, the resulting wall or floor can be made to have a particular texture, color, or other property. In principle, it is the same with giant molecules. They consist of many identical small units that are attached to one another, not by nails or mortar but by chemical bonds. The nature of the small units used determines the properties of the giant molecules.

A large molecule that is made up of a repeating arrangement of small molecules is called a **polymer** (from the Greek, *poly* + *meros*, many parts). Each small molecule is called a **monomer** (*mono* = one). Polymers contain several hundred to several thousand monomers. Generally, the monomers resemble links in a chain. There are some important exceptions to this, but for the majority of polymers it is a correct description.

A question that naturally comes to mind is: How do small molecules combine in a regular way to produce the giant molecules? In the following two sections you will consider the answer to this question.

18-2 Condensation Polymerization

Nylon is a synthetic fiber with which you are probably familiar. Chemists invented it by using a knowledge of how different kinds of organic functional groups react with one another. It was the first entirely synthetic fiber.

In order to appreciate the making of nylon, you have to recall the type of compound called an amide, first discussed in Section 16-18. It contains the amide-bond linkage, where R represents H or a hydrocarbon group.

$$H_3C-CH_2-\underset{\underset{O}{\|}}{C}-\underset{\underset{}{|}}{\overset{\overset{H}{|}}{N}}-R$$

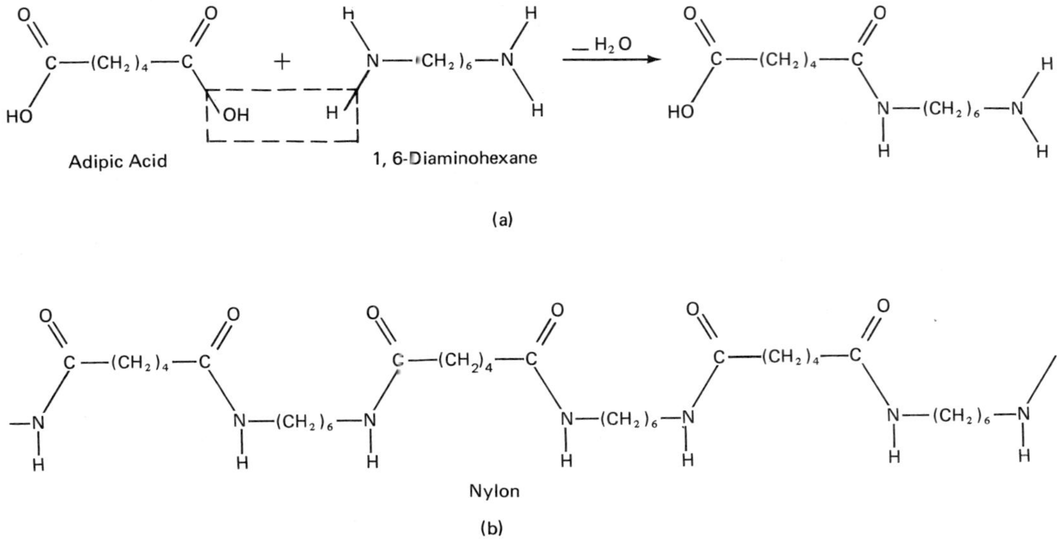

Figure 18–1 *The condensation of adipic acid with 1,6-diaminohexane leads first to a dimer (a) and eventually to a long-chain polymer (b) called nylon.*

You also learned previously that one way to make an amide is by reaction of an ester with an amine, as the following example shows.

$$C_2H_5COOCH_3 + H_2NC_3H_7 \longrightarrow$$
methyl propionate   n-propylamine

$$C_2H_5CONHC_3H_7 + CH_3OH$$
N-propylpropionamide   methyl alcohol

In this reaction methyl alcohol forms. Another closely related reaction that gives amides is that of a carboxylic acid with an amine. The following equation shows one example of this reaction.

$$C_2H_5COOH + H_2NC_3H_7 \longrightarrow$$
propionic acid   n-propylamine

$$C_2H_5CONHC_3H_7 + H_2O$$
N-propylpropionamide

The making of nylon depends on the latter type of reaction. In general, one group of molecules made up of a carbon chain with a carboxyl group at each end (diacid) reacts with another group of molecules made up of a carbon chain with an amine group at each end (diamine).

In particular, chemists react adipic acid

$$HOOC-CH_2-CH_2-CH_2-CH_2-COOH$$

with 1,6-diaminohexane.

$$H_2N-CH_2-CH_2-CH_2-CH_2-CH_2-CH_2-NH_2$$

This reaction must be repeated many times so that the monomers can "link up" to form a polymeric chain, as seen in Figure 18-1. Every time that the molecules react, water is eliminated between the acid and amine functional groups. An amide-bond linkage results at the ends of the molecules. For this

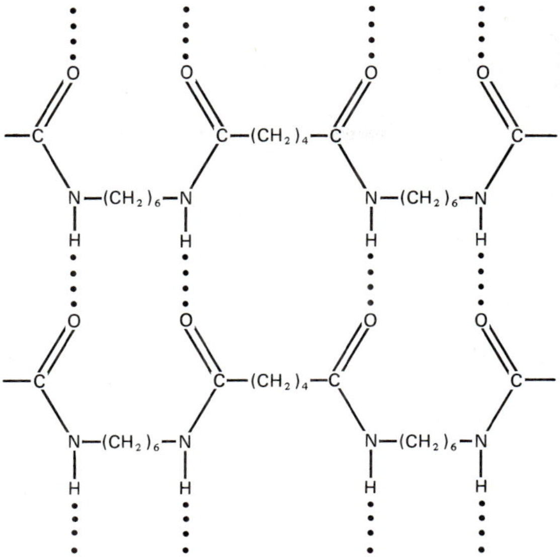

Figure 18-2 *The fibrous character and strength of nylon are due to the formation of hydrogen bonds between parallel chains.*

reason, molecules with functional groups at both ends are needed for the reaction to occur.

The making of nylon depends on a type of reaction called **condensation polymerization**. In all such reactions monomer molecules combine with the loss of some simple molecule, usually water. Other polymeric amides, or *polyamides*, besides nylon can be formed by this type of reaction. These polyamides are formed simply by varying the diacid and diamine used. In this way, polyamides with different properties can be made.

It should be noted that nylon is a fibrous material because the long polyamide molecules are mostly parallel to each other. This parallel arrangement results from the formation of hydrogen bonds between the chains, as shown in Figure 18-2.

Up to now you have considered the condensation polymerization of diacids and diamines. This reaction produces amide-linked polymers. In a similar manner, diacids and dialcohols can react to form ester-linked polymers. This, too, occurs by condensation polymerization. In particular, when the diacid called *para*phthalic acid reacts with the dialcohol called ethylene glycol, very long *polyester* molecules form. This polyester substance is the well-known synthetic fiber called Dacron. Its formation is shown on the next page, in Figure 18-3.

18-3 Addition Polymerization

Besides undergoing condensation polymerization, small molecules can react to form giant molecules in a process called **addition polymerization**. The commonest type of addition polymerization reaction is the following one. Here, alkenes open up their double bonds to form new single bonds between the carbon atoms.

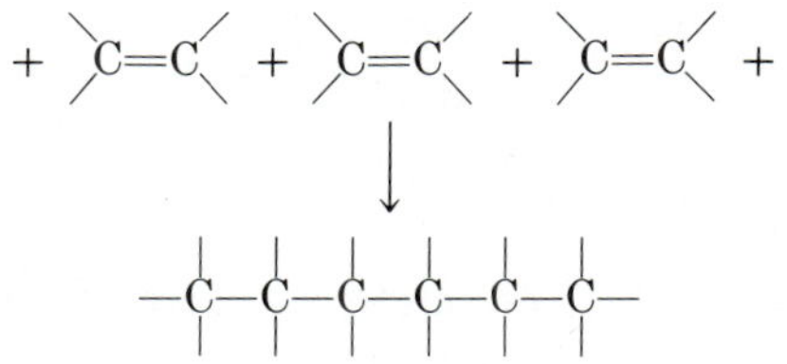

There is no elimination of atoms in this type of reaction. A great many of the most familiar and useful polymers are made in this way. They differ in the types of groups that are attached to the long chain of carbon atoms. When the attached groups are all hydrogen atoms, the polymer is called polyethylene.

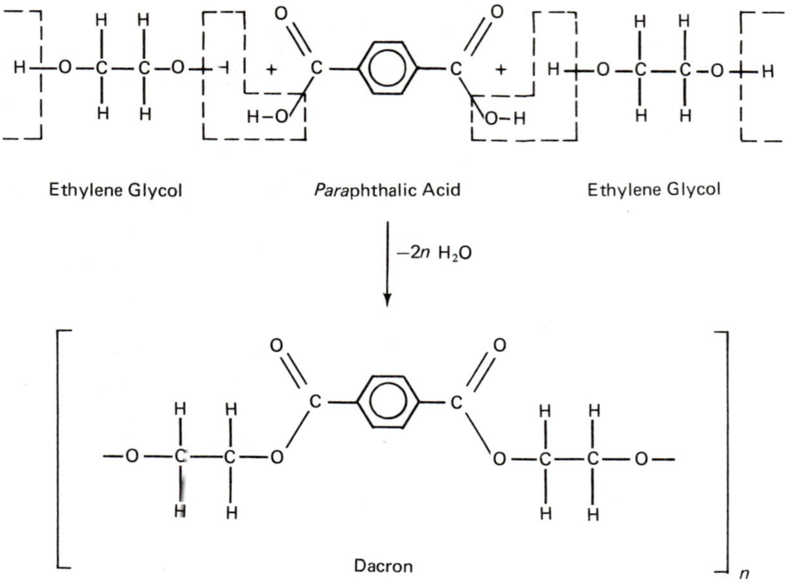

Figure 18-3 The condensation polymerization of ethylene glycol, a dialcohol, with paraphthalic acid, a diacid, leads to the polymer known commercially as Dacron.

TABLE 18-1 Some Familiar Addition Polymers

Monomer	Portion of Polymer Chain	Common Name
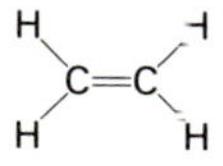 $H_2C=CH_2$	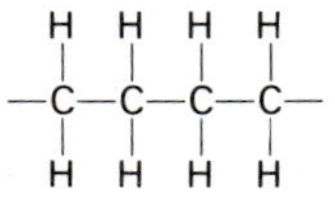 $-CH_2-CH_2-CH_2-CH_2-$	polyethylene
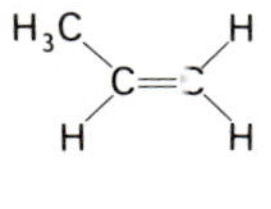 $H_3C-CH=CH_2$	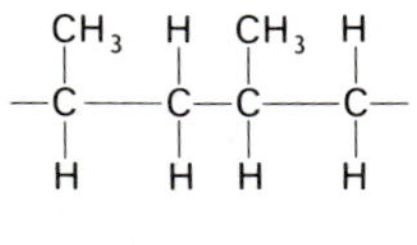 $-CH(CH_3)-CH_2-CH(CH_3)-CH_2-$	polypropylene
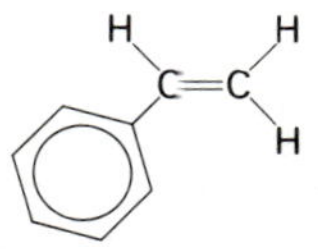 $C_6H_5-CH=CH_2$	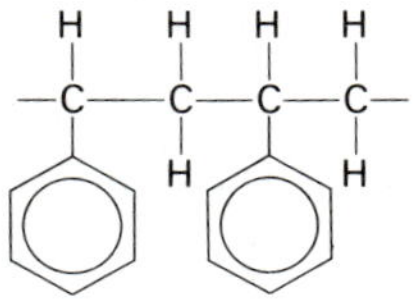 $-CH(C_6H_5)-CH_2-CH(C_6H_5)-CH_2-$	polystyrene, styrene
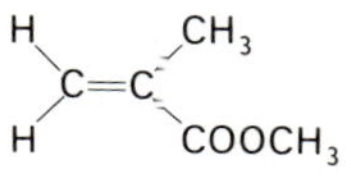 $H_2C=C(CH_3)COOCH_3$	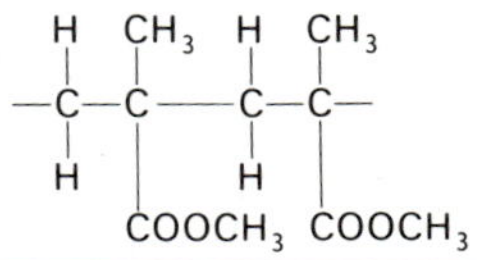 $-CH_2-C(CH_3)(COOCH_3)-CH_2-C(CH_3)(COOCH_3)-$	Lucite, Plexiglas

The small alkene known as ethylene, $H_2C{=}CH_2$, is the monomer that is used. If the monomer used is tetrafluoroethylene, $F_2C{=}CF_2$, the resulting polymer is the substance known as Teflon. Table 18-1 lists some of the better known polymers made by addition polymerization.

Many other synthetic polymers and other giant molecules are encountered every day in an industrialized society. Perhaps you can think of the names of some of them. As you do the several parts of the following experiment, you will become familiar, firsthand, with several polymers.

INVESTIGATION

18-4   The Preparation of Some Polymers

The production of many high quality, high molar mass polymers requires conditions of temperature and pressure control not easily achieved in the high school laboratory. The reactions that you will carry out in this experiment, however, do illustrate the formation of compounds typical of many polymers that have commercial applications.

PURPOSE

You will prepare condensation and addition polymers and will learn how cross-linking affects the properties of polymers.

MATERIALS

5 ml glycerol
3 g powdered phthalic anhydride
10 ml 40% formaldehyde solution
4 g adipic acid
3 ml ethylene glycol
10-15 ml methyl methacrylate
Beaker, 50-ml (or small tin can)
Glass stirring rod
Watch glass
Electric hot plate (or burner)
10 ml each solvent (ethyl acetate, methyl ethyl ketone, acetone)
10 ml saturated solution aniline hydrochloride in water
4 drops concentrated sulfuric acid
3-4 drops benzoyl peroxide in diethyl phthalate or butyl benzyl phthalate
20-30 ml Clear Cast polyester prepolymer
2-3 drops methyl ethyl ketone peroxide

Mortar and pestle
Ringstand and ring
Test tube
Balance
Small glass jar
Aluminum foil, heavy-duty
Paper cup, 3 oz
Rubber band
Piece of wood (or metal)

Figure 18-4

PROCEDURE

Part A. Preparation of a Glyptal Resin

The condensation of alcohols containing more than one —OH group and compounds called anhydrides leads to polyesters known as alkyd resins. These are used in making modern paints and enamels. A common member of this type is glyptal resin, which can be formed from glycerol and phthalic anhydride.

glycerol   phthalic anhydride

1. Place 2 g of glycerol and 3 g of powdered phthalic anhydride into a 50-ml beaker or a small tin can. Mix with a glass stirring rod. Cover with a watch glass and heat gently over an electric hot plate or *very* low burner flame. (Be sure to wear your safety glasses. The resin is flammable.) Keep heating until large bubbles form and the mixture puffs up. Allow the resin to cool.

2. Remove the resin from the container and grind it in a mortar, using a pestle, as shown in Figure 18-4. Try to dissolve some of it in a solvent suggested by your teacher. Do *not* heat over an open flame. When you have obtained a solution of the resin, pour some of it out on a piece of wood or metal and let it dry. In your notebook, note the nature of the residue. Suggest a practical use for this type of resin.

Part B. Preparation of an Amine-Aldehyde Type Polymer

In 1909, L. Baekeland first demonstrated the possibilities of forming plastics with formaldehyde, HCHO, by using aromatic compounds such as phenol, ⌬—OH. Such polymers are called Bakelite plastics. The first stage in the condensation reaction between phenol and formaldehyde is as follows. Note the elimination of water.

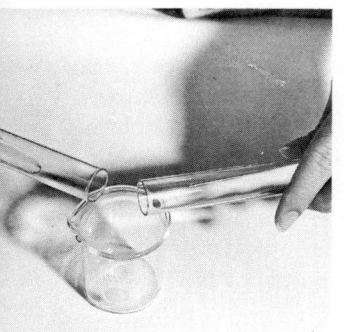

Figure 18-5

You will use an aromatic amine called aniline, $C_6H_5NH_2$, instead of phenol. Its reaction with formaldehyde is analogous to the phenol-formaldehyde condensation.

Measure 10 ml of 40% formaldehyde solution, called formalin, into one test tube and 10 ml of a saturated aqueous solution of aniline hydrochloride into another test tube. As shown in Figure 18-5, pour the two solutions simultaneously into a 50-ml beaker or a small tin can, in order to mix them. Note whether the reaction releases or absorbs heat. Examine the product and record its properties in your notebook.

Part C. Examination of the Effects of Cross-Linking

Neighboring chains of atoms, as in polymers, are often connected to each other by chemical bonds. This connection is called **cross-linking** or **cross-linkage.** The hydrogen bonds in Figure 18-2, for example, cross-link molecules of nylon.

Special Precautions. The heating of the mixtures in steps 2 and 3 should be done very carefully to minimize the formation of glycerol and ethylene glycol vapors.

1. Cut a piece of heavy-duty aluminum foil about 16 cm square. Attach an iron ring to a ringstand. Make a shallow container by placing the aluminum foil on the ring, pressing it down gently in the center and wrapping the excess foil around the ring to hold it in place. Be careful not to make any holes in the foil while putting it in place.

2. Add approximately 2 g of adipic acid to the center of the aluminum foil on the ring. Add about 40-50 drops of glycerol and 2 drops of concentrated sulfuric acid to the adipic acid. Mix gently with a stirring rod until a pasty mixture is achieved. Hold a burner below the aluminum foil and heat very carefully with a low flame for 4-5 minutes. Note all changes in the consistency of the mixture during the heating. Be careful not to scorch or char the product during the final stages of the heating. Allow the mixture to cool and examine the properties of the final product. Save the product for comparison with the product to be produced in step 3.

3. Place a fresh piece of heavy-duty aluminum foil on the iron ring. Add approximately 2 g of adipic acid, 40-50 drops of ethylene glycol, and 2 drops of concentrated sulfuric acid. Mix gently to form a pasty mixture. Heat carefully for several minutes and note all changes that take place. Allow the mixture to cool and compare it with the product formed in step 2.

Part D. Preparation of an Addition Polymer

1. Place 10 ml of methyl methacrylate in a small, expendable glass container (such as a baby food jar) and add several drops of benzoyl peroxide dissolved in diethyl phthalate or butyl benzyl phthalate. Cover it with a piece of aluminum foil, tucked in around the edges and held in place with a rubber band. Place the jar in a hot water bath and keep the temperature just below boiling for about 30 minutes or until polymerization is complete. Allow the polymer to cool. Record a description of its properties in your notebook.

2. Alternate procedure. Place 20–30 ml of commercial-type prepolymer, such as Clear Cast, obtainable from an art supply or hobby shop, in a small paper cup. Add 2-3 drops of methyl ethyl ketone peroxide and stir thoroughly with a glass stirring rod. Make a small cup-shaped mold from heavy-duty aluminum foil. Pour the mixture into the mold and let it stand in a ventilation hood or other well-ventilated place for twenty-four hours. Remove the aluminum foil and describe the properties of the product in your notebook.

   Colorful artistic objects can be made by this procedure by adding small amounts of dye to the mixture before stirring. The mixture may then be poured into molds carved in paraffin wax. Multicolored sheets of plastic can be created by pouring several portions of different colored mixtures together on a sheet of aluminum foil turned up slightly on the edges to prevent spilling.

RELATED PROBLEMS

1. The reaction between glycerol and phthalic anhydride to form a glyptal resin is somewhat more difficult to illustrate on paper than the reaction between glycerol and *ortho*phthalic acid, which forms essentially the same type of product.

$$\underset{\text{\textit{ortho}phthalic acid}}{\bigcirc\! \begin{array}{c} \text{—COOH} \\ \text{—COOH} \end{array}}$$

   Draw four structural formulas for glycerol molecules, spaced several inches apart at random locations on a piece of paper. Show how *ortho*phthalic acid molecules can link the glycerol molecules together by forming ester linkages in various directions.

2. Draw a structural formula for aniline and include all the H atoms on the benzene ring. Assume that the two hydrogen atoms adjacent

to the NH_2 group and the hydrogen atom directly opposite the NH_2 group will react with formaldehyde molecules. (See the analogous reaction, illustrated for the hydrogen atoms adjacent to the functional group in phenol, in Part B.) Show how the reaction of a number of aniline molecules and formaldehyde molecules can form a lattice-type structure.

3. How can you explain the difference between the adipic acid-ethylene glycol polymer and the adipic acid-glycerol polymer?

4. In the formation of the addition polymer in Part D, was there any other product formed besides the principal product? Would you expect the principal product to be more unsaturated or less unsaturated than the material you started with?

Summary

Plastics and synthetic fibers are made up of giant molecules called polymers. Polymers are made by joining together many small molecules, called monomers. This can be done in two basic ways. One method is called condensation polymerization. Here two different kinds of small molecules react with each other to eliminate water and link up into very long chains. The other method is called addition polymerization, which is the process by which the double bonds of small alkenes open up and new C—C bonds are formed between the monomers.

Self Check

1. Show with formulas how a diacid and a diamine combine to form a polyamide such as nylon.
2. Draw the polymer that can be made from the monomer $F_2C{=}CF_2$. Name the polymer. What is a common use for it?

The Giant Molecules of Nature

18-5 Proteins

Most of the substances of our bodies, including muscle, hair, nails, and skin, are made up of giant molecules called proteins. **Proteins** are a kind of polyamide. However, they differ from synthetic polyamides such as nylon in several important respects.

The small molecules from which proteins are built are called **amino acids**. They are so named because they contain an amino group ($-NH_2$) and a carboxyl group ($-COOH$). Nearly all of them have the following structure.

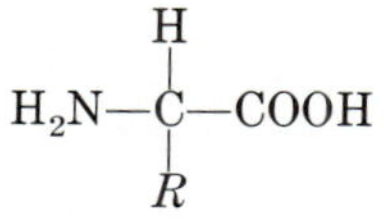

$$H_2N-\underset{R}{\overset{H}{C}}-COOH$$

Amino acids differ from each other in the nature of the —R group. There are approximately 20 different amino acids that make up nearly all the important proteins. There are, however, other less common amino acids. Several of the important amino acids are given in Table 18-2.

A protein is formed by the condensation polymerization of many amino acids. Note the presence of the amide groups, as indicated in Figure 18-6. In general, all protein molecules contain all, or nearly all, of the 20 amino acids. The exact order in which they occur determines the identity of a protein. Now, think about the possibilities presented by the availability of 20 different amino acids in a molecule that may contain 100 to 1000 of them. You can then appreciate why so many different proteins, capable of serving so many different purposes in the body, can be made. In a protein molecule containing 100 amino acids, with a choice of 20 possibilities for each place in the chain 20^{100} different molecules can be formed. This is an inconceivably large number, and only *one* of the combinations is found in a particular protein.

TABLE 18-2 Some Important Amino Acids

Formula	Name
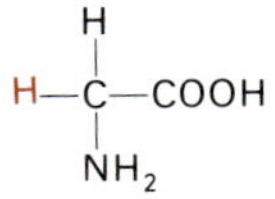 H—C(H)(NH$_2$)—COOH	glycine
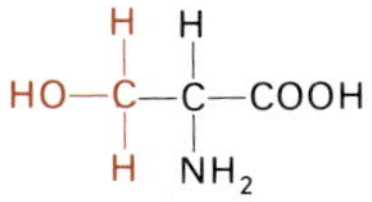 HO—C(H)(H)—C(H)(NH$_2$)—COOH	serine
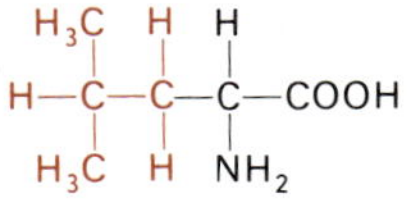 (H$_3$C)$_2$CH—CH$_2$—C(H)(NH$_2$)—COOH	leucine
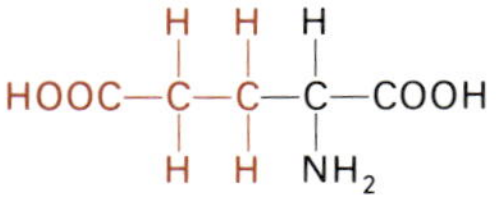 HOOC—C(H)(H)—C(H)(H)—C(H)(NH$_2$)—COOH	glutamic acid
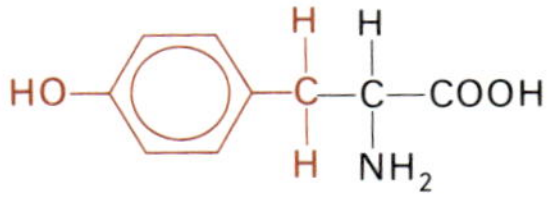 HO—C$_6$H$_4$—C(H)(H)—C(H)(NH$_2$)—COOH	tyrosine
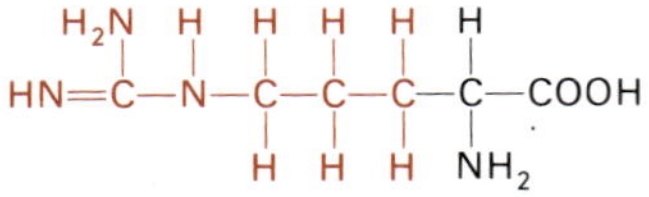 H$_2$N—C(=NH)—NH—C(H)(H)—C(H)(H)—C(H)(H)—C(H)(NH$_2$)—COOH	arginine

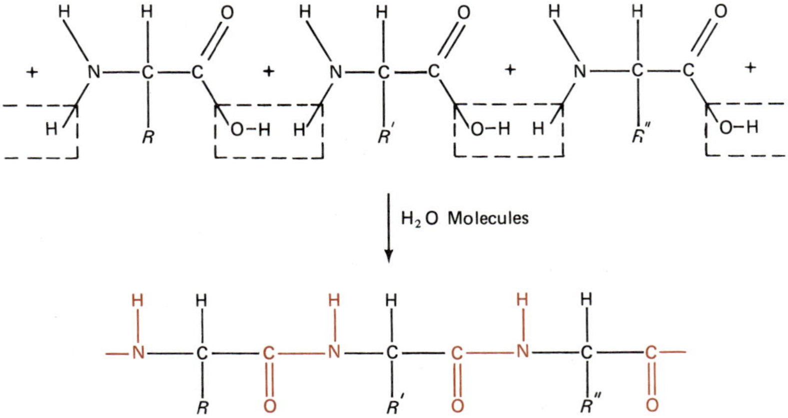

Figure 18-6 *Proteins are condensation polymers formed by head-to-tail joining of amino acids, as shown.*

Choh Hao Li (1913–)

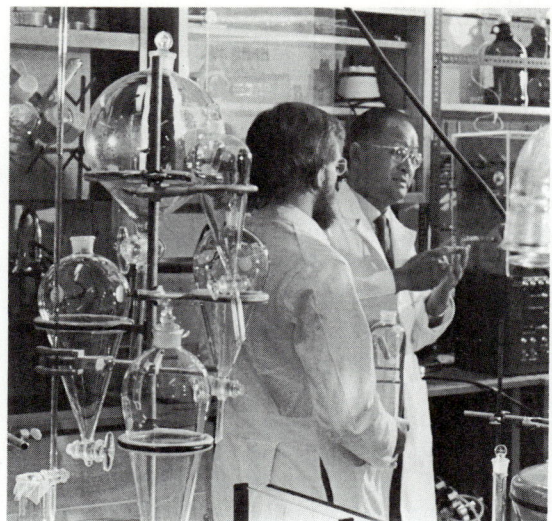

CHOH HAO LI has spent most of his scientific career examining the hormones of the pituitary gland. Li received his bachelor's degree in chemistry from the University of Nanking in China in 1933. Two years later he came to the United States. He earned his Ph.D. at the University of California at Berkeley in 1938 and joined the faculty there. By 1950, he had become a full professor and Director of the Hormone Research Laboratory. In 1955, he became a naturalized citizen of the United States.

How could anybody spend so many years working on only one gland? The answer lies in the complexity of the organ and the way it functions. The pituitary gland is located in the center of the head and controls many other glands in the body. Each of the three lobes of the pituitary has specialized functions and produces different hormones.

Li and his colleagues at the University of California isolated several hormones from the pituitary gland and studied their structures and functions. They discovered that certain hormones stimulated the thyroid gland, controlled the production of sex hormones, directed the functioning of the adrenal cortex, and stimulated the production of milk. They found that others controlled blood pressure, the balance of water in the body, and the rate of metabolism in cells.

One of the better-known hormones isolated by Li is ACTH, short for adrenocorticotrophic hormone. When administered to patients suffering from rheumatoid arthritis, ACTH stimulates the adrenal cortex to produce cortisone, which brings significant relief from the pain caused by the disease.

Li found that some of the pituitary hormones are far more complicated than the hormones produced in other parts of the body. Using methods developed by Nobel prizewinner Frederick Sanger of Cambridge, Li and his group systematically began to take the hormone molecules apart, piece by piece, examining the structures of the fragments.

By 1965, Li's group had determined that ACTH is a complicated protein composed of 39 amino acid groups arranged in a specific order. In the same year, the researchers discovered that the human growth hormone, which can produce giants or midgets depending on its amount in the body, is a protein with a molecule containing 256 amino acid groups.

With so much of the groundwork done by Li and his staff, many researchers are studying the mechanisms by which the pituitary hormones exert their controlling actions. Such knowledge will help physicians in their search for better ways to insure the best functioning of the human body and to treat disorders that arise.

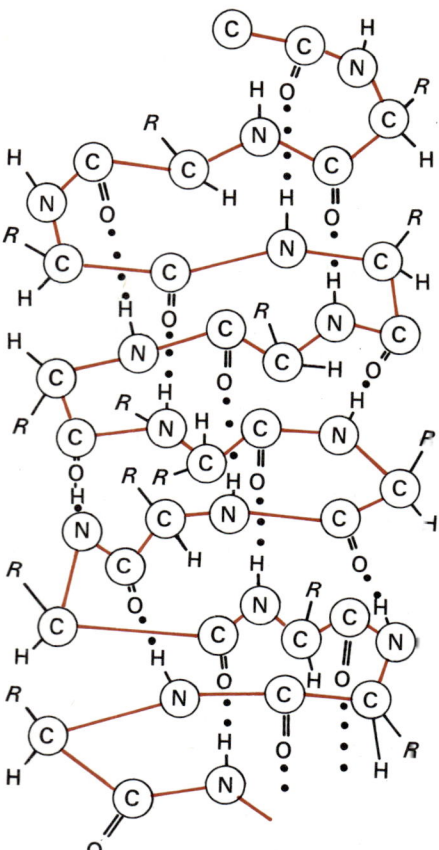

Figure 18-7 *A protein chain, held rigid by hydrogen bonds between N—H and C=O groups. The bonds shown in color form a spiral pattern.*

acids in their chains. (Actually, they differ in the order of amino acid residues. The residues are the pieces left after the water molecules have been removed.) One of the main reasons why molecules with different sequences have different properties is that they arrange themselves differently in space. The shape of a protein molecule is strongly influenced by the formation of hydrogen bonds within the molecule. Figure 18-7 shows one of the important shapes that protein molecules, or portions of protein molecules, often adopt. As you can see, the protein chain winds itself into a helix, which is held together by many hydrogen bonds. A helix resembles the shape of a spiral staircase. When proteins are heated, structures such as this helix break apart because the hydrogen bonds break. Thus meat, which contains protein, becomes tender when it is cooked, and gelatin melts when warmed and "sets" again when cooled.

Figure 18-8 shows another rigid arrangement, called a pleated sheet, found in protein molecules. It results from the formation of hydrogen bonds. The pleated sheet is the most prominent arrangement found in fibers of silk, giving silk its fibrous character. Does this remind you of nylon? It should. Though there are many differences in detail, both silk and nylon have separate strands that are connected by hydrogen bonds. Therefore, nylon has many properties similar to those of silk.

The way in which an organism makes certain that precisely the one right combination is always made for each of the thousands of proteins needed in the body is in itself a fascinating problem. Chemists and biologists are now beginning to understand how it is done. Research on this problem is one of the most challenging areas in science today.

As you have just learned, protein molecules differ in the order or sequence of the amino

18-6 Starches and Cellulose

Cellulose is the important structural material of plants. It gives stiffness and shape to plants of all kinds. It is also the major constituent of wood. Starch in various forms occurs in both plants and animals. However, the word "starch" is usually used to refer to the types found in plants. Potatoes are largely made up of granules of starch.

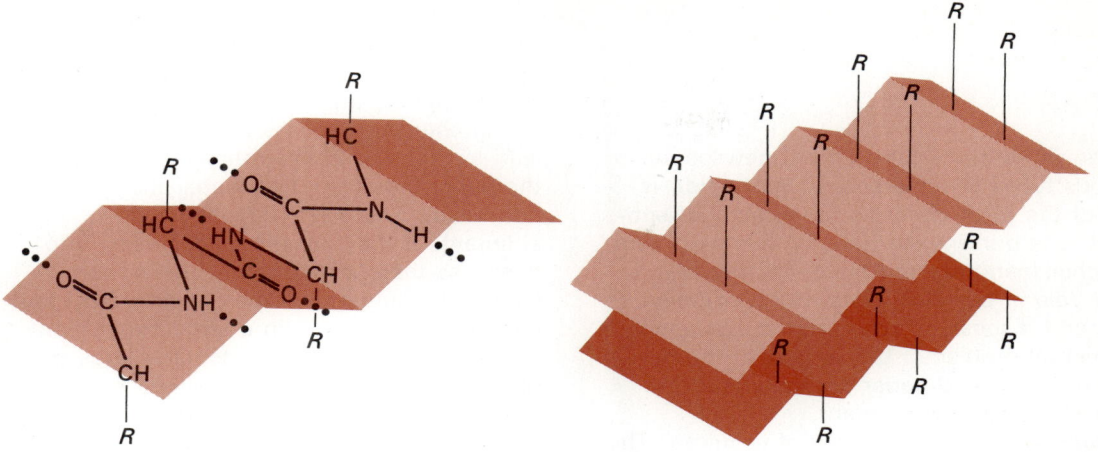

Figure 18-8 *Another way in which protein chains can be arranged is in parallel zig-zag rows to form "pleated sheets," again held rigid by hydrogen bonds.*

The monomer of which both cellulose and starch are made is a cyclic molecule called glucose. Its structure is as follows.

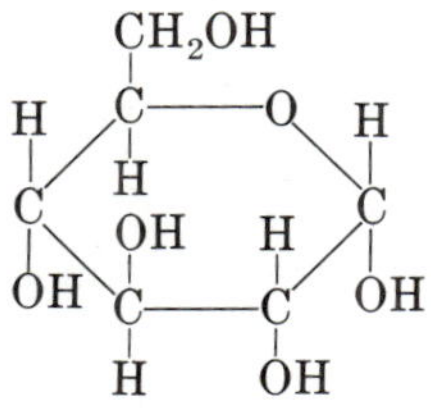

Glucose is also an important sugar. Glucose molecules can be polymerized by a condensation process in which water is eliminated. In this way, the giant molecules of cellulose and starch are formed. Starch and cellulose differ because they have different oxygen linkages between the glucose molecules. This is shown in Figure 18-9. You may have to look carefully to see that there is a difference. It may seem like a pretty small one, but it

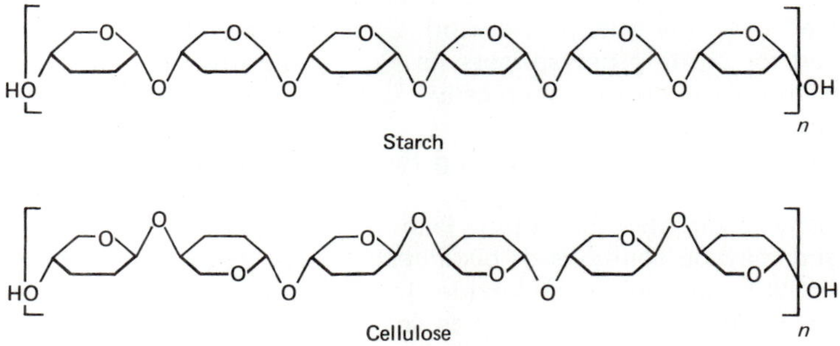

Figure 18-9 *Both starch and cellulose are polymers of glucose. They differ only in the arrangement about the connecting oxygen atoms.*

Chemistry and Our Environment

PAPER

Paper, in its many forms, from newspaper to various types of paperboard (cardboard), is one of the most common manufactured substances in our environment. In common with all other manufactured materials that play a large role in our daily lives, the production, use, and disposal of paper can cause environmental problems. Let us look at the nature of paper and paper products and how they affect our environment.

Paper is composed mainly of cellulose. The source of the cellulose is plants, especially trees, which supply about 90 percent of the cellulose in paper. Paper has its first environmental impact at this stage. In the United States and Canada alone, over a billion trees are cut each year for papermaking. It is important that this extensive cutting be done so as to avoid damage to the land through erosion. It is also necessary to see that forest reserves are maintained by replanting.

Wood typically consists of about 60 percent cellulose, with the remainder being mostly a complex polymer called *lignin*. The cellulose is present in wood as fibers, which are bound together by the lignin. In order to make paper, it is necessary to break down the wood into pulp. *Pulp* is a kind of soup of cellulose fibers in water. The cheapest way to produce it is to grind the wood mechanically. In this way the lignin is not removed and the paper produced is of low quality. Newsprint, the paper used for printing newspapers, is made almost entirely of this mechanically-produced pulp.

Much better pulp is made by chopping the wood into small chips and then treating these chemically to dissolve the lignin. The chemicals used may be solutions of bisulfites, such as $Mg(HSO_3)_2$, or sodium hydroxide. In order for these solutions to act rapidly as solvents for the lignin and other noncellulose components of the wood, they are boiled in vats. In some cases, sealed, high-pressure steam reactors are used.

The chemical processes for making pulp lead to the second environmental impact of the paper industry. Both water and air pollution result. The waste liquids are major water pollutants if they are dumped directly into rivers, as they unfortunately are in many cases. The sulfurous fumes (mainly SO_2) and particles emerging from a large paper plant can be serious air pollutants. One large paper plant can release a million pounds of SO_2 a year into the air. In recent years, the paper industry has begun to control polluting emissions, but there is still a long way to go.

The pulp is made into paper with large, continuously-operating machines. These form the pulp into thin layers, which are squeezed and dried. In most cases the paper is then treated with other materials to make it more suitable for its application. Paper intended for printing, other than newsprint, is coated with clay, or other white mineral matter, plus an adhesive. Paper intended for packaging foods is coated with wax or another water-repellent material.

Once the paper products have been made and used, the third environmental impact arises: waste paper. Waste paper makes up more than fifty percent of the trash produced in America. Just consider the trash produced in your own home, your school, or a typical local business. What can be done with all of this waste paper? In the past it has been either dumped or burned. But as the total volume has increased and the demand for paper has increased, another answer to the problem has begun to be adopted: recycling.

Every year, people in the United States consume over 60 million tons of paper and paperboard, and this is expected to double by 1985. Even if only a fraction is recycled, a considerable reduction in pollution (both by paper pulp plants and in the form of solid

The paper mill shown above has installed a primary waste water clarifier (left foreground) to reduce water pollution. The waste water is held in the basin while the settleable solids sink to the bottom. Afterwards, the solids are removed. Following primary treatment, the waste water receives secondary treatment.

waste-trash) can be made. Already roughly 20 percent of the paper produced is recycled. One ton of recycled paper conserves about 17 trees. At present, some 200 million trees each year are conserved by recycling.

Recycling requires large and expensive machinery. It must be designed to cope with the many other materials present in used paper products besides the cellulose fiber that it is desired to recover.

The paper trash is mixed with about 30 times its weight of steaming water and stirred in a huge tub with a device that resembles a giant eggbeater. The stirring force is so great that the paper is soon broken down to a slurry that is drained out the bottom through one-eighth inch holes. The stirring chamber is designed and equipped to remove a great many contaminants such as string, cloth, and pieces of metal. The slurry must then pass through several stages of mechanical and chemical purification. It is then used mainly to make paperboard.

Much of our waste paper is heavily contaminated by coating materials and ink. This makes recycling difficult. Methods are being developed to remove some of these contaminants. However, if the volume of recycling is to increase, it will be advisable to modify the present practices in manufacturing many forms of paper and paper products, so as to make recycling easier and cheaper.

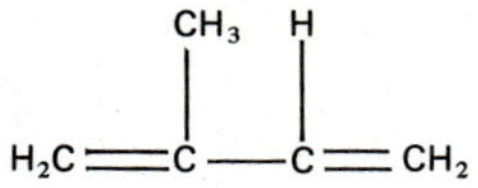

Figure 18-10 *Natural rubber consists of an addition polymer of isoprene called polyisoprene.*

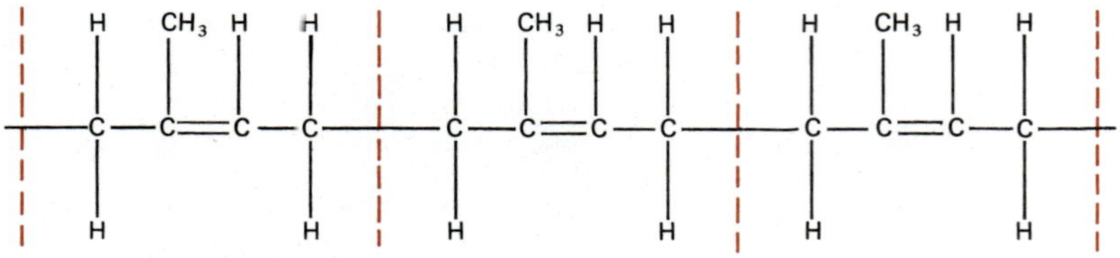

makes the difference between tasty French fries and the wood you sit on while you eat the French fries.

18-7 Rubber

Natural rubber is an elastic solid obtained chiefly from the milky juice of the rubber tree. An elastic material is one that can be bent, stretched, and compressed and then can recover its original shape when the force causing the distortion is removed. Natural rubber contains mainly a hydrocarbon polymer made from a monomer called isoprene. The polymer, called polyisoprene, has a molar mass of 300,000. The structures of both the monomer and the polymer are shown in Figure 18-10.

Polyisoprene itself is soft, sticky, and not very elastic. What gives rubber its elasticity is a process known as *vulcanization*. In this process one to eight percent by mass of sulfur is added to the polyisoprene. The mixture is then heated, causing the sulfur atoms to cross-link the polyisoprene chains, as the following shows.

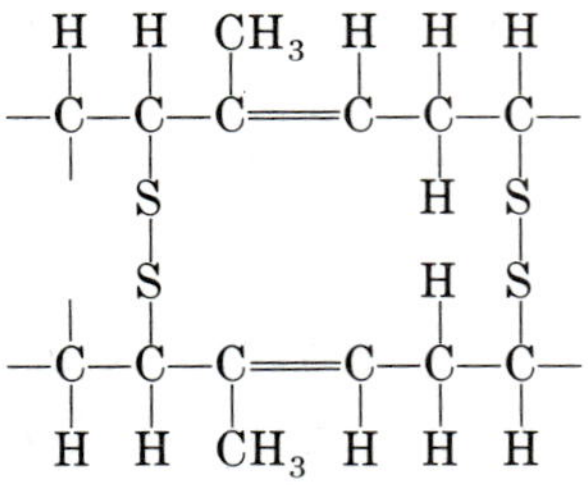

This leads to a stiffening of the substance. When vulcanized rubber is deformed, the sulfur cross-links do not permit much movement of the chains. In essence, the chains are "tied together." When the deforming force is released, they have a very strong tendency to return completely to their original positions. When 30 to 50 percent by mass of sulfur is used, very many cross-links form. This results in a material that cannot be

Wallace Hume Carothers (1896–1937)

THE ENORMOUS USE of synthetic polymers in textile fibers, films, plastics, and elastomers can be traced back to the pioneering research efforts of Wallace Carothers between 1928 and 1937.

After receiving his doctorate in chemistry at the University of Illinois in 1924, Carothers quickly established his reputation as a brilliant researcher, first at Illinois and then at Harvard University. In 1928 he was chosen to direct a program of basic research for the DuPont Company.

Having already become interested in giant molecules, he decided that the major efforts of his group would be dedicated to finding out all they could about the structure and synthesis of such compounds. The existence of large molecules with repeating units had been suspected as early as 1871, but little serious work had been done in this field.

Carothers first investigated addition polymers. In searching the literature, he discovered that Julius A. Nieuwland, a chemist at the University of Notre Dame and an expert on acetylene chemistry, had found out how to form large molecules from acetylene. Working with Nieuwland's cooperation, Carothers found that two molecules of acetylene could be combined to produce vinyl acetylene. When treated with hydrochloric acid, vinyl acetylene yielded a product known as chloroprene. Within a year or so, Carothers and his group were able to polymerize chloroprene, producing a new synthetic elastomer called neoprene.

Turning to condensation polymers, Carothers first worked with a number of polyesters and later with polyamides. The most famous of the polyamides was first named compound-66 because of the six carbon atoms in each of its starting molecules. This compound, soon to be named nylon, was discovered in 1934. It was very strong and became even stronger when its cold fibers were drawn out to several times their original length. First marketed in 1940, nylon was an instant success. Carothers, however, did not live to see how valuable a product he had invented. Today, varieties of nylon, polyesters, and many other condensation and addition polymers make up a major part of our organic chemical industry.

546 Chapter Eighteen

deformed at all. This material, called hard rubber, is used, for example, in battery cases.

Once chemists understood the molecular structure of natural rubber, they were able to invent various types of synthetic rubbers. Synthetic rubbers are elastic because their molecular structures resemble that of natural rubber. Chemists have been able to make small changes in the molecules that make up synthetic rubbers. In this way synthetic rubbers gain properties that are different from those of natural rubber. For example, natural rubber tends to soften and break down upon prolonged contact with gasoline. A synthetic rubber called *neoprene* is far more resistant to the action of gasoline. It is made from a monomer known as *chloroprene*. This monomer has chlorine atoms where isoprene has methyl groups. Because of its gasoline-resistant property, neoprene is used wherever contact with gasoline is expected.

18-8 The Nucleic Acids, DNA and RNA

Heredity is the transmission of characteristics from one generation to the next in both plants and animals. It is controlled by giant molecules called nucleic acids, so called because they were first found in the nuclei of cells. There are two types of nucleic acids. They are (1) *deoxyribonucleic acid*, DNA, and (2) *ribonucleic acid*, RNA. DNA is the substance that carries the coded messages that guide the development of each organism. The code itself is based on the sequence in which certain chemical groups occur along the chain of a giant molecule. RNA is very closely related to DNA and serves, among other functions, as a messenger in transmitting the coded messages. RNA differs from DNA, among other ways, in the presence of an additional —OH group in each sugar (ribose) unit. However, the following discussion of the nature of DNA and how it serves to carry genetic information applies to RNA as well as to DNA, though there are important differences. In Section 18-10 the differences between DNA and RNA will be further explained.

The DNA molecule consists of two parts. One part is the *backbone*. This is a very long chain of alternating deoxyribose sugar and phosphate units.

Compounds called bases make up the second part. One of the four bases shown in Figure 18-11 becomes attached to each deoxyribose unit. The base is attached by the reaction of one N—H bond with the —OH group of the deoxyribose unit, as shown in Figure 18-12. (You can find a discussion of bases in Chapter 23. However, you do not have to know their general properties for the present discussion of nucleic acids.)

How does the arrangement of these bases form a code for transmitting genetic information, the so-called *genetic code?* While much remains to be learned about this, scientists have discovered some of the most basic properties of the code. For example, a DNA molecule can give, in code, a statement of the order in which the amino acids are to be linked to produce a particular protein. It does this by assigning a particular sequence of three bases to stand for a particular amino

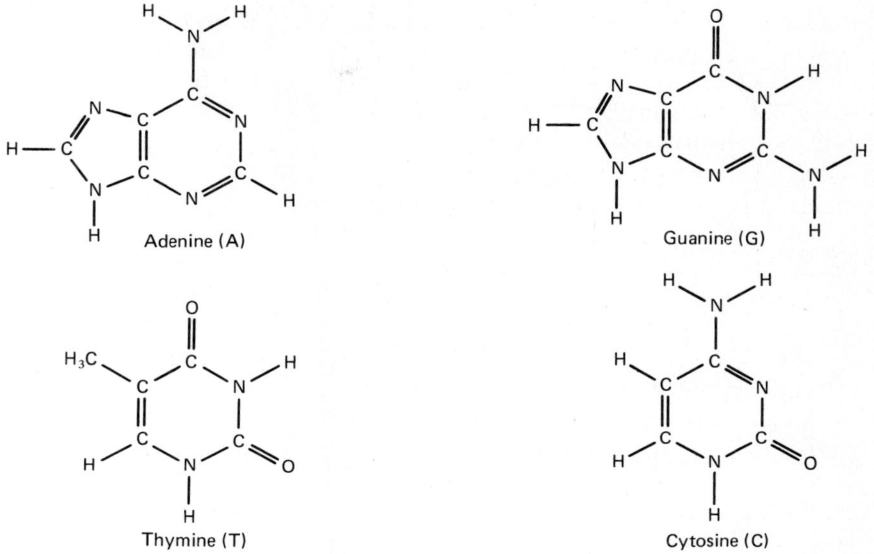

Figure 18-11 *The four bases found in DNA: adenine, thymine, guanine, and cytosine.*

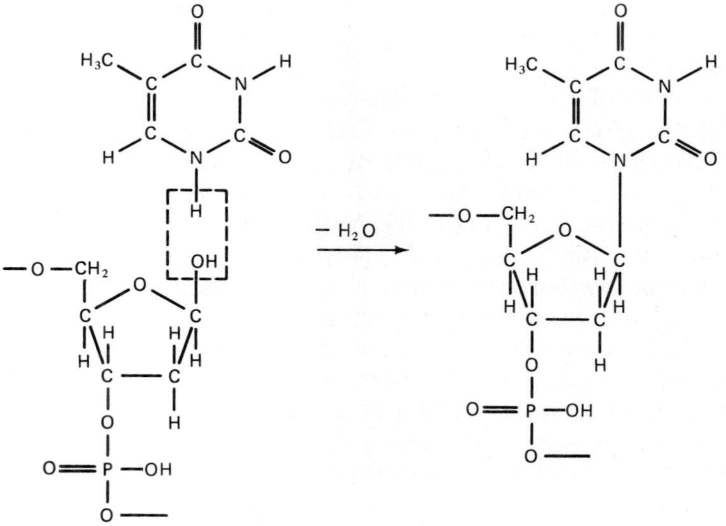

Figure 18-12 *A base group is attached to the DNA backbone as shown above.*

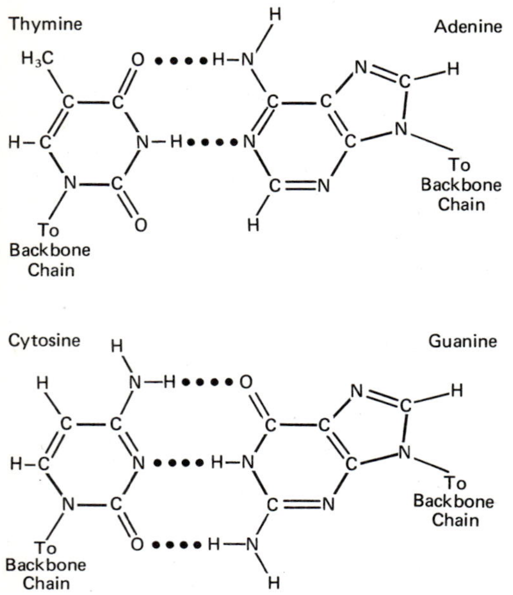

Figure 18-13 *Certain pairs of bases, thymine/adenine and cytosine/guanine, are particularly well matched and form base pairs, by means of hydrogen bonding, in DNA.*

acid. There are 64 different ways to make a sequence of three of the four bases, A, T, G, and C. The first base in the sequence may be any of the 4, so may the second, and so may the third. Hence, $4 \times 4 \times 4 = 64$. Not all of these sequences are used. In fact, only a little over 20 combinations are used to give code words for the 20 common amino acids. Also, a few of these combinations are used to indicate where reading of the code is meant to start and end, rather than as code words for an amino acid. Scientists now know the code words for most of the important amino acids.

You might now be wondering how the coded information is passed on from one cell to another. The answer lies in the structure of the DNA molecule, which is discussed in the following section.

18-9 The Double Helix and the Template Idea

The bases that occur in DNA are all capable of participating in the formation of hydrogen bonds. In fact, these bases have highly selective abilities to combine with each other. As you can see in Figure 18-13, thymine bonds to adenine, while cytosine bonds to guanine. These are the only bonding combinations that occur.

Because base pairs form, DNA molecules have a double helix structure. This is shown in diagram form in Figure 18-14a and in an actual molecular model in Figure 18-14b. In the diagram each band represents the backbone of alternating deoxyribose and phosphate units. The dots symbolize base pairs linked together by hydrogen bonds.

Because of the very selective *complementary* nature of the base pairing, the sequence of bases on each of the intertwined chains, or strands, restricts the sequence on the other. If we have, for example, the sequence ...TGATCCG... on one chain, the corresponding part of the other must run ...ACTAGGC.... The sequence of bases attached to each chain acts as a *template* (meaning "a pattern or guide") for the other.

Because of the template mechanism, duplicates of DNA molecules can be made. To do this, the two strands of one DNA molecule first unwind and separate. Then each one has built onto itself a new complementary strand that exactly duplicates and replaces the previous complementary strand. Meanwhile, the previous complementary strand is also getting a new matching partner built onto itself. Therefore, from one DNA molecule

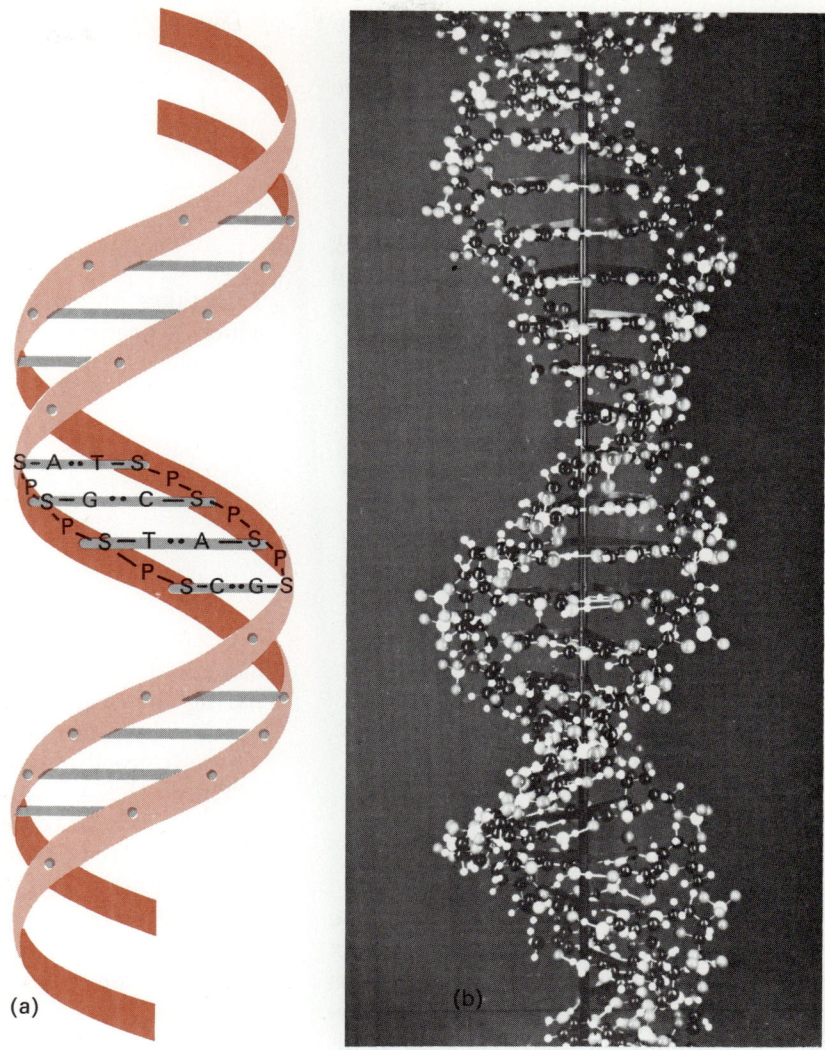

Figure 18-14 (a) A simplified model of the double-helix structure of DNA. The letters S and P stand for the sugar and phosphate groups in the backbone. The letters A, T, C, and G stand for the bases shown in Figure 18-11. (b) An actual atomic model of the DNA structure.

18-10 RNA

As was stated in Section 18-8, RNA is closely related to DNA. It differs in structure from DNA in three ways. First, it contains ribose sugar.

DNA, on the other hand, contains deoxyribose sugar.

Note that deoxyribose sugar contains one less oxygen than ribose. In fact, the prefix "deoxy-" indicates that a molecule contains less oxygen than another to which it is related. The backbone of RNA consists of alternating ribose and phosphate units. As in DNA, bases are attached to the backbone.

Second, the base uracil is found in RNA.

This takes the place of the base thymine, which is present in DNA.

Thymine has a methyl group where uracil has a hydrogen. This is seen clearly by comparing the diagram of thymine with that of uracil. Thymine bonds to adenine in DNA, whereas uracil bonds to adenine in RNA.

Third, RNA usually exists as single strands. This, of course, differs from the double-stranded structure of DNA.

As you have seen, a DNA strand can serve as a template for a complementary DNA strand, but not for one just like itself. However, a given DNA strand can serve as a template for an RNA strand. The RNA strand, in turn, may function as a template for making further copies of the first DNA strand.

Summary

Giant molecules are found in natural and synthetic rubbers. They also form an essential part of living things. The main types of giant molecules are proteins, starches and cellulose, and nucleic acids (DNA and RNA). Proteins are long chains of amino acid groups joined together by the formation of amide linkages. Starches and cellulose are long chains of glucose molecules linked together by the elimination of water molecules between pairs of —OH groups. Oxygen linkages result. Starches and cellulose differ in the detailed structure about each oxygen linkage. The nucleic acids are the molecules that carry the genetic code. They contain a backbone of alternating phosphate and ribose or deoxyribose units, with a base attached to each ribose or deoxyribose. The genetic code is based on the sequence of the bases along the backbone chain. The base sequence serves as a template for making new nucleic acid molecules with exactly the proper sequence of bases.

Self Check

1. Write the general formula for an amino acid. Then show how two of them unite to begin to form a protein molecule.
2. Draw a representative portion of the chain of glucose units that occurs in cellulose.

Silicates Are Inorganic Giant Molecules

18-11 Silicates ★

Silicates are substances containing silicon and oxygen combined into complex anions. The cations that accompany these anions are generally from the alkali and alkaline earth groups, although many important silicates contain transition metal cations as well.

The simplest type of silicate is an orthosilicate. It contains only individual SiO_4^{4-} tetrahedral anions. Orthosilicates are rare. The next most complicated type of silicate structure is one containing the anion $[O_3Si—O—SiO_3]^{6-}$. This is formed by two tetrahedra sharing one oxygen atom. (See Figure 18-15.) This relatively simple structure is also rare. The important types of silicate structure are much more elaborate. They consist of chains, bands, and sheets formed by SiO_4 tetrahedra that share two or more oxygen atoms.

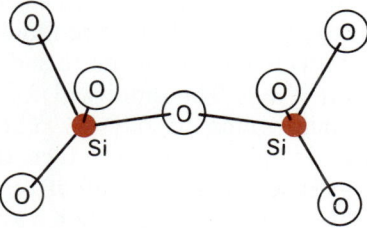

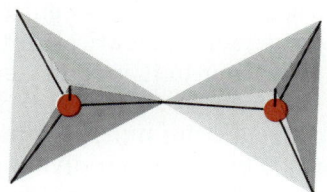

Figure 18-15 *When two SiO_4 tetrahedra share an oxygen atom, the disilicate anion, $Si_2O_7^{6-}$, is formed.*

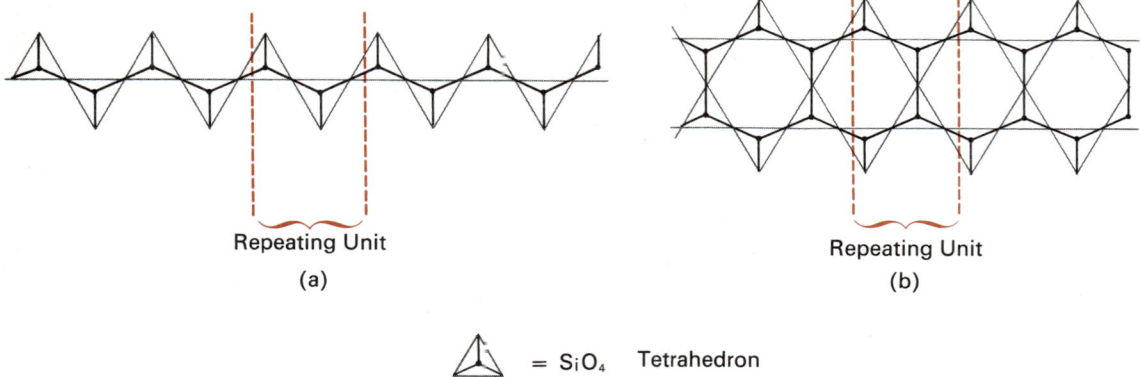

Figure 18-16 SiO_4 tetrahedra may join by sharing oxygen atoms to form (a) chain anions, $(SiO_3^{2-})_n$, and (b) band anions, $(Si_4O_{11}^{6-})_n$.

A silicate chain anion is shown in Figure 18-16a. Attached to each silicon atom there are two oxygen atoms that belong entirely to that silicon atom and two more that are shared with neighbors. A given silicon atom can be considered to have a half-share in each of the shared oxygen atoms. Hence the number of oxygen atoms per silicon atom in such a chain is $2 + \frac{1}{2} + \frac{1}{2} = 3$. The empirical formula of the chain anion is SiO_3^{2-}. The charge of $2-$ can be figured from the formal oxidation numbers (see Section 24-10) of $2-$ for oxygen and $4+$ for silicon. Another way to get this result is to consider the repeating unit in this chain. This is the smallest unit which can be chosen such that the entire structure is formed by joining identical units end to end. The repeating unit is indicated in Figure 18-16. Counting the number of Si and O atoms in the repeating unit, you find 2 Si and 6 O atoms. This gives, or dividing through by 2, SiO_3^{2-}. The charge is again figured by allowing $2-$ for each oxygen atom and $4+$ for the silicon atom.

An important form of band anion is shown in Figure 18-16b. The empirical formula here is most easily determined by counting up all the silicon and oxygen atoms in the repeating unit. This unit is indicated in Figure 18-16b. As you can see, there are 4 silicon atoms and 11 oxygen atoms between the vertical broken lines. Since 4 and 11 have no common divisor this leads directly to the empirical formula, $Si_4O_{11}^{6-}$.

Each band can be thought of as the result of combining two chains, with sharing of oxygen atoms. Suppose now you place bands side by side and connect them by further sharing of oxygen atoms. This results in the formation of a sheet anion as shown in Figure 18-17. The empirical formula of a sheet anion can be figured from the atom count in the repeating unit, which is outlined. Within this repeating unit are 4 Si atoms and 10 O atoms. After dividing through by 2, we get $Si_2O_5^{2-}$ as the empirical formula.

18-12 Asbestos, Mica, and Glass ★

Asbestos is the name given to silicate minerals that have a fibrous form. The majority

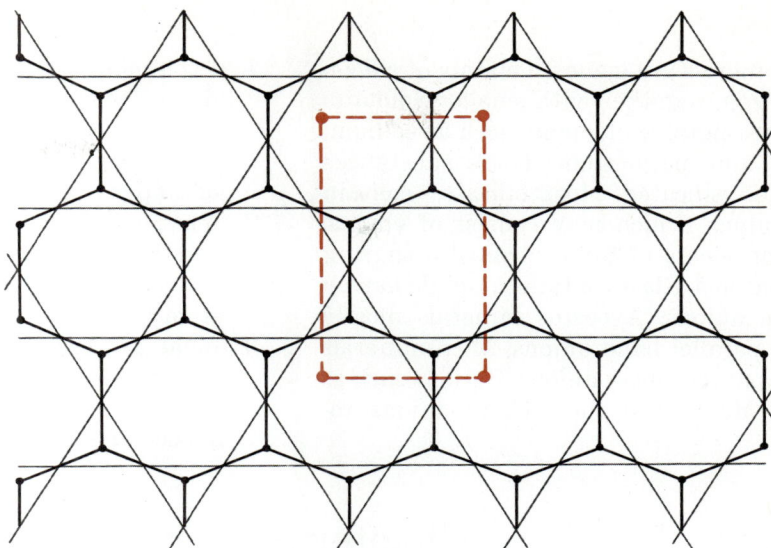

Figure 18-17 SiO_4 tetrahedra can link up to form a sheet anion. The formula of the repeating unit, outlined above, is $Si_4O_{10}^{4-}$.

of these minerals consist of band anions. Each band has a negative charge. The bands are held together in parallel bundles by positive ions, such as Na^+, Ca^{2+}, and Mg^{2+}, which lie between them in sufficient numbers to make the entire structure electrically neutral. The structure breaks most easily along the direction of the parallel bundles, thus accounting for the fibrous character of asbestos. Since asbestos, like all silicate minerals, already contains as much oxygen as the other elements present are capable of combining with, the process of combustion cannot occur. Thus, asbestos is fireproof. It can be woven in much the same way as organic fibers. In this way fireproof cloth can be made.

Mica is the general name for another class of silicate minerals. You have probably seen specimens of mica (like the one shown in Figure 17-19) and been astonished at how the sheets can be separated into thinner and thinner sheets, apparently indefinitely. Again, the logical explanation for the appearance and properties of the substance is found in the way it is built up from atoms. Micas are built of sheet silicate anions. The sheets are parallel and are held together by cations, such as Na^+ and Mg^{2+}, lying between them. Mica therefore breaks into sheets in planes parallel to the planes of the sheet anions.

Glass is one of the most important of all polymeric materials and is certainly the most important inorganic polymer. Glass consists of chain anions, together with enough cations, such as Na^+, to make the substance electrically neutral. In glass, however, the anions are not arranged in an orderly way. Instead, they are tangled up, like a bowl of well-stirred spaghetti. Thus glass has the mechanical characteristics of a very stiff liquid.

Summary

Many minerals are made up mainly of silicon and oxygen, together with smaller amounts of various metallic elements such as sodium, calcium, and magnesium. These substances are called silicates. Most silicates contain giant anions, which may consist of chains, bands, or sheets of SiO_4 tetrahedra sharing oxygen atoms. Glass contains a tangled array of chain anions. Asbestos minerals mostly contain parallel band anions. Mica minerals contain parallel sheet anions. Cations such as Na^+ or Mg^{2+} bind the silicate anions together.

Self Check

1. A common form of asbestos is called tremolite. The silicate part has the formula $Si_8O_{22}(OH)_2$. What is the charge on this unit? How many magnesium ions must accompany it?
2. What will be the empirical formula of the substance formed when SiO_4 tetrahedra are packed together so that every oxygen atom is shared?
3. What are the principal properties of asbestos, mica, and glass? How can you account for these properties, in terms of molecular structure?

Chapter Summary

Giant molecules, molecules of very great size and mass, are extremely important. Many occur naturally, while others are made by man to serve particular purposes.

Small molecules can be made to unite into long chains by several general processes. The resulting large molecules are called polymers. Molecules with double bonds can polymerize by having the double bonds open up so that new single bonds between molecules can then be formed. Polymers are also formed by condensation reactions. In these, a carboxyl group of one molecule reacts with an —OH or —NH_2 group of another to eliminate H_2O and form either an ester or an amide group. If molecules with two carboxyl groups (diacids) react with molecules having either two alcohol groups (dialcohols) or two amino groups (diamines), long molecules in which the two components alternate result.

In living organisms, the important types of giant molecules include proteins, starches and cellulose, and nucleic acids. Proteins are made up of long chains of amino acids linked together by the formation of amide groups. These groups form when the carboxyl group of one amino acid reacts with the amino group of another. Approximately twenty different amino acids commonly occur in proteins. The number of possible sequences in which they can be combined is enormous. As a result, many types of protein molecules exist. Starches and cellulose are both made up of long chains of glucose molecules linked together by loss of H_2O between two —OH groups. Oxygen linkages form between the molecules. The difference in properties between starch and cellulose results from the different spatial relationships in the two types of oxygen linkages they contain. DNA and RNA are responsible for the preservation and transmission of hereditary traits from one generation to the next. DNA is made up of long, double-helical chains of alternating phosphate groups and deoxyribose sugar groups to which a particular sequence of organic bases is attached. The information is recorded in a code based on the sequence of the base molecules. RNA differs from DNA in several structural ways.

Questions and Problems

1. What distinguishes polymers from other large molecules?
2. Illustrate by means of formulas how simple amides and esters differ from polyamides and polyesters.
3. How do hydrogen bonds contribute to the strength of nylon?
4. Show a structural formula for the compound formed when two molecules of ethylene react to form a four-carbon, linear molecule. Then show how this molecule can react with more ethylene to form longer molecules.
*5. Draw a section of a polypropylene chain formed from five propylene (propene) molecules.
6. Both nylon and protein molecules are polyamides. Show similarities and differences between these two types of substances.
7. Cellulose and starch have the same empirical formula. What accounts for the significant differences in the properties of these materials?
8. Show how two molecules of glucose with the empirical formula $C_6H_{12}O_6$ can form a molecule with the formula $C_{12}H_{22}O_{11}$. If glucose is the monomer for starch, why is the empirical formula for starch $(C_6H_{10}O_5)_n$, where n is a large number?
9. Give two examples of cross-linked molecules. Why is cross-linking important in the rubber industry?
10. Name two shapes that protein molecules can adopt. What type of bonding enables the molecules to adopt these shapes?
11. How is a duplicate of a DNA molecule made?
12. Show how the base groups bond to the sugar groups in DNA.
13. What is meant by the complementary nature of base pairing?
14. Suppose a DNA segment contains three of the base groups, say A, T, and G. How many different base sequences are possible? Show them.
15. How does RNA differ from DNA?

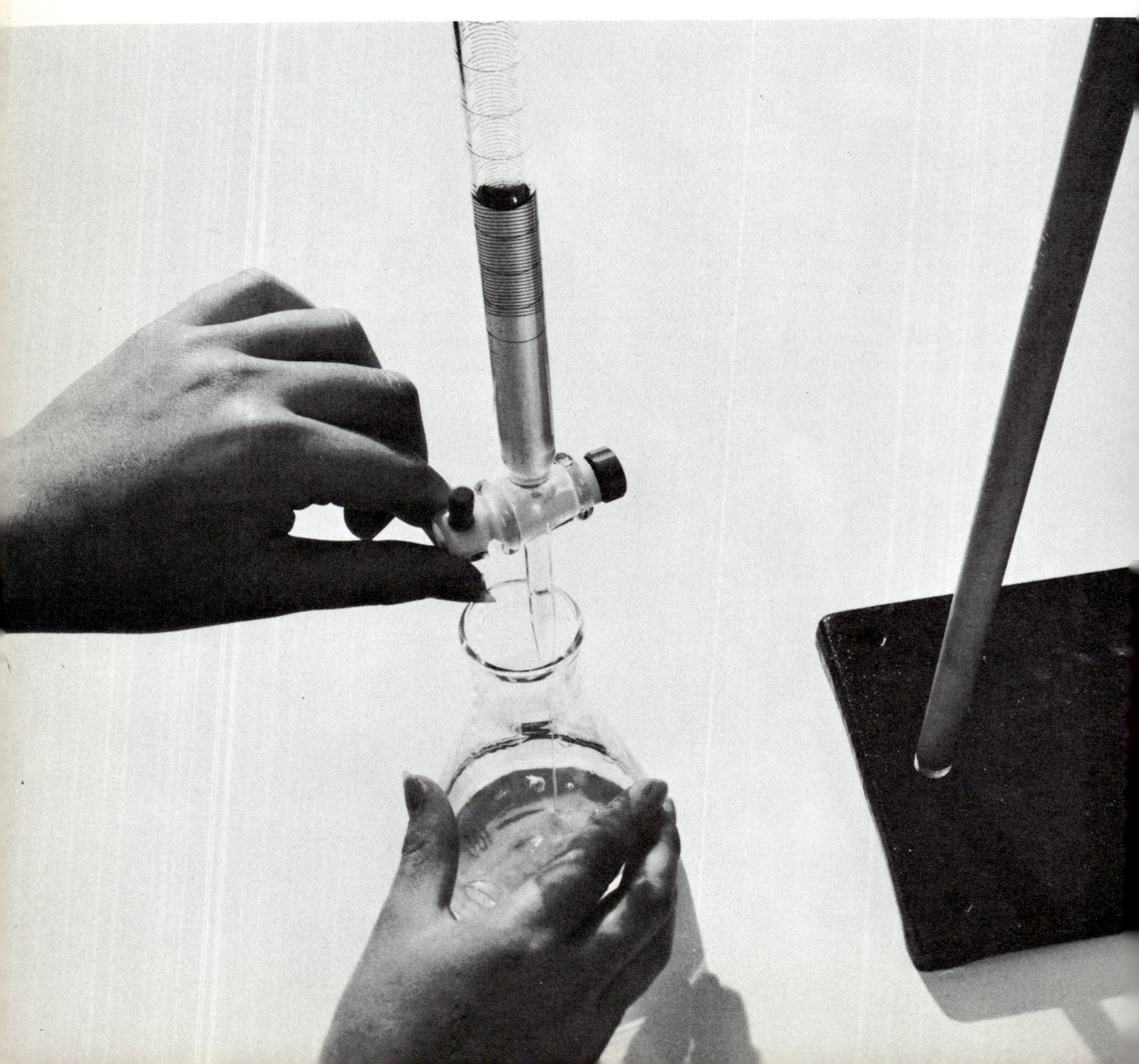

5

The Dynamics of Chemistry

Chemistry is concerned with changes and transformations. Some of these chemical changes are as powerful and spectacular as the making of iron and steel. Others are as quiet and familiar as those that occur when an egg is fried. As you know, learning the structures of atoms and molecules is a vital part of chemistry, and the understanding of chemical bonds still offers the research chemist problems worthy of all his intelligence. However, in the final analysis, the science of chemistry seeks to explain the *how* and *why* of *chemical reactions*. Reactions are the dynamics of chemistry.

Chemical changes go on all the time in the natural world. Chemists, in their laboratories, try to control certain reactions so that they can study them. They bring substances together, study the changes that occur when they react, and identify the products that are formed. By doing this, they learn to control reactions, so that the reactions can be put to use.

In this unit you will learn how chemists measure the energies involved in chemical changes and the rates of these changes. You will also discover how they study the factors that control the rates and products, and how chemical calculations are made.

19 Chemical Reactions and Calculations

Calculations are an important aspect of a chemist's work. Here, the chemist has his slide rule (left) close at hand.

Changes are taking place around you every day—wood burning, iron rusting, ice melting, organic matter decomposing, to name but a few. Are these merely phase changes or are they chemical changes in which entirely new substances are formed? When you melted candle wax, it appeared to change from an opaque solid to a clear liquid. Upon cooling, the liquid hardened into the original opaque solid wax. That was obviously a phase change. When you *burned* the wax, however, gases were given off. These gases turned out to be water vapor and carbon dioxide, and these materials have very different properties from those of the wax with which you began.

How would you express this chemical reaction qualitatively (in words) or quantitatively (in numbers)? In this chapter you will learn how to determine the numbers and kinds of atoms in a given substance and the ways in which the atoms in a substance are combined and rearranged when chemical changes occur. You will also learn the chemical shorthand used to describe the substances and the reactions involved. This study of the quantitative relationships implied by a chemical reaction is called *stoichiometry*.

Chemical Formulas and Equations

19-1 *Valence, the Combining Capacity of Elements*

Let us look at some of the formulas for compounds with which you are already familiar. In the case of water, H_2O, the formula indicates that two atoms of hydrogen have combined with one atom of oxygen. In the case of sodium chloride, NaCl, the sodium and chlorine have combined in a one-to-one ratio. In potassium hydroxide, KOH, there is one atom of potassium, one atom of oxygen, and one atom of hydrogen. Your studies of electron configurations and chemical bonding in Chapter 14 have provided an understanding of what actually determines the "combining capacity" of the elements. You have learned, for example, why one atom of the metal potassium combines with one atom of the halogen bromine to form KBr, while one atom of the metal calcium combines with two atoms of bromine to form $CaBr_2$. You have discovered that the combining capacity of the atom is in reality its bonding capacity. In simplest terms, bonding capacity involves the tendency of an atom to transfer its outer electrons to, or share them with, another atom.

A term widely used in referring to the combining capacities of atoms is *valence*. This term can be a most useful tool in predicting correct formulas for many of the simpler compounds. Valence represents the number of electrons an atom loses, gains, or shares when it forms chemical bonds with other atoms. Plus and minus signs are used to indicate valence. Plus indicates that an atom is mainly losing electrons, while minus indicates that it is mainly gaining electrons.

As stated before, valence can refer to atoms that primarily share electrons. Some scientists, however, prefer to apply the term only when discussing ions and their charges. When used this way, valence means "the charge on an ion." In this book, valence will be used with this meaning only.

TABLE 19-1 Names, Formulas, and Charges (Valences) of Some Common Ions

Positive Ions	Negative Ions
1+	**1−**
ammonium, NH_4^+	acetate, $C_2H_3O_2^-$
copper(I), Cu^+	bromide, Br^-
hydrogen, H^+	chlorate, ClO_3^-
lithium, Li^+	chloride, Cl^-
mercury(I), Hg_2^{2+}	fluoride, F^-
potassium, K^+	hydrogen carbonate, HCO_3^-
silver, Ag^+	hydrogen sulfate, HSO_4^-
sodium, Na^+	hydroxide, OH^-
2+	iodide, I^-
barium, Ba^{2+}	nitrate, NO_3^-
calcium, Ca^{2+}	nitrite, NO_2^-
chromium(II), Cr^{2+}	permanganate, MnO_4^-
copper(II), Cu^{2+}	**2−**
iron(II), Fe^{2+}	carbonate, CO_3^{2-}
lead(II), Pb^{2+}	chromate, CrO_4^{2-}
magnesium, Mg^{2+}	oxide, O^{2-}
manganese(II), Mn^{2+}	peroxide, O_2^{2-}
mercury(II), Hg^{2+}	sulfate, SO_4^{2-}
nickel(II), Ni^{2+}	sulfide, S^{2-}
zinc, Zn^{2+}	sulfite, SO_3^{2-}
3+	**3−**
aluminum, Al^{3+}	ferricyanide, $Fe(CN)_6^{3-}$
chromium(III), Cr^{3+}	phosphate, PO_4^{3-}
iron(III), Fe^{3+}	

In order to become familiar with the common ions and their charges, you should study carefully Table 19-1.

One of the first things that you should notice in the table is that the alkali metal ions have a charge of 1+. They have this charge because each of these elements has one electron in its outer (valence) orbital available for bonding.

Observe that the alkaline earth ions have charges of 2+. Each of these elements has two valence electrons available for bonding.

Look closely at the large number of polyatomic ions. Such ions were described in Section 17-11. Remember that each polyatomic ion is a very stable group of atoms that, as a group, bears a charge.

If you examine the table further, you can see that some elements have more than one charge, or valence. That is, they form ions that have several different charges. This indicates that the elements gain or lose different numbers of electrons. For example, copper forms two different ions, Cu^+ and Cu^{2+}. These two ions indicate that copper can lose either one or two electrons, respectively.

19-2 Using Valence Rules To Write Formulas

In writing the formulas for most simple compounds, the following rules are usually applied.

1. The element with the most positive valence is written first.
2. The element with the most negative valence is written last.
3. The net charge on the resulting compound must be zero. Thus, the charges indicated by the valence numbers must be balanced. To do this, subscripts are written to the right of the element or polyatomic ion.
4. The valence of a polyatomic ion must equal the algebraic sum of the charges

assigned to the individual atoms making up the polyatomic ion.

5. If a polyatomic ion appears more than once in the formula, it is enclosed in parentheses and the subscript is placed just outside to the right. Parentheses are never used around a single element.

The application of these rules can be illustrated by writing the formulas for some compounds with which you are already familiar. For example, to write the formula for water, write the element with the positive valence (hydrogen) first. Then write the element having the negative valence (oxygen). The formula so far is HO. Note, however, that the valence number for hydrogen is $1+$, while that for oxygen is $2-$. Two hydrogen atoms of valence $1+$ are required in the formula to balance the one oxygen atom of valence $2-$. Thus, the formula for this compound becomes the familiar H_2O.

In a similar fashion, you can write the formula for the compound formed between calcium and oxygen. Because its valence is positive, calcium is written first. Oxygen, with a negative valence, is written last. Since their valence numbers are equal, the formula is CaO.

You know at this point that the number and arrangement of the electrons in the valence orbitals of an atom account for its bonding capacity. The foregoing rules, however, provide a quick technique for writing formulas without investigating the electron structures involved in each and every case.

For the hydroxide ion, OH^-, the net charge is $1-$. This results from the fact that oxygen has a valence of $2-$, while hydrogen has a valence of $1+$.

19-3 Writing the Names of Compounds

It is useful to know a few rules about naming chemical compounds once you have written their correct formulas. Both names and formulas are useful in identifying compounds, as you can see in Figure 19-1.

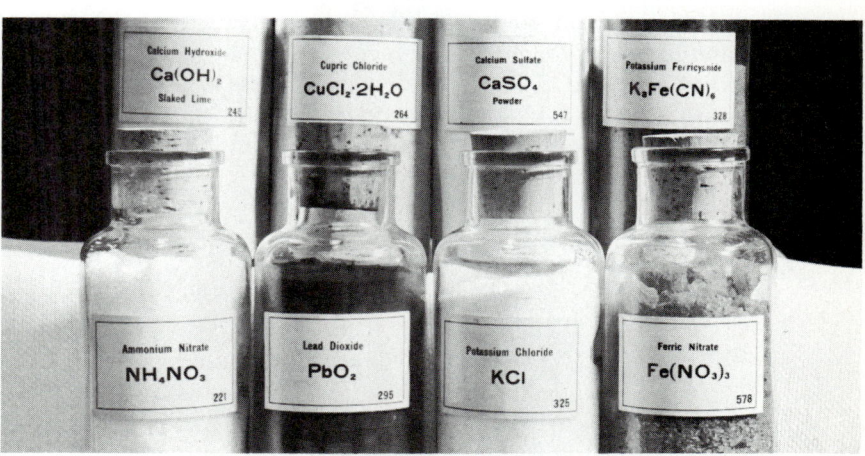

Figure 19-1 *Both names and formulas are useful in identifying chemical compounds.*

Many of the compounds discussed so far contain only two kinds of elements. They are called **binary** compounds. Most likely, you have noted that the names of binary compounds usually end in *-ide*. The name of the element with the negative valence is the one that ends in *-ide*.

NaCl	sodium chlor*ide*
CaCl$_2$	calcium chlor*ide*
CaO	calcium ox*ide*
Al$_2$O$_3$	aluminum ox*ide*

Notice that many, but not all, of the names of negative polyatomic ions end in *-ate*. Notice also that this ending is used when naming a compound containing a polyatomic ion.

NaNO$_3$	sodium nitr*ate*
NH$_4$NO$_3$	ammonium nitr*ate*
Ca$_3$(PO$_4$)$_2$	calcium phosph*ate*

It has been mentioned that, in forming compounds, several elements show more than one common valence. Such compounds are different substances and so have different properties and different formulas. Therefore, it is important for you to know how these compounds are named. Consider, for example, the reaction between mercury and oxygen. Using the previous rules, you can see that two formulas are possible.

Hg$_2$O	(mercury has a valence of 1+)
HgO	(mercury has a valence of 2+)

In the preferred system, the different valence states are identified by calling Hg$_2$O *mercury(I) oxide* and HgO *mercury(II) oxide*. The *I* and *II* indicate the valences of mercury. There is, however, another system commonly used. In this system, the lower valence of the metal is indicated by the ending *-ous* and the higher valence by the ending *-ic*. Thus, for the preceding compounds of mercury, the names would be as follows.

Hg$_2$O	mercur*ous* oxide
HgO	mercur*ic* oxide

As another illustration, the two chlorides of copper, CuCl and CuCl$_2$, can be named in either of the following ways.

CuCl	copper(I) chloride, or cuprous chloride
CuCl$_2$	copper(II) chloride, or cupric chloride

Note that the old Latin word *cuprum*, from which the symbol Cu is derived, is used to form the words *cupric* and *cuprous*.

Both types of names are found in books and are heard when chemists converse.

INVESTIGATION   *19-4 Behavior of Solid Iron in a Water Solution of Copper Sulfate*

When a chemist wants to produce a certain amount of a product, he should know how the quantities of his starting materials are related to the product he plans to make. A knowledge of the relationship between moles of reactants and moles of products makes it possible to mix the proper amounts without wasting chemicals.

PURPOSE

In this experiment you will react a known mass of iron with an excess of copper sulfate in water. You can then determine what relationship exists between the number of moles of iron used and the number of moles of the product formed in the reaction.

MATERIALS

1 gram iron filings or powder
2 beakers, 100-ml
Ringstand, ring
Wire gauze
5 grams copper sulfate crystals ($CuSO_4 \cdot 5H_2O$)
Burner
Balance, centigram
1-2 drops of detergent solution

PROCEDURE

1. Find the mass of a 100-ml beaker to the nearest 0.01 gram. Add 5.0-5.1 grams of copper sulfate crystals to the beaker on the balance. The amount does not have to be measured accurately but must be at least 5.0 grams.

2. Find the mass of an empty beaker to the nearest 0.01 gram. With the beaker still on the balance pan, add 0.95-1.05 grams of iron filings or powder, as shown in Figure 19-2. Find the total mass of the beaker and iron to the nearest 0.01 gram.

 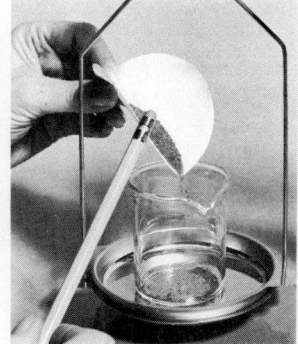

Figure 19-2 *Proper techniques for transferring solid chemicals include both slow rotation of the stock bottle (left) and gentle tapping of the filter paper (right). Each produces a smooth, constant transfer of material.*

Figure 19-3

3. Add 20-25 ml of water to the copper sulfate in the beaker. Heat the beaker and its contents until the copper sulfate dissolves and the mixture boils gently.

4. Remove the burner from under the beaker. Carefully pour the iron filings, a little at a time, from the beaker in which their mass was measured into the copper sulfate solution. After each addition of iron filings, wait for the reaction to subside before adding more.

5. After all the iron has been added to the solution, heat the solution again until it almost boils and keep it at that temperature for three to four minutes. Note the change in appearance of the substance in the beaker.

6. Remove the burner and allow the beaker to cool until it is comfortable to handle. Add 1-2 drops of a detergent solution to the mixture to prevent the solid particles from floating on the surface of the liquid.

7. Carefully decant the clear liquid, as in Figure 19-3, into another small beaker before discarding it. Be very careful not to pour off any of the solid particles. If any solid particles do get into the receiving beaker by mistake, pour off most of the remaining liquid from the receiving beaker. Then rinse the particles back into the original beaker with small amounts of demineralized water from a wash bottle, as shown in Figure 19-4. Decant again.

8. After as much liquid as possible has been decanted from the solid particles in the beaker, rinse the remaining particles several times with 10-ml portions of demineralized water. Each time swirl the particles with the water and decant the water as in step 7, making sure not to lose any of the particles.

9. After the final washing with water, place the beaker with its product in a low temperature oven (50-60°C) or under a heat lamp for several hours to dry it. Satisfactory results can be also obtained by drying in the air at room temperature overnight.

10. When the beaker and its contents are dry and cool, determine their mass to the nearest 0.01 gram.

DATA AND RESULTS

Your data table should include the following information.

Mass of beaker and iron filings	Mass of beaker and product
Mass of beaker alone	Mass of empty beaker
Mass of iron filings	Mass of product

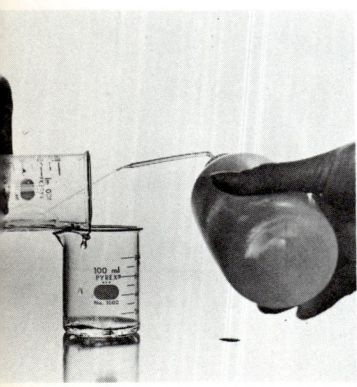

Figure 19-4

Calculations

1. Calculate the number of moles of iron that reacted with the copper sulfate. Remember that the mass of the substance divided by the mass per mole equals the number of moles.
2. Calculate the number of moles of the product (copper) obtained.
3. Determine the ratio of moles of copper to moles of iron involved in this reaction. Be sure to express your calculations using the correct number of significant figures.
4. In order to evaluate the results of this experiment, your teacher will collect the data obtained by other members of your class. When you have this compiled data, make a graph. Plot the number of individuals obtaining a given copper/iron ratio along the vertical axis. Plot the Cu/Fe ratio along the horizontal axis. The Cu/Fe ratio should be rounded off so that each division on the graph will represent values of ± 0.05. For example, values from 0.85 up to but not including 0.95 should be plotted as 0.9.
5. Considering only the middle two-thirds of the data plotted, what is the range of the values obtained for the Cu/Fe ratio? How does this compare with the uncertainty you considered justifiable from your measurements?

RELATED PROBLEMS

1. Using the results obtained in this experiment, supply the nearest whole number coefficients to complete the following statement.

One mole of iron (*solid*) + ? mole(s) of copper sulfate (*in water*) $\longrightarrow$? mole(s) of copper (*solid*) + ? mole(s) of iron sulfate (*in water*).

2. How many *atoms* of solid iron were involved in *your* experiment?
3. How many *atoms* of solid copper were involved in *your* experiment?
4. What is the relationship between the number of atoms of iron and the number of atoms of copper calculated in Problems 2 and 3?
5. When your reaction was complete, what evidence was there to indicate that you had actually started with an excess of copper sulfate?

19-5 How Chemical Equations Are Written

The process of writing chemical equations has been described in Section 5-13, and you have dealt with simple chemical equations quite a few times. Nevertheless, the study of this and the following chapters depends so strongly on your being able to write correct chemical equations with ease that more will be said about this very important subject. Part of what will be said will be purely review, but some new ideas will also be presented.

A chemical equation expresses in symbols what happens in a chemical reaction. On the left side of the equation are listed the substances that must be brought together to make the reaction occur. These substances are called **reactants.** On the right side are listed the substances formed by the reaction. These substances are known as **products.** The equation must be balanced. That is, there must be the same number of each kind of atom on each side. In chemical reactions, atoms are conserved. Mass is also conserved. The total mass of all the products equals the total mass of all the reactants.

To write a chemical equation, place the correct formula for each of the reactants to the left of the arrow and the correct formula for each of the products to the right of the arrow. For example, aluminum metal reacts with iron(III) oxide (ferric oxide) to form metallic iron and aluminum oxide. To arrive at the correct equation for this reaction, begin by writing the following.

$$Al + Fe_2O_3 \longrightarrow Fe + Al_2O_3$$

It is now necessary to decide what coefficient must be placed before each formula so that atoms are conserved. Oxygen atoms are already conserved, but iron atoms are not. A coefficient of 2 is required for both Al and Fe, making the complete, correct equation the following one.

$$2Al + Fe_2O_3 \longrightarrow 2Fe + Al_2O_3$$

How are equations involving *ionic* compounds written? In Chapter 17 you learned that ionic compounds *dissociate* (separate) into ions when they are dissolved in water.

In subsequent chapters you will learn about reactions between ions in solution. You should, therefore, become familiar with the best procedure for writing equations that describe ionic reactions.

Consider, for example, the reaction between aqueous solutions of silver nitrate and sodium chloride. An **aqueous** solution has water as the solvent. In each of these solutions the reactants entirely dissociate into ions and are written as follows.

Silver nitrate in aqueous
solution $\quad Ag^+ + NO_3^-$
Sodium chloride in aqueous
solution $\quad Na^+ + Cl^-$

The products of the reaction are silver chloride and sodium nitrate. Silver chloride is insoluble in water and thus is called a precipitate. A **precipitate** is a solid that separates from a solution. Sodium nitrate remains in solution, entirely dissociated into its ions. The products are written as follows.

Silver chloride $\quad AgCl$

Sodium nitrate in aqueous solution $\quad Na^+ + NO_3^-$

To begin writing the correct equation for the reaction, you therefore start with the following.

$$Ag^+ + NO_3^- + Na^+ + Cl^- \longrightarrow$$
$$AgCl + Na^+ + NO_3^-$$

You can see that this equation is already balanced. However, it is not in an entirely suitable form because there is no reason for including the formulas of ions that do not change. Since these formulas appear on both sides of the equation, including them makes the reaction look more complicated than it really is. Thus, it becomes difficult to focus attention on the essentials.

 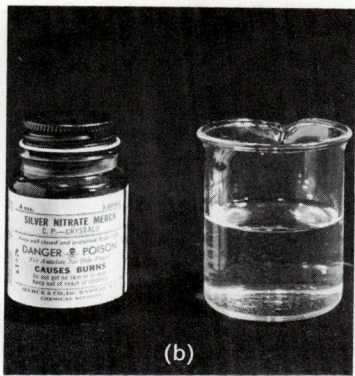 

Figure 19-5 *Mixing solutions of sodium chloride (a) and silver nitrate (b) produces a chemical reaction (c) in which an insoluble solid, silver chloride, settles to the bottom of a clear solution of sodium nitrate.*

In arriving at the best equation for an ionic reaction, you cross out those ions that appear on both sides of the equation in its initial form. In the example used above, the Na^+ and NO_3^- ions should be deleted, leaving the following equation.

$$Ag^+ + Cl^- \longrightarrow AgCl$$

This equation tells you what actually happens. One silver ion combines with one chloride ion to form silver chloride. The insoluble AgCl precipitates, as shown in Figure 19-5.

While it is not a universal rule, many chemists will indicate that a substance precipitates either by underlining the formula or by placing a small downward-pointing arrow immediately after it. Thus, the preceding equation could also be written in either of the ways shown in the next column.

$$Ag^+ + Cl^- \longrightarrow \underline{AgCl}$$
$$Ag^+ + Cl^- \longrightarrow \underline{AgCl}\downarrow$$

The same equation can also be written in the following way.

$$Ag^+(aq) + Cl^-(aq) \longrightarrow AgCl(s)$$

The symbol aq is shorthand notation for "in aqueous solution." Thus, $Ag^+(aq)$ indicates that silver ions are in aqueous solution. The symbol (s), which stands for "solid," indicates here that AgCl separates from solution as a solid. In other words, AgCl precipitates. This method of indicating precipitates and ions in aqueous solution will be used in the remainder of this book.

Summary

The combining capacity of an element is called its valence. Most elements have one or only a few common valences. By learning these, you can write the formulas for nearly all of the simple and common compounds without specifically considering the electron structures of the atoms. In the correct formula the net algebraic sum of all the valences is zero for neutral compounds or equal to the net charge for polyatomic ions.

The naming of most common compounds follows very simple rules. The metallic element (the one with the positive valence number) is given first. The name of the element with the negative valence number is usually in a form ending in -*ide*, as in oxide or chloride. For a number of important polyatomic ions, such as NH_4^+ (ammonium) and NO_3^- (nitrate), the names must be learned. Then they can be used to form the names of compounds in which they occur. For example, NH_4NO_3 is "ammonium nitrate."

Chemical equations are symbolic representations of chemical reactions. Both sides of the equation (reactants and products) must balance. That is, each side must show the same number of atoms, as well as equal masses. For reactions involving compounds that are entirely dissociated into ions, the charge should be indicated for each ion in the equation. The net charge on each side of the equation must be the same.

Self Check

1. Aluminum and oxygen combine to form aluminum oxide, with the formula Al_2O_3. Explain.
2. Write formulas for the two chlorides formed by mercury.
3. What charge does the $AlCl_4$ ion have?
4. Compounds of iron are frequently named by using the Latin *ferrum*, meaning "iron," with the -*ous* and -*ic* endings. Using this information, give two names for each of the following compounds of iron. (a) $FeCl_2$, (b) $FeCl_3$, (c) $FeSO_4$, (d) $Fe_2(SO_4)_3$.
5. Write balanced equations for the following reactions.
   (a) Zinc, Zn, reacts with an aqueous solution of copper sulfate, producing copper metal and a zinc sulfate solution.
   (b) An aqueous solution of iron(III) nitrate reacts with an aqueous solution of sodium hydroxide, producing iron(III) hydroxide as a precipitate.

Chemical Calculations

INVESTIGATION

19-6 *Mass Relationships in Chemical Changes*

Earlier in this course, you have determined changes in mass in chemical reactions. For example, you produced copper oxide, and you reacted solid iron with copper sulfate. In this experiment you will carry out several chemical changes, separate the products from the excess reactants, and measure the relations between the reactants and products at different steps of the process.

You will produce silver nitrate by reacting silver with nitric acid. After determining how much silver nitrate is formed, you will react it with sodium chloride to produce silver chloride. After washing and drying the silver chloride, you will find its mass and the mass of the other remaining materials. This will enable you to make a comparison of the moles of each of the substances involved in the process.

Before starting the experiment, examine the information required in the section on Data and Results. Organize a data sheet in your notebook to record this information. This should lessen the possibility of omitting important measurements.

PURPOSE

You will determine the mole relationships for the preceding series of reactions. You will also review and learn several important techniques, including decanting, filtering, washing, and drying.

MATERIALS

10 ml 6 M nitric acid
Beaker, 250-ml
Stirring rod
Wash bottle
Filter paper
Funnel and support
Ringstand, ring
Wire gauze
Burner

1.5–2.0 grams silver metal foil
2.0–2.5 grams sodium chloride (solid)
Beaker, 100-ml (or another 250-ml beaker)
Graduated cylinder, 10- or 25-ml

Special Precautions. Nitric acid is extremely corrosive to eyes, skin, and clothing. Be sure to wear safety glasses while handling it. If you spill it on your skin, wash it off at once with large amounts of water. Avoid inhaling the poisonous reddish-brown fumes of nitrogen dioxide, NO_2, which form when silver reacts with nitric acid. Do not touch silver nitrate. It reacts with your skin.

PROCEDURE

Part A. Preparation of Solid Silver Nitrate from Metallic Silver

1. Use a pencil to place your initials and the number 1 on the ground glass circle of a 250-ml beaker. Determine the mass of the beaker to the nearest 0.01 gram.

2. With the beaker still on the balance pan, add 1.5–2.0 grams of silver foil. Determine the mass of the beaker and the silver foil to the nearest 0.01 gram.

3. To the beaker containing the silver, add 10 ml of nitric acid, labeled 6 M HNO$_3$. Since the fumes that are produced are poisonous, this operation should be performed in a fume hood or near an open window. After the reaction has stopped, leave the beaker in the fume hood, or in a well-ventilated place, to be evaporated overnight.

4. After they have become dry, find the mass of the beaker and its contents (AgNO$_3$), to the nearest 0.01 gram. Remember the caution concerning the handling of silver nitrate.

Part B. The Effect of Adding a Solution of Sodium Chloride to a Solution of Silver Nitrate

1. Add about 15 ml of distilled water to the AgNO$_3$ in beaker 1. Stir until the solid fully dissolves.

2. Remove from the stock bottle about 2–2.5 grams (approximately 0.5 teaspoonful) of sodium chloride, NaCl. Gently rotate the bottle back and forth to pour out the approximate amount needed onto a piece of clean paper. (This technique was shown earlier, in Figure 19-2.)

3. Label a clean, dry 100-ml beaker as number 2 and find its mass to the nearest 0.01 g.

4. Adjust your balance so that it reads somewhere between 2 and 2.5 grams greater than the mass of beaker 2. Carefully add enough NaCl to make the balance pan drop. Discard any remaining salt. Determine the mass of the NaCl and beaker to the nearest 0.01 g.

5. Add about 15 ml of distilled water to the solid NaCl. Stir until the solid fully dissolves.

6. While briskly stirring the AgNO$_3$ solution in beaker 1, slowly add the NaCl solution. Note the result. The white solid produced is the compound silver chloride, AgCl. Rinse the empty beaker, number 2, with about 5 ml of distilled water from the wash bottle by directing the water around the inside of the beaker. Add the rinse water to the mixture in beaker 1. Rinse beaker 2 again with distilled water and this time discard the rinse water. The clean beaker will be used again in step 9.

7. Heat the resulting precipitate and the solution to boiling for about 2 minutes or until the solution becomes reasonably clear as the precipitate settles. Place a stirring rod in the beaker to help prevent unsteady boiling (bumping).

8. Determine the mass of a piece of filter paper to the nearest 0.01 gram. Fold it as shown in Figure 19-6. Fit it into a funnel and

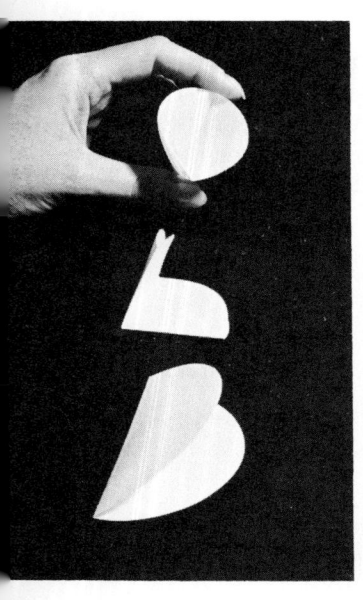

Figure 19-6

Figure 19-7

moisten the paper with some distilled water from a wash bottle. Set up the funnel for filtering as in Figure 19-7.

9. Place beaker 2 under the funnel. The tip of the funnel should touch the beaker so that a steady stream can run down the side. Decant the clear liquid from beaker 1, pouring it into the funnel along a glass rod. (See Figure 19-7.) A small amount of the precipitate may transfer to the filter paper, but try to keep most of it in the beaker, where it can be washed more readily.

10. Wash the precipitate in the beaker with about 15 ml of distilled water, stirring with a glass rod to aid the washing. Decant the wash water into the funnel. Repeat the washing procedure with another 15 ml of water. Decant the wash water again into the funnel.

11. After the filtration is complete, place the filter paper and any solid it contains in beaker 1, which contains the precipitate.

12. The filtrate (the solution that passed through the filter) in beaker 2 and the wet precipitate in beaker 1 should now be evaporated and dried. Put them in a place designated by your teacher for evaporation and drying overnight. Be sure that each beaker is numbered (1 or 2) and has your name or locker number on it.

13. Find the mass of each dry sample and record the masses. Save the silver chloride as directed by your teacher.

DATA AND RESULTS

Your data table should include the following.

Mass of beaker 1
Mass of beaker 1 and silver foil
Mass of beaker 1 and solid silver nitrate
Mass of beaker 1, filter paper, and solid silver chloride
Mass of beaker 2
Mass of beaker 2 and solid sodium chloride
Mass of a piece of filter paper
Mass of beaker 2 and solid residue

RELATED PROBLEMS

1. How many moles of silver did you start with? How many moles of dry silver nitrate were produced when the silver reacted with nitric acid? How many moles of solid silver chloride were finally produced?

2. Find the following ratios: moles Ag/moles $AgNO_3$; moles $AgNO_3$/moles AgCl; moles Ag/moles AgCl. Use the nearest whole number to express your results. If your results are not near to a whole number, how can you explain the difference? What can you conclude about the number of moles involved in these chemical changes?

3. Compare the sum of the masses of the $AgNO_3$ and NaCl used with the sum of the masses of the AgCl and the residue in beaker 2. Your conclusions will be more significant if based on class data compiled by your teacher. What is the significance of these results?

4. By examining your answer to Problem 3, predict what the residue in beaker 2 should be.

19-7 *Stoichiometry*

Many aspects of chemistry are quantitative. That is, the amounts or sizes of things are measurable and are related to one another in an exact, numerical way. One of the most important of these quantitative aspects of chemistry is called stoichiometry (*stoi*-kee-om-eh-tree). Chemists define **stoichiometry** as the quantitative relationships among reactants and products in a chemical reaction. A great deal of the fundamental knowledge of chemistry was built up by careful quantitative analyses of substances. Such analyses are possible only because of the exact stoichiometric relationships in reactions. From a commercial and industrial point of view, stoichiometry is enormously important. It enables chemists and engineers to calculate the actual amounts of reacting material required to produce a desired amount of a product.

Consider, for example, the chemical equation written and balanced earlier.

$$2Al(s) + Fe_2O_3(s) \longrightarrow Al_2O_3(s) + 2Fe(s)$$

This reaction does not occur unless the reactants are mixed and raised to a high temperature. Once begun, it proceeds vigorously.

The equation tells us that two moles of aluminum react with one mole of iron(III) oxide (ferric oxide) to form one mole of aluminum oxide and two moles of iron. How much iron is produced in this reaction from a given quantity of aluminum and iron oxide? To answer this question, you must make the following assumptions.

1. You must assume that the reaction as written is complete. That is, all of the materials on the left-hand side are reacting to form the products on the right-hand side, with no reverse reaction occurring. (There are reactions in which this is not a proper assumption. Such reactions and how to deal with them are the subject of Chapter 22.)

2. You must assume further, that only this one reaction, as expressed by the equation, takes place.

Problems in stoichiometry are generally one of the following types.

1. Given the available mass of one of the reactants, the *limiting reactant*, calculate how much of one or several of the products can be obtained. It is assumed that adequate quantities of all other necessary reactants are at hand.

2. How much of each of the reactants will be required to produce a specific quantity of one of the reaction products?

19-8 The Mole Method

The most efficient way of using a balanced chemical equation to calculate mass relationships in a chemical reaction is to express the quantities in *moles*. The "mole method" is an efficient way to solve problems in stoichiometry because the coefficients in a chemical equation show the number of moles of each substance involved in the reaction.

In order to find the mass of one mole of a substance, you need only add the molar masses of all of the atoms represented in its formula. For example, the mass of one mole of sulfuric acid, H_2SO_4, is found as follows.

$$
\begin{array}{l}
2 \times \text{molar mass of H atoms} = 2 \times 1.0 = 2.0 \text{ g} \\
1 \times \text{molar mass of S atoms} = 1 \times 32.1 = 32.1 \text{ g} \\
4 \times \text{molar mass of O atoms} = 4 \times 16.0 = 64.0 \text{ g} \\
\hline
\text{molar mass of } H_2SO_4 = 98.1 \text{ g}
\end{array}
$$

A step-by-step description of the mole method, for a given reaction, follows. Throughout the description, the *given* substance is the substance present in a specified amount. From the amount of the given substance, the mass of the *required* substance (or substances) is calculated. The reaction used in the description is the one that has been discussed in preceding sections.

$$2Al(s) + Fe_2O_3(s) \longrightarrow 2Fe(s) + Al_2O_3(s)$$

The given substance in the description is Fe_2O_3 and the required substance is iron. The amount of Fe_2O_3 to be used is 32.0 grams.

The main steps in the mole method follow.
1. Write a balanced equation.
2. Convert the mass of the given substance into moles of the given substance. This is done using the following relationship.

$$\text{Moles} = \frac{\text{mass, in grams}}{\text{molar mass, in grams}}$$

The molar mass of Fe_2O_3 is found as follows.

$$2 \times (55.9) + 3 \times (16.0) = 159.8$$

Thus,

$$32.0 \text{ g of } Fe_2O_3 \text{ is } \frac{32.0 \text{ g}}{159.8 \text{ g/mole}} = 0.200 \text{ mole.}$$

3. Inspect the equation for the reaction to determine how many moles of the required substance are produced from one mole of the given substance. In the reaction of aluminum with ferric oxide, one mole of Fe_2O_3 produces two moles of iron.
4. Determine, using a simple proportion, how many moles of the required substance can be produced by the actual number of moles of the given substance. For example, if only 0.200 mole of Fe_2O_3 were available, then 0.400 mole of Fe would be obtained.

$$\frac{2 \text{ moles Fe}}{1 \text{ mole } Fe_2O_3} = \frac{x}{0.200 \text{ mole } Fe_2O_3}$$

$$x = 0.400 \text{ mole Fe}$$

5. Convert the moles of required substance into the actual mass, or other desired unit (for example, volume at STP if it is a gas). Since the mass of Fe is 55.9 g/mole, 0.400 mole of Fe is 0.400 mole × 55.9 g/mole = 22.4 g.

19-9 Conversion Factors

The preceding problem was solved in a step-by-step way. You can solve it more simply by combining these steps, using conversion factors. A **conversion factor** is a ratio that relates one set of numbers or units to another. The following example shows how a conversion factor can be derived.

Example

How many dozens of donuts are 60 donuts?

$$\frac{12 \text{ donuts}}{60 \text{ donuts}} = \frac{1 \text{ dozen donuts}}{x}$$

$(x)(12 \text{ donuts}) = (60 \text{ donuts})(1 \text{ dozen donuts})$

$$x = (60 \text{ donuts})\left(\frac{1 \text{ dozen donuts}}{12 \text{ donuts}}\right)$$

$x = 5 \text{ dozen donuts}$

The ratio $\frac{1 \text{ dozen donuts}}{12 \text{ donuts}}$ is a conversion factor. It relates the unit "dozen" to a specific quantity of donuts. Conversion factors such as this one make solving problems easier by leading to the cancellation of units.

The preceding problem can just as easily be worked by using the conversion factor directly.

$$x = (60 \text{ donuts})\left(\frac{1 \text{ dozen donuts}}{12 \text{ donuts}}\right)$$

$x = 5 \text{ dozen donuts}$

The problem in Section 19-8 can be worked in a similar way, using conversion factors. In that problem you found the number of grams of Fe produced in the following reaction. Remember that 32.0 g of Fe_2O_3 were available.

$$2Al(s) + Fe_2O_3(s) \longrightarrow Al_2O_3(s) + 2Fe(s)$$

Using conversion factors, you can solve the problem in the following way.

Example

1. Moles of the given substance

$$= (32.0 \text{ g } Fe_2O_3)\left(\frac{1 \text{ mole } Fe_2O_3}{159.8 \text{ g } Fe_2O_3}\right)$$

$= 0.200 \text{ mole } Fe_2O_3$

2. Moles of the required substance

$$= (0.200 \text{ mole } Fe_2O_3)\left(\frac{2 \text{ moles Fe}}{1 \text{ mole } Fe_2O_3}\right)$$

$= 0.400 \text{ mole Fe}$

3. Mass of the required substance

$$= (0.400 \text{ mole Fe})\left(\frac{55.9 \text{ g Fe}}{1 \text{ mole Fe}}\right)$$

$= 22.4 \text{ g Fe}$

Combining all three steps, you get the following equation.

$$\text{Mass of Fe} = (32.0 \text{ g } Fe_2O_3)\left(\frac{1 \text{ mole } Fe_2O_3}{159.8 \text{ g } Fe_2O_3}\right)$$

$$\times \left(\frac{2 \text{ moles Fe}}{1 \text{ mole } Fe_2O_3}\right) \times \left(\frac{55.9 \text{ g Fe}}{1 \text{ mole Fe}}\right)$$

$= 22.4 \text{ g Fe}$

Notice the conversion to moles. This conversion shows that the mole method is used here, too. Notice also that the cancellation of units in the conversion-factor method leads to the desired units. This is, in essence, a check on your calculations. If the units in your answer are wrong, you have made a mistake somewhere in your work.

In the following sections, the problems will be worked without conversion factors at first. Each step in the solution will be explained verbally. This step-by-step method will be followed by the "shorthand" conversion factor method. In this way, you will become familiar with more than one way of solving stoichiometric problems.

19-10 The Manufacture of Sulfuric Acid: Mass-Mass Chemical Calculations

One of the most important commercially-produced chemicals is sulfuric acid, H_2SO_4. In fact, this product is so universally

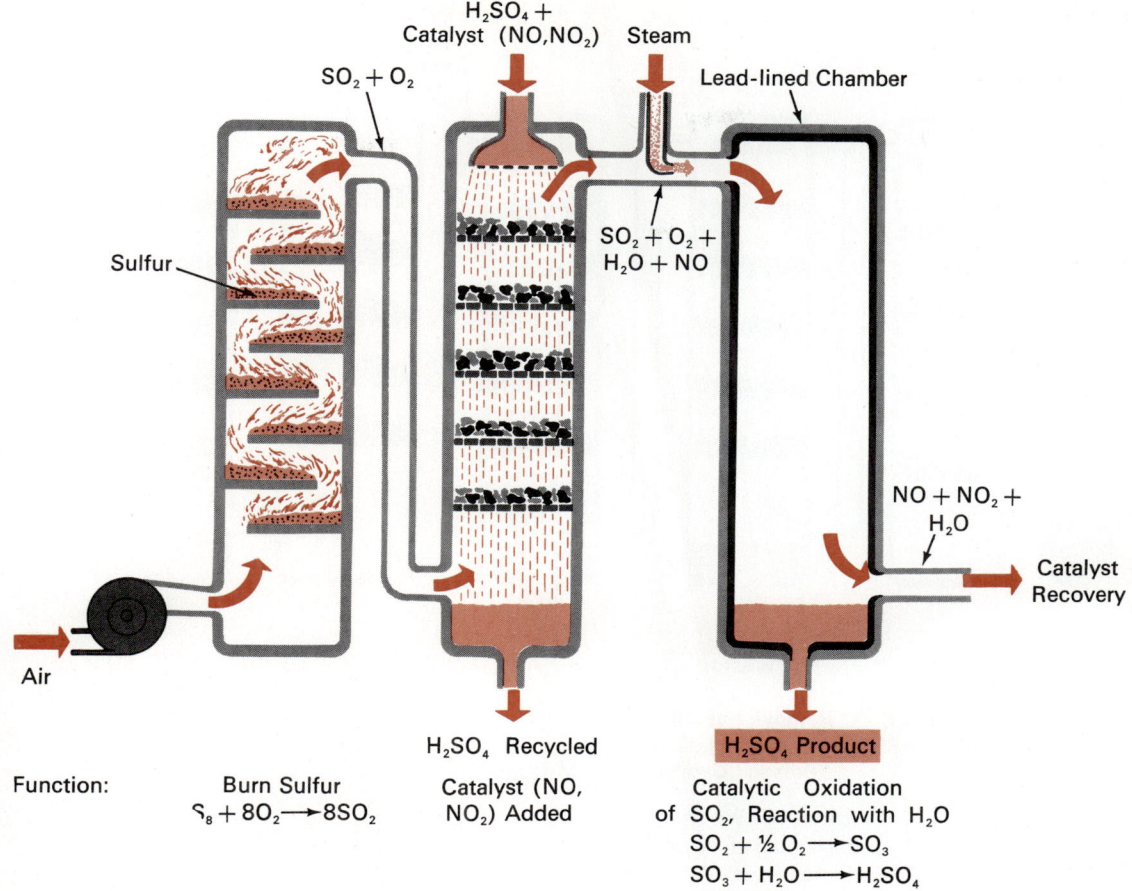

Figure 19-8 *In the lead chamber process for making sulfuric acid, a mixture of NO and NO_2 serves as a catalyst to oxidize SO_2 to SO_3.*

significant that the amount of H_2SO_4 used annually by a country can serve as a rough indication of its economic prosperity. Two important commercial processes have been developed for the production of sulfuric acid. These are the *lead chamber process* and the *contact process*. Most of the H_2SO_4 currently produced in this country is made by the contact process. This process results in a pure, highly concentrated product. The lead chamber process produces a less pure product that is suitable for some commercial uses. Figures 19-8 (above) and 19-9 (next page) show the general features of each process.

In both processes the preparation involves three reactions, each of which can be represented by a balanced equation, as follows.
1. Sulfur is burned in air to form sulfur dioxide.

$$S_8(s) + 8\ O_2(g) \longrightarrow 8\ SO_2(g)$$

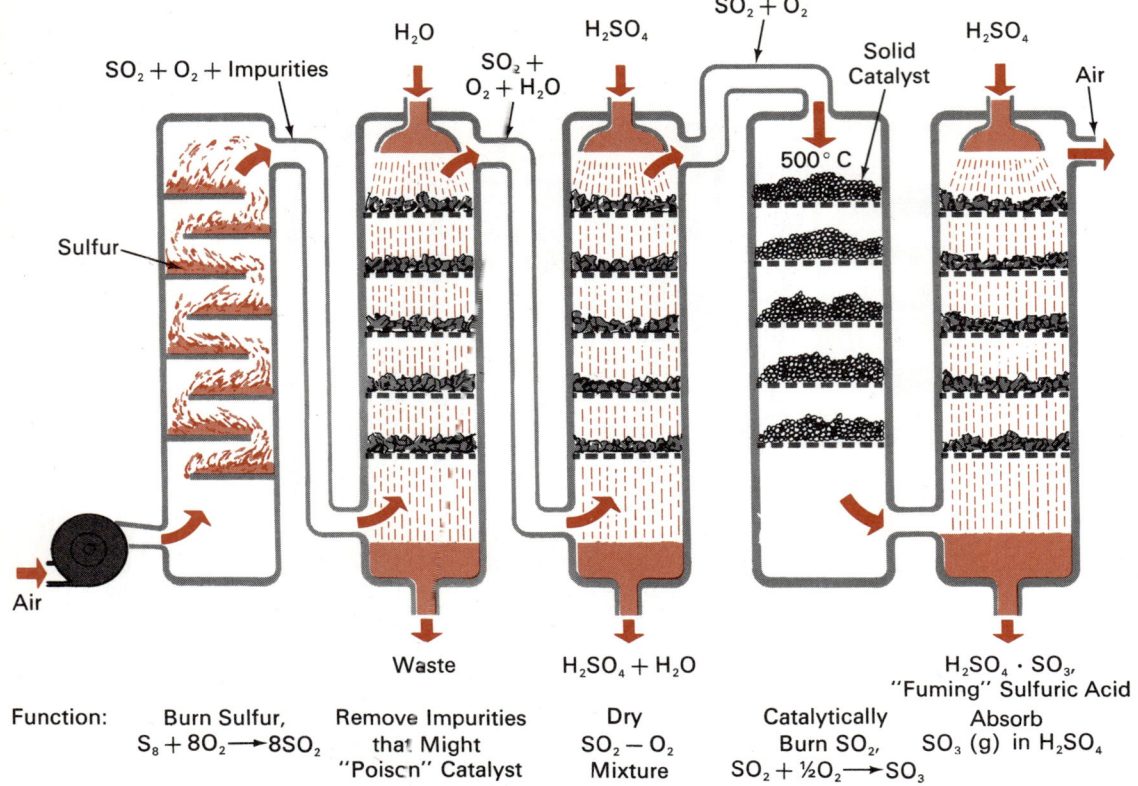

Figure 19-9 The contact process for making sulfuric acid uses a solid catalyst. It is a more costly process than is the lead chamber process, but it produces the more concentrated acid that is required for many purposes.

2. Sulfur dioxide gas reacts with oxygen to give sulfur trioxide. This reaction proceeds at a faster rate in the presence of a substance that is not changed in the course of the reaction. This substance is a "catalyst." Catalysts are discussed in more detail in Section 19-12 and in Chapter 21.

$$8\ SO_2(g) + 4\ O_2(g) \longrightarrow 8\ SO_3(g)$$

3. The sulfur trioxide is then dissolved in water to form sulfuric acid.

$$8\ SO_3(g) + 8\ H_2O(l) \longrightarrow 8\ H_2SO_4(l)$$

The overall reaction can be written by adding algebraically the equations representing the three steps just described. The resulting equation follows.

$$S_8(s) + 12\ O_2(g) + 8\ H_2O(l) \longrightarrow 8\ H_2SO_4(l)$$

Now the equations representing the reactions are balanced, a necessary first step in solving any problems concerning the stoichiometry of the process.

Suppose that you are a chemist at a plant. You are told to find out how much sulfuric acid will be produced by the burning of two kilograms of sulfur. What would you do? Here are the steps that you should follow, using the mole method.

1. Write the balanced equation for the reaction.

$$S_8(s) + 12O_2(g) + 8H_2O(l) \longrightarrow 8H_2SO_4(l)$$

2. Convert the amount of the given substance, solid sulfur, to the corresponding number of moles. The molar mass of S_8 is equal to the mass (32.1 g) of a mole of sulfur atoms times 8, or 256.8 g/mole. Hence 2 kg of sulfur equal 2000 g ÷ 256.8 g/mole = 7.79 moles of sulfur.
3. Determine the number of moles of the required substance formed from one mole of the given substance. The coefficients in the equation indicate that one mole of S_8 yields eight moles of H_2SO_4.
4. Determine by a proportion how many moles of H_2SO_4 are produced from the actual number of moles of S_8.

$$\frac{1 \text{ mole } S_8}{7.79 \text{ moles } S_8} = \frac{8 \text{ moles } H_2SO_4}{x}$$

$$x = 62.3 \text{ moles } H_2SO_4$$

5. Convert the moles of H_2SO_4 into the desired units. No particular units were specified. When this is the case, you usually should convert the number of moles of the required substance into the units of the given substance. In this example, convert the 62.3 moles of H_2SO_4 to kilograms. The mass of one mole of H_2SO_4 is 98.1 g. This means that 62.3 moles of H_2SO_4 must have a mass of 62.3 moles × 98.1 g/mole = 6112 g. A mass of 6112 g equals 6.11 kg. Therefore, 2 kg of S_8 produce 6.11 kg of H_2SO_4.

19-11 The Manufacture of H_2SO_4: Mass-Volume Calculations

When some or all of the reactants and products are gases, it is frequently desirable to determine the volumes of these substances rather than their masses. Relatively small masses of gaseous materials occupy rather large volumes. For this reason the chemist is usually more concerned with the volumes of the gaseous materials in the reaction than he is with their actual masses. To deal with such a problem, reconsider the relationship between the mass of one mole of a gas and the volume it occupies at STP.

One mole of a gas occupies 22.4 liters at 0°C (273 K) and one atmosphere of pressure. The volume of a fixed mass of gas at constant pressure changes in direct proportion to the change in absolute (Kelvin) temperature. This is one way of defining Charles' law, which was introduced in Section 6-3.

Suppose you want to determine the volume, in liters, of SO_2 formed in the reaction between sulfur and oxygen, at a temperature of 500°C and a pressure of one atmosphere. Suppose further that you burn two kilograms of sulfur. To solve this problem, you may use the mole method, taking the following steps.

1. Write the balanced equation.

$$S_8(s) + 8O_2(g) \longrightarrow 8SO_2(g)$$

2. Convert 2 kg of S_8 into moles. In Section 19-10 you found that this amount equals 7.79 moles of S_8.
3. Inspect the equation to see how many moles of SO_2 are produced from one mole

of S_8. The equation shows that 8 moles of SO_2 form for each mole of S_8 that reacts.
4. Using a simple proportion, find the actual number of moles of SO_2 that are produced.

$$\frac{1 \text{ mole } S_8}{7.79 \text{ moles } S_8} = \frac{8 \text{ moles } SO_2}{x}$$

$$x = 62.3 \text{ moles } SO_2$$

5. Convert 62.3 moles of SO_2 into liters at STP. You know that one mole of SO_2 at STP occupies 22.4 liters. Therefore, 62.3 moles of SO_2 will occupy 1400 liters at STP (62.3×22.4).
6. Find the volume of SO_2 in liters at 500°C (773 K) and one atmosphere. According to Charles' law, the volume at 273 K increases by 773/273 when the temperature is raised to 773 K. The final answer, then, is 3.96×10^3 liters (773/273 times 1400). You can work this problem by using conversion factors, as follows.

Volume of SO_2

$$= (2 \text{ kg } S_8)\left(\frac{1000 \text{ g } S_8}{1 \text{ kg } S_8}\right)\left(\frac{1 \text{ mole } S_8}{256.8 \text{ g } S_8}\right)$$

$$\times \left(\frac{8 \text{ moles } SO_2}{1 \text{ mole } S_8}\right)\left(\frac{22.4 \text{ liters } SO_2 \text{ STP}}{1 \text{ mole } SO_2}\right)$$

$$= 1400 \text{ liters } SO_2 \text{ at STP}$$

According to Charles' law, you again multiply this volume by 773/273 to get the final answer of 3.96×10^3 liters SO_2 at 773 K and one atmosphere.

The following example provides another illustration of mass-volume calculations. It shows the determination of the volume, in liters, of SO_2 produced, at 373 K and one atmosphere pressure, when enough S_8 burns to use up 64.0 g of oxygen.

Example

$$V_1 = (64.0 \text{ g } O_2)\left(\frac{1 \text{ mole } O_2}{32.0 \text{ g } O_2}\right)\left(\frac{8 \text{ mole } SO_2}{8 \text{ mole } O_2}\right)$$

$$\times \left(\frac{22.4 \text{ liters } SO_2 \text{ at STP}}{1 \text{ mole } SO_2}\right)$$

$$V_1 = 44.8 \text{ liters } SO_2 \text{ at STP}$$

$$V_2 = (44.8 \text{ liters } SO_2)\left(\frac{373}{273}\right)$$

$$V_2 = 61.2 \text{ liters } SO_2 \text{ at 373 K and 1 atmosphere}$$

19-12 The Manufacture of H_2SO_4: Volume-Volume Calculations

The two processes most widely used for the production of H_2SO_4 are, as mentioned earlier, the lead chamber process and the contact process. The most difficult step in the production of H_2SO_4 is the following one.

$$2 SO_2(g) + O_2(g) \longrightarrow 2 SO_3(g)$$

This reaction is extremely slow. The overall cost of the process would be very high were it not for the fact that any one of several catalysts greatly increase the rate of the reaction. A **catalyst** is a substance that alters the rate of a chemical reaction without itself undergoing change. Other than altering the rate of the reaction, a catalyst does not change the chemistry of the process. Therefore, the catalyst does not have to appear in the chemical equation for the reaction.

The two sulfuric acid processes, contact and lead chamber, differ chiefly in the catalyst used in converting SO_2 to SO_3. In the lead chamber process the reaction between SO_2 and O_2 is accelerated by mixing them in a lead-lined chamber with gaseous oxides of nitrogen, which serve as the catalyst. In the contact process catalysis occurs by the actual contact between the SO_2 and O_2 on the surface of a solid catalyst. Two catalysts used in the contact process are finely divided platinum, Pt, and vanadium pentoxide, V_2O_5.

At this stage in the production of sulfuric acid, the volume, in liters, of oxygen needed to react with a specified volume of SO_2 must be determined. For example, suppose that the volume of SO_2 is 112 liters at STP. You may then find the volume of oxygen in the following way.

1. Write the balanced equation.

$$2SO_2(g) + O_2(g) \longrightarrow 2SO_3(g)$$

2. Calculate the number of moles of SO_2. Since one mole of any gas will occupy 22.4 liters at STP, it follows that 112 liters must be the volume occupied by $112 \div 22.4$, or 5, moles.
3. Determine the number of moles of O_2 required to react with one mole of SO_2. The balanced equation indicates that two moles of SO_2 react with one mole of O_2. Thus, one-half mole of O_2 is required per mole of SO_2.
4. Determine the number of moles of O_2 that react. The five moles of SO_2 in this problem will require 5 times as much O_2 as one mole of SO_2. In other words, 2.5 moles of O_2 ($5 \times \frac{1}{2}$) are required.
5. Convert from moles of O_2 to liters of O_2. You know that one mole of O_2 occupies 22.4 liters at STP. Thus, 2.5 moles of O_2 occupy 2.5 moles $\times \dfrac{22.4 \text{ liters STP}}{\text{mole}} =$ 56.0 liters at STP.

The following example shows how this problem can be worked by the conversion-factor method.

Example

Volume of O_2

$$= (112 \text{ liters } SO_2 \text{ STP}) \left(\frac{1 \text{ mole } SO_2}{22.4 \text{ liters } SO_2 \text{ STP}} \right)$$

$$\times \left(\frac{1 \text{ mole } O_2}{2 \text{ moles } SO_2} \right) \left(\frac{22.4 \text{ liters } O_2 \text{ STP}}{1 \text{ mole } O_2} \right)$$

$$= 56.0 \text{ liters } O_2 \text{ at STP}$$

Calculations Using Volumes Only. You can work this problem directly in volume units, without bringing moles into the procedure. To show why this is so consider again the ideal gas law, $PV = RNT$. It was first discussed in Section 6-6.

Using this law, you can write the following equation for the pressure of SO_2.

$$P_1 = \frac{RN_1T_1}{V_1}$$

Similarly, the following equation may be written for the pressure of O_2.

$$P_2 = \frac{RN_2T_2}{V_2}$$

Because both gases are at the same pressure, the right-hand side of both equations can be set equal to each other.

$$\frac{RN_1T_1}{V_1} = \frac{RN_2T_2}{V_2}$$

The gases are also at the same temperature. Therefore, the following proportion results, by cancellation of equal variables.

$$\frac{N_1}{V_1} = \frac{N_2}{V_2}$$

For the specific chemical equation under discussion, when N_1 equals two moles, N_2 equals one mole. When you substitute these values into the preceding proportion, you get the following relationship.

$$\frac{2 \text{ moles}}{V_1} = \frac{1 \text{ mole}}{V_2}$$
$$2 V_2 = V_1$$

Thus, you see that, at the same temperature and pressure, the same ratio applies to both moles and volumes. Knowing this, the preceding example can be worked in the following way. Note that it is worked solely in volume units.

Volume of O_2

$= (112 \text{ liters SO}_2 \text{ STP}) \left(\dfrac{1 \text{ liter } O_2 \text{ STP}}{2 \text{ liters SO}_2 \text{ STP}} \right)$

$= 56.0$ liters O_2 at STP

19-13 Stoichiometry in an Organic Reaction ★

Among the many types of organic compounds discussed in Chapter 16 were esters and amides. You learned that esters could be transformed into amides by treating them with amines. A typical reaction of this type is the following one.

$$C_5H_{11}\overset{\overset{O}{\|}}{C}-O-CH_3 + C_4H_9NH_2 \longrightarrow$$

$$C_5H_{11}\overset{\overset{O}{\|}}{C}-\underset{\underset{H}{|}}{N}-C_4H_9 + CH_3OH$$

Here, the methyl ester of hexanoic acid reacts with butylamine to form the butyl amide of hexanoic acid and methanol.

Suppose a research chemist needed to prepare 50.0 g of the amide, using this reaction. How much of the ester must he start with to get the necessary 50.0 g of amide, assuming a complete reaction?

It is possible to make the reaction "go to completion," that is, to convert virtually all of the ester to the amide. To do this, the chemist uses a large amount of butylamine and heats the reaction mixture gently to drive off the methanol that is formed. Such techniques for producing a complete reaction are discussed further in Chapter 22.

The research chemist can solve his problem by using the mole method. The following steps in his solution are similar to those in the solutions to earlier problems in this chapter.

1. The chemist makes sure that the equation is balanced. It is balanced as it is now written.
2. He converts the desired mass of amide into moles. This is done as follows.

$$C_5H_{11}\overset{\overset{O}{\|}}{C}-\underset{\underset{H}{|}}{N}-C_4H_9 = C_{10}H_{21}NO$$

C: 10 × 12.0 = 120
H: 21 × 1.0 = 21.0
N: 1 × 14.0 = 14.0
O: 1 × 16.0 = 16.0
molar mass = 171.0

The number of moles of amide is 50.0 g ÷ 171 g/mole = 0.292 mole.

3. He determines by inspecting the equation that one mole of the ester yields one mole of the amide.
4. He then determines by a simple proportion how many moles of the ester are needed to produce the desired number of moles of the amide.

$$\frac{1 \text{ mole amide}}{0.292 \text{ mole amide}} = \frac{1 \text{ mole ester}}{x}$$

$$x = 0.292 \text{ mole ester}$$

Summary

Stoichiometry is the name for the quantitative relationships, including mass and volume relationships, among the reactants and products in a chemical reaction. Problems in stoichiometry generally involve calculations to obtain specific quantities of the reactants or the products of a chemical reaction. The most efficient way of calculating mass relationships is by the mole method. This method involves converting all given masses or volumes to moles. Using the molar ratios provided by the balanced chemical equations, the reactant needed or product obtainable is calculated. The answer in moles is then converted to grams, liters, or whatever other units may be desired.

5. Finally, he converts the moles of ester into grams. To do this, he must first calculate its molar mass as follows.

$$C_5H_{11}\overset{\overset{O}{\|}}{C}-O-CH_3 = C_7H_{14}O_2$$

C: 7 × 12.0 = 84.0
H: 14 × 1.0 = 14.0
O: 2 × 16.0 = 32.0
molar mass = 130.0

The number of grams of the ester required is therefore 0.292 mole × 130 g/mole = 38.0 g.

Using conversion factors, the research chemist can work the problem in the following way.

$$x = (50.0 \text{ g amide}) \left(\frac{1 \text{ mole amide}}{171 \text{ g amide}}\right)$$
$$\times \left(\frac{1 \text{ mole ester}}{1 \text{ mole amide}}\right) \left(\frac{130.0 \text{ g ester}}{1 \text{ mole ester}}\right)$$

$$x = 38.0 \text{ g ester}$$

Self Check

1. What is the molar mass of ferric oxide?
2. How many moles of Al are needed to produce 3 moles of Fe when Al reacts with Fe_2O_3?
3. Suppose that SO_2 and O_2 react at 773 K.
   (a) What volume would 112 liters of SO_2 at 273 K occupy at 773 K?
   (b) What volume would be occupied by the required 2.5 moles of O_2 at 773 K?

Chemistry and Our Environment

AUTOMOTIVE SMOG

In an earlier article (Chapter 6) we discussed one type of air pollution, which we called London smog. There is another type found in every major city where automobiles are in wide use. Ecologists often name it after the city in which it first became (and still is) a serious problem.

Los Angeles Smog. An automobile engine generates power by burning gasoline in its cylinders. Gasoline consists mainly of hydrocarbons such as octane, C_8H_{18}. The main thing that happens when octane, for example, is burned in an automobile engine is represented by the following equation.

$$C_8H_{18} + \tfrac{25}{2}O_2 \longrightarrow 8CO_2 + 9H_2O$$

If this were *all* that happened, there would be no problem. Water vapor is harmless, and so is carbon dioxide, in the quantities generated by automobiles. (In much larger quantities, CO_2 can be a problem.)

There are two troubles with automobile engines. First, the gas is not burned entirely to carbon dioxide and water. The combustion products, that is, the exhaust gases, contain some unburned hydrocarbons and some carbon monoxide, CO. Second, when the engine draws in air, in order to provide the oxygen needed to burn the gas, it also draws in nitrogen. A small amount of this nitrogen gets converted into two oxides: nitric oxide, NO, and nitrogen dioxide, NO_2. Therefore, in addition to carbon dioxide and water vapor, all automobile exhausts contain the following pollutants: carbon monoxide, unburned hydrocarbons, and oxides of nitrogen.

Carbon monoxide is a poison to all forms of animal life. It becomes attached to the hemoglobin molecules in the red blood cells. This prevents the cells from carrying out their normal function, which is to transport oxygen molecules from the lungs to body tissues. A person who is confined in a small, closed space, such as a garage, in which an automobile engine is running can be killed by the carbon monoxide produced. In areas with heavy automobile traffic, particularly in periods of static air patterns, carbon monoxide concentrations can become substantial. Long term exposure to high CO concentrations is probably harmful, even if there are no noticeable short term effects.

The danger from the unburned hydrocarbons and the oxides of nitrogen is much more certain. The oxides of nitrogen are themselves poisonous. Their more serious effect, however, is in promoting the generation of ozone, O_3, and various oxidation products of the unburned hydrocarbons. In the air, nitric oxide is converted into nitrogen dioxide.

$$2NO + O_2 \longrightarrow 2NO_2$$

The action of sunlight on NO_2 causes it to decompose as follows.

$$NO_2 \xrightarrow{sunlight} NO + O$$

The oxygen atoms thus produced react with oxygen molecules to produce ozone.

$$O + O_2 \longrightarrow O_3$$

Ozone is a form of oxygen that is extremely reactive. It is injurious in two ways. First, ozone reacts directly with the tissues of both plants and animals, damaging them severely. Many plants, including some food crops, are injured by as little as one milligram of ozone per ton of air. This problem is so severe in urbanized regions of Southern California that

The second way in which ozone is injurious is by oxidizing the unburned hydrocarbon molecules. A typical reaction follows.

$$CH_3CH=CH_2 + 2O_3 \longrightarrow CH_3CHO + CH_2O + 2O_2$$

Many of the compounds so produced are very irritating to the eyes and the respiratory tract.

What can be done about the pollution caused by automobiles? One thing would be to cut down on the use of automobiles by building clean, fast, and convenient public transportation. Another would be to improve engine design so that each car produces a smaller quantity of pollutants.

Still another important measure, one in which chemists must play the main role, is the development of effective *afterburners*. These are devices placed immediately after the exhaust manifold of the engine so that the hot exhaust gases pass over some kind of catalyst. The catalyst must promote reactions like the following ones, which convert the pollutants into harmless substances.

$$2NO_2 \longrightarrow N_2 + 2O_2$$
$$2NO \longrightarrow N_2 + O_2$$
$$2CO + O_2 \longrightarrow 2CO_2$$
$$\text{hydrocarbons} + O_2 \longrightarrow CO_2 + H_2O$$

The polluting exhausts contributed by these automobiles in Detroit help to obscure the setting sun.

The first two reactions remove the oxides of nitrogen, while the next two complete the burning of carbon monoxide and the hydrocarbons. At the present time, little progress has been made toward a suitable catalyst for the first two, but there has been some success with catalysts for the second two. Several metals and metal oxides dispersed on ceramic supports seem capable of doing the job. The length of time over which the catalyst will remain effective is still a problem, however.

certain vegetables can no longer be grown there and plant damage is widespread and clearly visible. Ozone is severely irritating to the bronchial tubes of animals.

Chapter Summary

Certain general rules concerning the capacity of each of the elements to combine with other elements are useful in writing chemical formulas. These rules are expressed by assigning valence numbers to the elements. It is to be emphasized that these valence numbers do not apply to every compound formed by an element. They do cover the vast majority of common compounds. By balancing valences so that, for neutral molecules, they equal zero, formulas can be written.

There are certain general rules for naming compounds. Binary compounds, which consist of only two elements, usually have names ending in -*ide*. Names ending in -*ite* or -*ate* are generally used for compounds containing negative polyatomic ions (for example, $CaSO_4$, calcium sulfate; $LiNO_2$, lithium nitrite). For metals with two common valences, the endings -*ous* and -*ic* are often used to indicate the lower and higher valences, respectively (for example, Fe^{2+}, ferrous; Fe^{3+}, ferric). In the preferred system, the valence states are shown by appropriate roman numerals.

Stoichiometry is a name used for the quantitative relationships that apply in chemical reactions. With the balanced equation for a reaction, its stoichiometry can be most conveniently worked out by using the mole method. In this method, given masses or volumes are first converted to moles. The number of moles of some other substance, or substances, corresponding to the known number of moles of the given substance, is then determined. Finally, this result is converted to any other desired units, such as grams, tons, or liters.

Questions and Problems

1. Write simple chemical formulas for the following compounds of sodium.
   (a) sodium chloride
   (b) sodium sulfide
   (c) sodium nitrate
   (d) sodium carbonate
   (e) sodium phosphate
2. Give the name and formula of the only common chemical radical discussed that shows a positive valence.
3. Write the formulas for compounds of iron and oxygen in the following cases: iron shows a valence of $3+$; iron shows a valence of $2+$. What are two possible names for each of the compounds formed?
4. Where would you look for information on the combining capacities of elements and radicals not listed in the text?
5. Name the following compounds: $NaNO_2$; KI; $(NH_4)_2SO_4$; $NaOH$; $AgNO_3$; $CuCl_2$ (give two names); K_3PO_4.
6. Write balanced chemical equations for the following reactions.
   (a) hydrogen + bromine $\longrightarrow$ hydrogen bromide
   (b) carbon + oxygen $\longrightarrow$ carbon dioxide
   (c) lithium + water $\longrightarrow$ lithium hydroxide + hydrogen
   (d) hydrogen + nitrogen $\longrightarrow$ ammonia gas
7. When iron rusts, it combines with oxygen of the air to form iron oxide, Fe_2O_3. Based on this information, which of the following is false?
   (a) The equation is $3O_2 + 4Fe \longrightarrow 2Fe_2O_3$.

(b) Five atoms are represented by the formula, Fe_2O_3.
(c) A mole of ferric oxide has a mass of 104 g.
(d) The mass of the reactants equals the mass of the products.
(e) Atoms are conserved.

8. A paraffin candle burns in air to form water and carbon dioxide. Although paraffin is composed of molecules of several sizes, the formula $C_{25}H_{52}$ will be used to represent all paraffin molecules. The (unbalanced) equation for the reaction follows.

$$C_{25}H_{52} + O_2 \longrightarrow H_2O + CO_2$$

(a) How many moles of oxygen are needed to burn one mole of paraffin?
(b) What is the mass of this oxygen?
(c) How many moles of oxygen are needed to burn 70.6 moles of paraffin?
(d) How many moles of water would be produced in (c)?

9. How many moles of oxygen, O_2, are required to produce 242 grams of magnesium oxide, by the following equation?

$$2Mg + O_2 \longrightarrow 2MgO$$

10. Sodium reacts with water according to the following equation.

$$2Na + 2H_2O \longrightarrow 2NaOH + H_2$$

(a) How many grams of sodium are needed to produce 1 mole of hydrogen gas?
(b) How many grams of sodium are needed to produce 4.0 liters of hydrogen gas at STP?

11. If a piece of sodium metal is lowered into a bottle of chlorine gas, a reaction takes place. Table salt, NaCl, is formed.

(a) Write the equation for the reaction.
(b) How many moles of NaCl could be formed from one mole of Na?
(c) How many moles of NaCl could be formed from 2.30 grams of Na?

12. Although sodium carbonate is needed in the manufacture of glass, very little is found in nature. It is made using two very abundant chemicals, calcium carbonate (marble) and sodium chloride (salt). The process involves many steps. The overall reaction follows.

$$CaCO_3 + 2NaCl \longrightarrow Na_2CO_3 + CaCl_2$$

(a) How many grams of sodium chloride react with 1.00 kg of calcium carbonate?
(b) How much sodium carbonate results?

13. The hourly energy requirements of an astronaut can be satisfied by the energy released when 34 grams of sucrose are burned in his body. How many grams of oxygen should a space capsule carry to meet this requirement, given the following (unbalanced) equation?

$$C_{12}H_{22}O_{11} + O_2 \longrightarrow CO_2 + H_2O$$

14. What mass of S_8 would have to be burned to produce 2.24×10^3 liters of SO_2 at STP?

15. In the preceding problem, what mass would be needed to produce the same volume (2.24×10^3 liters) at $400°C$? (Hint: First, calculate what this volume would be at STP.)

16. Graphite, a form of carbon, C, burns in air to produce the colorless gas, carbon dioxide. On the basis of this information:
(a) Write the equation for the reaction.
(b) Find out how many moles of carbon dioxide are produced when one mole

of graphite is burned. What is the mass (in grams) of this amount of carbon dioxide?

17. The reaction of sodium borohydride, $NaBH_4$, with water produces sodium dihydrogen borate, NaH_2BO_3, and hydrogen gas. Suppose you wish to inflate a life raft of 30-liter capacity with hydrogen, to a pressure of 5 atm at 20°C. How many grams of sodium borohydride would have to be used?

18. In the preceding problem, how many grams of water are used up?

19. The chlorine used to purify drinking water is made by electrolyzing molten NaCl to produce liquid sodium and gaseous chlorine.
    (a) How many grams of sodium chloride are needed to produce 355 grams of chlorine gas?
    (b) What volume would the chlorine occupy at STP?

20. Iron, Fe, burns in air to form a black, solid oxide, Fe_3O_4.
    (a) Write the equation for the reaction.
    (b) How many moles of oxygen gas are needed to burn one mole of iron?
    (c) How many grams of oxygen gas is that?
    (d) Can a piece of iron with a mass of 5.6 grams burn completely to Fe_3O_4 in a vessel containing 0.05 mole of O_2?

21. Kerosene, a mixture of hydrocarbons, contains a compound called decane, $C_{10}H_{22}$. Assuming that the kerosene is all $C_{10}H_{22}$ and that a stove burns 1.0 kg of kerosene per hour, answer the following questions.
    (a) How many liters (STP) of oxygen are needed per hour?
    (b) How many liters (STP) of carbon dioxide are produced per hour?

22. Hydrazine, N_2H_4, can be burned with oxygen to provide energy for rocket propulsion. The energy released is 150 kcal per mole of hydrazine burned. When a mole of hydrogen is burned, 57.8 kcal are released.
    (a) How much energy is released when 10.0 kg of hydrazine are burned?
    (b) Compare the energy released when the same mass of hydrogen (10.0 kg) is burned as a fuel instead.

23. If potassium chlorate, $KClO_3$, is heated gently, the crystals will melt. Further heating will decompose it to give oxygen gas and potassium chloride, KCl.
    (a) Write the equation for this "decomposition" reaction.
    (b) How many moles of $KClO_3$ are needed to give 1.5 moles of oxygen gas?
    (c) How many moles of KCl would be given by $\frac{1}{3}$ mole of $KClO_3$?
    (d) How many moles of oxygen gas would be produced by 122.6 grams of $KClO_3$?

24. Suppose that one gallon of gasoline is about 25 moles of octane, C_8H_{18}.
    (a) How many moles of oxygen must be used to burn this gasoline, assuming the only products are CO_2 and H_2O?
    (b) How many moles of carbon dioxide are formed?
    (c) What is the mass of this carbon dioxide? (Express your answer in kilograms.)
    (d) How many pounds of carbon dioxide are released into the atmosphere when your automobile consumes 10 gallons of gasoline? (The density of gasoline is 0.85 g per ml. One liter = 1.058 qt. One kilogram = 2.20 lb.)

25. The following reaction is involved in the production of iron from iron ore.

$$Fe_2O_3(s) + 3CO(g) \longrightarrow$$
$$2Fe(s) + 3CO_2(g) + 4.3 \text{ kcal of heat}$$

(a) How many grams of CO must react to release 13 kcal?
(b) How many liters of CO (STP) are needed to produce 1.0 kg of Fe?

26. More C_8H_{18} (a hydrocarbon desirable in gasoline) can be obtained from petroleum if the following reaction takes place.

$$C_{16}H_{32}(g) + 2H_2(g) \longrightarrow 2C_8H_{18}(g)$$

How many grams of C_8H_{18} can be made using 224 liters of H_2 at STP?

27. In a reaction with hydrochloric acid, HCl, how many grams of zinc metal would be needed to produce enough hydrogen gas to fill an 11.2-liter balloon at STP? What would be the volume of the balloon at 27°C and 680 Torr pressure?

$$Zn(s) + 2HCl(aq) \longrightarrow ZnCl_2(aq) + H_2(g)$$

28. The following reaction needs to be balanced.

$$NH_3(g) + O_2(g) \longrightarrow NO(g) + H_2O(g)$$

If 4.48 liters of ammonia gas, measured at STP, are used, how many liters of oxygen, measured at STP, will be needed to react with all the ammonia?

29. The following reaction (which needs to be balanced) is carried out with all gas volumes measured at the same pressure and temperature.

$$C_4H_{10}(g) + O_2(g) \longrightarrow CO_2(g) + H_2O(g)$$

(a) How many liters of oxygen are required to produce 2.0 liters of CO_2?
(b) If 15 liters of oxygen are used, how many liters of butane, C_4H_{10}, will be burned?
(c) If 8.0 liters each of oxygen and butane are mixed, how many liters of CO_2 will be produced? Assume that the reaction is as complete as possible.

30. Suppose 105 liters of NH_3 and 285 liters of O_2 are allowed to react until the following reaction (which needs to be balanced) is complete.

$$NH_3(g) + O_2(g) \longrightarrow H_2O(g) + NO_2(g)$$

The temperature and pressure are maintained constant at 200°C and 0.30 atm during all volume measurements. Measured at the stated conditions, which gas remains unreacted? What volume of this gas remains unreacted?

20 Energy Effects in Chemical Reactions

This fireworks display illustrates a spectacular release of energy (light and heat) in a chemical reaction. Many reactions proceed more slowly, but in all reactions a gain or loss of energy can be measured.

Most of the energy resources of the world involve the energy produced by chemical reactions. Thus, chemical reactions form an important part of chemistry.

How much energy is involved in a chemical reaction?

Where does this energy come from?

Where does it go when the reaction is complete?

In this chapter you will investigate the answers to these and related questions. You will begin to realize that although an understanding of atomic and molecular structure is important to chemists, it is often not an end in itself. Rather, reactions, which are chemical changes and transformations, are what the chemist wishes to understand and control as his final goal. You will also begin to appreciate more fully the value of applying a model drawn from experience at a macroscopic level to explain microscopic phenomena.

Heat Is Absorbed and Released in Chemical Reactions

20-1 A Practical Illustration

At a temperature of 600°C, steam passed over hot coal (coal is mostly carbon) reacts to give carbon monoxide and hydrogen.

$$H_2O(g) + C(s) \longrightarrow CO(g) + H_2(g) \quad (1)$$

This reaction is quite useful because the mixture of product gases, called *water gas*, is an excellent industrial fuel. In the commercial preparation of water gas, the chemical engineer must allow for the absorption of heat during the reaction. In fact, he must periodically turn off the steam and reheat the coal to keep the reaction going. When he measures the amount of heat absorbed by the system, he finds that 31.4 kcal of heat are absorbed per mole of carbon reacted. Since the heat is used up, as are the reactants, reaction (1) can be rewritten to show the heat on the left side of the equation, as follows.

$$H_2O(g) + C(s) + 31.4 \text{ kcal} \longrightarrow CO(g) + H_2(g) \quad (1a)$$

Now consider the mechanical engineer who is designing a boiler to be heated with water gas fuel. He is interested in burning the water gas. This involves the following two chemical combustion reactions.

$$CO(g) + \tfrac{1}{2}O_2(g) \longrightarrow CO_2(g) \quad (2)$$

and

$$H_2(g) + \tfrac{1}{2}O_2(g) \longrightarrow H_2O(g) \quad (3)$$

These reactions release heat; the engineer needs to know how much. Again, these amounts of heat can be measured. Once known, they can be added to reactions (2) and (3). Since heat, as well as the chemical product, is produced by the reaction, it appears on the right side of the equation.

$$CO(g) + \tfrac{1}{2}O_2(g) \longrightarrow CO_2(g) + 67.6 \text{ kcal} \quad (2a)$$

$$H_2(g) + \tfrac{1}{2}O_2(g) \longrightarrow H_2O(g) + 57.8 \text{ kcal} \quad (3)$$

Now consider the business manager. He thinks in terms of profits and losses. He is likely to observe that the consumption of coal and water to generate water gas is followed by the combustion of the water gas to form carbon dioxide and the same amount of water that was consumed. Without knowing much chemistry, he can see that what is finally accomplished is the combustion of coal to form carbon dioxide. The overall reaction is the sum of all the reactions involved.

In order to find the sum of a series of reactions, you simply write the reactions one under the other, in the order in which they occur. You then cross out all substances that appear in equal quantities on the right side of one equation and on the left side of another. In doing this, you are merely eliminating those substances that are produced by one reaction only to be entirely consumed in another. Such substances obviously cannot be either reactants or products in the overall reaction. In the case under discussion, you write reactions (1), (2), and (3) as follows, crossing out all substances that appear in equal quantities on both sides of the equation.

$$\cancel{H_2O(g)} + C(s) \longrightarrow \cancel{CO(g)} + \cancel{H_2(g)}$$
$$\cancel{CO(g)} + \tfrac{1}{2}O_2(g) \longrightarrow CO_2(g)$$
$$\cancel{H_2(g)} + \tfrac{1}{2}O_2(g) \longrightarrow \cancel{H_2O(g)}$$

Adding up the three reactions, you get the following overall reaction.

$$C(s) + O_2(g) \longrightarrow CO_2(g) \quad (4)$$

The business manager is frugal so he asks, "Why not burn the coal directly and save the cost of manufacturing the water gas?" The mechanical engineer is practical so he asks, "How much heat will the boiler receive if I use coal instead of water gas?" The chemical engineer goes to the laboratory to find the answers by measuring the heat released per mole of carbon burned in reaction (4). The laboratory result shows that this overall reaction releases 94.0 kcal/mole.

$$C(s) + O_2(g) \longrightarrow CO_2(g) + 94.0 \text{ kcal} \quad (4a)$$

The chemical engineer can now answer all the questions. If one mole of carbon is burned directly, 94.0 kcal of heat are released for the mechanical engineer. The same amount of coal converted into water gas releases the sum of the heats evolved in reactions (2a) and (3a).

$$67.6 \text{ kcal} + 57.8 \text{ kcal} = 125.4 \text{ kcal}$$

The mechanical engineer obviously has a better fuel in water gas than in coal.

With added respect, the business manager now asks the chemical engineer, "Where did this extra heat come from? Did we get something for nothing?" The answer to the second question is "No." The water gas releases more heat per mole of carbon because the chemical engineer put heat into reaction (1a). The business manager's ledger follows.

	Debits (Heat absorbed)	Credits (Heat released)
Reaction (1a)	31.4 kcal	—
(2a)	—	67.6 kcal
(3a)	—	57.8 kcal
Totals	31.4 kcal	125.4 kcal

Overall Reaction (4a) = 94.0 kcal (125.4 − 31.4 = 94.0)

20-2 Heat Content of a Substance

The example just given shows that the 31.4 kcal of heat absorbed during reaction (1a) was "stored" in the water gas. Furthermore, the amount of energy "stored" is definite. It does not change at the demand of the business manager or at the whim of the chemical engineer. How much energy is stored depends upon the reactants and products of the reaction. A fixed amount of energy, as heat, must be added to coal and steam to make a specified amount of carbon monoxide and hydrogen. This heat is retained by the CO and H_2 molecules. You might say that reaction (1a) increases the heat content of the reacting atoms by rearranging them to form the products.

The **heat content** is the energy stored in a substance during its formation. It is a fixed quantity of heat per mole of the substance. It is a characteristic property of the substance, just as the molar mass is a characteristic property. The concept of heat content provides a basis for a quantitative treatment of the energy changes in chemical reactions.

The amount of heat absorbed or released in a chemical reaction is called the **heat of reaction**. It equals the heat content of the products minus the heat content of the reactants. If more energy is stored in the reactants than in the products, heat will be released during a reaction. Conversely, heat will be absorbed if more energy is stored in the products than in the reactants.

Chemists symbolize heat content by H. Therefore, they indicate the heat of reaction by ΔH. Recall that the Greek letter Δ (delta) signifies "difference."

To gain a clear understanding of ΔH, consider again reaction (1).

$$H_2O(g) + C(s) \longrightarrow CO(g) + H_2(g) \quad (1)$$

Its heat of reaction is calculated as follows.

$$\Delta H = \begin{pmatrix} \text{heat content} \\ \text{of products} \end{pmatrix} - \begin{pmatrix} \text{heat content} \\ \text{of reactants} \end{pmatrix}$$
$$= (H_{CO} + H_{H_2}) - (H_{H_2O} + H_C)$$
$$\Delta H = H_{CO} + H_{H_2} - H_{H_2O} - H_C$$

Since the reaction consumes heat, the heat content of the products is higher, and ΔH is positive. You can express this information by writing the following.

$$H_2O(g) + C(s) \longrightarrow CO(g) + H_2(g)$$
$$\Delta H_1 = +31.4 \text{ kcal} \quad (1b)$$

Equation (1b) conveys exactly the same information as equation (1a).

Using the same procedure, you can find the heat of reaction for reaction (2), as the following shows.

$$\Delta H = \begin{pmatrix} \text{heat content} \\ \text{of products} \end{pmatrix} - \begin{pmatrix} \text{heat content} \\ \text{of reactants} \end{pmatrix}$$
$$= H_{CO_2} - (H_{CO} + H_{\frac{1}{2}O_2})$$
$$\Delta H = H_{CO_2} - H_{CO} - H_{\frac{1}{2}O_2}$$

In this reaction, heat is evolved. Hence, the heat content of the products is below that of the reactants, and ΔH must be negative.

$$CO(g) + \tfrac{1}{2}O_2(g) \longrightarrow CO_2(g)$$
$$\Delta H_2 = -67.6 \text{ kcal} \quad (2b)$$

This equation has exactly the same meaning as equation (2a).

Similarly, you can write equation (3) and its heat of reaction as follows.

$$H_2(g) + \tfrac{1}{2}O_2(g) \longrightarrow H_2O(g)$$
$$\Delta H_3 = -57.8 \text{ kcal} \quad (3b)$$

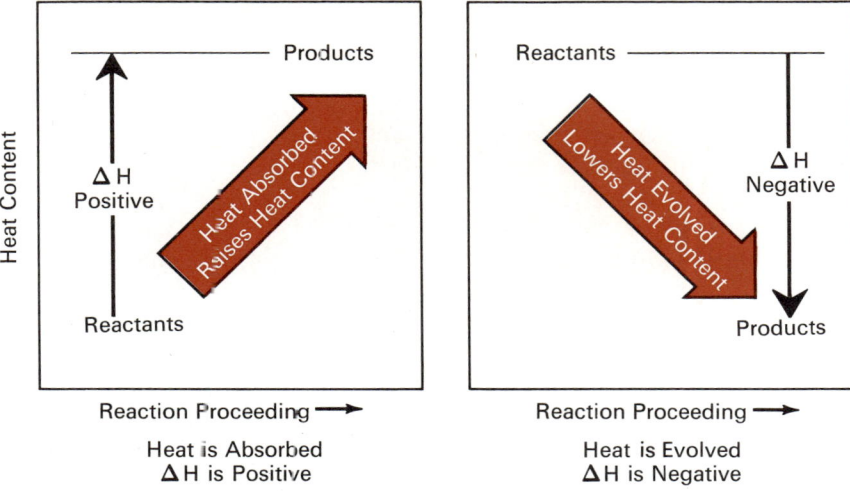

Figure 20-1 When heat is absorbed (left), ΔH is positive. When heat is given up (right), ΔH is negative.

Equations (3b) and (3a) give identical information.

You can see that the sign of ΔH is sensible. It is positive when the heat content rises (by heat absorption). It is negative when the heat content drops (by heat evolution). This is shown diagrammatically in Figure 20-1.

Reactions that absorb heat are known as **endothermic** reactions, while those that evolve heat are known as **exothermic** reactions. These two terms are derived by placing prefixes that mean "in" and "out," respectively, before the stem *thermic*, meaning "having to do with heat."

INVESTIGATION

20-3 Measuring the Heat of Reaction

A calorimeter is a device used for measuring the heat of reaction, ΔH. In this experiment you will use a 250-ml Erlenmeyer flask as a reaction vessel and as a simple calorimeter to measure the heat evolved or absorbed during several reactions.

PURPOSE

In this experiment you will measure and compare the heats of three reactions.

MATERIALS

4 g NaOH(s)
100 ml 0.5 M HCl
200 ml 0.25 M HCl
Erlenmeyer flask, 250-ml
Beaker, 250-ml
Graduated cylinder, 100-ml (or volumetric flask, 100-ml)
Pneumatic trough (or flat pan)

Thermometer, $-10°C$ to $110°C$
Balance, centigram
Glass stirring rod
Watch glass, small

PROCEDURE

Part A. Determination of Heat of Reaction (5)

Solid sodium hydroxide dissolves in water to form an aqueous solution of ions.

$$\text{NaOH}(s) \longrightarrow \text{Na}^+(aq) + \text{OH}^-(aq) + x_1 \text{ cal} \qquad \Delta H_5 = -x_1 \text{ cal} \qquad (5)$$

Figure 20-2

1. Determine the mass of a clean, dry 250-ml Erlenmeyer flask to the nearest gram.

2. To the nearest 1 ml, pour 200 ml of cool tap water into the flask. With a thermometer in the flask, stir carefully with a glass stirring rod until a constant temperature is reached (about room temperature). Record this temperature and all the following temperatures to the nearest 0.2°C.

3. Place a small watch glass on the balance pan and determine its mass. Add between 1.9 g and 2.1 g of sodium hydroxide, NaOH, pellets to the watch glass. Determine the mass of the compound to the nearest 0.01 g. (*Special precaution.* Do not touch the solid sodium hydroxide with your fingers. If this substance does get on your skin, wash it off with large amounts of water.) Since sodium hydroxide becomes moist in the open air, your teacher will give special instructions on finding the mass of a prescribed number of pellets.

4. Place the known amount of sodium hydroxide into the water in the Erlenmeyer flask. As shown in Figure 20-2, swirl the flask until the sodium hydroxide is dissolved. Place the thermometer in the flask. If the temperature rises, record the highest temperature reached. If it decreases, record the lowest temperature. Before proceeding to Part B, rinse the flask thoroughly with water.

Part B. Determination of Heat of Reaction (6)

Solid sodium hydroxide reacts with hydrochloric acid to form water and an aqueous solution of sodium chloride.

$$NaOH(s) + H^+(aq) + Cl^-(aq) \longrightarrow$$
$$H_2O + Na^+(aq) + Cl^-(aq) + x_2 \text{ cal} \qquad \Delta H_6 = -x_2 \text{ cal} \qquad (6)$$

1. Repeat the preceding four steps, except in step 2 use 200 ml of 0.25 M hydrochloric acid instead of tap water.
2. Rinse the 250-ml flask and proceed to Part C.

Part C. Determination of Heat of Reaction (7)

An aqueous solution of sodium hydroxide reacts with hydrochloric acid to form water and an aqueous solution of sodium chloride.

$$Na^+(aq) + OH^-(aq) + H^+(aq) + Cl^-(aq) \longrightarrow$$
$$H_2O + Na^+(aq) + Cl^-(aq) + x_3 \text{ cal} \qquad \Delta H_7 = -x_3 \text{ cal} \qquad (7)$$

1. Prepare 100 ml of 0.5 M sodium hydroxide solution by adding 2 g of solid sodium hydroxide to 100 ml of water in a 250-ml beaker. Stir to dissolve the sodium hydroxide. Place the beaker and its contents in a pan of cool water to bring its temperature to or slightly below room temperature.

2. Place 100 ml of 0.5 M hydrochloric acid in the Erlenmeyer flask. Be sure that the temperature of this solution is at or slightly below room temperature. You should check the temperature of each solution with a thermometer, being careful to rinse and dry it before transferring it from one solution to the other. Record the temperatures of both solutions just before proceeding to the next step.

3. Add the sodium hydroxide solution to the hydrochloric acid solution. Mix quickly and note the highest temperature reached.

DATA AND RESULTS

Record the following data in your notebook.

Part A.	Part B.
Mass of 250-ml flask	Mass of 250-ml flask
Volume of H_2O	Volume of 0.25 M HCl
Mass of NaOH	Mass of NaOH
Temperature of mixture	Temperature of mixture
Temperature of H_2O	Temperature of 0.25 M HCl

Part C.
Mass of 250-ml flask
Volume of 0.5 M HCl
Volume of 0.5 M NaOH

Temperature of mixture
Temperature of 0.5 M HCl
Temperature of 0.5 M NaOH

Calculations

You may assume that the heat of reaction will be used to change the temperature of the aqueous solution and of the glass container, and you may neglect other small losses to the surroundings. Recall that it takes 1.0 calorie to change the temperature of one gram of water one Celsius degree. It takes 0.2 calorie to change the temperature of one gram of glass one Celsius degree. Recall also that one milliliter of water has a mass of very nearly one gram. Thus, you can find the mass of water from its volume. From the change in the temperature and the mass of the reactants you can calculate the number of calories evolved or absorbed.

1. For each reaction calculate:
    (a) the change in temperature.
    (b) the amount of heat absorbed by the solution.
    (c) the amount of heat absorbed by the flask.
    (d) the total amount of heat absorbed.
    (e) the number of moles of NaOH in each reaction.
    (f) the total amount of heat per mole of NaOH.

2. Express your results as heats of reaction, ΔH_5, ΔH_6, and ΔH_7.

3. (a) Compare ΔH_6 with $(\Delta H_5 + \Delta H_7)$. Explain what you find.
   (b) Assuming ΔH_6 to be correct, calculate the percent error between ΔH_6 and $(\Delta H_5 + \Delta H_7)$.

RELATED PROBLEMS

1. Examine the equations for reactions (6) and (7). Find the net ionic equations for these reactions by subtracting from both sides of the equations those substances that appear on each side of the arrow in identical form. The resulting equations will then show only those substances actually involved in the reactions.

2. In reaction (5), ΔH_5 represents the heat of solution of NaOH(s). Look at the net ionic equations for reactions (6) and (7) and make a statement concerning the significance of ΔH_6 and ΔH_7.

3. Suppose you had used 4 g of NaOH(s) in reaction (5).
    (a) What would be the number of calories evolved?
    (b) What effect would this have on your calculation of ΔH_5, the heat evolved per mole of NaOH?

Germain Henri Hess (1802–1850)

GERMAIN HESS was trained as a physician, but he found time to study geology and chemistry, as well. After practicing medicine in Siberia for several years, he returned to St. Petersburg, his former home, and was soon appointed professor of chemistry at the Technological Institute.

Because of his interest in geology, much of Hess's early research was in the field of minerals. Late in the 1830's, however, he began to measure the heat produced when various substances react with each other. Antoine Lavoisier and P. S. de Laplace had done some work in this area about sixty years earlier. As a result of their observations, they had concluded that the quantity of heat required to decompose a compound into its elements is equal to the heat evolved when that compound is formed from its elements. This was the first general law worked out in the field of thermochemistry.

After spending several years in measuring the reaction heats of many substances, Hess reported a second law of thermochemistry. He stated that the heat change in a chemical reaction is the same whether it takes place in only one stage or in several stages. This meant that the net heat of a reaction depends only on the states of the reactants and products at the beginning and end of the changes and not on any intermediate steps through which they might pass. Hess's law makes it possible to add and subtract thermochemical equations, enabling chemists to calculate heats of reaction for reactions that are not easily performed.

The success of Hess's work in thermochemistry set the stage for the development of the science of chemical thermodynamics about thirty years later.

20-4 Additivity of Reaction Heats

Let us return to the water gas-fuel discussion. The overall reaction in this discussion can be written as follows.

$$C(s) + O_2(g) \longrightarrow CO_2(g)$$
$$\Delta H_4 = -94.0 \text{ kcal} \quad (4b)$$

Not only is equation (4b) equal to the sum of equations (1b), (2b), and (3b) in terms of atoms, but it is also equal in terms of heat.

$$\Delta H_4 = \Delta H_1 + \Delta H_2 + \Delta H_3$$
$$= 31.4 + (-67.6) + (-57.8)$$
$$= 31.4 - 67.6 - 57.8$$
$$\Delta H_4 = -94.0 \text{ kcal}$$

When a reaction can be expressed as the algebraic sum of a sequence of two or more other reactions, the heat of the reaction is the algebraic sum of the heats of the other reactions. This generalization has been found applicable to every reaction ever tested. Because it has been so widely tested, it is called a law—the **Law of Additivity of Reaction Heats.** (Because G. H. Hess first proposed this generalization, it is sometimes called *Hess's Law of Constant Heat Summation.*) Your data for the experiment in Section 20-3 should have provided evidence, within experimental error, of this law.

If you stop and think about it, you might realize that this law is really only a particular example of the Law of Conservation of Energy. Since energy cannot be created or destroyed, the same change in heat content must accompany both an overall reaction and the series of reactions that give the overall reaction.

TABLE 20-1 Heats of Formation of Various Compounds ($t = 25°C$, $p = 1$ atm.)

Elements		Compound		Heat of Reaction (kcal/mole of product)
		Formula	Name	
$H_2(g) + \frac{1}{2}O_2(g)$	$\rightarrow$	$H_2O(g)$	water vapor	-57.8
$H_2(g) + \frac{1}{2}O_2(g)$	$\rightarrow$	$H_2O(l)$	water	-68.3
$S(s) + O_2(g)$	$\rightarrow$	$SO_2(g)$	sulfur dioxide	-71.0
$H_2(g) + S(s) + 2O_2(g)$	$\rightarrow$	$H_2SO_4(l)$	sulfuric acid	-194.0
$\frac{1}{2}N_2(g) + \frac{1}{2}O_2(g)$	$\rightarrow$	$NO(g)$	nitric oxide	$+21.6$
$\frac{1}{2}N_2(g) + O_2(g)$	$\rightarrow$	$NO_2(g)$	nitrogen dioxide	$+8.1$
$\frac{1}{2}N_2(g) + \frac{3}{2}H_2(g)$	$\rightarrow$	$NH_3(g)$	ammonia	-11.0
$C(s) + \frac{1}{2}O_2(g)$	$\rightarrow$	$CO(g)$	carbon monoxide	-26.4
$C(s) + O_2(g)$	$\rightarrow$	$CO_2(g)$	carbon dioxide	-94.0
$2C(s) + 3H_2(g)$	$\rightarrow$	$C_2H_6(g)$	ethane	-20.2
$3C(s) + 4H_2(g)$	$\rightarrow$	$C_3H_8(g)$	propane	-24.8
$\frac{1}{2}H_2(g) + \frac{1}{2}I_2(g)$	$\rightarrow$	$HI(g)$	hydrogen iodide	$+6.2$

20-5 Heats of Formation

Table 20-1 shows the heats of a number of reactions. Despite the diversity of elements and compounds represented, thoughtful examination of the table will reveal a basic similarity in all the reactions. In every case the reactants are elements. In other words, every reaction in the table involves making a compound from the elements that constitute it.

The heat of such a reaction is called the **heat of formation** of the compound. The reactions used to define the heats of formation must be written so as to satisfy certain rules. The main rules are the following ones.

1. The reaction is carried out at 25°C.
2. The reaction is carried out at a pressure of 1 atmosphere.
3. Each of the elements is assumed to enter into the reaction in whatever phase (gas, liquid, or solid) is natural to it at 25°C and 1 atm pressure.

The reason that a table of heats of formation is useful is *not* because most of the heats of formation themselves are of interest as such. Instead, it is valuable because it provides a compact list of data from which the heats of an enormous number of chemical reactions can be calculated. The value of such a table will be shown in the next section.

20-6 Predicting the Heat of a Chemical Reaction

You can use the Law of Additivity of Reaction Heats and the data given in a table of heats of formation, such as Table 20-1, to calculate the heats of many chemical reactions. The only requirement is that you know the heat of formation for every compound that appears in the reaction. This will now be illustrated with some examples.

$$NO(g) + \tfrac{1}{2}O_2(g) \longrightarrow NO_2(g)$$
$$\Delta H_8 = ? \quad (8)$$

Since this reaction can be obtained by combining the following two reactions, which are found in Table 20-1, ΔH_8 can be predicted.

$$\tfrac{1}{2}N_2(g) + \tfrac{1}{2}O_2(g) \longrightarrow NO(g)$$
$$\Delta H_9 = +21.6 \text{ kcal} \quad (9)$$

$$\tfrac{1}{2}N_2(g) + O_2(g) \longrightarrow NO_2(g)$$
$$\Delta H_{10} = +8.1 \text{ kcal} \quad (10)$$

To obtain reaction (8), combine reactions (9) and (10). Since NO is a reactant in reaction (8), you will need the reverse of reaction (9). The heat of the reverse reaction is obtained merely by changing the algebraic sign of ΔH_9. From the table you know that 21.6 kcal of heat are absorbed when one mole of NO is formed. Thus, 21.6 kcal of heat will be released when one mole of NO is decomposed in the reverse reaction, as shown in reaction (11).

$$NO(g) \longrightarrow \tfrac{1}{2}N_2(g) + \tfrac{1}{2}O_2(g)$$
$$\Delta H_{11} = -21.6 \text{ kcal} \quad (11)$$

When reactions (10) and (11) are combined, the following reaction is found to result.

$$NO(g) + O_2(g) + \tfrac{1}{2}N_2(g) \longrightarrow$$
$$NO_2(g) + \tfrac{1}{2}O_2(g) + \tfrac{1}{2}N_2(g)$$
$$\Delta H_8 = -21.6 + 8.1$$

By subtracting like terms from both sides of the equation, you obtain the overall reaction.

$$NO(g) + \tfrac{1}{2}O_2(g) \longrightarrow NO_2(g)$$
$$\Delta H_8 = -13.5 \text{ kcal/mole NO} \quad (8a)$$

Now consider a more complicated example, the oxidation of ammonia, NH_3.

$$NH_3(g) + \tfrac{7}{4}O_2(g) \longrightarrow NO_2(g) + \tfrac{3}{2}H_2O(g)$$
$$\Delta H_{12} = ? \quad (12)$$

In this reaction the three compounds, $NH_3(g)$, $NO_2(g)$, and $H_2O(g)$, are all found in Table 20-1. Consequently, you can calculate ΔH_{12} by algebraically adding the following three equations.

$$NH_3(g) \longrightarrow \tfrac{1}{2}N_2(g) + \tfrac{3}{2}H_2(g) \quad \Delta H = +11.0 \text{ kcal}$$

$$\tfrac{1}{2}N_2(g) + O_2(g) \longrightarrow NO_2(g) \quad \Delta H = +8.1 \text{ kcal}$$

$$\tfrac{3}{2}H_2(g) + \tfrac{3}{4}O_2(g) \longrightarrow \tfrac{3}{2}H_2O(g) \quad \Delta H = -86.7 \text{ kcal}$$

You can obtain the last reaction and its ΔH by multiplying the reaction given in Table 20-1 for the formation of water vapor by $\tfrac{3}{2}$. The overall reaction that results from the algebraic addition of the three foregoing equations is reaction (12a).

$$NH_3(g) + \tfrac{7}{4}O_2(g) \longrightarrow NO_2(g) + \tfrac{3}{2}H_2O(g)$$
$$\Delta H_{12} = -67.6 \text{ kcal} \quad (12a)$$

When you wish to predict the heat of some reaction, it takes but a moment to decide whether Table 20-1 includes the necessary reactions. If every compound in the reaction of interest is in the compound column, then you can make the prediction. Of course, the list in Table 20-1 includes only a small fraction of the known values.

20-7 *Calorimetry*

The measurement of reaction heats is called **calorimetry,** a name related to a unit of heat, the calorie. You already have some experience in calorimetry in Section 20-3. Earlier, in Section 4-8, you measured the heat of combustion of a candle and the heat of solidification of wax.

In order to study calorimetry, you must have an understanding of a calorimeter, an apparatus for measuring heat. There are many types of calorimeters, each designed to measure the heats of different types of reactions. Figure 20-3 shows the general plan of a calorimeter that could be used in measuring the heat evolved during a combustion reaction. It could be applied to the combustion of a candle to yield a much more reliable answer than the one that you obtained in your experiment.

Essentially, the device consists of an insulated vessel containing a known mass of water. A known mass of the substance to be burned and an excess of oxygen are introduced under pressure into a reaction chamber placed in the vessel. The reaction mixture is ignited by means of an electrical resistance wire sealed into the reaction chamber. The heat produced by the reaction changes the temperature of the water, which is stirred to keep its temperature uniform. From this temperature change and from the amount of heat that has been found to be necessary to raise the temperature of the calorimeter and its contents by one degree, the heat of combustion per mole of substance burned can be calculated.

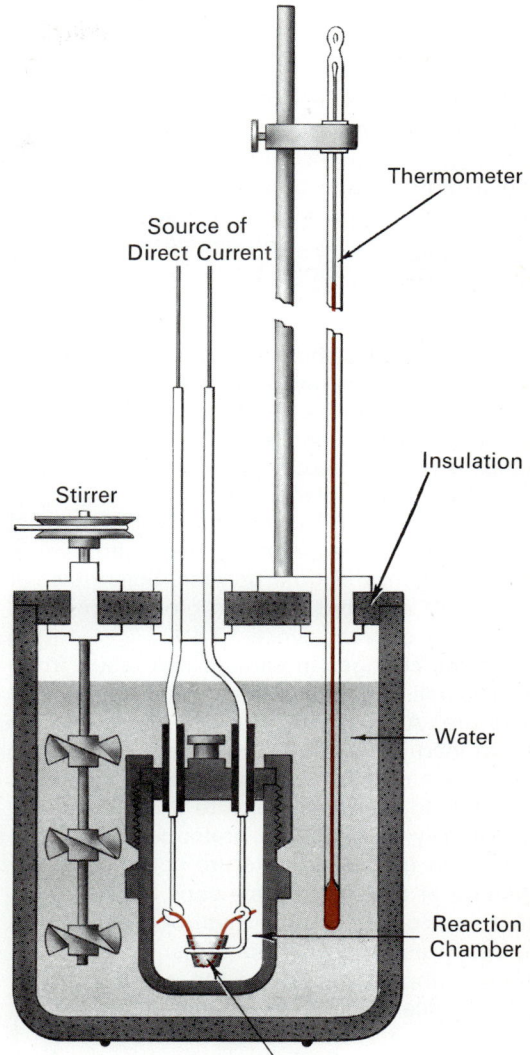

Figure 20-3 *A combustion calorimeter has insulated walls, and the combustion chamber within it is surrounded by a measurable volume of water. When the sample is burned, the water is stirred and its temperature rise is noted. From this it is possible to calculate the heat of combustion.*

Chemistry and Our Environment

CAN OUR ACTIVITIES CHANGE THE CLIMATE?

Our climate results from a very delicate balance of heat input and heat output. Heat comes to the earth's surface from the sun. Heat is also constantly being lost by radiation into outer space. The balance varies, of course, from place to place on the earth's surface. It is possible, however, to calculate an average temperature of the earth's surface, and to observe how this varies over the years. Naturally, the earth's temperature varies during the course of each year, as well as from place to place. This variation can also be averaged. Scientists thereby arrive at a mean global temperature, averaged over both location and time of year. Scientists have kept a record of mean global temperature, year by year, since 1880.

Drastic changes in climate can result from even small variations in the mean global temperature. A decrease of only a degree or two could lead to steady enlargement of the polar ice caps and eventually to growth of glaciers. This would ultimately lead to a new ice age. Conversely, a slight but prolonged increase in the mean global temperature could cause melting of the existing ice caps. Thawing of such ice could raise the level of the oceans as much as seven meters. This, in turn, would submerge many of the world's largest cities, since many of them are situated on the coasts, less than seven meters above present sea level.

Man's activities are now so extensive that they may be starting—or may have already started—to significantly affect the mean global temperature. There are two reasons for this. One is the introduction of particles into the air. The other is the introduction of carbon dioxide into the atmosphere. Let us see how each of these factors works.

A temperature inversion in Pennsylvania. The clouds were formed by fog condensing on pollutant particles. Trapped by a ceiling of warmer air above, they spill from an industrial area into an agricultural valley.

Man's activities at present are responsible for at least 10 percent and perhaps as much as 50 percent of the particles in the atmosphere. Of course, nature also puts particles into the air, as a result of winds blowing dust and soil, and as a result of volcanic eruptions. It may surprise you to learn that many of the particles introduced by man enter the atmosphere largely as gases, rather than directly as particles. Of course, some particles do enter directly as such. Much of the sulfur dioxide, SO_2, and oxides of nitrogen, mainly NO and NO_2, that enter the atmosphere eventually undergo reactions with other combustion products. This leads to the formation of sulfates and nitrates, such as Na_2SO_4 and $NaNO_3$, which form particles.

Records show that the number of particles in the atmosphere has been increasing every decade since the turn of the century. This is true not just over cities and industrial areas but over the earth as a whole. In Yellowstone National Park, a supposedly "clean" area, the dust content of the air has increased tenfold in the decade 1960–1970.

The effect of dust in the air is to scatter and deflect the sun's rays. The amount of heat reaching the earth's surface is thereby reduced. Not only do the dust particles themselves do this, but they lead to cloud formation. The clouds then shield the earth's surface from the sun's heat. Dust particles form nuclei or centers on which the condensation of water vapor in the air begins. This causes clouds and fog to develop where otherwise the air would remain clear. Studies of the climate records in many parts of the world indicate that most cities now have 25 percent fewer clear days each year than they did in 1900.

It has been estimated that a further 50 percent increase in the particle load of the atmosphere from man-made sources could lead directly to a 0.5 to 1.0°C drop in the mean global temperature. Indirectly, due to increased cloud formation, a further reduction of the same size might occur. A total drop of 2–3°C could have a catastrophic effect on climate all over the world.

Carbon dioxide in the air has a great effect on the earth's temperature. The sun's rays are not absorbed by the carbon dioxide. They strike the earth's surface, where they generate heat. Some of this heat tends to be emitted from the surface in the form of infrared rays. Carbon dioxide intercepts part of this infrared radiation emanating from the earth's surface and reflects it back. This is often called the "greenhouse" effect. The glass in a greenhouse allows the visible light from the sun to pass in, but turns back the infrared rays from inside. A greenhouse is thus heated by the sun's rays. The presence of CO_2 in our atmosphere makes the entire earth a kind of greenhouse.

In 1900 the level of carbon dioxide in the atmosphere was about 290 parts per million (ppm). By 1970 it had risen to 320 ppm, an 11 percent increase. The increase resulted from our continuously increasing combustion of fuels, coal and oil, to generate power. Nearly all of the carbon in the fuels we burn enters the atmosphere as CO_2. Of course, some of this is removed by green plants, which use CO_2 in photosynthesis, but there has been a net increase remaining in the atmosphere.

The continuously increasing amount of carbon dioxide in the atmosphere can lead to a significant increase in the mean global temperature. If the increase in CO_2 concentration continues at the present rate, the mean global temperature could increase by 0.5°C by the year 2000.

You will probably have noticed that increasing dust production tends to lower the earth's temperature, while increasing carbon dioxide production tends to raise it. Thus, the two factors, both of which result from our industrialized way of life, tend to cancel each other. This is fortunate. However, it would be more than foolish for us to just assume that there will be no long range problem. We know so little about the exact magnitudes of these effects that we cannot simply trust to cancellation. There is certainly no reason to expect the cancellation to be more than partial, anyway. Thus, it is important for scientists, including chemists, to learn much more about how the products of technology that enter the atmosphere affect the global heat balance.

Summary

All chemical reactions produce or consume some finite quantity of heat. This heat can be measured and expressed in kcal per mole of one reactant or product appearing in the equation. Each substance has a characteristic heat content, H. The heat of reaction, ΔH, is equal to the heat content of all the products minus the heat content of all the reactants. Heats of reaction can be added algebraically. Hence, if a reaction can be expressed as the algebraic sum of two or more other reactions, its ΔH is given by the algebraic sum of the ΔH's for these other reactions.

Heats of reaction are measured in a device called a calorimeter. Reactions that absorb heat are called endothermic reactions; for these ΔH is positive. Reactions that give off heat are called exothermic reactions; for these ΔH is negative.

The heat of reaction in which a single compound is formed from the elements that constitute it is called the heat of formation of that compound. A table of the heats of formation of many chemical compounds contains the data necessary to calculate the heats of reactions involving the compounds in the table.

Self Check

1. Predict the heat of the reaction

$$CO(g) + \tfrac{1}{2}O_2(g) \longrightarrow CO_2(g)$$

   from two reactions listed in Table 20-1. Compare your result with ΔH given in Equation (2b) in Section 20-2.
2. Explain what determines whether heat will be absorbed or evolved during a chemical reaction.

Molecules Store Energy

20-8 The Energy of a Molecule ★

As you have learned, chemists can account for the energy changes involved in a reaction by considering the heat contents of the reactants and products. These heat contents, of course, are assigned on the basis of heats of reaction that have been determined by experiment. How do chemists explain "heat content" on the microscopic level, that is, in terms of atoms and molecules? Chemists can account for the heat content of atoms and molecules in terms of the energy of motion (kinetic energy) and the energy of position (potential energy) of atoms and molecules.

Picture a molecule in terms of a model consisting of balls of proper relative masses hooked together by springs. The springs represent the bonds between the atoms. Suppose you start the springs vibrating and then toss the entire assembly through space in an end-over-end motion. There will be three kinds of kinetic energy associated with this model, as shown in Figure 20-4.

The model applies quite well to a molecule in the gaseous state. However, in the liquid state, and even more so in the solid state, all such motions are restricted. In the solid and liquid phases the chief indication of kinetic energy is motion of the molecule about a fixed point.

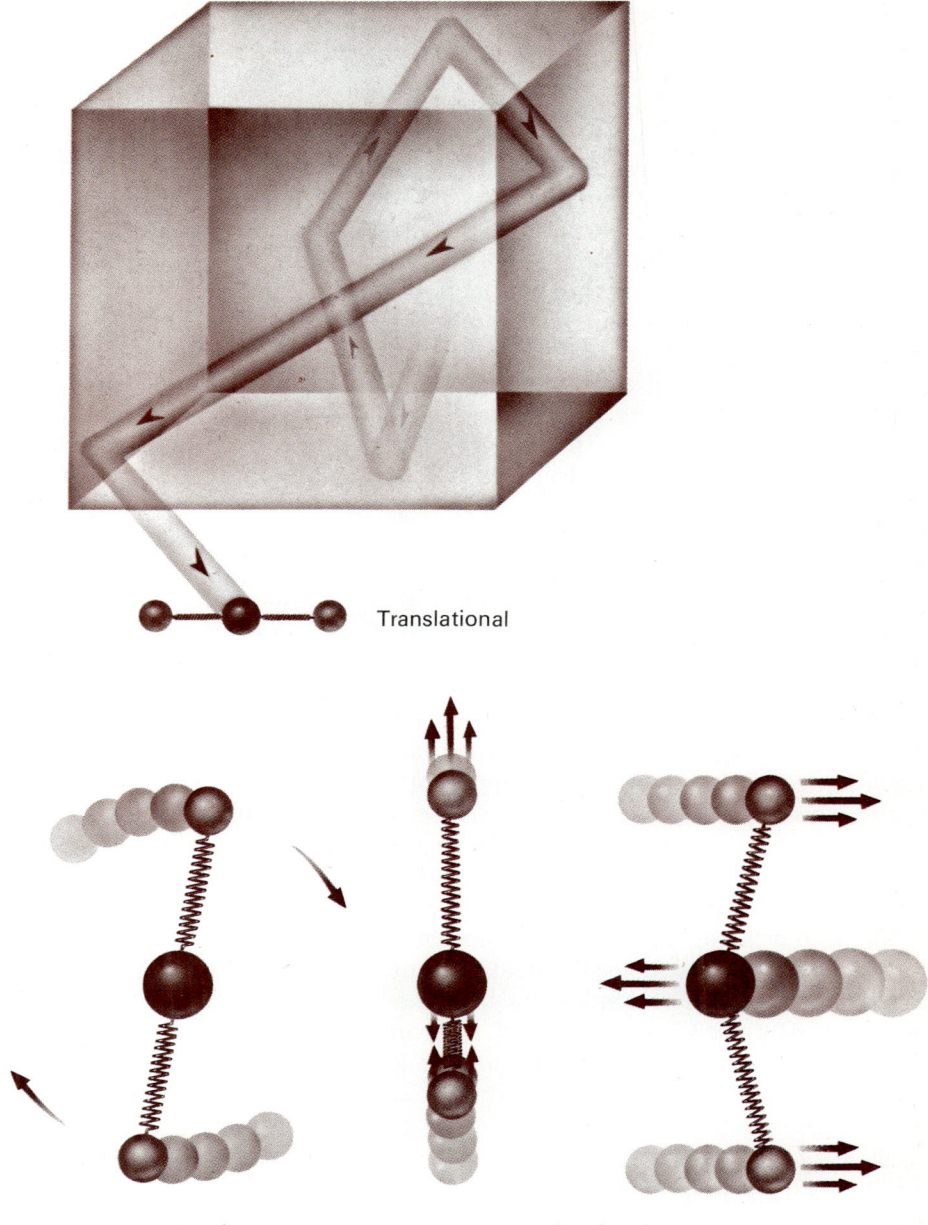

Figure 20-4 *The translational velocity of a gas molecule increases when it absorbs energy. At the same time, its constituent atoms can undergo rotational (left) and vibrational (center and right) motion, without breaking the springlike chemical bonds.*

In addition to the various kinds of kinetic energy given in Figure 20-4, there is also potential energy. This energy is related to the forces that act between molecules. These forces are attractive, having a very small average value in the gaseous state where the molecules are far apart. On the average, the forces are greater in the liquid state and are still greater in the solid state.

The chemical energy within the molecule is related to the forces that hold the atoms in the molecule together. As you know, this is referred to as chemical bond energy.

In addition, each atom has energy, some of which is associated with the electrons and some with the nucleus. The electrons possess kinetic energy. Because of their attraction to the nucleus and their repulsion from each other, they also possess potential energy. The algebraic sum of these energies represents the energy necessary to pull an electron away from an atom.

Finally, there is present within the nucleus of each atom a store of energy related to the forces holding the nuclear particles together. Each nucleus remains intact and apparently uninfluenced through chemical reactions. Therefore, this nuclear contribution to the heat content of molecules does not usually concern a chemist.

The sum of all of these forms of molecular energy makes up the molecular heat content. If you add together the molecular heat contents of 6.02×10^{23} molecules of a given kind, you obtain the molar heat content of that substance.

20-9 Energy Changes on Warming ★

Knowing what makes up the heat content of a substance, you can begin to visualize the effects brought on by warming the substance. If the temperature is low at first, the substance will be a solid. Warming the solid increases the kinetic energy of the motions of the molecules about their regular crystal positions. As the temperature rises, these motions disturb the regularity of the crystal more and more. Too much of this random movement leads to the destruction of the lattice. When the lattice is no longer stable, a phase change occurs. The solid melts.

In the liquid phase each molecule has considerably more freedom of movement, particularly for translation and rotation. Warming the liquid increases the amount of molecular movement. As always, kinetic energy provides a randomizing effect, tending to carry the molecules everywhere in the container. With the rising temperature, the energy of motion rises. More of the molecules are able to move away from the liquid region. Another phase change occurs. The liquid vaporizes.

Warming the substance sufficiently produces kinetic energies in vibration, rotation, and translation that are comparable to chemical bond energies. Then molecules begin to disintegrate. This is the reason that only the very simplest molecules, diatomic molecules, are found in the sun. The temperature there is so high (6000 K at the surface) that more complex molecules cannot survive.

Finally, if heating is continued still further, a temperature will ultimately be reached at which the kinetic energies are large enough to disrupt the nuclei. Then, *nuclear reactions* begin. The conditions in some stars are considered to be suitable for rapid nuclear reactions.

A comparison of the magnitudes of the energy effects concludes this study of molecular energy. Phase changes usually involve energies of several kilocalories per mole. Chemical reactions usually involve energies of twenty to several hundred kilocalories per mole. Thus, the energies involved in chemical reactions are usually 10 to 100 times larger than those involved in phase changes.

Summary

Energy that appears in many forms on the macroscopic level has its ultimate origin in the energy stored within atoms and molecules. A typical molecule has kinetic energy due to its translation (traveling) through space, its rotation (tumbling), and its vibration. It also has potential energy stored in the chemical bonds. Other amounts of energy are stored in the intermolecular interactions. As a substance is heated, all kinds of kinetic energy increase. Phase changes require increases of only a few kilocalories per mole, whereas chemical changes require increases of energy 10 to 100 times larger than this.

Self Check

1. What are three kinds of kinetic energy?
2. How does warming affect the lattice structure of crystals?

Chapter Summary

The absorption or evolution of heat in chemical reactions is best explained by the concept of heat content. Each substance has a fixed amount of stored energy, termed heat content, H. The heat of reaction, ΔH, is equal to the total heat content of the products minus the total heat content of the reactants. If ΔH is positive, the reaction consumes heat (is endothermic). If ΔH is negative, the reaction gives off heat (is exothermic). In an endothermic reaction the heat content of the products is greater than that of the reactants. In an exothermic reaction the opposite is true.

Chemists measure ΔH by using calorimeters and usually express the results in kilocalories per mole of reactant. You can calculate ΔH with the use of a table that gives the heats of formation of compounds involved in the reaction. You can make this calculation because heats of reaction can be added algebraically.

Questions and Problems

1. How is the Law of Conservation of Energy involved in the production, absorption, and measurement of heat in chemical reactions?
2. Explain why the ΔH of a reaction is negative when a reaction evolves heat.
3. Why is it considered desirable to go through the extra steps involved in producing water gas rather than to burn coal directly?
4. Rewrite the following equation in terms of one mole of hydrogen gas. Include the heat effect (ΔH, below) as a term in the equation.

$$\tfrac{1}{2}H_2(g) + \tfrac{1}{2}Br_2(l) \longrightarrow HBr(g)$$
$$\Delta H = -8.60 \text{ kcal/mole HBr}$$

5. Rewrite the following equation, using the ΔH notation for one mole of carbon.

$$3C(s) + 2Fe_2O_3(s) + 110.8 \text{ kcal} \longrightarrow 4Fe(s) + 3CO_2(g)$$

6. Which of the following reactions are endothermic?

   (a) $H_2(g) + \frac{1}{2}O_2(g) \longrightarrow H_2O(g)$
   $$\Delta H = -57.8 \text{ kcal}$$

   (b) $\frac{1}{2}N_2(g) + \frac{1}{2}O_2(g) \longrightarrow NO(g)$
   $$\Delta H = +21.6 \text{ kcal}$$

   (c) $\frac{1}{2}N_2(g) + O_2(g) + 8.1 \text{ kcal} \longrightarrow NO_2(g)$

   (d) $\frac{1}{2}N_2(g) + \frac{3}{2}H_2(g) \longrightarrow NH_3(g) + 11.0 \text{ kcal}$

   (e) $NH_3(g) \longrightarrow \frac{1}{2}N_2(g) + \frac{3}{2}H_2(g)$
   $$\Delta H = +11.0 \text{ kcal}$$

7. Consider the following equations.

   $C(diamond) + O_2(g) \longrightarrow CO_2(g)$
   $$\Delta H = -94.50 \text{ kcal}$$

   $C(graphite) + O_2(g) \longrightarrow CO_2(g)$
   $$\Delta H = -94.05 \text{ kcal}$$

   Find ΔH for the manufacture of diamond from graphite.

   $$C(graphite) \longrightarrow C(diamond)$$

   Is heat evolved or absorbed as graphite is converted to diamond?

8. The heat of combustion of methane, CH_4, is -210 kcal/mole.

   $CH_4(g) + 2O_2(g) \longrightarrow CO_2(g) + 2H_2O(g)$
   $$\Delta H = -210 \text{ kcal}$$

   Discuss why this fuel is better than water gas in terms of one mole of carbon atoms.

9. To change the temperature of a particular calorimeter and the water it contains by one Celsius degree requires 1550 calories. The complete combustion of 1.40 grams of ethylene gas, $C_2H_4(g)$, in the calorimeter causes a rise of 10.7°C. Find the heat of reaction per mole of ethylene.

10. The "thermite reaction" is spectacular and highly exothermic. It involves the reaction between Fe_2O_3 (ferric oxide) and metallic aluminum. The reaction produces white-hot, molten iron in a few seconds.

    $2Al(s) + \frac{3}{2}O_2(g) \longrightarrow Al_2O_3(s)$
    $$\Delta H_1 = -400 \text{ kcal/mole } Al_2O_3$$

    $2Fe(s) + \frac{3}{2}O_2(g) \longrightarrow Fe_2O_3(s)$
    $$\Delta H_2 = -200 \text{ kcal/mole } Fe_2O_3$$

    Determine the amount of heat liberated in the reaction of one mole of Fe_2O_3 with Al.

11. In the preceding problem, how much energy is released in the manufacture of 1.00 kg of iron by the "thermite reaction"?

12. How many grams of water could be heated from 0°C to 100°C by the heat liberated per mole of aluminum oxide formed by the "thermite reaction"?

13. How much energy is liberated when 0.100 mole of H_2 (at 25°C and one atmosphere) is combined with enough $O_2(g)$ to make liquid water at the same temperature and pressure?

14. Calculate the heat of burning ethane in oxygen to give CO_2 and water vapor.

15. How much energy is consumed in the decomposition of 5.0 g of $H_2O(l)$ at 25°C and one atmosphere into its gaseous elements at the same conditions of temperature and pressure?

16. Which would be the better fuel, nitric oxide, NO, or ammonia, NH_3, on the basis of the heat released per mole burned? Assume the products are $NO_2(g)$ and $NO_2(g) + H_2O(g)$, respectively.
17. What is the minimum energy required to produce sulfur dioxide from sulfuric acid?

$$H_2SO_4(l) \longrightarrow SO_2(g) + H_2O(g) + \tfrac{1}{2}O_2(g)$$

18. What is the minimum energy required to synthesize one mole of nitric oxide, NO, from the elements?
19. How many kilocalories will be liberated upon the combustion of 12.5 grams of ethylene, C_2H_4?

$$C_2H_4(g) + 3O_2(g) \longrightarrow 2CO_2(g) + 2H_2O(g)$$
$$\Delta H = -331.6 \text{ kcal/mole}$$

20. The heat of formation of sodium chloride, NaCl, is 98.3 kcal/mole. How much heat will be liberated upon the reaction of 92 grams of sodium with excess chlorine?
21. Consider the following three reactions.

$$Ca(s) + 2H_2O(l) \longrightarrow Ca^{2+}(aq) + 2OH^-(aq) + H_2(g) + 103 \text{ kcal}$$

$$CaO(s) + H_2O(l) \longrightarrow Ca^{2+}(aq) + 2OH^-(aq) + 19.5 \text{ kcal}$$

$$H_2(g) + \tfrac{1}{2}O_2(g) \longrightarrow H_2O(l) + 68.3 \text{ kcal}$$

Determine ΔH for the overall reaction:

$$Ca(s) + \tfrac{1}{2}O_2(g) \longrightarrow CaO(s).$$

21 The Rates of Chemical Reactions

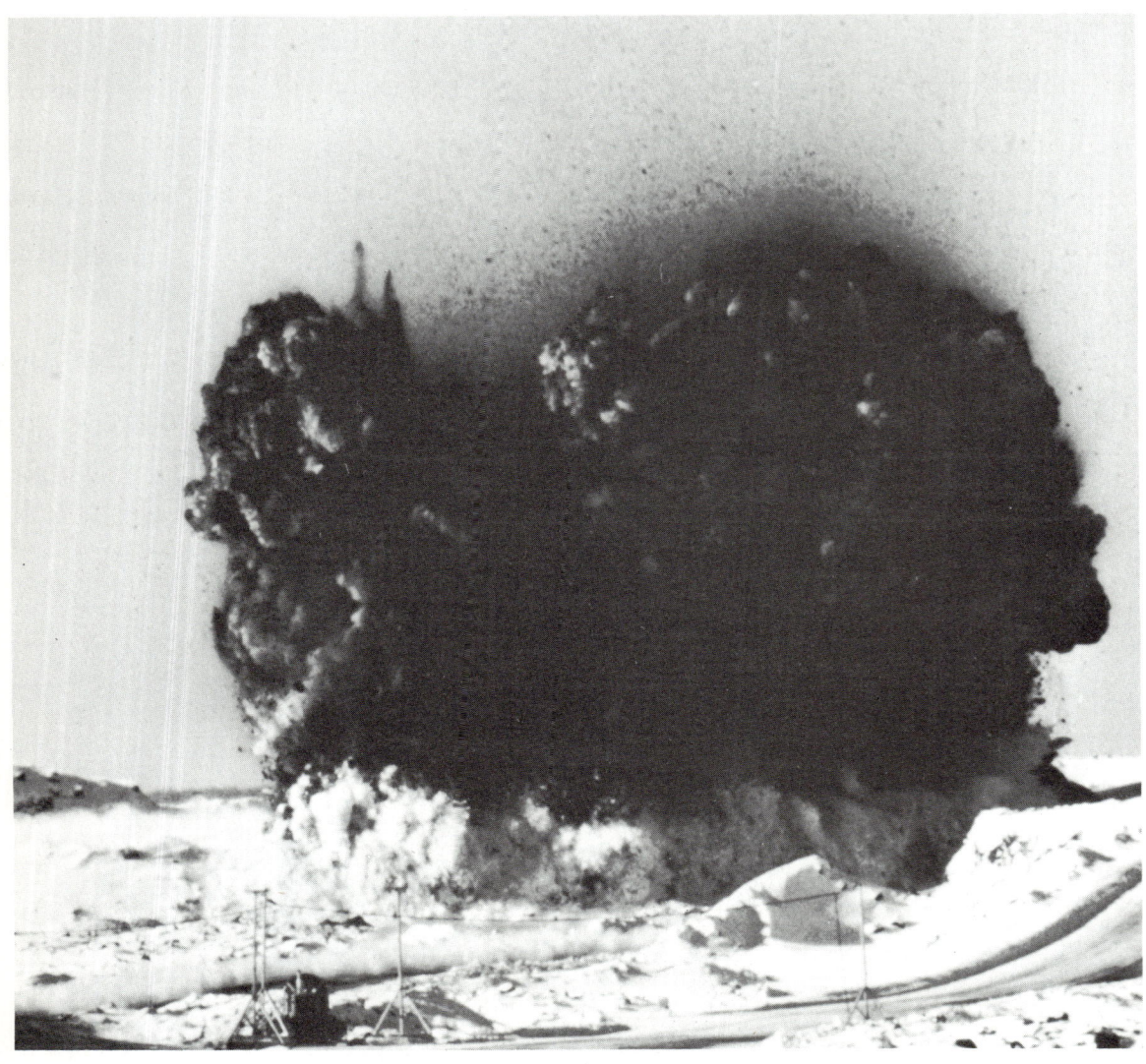

An explosion is a chemical reaction that takes place at a very rapid rate. Many other reactions proceed less rapidly.

You can leave an unlit candle in contact with air indefinitely without seeing any reaction, but hold a lighted match to the wick and you see one instantly. A mixture of household gas and air can be kept in a closed room indefinitely, but if a lighted cigarette is brought into the room there may be a violent explosion. A piece of iron reacts quite slowly with air (it rusts), while a piece of white phosphorus bursts into flame as soon as it is exposed to the atmosphere. All of these reactions involve oxygen from the air. Yet the times necessary for these reactions to take place differ greatly.

Why are there such great differences in reaction times? By learning the answers to such questions as this, chemists can bring reactions under control. This chapter will help you to understand why this is an important factor in chemical reactions.

Many Factors Determine Reaction Rates

21-1 Definition of the Rate of Reaction

What does the expression "rate of a reaction" mean? Consider as an example the reaction between carbon monoxide, CO, gas and nitrogen dioxide, NO_2. Chemical tests show that the products are carbon dioxide, CO_2, and nitric oxide, NO. The equation for the reaction follows.

$$CO + NO_2 \longrightarrow CO_2 + NO \qquad (1)$$

When you prepare a mixture of the reactants and then heat it to 200°C, you observe a gradual disappearance of the reddish-brown color of NO_2. A reaction is taking place. You can determine the amount of time that the reaction takes by measuring the change in color during a measured interval of time. Other gases in the reaction are colorless. Thus this color change indicates the quantity of NO_2 that has reacted in that time interval. The quantity of NO_2 divided by the time interval is one example of a rate of reaction.

$$\text{rate} = \frac{\text{quantity } NO_2 \text{ consumed}}{\text{time interval}}$$

$$= \text{quantity } NO_2 \text{ consumed per unit time}$$

Chemists call the study of reaction rates **chemical kinetics**.

From the reaction equation you know that CO must be consumed at exactly the same rate as NO_2. Therefore, you can also express the rate of reaction (1) in terms of the rate of CO consumption. Also, you can express the rate of the reaction equally well in terms of the appearance of either product, CO_2 or NO. Which compound is used depends upon the convenience of measurement. If you should measure the production of carbon dioxide, for example, you would express the rate in the following way.

$$\text{rate} = \frac{\text{quantity } CO_2 \text{ produced}}{\text{time interval}}$$

$$= \text{quantity } CO_2 \text{ produced per unit time}$$

If the substance is a gas, the quantity consumed or produced is conveniently expressed in partial pressure units. Concentration units are convenient if the reactant or product is in solution. The time measurement, furthermore, is expressed in whatever units fit the reaction. The units can be microseconds for the explosion of household gas and oxygen;

seconds or minutes for the burning of a candle; days for the rusting of iron; months for the rotting of wood.

21-2 The Nature of the Reacting Substances

Experimental studies show that the rate of a reaction depends on the nature of the reacting substances. The following reactions, all of which occur in water solutions, offer an interesting comparison.

slow

$$5C_2O_4^{2-}(aq) + 2MnO_4^-(aq) + 16H^+(aq) \longrightarrow 10CO_2(g) + 2Mn^{2+}(aq) + 8H_2O \quad (2)$$

very fast

$$5Fe^{2+}(aq) + MnO_4^-(aq) + 8H^+(aq) \longrightarrow 5Fe^{3+}(aq) + Mn^{2+}(aq) + 4H_2O \quad (3)$$

very fast

$$Fe^{2+}(aq) + Ce^{4+}(aq) \longrightarrow Fe^{3+}(aq) + Ce^{3+}(aq) \quad (4)$$

Both ferrous ion, $Fe^{2+}(aq)$, and oxalate ion, $C_2O_4^{2-}(aq)$, can decolorize a solution containing the permanganate ion at room temperature. Yet, there is a great contrast in the time required for the decoloration. (See Figure 21-1.) The difference lies in specific characteristics of $Fe^{2+}(aq)$ and $C_2O_4^{2-}(aq)$.

Ferrous ion, $Fe^{2+}(aq)$ is changed to $Fe^{3+}(aq)$ by reacting either with $MnO_4^-(aq)$ or with ceric ion, $Ce^{4+}(aq)$. One of these reactions is simple, and the other involves many molecules. Both of them, however, are rapid reactions.

As another illustration, consider the following two combustion reactions. Both of the reactions take place in the gas phase.

moderately rapid at 20°C

$$2NO + O_2 \longrightarrow 2NO_2 \quad (5)$$

extremely slow at 20°C

$$CH_4 + 2O_2 \longrightarrow CO_2 + 2H_2O \quad (6)$$

The oxidation of nitric oxide, NO, is a reaction involved in smog production. It is moderately rapid at normal temperatures. The oxidation of methane, CH_4, however, occurs so slowly at room temperature that for all practical purposes there is no reaction at all. Again, the difference in the reaction rates must depend upon specific characteristics of the reactants, NO and CH_4.

The determination of the molecular characteristics important to rate behavior is an interesting frontier of chemistry. Chemists have found that most, though not all, chemical reactions that involve the breaking of several chemical bonds and the formation of new chemical bonds tend to proceed slowly at room temperature. Reaction (2) is of this type. Many bonds must be broken in the five $C_2O_4^{2-}$ ions and the two MnO_4^- ions to form the ten CO_2 and two Mn^{2+}. This reaction might be expected to proceed slowly (as it does). Reaction (6) also involves breaking of bonds and forming of new bonds, and it is slow at room temperature. In contrast, reaction (3) is very rapid. Since it involves breaking of chemical bonds, it might be expected to proceed slowly. Apparently you cannot be certain in predicting that a reaction might be slow. Reaction (4) does not require bond breaking or bond formation. It might be expected to be rapid (as it is). A prediction of this type is usually reliable. Reaction (5) requires breaking of only one bond and the formation of two. It has a moderate reaction rate, rapid at high pressures and slow at low pressures.

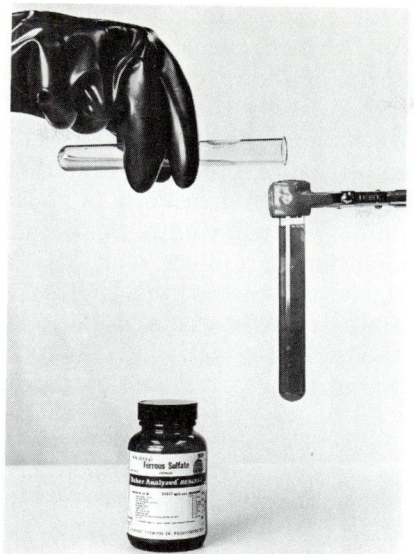 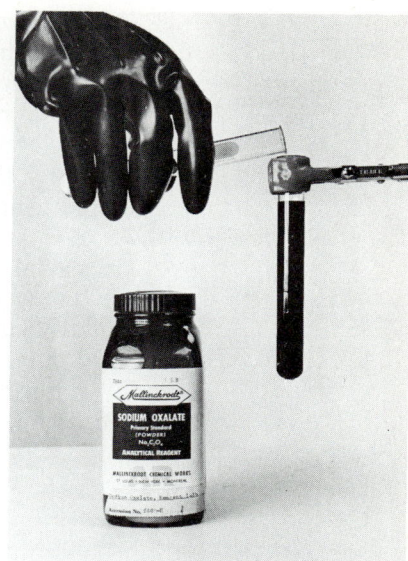

Figure 21-1 *The rate at which solutions containing MnO_4^- ions can be decolorized varies. The left photo shows that a potassium permanganate solution begins to lose its color almost immediately after the addition of Fe^{2+} (aq). Immediately after the addition of $C_2O_4^{2-}$ (aq), however, there is no appreciable decolorization (right photo).*

These and other examples lead to the following rules.
1. Reactions that do not involve bond rearrangements are usually rapid at room temperature.
2. Reactions in which bonds are broken tend to be slow at room temperature.

Little more can be said about how the nature of the reactants determines the reaction rate until you consider in detail how some reactions take place. For the time being, it will suffice to observe that this is an active field of study and that much remains to be learned.

INVESTIGATION

21-3 *A Study of Reaction Rates*

When a chemist performs a chemical reaction, he must consider a number of conditions that may affect the rate at which the reaction proceeds. Even though you might not be able to predict their effect, you could list a number of conditions that could possibly change the rate of a chemical reaction. Among these might be the rate of stirring, the size of the particles involved in the reaction, and the concentration of the reacting substances. Others might be the temperature at which the reaction is performed, the pressure on the mixture, and the size, shape, and material of the reaction container.

When you want to study the effect of certain controlled conditions on the rate of a reaction, it is helpful if you can conduct an experiment in which the completion of a reaction is marked by an easily detectible rapid change in color. Such a reaction is known as a *clock reaction*.

In this experiment you will study a clock reaction produced by mixing two solutions. Solution A, a dilute solution of potassium iodate, KIO_3, is the source of one of the reacting species, the iodate ion, $IO_3^-(aq)$. Solution B contains some starch and the other reacting species, the hydrogen sulfite ion, $HSO_3^-(aq)$.

The initial step in the reaction can be represented by the following equation.

$$IO_3^-(aq) + 3HSO_3^-(aq) \longrightarrow I^-(aq) + 3SO_4^{2-}(aq) + 3H^+(aq)$$

When the hydrogen sulfite ions, $HSO_3^-(aq)$, are used up, the iodide ions, $I^-(aq)$, react with the remaining iodate ions, $IO_3^-(aq)$, to produce iodine, $I_2(s)$.

$$5I^-(aq) + 6H^+(aq) + IO_3^-(aq) \longrightarrow 3I_2(s) + 3H_2O$$

The molecular iodine then combines with the starch from solution B to form a blue substance. The appearance of a blue color, then, shows that the reaction has begun to produce $I_2(s)$.

PURPOSE

You will investigate the role of the concentration of the reactants and the temperature of the reaction mixture in the reaction between the iodate ion and the hydrogen sulfite ion. By varying these factors you will determine what effect they may have on the rate of the reaction.

MATERIALS

2 test tubes, 18- × 150-mm
Beaker, 400-ml
Ringstand, ring
Wire gauze
Burner
Graduated cylinder, 10- or 25-ml
1 or 2 thermometers, −10°C to 110°C
Clock or watch with sweep second hand

Ice cubes
45 ml of solution A
70 ml of solution B
Distilled water

PROCEDURE

Part A. The Effect of Concentration Changes

To investigate the effect of the concentration of one of the reactants on the reaction time, you will prepare dilutions of solution A to vary the concentration of the $IO_3^-(aq)$ ion. In each case, the concentration of the $HSO_3^-(aq)$ ion will be kept constant and all the solutions should be at room temperature. If there is not sufficient time for you to investigate several concentrations on your own, your teacher will assign certain concentrations for you and your partner to do. By exchanging results with other members of the class you will be able to draw some conclusions concerning the effect of this variable.

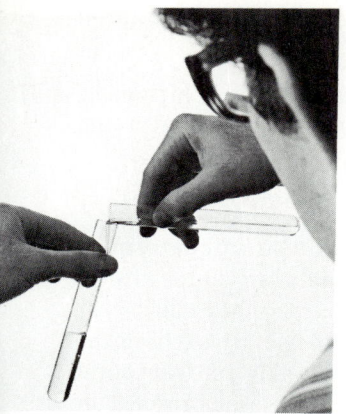

Figure 21-2

1. Use a clean graduated cylinder to measure 10.0 ml of solution A. Then pour the solution into a clean test tube. Rinse the graduated cylinder. In a similar manner place 10.0 ml of solution B into another test tube. If the solutions have been in the laboratory for some time you may assume that they are at room temperature. Otherwise, you should put the test tubes containing the solutions into a 250-ml beaker about two-thirds full of water at room temperature and let them stand for several minutes.

2. Using a watch with a second hand, record the time to the nearest second as you pour solution A into solution B, as shown in Figure 21-2. Then pour them back and forth *quickly* three times to obtain uniform mixing. Time should be recorded from the instant both solutions are in contact.

3. Watch the solution in the test tube carefully and record the time again at the first sign of a reaction.

4. Repeat the experiment to check your results, if directed to do so by your teacher.

5. Prepare different concentrations of the KIO_3 solution by diluting solution A as indicated in the following table. Do as many dilutions as directed by your teacher.

Solution A	Distilled Water
9.0 ml	1.0 ml
8.0 ml	2.0 ml
7.0 ml	3.0 ml
etc.	etc.

Note that the total volume is always 10.0 ml. Mix each of the diluted solutions well.

6. Repeat the timing procedure, adding one of the diluted solutions of KIO_3 to 10.0 ml of solution B. Both solutions should be at room temperature.

Part B. The Effect of Temperature

To investigate the effect of changes in temperature you will determine the time of this reaction at room temperature and at other temperatures within a range of $\pm 20°C$ of room temperature. Your teacher will assign particular temperatures for you and your partner to use. By exchanging results with other members of the class you will be able to draw some conclusions concerning the effect of temperature on the time of reaction.

1. Put 10.0 ml of a dilution of solution A into one test tube (18- × 150-mm) and 10.0 ml of solution B into another. (Use a dilution of solution A that you prepared in Part A, as your teacher directs.) These solutions must be brought to the desired temperature before they are mixed. Put both test tubes into a 250-ml beaker about two-thirds full of water at the temperature you were assigned to investigate. Let them stand for about ten minutes so the solutions will come to the temperature of the water bath. Keep the temperature of the water bath as nearly constant as you can. Heat it gently, or add hot water to it, if it begins to cool. If it starts to warm up, cold water or a small piece of ice can be added.

2. Using a watch with a second hand, record the time to the nearest second as you pour solution A into solution B. Then pour them back and forth *quickly* three times to obtain uniform mixing, as before. Time should be recorded from the instant both solutions are in contact.

3. Put the test tube back into the water bath and observe it carefully. Record the time again at the first sign of a reaction.

4. To check your results, repeat the experiment at the same temperature. Repeat it at other temperatures also, if directed by your teacher to do so.

RELATED PROBLEMS

1. The concentration of KIO_3 in solution A is 0.02 M. Calculate the number of moles of potassium iodate in each milliliter of solution A.

2. Calculate the initial molar concentration of KIO_3 in each of the mixtures of *A* plus *B* prepared in Part A.

3. Why is it important to keep the total volume at 10 ml when diluting solution *A*, in Part A?

4. Plot a graph of the concentration-time data. Plot time on the vertical axis and the concentration of the KIO_3 on the horizontal axis. Use data obtained by other members of the class also.

5. What generalizations can you make concerning the effect of varying the concentration on the time of the reaction?

6. How is the time of the reaction related to the rate of the reaction?

7. Plot a graph of the temperature-time data. Plot temperature on the horizontal axis and time on the vertical axis.

8. What general relationships can you derive from the temperature-time graph?

9. Make a prediction of the time of the reaction between undiluted solutions *A* and *B* at 0°C and at 50°C, assuming that the other variables in the experiment are kept constant.

21-4 Effect of Concentration on Reaction Rate

From here on you will study one reaction at a time. The nature of the reactants will remain constant, while the other factors that affect rates are considered. The first of these factors that you will study is concentration.

For many reactions, raising the concentration of a reactant increases the reaction rate. You have just observed this effect in your study of the reaction of $IO_3^-(aq)$ with $HSO_3^-(aq)$. Not infrequently, though, there will be no such effect. In this section you will learn why a rate usually, though not always, increases with increased reactant concentration. In Section 21-5 you will consider why some reactions proceed at a rate independent of the concentration of one or more reactants. Explanations for differences in reaction behavior are based on a model of the way chemical reactions take place on the molecular scale.

In the molecular view of matter, chemists assume that two molecules must come close together in order to react. Therefore, they postulate that chemical reactions depend upon collisions between the reacting particles. These particles may be atoms, molecules, or ions. This model of reaction-rate behavior is called the **collision theory**. It provides a successful basis for understanding the effect of concentration. In this connection, you know that an increase in the number of cars in motion on a highway leads to a higher rate of dented fenders. Similarly, increasing the number of particles in a given volume produces more frequent molecular

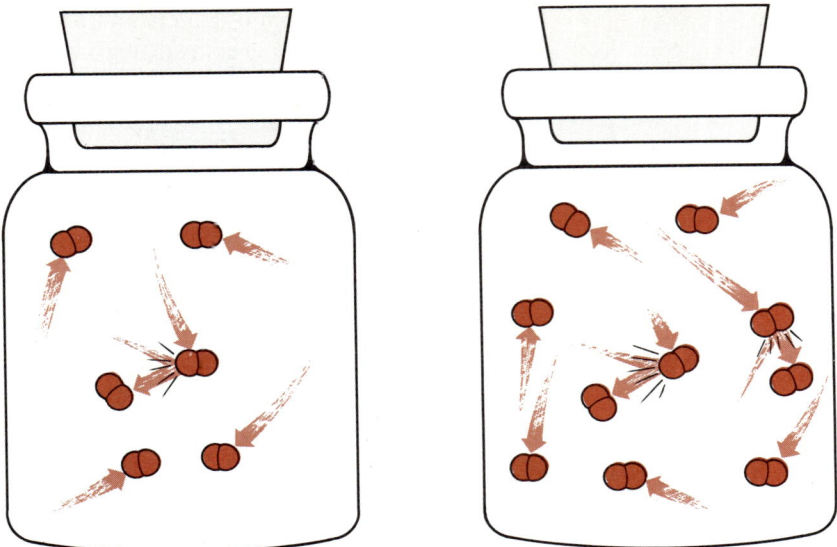

Figure 21-3 *The reaction rate in the container on the right is faster because the higher concentration of reactants in that container results in a greater number of collisions between molecules of these reactants.*

collisions. As Figure 21-3 indicates, the higher frequency of collisions results in a faster rate of reaction.

Consider a *homogeneous* system, that is, one in which all components are in the same phase, or state. A solution is a good example of such a "system." According to the collision theory, increasing the concentration of one or more reactants should result in an increase in the rate of the reaction. Lowering the concentration should have the opposite effect. This is exactly the behavior found in the reaction between $HSO_3^-(aq)$ and $IO_3^-(aq)$, when the concentrations are varied. In gases, which are also homogeneous systems, the concentration of an individual reactant can be raised by increasing the quantity of that substance. The concentrations of all gaseous components can be raised simultaneously by decreasing the volume that they occupy. Thus, decreasing the volume by compressing the gas increases the rates of the reactions taking place. Increasing the volume by expanding the gas has the opposite effect on concentrations. A decrease in reaction rates results.

In a *heterogeneous* reaction system, the components are in two or more different phases. The reaction below is a heterogeneous one.

$$\text{wood (}solid\text{)} + \text{oxygen (}gas\text{)} \xrightarrow{\text{burning}} \text{carbon dioxide (}gas\text{)} + \text{water (}gas\text{)}$$

In a system of this sort, the rate of the reaction depends upon the area of contact between the phases. For example, a log burns

Figure 21-4 *The wooden splint burns much more slowly in air (left), which is only twenty percent oxygen, than it does in pure, one hundred percent oxygen (right).*

in air at a relatively slow rate. If the amount of exposed surface of the wood is increased by reducing the log to splinters, the burning is much more rapid. If the wood is further reduced to fine sawdust that is then suspended in a current of air, combustion takes place so rapidly that there is an explosion. Where one of the reactants is a gas, as in the above example, the concentration of the gas is also a factor. A piece of wood burns much more rapidly in pure oxygen than it does in ordinary air. In the latter, oxygen makes up only about twenty per cent of the mixture. Figure 21-4 shows this difference in burning rates.

You can see that the collision theory provides a good explanation of reaction-rate behavior. It is quite reasonable that the reaction rate should depend upon collisions among the reactant molecules. In fact, it is so reasonable that one wonders why the concentrations of some reactants in some reactions do not affect the rate. The explanation is found in the detailed steps by which the reaction takes place.

21-5 *Reaction Mechanism*

It has been proposed that particles must collide for a chemical reaction to occur. The particles may be atoms, molecules, or ions. As a result of collisions, there can be rearrangements of atoms, electrons, and chemical bonds, resulting in the production of new molecular arrangements. For example, take another look at the reaction between Fe^{2+} and MnO_4^-.

$$5Fe^{2+}(aq) + MnO_4^-(aq) + 8H^+(aq) \longrightarrow$$
$$5Fe^{3+}(aq) + Mn^{2+}(aq) + 4H_2O \quad (7)$$

The equation indicates that one MnO_4^- ion, five Fe^{2+} ions, and eight H^+ ions must react with each other. If this reaction were to take place in a single step, these fourteen ions would have to collide with each other *simultaneously*. The probability of such an event occurring is extremely small. It is so small that a reaction that depended upon such a collision would proceed at an immeasurably slow rate.

In fact, chemists regard the simultaneous collision of even four molecules as an extremely improbable event if the molecules are at low concentration or if they are in the gas phase. The conclusion, therefore, is that a complex chemical reaction proceeding at a measurable rate probably takes place in a series of steps. The series of reaction steps is called the **reaction mechanism**.

Consider the following reaction, one that is reasonably rapid in the temperature range from 400°C to 600°C.

$$4HBr(g) + O_2(g) \longrightarrow 2H_2O(g) + 2Br_2(g) \quad (8)$$

According to the collision theory, increasing the partial pressure, and thus the concentration, of either the HBr or O_2 should speed up the reaction. Experiments show that this is the case. Quantitative studies of the rate of reaction (8) at various pressures and with various mixtures show that HBr and O_2 are equally effective in changing the reaction rate. However, this result raises a question. Observe that reaction (8) requires four molecules of HBr for every one molecule of O_2. Why, then, does a change in the HBr pressure have just the same effect as an equal change in the O_2 pressure?

The explanation is found by considering the details of the process by which reaction (8) occurs. The overall reaction brings together five molecules, four of HBr and one of O_2. The chance that five gaseous molecules will collide simultaneously is practically zero. For this reason, the reaction must occur in a sequence of simpler steps.

Reaction (8) has been studied very extensively in the laboratory. Chemists believe that the following series of reactions occurs.

slow

$$HBr + O_2 \longrightarrow HOOBr \quad (9)$$

fast

$$HOOBr + HBr \longrightarrow 2HOBr \quad (10)$$

fast

$$HOBr + HBr \longrightarrow H_2O + Br_2 \quad (11)$$

First, observe that adding reactions (9) and (10) plus twice reaction (11) gives the overall reaction (8). Next, you can see that each step in the sequence requires only two molecules to collide. Finally, the slow rate of reaction (9) explains why HBr and O_2 have the same effect on the reaction rate.

Reaction (9) is a "bottleneck." As fast as HOOBr is formed by this slow reaction, it is consumed in the rapid reaction (10). No matter how rapid reactions (10) and (11) are, they can produce H_2O and Br_2 only as fast as the slowest reaction in the sequence. Hence, the factors that determine the rate of reaction (9) determine the rate of the overall process.

The sequence of reactions (9), (10), and (11) is the reaction mechanism of the overall reaction (8). Because it is the slowest reaction in the mechanism, reaction (9) is the step that fixes the rate. The slowest reaction in a reaction mechanism is called the **rate-determining step**.

Two features of this example are rather common. First, none of the steps in the reac-

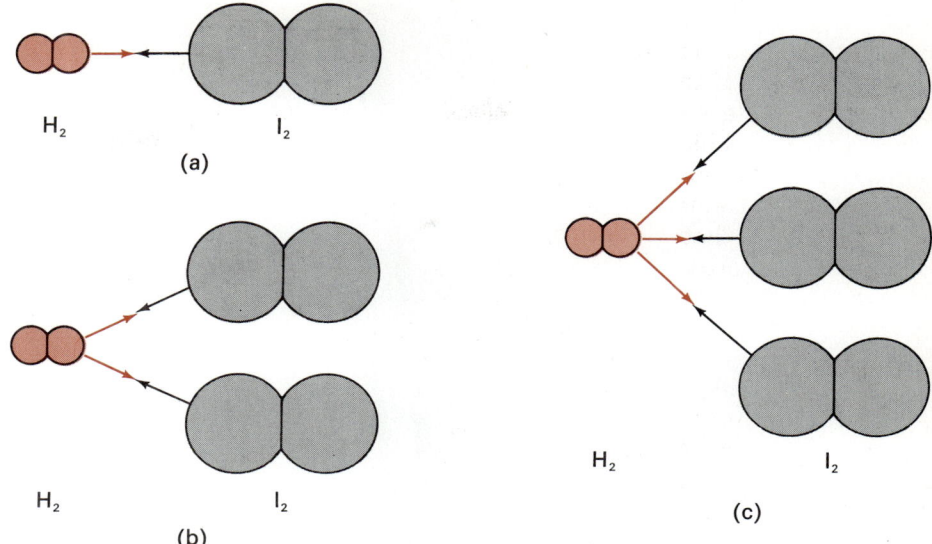

Figure 21–5 *For every collision of H_2 with I_2 that occurs at a given I_2 concentration (a), there will be twice as many if the I_2 concentration is doubled (b), three times as many if it is tripled (c), and so on.*

tion mechanism requires the collision of more than two particles. *Most chemical reactions proceed by sequences of steps, each involving only two-particle collisions.* Second, the overall or net reaction does not show the mechanism. In general, *the mechanism of a reaction cannot be deduced from the net equation for the reaction.* The various steps by which atoms are rearranged and recombined must be determined through experiment.

To determine a reaction mechanism by experiment, you must further develop your understanding of the collision model. So far you have used it only qualitatively. You can also use the model quantitatively. In this way you can tell *how much* a change in concentration will change the rate of a reaction.

Consider the following reaction between gaseous hydrogen, H_2, and gaseous iodine, I_2.

$$H_2(g) + I_2(g) \longrightarrow 2HI(g) \tag{12}$$

Each time a molecule of H_2 collides with an I_2 molecule, the reaction may occur. For a particular H_2 molecule, the frequency of these encounters is determined by how many I_2 molecules are present. Doubling the number of I_2 molecules per unit volume doubles the collisions. Tripling the number of I_2 molecules per unit volume triples the collisions. (See Figure 21-5.) Since the iodine partial pressure fixes the iodine concentration, the rate of the reaction is proportional to the iodine partial pressure.

$$\text{rate is proportional to} \begin{bmatrix} \text{iodine} \\ \text{partial} \\ \text{pressure} \end{bmatrix} \tag{13}$$

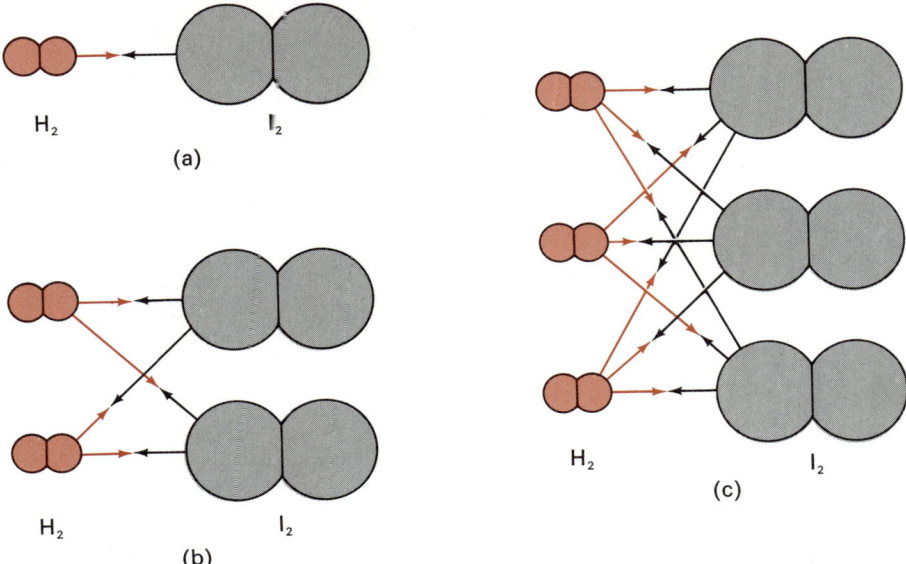

Figure 21-6 If the concentrations of both H_2 and I_2 are doubled from some initial values (a), the number of collisions will be increased by a factor of four (b). If both original concentrations are tripled, the number of collisions will increase by a factor of 9 (c).

In the same way, a particular iodine molecule must find a hydrogen molecule with which to react. The rate of the reaction is also proportional to the hydrogen partial pressure.

$$\text{rate is proportional to} \begin{bmatrix} \text{hydrogen} \\ \text{partial} \\ \text{pressure} \end{bmatrix} \quad (14)$$

In light of (13) and (14), the rate must be proportional to the *product* of the partial pressures of iodine and hydrogen. (See Figure 21-6.)

$$\text{rate is proportional to} \begin{bmatrix} \text{hydrogen} \\ \text{partial} \\ \text{pressure} \end{bmatrix} \times \begin{bmatrix} \text{iodine} \\ \text{partial} \\ \text{pressure} \end{bmatrix} \quad (15)$$

In symbols, you can write

$$\text{rate} = k[p_{H_2}] \times [p_{I_2}] \quad (16)$$

The term k is called the **rate constant**. It is a proportionality constant, just as G is in the Law of Gravitation (see Section 2-2). The rate constant has a particular value for this reaction at a particular temperature, but its value increases as the temperature increases.

21-6 Effect of Temperature on Reaction Rate

In the experiment in Section 21-3, you discovered that temperature has a marked effect upon the rate of chemical reactions. Raising the temperature speeded up the reaction between $IO_3^-(aq)$ and $HSO_3^-(aq)$. This effect is qualitatively the same as that observed in the reaction of a candle with air. The match lit the candle by raising its temperature at the wick. Once started, the reaction of combustion released enough heat to keep the temperature high and the reaction going at a reasonable rate. Raising the temperature speeded up the reaction. The same type of explanation applies to the explosion resulting from the introduction of a lighted cigarette into a kitchen filled with cooking gas and air. The temperature of the gas around the glowing tip of the cigarette is raised, speeding the reaction and liberating heat. This heat warms the nearby region even more and the reaction goes somewhat faster. This acceleration continues until finally, in a millisecond or so, it reaches an explosive rate. This explosive reaction is the most rapid reaction permitted by the collisional properties of the gas. Raising the temperature started it all by speeding up the reaction.

Increasing the temperature speeds up almost all reactions. The question may come to mind: "Why does a temperature rise speed up a reaction?" To answer this question, you will have to return once again to the collision theory.

Consider again the mixture of cooking gas, mostly CH_4, and air under normal conditions. Chemists have calculated that in the mixture a particular CH_4 molecule collides with an O_2 molecule about once every one-thousandth of a microsecond (10^{-9} second). This means that every second this CH_4 molecule encounters 10^9 O_2 molecules! Yet the reaction does not proceed noticeably. You might conclude either that most of the collisions are ineffective or that the collision theory is not a good explanation. You will see that the former is the case. The collision theory explains why most collisions are ineffective.

Chemists have learned that chemical reactions occur when collisions occur, but *only when the collision involves more than a certain amount of energy*. This can be better understood by returning to the analogy of cars bumping each other on a highway. In a line of heavy traffic one frequently receives gentle bumps from the car in front or from the car behind. No damage is done to the cars—only to tempers. Occasionally, though, a high-speed collision takes place. If this occurs with enough energy, a bumper may be knocked off a car, or a fender may be dented. High energy collisions also cause "molecular damage." The high energy molecular collisions that cause molecular damage are called *chemical reactions*. Just as a minimum amount of energy is required to break loose a bumper, a minimum amount of energy is required to cause a chemical reaction. You can think of this minimum amount of energy as "threshold energy." If there is more than this threshold energy, the reaction can occur. If there is less, it cannot. This topic will be more fully discussed in succeeding sections of this chapter.

Summary

The rate of a reaction is defined as the quantity of some reactant consumed or of some product produced per unit of time. The rate depends on many factors, one of which is the nature of the reactants. Different reactions occur with inherently different rates, even under similar conditions. In general, reactions that do not involve bond rearrangements are more rapid at normal temperatures than those that do, though there are exceptions. Reaction rates usually increase with increasing concentrations of reactants and with increasing temperature. The collision theory explains this. In general, reactions proceed in a series of steps. Each of the steps requires only two species to collide. The series of such steps that accounts for an overall reaction is called the mechanism of the reaction. The slowest step in the series determines the maximum possible speed for the overall reaction. This step is called the rate-determining step.

Self Check

1. Are any of the following three reactions likely to be extremely rapid at room temperature? Are any likely to be extremely slow at room temperature? Give the reasons for your answers.

   (a) $Cr^{2+}(aq) + Fe^{3+}(aq) \longrightarrow$
   $$Cr^{3+}(aq) + Fe^{2+}(aq)$$

   (b) $3Fe^{2+}(aq) + NO_3^-(aq) + 4H^+(aq) \longrightarrow$
   $$3Fe^{3+}(aq) + NO(g) + 2H_2O$$

   (c) $C_8H_{18}(l) + 12\frac{1}{2}O_2(g) \longrightarrow$
   $$8CO_2(g) + 9H_2O(g)$$

2. Imagine five people working together to wash a stack of very greasy dishes. The first two clear the table and hand the dishes to the third person, who washes them and hands them on. The last two dry and stack them. Which step is likely to be the rate-determining step? In light of your answer, discuss how the rate of the overall process would be affected if a sixth person joined the group (a) as a table clearer; (b) as a second dishwasher; (c) as a dish dryer.

Energy Plays an Important Role in Reaction Rates

21-7 Distribution of Kinetic Energies

The discussion of threshold energy raises the question, "What energies do molecules possess at a given temperature?" An experiment can provide the answer.

Figure 21-7 shows a device for measuring the distribution of atomic or molecular velocities. It consists of two discs, D_1 and D_2, in a vacuum chamber, rotating rapidly on a common axle. Inside the chamber is also an oven containing molten tin held at a controlled temperature. Vapor streams out of the small opening in the oven and strikes the rotating disc, D_1. When the disc has rotated to the position shown in Figure 21-7b, a small amount of gas has passed through the slot in disc D_1. A short time later (Figure 21-7c) the atoms of tin have traveled part of the way toward the second rotating disc. The fastest moving atoms have traveled farther than the others and are leading the way. The slowest moving atoms are beginning to lag behind. Still later (Figure 21-7d), the atoms have spread out in space even more. The fastest atoms have already reached the second rotating disc. As the atoms reach disc D_2, they condense at positions determined by how long each atom takes to travel from D_1 to D_2 and how fast D_2 is rotating.

As the slotted disc D_1 lets through burst after burst of tin atoms, a layer of tin builds up on the surface of disc D_2. The pattern of

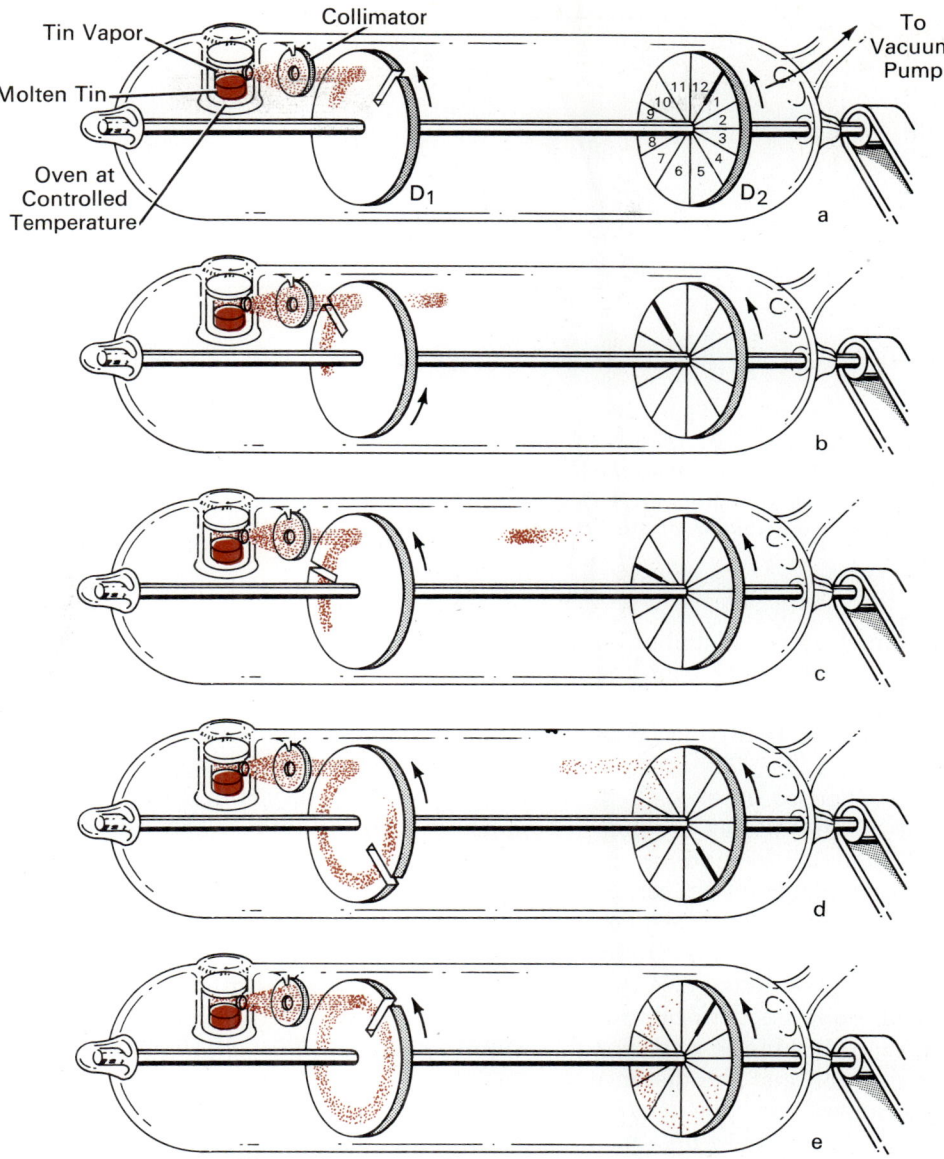

Figure 21-7 *These diagrams illustrate the operation, through one revolution of the discs, of a device for measuring the distribution of velocities of vaporized tin atoms.*

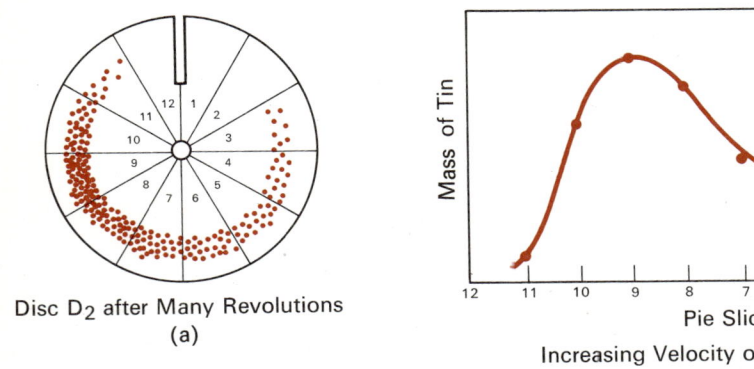

Figure 21-8 *The pattern formed by tin atoms after many revolutions of the disc D_2 (from the device in Figure 21-7) is shown in (a). A graph showing the distribution of tin by mass (b) indicates the distribution of atomic velocities.*

this layer is determined by the distribution of velocities of the atoms escaping from the oven at temperature T.

Figure 21-8a shows disc D_2 divided into sections, like slices of a pie. The fastest moving atoms are condensed on slices 3, 4, and 5. The slowest moving atoms are condensed on slices 10 and 11. By finding the mass of each slice, the amount of tin can be determined. Figure 21-8b shows the mass of tin plotted against the slice number to indicate the distribution of atomic velocities.

Clearly, this plot contains information about the distribution of kinetic energies. From the rate of rotation and the distance between discs, you can calculate the velocity necessary for an atom to condense on a particular slice. You can then calculate the kinetic energy of an atom from its atomic mass and its velocity.

There is still one very important point to be considered. As shown in Figure 21-7, the oven from which the tin vapor emerges is held at a "controlled temperature." This temperature must obviously be high enough to produce a stream of tin atoms that can be measured in the experiment. However, this still leaves a considerable range of temperatures over which the experiment may be performed. Suppose you perform the experiment at two different temperatures, at a lower temperature T_1 and at a higher one, T_2. How will the distribution of tin atoms on disc D_2 differ?

Figure 21-9 illustrates the difference. At temperature T_1 a few atoms have very low kinetic energies, and a few have very high kinetic energies. Most of them have intermediate kinetic energies, as shown by the solid curve. At the higher temperature, T_2, the energy distribution is that shown as a dashed curve. As can be seen, increasing the temperature causes a general shift of the distribution toward higher kinetic energies. At any particular value of kinetic energy, there are more molecules with energies equal to or above this value at the higher temperature, T_2, than at T_1.

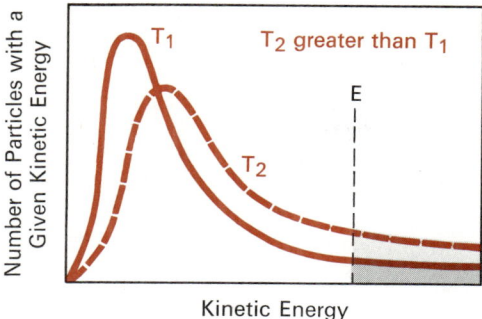

Figure 21–9 *An increase in temperature from T_1 to T_2 causes a shift in the kinetic energy distributions from a lower kinetic energy (solid line) toward a higher one (dashed line).*

These curves can be applied to the reaction rate problem. Suppose a reaction can occur only if two molecules collide with kinetic energy exceeding a certain threshold energy E, shown in Figure 21-9. At T_1 the darker shaded area is proportional to the number of the molecules possessing this much or more energy. Since only a small number of molecules have energy equal to or greater than E, few collisions are effective, and the reaction is slow. When the temperature is raised to T_2, the number of molecules with energy E or greater increases. Now the sum of the light and dark shaded areas is proportional to the number of molecules with energy equal to or greater than E. Only a small temperature change is needed to make a large change in the area to the right of energy E. Consequently, the reaction rate is very sensitive to change in temperature.

Raising the temperature should also increase the reaction rate by increasing the frequency of the collisions. This is true, but it is a very small effect compared with that caused by the increase in the number of molecules with sufficient energy to cause reaction.

21-8 Activation Energy

In the preceding section you learned that raising the temperature increases the number of molecules that have energy greater than the threshold energy, E. You can now consider the concept of the threshold energy itself.

Suppose you are a cross-country skier. Cross-country skiing is a test of both speed and endurance. In this sport you travel across natural terrain under your own power, going up some slopes and down others. Figure 21-10 (next page) shows a possible cross-country ski route.

Now imagine that a cross-country race is to be held over the terrain shown in Figure 21-10. The object is to get from the lodge at A to the finish line at B in the shortest time. On the way there are several ridges to cross, one of which is considerably higher than the others. It is clear that crossing this ridge is the crucial part of the race. Obviously, if you do not have enough energy to get over it, you will not finish the race. Of 100 skiers who start the race, perhaps only 60 will have enough energy to cross the big ridge. Once they have done this, they should have enough energy to cross the other much smaller ones and get to the end. Figure 21-11, on the next page, shows a graph of the route and makes clear the importance of the highest ridge. In essence, what determines whether you get to the finish line is whether you are energetic enough to get over the highest ridge.

You may be surprised to learn that there is a remarkably close analogy between this situation and the question of how fast a chemical reaction can proceed. As molecules

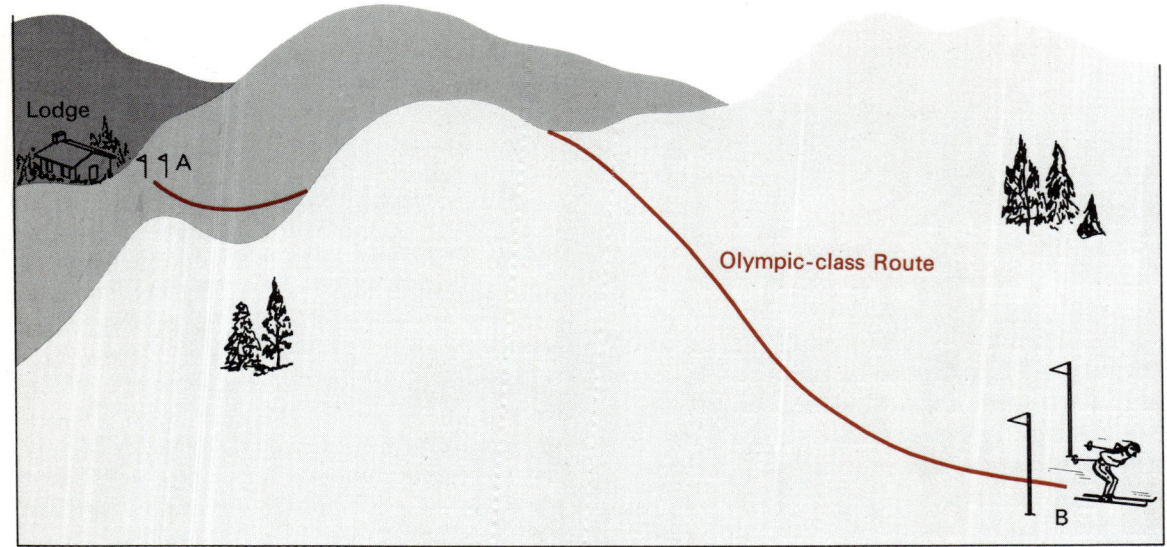

Figure 21-10 The Olympic-class route from the lodge (A) to the finish line (B) requires enough energy to get over a high ridge along the way.

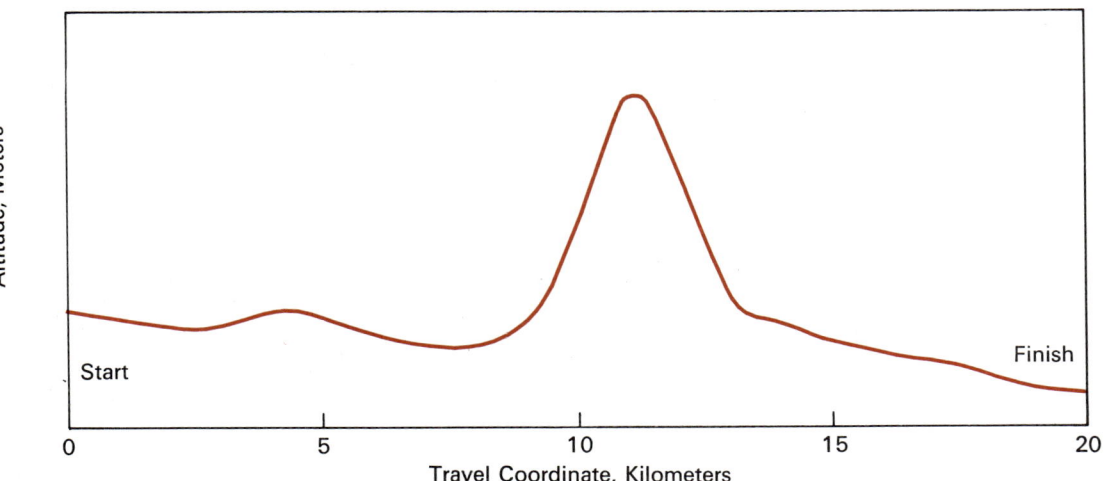

Figure 21-11 A graph of the skier's altitude as he travels the Olympic-class route shown in Figure 12-10 from the lodge to the finish line.

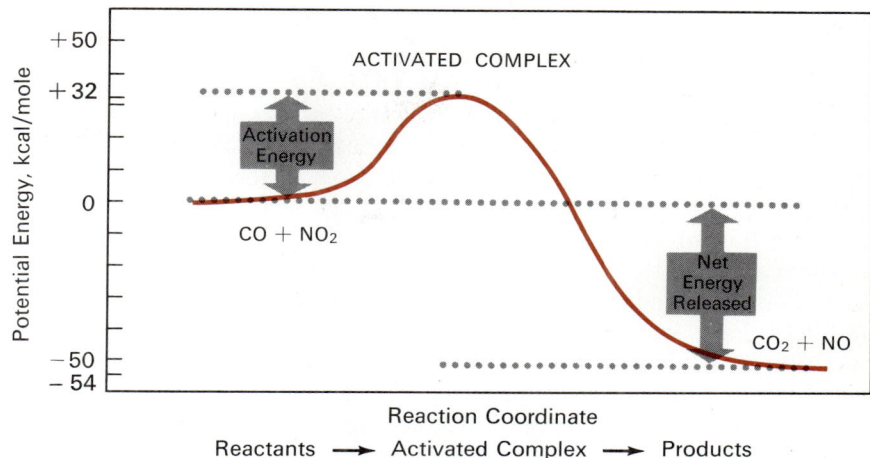

Figure 21–12 *In a chemical reaction, the reactants must have enough kinetic energy to reach the high potential energy (activated complex) required to complete the reaction. Attaining the activated complex is similar to reaching the top of the highest ridge in a cross-country ski race.*

collide and a reaction takes place, the atoms must momentarily take up bonding arrangements that are less stable than either those of the reactants or those of the products. Like the highest ridge, these high energy molecular arrangements place an energy barrier between reactants and products. Only if the colliding molecules have enough energy to surmount the barrier imposed by the unstable arrangements can a reaction take place. This barrier determines the "threshold energy," or minimum energy necessary for a reaction to occur.

This barrier is shown graphically in Figure 21-12. You can see that this figure is similar to Figure 21-11. Figure 21-12 applies to the reaction between carbon monoxide, CO, and nitrogen dioxide, NO_2, to produce carbon dioxide, CO_2, and nitric oxide, NO. The horizontal axis of the diagram, called the *reaction coordinate*, shows the progress of the reaction. Proceeding from left to right along this reaction coordinate signifies, first, that the CO and NO_2 molecules approach each other. Then they collide and go through intermediate processes of reaction that result in the formation of CO_2 and NO. The vertical axis represents the total potential energy of the system. Thus the curve provides a history of the potential energy change during a collision that results in a reaction. The energy needed to get over the barrier to reaction usually is provided by the kinetic energies of the colliding particles. As you know, it is the temperature that determines the kinetic energy.

Consider Figure 21-12 in more detail. From left to right along this curve, the following events occur. Along the flat region at the left, CO and NO_2 molecules approach each other. In this region they possess kinetic energy, and their total potential energy shows no change. The beginning of the rise in the curve signifies that the two molecules have come sufficiently close to have an effect on each other. During this approach, the molecules slow down as their kinetic energies furnish the potential energy to climb the curve. If they have sufficient kinetic energy, they can ascend the left side of the barrier all the way up to the summit. Attaining this point is interpreted as meaning that the CO and NO_2 have sufficient kinetic energy to overcome the mutually repulsive forces of their nuclei and negative electron clouds. Therefore, they can come very close to each other. Here at the summit the molecular arrangement is unstable with respect to either the forward reaction or the reverse reaction. The forward reaction produces CO_2 and NO, while the reverse reaction restores the molecules of CO and NO_2. The unstable, transitory arrangement is called the **activated complex**.

Once the activated complex has formed, there are the following two possibilities. (1) The activated complex may separate into the two original molecules, CO and NO_2. This is represented by moving down the left side of the curve. (2) The activated complex may separate into CO_2 and NO molecules. The latter possibility is represented by moving down the right side of the barrier. In the flat region at the right, CO_2 and NO have separated beyond the point of having any effect on each other. The potential energy of the activated complex has become kinetic energy again.

In the event that the CO and NO_2 molecules do not have sufficient energy to attain the summit, they reach a point only part of the way up the left side of the barrier. Repelling one another, they then separate again, going downhill to the left.

The difference between the high potential energy of the activated complex and the lower potential energy of the reactants is called the "activation energy." The **activation energy** is the energy necessary to transform the reactants into the activated complex. The transformation may involve weakening or breaking chemical bonds or forcing reactants close together in opposition to repulsive forces. It may also involve storing energy in a vibrating molecule so that it reacts on collision. Raising the temperature increases the reaction rate by increasing the number of colliding particles that have sufficient energy to form the activated complex.

The magnitude of the activation energy can be calculated from certain experimental results. It is only necessary to measure the changes in reaction rate caused by changes in temperature. However, the mathematics required go beyond what may generally be assumed in a high school course. For that reason, this subject must be left for the future.

21-9 Activation Energies and Heats of Reaction

You can determine the heat of reaction from Figure 21-12. The fact that the reactants have a higher total energy than the products have means that in the course of the reaction there will be a net release of energy. The reaction is exothermic. Figure 21-12 shows that the reaction releases 54 kcal of heat per mole of carbon monoxide consumed. This, of course, is the heat of reaction. The height of the activated complex has no effect on the net heat release. To get to the activated complex, you must put in an amount of energy equal to the activation energy. However, you get it all back on the way down the other side.

Now consider the reverse reaction, between CO_2 and NO to produce CO and NO_2.

$$CO_2 + NO \longrightarrow CO + NO_2 \qquad (17)$$

Henry Eyring (1901–)

HENRY EYRING ORIGINALLY intended to be a mining engineer. He received his degree in this subject at the University of Arizona, followed by a master's degree in metallurgy. Later, his interests changed, and he took his doctorate in physical chemistry, at the University of California.

Eyring has worked on an amazing variety of research problems. He applied quantum mechanics to chemical problems, and used the science of thermodynamics to explain differences in reaction rates. He developed theories of optical activity, worked in the field of mass spectroscopy, and studied the addition of dipoles and bond lengths in flexible polymers. He even probed into the mysteries of bioluminescence, the emission of visible light by living organisms such as fireflies.

In his work with reaction rates, Eyring proposed that most chemical reactions would not take place unless enough energy was put into the system to form an "activated complex" as an intermediate step in the reaction. From Eyring's theories it seemed like a simple problem to calculate the energy needed to form the activated complex and to make sure that the colliding molecules had enough energy to form the complex. In many cases in which the colliding molecules possessed the calculated energy, however, they failed to react as predicted. This led Eyring to work out the details of the *theory of absolute reaction rates* based on such fundamental physical properties as the dimensions, vibration frequencies, and masses of the reacting molecules.

Henry Eyring has taught and carried out research at a variety of institutions, including the Universities of Wisconsin and of Berlin, the University of California at Berkeley, Princeton University, and the University of Utah, where he became Dean of the Graduate School in 1946. A prolific writer, he has published over 375 scientific papers and co-authored five books. He was awarded the Gilbert N. Lewis Award by the American Chemical Society in 1963 for his many achievements in the field of chemistry.

Notice that the reaction begins at the lower energy shown on the *right* side of Figure 21-12 and ends at the higher energy shown on the *left* side of that figure. The difference in energy is the heat of this reaction. It is equal in magnitude to that of the reverse reaction, but is opposite in sign. This is an endothermic reaction. It absorbs 54 kcal of heat per mole of carbon monoxide produced.

Figure 12-12 contains one other interesting piece of information concerning the rate of the reverse reaction (17). The energy barrier confronting the colliding molecules of CO_2 and NO controls the reaction rate. You can see from the figure that the activation energy for this reaction is higher than that for the reaction studied earlier by exactly the heat of reaction. Therefore, at the same partial pressures and at any given temperature, the reaction between CO_2 and NO will be slower than that between CO and NO_2.

The relationship between activation energies for the forward and reverse reactions can be expressed mathematically. Activation energy is denoted by the symbol $\Delta H^{\ddagger}$ (read "delta-H-cross") with a subscript. The subscript R indicates that the reaction proceeds to the right, while the subscript L indicates that the reaction proceeds to the left. The heat of the reaction proceeding from left to right is denoted ΔH_R. Mathematically, ΔH_R is written as follows.

$$\Delta H_R = \Delta H_R^{\ddagger} - \Delta H_L^{\ddagger} \qquad (18)$$

As explained in Chapter 20, reactions that absorb heat are called endothermic. Their ΔH values are positive. Reactions that release heat are exothermic. Their ΔH's are negative. The activation process must always absorb energy, so all $\Delta H^{\ddagger}$'s are positive. The general equation (18) can be applied to this particular reaction in the following way. You can find ΔH_R for the reaction from left to right by direct measurement. As noted earlier it is -54 kcal/mole for this exothermic reaction. You can find $\Delta H_R^{\ddagger}$ by measuring how the reaction rate varies with temperature. Such a procedure yields a value of 32 kcal/mole. Using equation (18), you can write the following.

$$-54 \text{ kcal/mole} = 32 \text{ kcal/mole} - \Delta H_L^{\ddagger}$$

This becomes:

$$\Delta H_L^{\ddagger} = +86 \text{ kcal/mole}.$$

This is in accord with the way the curve in Figure 21-12 is drawn.

The relationships discussed so far can be summarized as follows.

$$CO + NO_2 \longrightarrow CO_2 + NO$$

$$\Delta H_R^{\ddagger} = +32 \text{ kcal/mole}$$
$$\Delta H_L^{\ddagger} = +86 \text{ kcal/mole}$$
$$\Delta H_R = -54 \text{ kcal/mole}$$

You can see that

$$-54 = 32 - 86$$

is in accordance with equation (18). This equation is important because it implies that it is necessary to know only two of the three quantities, $\Delta H_R^{\ddagger}$, $\Delta H_L^{\ddagger}$, and ΔH_R. The third can be calculated.

It is possible to show a potential energy diagram in terms of the reverse reaction, if desired. Such a diagram is shown in Figure 21-13. It is typical of an endothermic reaction. Note that it is simply the reverse of Figure 21-12.

21-10 Catalysts

Many reactions proceed quite slowly when the reactants are mixed. However, they can be made to take place much more rapidly by

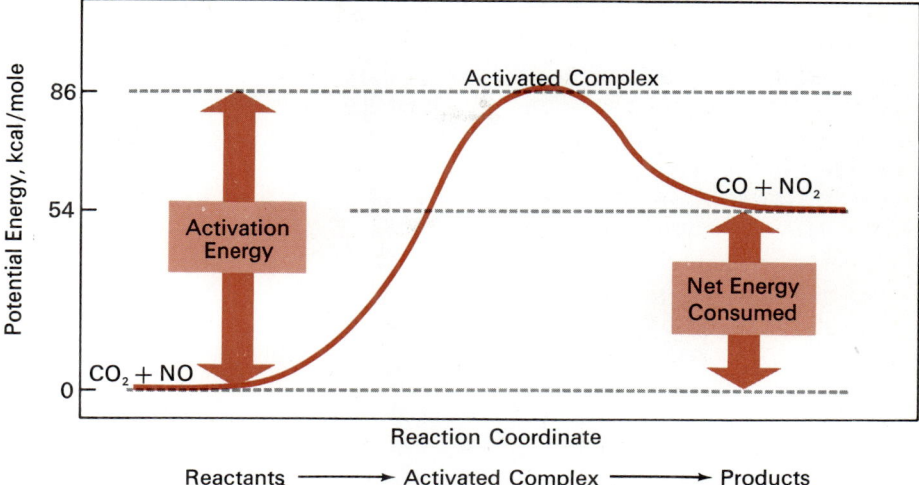

Figure 21-13 *In an endothermic reaction the potential energy of the products is higher than that of the reactants, while that of the activated complex is still higher.*

the introduction of other substances. As you may recall, these substances are called catalysts. (See Section 19-12.) Catalysts are not used up in a reaction.

The process of increasing a reaction rate through the use of a catalyst is referred to as **catalysis**. In Chapter 19 you learned about the use of a catalyst to accelerate the production of SO_3 from SO_2 in the manufacture of sulfuric acid.

The action of a catalyst can be explained in terms of the cross-country ski race. Suppose that the particular race route that you considered before is the most difficult one. This is the one used for Olympic-class competition. There is actually another way to get across the big ridge without having to climb quite so high. This other pathway, which is shown along with the first one in Figure 21-14 (next page), might be used in races at a lower level of competition. People with less energy could then make it over the big ridge by using the lower pass.

What would you expect to happen if the top skiers were allowed to use the second route, with the easier passage over the big ridge? More of them would make it over and finish the race. Instead of only the 60 percent who could finish the tougher route, 80 percent might finish the easier one.

A two-dimensional representation showing both cross-country routes is given in Figure 21-15 (next page). The crucial difference between the routes is clearly the altitude at which the high ridge must be crossed. Figure 21-16 (page 633) is an analogous representation for a chemical reaction. The higher curve shows the activation energy barrier that must be overcome for a reaction to take place. When a catalyst is added, there is a new reaction path, with a different activation energy barrier, as indicated by the lower curve.

The new reaction path corresponds to a new reaction mechanism. This mechanism permits the reaction to occur via a different

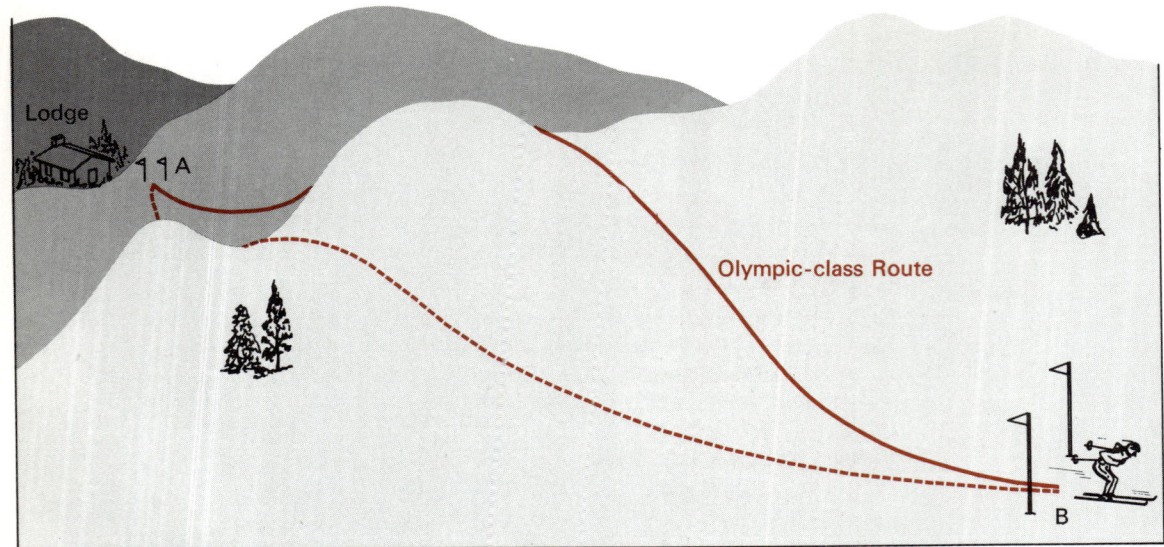

Figure 21–14 *A skier who does not have enough energy to surmount the ridge on the Olympic-class route can still get from A to B by using the alternate route. This route requires less energy, since it crosses the ridge at a lower point.*

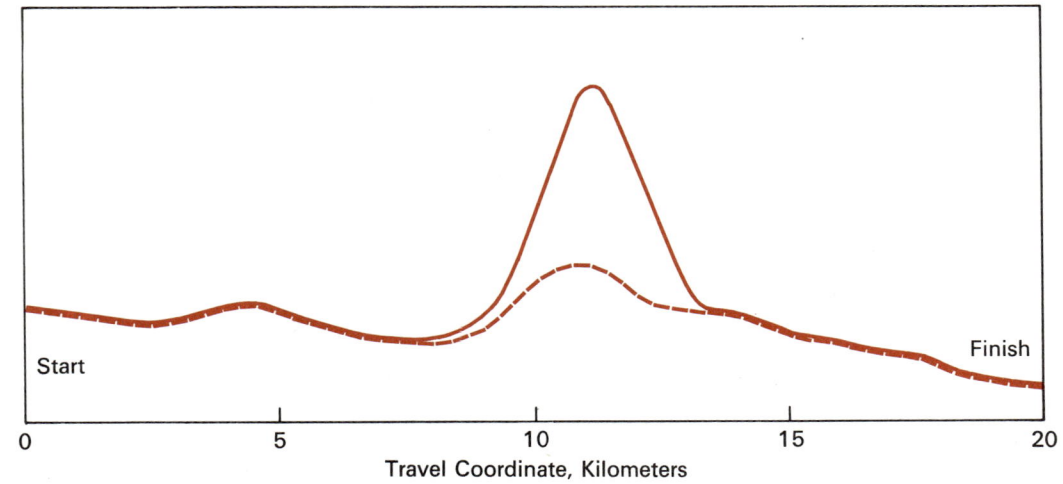

Figure 21–15 *The easier route does not require the skier to go over such a high point on the ridge. This is shown in the above diagram, which is a plot of altitude versus distance traveled, for the two routes.*

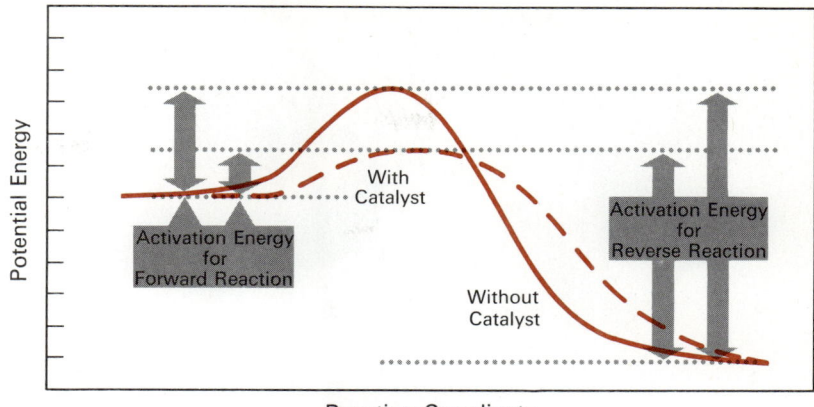

Figure 21-16 *The presence of a catalyst in a chemical reaction lowers the activation energy of the reaction in some way, thus reducing the kinetic energy necessary to complete the reaction.*

activated complex. Hence, more particles can get over the new, lower energy barrier, and the rate of reaction is increased. Note that the activation energy for the reverse reaction is lowered exactly the same amount as for the forward reaction. This accounts for the experimental fact that a catalyst for a reaction has an equal effect on the reverse reaction. That is, both reactions are speeded up by the same factor. If a catalyst doubles the rate in one direction, it also doubles the rate in the reverse direction.

In all cases of catalysis, the catalyst acts by inserting intermediate steps in a reaction. These steps would not occur without the catalyst. The catalyst itself must be regenerated in a subsequent step. An example is provided by the catalytic action of another acid on the decomposition of formic acid, HCOOH. (This effect is illustrated in Figure 23-1.)

The carbon atom in formic acid has attached to it a hydrogen atom, an oxygen atom, and an —OH group. Figure 21-17, on the next page, shows how this molecule might decompose. If the hydrogen atom attached to carbon migrates over to the —OH group, the carbon-oxygen bond can break to give a molecule of water and a molecule of carbon monoxide. This migration, which is shown in the center of the figure, requires a large amount of energy. This means there is a high activation energy, and the reaction occurs very slowly.

$$HCOOH \longrightarrow H_2O + CO \qquad (19)$$

If sulfuric acid, H_2SO_4, is added to an aqueous solution of formic acid, carbon monoxide bubbles out rapidly. This also occurs if phosphoric acid, H_3PO_4, is added instead. The common factor is that both of these acids release hydrogen ions, H^+. Yet careful analysis shows that the concentration of hydrogen ion is constant during the rapid decomposition of formic acid. It is not used up. Evidently, hydrogen ion acts as a catalyst in the decomposition of formic acid.

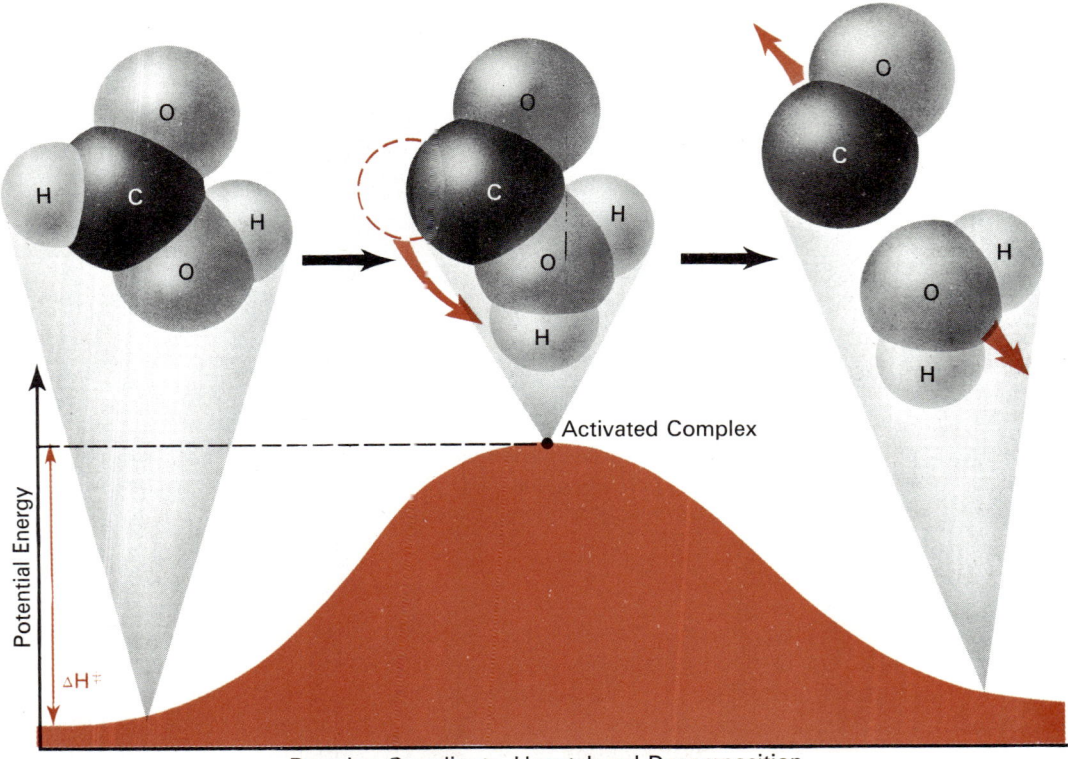

Figure 21–17 In the uncatalyzed decomposition of formic acid a large amount of energy is required for the hydrogen atom to migrate to the OH group, producing water and carbon monoxide.

Chemists have a rather clear picture of how H^+ catalyzes reaction (19). The availability of H^+ in the solution makes a new reaction mechanism available. The new mechanism begins with the addition of H^+ to formic acid, as shown in Figure 21-18. Thus, the catalyst is consumed at first, forming a new species, $(HCOOH_2)^+$. In this species one of the carbon-oxygen bonds is weakened. With only a small input of energy, the next reaction shown in Figure 21-18 can occur, producing $(HCO)^+$ and H_2O. Finally, $(HCO)^+$ decomposes to produce carbon monoxide, CO, and H^+. This last reaction of the sequence regenerates the catalyst, H^+.

Each of the steps in this new reaction mechanism is governed by the same principles that govern a simple reaction. Each reaction has an activation energy. The overall reaction has a potential energy diagram that is merely a composite of the simple energy curves of the succeeding steps.

The highest energy required in this new reaction path is only 18 kcal, much lower than

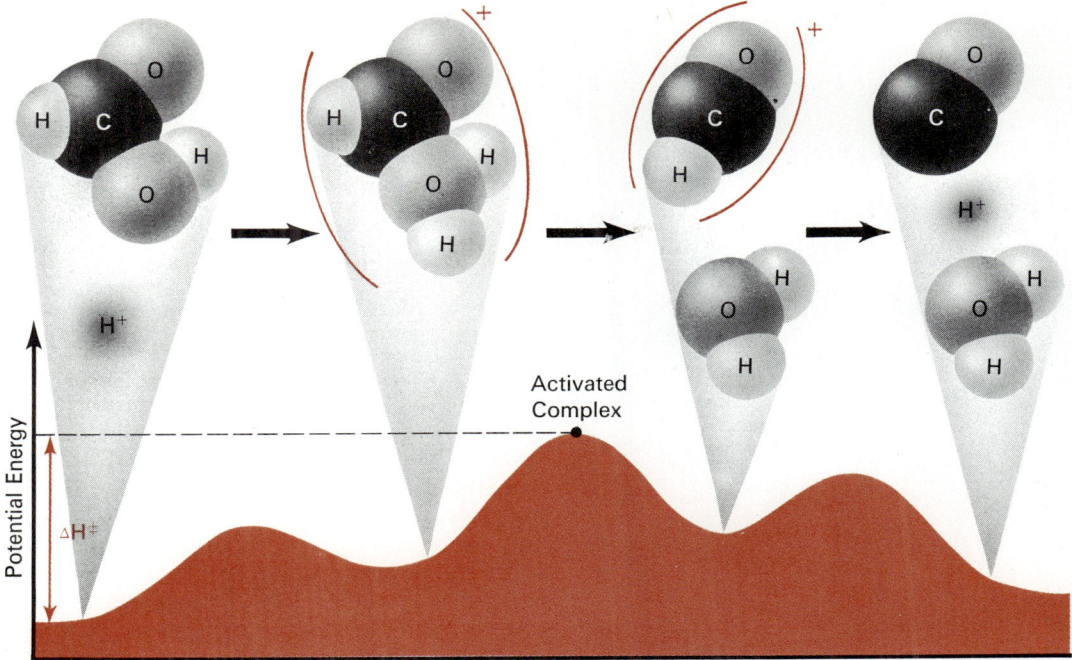

Figure 21-18 *The catalyzed reaction has a lower activation energy than the uncatalyzed one, since less energy is needed to produce the activated complex (which contains the catalyst, H^+).*

the activation energy shown in Figure 21-17 for the uncatalyzed reaction. Hence the rate of decomposition is faster when acid is present.

Notice that catalytic action does not *cause* the reaction. A catalyst simply speeds up a reaction that might take place in its absence.

In some cases the catalyst is a solid substance that can hold, or adsorb, reactant molecules on its surface. Two adsorbed molecules can make contact with each other more readily than molecules that are not adsorbed. In reactions involving gases, metals such as iron, nickel, platinum, and palladium act in this way. There is evidence that in some cases of adsorption, bonds of reactant particles are weakened or broken. Such action also helps the reaction to occur.

There is a very large group of catalysts called enzymes. An enzyme is a protein molecule that catalyzes reactions in living cells. (Proteins are discussed in Section 18-5). Among the best known examples of these are the digestive enzymes, such as ptyalin in saliva and pepsin in gastric juice. A common function of these two enzymes is to hasten the breakdown of large molecules into simpler molecules that can be utilized by body cells. In addition to these digestive enzymes, many other types of enzymes are

Edward Calvin Kendall
(1886–1972)

AFTER RECEIVING HIS DOCTORATE from Columbia University in 1910, Edward Kendall became interested in the thyroid gland. He suspected that this gland must produce a hormone that plays an important part in human growth and development. He decided to try to isolate and identify this substance.

Starting with thyroglobulin, a complex compound secreted by the thyroid gland, Kendall found that this material contained some protein and a functionally-active part, which he called thyroxine. Thyroxine proved to be one of the naturally-occurring amino acids, with four iodine atoms in its molecule. Further research proved that the hormone became increasingly inactive with the removal of one or more of the iodine atoms. The discovery of the function and structure of thyroxine made it possible to treat people suffering from damaged or poorly functioning thyroid glands.

Kendall then turned his attention to the adrenal glands, one of which is mounted on top of each of the two kidneys in the human body. An adrenal gland consists of an outer portion, or cortex, and an inner core. Each is actually a separate gland, producing different hormones. The inner core was shown to be relatively simple, producing only the hormone adrenalin. Adrenalin sets off a variety of alarm reactions when a person is subjected to great emotional stress.

The adrenal cortex, however, is far more complex. In the 1930's, Kendall discovered at least 28 different hormones produced by the cortex. Only four of these compounds showed significant physiological activity. One of these is the hormone known as cortisone, widely used for treating patients with rheumatoid arthritis. Since these compounds are produced in such minute quantities in animals, it has been estimated that 14,600 oxen would be required to produce enough cortisone to treat one patient for one year. Fortunately, Percy Julian's discovery of an inexpensive way to synthesize cortisone, in 1944, made cortisone therapy possible for many persons.

In 1950, Edward Kendall shared the Nobel Prize in Medicine and Physiology with P. S. Hench and T. Reichstein for their work in isolating and identifying a variety of hormones from the adrenal cortex.

found in living cells. Enzymes are discussed in more detail in the next section.

21-11 Enzymes, the Catalysts in Living Things ★

The processes of life involve thousands of chemical reactions. For example, both the replacement of cells and the digestion of food involve chemical reactions. Each of the individual chemical steps required to carry out these processes must take place smoothly, at the right time, at the right place, and at the temperature of the tissue in which it plays its role. Most of these reactions would not occur rapidly enough if the reactants were merely mixed (in a test tube, for example) at the temperature of the organism. The assistance of catalysts is required in almost every case. Thus all complex organisms, including you, contain thousands of these catalysts, known as **enzymes**.

Some of the major types of reactions catalyzed by enzymes follow.

1. *Hydrolysis of proteins*. Enzymes that catalyze this type of reaction are called *peptidases*. As you know, proteins consist of long chains of amino acids connected by amide linkages. These are called peptide chains. The hydrolysis reaction is represented generally by the following reaction.

$$\underset{\underset{H_2O}{\uparrow}}{-\overset{H}{\underset{H}{N}}-\overset{R}{\underset{}{C}}-\overset{O}{\underset{}{\overset{\|}{C}}}-\overset{H}{\underset{}{N}}-\overset{R'}{\underset{}{C}}-\overset{O}{\underset{}{\overset{\|}{C}}}-} \longrightarrow$$

$$-\overset{H}{\underset{H}{N}}-\overset{R}{\underset{}{C}}-\overset{O}{\underset{OH}{\overset{\|}{C}}} \; + \; H-\overset{H}{\underset{}{N}}-\overset{R'}{\underset{}{C}}-\overset{O}{\underset{}{\overset{\|}{C}}}-$$

Some enzymes, such as chymotrypsin, which is found in the pancreas, act on a wide variety of amide linkages. Chymotrypsin acts particularly on those linkages that are not near the ends of the protein chain. Other peptidases are specific for the last amide bond at either the carboxyl end of the chain (carboxypeptidases) or the amino end of the chain (aminopeptidases).

2. *Hydrolysis of esters*. Enzymes with this function are called *esterases*. The type of reaction they catalyze follows.

$$R-\overset{O}{\overset{\|}{C}}-O-R' \underset{\underset{H_2O}{\uparrow}}{\longrightarrow} R-\overset{O}{\overset{\|}{C}}-OH + H-O-R'$$

This reaction is similar in certain ways to the hydrolysis of amide links in proteins. It is not surprising then that some peptidases can also function as esterases.

3. *Reactions forming amide, ester, or other linkages*. Enzymes that catalyze such reactions are called *synthetases*. Some of these reactions are simply the reverse of the hydrolyses just mentioned. As the body digests food, peptidases and esterases are necessary to break down proteins and fats. However, in the building of new tissues, the synthetases are needed to form new protein and fat molecules from the components obtained by digestion.

4. *Transfer of functional groups*. Enzymes that assist in such reactions are called *transferases*. An important type of transfer reaction is the following, called a transamination.

$$R-\overset{NH_2}{\underset{H}{\overset{|}{C}}}-COOH \; + \; \overset{H}{\underset{O}{\diagdown}}C-R' \longrightarrow$$

$$R-\overset{O}{\overset{\|}{C}}-COOH + H_2NCH_2R'$$

There are many other types of enzyme-catalyzed reactions, which we are not able to discuss here. One more may be mentioned, however, because of its importance and simplicity. The following reaction is very fundamental in respiration.

$$CO_2 + H_2O = H_2CO_3$$

Left to itself, this reaction proceeds relatively slowly, in either direction. In order for respiration to take place satisfactorily, it is necessary for release of carbon dioxide (the reaction from right to left, as written) to be speeded up. There is an enzyme to do this, called *carbonic anhydrase*.

Now that you have a general idea about what enzymes do and how important they are, you might naturally ask, "What are they?" The answer, in a general way, is simple. All enzymes are protein molecules. Some are relatively small, with molar masses as low as about 10,000. Most are far larger, ranging up to molar masses of 50,000, 100,000, and even several million. Moreover, enzymes are proteins of the type called *globular*. This means that the long protein chain is coiled up to form a ball—a globule. An example is shown in Figure 21-19a. This is a space-filling model of an enzyme called *lysozyme*. Its function is to catalyze the hydrolytic breakdown of certain compounds that occur in cell walls. The first enzyme structure to be determined in detail was that of lysozyme. In the past few years, the structures of about a dozen more have been worked out. All of them are more or less globular.

How do enzymes work? What is it about these very large, complex molecules that gives them the ability to catalyze reactions? These questions still cannot be answered in complete atomic detail, and undoubtedly the details differ for each case. Still, in more general terms, chemists have learned a great deal about how enzymes work.

Like any catalyst, an enzyme has to do three things. (1) It must be able to combine with the substance or substances (called substrates) on which it is to act. This is called *substrate* binding. (2) The nature of its interaction with the substrate must be such that the reaction can proceed through a pathway with a lower activation energy than would otherwise be the case. In other words, the enzyme must make the activated complex easier to form. (3) The enzyme must bind the reaction product or products relatively weakly. The products must be released quickly so that the next substrate can be bound and activated.

Enzyme molecules must be large and complex to enable them to provide a place, called the *active site*, at which the substrate can be closely and accurately held and activated. The exact shape of the active site region of the enzyme must be designed to match, in a complementary way, the shape of the substrate molecule. Figure 21-19b shows how neatly the substrate on which lysozyme operates fits the active site part of the lysozyme surface.

From the above discussion, you can see that an enzyme can bind only a specific molecule, or a few very similar ones. This property of enzymes is called *specificity*. Specificity is important for two main reasons. First, if a particular enzyme is required to catalyze the reaction of only one molecule, or a few very similar ones, it can be highly specialized. It can therefore be extremely efficient. The enzyme is like a very complex precision tool ideally suited for one job and one job only. Second, this property helps preserve order in the operations of cell chemistry. To understand this, consider what would happen if one enzyme could catalyze the hydrolysis of all protein linkages. Chemical

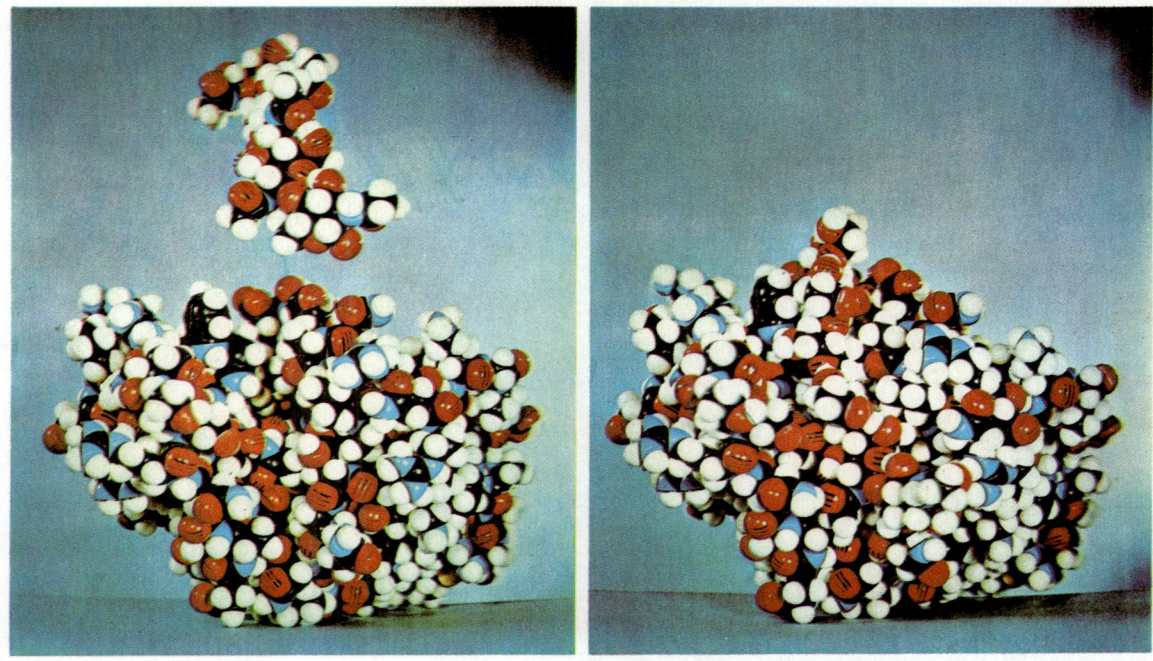

Figure 21-19 (a) *A space-filling model of the lysozyme molecule, together with a smaller substrate molecule with which the enzyme can interact.* (b) *During catalysis the substrate fits neatly into the active site part of the lysozyme surface.*

chaos would result, because the enzyme would go around catalyzing the destruction of other enzymes (which are proteins). It would also destroy the protein tissues making up parts of the digestive tract in which the enzyme must function.

The enormous range of enzyme structures and the operation of many thousands of enzymes in balance and harmony constitute a chemical "miracle" that scientists are trying to understand in more and more detail. The investigation of enzymes, their structures, and their activities constitutes one of the most exciting areas of research on the borderline between chemistry and biology today.

Summary

At constant temperature, the molecules in a sample of gas have different velocities and kinetic energies. As the temperature is raised, the fraction of molecules with high kinetic energy increases.

Only those collisions between molecules in which the total energy exceeds a certain threshold value can lead to a reaction. The threshold value is called the activation energy. This energy is required for the reacting molecules to combine and form the activated complex. The complex may then decompose to give the products or the reactants.

A catalyst makes a reaction go faster by providing a different pathway along which the activation energy is less than that in the uncatalyzed reaction. Hence, at any given temperature, there will be more molecules with enough energy to follow the catalyzed path than with the energy necessary to follow the uncatalyzed one.

Self Check

1. Do you expect reaction (17) to be faster or slower than reaction (19)? Explain why.
2. Describe what happens in transforming reactants into the activated complex.

Chapter Summary

Chemical reactions proceed at different rates, from the undetectably slow to the explosively fast, depending upon a number of factors. One factor is the particular substances that are reacting. Different reactions have different tendencies to occur. Thus they have inherently different rates even when factors such as temperature and concentration are as similar as possible. A second factor is the concentration (or partial pressure). For a particular reaction the rate will often, though not necessarily, be increased when the concentration of one of the reacting species is increased. A third factor is temperature. For virtually all reactions, the rate will increase with increasing temperature.

The collision theory of reactions explains the role of these factors. According to this theory, molecules must collide to react. Clearly, the more often they collide, the more often reaction will occur. This explains why increasing concentration or partial pressure of a reactant may increase the rate. Examination of the reaction mechanism may explain why an increase in concentration sometimes does not increase the rate, or increases it less than might be expected. Since collisions between more than two molecules at once are very rare, reactions in which many molecules are involved must occur by a series of subreactions. The rate of the entire reaction is governed by the rate of the slowest step in the mechanism. This step is called the rate-determining step. Only the concentrations of molecules that collide in this step affect the rate of the reaction.

The effect of temperature is explained by postulating that two molecules must collide with enough energy to form an activated complex. The complex can separate into the reactants or the products. The minimum energy required to form the activated complex is called the activation energy. Of the many collisions that occur, only a small fraction have energy equal to or greater than the activation energy. Hence, only a fraction of the collisions leads to reaction products. As the temperature is raised, the fraction of molecules with kinetic energy above a particular high value will increase. Therefore, the fraction of collisions energetic enough to produce the activated complex increases and the reaction goes faster.

A catalyst is a substance that is not consumed in the reaction. It makes possible a path or mechanism in which the highest activation energy required is lower than that in the uncatalyzed reaction. Hence, a catalyst increases the rate of a reaction at a given temperature.

Questions and Problems

1. The rate of movement of an automobile can be expressed in the units *miles per hour*. Describe various ways in which the rate of a chemical reaction can be expressed.
2. Consider two gases, A and B, in a container at room temperature. These two gases can undergo the following reaction.

$$A + B = C + D$$

   State and explain the effect on the reaction rate between these two gases if
   (a) the pressure is doubled.
   (b) the number of molecules of gas A is doubled.
   (c) the temperature is decreased at constant volume.
3. How does a homogeneous system differ from a heterogeneous one? Describe two situations in which reacting substances are mixed in a heterogeneous system. Do the same for two reacting substances mixed in a homogeneous system. Use systems different from those described in the book.
4. Which of the two homogeneous systems you described in Question 3 has the greater reaction rate? Which of the two heterogeneous systems will react faster? Explain your answers. Assume similar conditions within each pair.
5. Why does a log burn slowly in air, while sawdust made from the same tree burns explosively if scattered throughout the room and ignited?
6. In a collision of particles, what is the primary factor that determines whether a reaction will occur?
7. Based on what you have learned about reactions, reaction rates, and the collision theory, label the following statements true or false and explain your choice.
   (a) When two gases are mixed in a closed container, decreasing the volume of the container generally reduces the reaction rate.
   (b) A change in the concentration of only one of the reactants will increase the reaction rate of the gases in (a).
   (c) An increase in the number of collisions between molecules of reactants automatically increases reaction rate.
8. State three methods by which the pressure of a gaseous system can be increased.
9. Give several ways of increasing the rate of combustion in a candle flame. Explain why each way works.
10. Use the "threshold energy" concept to describe the life and death of an ordinary drinking glass. (Assume that it is dropped from time to time.)
11. What is the relationship between the rate at which a reaction proceeds at room temperature and the number of bond rearrangements involved in the reaction?

12. In your own words, describe what is meant by a reaction mechanism. Which step in a reaction mechanism is the rate-determining step?
13. A group of students is about to prepare 50 copies of a 10-page directory. The pages have been printed and are stacked in 10 piles, page by page. The pages must be (a) assembled in numerical order, (b) straightened into a neat pile and (c) stapled in sets. If three students work together, each executing a different operation, which would probably be the rate-controlling step? What would be the effect on the overall rate if the student doing the first step had ten new helpers? What if these ten helpers joined the student working on the second step? the third step?
14. Explain in terms of molecular collisions why an increase in the concentration of a reactant may increase the rate of reaction.
15. Methane reacts very slowly with oxygen. Increasing the concentration of methane does not speed up the reaction appreciably. Why? However, if a spark is provided in a methane-oxygen atmosphere, the greater the concentrations of reactants, the more rapid the reaction. Explain.
16. Do you expect the equation for the overall reaction

$$C_2H_4(g) + 3O_2(g) \longrightarrow 2CO_2(g) + 2H_2O(g)$$

to represent the mechanism by which ethylene, C_2H_4, burns? Explain your answer.
17. An increase in temperature of 10°C rarely doubles the average kinetic energy of particles. Therefore the number of collisions is not doubled. Yet this temperature increase may double the rate of a slow reaction. Explain.
18. In Figure 21-12, why is kinetic energy decreasing as NO_2 and CO go up the left side of the barrier? Why is kinetic energy increasing as CO_2 and NO go down the right side? Explain in terms of conservation of energy and in terms of what is happening to the various particles in relation to each other.
19. Considering that so little energy is required to convert graphite to diamond, why is the industrial production of diamonds so difficult?
20. Phosphorus, P_4, exposed to air burns spontaneously to give P_4O_{10}. ΔH for this reaction is -712 kilocalories per mole of P_4.
    (a) Draw an energy diagram for the net reaction. Explain the critical parts of the curve.
    (b) How much heat is produced when 12.4 grams of phosphorus burn?
21. Explain why a burning match will light a candle.
22. Draw an energy diagram for the burning of carbon when C is in the form of large chunks of coal.

$$C(s) + O_2(g) \longrightarrow CO_2(g)$$

Would the curve change if very fine carbon powder were used rather than chunks of coal? Explain.
23. Sketch a potential energy diagram to represent an endothermic reaction. Label parts of the curve representing the activated complex, the activation energy, and the net energy absorbed.
24. Give two factors that increase the rate of a reaction and explain why they cause the rate to increase.

25. In the Haber process, an important industrial method for manufacturing ammonia, the overall reaction is as follows.

$$N_2(g) + 3H_2(g) \longrightarrow 2NH_3(g) + 24 \text{ kcal}$$

A yield of approximately 98% can be obtained at 200°C and 1000 atm. The process makes use of a catalyst. The catalyst usually consists of finely divided mixed iron oxides containing small amounts of potassium oxide, K_2O, and aluminum oxide, Al_2O_3.
  (a) Is this reaction exothermic or endothermic?
  (b) Suggest a reason why this reaction is generally carried out at a temperature of 500°C and 350 atm although the yield under these circumstances is only about 30%.
  (c) What is ΔH for the reaction, in kcal/mole of $NH_3(g)$?
  (d) How many grams of hydrogen must react to form 1.60 grams of ammonia?
26. What is a catalyst? Describe the way in which catalysts function in a chemical reaction.
27. If a catalyst were added to a reacting system, what effect, if any, would this have on the kinetic energy distribution curve and on the threshold energy for the reaction?

22 Chemical Equilibrium

Opposing forces and opposing motions, illustrated by the motorcyclists, give an apparent condition of no change. In chemical processes, particles interact in a similar fashion to produce what chemists call a state of equilibrium.

In ordinary conversation the word equilibrium suggests a stable, or at least a steady, condition produced by balancing the forces that tend to act in different directions. Equilibrium has the same general meaning in chemistry, but there are some important distinctions.

So far it has been assumed that chemical reactions go to completion. That is, the proper proportion of reactants undergo reaction, leaving no trace of reactants and producing the full amount of each product indicated by the equation. While this might *appear* to be the case in many reactions, it is never actually true. Always, a state of equilibrium is reached in which some quantity (perhaps exceedingly small) of each of the reactants and products is still present. Although the reactants yield the products, to some extent the "products" react with one another to restore the original "reactants." What causes this "incompleteness" of all reactions?

Examining Some Qualitative Aspects of Equilibrium

22-1 Equilibrium in Chemical Reactions

In Chapter 21 you studied the rate of the reaction between CO and NO_2.

$$CO(g) + NO_2(g) \longrightarrow CO_2(g) + NO(g) \quad (1)$$

You also studied the rate of the reverse of reaction (1). This is the reaction between CO_2 and NO.

$$CO_2(g) + NO(g) \longrightarrow CO(g) + NO_2(g) \quad (2)$$

This raises the question: How does a reaction go both ways?

To begin with, it must be emphasized that this question must be discussed because this reaction *does* go both ways. If you mix $CO(g)$ and $NO_2(g)$, reaction (1) begins. As soon as it does, $CO_2(g)$ and $NO(g)$ form. As these products accumulate, reaction (2) begins, undoing reaction (1). The two reactions oppose each other. Which will win out?

By direct observation of the reddish-brown color of NO_2, you can see the progress of reaction (1). You can also carry out reaction (2), beginning with pure CO_2 and NO. In either case, the intensity of color due to NO_2 will at first change very noticeably. It fades in reaction (1) and builds up in reaction (2). However, after a time the color will stop changing. When this happens, the reaction is said to have reached a *state of equilibrium*.

22-2 Equilibrium in a Phase Change

The preceding example illustrates equilibrium in a chemical reaction. Another type of equilibrium important in chemistry is the type involved in phase changes. The equilibrium between a liquid and its vapor is an important example.

Consider what happens when you pour some liquid water at 20°C into a flask and then seal the flask. You find that some water molecules leave the liquid and enter the gas phase. In other words, they become water vapor. The partial pressure of water vapor in the flask rises. When it reaches 17.5 Torr, you observe no more change. The amount of excess liquid remains constant thereafter. Also, the partial pressure of water vapor in the flask remains at 17.5 Torr as long as the temperature is maintained at 20°C. This partial pressure is called the *vapor pressure* of water at 20°C. At this vapor pressure, liquid water and water vapor can coexist indefinitely at 20°C. This vapor pressure, 17.5 Torr, is the same whether air is present or not. It is a property of water. If a vacuum had existed within the flask, liquid would still have evaporated until the pressure rose from 0 Torr to 17.5 Torr. If the flask had originally contained dry air at a pressure of 750 Torr, liquid would have evaporated until the pressure rose from 750 Torr to 767.5 Torr. By subtracting the first pressure from the second, you can see that the partial pressure of water vapor again rises from 0 Torr to 17.5 Torr.

When the liquid is in contact with its vapor at the vapor pressure, the liquid and gas are

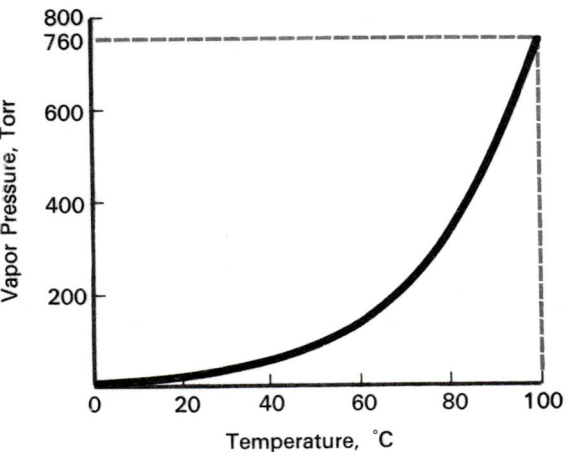

Figure 22–1 *The vapor pressure over liquid water rises with temperature. The shape of the curve is typical for all liquids.*

 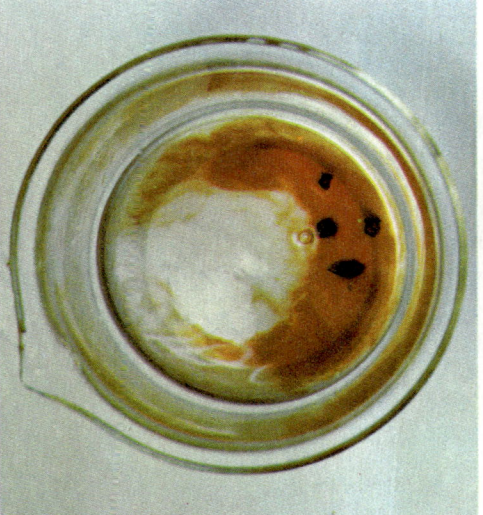 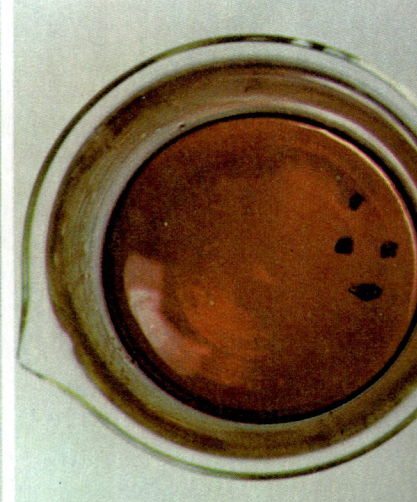

Figure 22–2 *(a) Iodine crystals have just been added to a water-alcohol solution. (b) After a time, some of the iodine has gone into solution. (c) Eventually no more iodine goes into solution, equilibrium having been attained.*

said to be at equilibrium. At equilibrium, no measurable changes are taking place.

The vapor pressure of water at 20°C is 17.5 Torr. At 40°C, the vapor pressure is 55.3 Torr; at 60°C, it is 149.4 Torr. These numbers indicate that the vapor pressure of water increases with increasing temperature. This is shown graphically in Figure 22-1.

Ethyl alcohol is also a liquid at room temperature. Its vapor pressure at 20°C is 44 Torr, higher than the vapor pressure of water at the same temperature. At 40°C, ethyl alcohol has a vapor pressure of 134 Torr. At 60°C, the vapor pressure is 352 Torr. Again the vapor pressure increases with increasing temperature. This is always so. *The vapor pressure of every liquid increases as the temperature is raised.*

22-3 The Boiling Point

At any temperature, molecules can pass from the liquid phase into the gas phase by leaving the surface of the liquid. What do you suppose happens when the pressure created by the vapor becomes so high that it is equal to the pressure exerted on the liquid by the atmosphere? In other words, what happens when the vapor pressure of the liquid can push as hard as the atmosphere pushes? The answer is that vapor forms right *in* the liquid, creating bubbles, and not just at the surface. When the vapor pressure of the liquid becomes large enough for this to happen, the liquid boils. The temperature at which the vapor pressure equals the atmospheric pressure is called the **boiling point.**

As you can see, the boiling point is fixed by the surrounding pressure. For example, if the surrounding pressure is 760 Torr, water boils at 100°C. This is the temperature at which the vapor pressure of water is 760 Torr, as seen in Figure 22-1. Ethyl alcohol achieves a vapor pressure of 760 Torr at 78.5°C. When the atmospheric pressure is 760 Torr, ethyl alcohol boils at 78.5°C. Thus the normal boiling points of water and ethyl alcohol are 100°C and 78.5°C, respectively. The **normal boiling point** of a liquid is the temperature at which the vapor pressure of that liquid is exactly one atmosphere, 760 Torr.

Suppose that the atmospheric pressure drops to 750 Torr, as it might just before a storm. Then bubbles of vapor can form anywhere in liquid water at a temperature of 99.6°C, since the vapor pressure of water is 750 Torr at 99.6°C. Thus water boils at 99.6°C when the surrounding pressure is 750 Torr.

22-4 Solubility Equilibrium

Figure 22-2 shows the addition of solid iodine to a mixture of water and alcohol. At first the liquid is colorless, but very quickly a reddish color appears near the solid. Stirring the liquid causes swirls of the reddish color to move out. The movement of the color indicates that the solid iodine is dissolving to become part of the liquid. As time passes, the liquid takes on an increasing color, and the pieces of solid iodine diminish in size. Finally, however, the color stops changing. The solid is still present, but the pieces of iodine no longer diminish in size. Since no more evidence of change can be detected, equilibrium is said to exist.

Many other solids will dissolve in water, or other liquids, until a state of equilibrium is reached. Equilibrium is attained when the dissolving of the solid ceases

22-5 Recognizing Equilibrium

Equilibrium situations such as those discussed in the preceding sections raise an interesting question. How do you recognize equilibrium?

Norbert Rillieux
(1806–1894)

NORBERT RILLIEUX was the son of Vincent Rillieux, a French sugar plantation owner in New Orleans, and his slave mistress, Vonstance Vivant. His father and mother lived together as man and wife and raised him as a free man with all the privileges accorded to a son of a wealthy citizen. He was educated both by private tutors and in local Catholic schools.

During his youth Rillieux was impressed by the frightfully hard work and inefficient methods used for extracting the juice from sugar cane and converting it into sugar. The juice from the crushed cane was heated in series of kettles and pans frequently referred to as the "Jamaica Train." Large numbers of slaves dipped the juice from one container to another with ladles as it became progressively thicker, until crystals formed in the last pan. The process was tedious and dangerous. Many slaves were severely burned when the hot liquid spilled on them. The final product was nothing like the pure white crystalline sugar to which we are accustomed.

When Norbert Rillieux was ready for college, he wanted to study engineering, but no college in the United States would admit him. As a result, his father sent him to L'Ecole Central in France, where he graduated in 1830. He was soon appointed as the youngest teacher ever to be employed by that school.

Rillieux's early work in engineering was focused mainly on improving the efficiency of steam engines. He knew that liquids boil at lower temperatures if the pressure on them is reduced. Remembering the inefficient process for making sugar, he thought of the great improvement that would result if the sugar juices could be boiled in a closed container under a partial vacuum. After several years' work, Rillieux invented the "multiple effect vacuum evaporator," which used the steam from one container to heat the contents of each container next in line.

Returning to the United States, Rillieux patented his invention in 1843. With the help and encouragement of several plantation owners, he was finally able to construct a dependable evaporator, and in 1846 he was granted another patent on his improved model.

Rillieux's evaporator revolutionized the sugar industry because it could produce thousands of times more sugar of higher quality than the old "Jamaica Train." Although slaves no longer had to endure the hazards of the old sugar-making process, the efficiency of Rillieux' evaporator created a demand for more sugar, and, hence, for more slaves to work in the cane fields.

In 1854, Rillieux returned to Paris, taking a teaching position at his old school, L'Ecole Central. He eventually became headmaster of the school and gained an outstanding reputation in the field of engineering.

If you recall what has been said about each of the examples of equilibrium so far considered, the answer to the question can be found. In each case equilibrium was reached when there appeared to be no further change. When observable properties, such as pressure, color, and amount of solid stopped changing, equilibrium had been attained. As you may recall from Chapter 5, properties such as these, that we can observe directly with the senses, are called macroscopic properties. Thus you can say that *equilibrium is characterized by constancy of macroscopic properties*.

In order to understand this better, consider the following reaction. Calcium carbonate, $CaCO_3$, decomposes upon heating to form carbon dioxide gas, CO_2, and calcium oxide (lime), CaO.

$$CaCO_3(s) \xrightarrow[\text{temp.}]{\text{high}} CaO(s) + CO_2(g) \qquad (3)$$

Figure 22-3 shows the steps in heating solid $CaCO_3$, initially under a vacuum, to 800°C. Decomposition begins according to reaction (3). Figure 22-3b shows the gas pressure rising. The pressure continues to rise until it reaches 190 Torr, as seen in Figure 22-3c. Thereafter, no further change is evident. Since there is no more evidence of change,

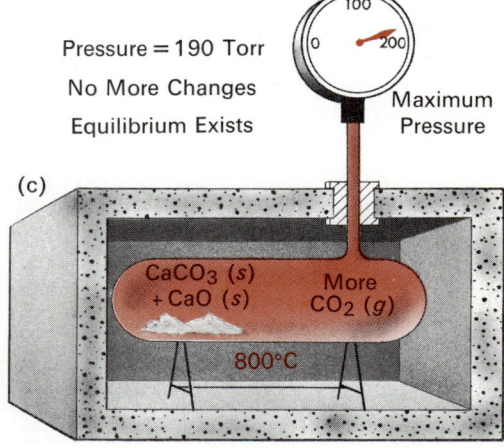

Figure 22-3 *Under a vacuum, $CaCO_3$ is heated to 800°C (a). It begins to decompose, as indicated by the rising pressure (b). After a time there is no evidence of further change (c) and equilibrium is said to exist.*

it is said that the system is at equilibrium. As before, equilibrium is characterized by constancy of macroscopic properties.

Such constancy is not the only requirement. Consider a laboratory burner flame, which can be thought of as a system. A **system** is any region and the material in it that you wish to consider. The flame system has a well-defined structure, an inner cone surrounded by a luminous region that does not appear to change. A temperature measurement made at a particular place in the flame shows that the temperature at that spot is constant. At another place in the flame the temperature might be different, but again it would be constant. It would not change with time. A measurement of the gas flow rate shows a constant movement of gas into the flame. Yet a laboratory burner flame is *not* at equilibrium, because chemical change is occurring. Methane, CH_4, and oxygen, O_2, are continuously fed into the flame, and carbon dioxide, CO_2, and water, H_2O, are continuously leaving. Substances are entering and leaving at all times. Such a system is called an open system. Furthermore, the temperature is not uniform throughout the flame system. Equilibrium can exist only in a *closed system* at a *uniform temperature*. A **closed system** contains a constant amount of matter.

It is now possible to give a complete statement about recognizing equilibrium. **Equilibrium** is constancy of macroscopic properties in a closed system at a uniform temperature.

Summary

A state of equilibrium can exist in chemical reactions, in a phase change, and in the formation of a solution. The equilibrium between a liquid and its vapor depends upon the vapor pressure of the liquid. The temperature at which the vapor pressure of a liquid equals the atmospheric pressure is the boiling point. Equilibrium is characterized by constancy of macroscopic properties in a closed system at a uniform temperature.

Self Check

1. What happens to the vapor pressure of a liquid when its temperature rises from 40°C to 60°C?
2. What is the difference between an open system and a closed system?
3. How can you recognize equilibrium? What two conditions are necessary in order for equilibrium to exist?
4. When does a liquid reach (a) its boiling point, (b) its normal boiling point?

Equilibrium Is Dynamic

22-6 *The Dynamic Nature of Equilibrium in Chemical Reactions*

The constancy of properties at equilibrium refers to macroscopic properties. Now you will look at equilibrium on the molecular level, as chemists picture it.

Suppose two identical bulbs are filled to equal pressures of nitrogen dioxide, NO_2. Bulb A is immersed in an ice bath and bulb B in boiling water, as shown in Figure 22-4. The gas in bulb A at 0°C has little color, while the gas in bulb B at 100°C is reddish-brown. Evidently the molecules responsible for the reddish-brown color are far more

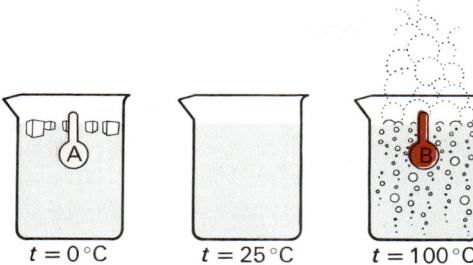

Figure 22-4 *The color of bulb A, in an ice bath, indicates that it contains mostly N_2O_4 molecules. The color of bulb B, at 100°C, indicates that NO_2 molecules predominate.*

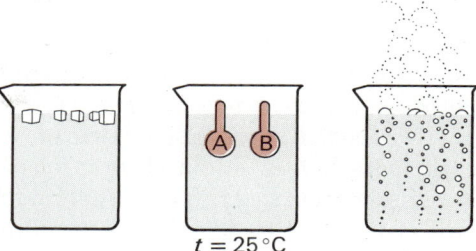

Figure 22-5 *At the same temperature (25°C) bulbs A and B both show the same color, indicating an equilibrium mixture of NO_2 and N_2O_4.*

abundant in the hot bulb than in the cold one. A variety of experiments shows that the cold bulb contains mostly N_2O_4 molecules. These same experiments show that the hot bulb contains mostly NO_2 molecules. The N_2O_4 molecules absorb no visible light, so the cold gas has little color. The NO_2 molecules do absorb some visible light, so the hot gas is reddish-brown.

Now these two bulbs are transferred to a bath at room temperature, as shown in Figure 22-5. Immediately the color begins to darken in bulb A. The darkening of the color shows that a chemical change has occurred, NO_2 molecules forming from N_2O_4.

$$N_2O_4(g) \longrightarrow 2NO_2(g) \qquad (4)$$

At the same time, the color in bulb B begins to pale, showing that a chemical change has occurred in this bulb as well. In bulb B, N_2O_4 molecules form from NO_2.

$$2NO_2(g) \longrightarrow N_2O_4(g) \qquad (5)$$

In each bulb the colors continue to change, bulb A becoming darker and bulb B becoming paler. Finally, as the two bulbs approach the same temperature, 25°C, the colors stop changing. A close examination shows that the two colors are now the same.

By direct observation you can watch the contents of these two bulbs approach the constancy of macroscopic properties (in this case, color) that indicates equilibrium. In bulb A equilibrium was approached by reaction (4). In bulb B it was approached by the opposite reaction, (5). Now it is clear why the color of each bulb stopped changing. The reaction between NO_2 and N_2O_4 can proceed in *both* directions, as the following indicates.

$$N_2O_4(g) \longrightarrow 2NO_2(g) \qquad (4)$$

$$N_2O_4(g) \longleftarrow 2NO_2(g) \qquad (5)$$

Since N_2O_4 molecules can dissociate, or split apart, in bulb A, they must also be able to dissociate in bulb B. Surely an N_2O_4 molecule does not act differently in bulb A (at 25°C) than it does in bulb B (at 25°C). The same must apply to the combination of two NO_2 molecules. If the reaction occurs in bulb B, then it must also occur in bulb A. The net change you see results from the changing NO_2 color. This net change represents the *difference* between the rate of production of

NO$_2$ by reaction (4) and the rate of loss of NO$_2$ by reaction (5). Macroscopic changes cease when these two rates are exactly equal.

In bulb A, equilibrium is approached from a lower temperature, which favors N$_2$O$_4$. Reaction (4) goes faster at first. However, as more and more NO$_2$ is produced, reaction (5) becomes faster and faster. When reaction (5) becomes just as fast as reaction (4), equilibrium is reached. Macroscopic properties no longer change even though both reactions still proceed in a state of balance. Thus you can replace the single arrow in the preceding reactions by a double arrow ($\rightleftharpoons$) or an equals sign ($=$) to show that equilibrium prevails. (As you learned earlier, the equals sign can also stand for a single arrow.)

$$N_2O_4(g) \rightleftharpoons 2NO_2(g)$$

or (6)

$$N_2O_4(g) = 2NO_2(g)$$

In bulb B equilibrium is approached from a higher temperature, which favors NO$_2$. Reaction (5) goes faster at first. Using the argument applied to bulb A, you can see that as time progresses, reaction (4) becomes more and more rapid as N$_2$O$_4$ is produced. Reaction (5), in turn, becomes slower as NO$_2$ is used up. Finally, when the rates become equal, equilibrium is reached. Thus the equilibrium expression (6) applies to bulb B as well as to bulb A.

You can now understand that for chemical reactions at equilibrium, microscopic processes continue. However, the processes balance, so no macroscopic changes occur.

22-7 The Dynamic Nature of a Phase Change Equilibrium

In Section 22-2 you learned that when liquid water at 20°C is placed in a flask and the flask is sealed, some water molecules leave the liquid and enter the gas phase. The partial pressure of water vapor rises as more and more water molecules become part of the gas. Finally, the pressure stops rising, and the partial pressure becomes constant. When the partial pressure of the vapor equals a constant value, the vapor pressure, equilibrium exists.

Water molecules from the liquid are still evaporating at equilibrium. Molecules in the liquid have no way of "knowing" that the partial pressure of the vapor is equal to the vapor pressure. In the gas phase, the randomly moving molecules continue to strike the surface of the liquid and condense. Equilibrium corresponds to a perfect balance between this continuing evaporation and condensation. At this point no net changes can be detected.

Figure 22-6 shows this schematically. In Figure 22-6a the partial pressure of the vapor is less than the equilibrium value. The vapor is said to be *unsaturated*. The rate of evaporation exceeds the rate of condensation until the partial pressure of the vapor equals the vapor pressure. In Figure 22-6b the partial pressure of the vapor equals the vapor pressure. In Figure 22-6c the partial pressure of the vapor is greater than the equilibrium value. This vapor is said to be *supersaturated*. Here condensation proceeds faster than evaporation until the excess of vapor has condensed.

As you have just seen, the partial pressure of the water vapor equals the equilibrium value, that is, the vapor pressure, in Figure 22-6b. Here the vapor is called *saturated* vapor. The *vapor pressure* corresponds to that concentration of vapor at which condensation and evaporation occur at exactly the same rate. Thus, at equilibrium microscopic processes continue. They balance, however, and no macroscopic changes occur.

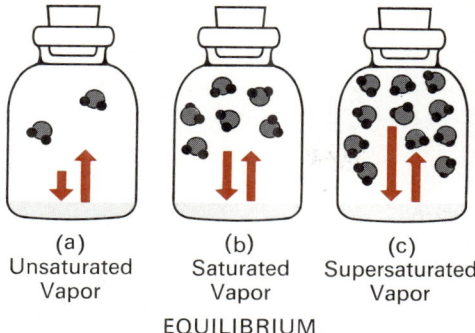

(a) Unsaturated Vapor (b) Saturated Vapor (c) Supersaturated Vapor

EQUILIBRIUM

Figure 22-6 *As the arrow sizes indicate, the rates of evaporation and condensation are equal in* (b), *while evaporation predominates in* (a), *and condensation predominates in* (c).

22-8 The Dynamic Nature of Solubility Equilibrium

You saw in Section 22-4 that solid iodine dissolves in an alcohol-water mixture until the solution is saturated. Then no more solid dissolves, and the color of the solution remains constant.

Among the molecules, however, there is still a great deal of activity. Iodine dissolves by the separation of surface layer molecules from the iodine crystals. The rate at which this occurs is fixed by the temperature and the stability of the crystal. Crystal stability tends to hold the molecules in the surface layer, while thermal agitation tends to dislodge the molecules from their lattice positions. As dissolving continues, the concentration of iodine molecules in the solution increases.

Occasionally a molecule moving about in the solution encounters the surface of an iodine crystal and lodges there. This addition to the crystal is called *precipitation*, or *crystallization*. It occurs more and more often as the concentration of iodine in solution increases.

Here you have two opposing processes. At a given temperature, molecules leave the surface of the crystal at a constant rate, tending to increase the concentration in solution. On the other hand, dissolved molecules are continually striking the surface and precipitating, tending to decrease the concentration of molecules in solution. What happens when the rate of return of molecules to the surface of the solid is just equal to the rate at which they are leaving the surface? No more net change will occur even though some molecules are continuously dissolving and others are continuously precipitating. As long as these two processes are in balance, the amount of iodine dissolved per unit volume of solution will be constant. The macroscopic property, the solubility, is now constant. The system is in solubility equilibrium. Chemists interpret this constancy as a balance between two opposing processes that continue at equilibrium. Again it must be emphasized that microscopic processes continue at equilibrium. However, they balance, so no macroscopic changes occur.

22-9 Direct Proof of the Dynamic Nature of Equilibrium

The concept that equilibrium is achieved by the *balancing*, not the *stopping*, of changes at the molecular level was accepted many years ago. Proof, however, has been possible only in recent decades. Direct evidence has been obtained by using *isotopic tracers*. As noted in Section 10-8, isotopes are atoms of the same element that have different numbers of neutrons. If a substance contains a small percentage of an isotope, the movement of this isotope can be detected by using a mass spectrograph, which was discussed in

Section 9-8. If the isotope is radioactive, it can be followed with a device for detecting radioactivity, such as a Geiger counter.

While normal oxygen consists mainly (99.76%) of ^{16}O, the isotope ^{17}O can be introduced at a level above normal (0.04%). Its presence can be detected by a mass spectrograph. Now, suppose you have the apparatus shown in Figure 22-7. One bulb contains ordinary water with an ^{17}O concentration of 0.04%. The other bulb contains water into which enough ^{17}O has been added to raise the concentration to 2%. The valve between the bulbs is kept closed until the temperature is uniform throughout and the equilibrium vapor pressure of water exists on both sides. What will happen when the valve is opened?

Since each of the two sides came to equilibrium at the same temperature, there will be no macroscopic changes in the system. However, the gas molecules do pass freely back and forth so that the traces of ^{17}O will be uniformly distributed throughout all the gas phase. The important point is that eventually the ^{17}O will be found at the same concentration in the two pools of liquid. This can only happen if molecules are continuously passing from the liquid phase to the gas phase and vice versa, even though equilibrium has been reached.

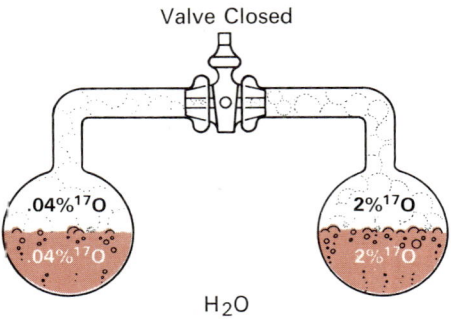

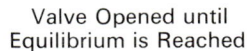

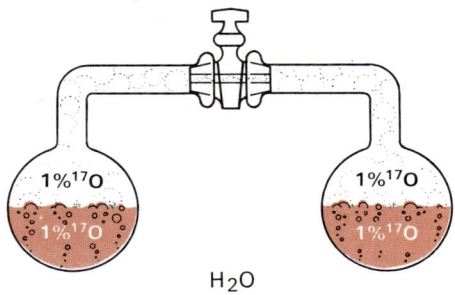

Figure 22-7 *The isolated bulbs (above) contain different amounts of ^{17}O, but when they are interconnected (below), the amounts of ^{17}O are found to be the same in each bulb at equilibrium. This could only happen if molecules pass continuously between the gas and liquid phases at equilibrium.*

Summary

Equilibrium in a chemical system is a state in which no macroscopic changes are occurring. On the molecular level, changes do occur, but they balance one another. When a chemical reaction reaches equilibrium, the quantities of the various compounds involved no longer change. The forward and reverse reactions still occur, but at equal rates. In the vaporization of a liquid, as well as in the dissolving of a solid in a liquid, there is again a continuous process of change at the molecular level. However, no macroscopic changes take place.

Self Check

1. What is the difference between unsaturated vapor and saturated vapor?
2. At equilibrium, what process opposes crystallization?

The State of Equilibrium Can Be Altered

22-10 The State of Equilibrium

In describing the equilibrium state for equation (6), it was not implied that the number of moles of N_2O_4 remaining was equal to half the number of moles of NO_2 produced. In other words, equation (6) gives us no information concerning the amount of nitrogen dioxide present at equilibrium. This is easily verified by raising the temperature of the water surrounding bulbs A and B about 10°C. The colors of the gases in both bulbs change to a new equilibrium color. This color corresponds to the presence of more NO_2. Yet the following expression is still applicable.

$$N_2O_4(g) \rightleftharpoons 2NO_2(g) \qquad (6)$$

What does equation (6) indicate, then? First, it indicates that equilibrium prevails; the $\rightleftharpoons$ sign tells you this. Next, it shows that two types of molecules are present, N_2O_4 and NO_2 molecules. Finally, it indicates that *during the approach* to equilibrium, *two* moles of NO_2 are produced (or consumed) for every *one* mole of N_2O_4 consumed (or formed). It does *not* tell you how rapidly the reaction moves to equilibrium. This is a matter of reaction rates. It also does *not* tell you whether at equilibrium there will be much or little NO_2 compared with the amount of N_2O_4.

To emphasize the latter point, consider another familiar reaction.

$$H_2O(g) \rightleftharpoons H_2(g) + \tfrac{1}{2}O_2(g) \qquad (7)$$

Until the necessary information is given, you cannot know how complete the decomposition of water is at equilibrium. All you can know is that for every mole of water that decomposes, one mole of hydrogen and one-half mole of oxygen will be obtained.

It has been experimentally determined that, in a closed vessel at 2273 K and a total pressure equal to one atmosphere at the time equilibrium is attained, 0.006 mole of water dissociates for every mole of water initially present. Therefore, $1 - 0.006 = 0.994$ mole of undecomposed water remains. From the coefficients in the balanced equation, you know that the products are 0.006 mole of H_2 and 0.003 mole of oxygen. This can be summarized as follows.

$$H_2O(g) \rightleftharpoons H_2(g) + \tfrac{1}{2}O_2(g)$$

Initial moles	1.	0	0
Moles present at equilibrium	0.994	0.006	0.003

In other words, not much water decomposes when the equilibrium state is attained at 2273 K.

Now what about approaching the equilibrium state when you start with hydrogen and oxygen? Assume you begin with one mole of hydrogen and one-half mole of oxygen. You allow the reaction to attain equilibrium at 2273 K and a total pressure of one atmosphere. At equilibrium you will, again, have 0.994 mole of water, 0.006 mole of H_2, and 0.003 mole of O_2. This can be summarized as follows.

$$H_2(g) + \tfrac{1}{2}O_2(g) \rightleftharpoons H_2O(g)$$

Initial moles	1	0.5	0
Moles present at equilibrium	0.006	0.003	0.994

If you start with hydrogen and oxygen, equilibrium is attained after most of the hydrogen and oxygen have united to form water. More important, the partial pressures at equilibrium are the same as those obtained in the reverse reaction. The equilibrium pressures are fixed by the temperature, the reactants and products, and the total pressure. They do not depend upon the direction from which equilibrium is approached. The balanced equation does not indicate the concentrations, or partial pressures, at equilibrium.

In many reactions the state of equilibrium lies very far to one side or the other. Thus, when equilibrium is attained in reaction (7), scarcely any H_2 or O_2 is present. For many practical purposes, the amounts present are negligible. Therefore, you might say that the reverse of reaction (7)

$$\tfrac{1}{2}O_2(g) + H_2(g) \longrightarrow H_2O(g)$$

is one that *goes to completion*. This expression is often used by chemists. However, it must be remembered that in a closed system no reaction can *literally* go to completion. Some finite quantity of each of the reactants must remain.

Of course, in an open system a reaction can literally go to completion. In this connection, recall reaction (3).

$$CaCO_3(s) \xrightarrow[\text{temp.}]{\text{high}} CaO(s) + CO_2(g) \qquad (3)$$

Suppose the vessel in the furnace is open so that the CO_2 produced can escape. In this case no equilibrium can be established because the pressure of CO_2 can never build up to its proper equilibrium value. Hence more and more $CaCO_3$ will decompose until literally all the $CaCO_3$ has been converted into CaO and CO_2. Remember that this is happening in an open system, where a substance is permitted to escape. In a closed system, where matter cannot escape, an equilibrium state will be reached.

22-11 Altering the State of Equilibrium

You have seen that the state of equilibrium for a system can be characterized by the relative amounts of products and reactants present. Consider again reaction (7), the decomposition of water. Any change in conditions that would cause more than 0.6 percent of the water to dissociate at equilibrium would alter the state of equilibrium for reaction (7).

$$H_2O(g) \rightleftharpoons H_2(g) + \tfrac{1}{2}O_2(g) \qquad (7)$$

Such a change would favor the formation of more hydrogen and oxygen.

What conditions might alter the equilibrium state? In order to answer this question, recall that concentration and temperature are two main factors that affect the rate of reaction. Equilibrium is attained when the rates of opposing reactions become equal. Any condition that changes the rate of only one of the reactions involved in the equilibrium should affect the conditions at equilibrium.

In order to understand the effect of concentration on equilibrium, consider the following reaction between ferric ion, Fe^{3+}, and thiocyanate ion, SCN^-.

$$Fe^{3+}(aq) + SCN^-(aq) \rightleftharpoons FeSCN^{2+}(aq) \qquad (8)$$

You can obtain visual evidence of the concentrations at equilibrium by observing the intensity of color. The intensity is deter-

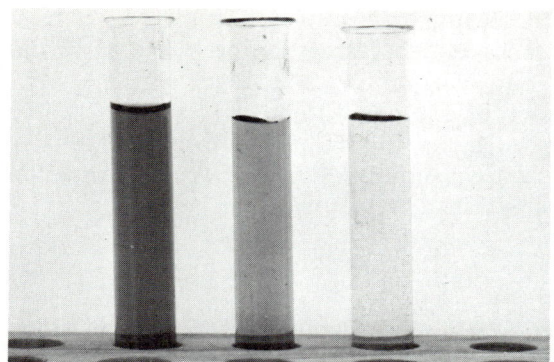

Figure 22-8 *These tubes all originally contained the same concentration of SCN^-(aq). The initial Fe^{3+}(aq) concentrations, however, were varied (decreasing from left to right). As you can see, the color intensities, due to the product, $FeSCN^{2+}$(aq), also decrease from left to right.*

librium concentrations are affected if the temperature is altered.

Now you know that a change in concentration or temperature alters the equilibrium state. Does a catalyst have the same effect? Although a catalyst increases the rate of a reaction, chemists have found experimentally that addition of a catalyst to a reaction at equilibrium does not alter the equilibrium state. Hence it must be true that any catalyst has the same effect on the rates of the forward and reverse reactions. You will recall that the effect of a catalyst on reaction rates can be discussed in terms of lowering the activation energy. This lowering is effective in increasing the rate in both directions, forward and reverse. For this reason, a catalyst produces no net change in the equilibrium concentrations even though the system may reach equilibrium much more rapidly than it did without the catalyst.

mined by the concentration of thiocyanato-iron(III) ion, $FeSCN^{2+}(aq)$. By using additional ferric ion and thiocyanate ion, you can change the concentration of one of the reactants in equation (8). As Figure 22-8 shows, the color of the solution darkens, indicating an increase in the amount of the colored ion, $FeSCN^{2+}$. Thus you have evidence that the equilibrium concentrations are affected if the concentrations of reactants or products are altered.

In order to understand the effect of temperature on equilibrium, consider reaction (6) again.

$$N_2O_4(g) \rightleftharpoons 2NO_2(g) \qquad (6)$$

You learned that the higher the temperature the greater the quantity of NO_2 that is produced. Therefore, you can see that the equi-

22-12 Predicting New Equilibrium Concentrations: Le Chatelier's Principle

The mere conclusion that this change or that change will in some way affect the equilibrium concentrations is not adequate to chemists. Among other things, they also want to know the *direction* of the effect (does it favor products or reactants?).

In this connection, the French chemist Henry Louis Le Chatelier (luh-shah-tuh-LYAY) sought regularities among a large amount of experimental data on equilibria. To summarize the regularities he found, he made the following generalization. *If an equilibrium system is subjected to a change, processes occur that tend to counteract partially the imposed change.* This generalization is known as **Le Chatelier's principle.**

How does this statement apply to the previous examples?

Consider first a change in concentration. As you know, the addition of $Fe^{3+}(aq)$ or $SCN^-(aq)$ to an equilibrium solution containing both these ions intensifies the color of the product in reaction (8).

$$Fe^{3+}(aq) + SCN^-(aq) \rightleftharpoons FeSCN^{2+}(aq) \quad (8)$$

A new state of equilibrium is attained in which more $FeSCN^{2+}$ is present. Increasing the concentration of, say, $SCN^-(aq)$ increases the concentration of the $FeSCN^{2+}$ ion. This agrees with Le Chatelier's principle. The change imposed on the system is an increase in the concentration of SCN^-. This change is counteracted in part by some Fe^{3+} and SCN^- ions reacting to form more $FeSCN^{2+}$. The same argument applies to an addition of Fe^{3+} ions. In each case, the formation of $FeSCN^{2+}$ uses up a portion of the added reactant. Thus, it partially counteracts the imposed change.

Consider now a change in pressure. You have learned that the concentration of all *gaseous* components can be altered by changing the pressure at which they are confined. In this connection, recall reaction (7).

$$H_2O(g) \rightleftharpoons H_2(g) + \tfrac{1}{2}O_2(g) \quad (7)$$

Suppose the total pressure is doubled. The volume then becomes much smaller. The total number of moles present per unit volume is greater than it was under the original equilibrium conditions. The change in concentration is counteracted in part by the combination of some hydrogen and oxygen to form gaseous water. This combination reduces the total number of moles present (one and one-half moles unite to form one mole). Hence, increasing the concentration of all components by increasing the pressure shifts the state of equilibrium in favor of the formation of gaseous water.

A change in total pressure does not always shift equilibrium. Consider the first reaction mentioned in this chapter.

$$CO(g) + NO_2(g) \longrightarrow CO_2(g) + NO(g) \quad (1)$$

If the pressure on a mixture of these four gases at equilibrium is increased, the gases are compressed to a smaller volume. Once again the concentrations are all increased. How does Le Chatelier's principle apply here? If the equilibrium state is altered to favor products, some CO and NO_2 molecules react to form an exactly equal number of molecules of CO_2 and NO. Since there is no change in the total number of molecules, the proposed change in the equilibrium does *not* partially reduce the pressure change. Le Chatelier's principle tells you that processes occur so as to "counteract partially the imposed change." Here neither a shift favoring products nor a shift favoring the reactants will counteract the imposed pressure change. Hence, Le Chatelier's principle tells you not to expect a change in the equilibrium state for reaction (1) when the pressure is altered. Experimentally, it is found that no change is observed. The equilibrium state is not affected by a pressure change for any such equilibrium gas mixture. It is not affected because the number of reacting molecules is the same as the number of product molecules in the balanced reaction.

It should be emphasized that the concentration of pure solids and liquids is not changed by altering the pressure. As you know, the concentration of gases is affected. For this reason, you should consider only gases, and not solids or liquids, when applying Le Chatelier's principle in connection with *pressure* changes. For example, consider the following reaction.

$$2HgO(s) \rightleftharpoons 2Hg(l) + O_2(g) \quad (9)$$

Is the forward or the reverse reaction favored by an increase in pressure? To answer this question, you must first observe that the only substance in the gas phase is oxygen gas. An increase in pressure affects the oxygen gas, but not the other substances. Thus, an increase in pressure favors the reverse reaction, which yields no moles of gas.

Now, what does Le Chatelier's principle say about a change in temperature? In order to answer this question, consider the heat effect of reaction (6). The decomposition of N_2O_4 is endothermic, as shown in reactions (6a) and (6b).

$$N_2O_4(g) + 14.1 \text{ kcal} \rightleftharpoons 2NO_2(g) \quad (6a)$$

$$N_2O_4(g) \rightleftharpoons 2NO_2(g)$$
$$\Delta H = +14.1 \text{ kcal} \quad (6b)$$

It was noted previously that warming a bulb containing NO_2 and N_2O_4 causes a shift of the equilibrium state in favor of the formation of NO_2 (the reddish-brown color grows darker). It is easy to see that this agrees with Le Chatelier's principle. A rise in temperature is caused by an input of heat. At the higher temperature, the equilibrium is changed to form more NO_2. The formation of NO_2 absorbs a portion of the heat that caused the temperature rise.

In this connection, recall that raising the temperature of liquid water raises its vapor pressure. This agrees with Le Chatelier's principle, since heat is absorbed as the liquid vaporizes. This absorption of heat partially counteracts the temperature rise.

22-13 Application of Equilibrium Principles: The Haber Process

Knowledge of chemical principles is rewarding in solving technical problems. Control of chemical reactions is the key. The large-scale commercial production of nitrogen compounds provides a practical example of the beneficial application of Le Chatelier's principle.

The commercial process involves the conversion of the inert nitrogen of the atmosphere into useful compounds, such as fertilizers, plastics, and dyes. The most difficult step in this process is reaction (10), which follows.

$$N_2(g) + 3H_2(g) \rightleftharpoons 2NH_3(g) + 22 \text{ kcal}$$

or

$$N_2(g) + 3H_2(g) \rightleftharpoons 2NH_3(g)$$
$$\Delta H = -22 \text{ kcal} \quad (10)$$

Can you predict the optimum conditions for a high yield of NH_3? Should the system be allowed to attain equilibrium at a low or high temperature? Application of Le Chatelier's principle suggests that the lower the temperature, the more the equilibrium state will favor the production of NH_3. Should low or high pressure be used? The production of NH_3 represents a decrease in total moles present from four to two. Le Chatelier's principle suggests use of pressure to increase concentration.

Chemistry and Our Environment

OZONE, LIFE ON EARTH, AND THE SST

You already know that the earth's atmosphere is about one-fifth oxygen, and that this oxygen is made up of diatomic molecules. There is also a more reactive form of oxygen, called *ozone,* in which the molecules are triatomic, O_3. The ozone molecules are more energetic than the ordinary O_2 molecules and the reaction

$$2O_3 = 3O_2$$

proceeds with release of energy. Even so ozone decomposes into ordinary oxygen only slowly unless the reaction is catalyzed.

Ozone plays an absolutely indispensable role in making life as we know it possible on this planet. Without ozone, little or none of the plant and animal life present today could survive. This is because there is a layer of ozone, as shown in the accompanying diagram, that lies between 10 and 30 miles above the earth's surface. The ozone in this layer is produced when oxygen molecules absorb strong ultraviolet radiation from the sun. Ozone molecules themselves absorb ultraviolet radiation very strongly. It is the ozone belt that keeps all but a tiny fraction of the ultraviolet radiation from the sun from reaching the earth's surface.

Ultraviolet radiation has so much energy that if the amount reaching the earth's surface were to increase very much, all plants except those living under water would be killed. Furthermore, all animals would be blinded, and, for man at least, skin cancer would probably become epidemic. The loss of land plants would cut off nearly all of our food supply at its source, thus leading to mass starvation. The introduction of any technology that might deplete the ozone belt to more than a very slight extent could well turn out to be race suicide.

The concentration of ozone in the ozone belt could be seriously depleted by the introduction of a substance that can increase the rate of decomposition of ozone to ordinary oxygen. One such substance is nitric oxide, NO, as shown by a young chemist named Harold S. Johnston in 1950. He worked out the following mechanism for the catalytic process.

$$O_3 + NO = NO_2 + O_2$$
$$NO_2 + O = NO + O_2$$

The oxygen atoms required in the second step come from the equilibrium involving O_3, O_2, and O that exists in the ozone belt.

$$O_3 = O_2 + O$$

Note that NO is regenerated in the catalytic process. Thus, a small amount of it can go around and around, decomposing many times its own mass of O_3.

How could significant amounts of NO get into the ozone layer? Although NO is formed as a pollutant near the earth's surface, by automobiles and some forms of combustion, very little of it is able to get into the stratosphere. The various temperature belts in the atmosphere oppose a flow in the upward

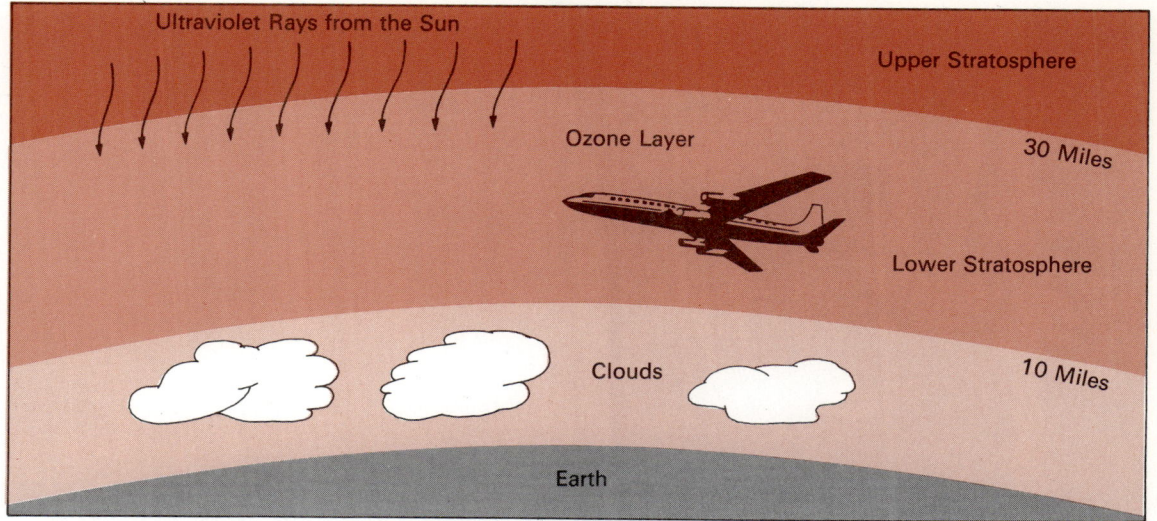

A layer of ozone lies between 10 and 30 miles above the earth's surface. The ozone belt keeps all but a tiny fraction of the sun's ultraviolet radiation from reaching the earth's surface.

direction. Even airplanes do not fly high enough to inject NO into the ozone layer—except for supersonic aircraft. These often fly in the stratosphere, and the supersonic transports, SST's, now being developed *must fly in this region to be economical.* They are being designed to do just that.

Until recently, the effect of the NO that these SST's will dump into the stratosphere was not considered to be serious. However, Professor Johnston has made extensive calculations, based on his and others' studies of the catalytic effect of NO on ozone decomposition. It is estimated that each SST would release about *one ton* of nitric oxide per hour into the stratosphere. With a fleet of 500 SST's, operating an average of seven hours a day (which is predicted for 1985 if present development continues), the ozone content of the stratosphere could be reduced by a half in only one year. If this were to happen, the effects might be worse than those of a nuclear war.

It must be emphasized that Johnston's calculations are not absolutely certain. Some of the reaction rates that enter into his calculations are not known with great accuracy. There is also some uncertainty in using laboratory findings, however accurate, to predict what will happen in the upper atmosphere. However, the calculations show that there is a real possibility that large scale use of SST's could cause a catastrophe. The most serious aspect of this situation is that if this catastrophe should occur, its effects would not be reversible. If the ozone concentration were lowered significantly, merely stopping further use of the SST's would not restore it.

Harold S. Johnston (1920–   )

WHAT GOOD IS fundamental research? This question is often asked by nonscientists. They see scientists in universities and research laboratories spending time and money to study problems that have to do with the basic properties and behavior of matter. Such problems appear to lack any direct bearing on everyday life.

Until very recently, those who ask this question might have pointed to the work of Harold S. Johnston and his students at the University of California as an example of "useless basic research." For years, Johnston has painstakingly studied the way in which reactions between relatively simple molecules occur in the gas phase. He has been able to work out the mechanisms of many such reactions. That is, he has discovered what the individual steps are, how fast they occur, and how added gases may influence these rates.

The last type of problem—how one substance can be a catalyst for some reaction of another—was first tackled by Johnston when he was a young faculty member at Stanford University. The particular reaction he chose to study was the breakdown of ozone into ordinary oxygen, under the influence of nitric oxide as a catalyst. He chose this problem because it seemed to him that the reaction was a very basic one and that if we could understand the processes involved in this reaction, we would gain an increased understanding of how chemical reactions occur. Indeed, there could not have been any other reason for studying this particular reaction at the time Johnston did so. No one could *then* see any practical application for the results. Today it seems possible that Johnston's careful work, his "useless basic research," may be instrumental in preserving life on this planet. Harold Johnston's career is surely a shining example of the fact that basic research may in the long run be the most *useful* research there is.

What about feasibility? Since reaction rates are slow at low temperatures, a compromise is necessary. Low temperature is required for a desirable equilibrium state, and high temperature is necessary for a satisfactory rate. The compromise used industrially involves an intermediate temperature around 500°C. Even then the success of the process depends upon the presence of a suitable catalyst to achieve a reasonable reaction rate.

With regard to pressure, another compromise is needed because it is expensive to build high-pressure equipment. A pressure of about 350 atmospheres is actually used. Under these conditions, 350 atmospheres and 500°C, only about 30 percent of the reactants are converted to NH_3. The NH_3 is removed from the mixture by liquefying it under conditions at which N_2 and H_2 remain as gases. The unreacted N_2 and H_2 are then used again. Because of this, the total percent conversion to ammonia is very high. Figure 22-9 shows the steps in this process.

Prior to World War I the principal sources of nitrogen compounds were some nitrate deposits in Chile. This situation changed when Fritz Haber, a German chemist, successfully developed the process just described. Thus it is called the *Haber process*. It allows chemists to use the almost unlimited supply of nitrogen in the atmosphere as a source of nitrogen compounds.

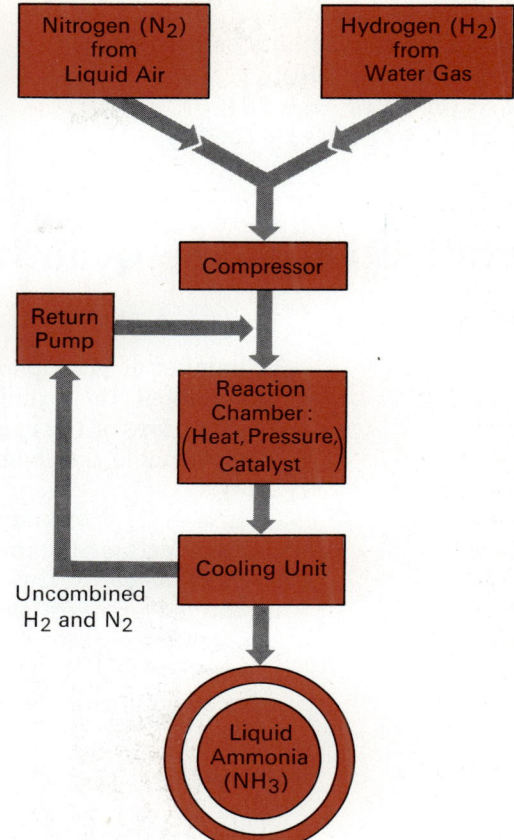

Figure 22-9 *Steps in the Haber process. Gaseous ammonia is formed in the reaction chamber and is then cooled to produce liquid ammonia. The remaining reactants are recycled.*

Summary

The state of equilibrium can be altered by changes in concentration and temperature. Le Chatelier's principle allows chemists to predict the direction of change of the equilibrium state. The principle states that if an equilibrium system is subjected to a change, processes occur that tend to counteract partially the imposed change. Although a catalyst increases the rate of a reaction, it does not alter the equilibrium state.

Self Check

1. Does Le Chatelier's principle predict a change of equilibrium concentrations for the following reactions when the mixture

of gases is compressed? If so, does the change favor reactants or products?
(a) $N_2O_4(g) \rightleftharpoons 2NO_2(g)$
(b) $H_2(g) + I_2(g) \rightleftharpoons 2HI(g)$
(c) $N_2(g) + 3H_2(g) \rightleftharpoons 2NH_3(g)$

2. Can a reaction in a closed system literally go to completion? Explain.
3. Why does the addition of a catalyst to a reaction at equilibrium not alter the equilibrium state?

Investigation of the Quantitative Aspects of Equilibrium

INVESTIGATION

22-14 A Quantitative Approach to Chemical Equilibrium

A number of principles permit the chemist to make qualitative predictions about the equilibrium state. Most of these principles rely on observations of the macroscopic state of affairs in reactions. Hence they do not enable the chemist to understand, predict, and control reactions fully.

It is helpful to know, for example, that raising the pressure favors the production of ammonia in its commercial preparation, as mentioned in Section 22-13. However, the chemist wants to know precisely how great the pressure should be in order to obtain the highest yield of ammonia. In other words, the chemist wants to know exactly how to control a reaction. To do this, he needs to have quantitative information about equilibrium conditions.

PURPOSE

This experiment introduces you to some quantitative aspects of equilibrium conditions. You will examine quantitatively reaction (8), which you have already studied qualitatively.

$$Fe^{3+}(aq) + SCN^-(aq) \rightleftharpoons FeSCN^{2+}(aq) \tag{8}$$

By adding varying concentrations of $Fe^{3+}(aq)$ to solutions having a constant concentration of $SCN^-(aq)$, you will examine the effects of these variations on the equilibrium concentrations of the reactants and products.

MATERIALS

25 ml 0.002 M KSCN
15 ml 0.2 M Fe(NO$_3$)$_3$
Distilled water
5 test tubes, 13- × 100-mm
Graduated cylinder, 25-ml

Beaker, 250-ml
Glass stirring rod
White paper
Centimeter ruler
Medicine dropper

PROCEDURE

You will determine the concentrations colorimetrically, that is, by comparing colors. You may have noticed that the color intensity of a colored liquid, such as iced tea, inside a glass is different when viewed through the sides of the glass than when viewed directly from the top. The color intensity depends on the concentration of the liquid *and* on the depth of the solution. Thus a 1-cm depth of $1\,M$ colored solution will appear to have the same color intensity as a 2-cm depth of a $0.5\,M$ solution. (See Figure 22-10.) The ratio of the concentrations is the inverse of the ratio of the depths. Using this relationship, you can compare the concentration of two solutions by altering their relative depths until the color intensity appears to be the same. This procedure gives relative values for the concentrations. To get absolute values, you must use a standard solution of known concentration. In step 1 you will

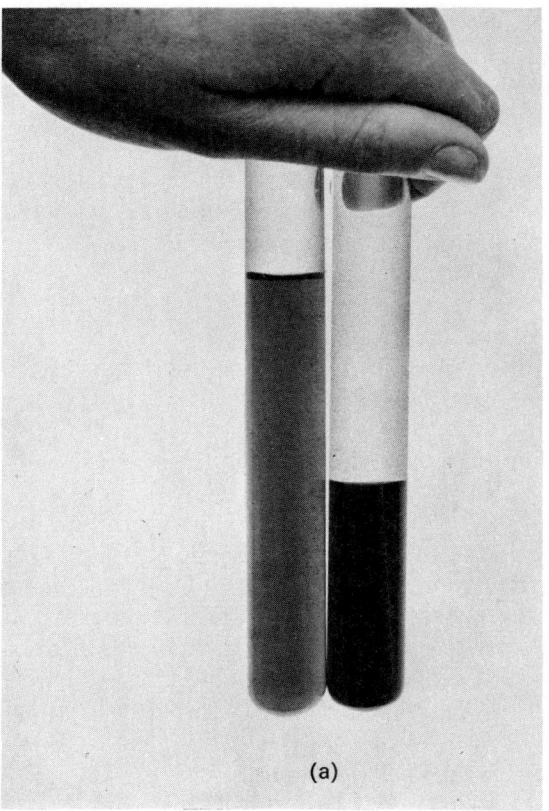

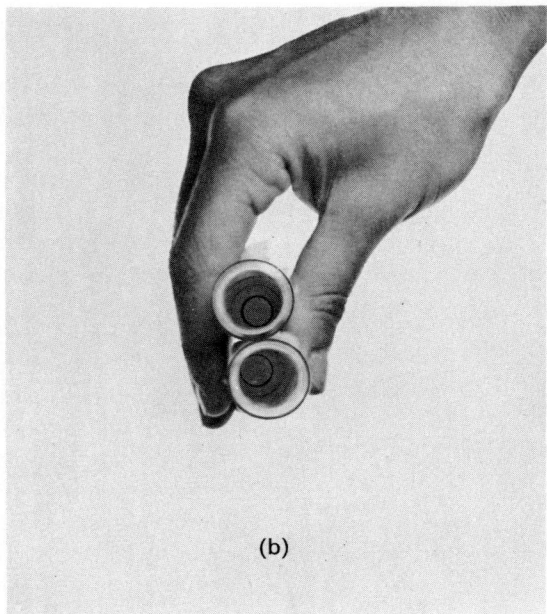

Figure 22-10 *The solution on the right in (a) has a concentration twice as great, and a volume half as great, as that of the other solution. When viewed from the top (b) the two solutions show equal color intensities.*

prepare a standard solution. You will start with a known, low concentration of thiocyanate ion, $SCN^-(aq)$, and will add a large excess of ferric ion, $Fe^{3+}(aq)$.

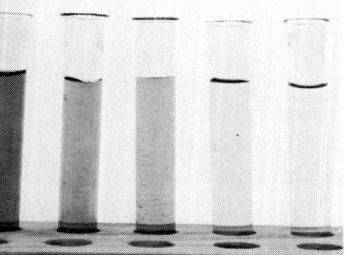

Figure 22-11

1. Line up five clean test tubes of the same diameter and label them 1, 2, 3, 4, 5. Add 5.0 ml of a 0.002 M solution of potassium thiocyanate, KSCN, to each of these five test tubes. To test tube 1 add 5.0 ml of a 0.2 M solution of ferric nitrate, $Fe(NO_3)_3$. This tube will be used as the standard.

2. Measure 10.0 ml of 0.2 M $Fe(NO_3)_3$ in your graduated cylinder and fill to the 25.0-ml mark with distilled water. Pour the solution into a clean, dry beaker and mix the solution with a glass stirring rod. Measure 5.0 ml of this solution and pour it into test tube 2. Save the remainder of the $Fe(NO_3)_3$ solution for use in step 3. Calculate the concentration of this solution as part of your prelab preparation.

3. Pour 10.0 ml of the solution prepared in step 2 into your graduated cylinder. Discard the remainder. Fill the cylinder to the 25.0-ml mark with distilled water and pour the solution into a clean, dry beaker to mix. Pour 5.0 ml of this solution into test tube 3.

4. Continue diluting in this manner until you have 5.0 ml of successively more dilute solution in each test tube. Your solutions should resemble those in Figure 22-11. Calculate the concentration of each of the solutions as part of your prelab preparation.

5. To determine the concentration of the thiocyanatoiron(III) ion, $FeSCN^{2+}(aq)$, compare the solutions in each of the test tubes with the standard (test tube 1). To help you do this, wrap a strip of paper around test tubes 1 and 2 to exclude light from the side. Holding the tubes over diffused light, look vertically down through the solutions, as shown in Figure 22-10b. If the color intensities appear the same, measure the depth of each solution to the nearest millimeter and record this. If the color intensities do not appear the same, remove some of the solution from the standard tube with a medicine dropper until the color intensities are the same. Put the portion you removed into a clean, dry beaker, since you may have to use it later. You may find it easier to match solutions by removing more standard than seems necessary and then replacing part of it drop by drop. When the color intensities are the same in each test tube, measure the depth of both solutions to the nearest millimeter. Repeat the procedure, comparing test tubes 1 and 3, 1 and 4, and 1 and 5.

DATA AND RESULTS

Make a table in your notebook for recording the depths of solutions in each of the four color comparisons.

Depth of solution in the standard test tube (1)
Depth of solution in the other test tubes (2, 3, 4, or 5)

Calculations

Assume that the ferric nitrate and the potassium thiocyanate exist in their respective solutions entirely as ions and that in the standard test tube, (1), essentially all of the thiocyanate ions have reacted to form $Fe(SCN)^{2+}(aq)$. Thus, you may assume that the equilibrium concentration of the $FeSCN^{2+}(aq)$ ion in the standard test tube is essentially the same as the concentration of the $SCN^-(aq)$ ion with which you started. Remember that *both* solutions, $Fe(NO_3)_3(aq)$ and $KSCN(aq)$, are diluted on mixing.

The symbol [ ] is a shorthand notation for concentration. Here it will be used to represent the equilibrium concentration in moles per liter of the formula appearing within the brackets. For example, the notation $[Fe^{3+}]$ stands for the equilibrium concentration of the ferric ion, $Fe^{3+}(aq)$, in moles per liter.

Do all of the calculations for each test tube (numbers 2 through 5) as follows.

1. (a) Calculate the ratio of depths in the color comparison. For test tube 2 this ratio is as follows.

$$\text{Ratio of depths} = \frac{\text{depth of standard matched with test tube 2}}{\text{depth of liquid in test tube 2}}$$

   (b) From this ratio, calculate the equilibrium concentration of thiocyanatoiron(III) ion, $[FeSCN^{2+}]$.

$$[FeSCN^{2+}] = \text{ratio of depths} \times \text{concentration of standard}$$

2. From your dilution data calculate the *initial* concentration of $Fe^{3+}(aq)$ ion.

3. Calculate the equilibrium concentration of $Fe^{3+}(aq)$ ion, $[Fe^{3+}]$, by subtracting the equilibrium concentration of the $FeSCN^{2+}(aq)$ ion from the initial concentration of the $Fe^{3+}(aq)$ ion.

4. Calculate the equilibrium concentration of the $SCN^-(aq)$ ion, $[SCN^-]$, in the same manner. In this case, subtract the equilibrium concentration of the $FeSCN^{2+}(aq)$ ion from the initial concentration of the $SCN^-(aq)$ ion.

5. Try to find some constant numerical relationship among the equilibrium concentrations of the ions in each test tube. To do this, multiply and divide the values obtained in each test tube in various ways. For example, for each test tube calculate the following.

(a) $[Fe^{3+}][FeSCN^{2+}][SCN^-]$

(b) $\dfrac{[Fe^{3+}][FeSCN^{2+}]}{[SCN^-]}$

(c) $\dfrac{[FeSCN^{2+}]}{[Fe^{3+}][SCN^-]}$

RELATED PROBLEMS

1. In Calculation 5, which of the combinations of concentrations—(a), (b), or (c)—gives the most constant numerical value?

2. The combination that gives the most constant numerical value is called the *equilibrium constant*. Restate this constant in words, using the terms *reactants* and *products*.

3. Explain why such a constant numerical relationship might exist.

22-15 The Equilibrium Constant for a Reaction

You have now made a colorimetric determination in the laboratory, as described in the preceding section. For reaction (8), you measured the concentration of the product, $FeSCN^{2+}$, in solutions containing ferric and thiocyanate ions, Fe^{3+} and SCN^-. (Remember that the square bracket notation [ ] is used to indicate concentration. Thus $[Fe^{3+}]$ is read "the ferric ion concentration.")

$$Fe^{3+}(aq) + SCN^-(aq) \rightleftharpoons FeSCN^{2+}(aq) \quad (8)$$

From $[FeSCN^{2+}]$ and the initial values of $[Fe^{3+}]$ and $[SCN^-]$, you calculated $[Fe^{3+}]$ and $[SCN^-]$ at equilibrium. You then made calculations for various combinations of these values to determine which ratio came closest to a constant value.

Colorimetric analysis based on human vision is not very exact. The data in Table 22-1 are more exact. The table contains various equilibrium concentrations for the reactants and products in reaction (11).

$$2HI(g) \rightleftharpoons H_2(g) + I_2(g) \quad (11)$$

The data are expressed in concentrations, although pressure units are more usual for a reaction involving gases.

What results do these data give? In Table 22-2 the experimental values of the following ratio are given for each of five measurements.

$$\dfrac{[H_2][I_2]}{[HI]} \quad (12)$$

Although the equilibrium concentrations are accurate, the five values for this ratio are far

TABLE 22-1 Equilibrium Concentrations of Hydrogen, Iodine, and Hydrogen Iodide at 698.6 K*

	$[H_2]$ (moles/liter)	$[I_2]$ (moles/liter)	$[HI]$ (moles/liter)
1	1.8313×10^{-3}	3.1292×10^{-3}	17.671×10^{-3}
2	2.9070×10^{-3}	1.7069×10^{-3}	16.482×10^{-3}
3	4.5647×10^{-3}	0.7378×10^{-3}	13.544×10^{-3}
4	0.4789×10^{-3}	0.4789×10^{-3}	3.531×10^{-3}
5	1.1409×10^{-3}	1.1409×10^{-3}	8.410×10^{-3}

*Values above the middle horizontal line were obtained by heating hydrogen and iodine together. Values below the line were obtained by heating pure hydrogen iodide.

from constant. Now consider the following ratio.

$$\frac{[H_2][I_2]}{[HI]^2} \quad (13)$$

Calculations of this ratio are summarized in Table 22-3. The results imply that with a fair degree of reliability, ratio (13) equals a constant. The following indicates this quite clearly.

$$\frac{[H_2][I_2]}{[HI]^2} = \text{a constant}$$
$$= 1.835 \times 10^{-2} \text{ at } 698.6 \text{ K} \quad (14)$$

TABLE 22-2 Values of $\frac{[H_2][I_2]}{[HI]}$ (Calculated from Data in Table 22-1)

Experiment	$\frac{[H_2][I_2]}{[HI]}$
1	32.429×10^{-5}
2	30.105×10^{-5}
3	24.866×10^{-5}
4	6.495×10^{-5}
5	15.477×10^{-5}

TABLE 22-3 Values of $\frac{[H_2][I_2]}{[HI]^2}$ (Calculated from Data in Table 22-1)

Experiment	$\frac{[H_2][I_2]}{[HI]^2}$
1	1.8351×10^{-2}
2	1.8265×10^{-2}
3	1.8359×10^{-2}
4	1.8390×10^{-2}
5	1.8403×10^{-2}
Average	1.835×10^{-2}

Look at this ratio in terms of reaction (11). Ratio (13) is the product of the equilibrium concentrations of the substances produced in the reaction divided by the square of the concentration of the reacting substance. In this ratio, notice that the power to which the concentration of each substance is raised is equal to its coefficient in reaction (11).

22-16 The Law of Chemical Equilibrium

The following summarizes what you have learned. For the reaction

$$Fe^{3+}(aq) + SCN^-(aq) \rightleftharpoons FeSCN^{2+}(aq) \quad (8)$$

you found (or should have found) experimentally that the concentrations of the ions involved have the following simple relationship.

$$\frac{[FeSCN^{2+}]}{[Fe^{3+}][SCN^-]} = \text{a constant} \quad (15)$$

Then you considered precise equilibrium data for reaction (11).

$$2HI(g) \rightleftharpoons H_2(g) + I_2(g) \quad (11)$$

The concentrations of the molecules appearing in reaction (11) were found to have the following simple relationship.

$$\frac{[H_2][I_2]}{[HI]^2} = \text{a constant} \quad (14)$$

In each of these simple relationships, (14) and (15), the concentrations of the products appear in the numerator and the concentrations of the reactants appear in the denominator. In reaction (11), two molecules of hydrogen iodide react. This influences expression (14) because it is necessary to square the concentration of hydrogen iodide, HI, in order to obtain a constant ratio.

These observations and many others like them lead to the generalization known as the **Law of Chemical Equilibrium.** For a reaction

$$aA + bB + \cdots \rightleftharpoons eE + fF + \cdots \quad (16)$$

when equilibrium exists, there will be a simple relation between the concentrations of products and the concentrations of reactants. For simplicity, the following equation is written in terms of only two products (E and F) and two reactants (A and B).

$$\frac{[E]^e[F]^f}{[A]^a[B]^b} = K = \text{a constant at a constant temperature} \quad (17)$$

Look carefully at the generalized equation, (17). You see that the numerator is the product of the equilibrium concentrations of the substances formed in the chemical reaction, each concentration raised to the appropriate power. The denominator is the product of the equilibrium concentrations of the reacting substances, again with each concentration raised to the appropriate power. The quotient remains constant at any fixed temperature. The constant K is called the equilibrium constant. This constant, and hence the Law of Chemical Equilibrium, applies to every chemical reaction. Using any chemical equation, you can immediately write an expression in terms of the concentrations of reactants and products, that will be constant at any given temperature.

In Table 22-4 some reactions are listed, along with the equilibrium law relation of concentrations and the numerical values of the equilibrium constants. Look closely at the table. Verify the forms of the equilibrium law relation among the concentrations. The very first relation has an unexpected form. For reaction (18)

$$Cu(s) + 2Ag^+(aq) \rightleftharpoons Cu^{2+}(aq) + 2Ag(s) \quad (18)$$

you do not find the following expression.

$$\frac{[Cu^{2+}][Ag]^2}{[Ag^+]^2[Cu]} = K \quad (19)$$

Rather, you find the following expression, equation (20).

$$\frac{[Cu^{2+}]}{[Ag^+]^2} = K \quad (20)$$

You find equation (20) because the concentrations of solid copper and solid silver are incorporated into the equilibrium constant. The concentration of solid copper is fixed by the density of the metal. It cannot be altered either by the chemist or by the progress of the reaction. The same is true of the concentration of solid silver. Neither of these concentrations varies no matter how much solid is added. For this reason, there is no need to write them each time an equilibrium calculation is made. Equation (20) is sufficient. Similarly, the equilibrium law relationships for the solubilities of AgCl(s) and AgI(s) do not include the concentrations of the solid salts.

TABLE 22-4 Some Equilibrium Constants

Reaction	Equilibrium Law Relation	K at Stated Temperature
$Cu(s) + 2Ag^+(aq) \rightleftharpoons Cu^{2+}(aq) + 2Ag(s)$	$K = \dfrac{[Cu^{2+}]}{[Ag^+]^2}$	2×10^{15} at 25°C
$Ag^+(aq) + 2NH_3(aq) \rightleftharpoons Ag(NH_3)_2^+(aq)$	$K = \dfrac{[Ag(NH_3)_2^+]}{[Ag^+][NH_3]^2}$	1.7×10^7 at 25°C
$N_2O_4(g) \rightleftharpoons 2NO_2(g)$	$K = \dfrac{[NO_2]^2}{[N_2O_4]}$	0.87 at 55°C
$2HI(g) \rightleftharpoons H_2(g) + I_2(g)$	$K = \dfrac{[H_2][I_2]}{[HI]^2}$	0.018 at 423°C
$HSO_4^-(aq) \rightleftharpoons H^+(aq) + SO_4^{2-}(aq)$	$K = \dfrac{[H^+][SO_4^{2-}]}{[HSO_4^-]}$	0.013 at 25°C
$CH_3COOH(aq) \rightleftharpoons H^+(aq) + CH_3COO^-(aq)$	$K = \dfrac{[H^+][CH_3COO^-]}{[CH_3COOH]}$	1.8×10^{-5} at 25°C
$AgCl(s) \rightleftharpoons Ag^+(aq) + Cl^-(aq)$	$K = [Ag^+][Cl^-]$	1.7×10^{-10} at 25°C
$AgI(s) \rightleftharpoons Ag^+(aq) + I^-(aq)$	$K = [Ag^+][I^-]$	10^{-16} at 25°C

Josiah Willard Gibbs (1839–1903)

J. W. GIBBS WAS A PROFESSOR of mathematical physics at Yale University for most of his life. He is best known for a series of papers that he published in a rather obscure scientific journal, *Transactions of the Connecticut Academy of Sciences,* in the years 1876–1878.

Earlier in the century, in Europe, such noted scientists as Carnot, Joule, Hemholtz, and Kelvin had made extensive studies in the field of thermodynamics. Thermodynamics is concerned with the relationships between heat and mechanical work. Their work dealt primarily with heat engines and made no mention of applications in the field of chemistry. Hess, who had measured the heats of reaction of many substances, did predict in 1840 that thermochemistry held the answers to many questions in chemistry.

Gibbs published his ideas showing the relationships between thermodynamics and chemical reactions, but few American chemists at the time could understand the complicated mathematics that he used. His work went unnoticed until 1892, when it was finally translated into German and published in European periodicals. In Europe his theories won immediate acclaim. A century of further scientific work has proved the thoroughness of Gibbs' treatment of the complex subject of chemical thermodynamics. Very little could be added to his work. He was the first to develop the concepts of "free energy" and "chemical potential" to help to explain the driving force behind chemical reactions.

An important application of Gibbs' theories came in the field of equilibria involving systems composed of several chemical components and more than one phase. By applying thermodynamic considerations, Gibbs developed a simple equation showing how the number of components and the number of phases in the system affected its "degrees of freedom." The degrees of freedom are the number of ways the temperature, pressure, and concentration can be varied while still maintaining an equilibrium.

As a result of his brilliant work, Gibbs was awarded the Copley Medal of the Royal Society in London and, in 1950, was elected to the Hall of Fame for Great Americans.

INVESTIGATION *22-17 Determination of the Solubility Product of Silver Acetate*

In a saturated solution containing ions in equilibrium with a solid, the rate at which ions are leaving the solid crystal is equal to the rate at which they are returning to the crystal. In one such solution, that of silver acetate, the following equation applies.

$$AgCH_3COO(s) \rightleftharpoons Ag^+(aq) + CH_3COO^-(aq)$$

The concentrations of the ionic products, silver ions, $Ag^+(aq)$, and acetate ions, $CH_3COO^-(aq)$, determine the equilibrium solubility. The equilibrium constant, called the *solubility product,* can be determined by obtaining the product of the equilibrium concentrations of the ions.

$$\text{Solubility product, } K_{sp} = [Ag^+][CH_3COO^-]$$

Note that the "concentration" of the solid silver acetate, $AgCH_3COO(s)$ does not appear in the equilibrium expression, since it does not change.

Because the solubility of silver acetate varies with the temperature, the solubility product is also dependent on the temperature.

PURPOSE

You will determine the equilibrium concentration of the silver ions in a saturated solution of silver acetate at room temperature. Since the acetate ion concentration is the same as the silver ion concentration, you will be able to calculate the K_{sp}.

MATERIALS

Distilled water
30 cm copper wire
Acetone
Emery cloth (or steel wool)
100 ml saturated $AgCH_3COO(aq)$ (one of three solutions, each prepared in a different way)

2 wash bottles
Graduated cylinder, 100-ml
Beaker, 250-ml
Balance, centigram

PROCEDURE

1. In a graduated cylinder carefully measure 100 ml of one of the three saturated silver acetate solutions. Record the number of the sample taken and the way it was prepared. Pour the solution into a clean, dry 250-ml beaker.

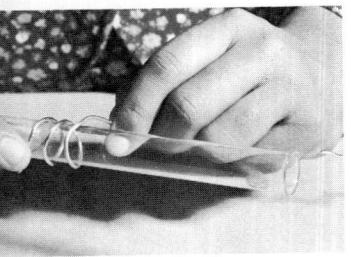

Figure 22-12

2. Obtain a 30-cm length of copper wire. Clean the wire with emery cloth. Then wind the wire into a loose coil around a test tube, as shown in Figure 22-12.

3. Find the mass of the copper coil to the nearest 0.01 g, and place it into the beaker containing the saturated silver acetate solution. Allow the solution and coil to stand overnight, so that all the silver ions will have an opportunity to react.

4. Shake the silver crystals free from the copper wire into the beaker. Wash the wire in a stream of water from the tap. Then rinse the wire in acetone. When the wire is dry, find its mass.

5. Decant the solution off the silver crystals and rinse them with distilled water and a wash bottle. Place the silver in a container designated by your teacher, so the silver can be used again.

DATA AND RESULTS

Record the mass of the copper coil both before and after it reacts.

Calculations

1. Calculate the number of moles of Cu(s) that reacted with the silver ions.

2. If one mole of Cu(s) reacts with two moles of $Ag^+(aq)$, how many moles of silver ions, $Ag^+(aq)$, are present in the 100-ml sample? What is the concentration of the silver ions in moles per liter?

3. What is the concentration of the acetate ions in moles per liter?

4. Calculate the value of the K_{sp} for silver acetate at room temperature.

5. Compare your results with those of other students who used the saturated solutions of silver acetate that contained different amounts of the solid, $AgCH_3COO(s)$. What do you conclude about the comparative results?

RELATED PROBLEMS

1. Using the procedure in this experiment or a different one, propose another method for determining the concentration of Ag^+ (aq) in the saturated solution.

2. Suppose that some sodium acetate, $NaCH_3COO(s)$, is added to a saturated solution of silver acetate that is in equilibrium with some $AgCH_3COO(s)$. After the sodium acetate has dissolved, what will be

the effect of the increased concentration of the acetate ion on the following equilibrium?

$$AgCH_3COO(s) \rightleftharpoons Ag^+(aq) + CH_3COO^-(aq)$$

3. Calculate the $Ag^+(aq)$ ion concentration when the acetate ion concentration in the solution in Problem 2 is $1.0\ M$. Is this result in agreement with the prediction you made in Problem 2?

22-18 Quantitative Treatment of Solubility Equilibrium

The Law of Chemical Equilibrium applies to solubility equilibrium as well as to equilibrium in chemical reactions. In this connection, you can regard the dissolving of a solid substance as a "reaction." Thus you can write equation (21).

$$\text{solid} = \text{dissolved substance} \quad (21)$$

The appropriate equilibrium constant is the following one.

$$K = \frac{[\text{dissolved substance}]}{[\text{solid}]} \quad (22)$$

Since the concentration of the solid is itself a constant for each substance, the equilibrium constant for a solubility equilibrium always takes the following simple form.

$$K = [\text{dissolved substance}] \quad (23)$$

Sometimes the dissolved substance consists entirely of molecules of one type as, for example, when iodine is dissolved in a water-alcohol solvent. In such cases as this, the equilibrium constant has a very simple interpretation. It is nothing more than the solubility, in moles per liter, of the solute in the particular solvent at a specified temperature.

The situation is more complicated for substances that dissolve to form ions. As you know, these solutes are called electrolytes, and their solutions are called electrolytic solutions. The dissolving of electrolytes in water was discussed in Section 17-18. Consider now the dissolving of silver chloride, AgCl.

When silver chloride dissolves in water, no molecules of silver chloride, AgCl, are present. Instead, silver ions, $Ag^+(aq)$, and chloride ions, $Cl^-(aq)$, are found in the solution.

$$AgCl(s) \rightleftharpoons Ag^+(aq) + Cl^-(aq) \quad (24)$$

The concentrations of these ions determine the equilibrium solubility. Equation (25) describes the equilibrium situation.

$$K = \text{a constant} = [Ag^+][Cl^-] \quad (25)$$

Remember that the "concentration" of a solid does not appear in an equilibrium expression, such as (25), because the concentration does not vary.

In another example, consider the solubility of lead chloride, $PbCl_2$.

$$PbCl_2(s) \rightleftharpoons Pb^{2+}(aq) + 2Cl^-(aq) \quad (26)$$

At equilibrium, the following expression holds.

$$K = \text{a constant} = [Pb^{2+}][Cl^-]^2 \quad (27)$$

As you learned in the preceding experiment, such solubility equilibrium constants are given a special name, the "solubility product," with the symbol K_{sp}. The **solubility product** of an electrolyte is the product of the concentrations of its ions in a saturated solution, each concentration raised to the appropriate power. A **saturated solution** is a solution containing the maximum amount of solute that will dissolve at a given temperature. A low value of the K_{sp} means the concentrations of ions are low at equilibrium. Hence the solubility must be low. Table 22-5 lists solubility products for some common compounds.

22-19 Calculation of the Solubility from the Solubility Product

The solubility product can be determined from measurements of the solubility. In turn, it can be used as a basis for calculations of solubility. Suppose you want to know how much cuprous chloride, CuCl, will dissolve in one liter of water. In order to determine this, you begin by writing the balanced equation for the reaction.

$$\text{CuCl}(s) \rightleftharpoons \text{Cu}^+(aq) + \text{Cl}^-(aq) \quad (28)$$

From this equation, you can write the equilibrium expression.

$$K_{sp} = [\text{Cu}^+][\text{Cl}^-] \quad (29)$$

You can find the numerical value of the K_{sp} at 25°C in Table 22-5.

$$K_{sp} = 3.2 \times 10^{-7} = [\text{Cu}^+][\text{Cl}^-] \quad (30)$$

Expression (30) indicates that cuprous chloride dissolves until the molar concentrations of cuprous ion and chloride ion rise enough to make their product equal to 3.2×10^{-7}.

Now suppose you designate the solubility of cuprous chloride in water by the symbol s. This symbol represents the moles of solid cuprous chloride that dissolve in one liter of water. Using equation (28), you see that s moles of solid cuprous chloride produce s moles of cuprous ion, $\text{Cu}^+(aq)$, and s moles of chloride ion, $\text{Cl}^-(aq)$. Hence these concentrations must be equal.

$$[\text{Cu}^+] = [\text{Cl}^-] = s \text{ moles/liter}$$
$$= \text{moles CuCl dissolved} \quad (31)$$

Substituting expression (31) into equation (30), you have the following.

$$K_{sp} = 3.2 \times 10^{-7} = s \times s = s^2$$
$$s^2 = 3.2 \times 10^{-7} = 32 \times 10^{-8}$$
$$s = \sqrt{32 \times 10^{-8}} = 5.7 \times 10^{-4} \text{ mole/liter}$$
$$s = 0.00057 \, M \quad (32)$$

22-20 Will a Precipitate Form?

Many electrolytes have only slight solubility in water. Chemists often face a problem concerning the solubility behavior of electrolytes of low solubility. Calcium sulfate, CaSO_4, provides a good illustration of this. Suppose that two solutions, one containing sulfate ions and the other containing calcium ions, are mixed. Then the two components of a calcium sulfate solution are present together in the same solution. They may not be present in equal concentrations, as they must be when calcium sulfate itself is dissolved in water. The question that presents itself is: For given concentrations of sulfate and calcium ions, will a calcium sulfate precipitate form? (A precipitate of calcium sulfate is

TABLE 22-5 Some Solubility Products at Room Temperature (25°C)

Compound	Solubility Product	Compound	Solubility Product
TlCl	1.9×10^{-4}	$SrCrO_4$	3.6×10^{-5}
CuCl	3.2×10^{-7}	$BaCrO_4$	8.5×10^{-11}
AgCl	1.7×10^{-10}	$PbCrO_4$	2.0×10^{-14}
TlBr	3.6×10^{-6}	$CaSO_4$	2.4×10^{-5}
CuBr	5.9×10^{-9}	$SrSO_4$	7.6×10^{-7}
AgBr	5.0×10^{-13}	$PbSO_4$	1.3×10^{-8}
TlI	8.9×10^{-8}	$BaSO_4$	1.5×10^{-9}
CuI	1.1×10^{-12}	$RaSO_4$	3.8×10^{-11}
AgI	8.5×10^{-17}	$AgBrO_3$	5.4×10^{-5}
		$AgIO_3$	3.1×10^{-8}

shown in Figure 22-13.) The solubility product allows you to answer the question with confidence.

For example, consider the following two reactions.
1. Equal volumes of 0.02 M $CaCl_2$ and 0.0004 M Na_2SO_4 are mixed.
2. Equal volumes of 0.08 M $CaCl_2$ and 0.02 M Na_2SO_4 are mixed.

Will a precipitate form in either reaction? in both?

In order to answer these questions, you first write the balanced equation for the reaction of calcium sulfate dissolving in water. Then you write the equilibrium law relation.

$$CaSO_4(s) \rightleftharpoons Ca^{2+}(aq) + SO_4^{2-}(aq) \quad (33)$$
$$K_{sp} = [Ca^{2+}][SO_4^{2-}] \quad (34)$$

Next you find the concentration of each ion. After equal volumes are mixed, each ion is present in twice as much solvent. Therefore, the concentration of each ion is only half as

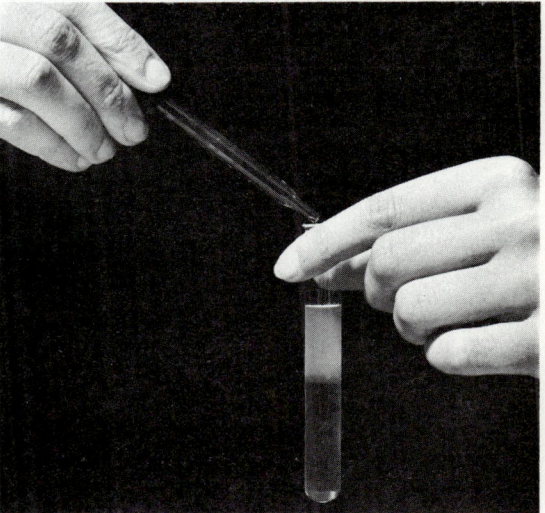

Figure 22-13 Predicting whether a precipitate, such as that of calcium sulfate (above), will form requires a knowledge of concentrations and solubility products.

great as it was before mixing. The solubility calculations for the first reaction follow.

$$[Ca^{2+}] = \frac{0.02\ M}{2} = 0.01\ M = 1 \times 10^{-2}\ M$$

$$[SO_4^{2-}] = \frac{0.0004\ M}{2} = 0.0002\ M$$
$$= 2 \times 10^{-4}\ M$$

If the product of the preceding ion concentrations is greater than the K_{sp}, precipitation will occur. If it is less, there will be no precipitation. In order to understand why this is so, recall that the K_{sp} describes a saturated solution. A saturated solution is a solution containing the *maximum* amount of solute that will dissolve at a given temperature. Thus, a product of ion concentrations greater than the K_{sp} indicates that the solvent cannot dissolve all of the ions. A product of ion concentrations less than the K_{sp} indicates that the solvent can dissolve additional ions.

With this knowledge in mind, you can interpret the calculation that follows.

$$[Ca^{2+}][SO_4^{2-}] = (1 \times 10^{-2})(2 \times 10^{-4})$$
$$= 2 \times 10^{-6}$$

This product, 2×10^{-6}, is *less* than the K_{sp}, which equals 2.4×10^{-5}. Thus, a $CaSO_4$ precipitate will *not* occur in the first reaction.

The calculations for the second reaction are done in the same way.

$$[Ca^{2+}] = \frac{0.08\ M}{2} = 0.04\ M = 4 \times 10^{-2}\ M$$

$$[SO_4^{2-}] = \frac{0.02\ M}{2} = 0.01\ M = 1 \times 10^{-2}\ M$$

Again you compute the ion product.

$$[Ca^{2+}][SO_4^{2-}] = (4 \times 10^{-2}) \times (1 \times 10^{-2})$$
$$= 4 \times 10^{-4}$$

This time the ion product, 4×10^{-4}, is *greater* than the K_{sp}, which equals 2.4×10^{-5}. Therefore, a $CaSO_4$ precipitate *does* form. Solid calcium sulfate will continue to form, lowering the concentrations of $Ca^{2+}(aq)$ and $SO_4^{2-}(aq)$ until they are low enough that the ion product equals the K_{sp}. Then equilibrium will exist, and no more precipitation will occur.

22-21 The Law of Chemical Equilibrium Derived from Rates of Opposing Processes ★

Chemists picture equilibrium as a dynamic balance between opposing processes. An understanding of the Law of Chemical Equilibrium can be built upon this basis.

Consider the reaction between nitric oxide, NO, and oxygen gas, O_2.

$$2NO(g) + O_2(g) \longrightarrow 2NO_2(g) \qquad (35)$$

The reaction to the right (R) proceeds with a rate that is found experimentally to depend upon the concentrations of the reactants as follows.

$$(\text{rate})_R = k_R[NO]^2[O_2] \qquad (36)$$

Remember that k is the symbol for a rate constant, which was introduced in Section 21-5. Chemists have also studied the reverse reaction, to the left (L).

$$2NO_2(g) \longrightarrow 2NO(g) + O_2(g) \qquad (37)$$

The rate of this reaction depends upon the concentration of the product as follows.

$$(\text{rate})_L = k_L[NO_2]^2 \qquad (38)$$

Macroscopic chemical changes cease when the rate of reaction (35) is exactly equal to that of reaction (37). When this is so, expressions (36) and (38) can be equated.

$$(\text{rate})_R = (\text{rate})_L \tag{39}$$

$$k_R[NO]^2[O_2] = k_L[NO_2]^2 \tag{40}$$

By algebraic rearrangement, equation (40) can be written as follows.

$$\frac{k_R}{k_L} = \frac{[NO_2]^2}{[NO]^2[O_2]} \tag{41}$$

Since both k_R and k_L are constants at a given temperature, their ratio is constant. Hence equation (41) is an example of the equilibrium law. In every reaction that has been sufficiently studied, this same result is obtained. Therefore, chemists have confidence in the molecular view of equilibrium in chemical reactions. According to this view, a dynamic balance between opposing reactions exists.

Solubility equilibrium is similar to equilibrium in chemical reactions. At solubility equilibrium the rates of dissolving and precipitation must be equal.

One of the factors that influences the rate of dissolving a solid, such as iodine, is the area, A, of the crystal surface that makes contact with the liquid. If many crystals (with large A) are dissolving simultaneously, the rate of dissolving is faster than if only a few crystals (with small A) are in the solvent. The rate of dissolving is proportional to this liquid-solid surface area, A.

A molecule of iodine, for example, is more stable in the crystal than in the solution. For this reason, the potential energy must rise as a molecule leaves the crystal. Here the principles that govern the rates of reaction are operating. Presumably there is an activated complex for the process. The rate at which molecules, passing over the energy barrier, leave a square centimeter of surface is determined by the height of the barrier and by the temperature. You can call this rate k_d. Changing the temperature does not affect the activated complex. However, the molecular energy distribution is altered. (See Figure 21-9.) Hence k_d is a function of temperature, as the following equation shows.

$$\text{rate of dissolving} = \begin{pmatrix} \text{surface} \\ \text{area} \end{pmatrix} \times \begin{pmatrix} \text{rate molecules} \\ \text{leave a square} \\ \text{centimeter of} \\ \text{crystal surface} \end{pmatrix}$$

$$\text{rate of dissolving} = A \times k_d \tag{42}$$

These two factors are depicted in Figure 22-14, on the next page.

The rate of precipitation, on the other hand, is the rate at which molecules return to the surface and fit into the crystal lattice. To do this, the molecules in solution must first strike the crystal surface. Again, the larger the surface, the more frequently will dissolved molecules encounter a piece of crystal. The rate of precipitation, then, is proportional to A.

In addition, the rate at which molecules strike the surface depends upon how many molecules there are per unit volume of solution. As the concentration rises, more and more molecules strike the surface per unit time. Thus, the rate of precipitation is proportional to the iodine concentration, $[I_2]$.

The final factor is, again, the rate at which molecules can pass over the energy barrier, the activated complex for precipitation. This rate constant, k_p, is determined by the temperature and the height of the energy barrier to precipitation.

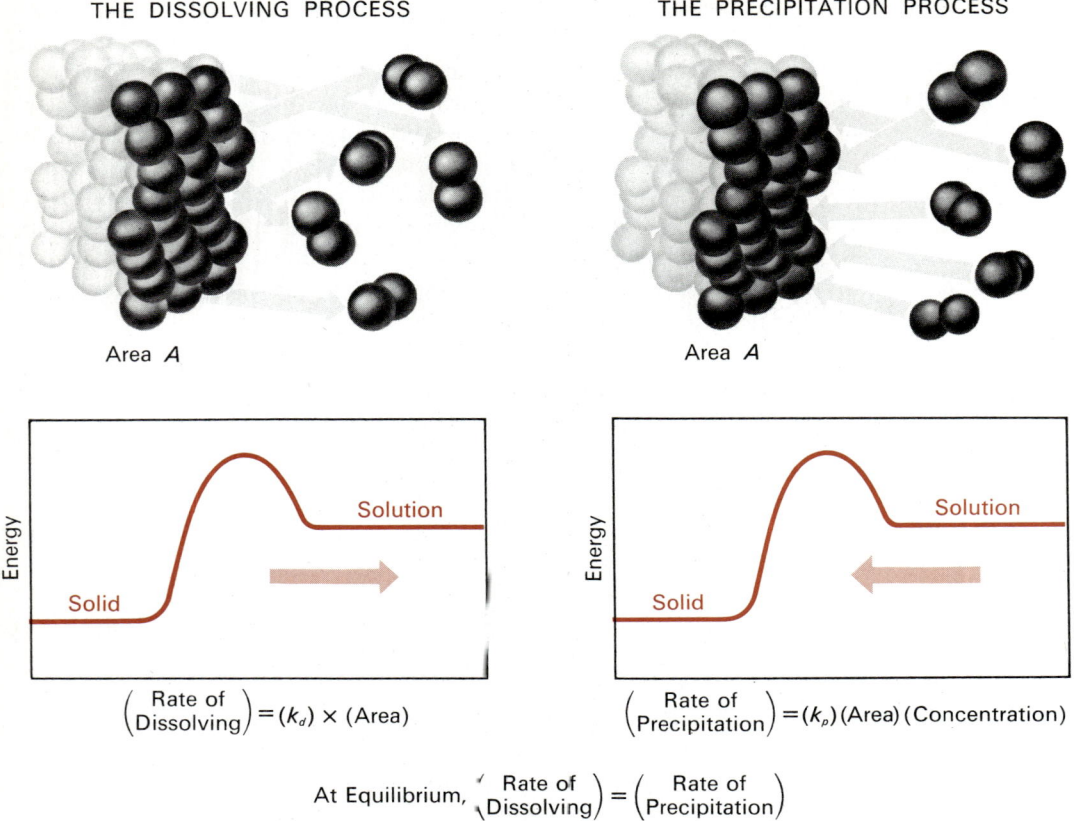

Figure 22-14 As depicted above, dissolving and precipitation are taking place on the microscopic scale at equal rates when equilibrium has been established.

You can see that three factors determine the rate of precipitation. These factors are depicted in Figure 22-14.

$$\text{rate of precipitation} = \begin{pmatrix} \text{surface} \\ \text{area} \end{pmatrix} \times \begin{pmatrix} \text{concentration} \\ \text{of dissolved } I_2 \end{pmatrix}$$

$$\times \begin{pmatrix} \text{rate dissolved molecules} \\ \text{pass over energy barrier} \end{pmatrix}$$

$$= (A) \times [I_2] \times (k_p) \quad (43)$$

At equilibrium you can equate (42) and (43).

rate of dissolving = rate of precipitation

$$A \times k_d = A \times [I_2] \times k_p \quad (44)$$

The area of contact, A, appears both on the left and on the right of expression (44). Hence, you can cancel it out. Dividing both sides of (44) by k_p, you obtain the following equation.

$$\frac{k_d}{k_p} = [I_2] \qquad (45)$$

Since k_d and k_p each depend upon temperature, their ratio depends upon temperature. Otherwise, each is constant. Thus you can write equation (46), in which $K = k_d/k_p$.

$$K = [I_2] \qquad (46)$$

By expressing the dynamic balance between the rates of dissolving and precipitating you obtain equation (46). The concentration of I_2 at equilibrium is a constant, fixed by the temperature. This constant is equal to the solubility of I_2, expressed in moles per liter.

22-22 Factors That Determine Equilibrium ★

You have learned much about equilibrium. You have discovered that although it is characterized by constancy of macroscopic properties, molecular processes continue in a state of balance. You know that at equilibrium every reaction that takes place does so at the same reaction rate as its reverse reaction.

You have gone further and discovered that the equilibrium conditions imply a constant relationship among the concentrations of reactants and products. This relationship is called the Law of Chemical Equilibrium. Using this law, the conditions at equilibrium can be expressed in terms of the equilibrium constant, K.

Despite this detailed familiarity with equilibrium, you have not considered one aspect at all. What determines the equilibrium constant? Why does one reaction favor reactants and another reaction favor products? What factors cause sodium chloride to have a large solubility in water? Why does equilibrium favor the reaction of oxygen with iron to form Fe_2O_3 (rust) but not the reaction of oxygen with gold? Scientists cannot resist wondering what factors determine the conditions at equilibrium.

As emphasized in Chapter 1, science is not merely a method of collecting facts or of discovering relations among facts. It is also—and this is in a sense its chief delight—an activity involving the search for explanations of the facts and of the patterns and relations observed.

An explanation is a likeness that connects the system under study with a model, that is, with a more familiar system. For example, consider the one shown on the next page (Figure 22-15). Here you see a golf bag that was thrown into the rear of a station wagon. The ball pocket was open, so all of the golf balls have spilled out onto the floor. Because this floor has a step in it, the golf balls on the upper level possess some potential energy, energy of position. The golf balls tend to roll to the lower level spontaneously, as shown in Figure 22-16 (next page). As a golf ball does this, its potential energy becomes kinetic energy, energy of motion. Finally, this kinetic energy is dissipated into heat. Now the golf balls lie at rest at the lower level of the floor.

This situation is similar to the chemical change in a spontaneous, exothermic reaction. The reactants of high heat content react spontaneously to form products of lower heat content. As each molecular reaction occurs, the excess heat content becomes kinetic energy. The product molecules separate from each other with high kinetic energy. As they collide with other molecules, this energy is dissipated into heat.

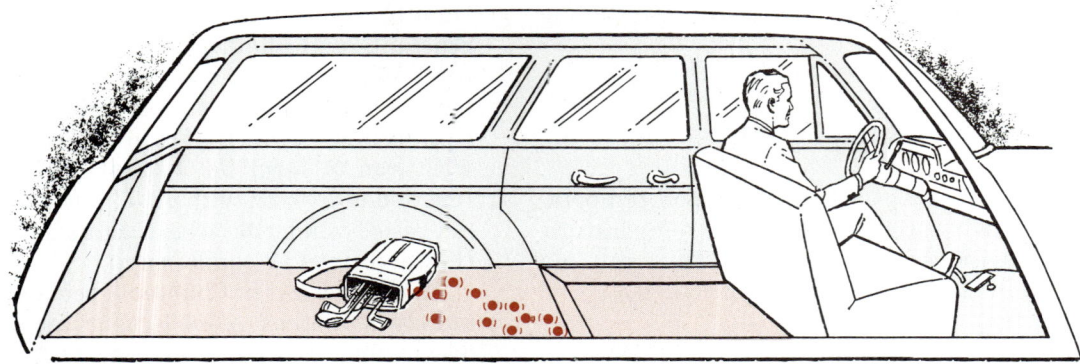

Figure 22-15 Loose golf balls on the floor of a station wagon are a useful model for the understanding of spontaneous exothermic reactions. Golf balls on the upper surface can roll to the lower potential energy surface. This is analogous to reactants forming products that have lower potential energy.

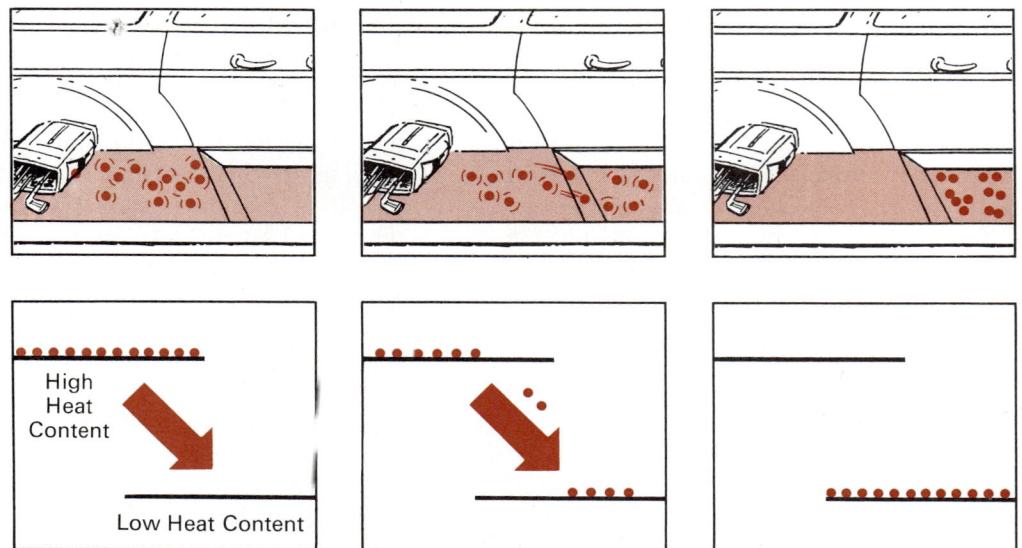

Figure 22-16 The golf ball model has many close parallels with spontaneous reaction heat diagrams. Initially, the golf balls have high potential energy and the reactants have a high heat content. In both cases, energy is converted to kinetic energy, then to heat, until the final state is reached. Both changes are spontaneous.

Figure 22-16 compares a chemical reaction with the rolling golf ball situation. This comparison can be summarized as follows.

1. Each system has two states.

	Initial State	Final State
Golf balls	on upper level ⟶	on lower level
Reaction	reactants ⟶	products

2. The potential energy of the initial state is higher than the potential energy of the final state.

	Initial State	Final State
Golf balls	high potential energy ⟶	low potential energy
Reaction	high heat content ⟶	low heat content

3. As the change from initial state to final state proceeds, the form of energy changes.

	Initial State	Final State
Golf balls	potential energy ⟶	kinetic energy, then heat
Reaction	heat content ⟶	molecular kinetic energy, then heat

4. The changes from initial to final state proceed spontaneously toward lowest potential energy, the direction corresponding to "rolling downhill."

	Initial State ⟶ Final State
Golf balls	spontaneous
Reaction	spontaneous

Having established these similarities, a possible generalization can be proposed.

Since golf balls always roll downhill spontaneously

Perhaps reactions always proceed spontaneously in the direction toward minimum energy.

From this generalization you might expect that a reaction tends to proceed spontaneously only if the products have lower energy than the reactants. This expectation agrees with some experimental evidence, especially for those reactions that release a large amount of heat.

There are, however, two basic and serious difficulties with this proposed explanation. (1) Some *endothermic* reactions proceed spontaneously. One example is the evaporation of a liquid. Water evaporates spontaneously, absorbing heat as it does so. It does not "roll downhill" energetically. When ammonium chloride dissolves in water, the solution becomes cooler. Again heat is absorbed. Yet the ammonium chloride dissolves. (2) Another difficulty is that spontaneous chemical reactions do not go to completion. Even if a spontaneous reaction is exothermic, it proceeds only until it reaches equilibrium. However, in the golf ball analogy, "equilibrium" is reached when *all* of the golf balls are on the lower level. Thus we might expect that an exothermic reaction would proceed until *all* of the reactants were converted to products, not to a dynamic equilibrium.

Because of these failures, our proposed explanation needs to be altered, and another analogy must be found. The analogy must correspond better with the behavior of chemical reactions. How can the golf ball analogy be altered to bring it into better agreement with experimental evidence? Another factor could be introduced.

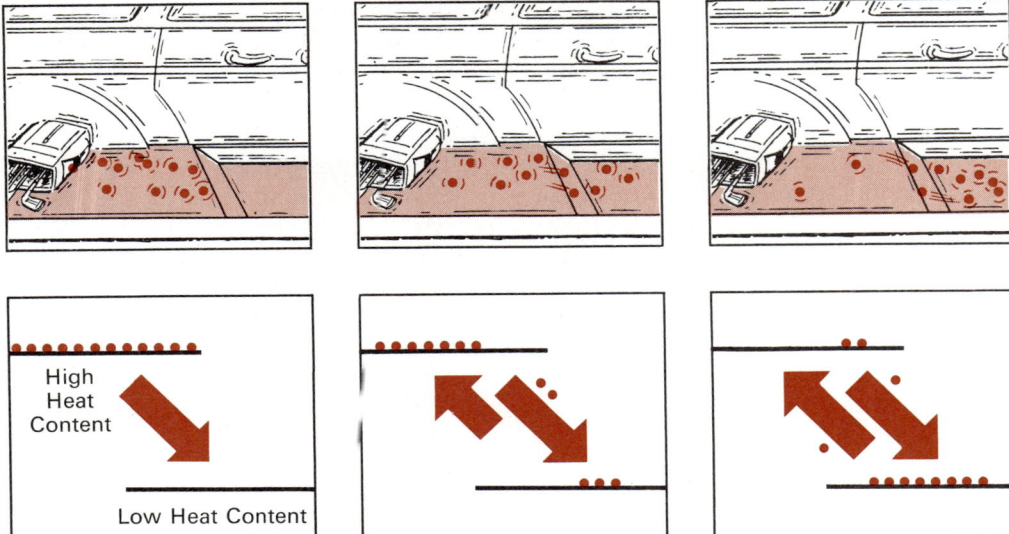

Figure 22-17 *The golf ball model provides further answers if it is assumed the wagon is being driven over a very bumpy road. Through collisions, golf balls are able to reach the higher potential energy on the upper surface. When just as many golf balls are rolling up as are rolling down, equilibrium exists.*

Consider how the golf ball situation shown in Figure 22-16 might change when the station wagon is driven over a bumpy road. Now the golf balls are shaken and jostled about. They roll around and collide with each other. Every now and then one of the golf balls even accumulates enough energy through collisions to return to the upper level of the station wagon floor. Of course, any golf ball that is bounced up tends to roll back down to the lower level a little later. As this bumpy ride continues, a state is reached in which golf balls are being jostled up to the higher level at the same rate they are rolling back down to the lower level. Since the rate of bouncing up equals the rate of rolling down, a dynamic balance exists, as indicated in Figure 22-17.

This analogy solves the problem. The bumpy road model contains a new feature that provides a basis for expecting "reaction" in the endothermic direction. Some golf balls roll uphill if they are shaken hard enough. The tendency to roll back down will always keep them coming back to the lower level. Finally, when the rate of rolling down equals the rate of jostling up, equilibrium will be reached.

What happens when the road becomes smoother? The jostling-up reaction is less favored. The equilibrium conditions change in favor of the golf balls at the lower level.

Returning to the chemical reaction, what feature in a reacting chemical system corresponds to the bumpy road in the analogy? It is the *temperature*. At any temperature except absolute zero there is a constant random jostling of the molecules. Some molecules have low kinetic energies, and others have high kinetic energies, as was described in

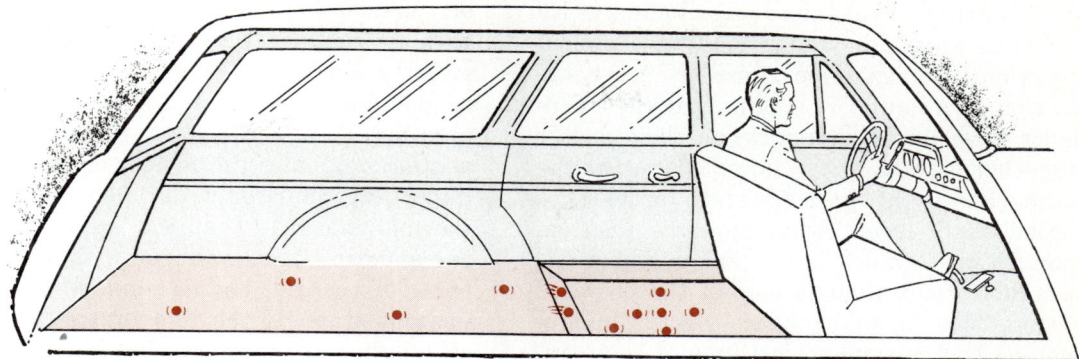

Figure 22-18 *With the golf bag removed, there is more space on the upper level, and more golf balls can remain at a higher energy. This "available space" is analogous to the gas phase, in which molecules can exist at equilibrium despite their higher kinetic energy.*

Section 21-7. Some of the molecules will occasionally accumulate enough energy to "roll uphill" to less stable molecular forms. The higher the kinetic energy, the greater the number of molecules that will "roll uphill."

On the one hand, molecular changes will take place in the direction of minimum energy. On the other hand, random jostlings, or energy transfers, will restore molecules to a higher energy state. When these opposing processes occur at the same rate, a state of dynamic equilibrium exists.

This analogy should aid you in understanding chemical reactions and equilibrium. The following features of chemical reactions should now be clear.

1. Chemical reactions spontaneously approach the equilibrium state.
2. One factor that fixes the equilibrium state is energy. Equilibrium tends to favor the state of lowest energy.
3. The other factor that fixes the equilibrium state is randomness caused by the temperature. Equilibrium tends to favor the state of greatest randomness.
4. The equilibrium state is a compromise between these two factors, minimum energy and maximum randomness. At very low temperatures energy tends to be the more important factor. Here equilibrium favors the molecular substances having the lowest heat content. At very high temperatures randomness becomes more important. In this case equilibrium favors a random distribution among reactants and products without regard for energy differences.

This analogy can be stretched one point further. You may have wondered whether the relative area of the upper floor level compared with that of the lower floor level may partially determine the distribution of golf balls. After all, if the "available space" on the upper level area is small, as in Figure 22-15, few golf balls are likely to remain there. In contrast, Figure 22-18 shows that, when the golf bag has been removed, the upper level has much more area. Golf balls reaching the upper level have a great deal of space. They can roll around longer on this level before returning to the lower level. By enlarging the upper level, the fraction of golf balls occupying the upper level at "equilibrium" is increased.

This extension of the analogy increases its value as an aid in understanding chemical reactions. For example, consider the vaporization of a liquid. It is true that the molecules have lower energy when they cluster together tightly in the liquid state. On the other hand, the gaseous state provides a broad upper level. Every molecule that vaporizes has a much larger amount of space available to it than it had in the crowded liquid. This "available space" factor, accompanied by the random jostlings of temperature to overcome the potential energy difference, aids vaporization.

Now recall what happens as a solid substance is warmed from a very low to a very high temperature. As the temperature of the solid is raised, the lower energy of the regular solid becomes unimportant compared with the random thermal energies. The solid melts, giving up its energy stability for the randomness of the liquid state. If the temperature is raised higher, the energies of attraction among the molecules become unimportant compared with the higher random thermal energies. Then the liquid vaporizes, surrendering the lower potential energy in favor of the still higher randomness of the gaseous state. If the temperature is raised still higher, the energies that hold molecules together begin to become unimportant compared with the random thermal energies. Finally, at extremely high energies, molecules no longer exist. All is chaos. This is the chemical situation that exists at the very high temperatures within the sun.

Summary

The position of an equilibrium at any specified temperature may be stated quantitatively by the Law of Chemical Equilibrium. This law relates the concentrations of the products to those of the reactants. The numerator of this expression contains the product of the concentrations of the substances formed in a reaction, each concentration raised to the appropriate power. The denominator is similarly formed, using all the substances on the left side. The Law of Chemical Equilibrium states that the resulting quotient is a constant at any fixed temperature.

The Law of Chemical Equilibrium applies to solubility equilibrium as well as to reaction equilibrium. For electrolyte solutions the solubility equilibrium constant is called the solubility product. It can be used to calculate whether a substance will precipitate when solutions containing its component ions are mixed.

Self Check

1. State in words the Law of Chemical Equilibrium.
2. Write the expressions for the equilibrium constants of the following reactions.
   (a) $Cu(s) + 2Ag^+(aq) \rightleftharpoons Cu^{2+}(aq) + 2Ag(s)$
   (b) $H_2O_2(g) \rightleftharpoons H_2O(g) + \frac{1}{2}O_2(g)$

Chapter Summary

All processes that occur in a closed system at a uniform temperature eventually reach a state of equilibrium. This state is characterized by constancy of the macroscopic properties of the system. However, the equilibrium is dynamic, for changes continue to occur at the molecular level. The rates of the opposing reactions or processes are equal and so cancel out one another. Thus no chemical reaction in a closed system ever truly goes to completion. Some definite concentration, however small, of each of the reactants must remain when equilibrium is attained. In an open system a reaction may go to completion due to the escape or withdrawal of one or more of the products.

The rate at which different reactions or systems come to equilibrium is highly variable. A catalyst speeds up both the forward and reverse rates to the same extent. Therefore, it merely increases the rate at which a reaction reaches equilibrium. It does not influence the position of that equilibrium.

The qualitative effects of changes in pressures, temperatures, and concentrations on the position of equilibrium can be predicted from Le Chatelier's principle. This principle states that in a system at equilibrium, processes will occur that tend to counteract any imposed change in temperature, pressure, or concentration.

The position of equilibrium can be quantitatively specified at any given temperature by the numerical value of the equilibrium constant of the reaction. For the reaction

$$aA + bB + cC = dD + eE + fF$$

the equilibrium constant is given as follows.

$$K = \frac{[D]^d[E]^e[F]^f}{[A]^a[B]^b[C]^c}$$

This is the Law of Chemical Equilibrium.

Solubility equilibrium also obeys this law. For simple molecular solutes the equilibrium constant is identical to the solubility in moles per liter. For ionic solutions, the equilibrium expression is called the solubility product.

Questions and Problems

1. When are a liquid and its vapor said to be in equilibrium?
2. Describe an example that illustrates the statement that "equilibrium is characterized by constancy of macroscopic properties."
3. What, specifically, is "equal" in a chemical reaction that has attained a state of equilibrium?
4. What is meant when a reaction is said to "go to completion"? Is this more likely to happen in a closed system or an open system? Explain.
5. Will one drop of water establish a state of vapor pressure equilibrium when placed in a closed bottle? Explain.
6. Why are chemical equilibria referred to as "dynamic"?
7. What effect does a catalyst have on the equilibrium conditions of a chemical system?
8. What do the following experiments, done at 25°C, indicate about equilibrium?

   (a) One liter of water is added, a few milliliters at a time, to a kilogram of salt. Only part of the salt dissolves.
   (b) A large saltshaker containing one kilogram of salt is gradually emptied into one liter of water. The same amount of salt dissolves as in (a).
9. In any discussion of chemical equilibrium, why are concentrations always expressed in moles per unit volume rather than in grams per unit volume?
10. The following chemical equation represents the reaction between hydrogen and chlorine to form hydrogen chloride.

    $$H_2(g) + Cl_2(g) \rightleftharpoons 2HCl(g) + 44.0 \text{ kcal}$$

    (a) List four important facts indicated by this equation.
    (b) What are three important areas of interest concerning this reaction for which no information is indicated?

11. Consider the phase change described by the following equation.

$$\text{heat} + H_2O(l) \rightleftharpoons H_2O(g)$$

   (a) What will be the effect of reducing the volume, thus increasing the pressure?
   (b) What will be the effect of an increase in temperature?
   (c) What will be the effect of injecting some steam into the closed system?

12. Consider the following reaction.

$$4HCl(g) + O_2(g) \rightleftharpoons 2H_2O(g) + 2Cl_2(g) + 27 \text{ kcal}$$

   What effect would the following changes have on the equilibrium concentration of Cl_2? Give reasons for your answers.
   (a) Increasing the temperature
   (b) Decreasing the total pressure
   (c) Increasing the concentration of O_2
   (d) Increasing the volume of the reaction chamber
   (e) Adding a catalyst

13. Write the equation for the dissociation of $HI(g)$ into its elements.
   (a) Will HI dissociate to a greater or lesser extent as the temperature is increased? $\Delta H = -6.2$ kcal/mole $HI(g)$.
   (b) How many grams of iodine are present at equilibrium if 0.050 mole of HI has dissociated?

14. Consider the following two separate closed systems at equilibrium.
   (a) HI and the elements from which it is formed
   (b) H_2S and the elements from which it is formed

   What would happen in each system if the total pressure were increased? Assume that all reactants and products are gases.

15. Each of the following systems is at equilibrium. In each case, would the equilibrium concentration of each substance increase, decrease, or show no change when the indicated reagent is added to the system?
   (a) Reaction: $C_2H_6(g) \rightleftharpoons H_2(g) + C_2H_4(g)$
       Added reagent: $H_2(g)$
   (b) Reaction: $Cu^{2+}(aq) + 4NH_3(g) \rightleftharpoons Cu(NH_3)_4^{2+}(aq)$
       Added reagent: $CuSO_4(s)$
   (c) Reaction: $Ag^+(aq) + Cl^-(aq) \rightleftharpoons AgCl(s)$
       Added reagent: $AgCl(s)$
   (d) Reaction: $PbSO_4(s) + H^+(aq) \rightleftharpoons Pb^{2+}(aq) + HSO_4^-(aq)$
       Added reagent: $Pb(NO_3)_2(s)$

16. Consider the following equilibrium.

$$SO_2(g) + \tfrac{1}{2}O_2(g) \rightleftharpoons SO_3(g) + 23 \text{ kcal}$$

   (a) What conditions favor a high equilibrium concentration of SO_3?
   (b) How many grams of oxygen gas are needed to form 1.00 gram of SO_3?

17. The following reaction takes place in a closed system at 450°C. Its equilibrium constant is 50.0 at this temperature.

$$H_2(g) + I_2(g) \rightleftharpoons 2HI(g)$$

   What is the equilibrium constant of the reverse reaction at 450°C?

18. Write the equilibrium law relation for each of the following reactions.
   (a) $N_2(g) + 3H_2(g) \rightleftharpoons 2NH_3(g)$
   (b) $CO(g) + NO_2(g) \rightleftharpoons CO_2(g) + NO(g)$
   (c) $Zn(s) + 2Ag^+(aq) \rightleftharpoons Zn^{2+}(aq) + 2Ag(s)$
   (d) $PbI_2(s) \rightleftharpoons Pb^{2+}(aq) + 2I^-(aq)$
   (e) $CN^-(aq) + H_2O(l) \rightleftharpoons HCN(aq) + OH^-(aq)$

19. Using the equilibrium constants given, state which of the following reactions goes farthest toward completion. Explain.
   (a) $CH_3COOH(aq) \rightleftharpoons H^+(aq) + CH_3COO^-(aq)$
       $K = 1.8 \times 10^{-5}$

(b) $CdS(s) \rightleftharpoons Cd^{2+}(aq) + S^{2-}(aq)$
$K = 7.1 \times 10^{-28}$
(c) $H^+(aq) + HS^-(aq) \rightleftharpoons H_2S(aq)$
$K = 1.0 \times 10^7$

20. Consider the following reaction.

$$2HI(g) \rightleftharpoons H_2(g) + I_2(g)$$

At 448°C the partial pressures of the gases at equilibrium are as follows.

$HI = 4 \times 10^{-3}$ atm
$H_2 = 7.5 \times 10^{-3}$ atm
$I_2 = 4.3 \times 10^{-5}$ atm

What is the equilibrium constant for this reaction?

21. Sugar is added to a cup of coffee until no more sugar will dissolve. Does addition of another spoonful of sugar increase the rate at which the sugar molecules leave the crystal phase and enter the liquid phase? Will the sweetness of the liquid be increased by this addition? Explain.

22. How does an aqueous iodine solution differ from an aqueous salt solution?

23. Assume that the following compounds dissolve in water to form separate ions in solution. Write the formulas and names for the ions that can be expected.
(a) HI (d) $Ba(OH)_2$
(b) $CaCl_2$ (e) KNO_3
(c) Na_2CO_3 (f) NH_4Cl

24. Write the solubility product expression for an aqueous solution of each of the following: (a) calcium carbonate, (b) silver sulfide, (c) aluminum hydroxide.

25. The solubility product of AgCl is 1.4×10^{-4} at 100°C. Calculate the solubility, in moles per liter, of silver chloride in boiling water.

26. Experiments show that 0.0059 gram of $SrCO_3$ will dissolve in 1.0 liter of water at 25°C. What is K_{sp} for $SrCO_3$?

27. Write the solubility product expression for each of the following reactions.
(a) $BaSO_4(s) \rightleftharpoons Ba^{2+}(aq) + SO_4^{2-}(aq)$
(b) $Zn(OH)_2(s) \rightleftharpoons Zn^{2+}(aq) + 2OH^-(aq)$
(c) $Ca_3(PO_4)_2(s) \rightleftharpoons 3Ca^{2+}(aq) + 2PO_4^{3-}(aq)$

28. How many milligrams of silver bromide dissolve in 20 liters of water at 25°C? Use the data given in Table 22-5.

29. To one liter of 0.001 M H_2SO_4 is added 0.002 mole of solid $Pb(NO_3)_2$. As the lead nitrate dissolves, will lead sulfate precipitate? Use data from Table 22-5.

30. Suppose 10 ml of 1.0 M $AgNO_3$ are diluted to one liter with tap water. If the chloride concentration in the tap water is about 10^{-5} M, will a precipitate of silver chloride form? Use the data given in Table 22-5.

31. Will a precipitate exist at equilibrium if 0.5 liter of a 2×10^{-3} M $AlCl_3$ solution and 0.5 liter of a 4×10^{-2} M NaOH solution are mixed and diluted to 10^3 liters with water at room temperature? The K_{sp} for $Al(OH)_3$ is 5×10^{-33}.

32. The solubility of silver chloride is so low that all but a negligible amount of it is precipitated when excess sodium chloride solution is added to silver nitrate solution. How much precipitate forms when 100 ml of 0.5 M NaCl are added to 50.0 ml of 0.1 M $AgNO_3$?

33. Recall the factors that determine the rate of dissolving. Propose two methods for increasing the rate at which iodine dissolves in water.

34. Compare the dissolving rate in water of a single crystal of salt having a mass of one gram with that of finely divided grains of salt with a total mass of one gram.

35. What factor that is not involved in determining the rate of dissolving must be considered when determining the rate of precipitation? Why is it considered?

23 Acids and Bases

The acid or base content of a solution can readily be measured with the simple equipment being used in the above photo.

Acids and bases have chemical "personalities" familiar to you from your everyday experience. These personalities help in identifying an acid or a base in a general way. Acids, for example, corrode metal and taste sour. Vinegar and lemon juice both contain acids. The familiar corrosion that takes place around the terminals of a car storage battery is a result of the action of the sulfuric acid that the battery contains.

Bases, such as lye, ammonia, and washing soda, are also corrosive. When even a dilute solution of a base is rubbed between the fingers, it feels slippery. Both acids and bases are key chemicals in all the chemistry that occurs in water.

As soon as you begin to investigate these two distinctive types of compounds, many interesting and fundamental questions come up. How are acids and bases related to water? Are there certain atoms or groups of atoms common to all, or most, of the acids and bases that could account for their general similarities? Do acids and bases reinforce or cancel each other's properties? In this chapter such questions will be answered.

Water Itself Is an Electrolyte

23-1 Strong and Weak Electrolytes

Electrolytes were discussed in Chapters 17 and 22. They are substances that dissolve in water to give solutions containing ions. Such solutions conduct electricity. Examples are solutions of the electrolytes $HCl(g)$ and $NaOH(s)$.

$$HCl(g) + water \longrightarrow H^+(aq) + Cl^-(aq) \quad (1)$$

$$NaOH(s) + water \longrightarrow Na^+(aq) + OH^-(aq) \quad (2)$$

According to reaction (1), when HCl gas dissolves in water, all of the HCl molecules break up, or dissociate, into ions, $H^+(aq)$ and $Cl^-(aq)$. There is no experimental evidence for the presence of molecules of HCl in aqueous hydrochloric acid solutions. Therefore, chemists consider the dissociation to be complete. In a similar way, there is no evidence for the presence of molecules of NaOH in aqueous solution. Apparently the sodium hydroxide crystal breaks up completely into sodium ions, $Na^+(aq)$, and hydroxide ions, $OH^-(aq)$. A substance that dissolves and exclusively gives ions is called a **strong electrolyte**.

Not all substances that form conducting solutions dissociate so completely. For example, vinegar is simply an aqueous solution of acetic acid. Such a solution conducts electric current, showing that ions are present.

$$CH_3COOH(aq) \rightleftharpoons H^+(aq) + CH_3COO^-(aq) \quad (3)$$

However, the conductivity of a $0.1\,M$ acetic acid solution is much lower than that of a $0.1\,M$ hydrogen chloride solution. This and other experiments show that only a small fraction of the dissolved acetic acid has formed ions. Such a substance, which dissolves but only partially dissociates into ions, is called a **weak electrolyte**.

23-2 Water As a Weak Electrolyte

Pure water does not conduct electric current readily. Yet an extremely sensitive meter shows that the purest water obtainable has a very small degree of conductivity. Thus, water must dissociate to a very small extent, forming ions. The ions are the hydrogen ion, $H^+(aq)$, and the hydroxide ion, $OH^-(aq)$.

$$H_2O(l) \rightleftharpoons H^+(aq) + OH^-(aq) \quad (4)$$

This is an equilibrium involving three species in the liquid phase, H_2O, $H^+(aq)$, and $OH^-(aq)$. For reaction (4) the equilibrium law relation is the following one.

$$K = \frac{[H^+][OH^-]}{[H_2O]} \quad (5)$$

The concentration of H_2O molecules in water solutions is essentially 55.5 M. Since a liter of water at room temperature has a mass of 1000 g, there are $\frac{1000 \text{ g}}{18.0 \text{ g/mole}}$, or 55.5 moles of H_2O per liter of water. A dilute aqueous solution has essentially the same concentration of H_2O molecules. Since so few ions are formed, the concentration of H_2O molecules is virtually constant. Consequently, expression (5) is usually simplified by incorporating the factor 55.5 into the constant. When this is done, chemists usually label the constant K_w to indicate that it includes the factor $[H_2O]$. Thus, K_w is defined as the product of $[H^+]$ and $[OH^-]$.

$$K_w = [H^+][OH^-] \quad (6)$$

You can easily see how this product is related to K. Multiply both sides of equation (5) by $[H_2O]$ to give: $[H_2O] \times K = [H^+][OH^-]$. From this equation, the following relationship is readily shown.

$$K_w = [H^+][OH^-] = 55.5 \times K \quad (6a)$$

K_w is called a **dissociation constant.** The magnitude of K_w at 25°C is the following.

$$K_w = 1.00 \times 10^{-14} \text{ mole}^2/\text{liter}^2 \quad (7)$$

Like all equilibrium constants, its value changes with temperature. However, nearly all calculations made on water solutions are for 25°C. This is the only temperature we shall be concerned with in this book. The value of K_w, 10^{-14}, is a number worth remembering, because water is such an important solvent. Knowing the value of K_w, you can calculate ion concentrations. The same methods that you used in doing solubility calculations apply.

In reaction (4) there is one $H^+(aq)$ ion formed for every $OH^-(aq)$ ion. Hence, in pure water, where the only source of ions is reaction (4), the concentrations of $H^+(aq)$ and $OH^-(aq)$ must be equal.

$$[OH^-] = [H^+] \quad (8)$$

Substituting equation (8) into the equilibrium expression (6), you can calculate the concentrations of the two types of ions.

$$K_w = [H^+][OH^-] = [H^+][H^+]$$
$$= [H^+]^2$$
$$[H^+] = \sqrt{K_w} = \sqrt{1.00 \times 10^{-14}}$$
$$[H^+] = 1.00 \times 10^{-7} M \quad (9)$$

Using equation (8), you also can write the following.

$$[OH^-] = [H^+] = 1.00 \times 10^{-7} M \quad (10)$$

Now you can understand the low conductivity of pure water. Though water dissociates into ions, $H^+(aq)$ and $OH^-(aq)$, it does so only to a very slight extent. At equilibrium, the ion concentrations are only $10^{-7} M$. Therefore, water is a very weak electrolyte.

23-3 Changing the Concentrations of H^+ and OH^- Ions

You have seen that the concentrations of H^+ and OH^- ions are each equal to $10^{-7}\ M$ in pure water at 25°C. Suppose you add to water some electrolyte that supplies H^+ or OH^- ions. What will this do to the concentrations of the two ions? Suppose, for example, that you add some HCl.

HCl is a strong electrolyte, dissolving to give the ions $H^+(aq)$ and $Cl^-(aq)$. Hydrogen chloride adds $H^+(aq)$ but not $OH^-(aq)$ to the solution. The concentrations $[H^+]$ and $[OH^-]$ are then no longer equal. However they are still related by equilibrium relationship (6).

$$K_w = [H^+][OH^-] \qquad (6)$$

or,

$$[OH^-] = \frac{K_w}{[H^+]} \qquad (11)$$

Expression (11) shows that as the concentration of H^+ rises, the concentration of OH^- decreases.

Suppose, on the other hand, that sodium hydroxide, NaOH, is added to pure water. Sodium hydroxide is also a strong electrolyte, adding OH^- ions to the solution. Now you can rewrite equation (6) in the following form.

$$[H^+] = \frac{K_w}{[OH^-]} \qquad (12)$$

You can see that raising the concentration of OH^- lowers the concentration of H^+.

The important thing to understand is that equilibrium equation (6) must always be obeyed. The concentration of one of the ions, H^+ or OH^-, can only be increased if the concentration of the other is simultaneously decreased. Moreover, there is an exact, quantitative relationship. The product of the two concentrations must have the same value, K_w, regardless of how much of one or the other of the two ions is added to the solution.

Consider an example. Suppose 0.10 mole of hydrogen chloride is dissolved in 1.0 liter of water. Since HCl is a strong electrolyte, 0.10 mole of HCl forms 0.10 mole of $H^+(aq)$ ions—and 0.10 mole of $Cl^-(aq)$ ions—in the one liter volume. The total number of H^+ ions present will be greater than 0.10 mole because dissociation of water molecules also supplies these ions. However, you can quickly see that this contribution is so small in comparison to the quantity supplied by the HCl that it can be ignored. To understand this, recall that in pure water the concentration of H^+ ions supplied by dissociation of H_2O molecules is only $10^{-7}\ M$. The very large concentration of H^+ ions supplied by the added HCl means that $[OH^-]$ must go down in order to keep $[H^+][OH^-] = K_w$. With this addition of HCl, even fewer molecules of water will dissociate. Therefore, even fewer than 10^{-7} moles per liter of H^+ ions will be supplied by the water itself. Even a concentration of $10^{-7}\ M$ ($0.0000001\ M$) is so small in comparison to $0.10\ M$ that you can ignore it. It is only 0.0001 percent of the total. It is therefore an excellent approximation to say that in a $0.10\ M$ solution of HCl, the concentration of H^+ ions is $0.10\ M$.

If $[H^+]$ is $0.10\ M$, expression (11) allows you to calculate $[OH^-]$.

$$[OH^-] = \frac{K_w}{[H^+]} = \frac{1.00 \times 10^{-14}}{0.10}$$

$$= \frac{1.00 \times 10^{-14}}{1.0 \times 10^{-1}}$$

$$[OH^-] = 1.0 \times 10^{-13}$$

Adding the 0.10 mole of HCl to 1.0 liter of water lowers the OH^- concentration from

10^{-7} M, its value in pure water, to 10^{-13} M. In other words, [OH$^-$] changes by a factor of a million (10^6).

It should be clear that if 0.10 mole of NaOH is dissolved in 1.0 liter of water, [OH$^-$] = 0.10 M and the H$^+$ concentration will be reduced, by a factor of a million, to 1.0×10^{-13} M.

$$[\text{H}^+] = \frac{1.00 \times 10^{-14}}{0.10} = 1.0 \times 10^{-13}$$

The ease with which chemists can control and vary the concentrations of H$^+$(aq) and OH$^-$(aq) would be only a curiosity but for one fact. The H$^+$(aq) and OH$^-$(aq) ions take part in many important reactions that occur in aqueous solutions. Thus, if H$^+$(aq) is a reactant or a product in a reaction, the variation of the concentration of hydrogen ion by a factor of 10^6 can have an enormous effect. Such a change in a reaction at equilibrium requires a large change in the concentrations of all other reactants and products so that the equilibrium law may continue to be satisfied. Furthermore, there are many reactions for which either the hydrogen ion or the hydroxide ion acts as a catalyst. For example, the catalysis of the decomposition of formic acid by sulfuric acid was discussed in Section 21-10. Formic acid is reasonably stable until the hydrogen ion concentration is raised. Then the rate of the decomposition reaction becomes very rapid, as shown in Figure 23-1.

By affecting equilibrium conditions and by changing reaction rates, the concentrations of H$^+$(aq) and OH$^-$(aq) help chemists control reactions in aqueous solutions.

23-4 Acidic, Basic, and Neutral Solutions; the Definition of pH

In pure water [H$^+$] = [OH$^-$]. The concentrations of these ions will remain equal even

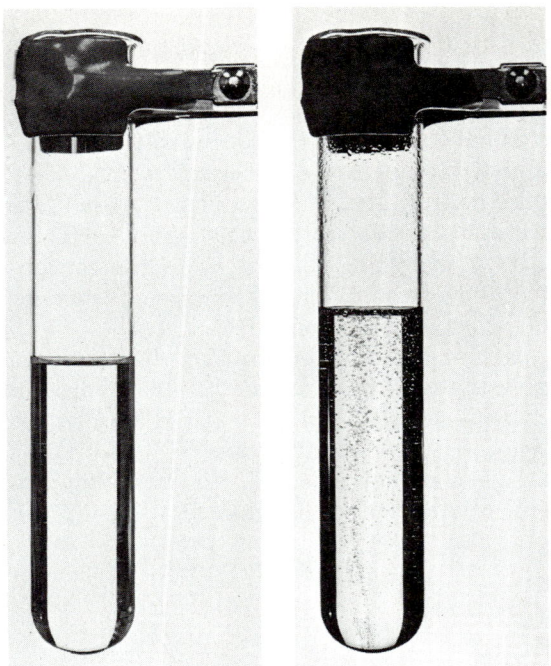

Figure 23-1 *The left photo shows formic acid to which no additional H$^+$ ions have been added. The addition of some sulfuric acid causes the formic acid to decompose rapidly* (right photo).

if solutes are dissolved in the water, provided the solutes do not supply either H$^+$ or OH$^-$ ions or do not react with these ions. Pure water itself, or any solution in which [H$^+$] = [OH$^-$], is said to be neutral.

An **acidic** solution is one in which [H$^+$] exceeds [OH$^-$]. In an acidic solution at 25°C, [H$^+$] is greater than 10^{-7} M, and [OH$^-$] is less than 10^{-7} M.

A **basic** solution is one in which [OH$^-$] is greater than [H$^+$]. In a basic solution at 25°C, [OH$^-$] is greater than 10^{-7} M, while [H$^+$] is less than 10^{-7} M.

The acidity of solutions must be referred to frequently. Therefore, it is useful to have a way of expressing acidity that is more compact than the numerical value of [H$^+$], which is usually a figure such as 1.00×10^{-4} mole/liter. Such a compact notation is provided by the pH scale.

The expression pH is defined by the following equation.

$$p\text{H} = -\log_{10}[\text{H}^+] \qquad (13)$$

Pure water at 25°C has [H$^+$] = 1.00×10^{-7} mole/liter. Hence, the pH of a neutral solution is equal to

$-\log_{10}(1.00 \times 10^{-7})$
$\qquad = -\log_{10}(1.00) - \log_{10}10^{-7}.$

Since $-\log_{10}(1.00) = 0$, then

$p\text{H} = -\log_{10}[10^{-7}] = -(-7) = 7.$

A solution that has a pH less than 7 is *acidic*. For instance, a 0.001 M solution of HCl has [H$^+$] = 10^{-3} mole per liter. You would calculate its pH as follows.

$$p\text{H} = -\log_{10}[10^{-3}] = -(-3) = 3$$

On the other hand, a basic solution has a pH higher than 7. For example, a 0.001 M solution of NaOH has a pH given by the following equations.

$$[\text{H}^+] = \frac{K_w}{[\text{OH}^-]} \qquad (12)$$

$$[\text{H}^+] = \frac{10^{-14}}{[\text{OH}^-]} = \frac{10^{-14}}{10^{-3}} = 10^{-11}$$
$$p\text{H} = -\log_{10}[10^{-11}] = -(-11) = 11$$

The symbol pH was first used by the Danish chemist Sörenson. The p stands for the Danish word *potenz* (power) and the H stands for the hydrogen ion. A common method of

Figure 23-2 *A pH meter, such as the one shown above, is commonly used to find the concentration of hydrogen ions (pH) in a solution.*

finding the pH of a solution is to use a pH meter such as the one shown in Figure 23-2.

23-5 *The Change of K_w with Temperature* ★

Experiments show that reaction (4) absorbs energy, as the following equations show.

H$_2$O(l) + 13.68 kcal $\longrightarrow$
$\qquad\qquad$ H$^+$(aq) + OH$^-$(aq) $\quad$ (14)

or,

H$_2$O(l) $\longrightarrow$ H$^+$(aq) + OH$^-$(aq)
$\qquad\qquad\qquad \Delta H = 18.68$ kcal $\quad$ (15)

As reactions (14) and (15) show, it takes 13.68 kcal to dissociate one mole of water into its ions.

This heat effect can be used in predicting how K_w changes with temperature. Le Chatelier's principle indicates that increased temperature will shift the equilibrium state toward larger concentrations of the ions (so as to absorb heat). Hence, K_w increases when the temperature is raised. Experimental values for K_w at various temperatures are given in Table 23-1.

TABLE 23-1 Values of K_w at Various Temperatures

Temperature (°C)	K_w
0	0.114×10^{-14}
10	0.295×10^{-14}
20	0.676×10^{-14}
25	1.00×10^{-14}
60	9.55×10^{-14}

Summary

There are two classes of electrolytes. Strong ones are completely dissociated into ions in aqueous solution, while weak ones are only partially dissociated. Water itself is a weak electrolyte. At 25°C, its dissociation constant, K_w, equals 1.00×10^{-14} mole²/liter². Although this dissociation is slight, it has an enormous influence on many chemical processes in aqueous solutions. Even a low concentration of a substance containing H^+ ions alters the concentration of the OH^- ion by many powers of ten. Addition of OH^- ions has a similarly large effect upon the H^+ concentration.

Pure water and all solutions in which $[H^+] = [OH^-]$ are said to be neutral, and have $pH = 7$. Solutions in which $[H^+]$ is greater than $[OH^-]$ are called acidic and have pH less than 7. Solutions in which $[OH^-]$ is greater than $[H^+]$ are called basic and have pH greater than 7.

Self Check

1. How do strong electrolytes differ from weak electrolytes?
2. If you calculate the product of $[H^+]$ and $[OH^-]$ of any aqueous solution, what result does it always yield at 25°C?
3. What happens to the H^+ ion concentration of an aqueous solution when an additional source of OH^- ions is added?
4. What are the pH values of solutions, at 25°C, with the following concentrations?
   (a) $[H^+] = 10^{-3}$ M
   (b) $[OH^-] = 10^{-4}$ M

Acids and Bases Have Characteristic Properties

23-6 Properties of Aqueous Solutions of Acids

Chemists recognized long ago the significant influence that $H^+(aq)$ and $OH^-(aq)$ exert on aqueous solutions. Consequently, they have long identified by the class names "acids" and "bases" those substances that change the concentrations of these two ions.

Consider the following compounds.

HCl	hydrochloric acid (or hydrogen chloride)
HNO_3	nitric acid
CH_3COOH	acetic acid
H_2SO_4	sulfuric acid
H_3PO_4	phosphoric acid

Each of these five compounds is called an acid. They all share this name because they have the following important properties in common.

Hydrogen Content. Each of these compounds contains hydrogen.

Electrical Conductivity. Each of these compounds dissolves in water to form solutions that conduct electricity. Ions are present in their aqueous solutions.

Liberation of Hydrogen Gas. The aqueous solutions of each compound produce hydrogen gas, H_2, when zinc metal is added.

Color of Litmus. Litmus, a dye, turns red in color when it is placed in these aqueous solutions.

Taste. The dilute aqueous solutions of each compound taste sour. Many chemicals, including some acids and some bases, are poisons. Therefore, you should *not* use taste as a voluntary means of deciding if an unidentified solution is acidic or basic.

Because the aqueous solutions of these compounds have the foregoing properties in common, the compounds are conveniently classed together and identified as acids. In fact, these properties constitute the simplest definition of an acid. They provide a basis for deciding whether a compound should be classified as an acid. Some familiar acids are shown in Figure 23-3.

What is the common factor that makes these different substances behave in the same ways? In water they all form conducting solutions. Therefore they all form ions in water. Each substance contains hydrogen, and each reacts with zinc metal to produce hydrogen gas. Perhaps all of these aqueous solutions contain the *same* ion, and this ion accounts for the formation of $H_2(g)$. Based on the properties of acids, it is reasonable to propose that the common ion is $H^+(aq)$. Then, one definition of an acid is the following one. *A substance has the properties of an acid if it can release hydrogen ions.*

Figure 23-3 *Such familiar household substances as those shown above have certain properties that are common to all acids.*

23-7 Properties of Aqueous Solutions of Bases

Consider the following compounds.

NaOH	sodium hydroxide
KOH	potassium hydroxide
$Mg(OH)_2$	magnesium hydroxide
Na_2CO_3	sodium carbonate
NH_3	ammonia

Each of these five compounds is called a base. They all share this name because they have the following important properties in common.

Electrical Conductivity. Like acids, these compounds dissolve in water to form conducting solutions. Ions are present in an aqueous solution of a base.

Reaction with Acids. When equal quantities of acids and bases are combined, the base destroys all identifying properties of the acid solution except electrical conductivity. Of course, you could also say that an acid destroys all identifying properties of a basic solution except electrical conductivity.

Color of Litmus. Litmus turns blue in color when it is placed in an aqueous solution of any of these compounds.

Taste. The dilute aqueous solutions taste bitter. Remember that you should *not* use taste as a means of identifying a solution.

Feel. The aqueous solutions feel "slippery."

Again, these properties constitute the simplest definition of a base. They provide a basis for deciding whether a compound should be classified as a base. Some familiar bases are shown in Figure 23-4.

23-8 An Explanation of the Properties of Bases

Employing the same argument used for acids, a common factor can be sought that accounts for the similarities of bases. Because of the electrical conductivity, you might look for an ion. Because of the ability to counteract the properties of acids, such an ion ought to be able to remove $H^+(aq)$ from solution. Recall that $H^+(aq)$ accounts for the properties of acids.

Sodium hydroxide, NaOH, when dissolved in water, gives a solution with the properties of a base. The hydroxides of other elements from the left side of the periodic table behave in the same way. Perhaps they dissolve to form ions as shown in the following equations.

$$NaOH(s) \longrightarrow Na^+(aq) + OH^-(aq) \quad (16)$$

Figure 23-4 *Such familiar household substances as those shown above have certain properties that are common to all bases.*

$$KOH(s) \longrightarrow K^+(aq) + OH^-(aq) \quad (17)$$

$$Mg(OH)_2(s) \rightleftharpoons Mg^{2+}(aq) + 2OH^-(aq) \quad (18)$$

$$Ca(OH)_2(s) \rightleftharpoons Ca^{2+}(aq) + 2OH^-(aq) \quad (19)$$

The hydroxide ion, $OH^-(aq)$, could react with hydrogen ion to account for the second property of bases, the removal of acid properties, as seen in equation (20).

$$OH^-(aq) + H^+(aq) \rightleftharpoons H_2O \quad (20)$$

The similarities among the hydroxides are obvious. Now compare sodium carbonate and ammonia. Sodium carbonate, Na_2CO_3, dissolves in water to give a solution with the properties that identify a base. Quantitative studies of the solubilities of carbonates show that carbonate ion, CO_3^{2-}, can react with water in the following way.

$$Na_2CO_3(s) \longrightarrow 2Na^+(aq) + CO_3^{2-}(aq) \quad (21)$$

$$CO_3^{2-}(aq) + H_2O \rightleftharpoons HCO_3^-(aq) + OH^-(aq) \quad (22)$$

Reaction (22) indicates that the presence of carbonate ion in water increases the OH^- concentration. The OH^- ion is present in the solutions of NaOH, KOH, $Mg(OH)_2$, and $Ca(OH)_2$.

Reaction (22) also provides a basis for understanding the removal of acid properties. If CO_3^{2-} readily forms bicarbonate ion, HCO_3^-, the following reaction is likely to occur.

$$CO_3^{2-}(aq) + H^+(aq) \rightleftharpoons HCO_3^-(aq) \quad (23)$$

The production of $HCO_3^-(aq)$ raises $[OH^-]$, since $[H^+]$ is decreased in the process.

You now can postulate that $OH^-(aq)$ accounts for the slippery "feel" and bitter taste of the basic solutions. The stability of the bicarbonate ion in reaction (23) also explains the removal of acid properties, because $H^+(aq)$ is removed from solution.

In Section 23-7, the fifth substance listed as a base is ammonia, NH_3. Ammonia readily forms the ammonium ion, NH_4^+. Ammonia can react with both water and the hydrogen ion.

$$NH_3(aq) + H_2O \rightleftharpoons NH_4^+(aq) + OH^-(aq) \quad (24)$$

Summary

An acid is a hydrogen-containing substance that has the following properties when dissolved in water. It is an electrical conductor; it reacts with zinc to give $H_2(g)$; it turns litmus red; it tastes sour. All of these prop-

$$NH_3(aq) + H^+(aq) \rightleftharpoons NH_4^+(aq) \quad (25)$$

The formation of NH_4^+ accounts for ammonia having the properties of a base. Reaction (24) produces hydroxide ion. According to the preceding postulate, this ion accounts for the taste and "feel" properties of solutions of bases. Reaction (25) shows how ammonia can act to destroy the acid properties of a solution containing hydrogen ions, $H^+(aq)$.

Investigation of the reactions of other compounds that have the properties of a base shows that each compound can produce hydroxide ions in water. The OH^- ions may be produced directly, as in equations (16)–(19). They may be produced indirectly, through reaction with water, as in equations (22) and (24).

Any substance that can produce hydroxide ions in water also can combine with hydrogen ions.

$$OH^-(aq) + H^+(aq) \rightleftharpoons H_2O \quad (20)$$

$$CO_3^{2-}(aq) + H^+(aq) \rightleftharpoons HCO_3^-(aq) \quad (23)$$

$$NH_3(aq) + H^+(aq) \rightleftharpoons NH_4^+(aq) \quad (25)$$

Production of $OH^-(aq)$ and reaction with $H^+(aq)$ occur together in aqueous solutions. Therefore, you can describe a base *either* as a substance that produces $OH^-(aq)$ *or* as a substance that can react with $H^+(aq)$. In solvents other than water, the latter description is more generally useful. Therefore, *a substance has the properties of a base if it can combine with hydrogen ions.*

erties may be attributed to the fact that an acid produces hydrogen ions, H^+.

A base is a substance that has the following properties when dissolved in water. It is an electrical conductor; it reacts with acids, cancelling their acidic properties; it turns litmus blue; it tastes bitter; it feels slippery. A substance is a base if it can combine with hydrogen ions.

Self Check

1. What ion is characteristic of an acid?
2. What ion is characteristic of a base?

Finding Amounts of Acid and Base

INVESTIGATION

23-9 Determining the Hydrogen Ion Concentration of Solutions, Using Indicators

Many chemical reactions require an accurate measurement and careful control of the hydrogen ion concentrations in their solutions if they are to be carried out properly. Some plants grow well in soils that are slightly acid while others grow well in slightly basic soils. If we can measure the concentration of hydrogen ions in the moisture in the ground, we can determine which chemicals to add to the soil to provide the best growing conditions.

Electrically-operated pH meters can be used for determining the hydrogen ion concentrations of solutions with a high degree of accuracy. (See Figure 23-2.) When such instruments are not available, acid-base indicators can be used to provide a good estimate of the concentration of hydrogen ions.

Acid-base indicators are dyes. Their colors depend on the hydrogen ion concentration, $[H^+]$, of a solution. There are many suitable dyes, and each changes color over a particular range of hydrogen ion concentration, as shown in Figure 23-5. You are already familiar with one dye, litmus, which is red when $[H^+] = 10^{-6}$ M or greater and blue when $[H^+] = 10^{-8}$ M or less. For this experiment you will use two indicators that have color changes in the acid range, methyl orange and orange IV, and two indicators for the basic range, indigo carmine and alizarin yellow R.

PURPOSE

Using a series of prepared solutions of known concentrations as color standards, you will determine the $[H^+]$ of an "unknown" solution by comparing the color of the unknown solution with the colors of the standard solutions. You will also determine the hydrogen ion concentration of a solution of a weak acid and calculate its equilibrium constant. You will also use an indicator to compare the volumes of 0.1 M

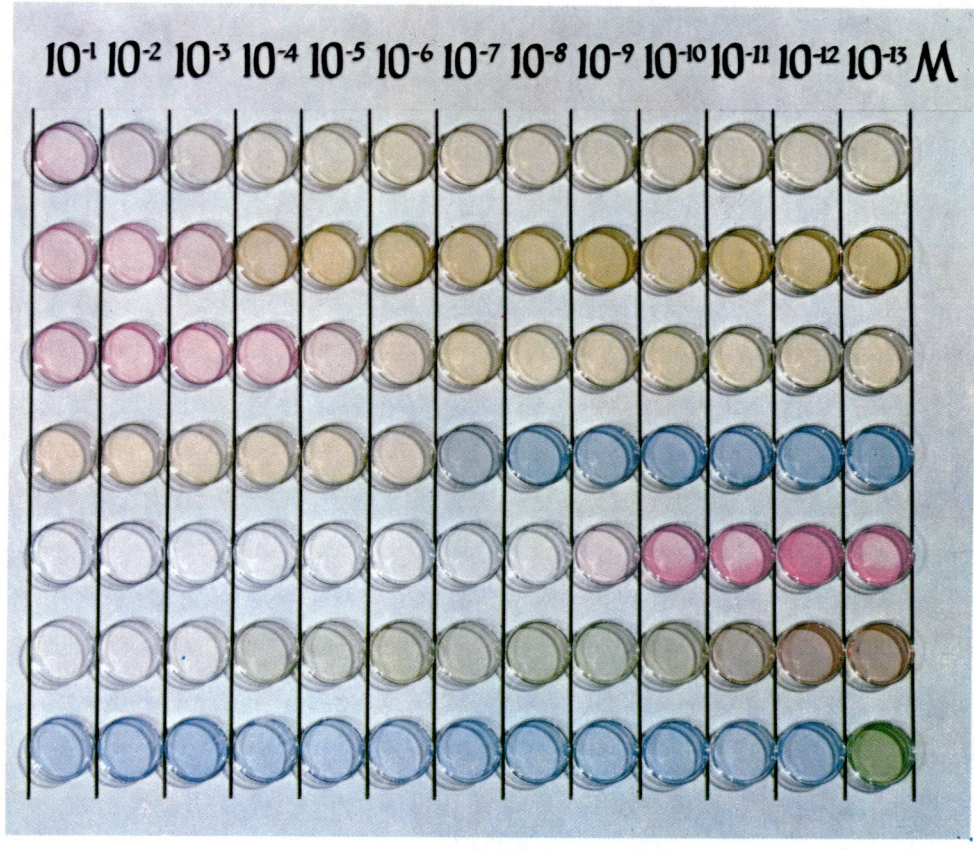

Figure 23–5 *Indicators show characteristic color variations with changing hydrogen ion concentrations. From the top to the bottom row these indicators are: orange IV, methyl orange, methyl red, bromthymol blue, phenolphthalein, alizarin yellow R, and indigo carmine.*

NaOH required to react with equal volumes of 0.1 M HCl and 0.1 M acetic acid, CH_3COOH.

MATERIALS

Litmus paper, red and blue
6 ml 0.1 M HCl
7 ml 0.1 M NaOH
5 ml "unknown" solution
6 ml 0.1 M acetic acid
5 ml 1.0 M acetic acid

Graduated cylinder, 10-ml
Medicine dropper
22 test tubes, 13- × 100-mm
Marking pencils or labels
Distilled water

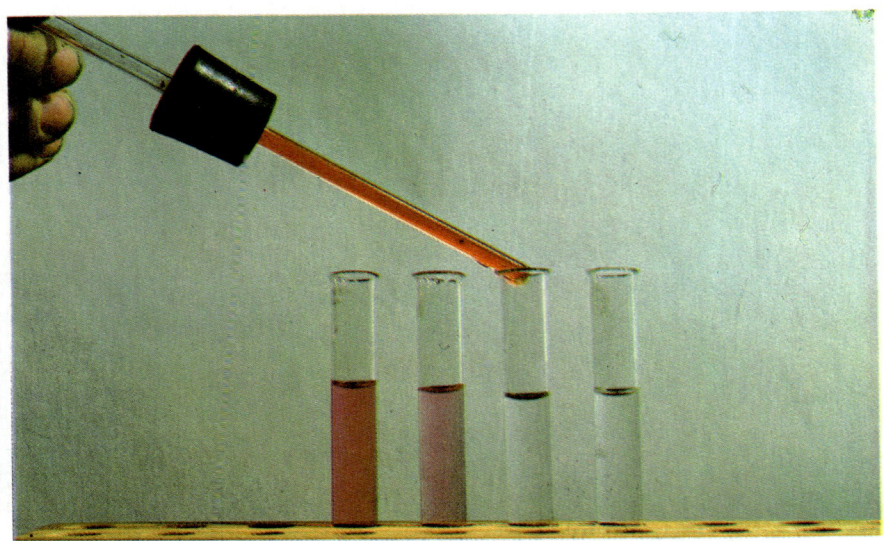

Figure 23-6 *Methyl orange indicator is being added to a series of "standard" solutions. The colors of these solutions can be used to estimate [H⁺] in other solutions that contain methyl orange and have [H⁺] within the same range.*

Indicator solutions in dropper bottles
    Orange IV   Alizarin yellow R
    Methyl orange   Phenolphthalein
    Indigo carmine

PROCEDURE

You should work in pairs to do this experiment. One student should prepare the standard solutions in the acid range. The other student should do the same for the basic range. Label the test tubes so that each of you can use the standards to determine [H⁺] of an "unknown" solution in Part C. You may work as partners on Parts D and E but each of you should keep individual data sheets.

Part A. The Preparation of Standard Solutions in the Acid Range:

$$[\text{H}^+] = 10^{-1}\,M \text{ to } 10^{-4}\,M$$

1. Obtain about 5 ml of 0.1 M HCl in a clean, dry 13- × 100-mm test tube. Label this test tube [H⁺] = 0.1 M. Since hydrochloric acid is a

strong acid it can be assumed to be completely ionized in this dilute solution.

2. Prepare some 0.01 M HCl by diluting one volume of 0.1 M HCl with nine volumes of distilled water. Use a calibrated pipette or medicine dropper. Since a final volume of 5 ml is desired, use 0.5 ml of 0.1 M HCl and 4.5 ml of distilled water. Thoroughly mix this solution and label it $[H^+] = 0.01\ M$ or $10^{-2}\ M$.

3. In a similar manner, prepare 5 ml of 0.001 M HCl by diluting some of the standard solution prepared in step 2. Mix it thoroughly and label it $[H^+] = 0.001\ M$ or $10^{-3}\ M$.

4. Finally, prepare 5 ml of 0.0001 M HCl by diluting some of the standard solution prepared in step 3. Mix it thoroughly and label it $[H^+] = 0.0001\ M$ or $10^{-4}\ M$.

5. Pour half of each of the standard solutions into clean test tubes so you will have two sets of standards. Label the new test tubes to correspond with each standard. For one set add one drop of the orange IV solution to each of the four test tubes. For the other set add one drop of methyl orange solution to each test tube (Figure 23-6).

6. Record the colors observed in each of the solutions for the various H^+ ion concentrations. Retain these standards for Parts C and D.

Part B. The Preparation of Standard Solutions in the Basic Range:

$$[OH^-] = 10^{-1}\ M \text{ to } 10^{-4}\ M$$

Follow the directions of Part A (steps 1-4) using 0.1 M NaOH instead of 0.1 M HCl to prepare solutions in the basic range.

Since sodium hydroxide is a strong electrolyte, a dilute solution of NaOH can be considered to be completely ionized. Thus, the hydroxide ion concentration of 0.1 M NaOH is 0.1 M or $10^{-1}\ M$.

Label the solutions prepared as $[OH^-] = 10^{-2}\ M$, $[OH^-] = 10^{-3}\ M$, and $[OH^-] = 10^{-4}\ M$, respectively.

1. Divide the solutions to obtain two sets of the four standards as outlined in Part A, step 5. For one set, add a drop of the indigo carmine solution to each of the four test tubes. For the other set add a drop of alizarin yellow R solution to each test tube.

2. Record the colors observed in each of the solutions for the various OH^- ion concentrations. Retain these standards for use in Part C.

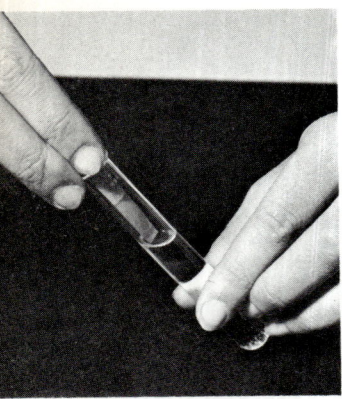

Figure 23-7

Part C. The Determination of the Hydrogen Ion Concentration of an "Unknown" Aqueous Solution (to be done individually)

1. Obtain about 5 ml of an "unknown" solution, in a clean, dry test tube, from your teacher. Test a portion of it with litmus paper, as in Figure 23-7, to determine whether [H$^+$] is in the range greater than 10^{-7} M or lower than 10^{-7} M.

2. Place about 2 ml of the "unknown" solution into each of two small test tubes. If the solution is in the acid range, add a drop of orange IV solution to one test tube and a drop of methyl orange to the other test tube. If the solution is in the basic range, use the indicators indigo carmine and alizarin yellow R.

3. Compare the colors with those of the standards prepared by either you or your partner. Record the hydrogen ion concentration of the "unknown" solution.

Part D. The Determination of the Hydrogen Ion Concentration of a Solution of a Weak Acid: Acetic Acid, CH$_3$COOH

1. Obtain about 5 ml of a solution of acetic acid in a small, clean, dry test tube. Note and record the concentration. Some students will use a 0.1 M solution and others will use a 1.0 M solution.

2. Place about 2 ml of the acetic acid solution assigned to you into each of two small clean test tubes. Add a drop of orange IV solution to one test tube, and a drop of methyl orange solution to the other.

3. Compare the colors with those of the standards. Estimate [H$^+$] of the acetic acid solution as accurately as you can. Record the H$^+$(aq) concentration. You will use this to calculate the equilibrium constant for dissociation of acetic acid.

Part E. The Determination of the Volume of 0.1 M NaOH Required to React with Equal Volumes of 0.1 M HCl and of 0.1 M CH$_3$COOH

1. Use a clean, calibrated medicine dropper to measure out 1.0 ml of 0.1 M HCl into a clean small test tube. To calibrate a medicine dropper, fill it with water and then count the number of drops needed to deliver 1.0 ml of liquid into a clean dry 10-ml graduated cylinder. Be sure that the bottom of the meniscus in the cylinder is exactly even with the 1.0 ml mark. Record the number of drops per milliliter for this medicine dropper.

2. Rinse the medicine dropper thoroughly and measure out 1.0 ml of 0.1 M CH$_3$COOH into another test tube.

3. Add a drop of the indicator solution phenolphthalein to each test tube. Phenolphthalein is a dye that is colorless in the range over which

$[H^+] = 10^{-1}$ M to 10^{-9} M. It turns pink when $[H^+]$ is about 10^{-9} M, as shown in Figure 23-5.

4. Rinse the medicine dropper thoroughly and fill it with 0.1 M NaOH. Add this solution of base, drop by drop, to each acid solution until the indicator turns to a pink color that does not disappear on mixing. Record the number of drops of 0.1 M NaOH required in each case.

DATA AND RESULTS

Calculations

Calculate the equilibrium constant for the equilibrium involving the aqueous solution of acetic acid.

$$CH_3COOH \rightleftharpoons H^+(aq) + CH_3COO^-(aq)$$

Use the value for $[H^+]$ determined in Part D, step 3. You may assume that the concentration of the acetate ion, $[CH_3COO^-]$, is equal to $[H^+]$ and that the concentration of acetic acid, CH_3COOH, is essentially the concentration of the solution you used, 0.1 M or 1.0 M.

RELATED PROBLEMS

1. Predict qualitatively the effect of each of the following experiments on the acetic acid equilibrium. Does $[H^+]$ increase or decrease?
   (a) Some of the salt sodium acetate, which produces the ions $Na^+(aq)$ and $CH_3COO^-(aq)$, is dissolved in the 0.1 M acetic acid solution.
   (b) Some sodium hydroxide solution is added, drop by drop, to the 0.1 M CH_3COOH.

2. How do you explain the results noted in Part E, when you compared the volume of 0.1 M NaOH required to react with equal volumes of 0.1 M HCl and 0.1 M CH_3COOH?

3. The $H^+(aq)$ concentration of a 1 M solution of benzoic acid is 8×10^{-3} M.
   (a) What percent of the benzoic acid, C_6H_5COOH, is ionized in this aqueous solution?
   (b) Predict what volume of 1 M NaOH would be required to react with 10 ml of 1 M C_6H_5COOH. Assume that the reaction will be carried to the point at which phenophthalein turns pink, as in Part E.

4. Calculate the $H^+(aq)$ concentration of each of the solutions prepared in Part B. Use the following relation.

$$[H^+][OH^-] = 10^{-14}$$

Chemistry and Our Environment
ACID RAIN AND SNOW

You have probably heard the expression "pure as the driven snow." Sad to say, this may no longer be a good simile. There is mounting evidence that snow—and rain—are no longer very pure. Over the past 15 years rain has been getting more and more acidic. This trend seems linked to the increasing quantities of gaseous air pollutants, mainly sulfur and nitrogen oxides.

As you know, pure water at 25°C has a pH of 7.0. The concentration of H^+ and OH^- are each equal to 10^{-7} M. Rainwater always has been somewhat acidic because of the carbon dioxide in the atmosphere. In the absence of pollution, a pH of about 5.7 is normal for rainwater.

Recent studies, particularly in Europe, have revealed that rain and snow now commonly have pH's between 5 and 3. Values as low as 2.8 have been recorded. Remember that a pH of 3 means a 0.001 M solution of hydrogen ions. Most chemists would call that a dilute acid.

The causes of this increased acidity are mainly sulfuric and nitric acids, H_2SO_4 and HNO_3, in the rain and snow. These acids are the final, water-soluble products of the SO_2 and nitrogen oxides, NO and NO_2, that enter the atmosphere in great quantities from power plants, automobile exhausts, jet planes, and various industrial off-gases. In the United States alone about 32 million tons each of sulfur dioxide and oxides of nitrogen are emitted into the air each year. The rest of the world must add at least this much again.

The results of this air pollution are obvious in terms of smog and other direct effects. It is shocking, but important, to realize that these forms of pollution can bedevil us in more subtle ways, such as acid rain. Acid rain is not pure water containing a little dissolved CO_2, but instead a not-so-dilute solution of sulfuric and nitric acids.

What are the effects of acid rain? We are just beginning to understand the full implications, but it is clear that they cannot be good. Undoubtedly buildings, machinery, and other man-made objects will suffer some degree of corrosion from prolonged exposure to dilute acids. More important, however, will be complex, long-range effects on nature.

An increased input of acid into soil will almost certainly lead to increased rates of leaching of vital nutrients, such as calcium. Direct adverse effects on vegetation seem likely, but studies to evaluate this are still lacking. Effects of the increasing acidity of ground water on aquatic life are already known to be very serious. A study in Norway has shown that acid rain has so altered the pH of certain streams that salmon eggs can no longer develop. The salmon runs have been eliminated in certain streams in southern Norway.

The problem of acid rain is an especially good example of the international character of environmental problems. Air pollution and its various effects, such as acid rain and snow, are equivalent to an unintended form of chemical warfare!

23-10 HCl and NaOH in the Same Solution

You have not considered what happens when HCl and NaOH are dissolved in the same solution. Before you consider this, recall what happens when HCl and NaOH are dissolved in separate solutions.

If, for example, 0.10 mole of HCl is added to a liter of water, a solution is formed in which $[H^+]$ is 0.10 M. However, the relationship given by equation (6) must hold.

$$K_w = 10^{-14} = [H^+][OH^-] \qquad (6)$$

Because of this, the value of $[OH^-]$ is reduced to $10^{-13}\ M$.

You may ask: What happened to OH^- ions that were lost in order to reduce $[OH^-]$ from $10^{-7}\ M$ to $10^{-13}\ M$? The answer is that they combined with H^+ ions to form water molecules. You may then ask another question. If some H^+ ions have been consumed by reacting with OH^- ions to form water, is there not *less* than 0.10 mole per liter of H^+ ions? The answer is "No"!

In reducing $[OH^-]$ from $10^{-7}\ M$ to $10^{-13}\ M$, essentially all of the initial 10^{-7} mole/liter of $OH^-(aq)$ combines with $H^+(aq)$ to form H_2O. Note that 10^{-13} is only one millionth of 10^{-7} and so is negligible in comparison. Initially, 10^{-7} mole per liter of H^+ ions from the dissociated water was also present. As you know, equal numbers of H^+ and OH^- ions must combine to form H_2O molecules. This means that the concentration of H^+ ions from the dissociated water decreases from $10^{-7}\ M$ to $10^{-13}\ M$, as does the OH^- concentration. The remaining, uncombined H^+ ions from the dissociated water are negligible. Hence, the actual concentration of H^+ ions after equilibrium has been attained is equal, to a *very* high degree of accuracy, to the concentration of H^+ ions from dissociated HCl. This concentration is 0.10 M. (Of course, there is no difference between the H^+ ions of dissociated water and those of dissociated HCl. They are discussed as if they are different simply to make the preceding explanation as clear as possible.)

Similarly, if 0.10 mole of NaOH is added to a liter of water, $[OH^-]$ becomes 0.10 M to a *very* high degree of accuracy. The value of $[H^+]$ is reduced to $10^{-13}\ M$.

$$\frac{10^{-14}}{10^{-1}} = 10^{-13}$$

Now suppose both HCl and NaOH are added to the same solution. To be specific, suppose both 0.10 mole of HCl and 0.10 mole of NaOH are added to the same liter of water. Clearly, if each solute behaved as it would alone, $[H^+]$ would equal 0.10 M and $[OH^-]$ would equal 0.10 M. Therefore, the following equation would be true.

$$[H^+][OH^-] = (0.10)(0.10) = 10^{-2}$$

This, however, could not be true because it violates the equilibrium law. The product $[H^+][OH^-]$ must equal, *and cannot exceed*, the value 10^{-14}. This requirement can be met only if enough pairs of H^+ and OH^- ions combine to form H_2O molecules.

In order to form H_2O, equal numbers of H^+ and OH^- ions must combine. Therefore, $[H^+]$ and $[OH^-]$ remain equal as H_2O molecules are being formed. They will still be equal when equilibrium is attained. This means that the concentration of each ion must be equal to 10^{-7}. This, of course, is the value for the concentration of each ion in pure water. Thus equation (10) applies.

$$[OH^-] = [H^+] = 1.00 \times 10^{-7}\ M \qquad (10)$$

The two ion concentrations are the same both before and after the addition of HCl and NaOH. This indicates that *all* of the H^+ and OH^- ions from these compounds must combine to form H_2O. From reaction (20), you can see that 0.10 mole of $H^+(aq)$ and 0.10 mole of $OH^-(aq)$ yield 0.10 mole of water.

$$H^+(aq) + OH^-(aq) \rightleftharpoons H_2O(l) \quad (20)$$

Now suppose that 0.20 mole of HCl and 0.10 mole of NaOH are added to a liter of water. Again, if each solute behaved independently of the other, the equilibrium law would be violated, since the following would be true.

$$[H^+][OH^-] = (0.20)(0.10) = 2 \times 10^{-2}$$

The only way the equilibrium law can be satisfied is for enough pairs of $H^+(aq)$ and $OH^-(aq)$ ions to combine, forming H_2O molecules. Then the product of the concentrations of the remaining uncombined H^+ and OH^- ions would be 10^{-14}.

Notice that more H^+ ions than OH^- ions are present in the solution. The formation of H_2O in reaction (20) thus consumes almost all of the 0.10 mole of OH^- ions from NaOH. Essentially 0.10 mole of H^+ ions from HCl combine with 0.10 mole of OH^- ions to form 0.10 mole of H_2O. This means that 0.10 mole of H^+ ions from HCl remains uncombined. Thus, $[OH^-] = 1.0 \times 10^{-13}$.

$$[OH^-] = \frac{K_w}{[H^+]} = \frac{1.00 \times 10^{-14}}{0.10}$$
$$= 1.0 \times 10^{-13}$$

Although 10^{-13} mole of OH^- ions remains in order to satisfy the equilibrium law this is only a million-millionth of the 0.10 mole initially put in. Chemists say that the base has been, for practical purposes, completely *neutralized* by the excess acid. Had acid been added to an excess of base, the acid would have been completely neutralized by the base.

Although the combining of $H^+(aq)$ and $OH^-(aq)$ to form H_2O is the only reaction that takes place, something else happens when an acid and a base react. The acid contains a negative ion to go with its H^+ ion in solution and the OH^- ion of the base is accompanied by a positive ion. After the H^+ and OH^- ions have neutralized each other, these other ions remain in the solution. For instance, the full equation for the neutralization of HCl solution by NaOH solution is as follows.

$$Na^+(aq) + OH^-(aq) + H^+(aq) + Cl^-(aq) \rightleftharpoons$$
$$H_2O + Na^+(aq) + Cl^-(aq)$$

If such a solution were to be evaporated, crystalline sodium chloride, NaCl, would be recovered. (See Figure 23-8.) NaCl is, of course, the compound commonly called "salt."

In a more technical sense, NaCl is only one of many salts. **Salts** are compounds formed when an acid and a base neutralize each other. The salt obtained contains the positive ion of the base and the negative ion of the acid. Another example of the formation of a salt by neutralization is the reaction between sulfuric acid, H_2SO_4, and calcium hydroxide, $Ca(OH)_2$. The product of the reaction is a solution of the salt calcium sulfate, $CaSO_4$, as shown by the following reaction.

$$2H^+(aq) + SO_4^{2-}(aq) + Ca^{2+}(aq) + 2OH^-(aq)$$
$$\longrightarrow 2H_2O + Ca^{2+}(aq) + SO_4^{2-}(aq)$$

Now a definition of neutralization can be given. **Neutralization** is a reaction in which the H^+ ion of an acid and the OH^- ion of a base unite to form water. The other ions that remain are the same as a solution of a salt. You can also think of neutralization in terms of the following general reaction.

$$\text{acid} + \text{base} \rightleftharpoons \text{salt} + \text{water}$$

23-11 The Process of Titration in Theory

Titration is the name given to the process by which the volume of a solution needed to react with a given mass or volume of a sample is determined. This process is commonly used in the quantitative study of acid-base reactions.

In *titrating* an acid with a base, or vice versa, a solution of one is added little by little to a solution of the other until the amounts of $H^+(aq)$ and $OH^-(aq)$ are equal. Consider, for example, a titration of HCl and NaOH solutions. The resulting solution will have $[H^+] = [OH^-] = 10^{-7} M$ when equivalent amounts of acid and base have been mixed. This is the *point of equivalence*, or *end point*, in the titration.

How do you know when the point of equivalence has been reached? One of the simplest ways is to add litmus, or some other acid-base indicator. Litmus is red in a solution of an acid and blue in a solution of a base. If you begin with an acid solution containing litmus, the solution will be red. As NaOH solution is slowly added, the color will remain red until just enough base has been added to neutralize all the acid. When the tiniest bit more base is added, the color will change to blue. By watching very closely, it is possible to see, to the nearest *drop* of base, when this color change occurs.

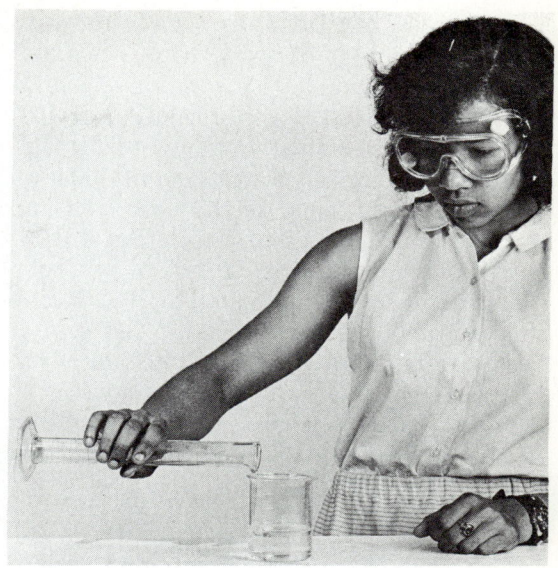

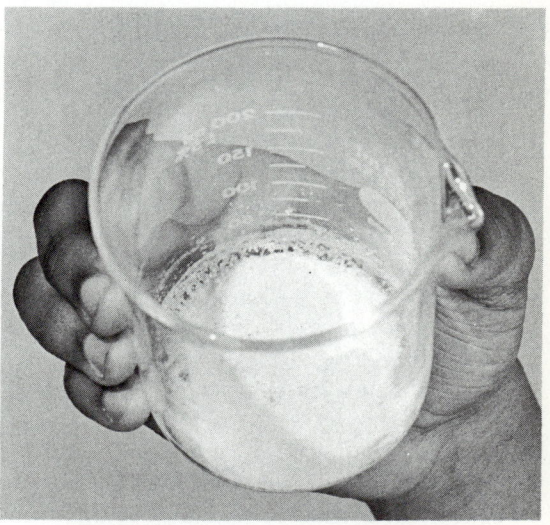

Figure 23-8 *When NaOH(aq) and HCl(aq) are mixed (top photo), a neutralization reaction takes place. One of the products of the reaction is sodium chloride, NaCl, which remains as a solid when the water has been evaporated away (bottom photo).*

The process of titration is used to compare the concentrations of solutions of acids and bases. Suppose you have a solution of NaOH known to be 0.100 M and a solution of HCl of unknown concentration. Suppose, further, that by carrying out a titration of 100.0 ml of the acid solution with the base, you find that 75.0 ml of the NaOH solution are required to neutralize the 100.0 ml of acid solution. This means that 75.0 ml of NaOH solution contain as many moles of $OH^-(aq)$ as there are moles of $H^+(aq)$ in 100.0 ml of the HCl solution.

Since there is 0.100 mole of NaOH in a liter of 0.100 M NaOH solution, there must be 0.00750 mole of NaOH in 75.0 ml of 0.100 M NaOH solution.

$$0.100 \text{ mole} \left(\frac{75.0 \text{ ml}}{1000 \text{ ml}}\right) = 0.00750 \text{ mole}$$

Therefore, 0.00750, or 7.5×10^{-3}, mole of HCl must be present in 100.0 ml of the HCl solution. It follows that in a liter of the HCl solution there must be 7.5×10^{-2} mole of HCl.

$$7.50 \times 10^{-3} \text{ mole} \left(\frac{1000 \text{ ml}}{100.0 \text{ ml}}\right) = 7.50 \times 10^{-2} \text{ mole}$$

The concentration of the HCl solution is thus 0.0750 M.

Care of Burets

1. *Cleaning*

   Place a few milliliters of a detergent solution into a buret, then use a buret brush to clean the inside surface, as in Figure 23–9. Rinse well, first with tap water and then with distilled water. After draining the buret, note if there are any droplets still adhering to the sides of the tube. If there are, the glass is not thoroughly cleaned and should be rewashed. When glass is clean, water wets it evenly.

2. *Preparation for Use*

   After cleaning a buret, add 5–10 ml of the solution that is to be used in that buret. Let several milliliters of solution flow through the tip. Turn the buret to a horizontal position and, with a rotary motion, slowly pour the rest of the solution out of the top. Make sure that the solution wets the inside completely. For a more complete rinsing, repeat the procedure.

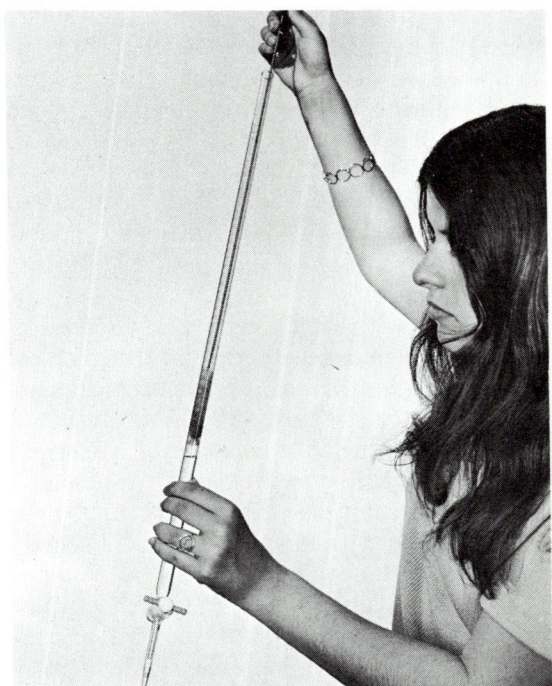

Figure 23–9

Acids and Bases 711

INVESTIGATION   *23-12 The Process of Titration in Practice*

The titration process is one of the most useful techniques available to the analytical chemist. Once "standard" solutions have been prepared, it is possible to use this method to determine quickly and accurately the composition of many different types of liquids and solids. Titration is one form of *volumetric* analysis, used to compare relative amounts of reactants and products by measuring the volumes and concentrations of reacting substances. *Gravimetric* analysis is used to find a relationship between reactants and products by measuring the masses of the products formed in the reaction.

In order for a titration to be successful, the reaction between the substances being measured must take place rapidly. In addition, there must be a sudden and easily observable change in the properties of the mixture when equivalent amounts of the reactants have been mixed, that is, at the *end point*. This point is frequently determined with the

---

Fill the buret to the top with the solution to be used. Permit air bubbles in the tip to escape by turning the tip upward, as shown in Figure 23–10. Let solution flow from the tip until the bottom of the meniscus is at zero or below.

If a drop hangs on the tip before you start a titration, discard it by touching it to a beaker. However, a drop formed during a titration must be caught by touching it to the side of the container being used and rinsing it into the container with distilled water.

3. *Reading the Volume*
When reading the volume on the buret, be sure to have your eye level with the bottom of the meniscus, and read the volume carefully at the bottom of the curve. In each titration use an absolute minimum of 10 ml of each solution to attain a precision of 1 percent.

4. *Care After Use*
Drain and rinse the buret several times with tap water, then, as a final rinse, use

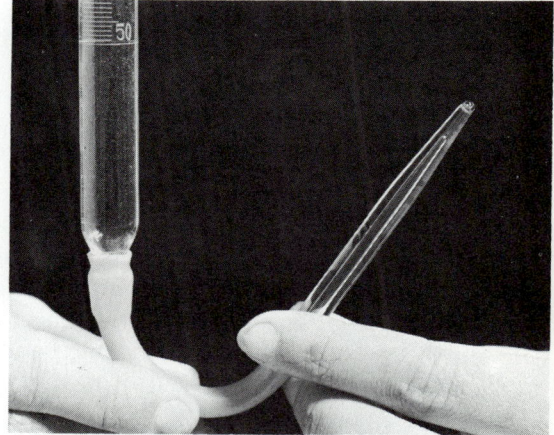

Figure 23–10

distilled water. Glass reacts with basic solutions, so take special care in rinsing a buret that has had such solutions in it. A rinse with dilute acid after one water rinse will help assure that the base is removed. Follow this with the water rinses.

use of a suitable indicator. Some titrations may depend on a change in the color or cloudiness of the mixture itself, or on a sudden change in the electrical conductivity of the mixture, to note the end point.

PURPOSE

In this experiment you will first "standardize" a sodium hydroxide solution by titrating it with hydrochloric acid of known concentration. Then you will use this standard base solution to titrate a known mass of an unknown solid acid. Phenolphthalein will be used as the indicator in both cases.

Figure 23-11

MATERIALS

Ringstand
Erlenmeyer flask, 250-ml
Wash bottle with distilled water
Detergent
2 burets, 50-ml (or 1 buret and a 10-ml pipet)
2 single, or 1 double, buret or utility clamp
2 beakers, 250-ml (or 1 250- and 1 400-ml beaker)
Few drops phenolphthalein solution

Buret brushes
75 to 100 ml of standard acid
100 to 150 ml of unknown base
1 to 2 grams unknown solid acid

PROCEDURE

Part A. Standardization of the Base Solution

1. Obtain two burets. Consult the section on the care of burets. Clean the burets and rinse one of them with 10 ml of the standard hydrochloric acid. Fill this buret with the standard acid. Rinse the other with 10 ml of the sodium hydroxide solution you are instructed to use. Fill this buret with the base. If there are not enough burets for each student, your teacher will demonstrate how to use a pipet to measure the volume of acid.

2. Record the volume in each buret by reading the bottom of each meniscus to the nearest 0.1 ml. Let about 10 ml of hydrochloric acid flow into a clean 250-ml Erlenmeyer flask, as in Figure 23-11. Add about 15 ml of distilled water and 3 drops of phenolphthalein.

3. Hold the neck of the Erlenmeyer flask with one hand and manipulate the buret with the other. As you add the sodium hydroxide, gently swirl the flask so the solutions will become mixed. (See Figure 23-12.) Continue adding sodium hydroxide until the first faint pink color develops. If the color disappears upon mixing the solution, add more sodium hydroxide, drop by drop, until a persistent pink color is obtained. If you go beyond this end point you may add a few drops

Figure 23-12

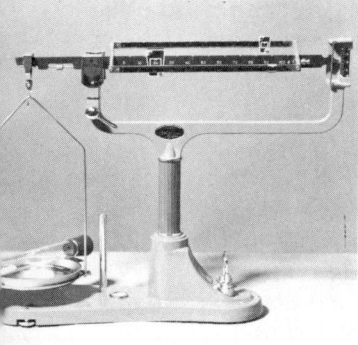

Figure 23-13

Figure 23-14

of acid, and then complete the titration with a few more drops of sodium hydroxide. (Take care not to go beyond the last calibration marks on the buret.) Record the volume reading, at the bottom of the meniscus, of each buret. Rinse the Erlenmeyer flask with distilled water before repeating the titration.

4. Perform at least one more titration. Refill the burets, if necessary. Continue until you obtain ratios of volume of acid to volume of base that agree to 1 or 2 percent, if your teacher so directs.

Part B. Titration of an Unknown Acid

1. Obtain a solid unknown acid from your teacher. Find the mass of the vial or test tube with the sample in it, to the nearest 0.01 g (Figure 23-13). Place a suitable amount (as directed by your teacher) of the solid acid into a clean flask. The transfer can be made by tapping the test tube gently, as shown in Figure 23-14. Then find the mass of the vial and contents again. Dissolve the sample in 50 ml of distilled water and add three drops of phenolphthalein. If all of the acid does not dissolve at this point, it will dissolve later during the titration when the acid will be converted to the more soluble sodium salt.

2. Refill the proper buret with some of the solution of base used previously and record the initial reading. Add the base to the acid solution until the first persistent, faint pink color appears. Be careful not to overrun the end point. (If you pass the end point, add a little more of the solid acid and reweigh the vial. Be sure to include the mass of any solid acid added to the mass of your sample. Retitrate to the end point and record the final buret reading.)

3. Repeat the titration with a similar sample. Use the knowledge you gained in the first titration. For example, assume that you used 40 ml of base to titrate a certain mass of acid, and that you have almost the same mass of acid for the second trial. You can then run 35 ml of base into the flask rapidly and complete the last part of the titration cautiously.

Part C. Optional Titration

If time permits, you may bring from home commercially available household acidic or basic substances to titrate with either your standard acid or base. Examples of items readily available are lemon juice, vinegar, household ammonia, and washing powders. Possible determinations include the percent of acetic acid, CH_3COOH, in vinegar; percent citric acid, $C_6H_8O_7$, in lemon juice; and percent ammonia, NH_3, in household ammonia. Your teacher will tell you how large a sample to use and which indicator is appropriate.

Since aspirin is an acid, acetylsalicylic acid, you might want to compare the amount of aspirin in a tablet of the least expensive and the most expensive brands. Dissolve the aspirin tablet in a mixture of warm water and alcohol. Titrate it with your standard sodium hydroxide solution, using phenolphthalein as an indicator.

DATA AND RESULTS

Calculations

1. From the concentration given and volume used, calculate the number of moles of hydrochloric acid involved in each titration of Part A. For example, if 10 ml of $0.05\ M$ HCl are used, the number of moles involved in the reaction is obtained as follows.

$$\frac{10\ \text{ml}}{1000\ \text{ml/liter}} \times \frac{0.05\ \text{mole}}{\text{liter}} = 0.0005\ \text{mole}$$

2. From the equation for the reaction, how many moles of base are used per mole of acid in Part A?

3. Using the relationship in calculation 2, find the moles of base used in the titration.

4. Calculate the molarity of the base.

5. From the mass of the solid acid and the volumes involved in its titration, calculate the mass of the solid unknown acid that will react with one mole of the base.

6. The formula of the acid and the equation for the reaction will be given you by your teacher. Calculate the theoretical value for the mass of the acid that will react with one mole of the base.

7. Determine the percentage error, using the value found in calculation 6 as the accepted value.

Summary

When H^+ and OH^- ions are added to the same solution, they must combine with one another to form H_2O until their concentrations are low enough to satisfy the equilibrium law for water, namely, $K_w = 10^{-14} = [H^+][OH^-]$. Neutralization is a reaction in which an acid and base combine to yield salt and water. The process by which a solution of an acid (or a base) is added to a solution of a base (or an acid) until a neutral solution is formed is called titration. By means of titration, the concentration of one solution can be measured if the concentration of the other is known. The point at which $[H^+] = [OH^-]$ can be shown by the color change of an acid-base indicator.

Self Check

1. If one liter of a 0.120 M solution of HCl is added to one liter of a 0.140 M solution of NaOH, show that [OH$^-$] in the resulting solution is 0.01 M.

2. How many ml of a 0.05 M HCl solution would be required to neutralize 100 ml of a 0.15 M solution of NaOH?

Acids Vary in Strength

23-13 Weak Acids

In Section 23-1 you learned that electrolytes could be classified as strong if they are completely dissociated in solution and weak if they are only partially dissociated. Since acids are electrolytes, the terms weak and strong can be applied especially to them. Thus there are *strong acids,* such as HCl and HNO$_3$, which are fully dissociated. They give a mole of H$^+$(aq) per mole of acid. This simple behavior needs no further comment. Many *weak acids,* however, such as acetic acid, CH$_3$COOH, or hydrofluoric acid, HF, dissociate to a slight extent only.

The dissociation of weak acids can be discussed in terms of equilibrium expressions and equilibrium constants. For example, consider the difference in the behavior of the two weak acids, acetic acid and hydrofluoric acid.

$$\text{CH}_3\text{COOH}(aq) \rightleftharpoons \text{H}^+(aq) + \text{CH}_3\text{COO}^-(aq) \quad (26)$$

$$\text{HF}(aq) \rightleftharpoons \text{H}^+(aq) + \text{F}^-(aq) \quad (27)$$

Measurements of the electrical conductivities of 0.10 M solutions of these two acids show that there are more ions present in the HF solution than in the acetic acid solution. Thus, acetic acid is a weaker acid than HF. You can see this quantitatively in terms of the equilibrium constants, equations (28) and (29), for reactions (26) and (27).

$$K_{\text{CH}_3\text{COOH}} = \frac{[\text{H}^+][\text{CH}_3\text{COO}^-]}{[\text{CH}_3\text{COOH}]}$$
$$= 1.8 \times 10^{-5} \quad (28)$$

$$K_{\text{HF}} = \frac{[\text{H}^+][\text{F}^-]}{[\text{HF}]} = 6.7 \times 10^{-4} \quad (29)$$

Since K_{HF} is a larger number than $K_{\text{CH}_3\text{COOH}}$, hydrofluoric acid dissociates in water to a larger extent than does acetic acid. Though HF is a weak acid (only partially dissociated), it is a stronger acid than is acetic acid.

You can express these ideas in terms of a general acid, HB. The acidic nature of HB is a result of its ability to release hydrogen ions.

$$\text{HB}(aq) \rightleftharpoons \text{H}^+(aq) + \text{B}^-(aq) \quad (30)$$

The equilibrium constant for reaction (30), K_A, measures quantitatively the ease with which HB releases H$^+$ ions. The symbol K_A stands for the *acid equilibrium constant.*

$$K_A = \frac{[\text{H}^+][\text{B}^-]}{[\text{HB}]} \quad (31)$$

Tables of K_A furnish a quantitative measure of acid strengths for comparing different acids and predicting their properties. Values

TABLE 23-2 Relative Strengths of Acids in Aqueous Solution at Room Temperature

$$K_A = \frac{[H^+][B^-]}{[HB]}$$

Acid	Strength	Reaction	K_A
HCl	very strong	$HCl(aq) \rightarrow H^+(aq) + Cl^-(aq)$	very large
HNO_3		$HNO_3(aq) \rightarrow H^+(aq) + NO_3^-(aq)$	very large
H_2SO_4	very strong	$H_2SO_4(aq) \rightarrow H^+(aq) + HSO_4^-(aq)$	large
HSO_4^-	strong	$HSO_4^-(aq) \rightleftharpoons H^+(aq) + SO_4^{2-}(aq)$	1.3×10^{-2}
HF	weak	$HF(aq) \rightleftharpoons H^+(aq) + F^-(aq)$	6.7×10^{-4}
C_6H_5COOH		$C_6H_5COOH(aq) \rightleftharpoons H^+(aq) + C_6H_5COO^-(aq)$	6.6×10^{-5}
CH_3COOH		$CH_3COOH(aq) \rightleftharpoons H^+(aq) + CH_3COO^-(aq)$	1.8×10^{-5}
HNO_2		$HNO_2(aq) \rightleftharpoons H^+(aq) + NO_2^-(aq)$	6.0×10^{-6}
H_2CO_3 $(CO_2 + H_2O)$		$H_2CO_3(aq) \rightleftharpoons H^+(aq) + HCO_3^-(aq)$	4.4×10^{-7}
H_2S	weak	$H_2S(aq) \rightleftharpoons H^+(aq) + HS^-(aq)$	1.0×10^{-7}
NH_4^+		$NH_4^+(aq) \rightleftharpoons H^+(aq) + NH_3(aq)$	5.7×10^{-10}
HCO_3^-		$HCO_3^-(aq) \rightleftharpoons H^+(aq) + CO_3^{2-}(aq)$	4.7×10^{-11}
H_2O	very weak	$H_2O(aq) \rightleftharpoons H^+(aq) + OH^-(aq)$	$1.8 \times 10^{-16*}$

*The acid equilibrium constant, K_A, for water equals $\frac{K_w}{[H_2O]} = \frac{1.00 \times 10^{-14}}{55.5}$

of K_A for several acids are given in Table 23-2.

In Table 23-2 you can see that the equilibrium view of acid strengths shows that water itself is a weak acid. It can release hydrogen ions. The extent to which it does so is indicated in its equilibrium constant, just as for the other acids.

Because the acidity of a solution can greatly affect many chemical reactions, it is important to be able to determine and control the hydrogen ion concentration. This control is obtained through application of the equilibrium law. Two types of calculations based on this law will be shown, using benzoic acid, C_6H_5COOH, as an example. In the first calculation K_A will be determined from experimental data. In the second, $[H^+]$ will be found by using K_A.

23-14 Determination of K_A

To apply the equilibrium law to acid solutions, a chemist must know the numerical value of the acid equilibrium constant, K_A. Experiments that provide this information require the measurement of hydrogen ion concentration. Acid-sensitive indicators, such as litmus, offer the easiest estimate of $[H^+]$.

A typical example follows. Benzoic acid, C_6H_5COOH, is a solid substance with only moderate solubility in water. The aqueous solution conducts electric current and has the other properties of an acid listed in Section 23-6. Its acidic behavior and its equilibrium relation are shown in equations (32) and (33), respectively.

$$C_6H_5COOH(aq) \rightleftharpoons H^+(aq) + C_6H_5COO^-(aq) \quad (32)$$

$$K_A = \frac{[H^+][C_6H_5COO^-]}{[C_6H_5COOH]} \quad (33)$$

The following experiment was performed to determine the acid equilibrium constant in equation (33). A 1.22-gram sample of benzoic acid was dissolved in 1.00 liter of water at 25°C. Using suitable indicators, the concentration of $H^+(aq)$ was estimated to be 8×10^{-4} M.

To make use of these data, all quantities must first be expressed in terms of moles. The molar mass of benzoic acid, C_6H_5COOH, is 122 grams/mole. Hence, the following can be written.

$$\text{Number of moles, } C_6H_5COOH = \frac{1.22 \text{ g}}{122 \text{ g/mole}}$$
$$= 0.010 \text{ mole}$$

Finally, since this number of moles was dissolved in one liter, the concentration is 0.010 mole per liter, or 1.00×10^{-2} M. However, the concentration of C_6H_5COOH as such cannot be quite this great because some of it dissociates. It is known qualitatively that benzoic acid is a fairly weak acid. Thus, you can make the assumption (or approximation) that the fraction that dissociates is small. Thus, the actual concentration will be given fairly accurately by equation (34).

$$[C_6H_5COOH] = 1.00 \times 10^{-2} \quad (34)$$

Careful measurement indicates that

$$[H^+] = 8 \times 10^{-4} \text{ } M. \quad (35)$$

Since only about 8 percent of the acid does dissociate, such an approximation is fairly good.

Now two of the concentrations in expression (33) are known. To complete the calculation, the concentration of benzoate ion, $[C_6H_5COO^-]$, must be found. Since the benzoic acid was dissolved in pure water, the only source of $C_6H_5COO^-$ is reaction (32). This is also the source of $H^+(aq)$. Because these two ions are both produced only by reaction (32), their concentrations must be equal in this solution.

$$[H^+] = [C_6H_5COO^-]$$

Since $[H^+] = 8 \times 10^{-4}$, then the following equation holds true.

$$[C_6H_5COO^-] = 8 \times 10^{-4} \quad (36)$$

Now you can complete the calculation by substituting equations (34), (35), and (36) into expression (33).

$$K_A = \frac{[H^+][C_6H_5COO^-]}{[C_6H_5COOH]}$$
$$= \frac{[8 \times 10^{-4}][8 \times 10^{-4}]}{[1.0 \times 10^{-2}]}$$
$$= \frac{64 \times 10^{-8}}{1.0 \times 10^{-2}}$$
$$K_A = 64 \times 10^{-6} = 6.4 \times 10^{-5}$$

Although the value of $[C_6H_5COOH]$ was approximated, the calculated value for K_A is very close to that given in Table 23-2.

23-15 Calculation of $[H^+]$

When dealing with weak acids and their salts, chemists often must calculate $[H^+]$ of an acidic solution by using K_A. This type of calculation is illustrated by the following example.

Suppose you need to know $[H^+]$ in a solution containing both $0.010\ M$ benzoic acid, C_6H_5COOH, and $0.030\ M$ sodium benzoate, C_6H_5COONa. Of course, you could go to the laboratory and proceed to investigate the colors of indicator dyes placed in the solution. However, it is possible (and easier) to calculate the desired information by using the value of K_A for benzoic acid. To work this problem, you will use the accurate value of K_A given in Table 23-2 rather than the approximate one calculated in the preceding section. Sodium benzoate is a strong electrolyte. Its aqueous solutions are found to contain sodium ions, $Na^+(aq)$, and benzoate ions, $C_6H_5COO^-(aq)$. Hence, the equilibrium involved is the same as before.

$$C_6H_5COOH(aq) \rightleftharpoons H^+(aq) + C_6H_5COO^-(aq) \quad (32)$$

At equilibrium, the concentrations must be in accord with the equilibrium expression.

$$K_A = \frac{[H^+][C_6H_5COO^-]}{[C_6H_5COOH]} = 6.6 \times 10^{-5} \quad (37)$$

The small value of this constant indicates that the fraction of benzoic acid that dissolves is small. Consequently, the concentration of the undissociated acid is almost equal to that of the dissociated acid. Therefore, you can consider $[C_6H_5COOH] = 0.010\ M$, as in equation (34). Similarly, the amount of $C_6H_5COO^-$ ion contributed by dissociation is small in comparison to the concentration of $C_6H_5COO^-$ from the added C_6H_5COONa. For this reason, you can consider $[C_6H_5COO^-] = 0.030\ M$. You can now find $[H^+]$ from equation (38).

$$K_A = \frac{[H^+][C_6H_5COO^-]}{[C_6H_5COOH]} = \frac{[H^+](0.030)}{(0.010)} \quad (38)$$

Rearranging equation (38), you obtain the following.

$$[H^+] = K_A \times \frac{(0.010)}{(0.030)}$$
$$= 6.6 \times 10^{-5} \times \frac{1.010}{0.030}$$
$$[H^+] = 2.2 \times 10^{-5}$$

The calculation cannot be considered complete until the assumptions are checked. Was it reasonable to assume that the concentrations of benzoate ion and benzoic acid were not changed by reaction (32)? To decide, compare the magnitude of $[H^+]$, $2.2 \times 10^{-5}\ M$, with the benzoate ion and benzoic acid concentrations. Such comparison shows that $[C_6H_5COOH] = 0.010\ M$ is about 500 times larger than the H^+ ion concentration, $2.2 \times 10^{-5}\ M$. The same argument applies to $[C_6H_5COO^-]$. The assumptions, therefore, are valid.

23-16 Competition for H^+ Among Weak Acids ★

The properties of acids have been explained in terms of their abilities to release hydrogen ions, $H^+(aq)$. From this point of view acetic acid is a weak acid because of the slight extent to which reaction (26) releases $H^+(aq)$.

$$CH_3COOH(aq) \rightleftharpoons H^+(aq) + CH_3COO^-(aq) \quad (26)$$

The properties of bases have been explained in terms of their abilities to *react* with hydrogen ion. Thus, ammonia is a base because it can react as shown in equation (25).

$$NH_3(aq) + H^+(aq) \rightleftharpoons NH_4^+(aq) \quad (25)$$

Now consider the result of mixing aqueous solutions of acetic acid and ammonia. You can think of the reaction that occurs as a sequence of reactions.

$$CH_3COOH(aq) \rightleftharpoons H^+(aq) + CH_3COO^-(aq)$$

$$NH_3(aq) + H^+(aq) \rightleftharpoons NH_4^+(aq)$$

The net reaction is reaction (39).

$$CH_3COOH(aq) + NH_3(aq) \rightleftharpoons$$
$$CH_3COO^-(aq) + NH_4^+(aq) \quad (39)$$

Practically speaking, the result of reactions (26) and (25) is reaction (39). In reaction (26) acetic acid releases hydrogen ions to form $H^+(aq)$, while in reaction (39) it releases hydrogen ions to NH_3 to form NH_4^+. In other words, ammonia acts as a base in reaction (39) by reacting with the hydrogen ion released by acetic acid. Thus reaction (39) is an acid-base reaction, though the net reaction does not show $H^+(aq)$ explicitly.

By taking one more step, you can view acid-base reactions in a broader sense. Suppose aqueous solutions of ammonium chloride, NH_4Cl, and sodium acetate, CH_3COONa, are mixed. You can tell by the odor that ammonia has been formed. The reaction that occurs is the following one.

$$NH_4^+(aq) + CH_3COO^-(aq) \rightleftharpoons$$
$$CH_3COOH(aq) + NH_3(aq) \quad (40)$$

Reaction (40) is just the reverse of reaction (39). Inspection reveals that reaction (40) is *also* an acid-base reaction. Once again there is an acid that releases H^+ and a base that accepts H^+. The acid is NH_4^+, and the base is CH_3COO^-. The net effect of the reaction is, once more, the transfer of a hydrogen ion from one species to another. The acid-base reaction between acetic acid and ammonia gave two products, one an acid, NH_4^+, and one a base, CH_3COO^-. A little thought will convince you that *every* acid-base reaction does so. The transfer of a hydrogen ion from an acid to a base necessarily implies that it *might* be transferred back. Transferring it back (the reverse reaction) is just as much a hydrogen ion transfer (hence an acid-base reaction) as is the original transfer.

Notice that in the preceding reactions a hydrogen ion is transferred from an acid to a base without specifically involving H^+ ions from dissociated water. A hydrogen ion, H^+, is nothing more than a proton. Consequently, you can now define a more general view of acid-base reactions in terms of *proton transfer*. The main value of this view is that it is applicable to a wider range of chemical systems, including nonaqueous systems.

This view of the acid-base type of reaction can be generalized as follows.

$$\underset{\text{an acid}}{CH_3COOH} + \underset{\text{a base}}{NH_3} \rightleftharpoons$$
$$\underset{\text{an acid}}{NH_4^+} + \underset{\text{a base}}{CH_3COO^-} \quad (39)$$

The acetic acid reacts as an acid, giving up its proton, to form acetate ion, CH_3COO^-, a substance that can act as a base. You can write expression (39) in a general form, as follows.

$$HB_1 + B_2 \rightleftharpoons HB_2 + B_1 \quad (41)$$
$$Acid_1 + Base_2 \rightleftharpoons Acid_2 + Base_1$$

You see that *an acid and a base react, through proton transfer, to form another acid and base.*

Gilbert Newton Lewis (1875–1946)

ONE OF GILBERT LEWIS' first positions, after having received his doctorate at Harvard University, was as supervisor of weights and measures and chemist in the Philippine Bureau of Science. Later he spent seven years at the Massachusetts Institute of Technology, where he became a professor. Then he went to the University of California at Berkeley. There he eventually became chairman of the chemistry department, which he transformed into an outstanding one.

In 1916, he proposed the idea that covalent bonds were formed by pairs of electrons shared between bonded atoms. This idea led to the use of the now familiar electron dot symbols representing the bonding in compounds.

For many years chemists had tried to find an adequate definition for acids and bases. Arrhenius had stated in 1884 that acids were compounds that formed hydrogen ions in solution, and that bases formed hydroxide ions. Many scientists felt that a broader definition was necessary.

In 1923, J. M. Brønsted and T. M. Lowry independently proposed the idea that acids are compounds that can donate a proton in a reaction, while bases are compounds that can accept a proton. Lewis had another idea. In order to include a wide variety of compounds in his very broad theory, Lewis suggested that an acid is a substance capable of receiving a pair of electrons to form a bond, while a base is a substance that can furnish a pair of electrons to form a bond. Lewis' theory did not exclude those compounds included in the theories of Arrhenius and of Brønsted and Lowry, but broadened the concept of acids and bases to include many compounds that did not contain hydrogen or the hydroxide group.

A classic example of Lewis' theory is the combination of BF_3 and NH_3 to form F_3BNH_3. An examination of BF_3 shows that it has an empty orbital that can accept two more electrons in its valence shell, while the NH_3 molecule has an unshared pair of electrons that it can donate to the BF_3 molecule. In this case, the BF_3 molecule is classified as an acid, while the NH_3 molecule is considered to be a base.

In his later work, Lewis was the first scientist to synthesize heavy water. He also proposed a theory of excited electron states that helped explain such phenomena as color, fluorescence, and phosphorescence in organic compounds. Although Gilbert Lewis was never awarded a Nobel prize for his work, a number of chemists who studied under him at the University of California between 1912 and 1942 have received this high honor.

Chemists use this more general view to discuss the strengths of acids. In reaction (41), the proton transfer implies that the chemical bond in HB_1 must be broken and the chemical bond in HB_2 must be formed. If the HB_1 bond is easily broken, then HB_1 will be a strong acid. Then equilibrium will tend to favor a proton transfer from B_1 to some other base, B_2. If, on the other hand, the HB_1 bond is extremely stable, then this substance will be a weak acid. Equilibrium will tend to favor a proton transfer from some other acid, HB_2, to base B_1, forming the stable HB_1 bond.

23-17 Hydrogen Ion in the Proton Transfer Theory of Acids ★

In the proton-transfer view of acid-base reactions, an acid and a base react to form another acid and another base. Let us see how this theory encompasses the elementary reaction between $H^+(aq)$ and $OH^-(aq)$ and the dissociation of acetic acid.

$$H^+(aq) + OH^-(aq) \rightleftharpoons H_2O$$

$$CH_3COOH(aq) \rightleftharpoons H^+(aq) + CH_3COO^-(aq)$$

It does so by making a specific assumption about the nature of the species $H^+(aq)$. It is considered to have the molecular formula $H_3O^+(aq)$. Thus, when HCl dissolves in water, the reaction is written as follows.

$$HCl(g) + H_2O \rightleftharpoons H_3O^+(aq) + Cl^-(aq)$$

Whenever $H^+(aq)$ might appear in an equation for a reaction, it is replaced by the **hydronium** ion, H_3O^+, and a molecule of water is added to the other side of the equation. The ionization of acetic acid can then be written as follows.

$$CH_3COOH(aq) + H_2O \rightleftharpoons H_3O^+(aq) + CH_3COO^-(aq)$$

Now the dissociation of acetic acid can be regarded as an acid-base reaction. The acid CH_3COOH transfers a proton to the base H_2O forming the acid H_3O^+ and the base CH_3COO^-.

The formation of water from its ions can be shown as follows.

$$H_3O^+(aq) + OH^-(aq) \rightleftharpoons H_2O + H_2O$$

In the preceding reaction the acid H_3O^+ transfers a proton to the base OH^-, forming an acid, H_2O, and a base, H_2O. Within the proton transfer theory, the molecule H_2O must be assigned the properties of an acid and, as well, those of a base.

The designation of the species $H^+(aq)$ as the hydronium ion, H_3O^+, is not necessitated by experimental evidence. There is no evidence that directly proves the unique existence of this ion, H_3O^+, in dilute aqueous solutions. The fact that water is a polar molecule, however, makes it difficult to believe that a positively-charged proton could exist in an aqueous solution without attaching itself to the negatively charged end of a water molecule. This assumption amply justifies its use as an aid in correlating acid-base behavior.

Summary

Acids that are not fully dissociated in solution are called weak acids. They are, of course, simply a special type of weak electrolyte. Their behavior can be understood quantitatively in terms of the acid equilibrium constant, K_A. The relative strengths of acids can be expressed and compared quantitatively using a Table of K_A's. Acid strength is defined as the ability to release H^+ ions. The K_A can be determined from experimental data. Once the K_A is known for a particular acid, the $[H^+]$ for any solution containing this acid can be calculated.

Self Check

1. Which of the following acids is the strongest acid and which is the weakest?

   (a) Nitrous acid, HNO_2 $K_{HNO_2} = 5.1 \times 10^{-4}$
   (b) Sulfurous acid, H_2SO_3
       $K_{H_2SO_3} = 1.7 \times 10^{-2}$
   (c) Phosphoric acid, H_3PO_4
       $K_{H_3PO_4} = 7.1 \times 10^{-3}$

2. Calculate $[H^+]$ when a solution of nitrous acid contains equal concentrations of HNO_2 and NO_2^-.

3. Write an equation that shows the ionization of formic acid, HCOOH, in aqueous solution, assuming the formation of hydronium ions.

Chapter Summary

There are two ways to define an acid in aqueous solution. One is directly in terms of its observable properties. (1) It contains hydrogen. (2) Its solution conducts electricity. (3) Various metals such as zinc will react with it, liberating hydrogen. (4) It turns litmus red. (5) It has a sour taste.

From experiment we know that all of these properties are due to the presence of $H^+(aq)$ in the solution. Therefore, a second definition is that an acid is a substance that supplies $H^+(aq)$ when dissolved in water.

A base may be defined as a substance that neutralizes an acid. That is, it reacts with and removes $H^+(aq)$ from acid solution. The most common and characteristic ion of bases is $OH^-(aq)$. Bases have as their other main properties the following. (1) Their solutions conduct electricity. (2) They turn litmus blue. (3) They taste bitter. (4) Their aqueous solutions feel slippery.

In order to understand the behavior of acids and bases in aqueous solution, consider the following equilibrium.

$$H_2O(l) \rightleftharpoons H^+(aq) + OH^-(aq)$$
$$K_w = 10^{-14} \text{ at } 25°C$$

When both an acid and a base are added to the same solution, it is this equilibrium that determines the result. The H^+ and OH^- ions must react with one another until the product of their concentrations is low enough to satisfy the equilibrium constant. The reaction of a base with an acid to produce water and a salt is called neutralization. The process by which a solution of an acid (or a base) is carefully added to a solution of a base (or an acid) until neutralization occurs is called titration.

The pH is a convenient way to indicate the acidity of a solution. It is defined as $-\log_{10}[H^+]$. Acid solutions have pH less than seven. Basic solutions have pH greater than seven.

Many acids are weak. They do not fully dissociate into $H^+(aq)$ and a negative ion. The behavior of their solutions is governed by an equilibrium constant. For an acid HB, the equilibrium law takes the following form.

$$K_A = \frac{[H^+][B^-]}{[HB]}$$

Questions and Problems

1. Write the names and formulas of three important acids.
2. Why is an acid called a proton donor? What would be a somewhat comparable name for a base?
3. Briefly discuss the pH scale as a device for indicating the acidity or basicity of an aqueous solution.
4. If the pH of a solution is 5, what is $[H^+]$? Is the solution acidic or basic?
5. What is the pH of a solution in which $[H^+] = 10^{-8}$? Is the solution acidic or basic?
6. Vinegar, lemon juice, and curdled milk all taste sour. What other properties would you expect them to have in common?
7. (a) Given the following acids: ammonium ion, NH_4^+ (in an NH_4Cl solution); bisulfate ion, HSO_4^- (in a $KHSO_4$ solution); hydrogen sulfide, H_2S. Which is the strongest acid? Which is the weakest?
   (b) If 0.1 M solutions are made of NH_4Cl, $KHSO_4$, and H_2S, in which will $[H^+]$ be highest and in which will it be lowest?
8. What is meant by neutralization? Illustrate it with an equation.
9. What characteristic makes an acid strong or weak? Define a strong acid and name one.
10. Describe the process of titration. For what purpose is it used?
11. Give the names and formulas of three hydrogen-containing compounds that are not classified as acids. For each compound state one or more properties of acids that it does not possess.
12. As a solution of barium hydroxide is mixed with a solution of sulfuric acid, a white precipitate forms and the electrical conductivity decreases markedly. Write equations for the reactions that occur and account for the conductivity change.
13. An acid is a substance, HB, that can form $H^+(aq)$ in the equilibrium:

    $$HB(aq) \rightleftharpoons H^+(aq) + B^-(aq).$$

    (a) Does equilibrium favor reactants or products for a strong acid?
    (b) Does equilibrium favor reactants or products for a very weak acid?
    (c) If acid HB_1 is stronger than acid HB_2, is K_1 a larger or smaller number than K_2?

    $$K_1 = \frac{[H^+][B_1^-]}{[HB_1]}; \quad K_2 = \frac{[H^+][B_2^-]}{[HB_2]}$$

14. What is the net ionic equation for the neutralization reaction?
15. What is the concentration of $H^+(aq)$ in an aqueous solution in which $[OH^-] = 1.0 \times 10^{-3}\ M$?
16. Nitric acid is a very strong acid. What is $[H^+]$ in a 0.050 M HNO_3 solution?
17. If 23 grams of formic acid, HCOOH, are dissolved in 10.0 liters of water at 20°C, the $[H^+]$ is found to be $3.0 \times 10^{-3}\ M$. Calculate K_A.
18. A chemist dissolved 25 grams of CH_3COOH in enough water to make one liter of solution. What is the concentration of $H^+(aq)$? Refer to Table 23-2, and assume a negligible change in CH_3COOH because of dissociation.
19. Show that the addition of 0.010 mole of solid NaOH to 1.0 liter of water reduces the concentration of $H^+(aq)$ to $1.0 \times 10^{-12}\ M$.
20. Suppose that 3.65 grams of HCl are dissolved in 10.0 liters of water. What is $[H^+]$? Show that $[OH^-] = 1.00 \times 10^{-12}\ M$.

21. Suppose that 0.099 mole of solid NaOH is added to 0.100 liter of 1.00 M HCl.
    (a) How many more moles of HCl are present in the solution than moles of NaOH?
    (b) From the excess number of moles and the volume, calculate the concentration of excess $H^+(aq)$.
    (c) Calculate the excess concentration of $H^+(aq)$ from the difference between the initial concentrations of HCl and NaOH.
    (d) Calculate the concentration of $OH^-(aq)$ at equilibrium.

22. 100 ml of the HCl solution described in Question 21 is diluted with water to 1.00 liter. What is the concentration of $H^+(aq)$? What is $[OH^-]$ in this solution?

23. An eyedropper is calibrated by counting the number of drops required to deliver 1.0 ml. Twenty drops are required.
    (a) What is the volume of one drop?
    (b) Suppose one such drop of 0.20 M HCl is added to 100 ml of water. What is $[H^+]$?
    (c) By what factor did $[H^+]$ change when the one drop was added?

24. Suppose drops (from the same eyedropper) of 0.10 M NaOH are added, one at a time, to the 100 ml of HCl in Question 23b.
    (a) What will be $[H^+]$ after one drop is added?
    (b) What will be $[H^+]$ after two drops are added?
    (c) What will be $[H^+]$ after three drops are added?

25. Calculate $[H^+]$ and $[OH^-]$ in a solution made by mixing 50.0 ml of 0.200 M HCl and 49.0 ml of 0.200 M NaOH.

26. Calculate $[H^+]$ and $[OH^-]$ in a solution made by mixing 50.0 ml of 0.200 M HCl and 49.9 ml of 0.200 M NaOH.

27. How much more 0.200 M NaOH solution need be added to the solution in Question 26 to change $[H^+]$ to 10^{-7} M?

28. A 0.25 M solution of benzoic acid (call it HB) is found to have the hydrogen ion concentration: $[H^+] = 4 \times 10^{-3}$ M.
    (a) Assuming the simple reaction $HB(aq) \rightleftharpoons H^+(aq) + B^-(aq)$, calculate K_A for benzoic acid.
    (b) Compare the values of $[HB]$, $[H^+]$, $[B^-]$, and K_A used in this problem to the corresponding quantities in the benzoic acid calculation presented in the text.

29. In the manufacture of a certain dye, it is necessary to know the amount of hydrochloric acid in the reaction mixture during various steps of the process. If a 2.00-gram sample of the mixture is neutralized when titrated with 22.0 ml of 0.1 M NaOH, what percent of the sample is hydrochloric acid?

30. A 0.150-gram sample of a pure acid is neutralized when titrated with 9.85 ml of 0.200 M NaOH. How many grams are in one mole of the acid?

31. What is the hydronium ion? What is its significance in describing the dissociation of acetic acid as an acid-base reaction?

32. When sodium acetate, CH_3COONa, is added to an aqueous solution of hydrogen fluoride, HF, a reaction occurs in which the weak acid HF loses H^+.
    (a) Write the equation for the reaction.
    (b) What weak base is competing with F^- for H^+?

33. (a) Write the equation that shows the acid-base reaction between hydrogen sulfide, H_2S, and carbonate ion, CO_3^{2-}.
    (b) What are the two bases competing for H^+?

(c) From the values of K_A in Table 23-2 for the two corresponding acids, predict whether the equilibrium favors reactants or products.
34. Write the equation for the reaction between each of the following acid-base pairs.
    (a) $HNO_2(aq) + NH_3(aq)$
    (b) $NH_4^+(aq) + F^-(aq)$
    (c) $C_6H_5COOH(aq) + CH_3COO^-(aq)$
35. For each reaction in Question 34, predict whether reactants or products are favored (using the values of K_A given in Table 23-2).
36. Write the equation for the reaction between each of the following acid-base pairs.
    (a) $HNO_2(aq) + HCO_3^-(aq)$
    (b) $H_2CO_3(aq) + NO_2^-(aq)$
    (c) $H_2CO_3(aq) + CO_3^{2-}(aq)$
37. For each reaction in Question 36, predict whether reactants or products are favored (using the values of K_A given in Table 23-2).

24 Oxidation-Reduction Reactions

These scientists are assembling a lithium/selenium electrochemical cell. The energy produced by electrochemical cells results from oxidation-reduction reactions.

One of the important classes of chemical reactions takes place every time you snap on a flashlight. Chemists term it an oxidation-reduction reaction—redox for short. A thorough understanding of redox reactions will enable you to understand devices such as batteries.

How does a battery operate? Why does it produce an electric current? Why does a battery run down? In this chapter you will find the answers to questions like these. You will also be able to understand a great number of other chemical processes that are similar to those in a battery. Each of these processes makes use of molecules or ions that transfer electrons to other molecules or ions. The transfer of electrons lies at the heart of redox reactions.

Since an actual investigation of a chemical process is one of the best ways to learn about it, this chapter begins with an experiment involving an electrochemical reaction. This is a reaction in which chemical and electrical changes occur together.

Understanding Redox Reactions

INVESTIGATION

24-1 The Chemistry of an Electrochemical Cell

In 1807 Sir Humphry Davy isolated the element potassium from one of its compounds electrochemically. He passed an electric current through potassium hydroxide. Since then, electrochemistry has been used in a variety of applications, many of which depend on the use of a simple electrochemical cell.

PURPOSE

In this experiment you will set up and operate an electrochemical cell. Then you will determine the products of the reaction that takes place while the cell is operating and find the ratio of these products to each other.

MATERIALS

200 ml 0.5 M $AgNO_3(aq)$
200 ml 0.5 M $CuSO_4(aq)$
50 ml 0.5 M $NH_4NO_3(aq)$
Silver strip, 2- × 10-cm
Copper strip, 2- × 10-cm
30 ml distilled water
Filter paper
Funnel
Balance, centigram
2 beakers, 250-ml
Ammeter
Glass wool, porous
Steel wool
Wash bottle
Erlenmeyer flask, 125-ml
Ringstand, ring
U-tube, 8-10 cm on each side
Filter rack
2 insulated connecting wires with alligator clips, 18-inch (46-cm)

Special Precautions. Use gloves when handling glass wool to avoid glass splinters in your fingers. Keep silver nitrate solution off your skin. It produces stains that are extremely difficult to remove. If the solution does get on your skin, wash it off at once with large amounts of water. To prevent stains on desks and floors, wipe up any spills immediately and rinse with water.

PROCEDURE

Part A. Preparation and Operation of the Electrochemical Cell

1. Fill one beaker approximately two-thirds full with the $AgNO_3$ solution. Carefully determine the mass of the strip of silver to the nearest 0.01 g. Place the silver strip in the beaker so that the lower end extends well below the surface of the solution, as in Figure 24-1.

2. Fill the other beaker approximately two-thirds full with the $CuSO_4$ solution. Clean the surface of the copper strip with a piece of steel wool. Determine the mass of the strip to the nearest 0.01 g. Immerse the copper strip in the beaker of $CuSO_4$ and position it as you did the silver strip in the first beaker.

3. With wire and its attached alligator clips, connect the silver strip to one terminal of the ammeter. With another wire and its alligator clips, connect the copper strip to the other terminal of the ammeter, as shown in Figure 24-1. (An ammeter measures electric current.) In your notebook record what happens to the needle of the ammeter.

4. Using an Erlenmeyer flask and a funnel, fill a U-shaped glass tube with NH_4NO_3 solution. (See Figure 24-2.) Plug the ends of the tube with porous glass wool. Use this device, known as a "salt bridge," to connect

Figure 24-1

Figure 24-2

the beakers containing the two solutions, as shown in Figure 24-3. Watch the needle of the ammeter and record your observations.

5. Remove the alligator clips from the meter terminals. Reverse the connections, placing the wire from the copper strip on the terminal that was previously connected to the silver strip and vice versa. Observe the meter and record your observation.

6. While the cell is operating, observe what is taking place at the metal surfaces that are in the solutions. Allow the cell to operate for ten or fifteen minutes.

Part B. Measuring the Products of the Electrochemical Reaction

1. Remove the salt bridge and disconnect the wires from the copper and silver strips. Look at the metal strips and record any changes that you see in them.

2. Remove the copper strip from the solution of $CuSO_4$ and rinse it off with distilled water from a wash bottle.

3. Carefully remove the silver strip from the $AgNO_3$ solution. Use distilled water to rinse into the solution any loose particles from the strip.

4. Determine the mass of a piece of filter paper to the nearest 0.01 g. Using the technique shown in Figure 19-7, filter the $AgNO_3$ solution through filter paper. Rinse the deposit on the filter paper with distilled water to remove any of the remaining solution from the deposit and the paper. Place the copper and silver strips along with the filter paper and its contents on a piece of clean paper towel. Set them aside in a safe place to dry.

5. When both pieces of metal and the filter paper are completely dry, determine the mass of the copper strip. Record the results in your notebook. Then place the piece of silver along with the filter paper and its contents on the balance pan and determine their combined masses. Find the mass of the silver alone by subtracting the mass of the filter paper from this result.

DATA AND RESULTS

Be sure to record all of the following measurements in your notebook.

Original mass of copper strip
Final mass of copper strip
Original mass of silver strip
Final mass of silver strip, filter paper, and deposit
Mass of filter paper

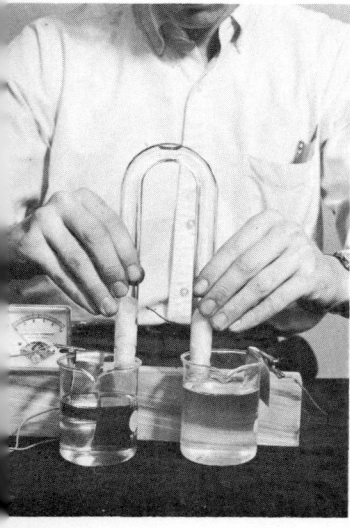

Figure 24-3

Calculations

1. Determine the change in mass, in grams, of the strip of copper metal. Label this as mass gained or lost.

2. Repeat calculation 1 to determine the change in mass of the silver.

3. Convert the mass gained or lost by each metal in this reaction to the number of *moles* gained or lost.

4. Determine the relationship of the number of moles of silver gained or lost to the number of moles of copper gained or lost.

RELATED PROBLEMS

1. Did the meter reading change after the salt bridge was included as part of the cell?

2. What function does the salt bridge perform in the operation of this electrochemical cell?

3. What changes, if any, were caused by reversing the terminal connections to the meter? What conclusion did you draw from this observation?

4. After the cell had been operating for some time, what changes in the two pieces of metal became apparent? In your own words, describe the processes that caused these changes in the cell.

5. Explain the relationship that you determined in calculation 4.

24-2 *Examination of the Electrochemical Experiment*

You have just completed a typical electrochemical experiment. Now consider it in detail. Suppose you determined that the Cu strip lost 1.27 grams and that the Ag strip gained 4.32 grams. If you convert the mass lost or gained by each metal strip in this reaction to the number of moles of metal lost or gained, the following interesting relationships become apparent.

$$\frac{\text{mass of Cu lost}}{\text{molar mass of Cu}} = \frac{1.27 \text{ g}}{63.5 \frac{\text{g}}{\text{mole}}} = 0.020 \text{ mole}$$

$$\frac{\text{mass of Ag gained}}{\text{molar mass of Ag}} = \frac{4.32 \text{ g}}{108 \frac{\text{g}}{\text{mole}}} = 0.040 \text{ mole}$$

Clearly, for every mole of Cu that dissolves in the reaction, two moles of Ag are deposited on the Ag strip. Chemical equations for the processes occurring in each of the beakers can be written. In the beaker containing the Cu strip, Cu atoms, symbolized by Cu(*s*), leave the strip and go into solution. In solution they are present as ions, $Cu^{2+}(aq)$, which are surrounded by water molecules. Two electrons, symbolized by $2e^-$, are left behind in the solid copper. The appropriate chemical equation is the following one.

$$Cu(s) \rightleftharpoons Cu^{2+}(aq) + 2e^- \qquad (1)$$

The electrons pass through the Cu strip, through the ammeter, and to the Ag strip. There they combine with Ag^+ ions in the $AgNO_3$ solution that are at the surface of the metal, forming neutral Ag atoms that deposit on the Ag strip.

$$Ag^+(aq) + e^- \rightleftharpoons Ag(s) \qquad (2)$$

Every Cu atom that dissolves, forming a Cu^{2+} ion, releases two electrons. Thus every atom of Cu that dissolves furnishes enough electrons to deposit two Ag^+ ions from the solution as Ag atoms. In other words, for every mole of Cu that dissolves in the reaction, two moles of Ag are deposited on the Ag strip.

Notice that equations (1) and (2) are a little different from those previously used, in that electrons, as well as atoms, molecules, and ions, appear in them.

The experiment that is being described has the main features of an **electrochemical cell**. The main parts of such a cell are as follows.

1. A reaction that provides electrons
2. A reaction that accepts electrons

Since the overall chemical process in the electrochemical cell has two separate parts, each part, or reaction, is called a **half-cell reaction**.

The half-cell reaction occurring in the Cu-$CuSO_4$ beaker is reaction (1).

$$Cu(s) \rightleftharpoons Cu^{2+}(aq) + 2e^- \qquad (1)$$

The half-cell reaction occurring in the Ag-$AgNO_3$ beaker is reaction (2a).

$$2Ag^+(aq) + 2e^- \rightleftharpoons 2Ag(s) \qquad (2a)$$

You find the overall reaction by adding the equations for the half-cell reactions.

$$Cu(s) \rightleftharpoons Cu^{2+}(aq) + 2e^- \qquad (1)$$
$$2Ag^+(aq) + 2e^- \rightleftharpoons 2Ag(s) \qquad (2a)$$
$$\overline{Cu(s) + 2Ag^+(aq) \rightleftharpoons Cu^{2+}(aq) + 2Ag(s)} \qquad (3)$$

In the equation for the overall chemical reaction for this cell, no electrons appear. Thus, the Ag^+ ions used up as many electrons as were furnished by the Cu atoms that dissolved. Before the half-cell reactions may be added, their coefficients must be adjusted to show this. Since each Cu atom furnished 2 electrons when it dissolved to form Cu^{2+} ions, the equation for the reaction in the Ag-$AgNO_3$ half-cell needs to be written as in equation (2a). When the two half-cell reactions are combined, the loss of electrons is equal to the gain of electrons. This results in the equation for the entire reaction, equation (3), in which no electrons appear as such.

Some terminology very useful in describing the reactions that occur in electrochemical cells must now be given.

1. The half-cell reaction in which electrons are given up (lost) is called **oxidation**. This occurred in the Cu-$CuSO_4$ beaker, in which Cu atoms dissolved, giving up two electrons.

$$Cu(s) \longrightarrow Cu^{2+}(aq) + 2e^- \qquad (1)$$

The Cu atoms are said to be *oxidized*.

2. The half-cell reaction in which electrons are taken up (gained) is called **reduction**. This occurred in the Ag-$AgNO_3$ beaker, in which Ag^+ ions gained electrons and became Ag atoms.

$$2Ag^+(aq) + 2e^- \longrightarrow 2Ag(s) \qquad (2a)$$

The Ag^+ ions are said to be *reduced*.

3. The substance that brings about the reduction of another is called a **reducing agent**. The substance that brings about the oxidation of another is called an **oxidizing agent**. In the experiment under discussion, Cu(s) is a reducing agent, while $Ag^+(aq)$ is an oxidizing agent. You should note that a reducing agent is oxidized and that an oxidizing agent is reduced. The overall reaction shown by combining the two half-cell reactions is called an **oxidation-reduction** or **redox** reaction. The term "redox" is such a convenient substitute for the more proper term, oxidation-reduction, that it is now an accepted scientific term.

24-3 Oxidation-Reduction Reactions in a Beaker

If a piece of Cu metal is immersed in a beaker containing a solution of $AgNO_3$, an interesting phenomenon can be observed. The copper begins to have flaky crystals of silver metal clinging to it as shown in Figure 24-4. The clear, colorless solution of $AgNO_3$ begins to turn blue, indicating the presence of dissolved Cu^{2+} ions.

Closer investigation shows that the reactions occurring here are the same ones described in Section 24-2. It appears that the conducting wires, salt bridge, and meter are not necessary for the reaction between Cu atoms and Ag^+ ions. In this case, exactly as in the electrochemical-cell experiment, the Cu atoms dissolve to form Cu^{2+} ions and liberate two electrons per atom.

$$Cu(s) \rightleftharpoons Cu^{2+}(aq) + 2e^-$$

The Ag^+ ions in solution combine with these electrons and deposit as neutral Ag atoms.

$$2Ag^+(aq) + 2e^- \rightleftharpoons 2Ag(s)$$

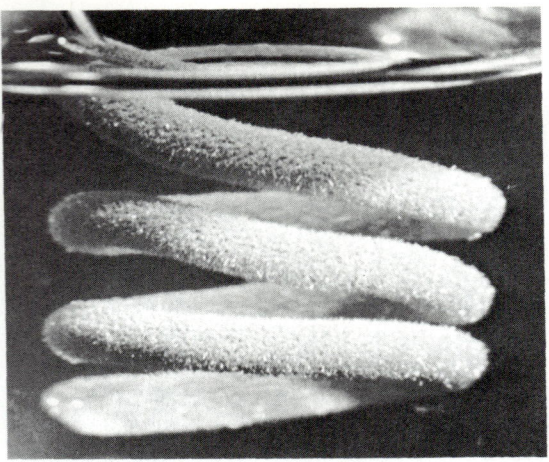

Figure 24-4 *Crystals of silver metal coat a copper coil when the coil is placed in a solution of silver nitrate.*

Suppose that you determine the loss of mass of copper and the mass of silver deposited on the copper, as you did in the experiment. You would find that the ratio of one mole of copper dissolved to two moles of silver deposited still holds. Apparently the transfer of electrons can and does take place directly between the metal and the free-moving ions in the solution in which it is suspended.

24-4 The Function of the Salt Bridge

In the preceding section you learned that a salt bridge is not needed for the reaction between Cu atoms and Ag^+ ions. Why, then, was one required in the experiment in Section 24-1? The answer involves charges within solutions.

As new Cu^{2+} ions are introduced from the copper strip into the copper sulfate solution, the solution tends to acquire an excess positive charge. At the same time, the electrons reaching the silver nitrate solution remove Ag^+ ions from it. An excess of negative ions tends to be created there. This buildup of excess positive and negative charges in the two beakers cannot proceed very far. Indeed, only a very small amount of current can flow before the initial tendency for the process to occur is canceled by the opposing effect of the excesses of charge. In order for the process to continue, the negative and positive charges in the two solutions must be kept in balance. Thus, ions must be able to flow from one beaker to the other, without simultaneously bringing the Ag^+ ions into direct contact with the Cu strip. The salt bridge permits NO_3^- ions to leave the $AgNO_3$ solution and pass into the $CuSO_4$ solution. This prevents the accumulation of extra negative ions in the $AgNO_3$ solution. It also provides negative ions to balance the charges on the new Cu^{2+} ions in the $CuSO_4$ beaker. In this way electrical neutrality is maintained within each of the beakers by the transfer of ions, mainly NO_3^- ions. These ions are not themselves involved in the redox reaction that generates the current.

As you know, redox reactions can always be carried out simply as a direct chemical reaction. To do this, you simply bring the substance to be oxidized (Cu metal, for example) and the substance to be reduced (Ag^+ ions, for example) into direct contact. Electrons then pass directly "on contact" from one to the other. If you want to use the reaction to generate an electric current, you must prevent this direct contact. Therefore, you must lower each metal into a solution of its own ions and place it in a separate beaker. To enable electrons to get from the metal being oxidized to the ions of the one being reduced, you must connect the two pieces of metal with a conductor of electrons, a wire. Equally important, to allow some ion or ions to move from one beaker to the other in order to keep the positive and negative charges in each solution in balance, you must use a salt bridge.

Figure 24-5 *When zinc metal is added to sulfuric acid, the zinc dissolves, and hydrogen gas is produced.*

24-5 Some Other Redox Reactions

In the chapter on acids and bases, the reaction of acid solutions with some metals such as zinc was mentioned as being an identifying property of an acid. Let us now examine this type of reaction more closely. If zinc metal is added to sulfuric acid, it dissolves, releasing hydrogen gas. This is shown in Figure 24-5. The beginning student might write the following reaction.

$$Zn(s) + H_2SO_4(aq) \rightleftharpoons ZnSO_4(aq) + H_2(g)$$

Is this what really happens in the reaction? You know that Zn reacts with any acid solution to release hydrogen. You know also that acid solutions have $H^+(aq)$ ions present, and that these ions account for the acidic properties of the solutions. Thus you can write the preceding reaction as follows.

$$Zn(s) + 2H^+(aq) \rightleftharpoons Zn^{2+}(aq) + H_2(g)$$

The Zn atoms apparently dissolve in the acid solution, releasing electrons. The H^+ ions gain these electrons, becoming neutral H atoms. These atoms combine immediately into H_2 molecules and escape as bubbles of hydrogen gas. This, then, is another example of an oxidation-reduction reaction. Electrons are lost by the Zn metal (oxidation), and electrons are gained by the H^+ ions (reduction). The equations for the half-cell reactions are as follows.

$$Zn(s) \rightleftharpoons Zn^{2+}(aq) + 2e^-$$
$$2H^+(aq) + 2e^- \rightleftharpoons H_2(g)$$

Adding, you get the following overall reaction.

$$Zn(s) + 2H^+(aq) \rightleftharpoons Zn^{2+}(aq) + H_2(g)$$

Earlier it was stated that acids react with *some* metals such as Zn to release hydrogen. Further investigation shows that many other metals react with acids just as zinc does. Some of them are iron, nickel, aluminum, and magnesium. Many others, however, show no reaction with acids. These include copper, silver, gold, and mercury. In the next section you will begin to understand how you can predict whether these and other types of redox reactions will occur.

24-6 Competition for Electrons

If a strip of zinc metal is placed in a copper nitrate solution, a reddish, metallic coating of copper forms on the zinc. Here there seems to be a redox reaction in which the zinc is losing electrons and the copper ions are gaining electrons.

$$Zn(s) + Cu^{2+}(aq) \rightleftharpoons Cu(s) + Zn^{2+}(aq)$$

The blue color in the solution (due to the presence of Cu^{2+} ions) gradually fades, indicating that Cu^{2+} ions are used up as this reaction proceeds. Furthermore, simple chemical tests show the presence of Zn^{2+} ions in the solution. These ions were not present in the solution before the reaction began. The *oxidation* in this example is the *loss of electrons* by the Zn metal to form Zn^{2+} ions.

$$Zn(s) \longrightarrow Zn^{2+}(aq) + 2e^-$$

The *reduction* is the *gain of electrons* by the Cu^{2+} ions to form pure $Cu(s)$.

$$Cu^{2+}(aq) + 2e^- \longrightarrow Cu(s)$$

As you learned in Section 24-2, reactions of this type cannot occur separately and are known as half-cell reactions. In the cell that you used in the electrochemical experiment, no current flowed in the circuit until the two half-cells had been connected by means of wires and a salt bridge.

It is of further importance to note that if you put a strip of copper metal in a zinc nitrate solution, no detectable reaction takes place. Apparently the equilibrium condition greatly favors the products in this reaction, namely $Cu(s)$ and $Zn^{2+}(aq)$.

If you continue the investigation of the reaction between various metals and the aqueous solutions of other metallic ions, you will find additional results similar to those in the reactions just presented. You saw that copper tended to give up electrons and go into solution as Cu^{2+} ions when placed in contact with Ag^+ ions. However, it did not tend to give up electrons when placed in contact with Zn^{2+} ions. Instead, Cu^{2+} ions placed in contact with $Zn(s)$ were reduced to $Cu(s)$. Evidently a very real competition for electrons exists in these reactions, and some reactants compete more successfully than others.

In fact, the half-cell reactions examined thus far can be arranged in decreasing order of the tendency to gain electrons. Recall that in the experiment in Section 24-1, the Ag^+ ion receives an electron from Cu. Therefore, the Ag^+ ion has a greater tendency than the Cu^{2+} ion to gain electrons. The two equations representing the half-cell reactions in the experiment can be written with the Ag^+ ion, which gains electrons more readily, above the Cu^{2+} ion.

$$Ag^+(aq) + e^- \longrightarrow Ag(s)$$
$$Cu^{2+}(aq) + 2e^- \longrightarrow Cu(s)$$

Since the Cu^{2+} ion readily accepts electrons from zinc, the Zn^{2+} ion is placed below the Cu^{2+} ion.

$$Ag^+(aq) + e^- \longrightarrow Ag(s)$$
$$Cu^{2+}(aq) + 2e^- \longrightarrow Cu(s)$$
$$Zn^{2+}(aq) + 2e^- \longrightarrow Zn(s)$$

This list leads you to predict that since the Cu^{2+} ion has a greater tendency than the Zn^{2+} ion to gain electrons, the Ag^+ ion should receive an electron from zinc. To test this prediction, you might place a zinc metal strip in a silver nitrate solution. If you did so, you would find that silver metal deposits on the zinc and that zinc ions appear in solution.

Recall that zinc reacts readily with acids to release hydrogen gas and that copper does not react with acids in a similar fashion. This means that the H^+ ion accepts electrons more readily than the Zn^{2+} ion, but not as readily as the Cu^{2+} ion. Thus the H^+ ion is placed above the Zn^{2+} ion and below the Cu^{2+} ion in the list. With this addition, the list now shows the tendency to gain electrons in all the reactions studied.

$$Ag^+(aq) + e^- \longrightarrow Ag(s)$$
$$Cu^{2+}(aq) + 2e^- \longrightarrow Cu(s)$$
$$2H^+(aq) + 2e^- \longrightarrow H_2(g)$$
$$Zn^{2+}(aq) + 2e^- \longrightarrow Zn(s)$$

Through careful laboratory investigation, you can expand this list to include a great many half-cell reduction reactions. Notice that any half-cell reaction in such a list can be combined with the reverse of one of the other half-cell reactions to produce a possible chemical reaction. Further study of the data from electrochemical cells will give **quantitative** information regarding these reactions.

Summary

Oxidation-reduction (redox) reactions are those in which one substance loses electrons (is oxidized), while another gains electrons (is reduced). Redox reactions may be carried out directly, in a beaker, or indirectly, in the form of an electrochemical cell that generates an electric current. In the direct reaction, the substances being oxidized and reduced come into direct contact and pass electrons from one to the other. In the electrochemical cell, electrons are passed from one reactant to the other through a conductor, a wire. In order to maintain electrical neutrality in the two solutions where ions are being made or consumed, the two solutions must be connected by a salt bridge. The salt bridge allows ions that are not directly involved in the redox reaction to pass from one reaction vessel to another. The oxidation and reduction parts of the overall reaction can be written separately and are called half-cell reactions. The half-cell reactions can be arranged in order of their tendency to proceed. Then, any two of them (one written in reverse) can be combined to give a possible chemical reaction.

Self Check

1. When is a chemical reaction considered to be a redox reaction?
2. In a redox reaction, what substances are oxidized? reduced?
3. Rewrite the equation below to show the true condition of the electrolytes and state what is oxidized and what is reduced.

$$Mg(s) + 2HCl(aq) \longrightarrow MgCl_2(aq) + H_2(g)$$

How Oxidation-Reduction Reactions Can Be Predicted

24-7 Measuring Cell Potentials

Simple experiments can provide information that will enable you to predict whether a given oxidation-reduction reaction will occur. In these experiments an instrument called a *voltmeter* is used. In the experiment in Section 24-1, you used another type of instrument, an ammeter. An ammeter measures the amount of electric current in a circuit. A voltmeter, on the other hand, measures the capacity, or **potential,** of the cell to push electrons through an external circuit. These electrons usually perform some useful work, such as producing light or heat. The electrons give up energy when performing work, and hence their potential energy is lowered as they go through the circuit. The voltmeter measures the cell potential, which is expressed in the familiar unit *volts*.

By placing a voltmeter across the terminals of each of several different electrochemical cells, you can read the cell potentials on the meter. Such readings indicate that the potential varies from cell to cell. Apparently, then, the capacity of different cells to do electrical work is measurably different.

Now consider some of the voltages produced by different cells, including some described earlier. Because the concentration affects the equilibrium conditions of chemical reactions, the concentration of all solutions used in the cells will be the same, 1.0 M.

If you construct a cell involving a Zn^{2+}-Zn half-cell connected by a salt bridge to a Ni^{2+}-Ni half-cell, as in Figure 24-6a, the voltmeter will indicate that electrons tend to flow from the Zn^{2+}-Zn half-cell (high potential side) to the Ni^{2+}-Ni half-cell (low potential side). The voltage produced by such a cell is found to be 0.5 volt.

If you connect a Zn^{2+}-Zn half-cell to a Ag^+-Ag half-cell, as in Figure 24-6b, the voltmeter will indicate that the electrons tend to flow from the Zn^{2+}-Zn to the Ag^+-Ag half-cell, and the potential will be 1.5 volts. The voltmeter needle will tend to be deflected in the *opposite* direction from that observed in the Ni-Zn cell. In the first cell the electrons will tend to go through the circuit in one direction, backed by a "push," or "force," of 0.5 volt. In the second cell the electrons will tend to be pushed in the opposite direction with a force of 1.5 volts.

The question arises: What will happen if the two cells are connected in opposition to each other, as in Figure 24-6c? If they are connected in this way, the voltmeter will indicate the tendency of the electrons to flow in the same direction as they did in the cell

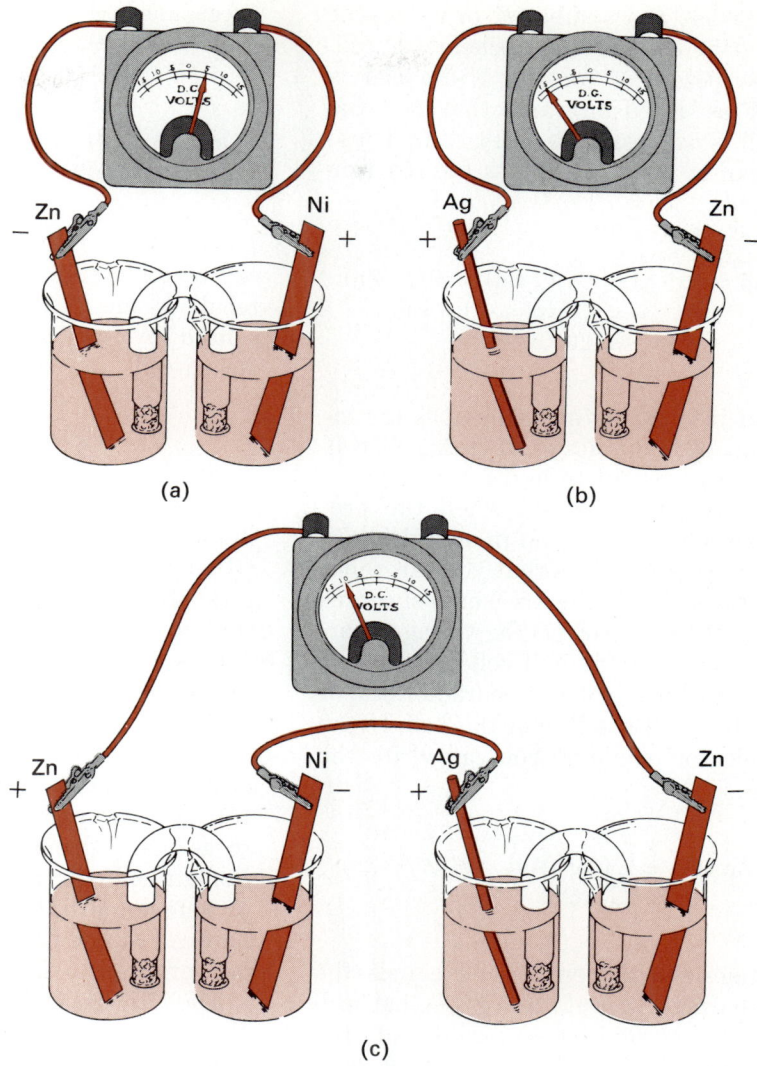

Figure 24-6 *When two cells, (a) and (b), that have opposite voltages are connected as shown in (c), the net voltage is equal to the difference and is in the direction of the larger one.*

with the higher voltage. Furthermore, the resulting overall potential will be 1.0 volt. This voltage is equal to the 1.5 volts of the original Ag-Zn cell minus the 0.5 volt of the original Ni-Zn cell. The reaction that produces the higher voltage will prevail, and the overall reaction will proceed in the direction of the prevailing reaction.

Chemists use the symbol E to represent the potential of an electrochemical cell. Hence the potential for the Ni-Zn cell is expressed as E(Ni-Zn) and that for the Ag-Zn cell is expressed as E(Ag-Zn). These symbols can be used to express the reaction just studied.

$$E\text{(overall reaction)} = E\text{(Ag-Zn)} - E\text{(Ni-Zn)}$$
$$= 1.5 \text{ volts} - 0.5 \text{ volt}$$
$$= 1.0 \text{ volt}$$

Another interesting experiment is the investigation of the voltage of a Ag-Ni cell. The voltage turns out to be 1.0 volt, exactly the same as that obtained when the preceding two cells were set up in opposition to each other. This suggests that you can think of the voltage of any electrochemical cell as consisting of two parts: (1) the voltage characteristic of one of the half-cell reactions of the cell and (2) the voltage characteristic of the other half-cell reaction of this cell. In the Ag-Zn cell, for example, you can write the following.

$$E\text{(Ag-Zn cell)} = E(\text{Ag}^+\text{-Ag}) - E(\text{Zn}^{2+}\text{-Zn})$$
$$= 1.5 \text{ volts}$$

The voltage for the entire cell equals the voltage characteristic of the Ag^+-Ag half-cell minus the voltage characteristic of the Zn^{2+}-Zn half-cell. Likewise for the Ni-Zn cell, the following equation applies.

$$E\text{(Ni-Zn cell)} = E(\text{Ni}^{2+}\text{-Ni}) - E(\text{Zn}^{2+}\text{-Zn})$$
$$= 0.5 \text{ volt}$$

When the preceding two electrochemical cells are placed in opposition, the voltage of the resultant cell can be found as follows.

E(combined cell)
$$= E\text{(Ag-Zn cell)} - E\text{(Ni-Zn cell)}$$
$$= [E(\text{Ag}^+\text{-Ag}) - E(\text{Zn}^{2+}\text{-Zn})]$$
$$\quad - [E(\text{Ni}^{2+}\text{-Ni}) - E(\text{Zn}^{2+}\text{-Zn})]$$
$$= E(\text{Ag}^+\text{-Ag}) - E(\text{Ni}^{2+}\text{-Ni})$$
$$= 1.0 \text{ volt}$$

Observe that the E for the Zn^{2+}-Zn half-cell reaction subtracts out and that the net result is the sum of the potentials represented by the half-cell reactions for the Ag-Ni cell.

24-8 Measuring Half-Cell Potentials

The cell potentials just studied indicate that it would be desirable to know the actual contribution that each half-cell makes to the total potential of a cell. However, a voltage that you can measure is always the *difference* between two half-cell potentials. You can never isolate just one half-cell and measure its potential.

Scientists have solved this dilemma by choosing one particular half-cell and arbitrarily assigning it a potential of 0.00 (zero to as many decimal places as you like) for its half-cell reaction. It is called a **standard** or **reference half-cell.** The voltage of any half-cell can be measured when that half-cell is connected to the standard half-cell.

The 2H^+-H_2 half-cell reaction is used as the standard. Its potential is assigned the value of 0.00 volt.

$$2\text{H}^+(aq) + 2e^- \longrightarrow \text{H}_2(g)$$
$$E° = 0.00 \text{ volt}$$

It has already been mentioned that if the concentration of the solution used in a half-cell reaction is varied, the voltage of the cell varies. Since the voltage of the cell is a measure of the tendency of the reaction to pro-

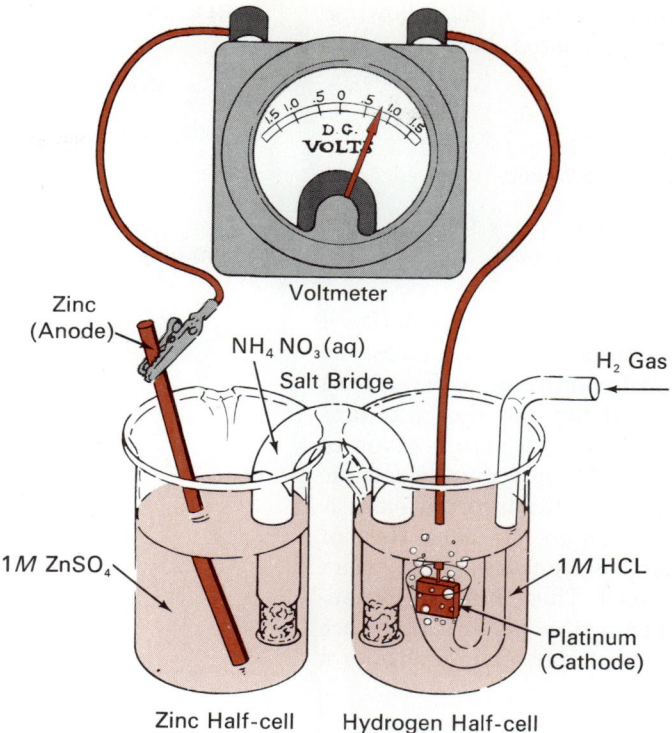

Figure 24–7 The potential of the Zn^{2+}-Zn half-cell relative to that of $2H^+$-H_2, which is taken as zero, can be measured directly, as shown in this experiment.

ceed, you could expect the equilibrium to be affected by a change in concentration of either reactants or products (Le Chatelier's principle). If the measurements of half-cell potentials are to have any real value, they must all be measured under some specific or *standard* conditions. In fact, the superscript zero in the symbol $E°$ refers to a *standard state*. The symbol indicates that this is the voltage to expect when all reactants are in their standard states. Thus, the value of $E°$ is the potential of any half-cell reaction occurring between substances in their standard states.

The standard state for gases has been taken to be a pressure of 760 Torr (one atmosphere) and a temperature of 25°C. The standard state for ions in solution is considered to be a concentration of $1\,M$. The standard state for pure substances is taken to be the form in which they normally exist at 25°C.

When the $2H^+$-H_2 half-cell is used as the standard and its standard potential, $E°$, is taken as 0.00 volt, the potentials of a great many half-cells can be measured. (See Figure 24–7.) Chemists have already measured these potentials. Thus, when measured at 25°C against the standard half-cell, the following potentials for the indicated half-cell reactions were found.

$$Zn^{2+}(aq, 1\,M) + 2e^- \longrightarrow Zn(s)$$
$$E° = -0.76 \text{ volt}$$

$$Cu^{2+}(aq, 1\,M) + 2e^- \longrightarrow Cu(s)$$
$$E° = +0.34 \text{ volt}$$

As stated before, the $E°$ for any half-cell is a measure of its tendency to gain electrons under the standard conditions as compared to the similar tendency of the standard $(2H^+\text{-}H_2)$ half-cell. A *positive* sign is given to potentials of half-cells that gain electrons *more* readily than the standard half-cell. A *negative* sign is given to the potentials of half-cells that gain electrons *less* readily than the standard half-cell. These half-cell potentials are called **reduction potentials**, since they are a measure of the tendency to gain electrons.

24-9 Using Reduction-Potential Tables

Some important and useful reduction potentials are given in Table 24-1. Of course, this is only a partial list of the reduction potentials that have been carefully and accurately measured. The standard conditions described earlier are assumed for all of the reactions listed. Now compare the order of the half-cell reactions listed in this table with that of the last list of half-cell reactions you prepared in Section 24-6. You will find that the equations in both lists are in the same order.

The higher (more positive) the value of $E°$ for a given half-cell reaction, the greater is the tendency to gain electrons. Therefore, you can predict that in the reaction between any two half-cells, electrons will go from the half-cell with the lower $E°$ to the half-cell

TABLE 24-1 Some Standard Reduction Potentials

Half-Cell Reaction	$E°$ (volts)
$MnO_4^- + 8H^+ + 5e^- \longrightarrow Mn^{2+} + 4H_2O$	+1.52
$Cl_2(g) + 2e^- \longrightarrow 2Cl^-$	+1.36
$MnO_2(s) + 4H^+ + 2e^- \longrightarrow Mn^{2+} + 2H_2O$	+1.28
$Br_2(l) + 2e^- \longrightarrow 2Br^-$	+1.06
$Ag^+ + e^- \longrightarrow Ag(s)$	+0.80
$I_2(s) + 2e^- \longrightarrow 2I^-$	+0.53
$Cu^{2+} + 2e^- \longrightarrow Cu(s)$	+0.34
$2H^+ + 2e^- \longrightarrow H_2(g)$	0.00
$Ni^{2+} + 2e^- \longrightarrow Ni(s)$	-0.25
$Co^{2+} + 2e^- \longrightarrow Co(s)$	-0.28
$Zn^{2+} + 2e^- \longrightarrow Zn(s)$	-0.76

with the higher $E°$. Further, you can calculate the net $E°$ for any electrochemical cell reaction by subtracting the lower $E°$ from the higher $E°$ for the half-cells involved.

The ideas developed in electrochemical reactions can be applied to reactions that occur in a beaker. Therefore, *chemists use half-cell potentials to predict what chemical reactions can occur spontaneously.*

Suppose a chemist wants to know whether Zn will be oxidized when it is placed in contact with a solution of nickel sulfate. To answer this question, he can use the values of $E°$. The half-cell potential for Zn^{2+}-Zn is -0.76 volt, which is less than that for Ni^{2+}-Ni (-0.25 volt). Because Ni^{2+}-Ni has a higher $E°$ value than that of Zn^{2+}-Zn, Ni^{2+} has a greater tendency to gain electrons than does Zn^{2+}. Therefore, Zn can transfer electrons to Ni^{2+}. The chemist predicts: Zn will react with Ni^{2+}, Zn being oxidized and Ni^{2+} being reduced.

Suppose the question is whether silver will be oxidized if it is immersed in copper sulfate. The half-cell potential for Ag^+-Ag is $+0.80$ volt and that for Cu^{2+}-Cu is $+0.34$ volt. The first value, $+0.80$ volt, is more positive than the second, $+0.34$ volt. The higher Ag^+-Ag $E°$ indicates that Ag^+ has a greater tendency to gain electrons than does Cu^{2+}. The oxidation of silver will *not* tend to proceed spontaneously. Silver will not be oxidized to an appreciable extent in copper sulfate.

You can generalize further, using Table 24-1. A substance on the left of the arrow reacts by gaining electrons. A substance on the right reacts by losing electrons. The following rules can be stated.

1. An oxidation-reduction reaction must involve a substance from the left side of the half-cell reactions in the table (something that can be reduced) and a substance from the right side (something that can be oxidized).

2. A substance on the left side of the half-cell reactions tends to react spontaneously with any substance on the right side that is at a *lower* position in the table. Applying these rules, you can predict the following results, among others. The Cu^{2+} ion can be reduced to Cu(s) by Ni(s) or Zn(s), but not by bromide ion, Br^-.

You must always remember that the reliability of such predictions based on Table 24-1 (or others like it) depends upon standard conditions being maintained. If the $E°$ for the overall reaction under consideration is *very small*, even small deviations from standard conditions that are not taken into account may throw predictions off. Furthermore, any predictions based on the table alone do not anticipate any reactions that may be exceedingly *fast* or *slow*. Thus, the use of a reduction-potential table does not exclude the necessity for further experimentation. It does, however, provide an excellent basis for eliminating from consideration many reactions that cannot take place.

Summary

You can predict whether a given oxidation-reduction reaction will actually occur (and if so, in which direction) by using a table of half-cell potentials such as Table 24-1 or the more extensive listing given in Appendix 2. These potentials are obtained by measuring the potentials of electrochemical cells, each of which is made up of two half-cells. Then, by arbitrarily assigning a voltage of zero to one such half-cell, namely the $2H^+$-H_2 half-cell, the values of all others may be calculated. Those with positive values have a greater tendency to accept electrons than the $2H^+$-H_2 half-cell. Those with negative values have less of such a tendency. The calculated values are called $E°$ values. The superscript zero means that they apply under standard conditions.

Self Check

1. Use the values of $E°$ to predict whether cobalt metal will tend to dissolve in a $1\,M$ solution of acid, H^+. Then predict whether cobalt metal will tend to dissolve in a $1\,M$ solution of zinc sulfate.
2. Use Table 24-1 to decide which of the following substances tend to oxidize the bromide ion: $Cl_2(g)$; H^+; Ni^{2+}.
3. Use the same table to decide which of the following substances tend to reduce $Br_2(l)$: Cl^-; $H_2(g)$; Ni(s); Mn^{2+}.

Balancing Equations for Oxidation-Reduction Reactions

24-10 Oxidation Numbers

In oxidation-reduction reactions such as those you have studied, there is always a transfer of electrons. The number of electrons released by the oxidation reaction must, of course, always equal the number gained in the resulting reduction. In attempting to balance an equation representing this type of reaction, you must be certain that both sides of the equation are balanced with respect to charge as well as to numbers of individual atoms.

You know that whenever you have unequal numbers of atoms or charges on either side of a chemical equation, you can make them equal by using appropriate numbers as coefficients for the reactants and products. In the discussion in Section 24-2 of the reaction between Cu atoms and Ag^+ ions, recall that as each atom of copper was oxidized to Cu^{2+} ions, *two electrons* were released.

$$Cu(s) \longrightarrow Cu^{2+}(aq) + 2e^-$$

Subsequently, as each ion of silver was reduced to a silver atom, one electron was gained.

$$Ag^+(aq) + e^- \longrightarrow Ag(s)$$

Obviously each Cu atom that was oxidized released enough electrons to reduce two Ag^+ ions, since each Ag^+ ion has a charge of $1+$ and needs only one electron to reduce it to a neutral silver atom. Thus, the reaction between Cu and Ag^+ ions, first written as

$$Cu(s) + Ag^+(aq) \longrightarrow Cu^{2+}(aq) + Ag(s)$$

is not balanced with respect to charges since there is a net positive charge of one on the left-hand side and of two on the right-hand side. Proper coefficients can balance this equation.

$$Cu(s) + 2Ag^+(aq) \longrightarrow Cu^{2+}(aq) + 2Ag(s)$$

The coefficient "2" in front of both the Ag^+ ion and the Ag atom brings the equation into balance, both with respect to atoms and to charge. The gain and loss of electrons, and consequently the coefficients needed to bring the equation into balance, are easily observed in a relatively simple redox equation such as the one just described. Many oxidation-reduction reactions, however, are considerably more complex than the relatively straightforward exchange of electrons between the atom of one element and the ion of another.

For example, in reactions involving polyatomic ions, which contain atoms of two or more elements, you can arbitrarily assume that the charges on the hydrogen and oxygen atoms in any such ion are $1+$ and $2-$ respectively. Hence, you can assign a charge to the remaining atom, or atoms, such that the sum of all the charges thus represented is equal to the net charge on the polyatomic ion. For instance, the MnO_4^- (permanganate) ion has a net charge of $1-$, as shown. Now, you can assume that each oxygen atom has a characteristic charge of $2-$, making a total negative charge of $8-$ for the four oxygen atoms in this ion. You can then assign a charge of $7+$ to the Mn atom, making the overall charge on the ion $(8-) + (7+)$, or $1-$.

The charge assigned to the Mn atom by this technique is called its **oxidation number**. It is a fictitious charge, as are all the oxidation numbers assigned to various atoms in ions by this method. However, it is useful in helping to keep track of the gain and loss of electrons.

Oxidation numbers are assigned to the atoms of polyatomic molecules in the same way that they are assigned to the atoms of polyatomic ions. Consider, for example, an equation representing an oxidation-reduction reaction found in Chapter 19.

$$2SO_2(g) + O_2(g) \rightleftharpoons 2SO_3(g)$$

This is another example of an oxidation-reduction reaction in which the transfer of electrons is not so readily apparent as it was in some of the half-cell reactions. Applying the same assumption used previously, you can call the charge on the oxygen atoms in these molecules $2-$. You can then readily assign oxidation numbers of $4+$ and $6+$ to the sulfur atoms in SO_2 and SO_3, respectively.

24-11 Using Oxidation Numbers

Oxidation numbers are used as a device for assigning a fictitious charge to an atom in an ion or molecule. According to this scheme, all oxidation-reduction reactions involve changes of the oxidation numbers. In order for charge to be conserved, there must be a balance between changes of oxidation number. The algebraic sum of all changes must equal zero. Consequently, oxidation numbers provide a basis for balancing redox equations.

For the present, the discussion will be limited to monatomic ions and to polyatomic ions and molecules containing hydrogen and/or oxygen along with the element to which an oxidation number is assigned. The rules to be used are the following ones.

1. The oxidation number of a monatomic ion is equal to the charge on the ion.
2. The oxidation number of any substance in the elementary state is zero.
3. The oxidation number of hydrogen is taken to be $1+$ (except in H_2, which is the elementary state, and in hydrides, which are compounds, such as NaH, containing hydrogen and a metal).
4. The oxidation number of oxygen is taken to be $2-$ (except in O_2; ozone, O_3; and peroxides, such as H_2O_2 and Na_2O_2).
5. The other oxidation numbers are selected to make the algebraic sum of the oxidation numbers equal to the net charge on the molecule or ion.
6. Reactions must be so balanced that the net change of oxidation numbers is zero. (This last rule is the result of the conservation of charge.)

The first step in balancing a reaction by the oxidation-number method (or by any other method) is to determine the products. Experiment, of course, provides the answer. For example, $H_2S(g)$ reacts with $MnO_4^-(aq)$ to give $S(s)$ and $Mn^{2+}(aq)$.

$$\begin{array}{cccc} MnO_4^-(aq) + & H_2S(g) & \text{gives} & S(s) + Mn^{2+}(aq) \\ 7+ & 2- & & 0 \quad\quad 2+ \end{array}$$

Changes in oxidation numbers are found.

$$Mn\ 7+ \xrightarrow{+(5-)} 2+$$
$$S\ 2- \xrightarrow{-(2-)} 0$$

If the gain in oxidation number by sulfur is to equal the loss by manganese, then five atoms of sulfur must react with two atoms of manganese.

$$2MnO_4^-(aq) + 5H_2S(g) \text{ gives } 5S(s) + 2Mn^{2+}(aq)$$
$$2(7+) \xrightarrow{(+2)(5-)\ =\ +(10-)} 2(2+)$$
$$5(2-) \xrightarrow{(-5)(2-)\ =\ -(10-)} 5(0)$$

From this point on, the balancing procedure that you learned in Chapter 19 is used. Once you have set the ratio between the atoms directly involved in the redox process, it must be altered.

The next step is to ensure the conservation of oxygen atoms. Eight oxygen atoms are on the left. Thus, eight molecules of H_2O must be added to the right. You can add water molecules to either side of the equation because the reaction occurs in aqueous solution. When you conserve oxygen atoms, you get the following reaction.

$$2MnO_4^-(aq) + 5H_2S(g) \text{ gives } 5S(s) + 2Mn^{2+}(aq) + 8H_2O$$

This is still not balanced, so next we must ensure conservation of hydrogen atoms. On the left, there are 10 hydrogen atoms (in 5 H_2S) and 16 on the right (in 8 H_2O). We will assume that the solution under discussion is either neutral or acidic. For such a solution, chemists assume that any hydrogen atoms that are needed to balance an equation are provided by $H^+(aq)$. (If it were a basic solution, chemists would assume that OH^- ions were available.)

$$2MnO_4^-(aq) + 5H_2S(g) + 6H^+(aq) \rightleftharpoons 2Mn^{2+}(aq) + 5S(s) + 8H_2O$$

The equation should now be balanced, but experience shows that a check should *always* be made on the basis of charge balance. If the equation is really balanced, the net charges on each side will be equal.

$$2(1-) + 5(0) + 6(1+) \stackrel{?}{=} 2(2+) + 5(0) + 8(0)$$
$$2- \qquad\qquad 6+ \stackrel{?}{=} 4+$$
$$4+ = 4+$$

24-12 Using Half-Cell Reactions to Balance Equations

The technique here consists of selecting two balanced half-cell reactions that can be added to give the full reaction in question. If the two half-cell reactions involve different numbers of electrons, it is necessary to multiply each reaction by a suitable coefficient so that each of them involves the same number of electrons. Then when they are added together, one as an oxidation and the other as a reduction, the same number of electrons will appear on each side and will cancel out.

As an example, consider the same reaction treated in the preceding section.

$$MnO_4^-(aq) + H_2S(g) \text{ gives } S(s) + Mn^{2+}(aq) \quad (4)$$

The two half-cell reactions needed are the following ones.

$$MnO_4^-(aq) + 8H^+(aq) + 5e^- \longrightarrow Mn^{2+}(aq) + 4H_2O$$

$$S(s) + 2H^+(aq) + 2e^- \longrightarrow H_2S(g)$$

Both of these half-cell reactions can be found, balanced, in the table of half-cell potentials given in Appendix 2. However, if there were no such table, the half-cell reactions themselves could be balanced by using, in a modified form, the oxidation-number technique described in the previous section. For example, suppose you begin with the statement.

$$MnO_4^- \text{ goes to } Mn^{2+}$$
$$7+ \qquad\qquad 2+$$

where the oxidation numbers have been written under the Mn symbols. Now, the easiest procedure is first to balance electrons.

Since Mn with the 7+ oxidation number has has five less electrons than Mn^{2+}, you must add five electrons to the side of the equation on which MnO_4^- appears.

$$5e^- + MnO_4^-(aq) \text{ goes to } Mn^{2+}(aq)$$

Now you must balance oxygen atoms by adding 4 H_2O to the right side. After you have done this, you must add 8 H^+ to the left to balance the hydrogen atoms. This results in the conservation of all atoms and electrons so that the result should now be a balanced equation.

$$5e^- + MnO_4^-(aq) + 8H^+(aq) \rightleftharpoons Mn^{2+}(aq) + 4H_2O$$

Again, as a check, make sure that charge is conserved.

$$5(1-) + (1-) + 8(1+) \stackrel{?}{=} (2+) + 4(0)$$
$$5- \quad 1- \quad 8+ \stackrel{?}{=} 2+$$
$$2+ = 2+$$

Once you have two balanced half-cell reactions (either by taking them from a table or by balancing them yourself), the problem of balancing the equation for the entire reaction is almost solved. In the present case, one of the half-reactions involves five electrons, while the other involves two electrons. Hence, you must multiply the first one by two and the second one by five.

$$2MnO_4^-(aq) + 16H^+(aq) + 10e^- \longrightarrow$$
$$2Mn^{2+}(aq) + 8H_2O \quad (5)$$

$$5S(s) + 10H^+(aq) + 10e^- \longrightarrow 5H_2S(g) \quad (6)$$

Now, in order to get an equation based on reaction (4), you should carry out the following steps.
1. Leave equation (5) as it is.
2. Write equation (6) in reverse.
3. Add equations (5) and (6).
4. Eliminate the hydrogen ions, electrons, and water molecules that appear on both sides of the equation.

These steps, illustrated below, give an equation that is balanced and identical to the equation obtained in the previous section, in which the oxidation-number technique was used directly on the entire equation.

1. $\quad 2MnO_4^-(aq) + 16H^+(aq) + 10e^- \longrightarrow 2Mn^{2+}(aq) + 8H_2O$
2. $\quad 5H_2S(g) \longrightarrow 5S(s) + 10H^+(aq) + 10e^-$
3. $\quad 2MnO_4^-(aq) + 5H_2S(g) + 16H^+(aq) + 10e^- \longrightarrow 2Mn^{2+}(aq) + 5S(s) + 8H_2O + 10H^+(aq) + 10e^-$
4. $\quad 2MnO_4^-(aq) + 5H_2S(g) + 6H^+(aq) \longrightarrow 2Mn^{2+}(aq) + 5S(s) + 8H_2O$

Summary

Redox reactions can be balanced by either of two methods. One method is as follows. (1) Using an expression for the entire reaction, oxidation numbers are assigned to all atoms. (2) The atoms with oxidation numbers that change are given coefficients such that the total increase in oxidation number for one type of atom is equal to the total decrease for the other type. (3) Keeping these numbers, or their ratio, fixed, the remainder of the equation is balanced as usual. As is necessary, H_2O, $H^+(aq)$, or $OH^-(aq)$ may be added to either side of the equation if the reaction occurs in an aqueous solution. Finally, a check is made to see that the charges are balanced.

The second method is as follows. (1) Two balanced half-cell reactions are added after each has been multiplied by the appropriate figure to make the number of electrons involved in each the same. (2) The electrons, hydrogen ions, and water molecules that appear on both sides of the equation are then eliminated.

Self Check

1. Show that the oxidation number of nitrogen is $5+$ in both NO_3^- and N_2O_5.
2. Explain how oxidation numbers provide chemists with a means of balancing oxidation-reduction (redox) reactions.

Useful Electrochemistry

24-13   Batteries ★

In the experiment in Section 24-1, you saw that an electric current was generated by two half-cell reactions. You also saw that an external circuit, which included a wire and an ammeter, was used. Such an external circuit may, for example, include a light bulb, a motor, a transistor radio, or the ignition circuit of an automobile. The electrochemical cells used to generate electricity for consumption in these devices are commonly called **batteries.**

Batteries in practical use today utilize different reactions from the ones you have just studied, but exactly the same principles are involved. Therefore, you are now in a position to examine the two most common types of batteries and understand how they work.

One type of battery is the *common dry cell.* This is the familiar type of battery used in flashlights, many transistor radios and other devices requiring fairly small currents. In this type of battery, the charge runs down and cannot be restored. It is different from the rechargeable type of battery you find in automobiles. As shown in Figure 24-8, the dry cell is encased in a zinc cylinder and has a carbon rod at the center.

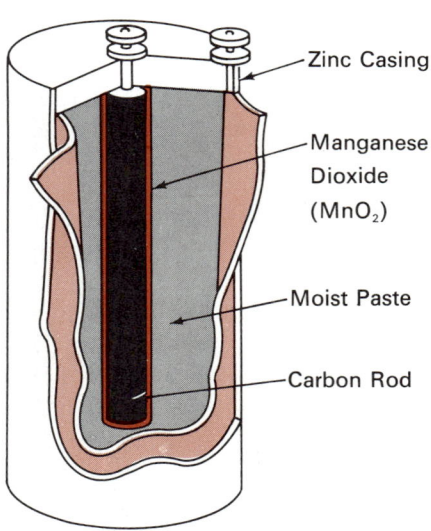

Figure 24–8 *The basic structure of an ordinary "dry cell" type of battery is shown in this cutaway drawing.*

The reaction that produces electrons occurs at the zinc electrode, which surrounds the entire cell.

$$Zn(s) \longrightarrow Zn^{2+}(aq) + 2e^-$$

The reaction that takes up electrons occurs at the carbon rod and is described by the following equation.

$$2NH_4^+(aq) + 2MnO_2(s) + 2e^- \longrightarrow 2MnO(OH)(s) + 2NH_3(aq)$$

The electrolyte in the cell is a moist, but not actually liquid paste of ammonium chloride, water, and an inert substance. Because of its pasty nature, the presence of a porous sheet, and the insolubility of the MnO_2, no direct chemical reaction between Zn and MnO_2 occurs.

The other common type of battery is the *lead-sulfuric acid battery*, found in automobiles. This battery is of the type called the *storage cell*, since it can be returned to its original state after current has been drawn from it. This is done by applying a voltage across its electrodes in the opposite direction, thus *recharging* it. (Section 24-14 has a discussion of this type of process, termed *electrolysis*.) The electrodes in this battery are lead and lead dioxide plates immersed in a sulfuric acid electrolyte.

In the lead battery, the two reactions that produce and consume electrons are, respectively, the following ones.

$$Pb(s) + SO_4^{2-}(aq) \longrightarrow PbSO_4(s) + 2e^-$$

$$PbO_2(s) + SO_4^{2-}(aq) + 4H^+(aq) + 2e^- \longrightarrow PbSO_4(s) + 2H_2O$$

Each of these half-cell reactions produces the same insoluble substance, $PbSO_4(s)$, which adheres to the plates. As the cell discharges, generating current, acid is consumed and water is produced. Thus the extent of discharge of the battery can be measured using a hydrometer, which was discussed in Section 4-4. The more dilute the acid becomes, the lower is the density of the liquid.

24-14 Using a Voltage to Drive a Reaction ★

So far, this chapter has dealt with reactions that proceed spontaneously. However, the same ideas and names are applied to reactions that are forced to take place against their natural tendency. In these reactions energy is supplied by an externally-applied electric current. Such a process is termed **electrolysis,** or "separation by electricity." Though you might not know it, you dealt with electrolysis in the discussion of the electrical conductivity of an ionic solution.

A distinguishing property of ionic solutions is electrical conductivity, just as it is a distinguishing property of metals. However, the current-carrying mechanism in ionic solutions differs from that in metals. Electrons flow through a wire without changing the metal chemically. In contrast, the movement of electric charge through an aqueous solution of an electrolyte causes significant chemical changes.

Consider, for example, the behavior of an aqueous hydrogen iodide solution during conduction (Figure 24-9, next page). The two carbon electrodes, or rods, are connected by wires to the terminals of a two-volt battery. Electrons flow from the battery through the left carbon electrode, entering the solution. An equal number of electrons leave the solution through the right carbon electrode to

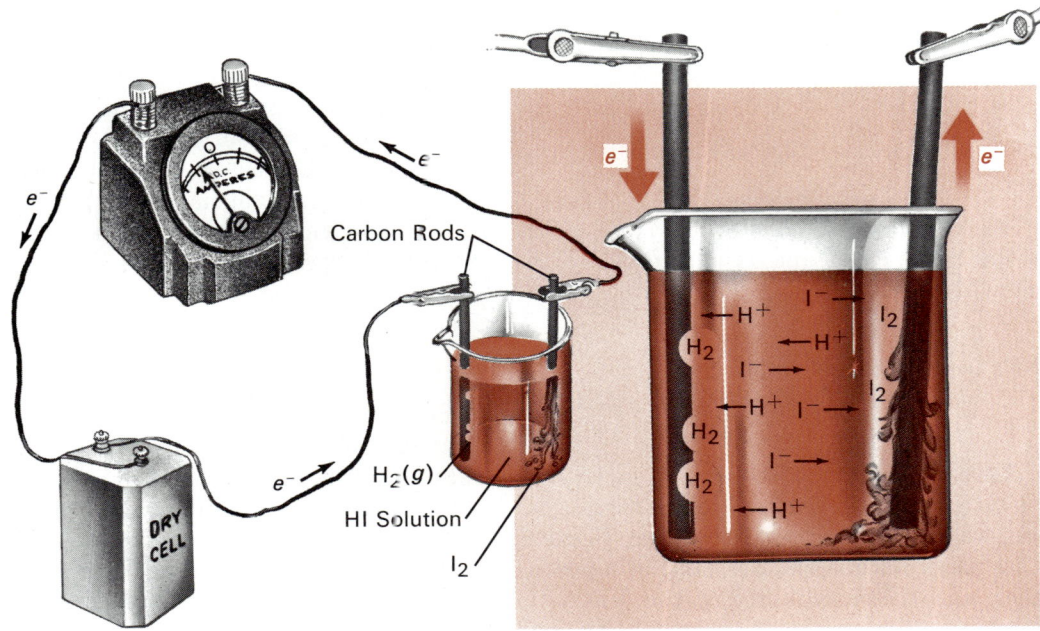

Figure 24-9 *Energy supplied by a dry cell drives a current through a solution of H^+ and I^- ions, producing chemical changes that would not occur spontaneously. The process is called electrolysis.*

return to the battery. The hydrogen ion, H^+, has the ability to accept an electron at the left electrode, where electrons are in excess. The H^+ ion is changed chemically to a neutral atom. An iodide ion, I^-, has one excess electron that can be released at the right electrode. The I^- ion changes chemically to a neutral atom. The net result of these two occurrences is the process called electrolysis.

The following steps sum up the process during the movement of one electron through the entire circuit shown in Figure 24-9.

1. An electron leaves the left carbon rod. Another electron shows up at the right carbon rod.
2. One $H^+(aq)$ ion and one $I^-(aq)$ ion vanish. One H atom and one I atom form.
3. As this process continues to take place, the

H^+ ions in the solution at the left and the I^- ions at the right tend to be used up. Since only positive ions are used up at the left, the remaining negative ions are repelled from this region and are attracted toward the solution at the right, where positive ions are plentiful. Here negative ions are used up, so the positive ions are repelled and are attracted toward the left, where negative ions are plentiful. Iodide ions, I^-, move from left to right through the solution, carrying a negative charge. At the same time, hydrogen ions, H^+, move from right to left through the solution, carrying a positive charge. This drift of the ions through the solution, positive ions in one direction and negative ions in the other, explains the conduction in aqueous solutions.

4. The battery performs work in forcing current to flow through the solution and in causing chemical changes to occur that would not proceed spontaneously. The net reaction is as follows.

$$2H^+(aq) + 2I^-(aq) \longrightarrow H_2(g) + I_2(s)$$

This is an oxidation-reduction reaction and is readily separated into two half-cell reactions.

Reaction	$E°$(V)
$2H^+(aq) + 2e^- = H_2(g)$	0.00
$2I^-(aq) = I_2(s) + 2e^-$	-0.53
$2H^+(aq) + 2I^-(aq) = H_2(g) + I_2(s)$	-0.53

The negative value of $E° = -0.53$ volt indicates that the reaction will not occur spontaneously as written. This voltage further indicates that electrolysis will occur only if a cell with a voltage exceeding 0.53 volt is placed in the external circuit. Then current can be forced to flow in opposition to the spontaneous direction.

When electrolysis is carried out, a voltage is applied to a solution by attaching two electrodes, one to the positive terminal and one to the negative terminal, to an external source of current such as a battery. The electrode attached to the positive terminal is the one at which something in the solution is oxidized. It is called the **anode** (it attracts anions). The negative electrode, at which reduction occurs, is called the **cathode** (it attracts cations).

24-15 *The Process of Electroplating* ★

A very important use of the process of electrolysis reactions is in electroplating. This is the process of depositing a coating of metal, from a solution containing its ions, upon some other metal object.

A common example of this is copper plating. The solution often used contains copper sulfate, that is, Cu^{2+} ions and SO_4^{2-} ions. The object on which a coating of copper is to be deposited is suspended in the solution and attached to the negative terminal of a battery or other similar energy source. The object thus becomes the cathode. The anode consists of a rod of pure copper. The following reactions then occur.

At the anode $Cu(s) \longrightarrow Cu^{2+}(aq) + 2e^-$
At the cathode $Cu^{2+}(aq) + 2e^- \longrightarrow Cu(s)$

Thus copper goes into solution at the anode and is "plated out" at the cathode, as shown in Figure 24-10. The "plating bath" simply

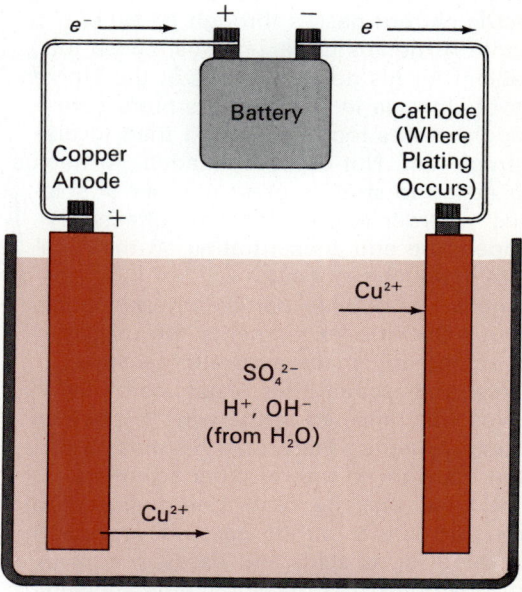

Figure 24-10 *A source of current, such as a battery, causes copper to go into solution as Cu^{2+} ions at the anode, while Cu^{2+} ions from the solution are electroplated at the cathode.*

Svante August Arrhenius (1859–1927)

WHAT HAPPENS IN A SOLUTION when an electric current passes through it? When Svante Arrhenius (uh-RAI-nee-us) chose this problem for his doctoral thesis at the University of Uppsala in Sweden, his professors thought it was too difficult and tried to discourage him. Not easily dissuaded, Arrhenius proceeded to study a great number of solutions, in order to determine the effects of temperature and concentration on their ability to conduct electricity.

There was nothing particularly spectacular about Arrhenius' experiments. He used simple equipment, made accurate measurements, and recorded his results carefully. His genius was reflected in the way he planned his experiments, interpreted his data, and related his work to that of other scientists.

He knew that the heat of neutralization of strong acids with strong bases is very nearly the same for all acids and bases in equivalent quantities. His measurements showed that these compounds were good conductors of electricity when they were dissolved in water. Arrhenius also discovered that acids and bases that were poor conductors of electricity in concentrated aqueous solutions became better conductors when they were diluted.

As a result of his experiments, Arrhenius proposed that solute molecules dissociate into tiny particles, called ions, which carry electric current through solutions. He interpreted the similarity in the heats of neutralization of acids and bases to indicate that all acids must form hydrogen ions and all bases must form hydroxide ions. He predicted that the chemical activity of aqueous solutions of acids and bases should be proportional to their conductivity.

Arrhenius' conclusions were so radical that his professors accepted his thesis with great reluctance. During the next few years he conducted many experiments to verify his theories, but doubt persisted. Eventually, his experiments and conclusions were confirmed by other noted scientists, and he was awarded the Nobel Prize for Chemistry in 1903.

serves as a means of transferring copper from a rod of the pure metal to the surface of an object where it is required.

Chromium plating is carried out in a slightly different way. Here the "bath" is a highly ionized solution containing chromate ions, CrO_4^{2-}. The object to be plated is made the cathode, and chromium metal deposits there in the following reaction.

$$CrO_4^{2-}(aq) + 8H^+(aq) + 6e^- \longrightarrow Cr(s) + 4H_2O$$

At the anode, the oxidation reaction is as follows.

$$2H_2O \longrightarrow O_2(g) + 4H^+(aq) + 4e^-$$

The supply of chromium in the bath is maintained by the addition of CrO_3, which dissolves according to the following reaction.

$$CrO_3(s) + H_2O = 2H^+(aq) + CrO_4^{2-}(aq)$$

Summary

Spontaneous redox reactions can be used to generate electric current for useful purposes. Devices in which this is done are called batteries. In the common dry cell the reaction that produces electrons occurs at the zinc electrode. The reaction that takes up electrons occurs at the carbon electrode. In the lead-sulfuric acid battery, a reaction involving metallic lead produces electrons, while a reaction using PbO_2 consumes electrons.

Many chemical reactions do not proceed spontaneously, but require some external energy to be added in order to make the reaction proceed. When this external energy is an electric current, the resultant reaction is a part of the process called electrolysis.

Electrolysis can be used to produce many substances, especially metals. When the process of electrolysis is used specifically to deposit a metal from a solution containing its ions onto another metal object, the process is called electroplating.

Self Check

1. Write the half-cell reactions that occur when a lead-sulfuric acid battery is discharging (delivering current). Write the reactions that occur when it is being recharged.
2. How does a common dry cell differ from a lead-sulfuric acid battery?

Chapter Summary

Oxidation-reduction (redox) reactions are reactions that involve the gain and loss of electrons. The part of such a reaction involving a loss of electrons is called oxidation. The part involving the gain of electrons is called reduction. Redox reactions can take place either directly, in a reaction, or indirectly in the form of an electrochemical cell that generates an electric current.

You can predict whether redox reactions will actually take place, and in which directions they will proceed, by consulting a table of half-cell potentials. Under standard conditions, the values in the table are called $E°$

values. They are based on a value of zero, which has been arbitrarily assigned to the $2H^+$-H_2 half-cell.

The balancing of redox reactions can be accomplished by either of two methods. These methods involve (1) using an expression for the overall reaction, or (2) balancing equations of half-cell reactions and then eliminating any electrons, hydrogen ions, and water molecules that appear on both sides of the overall equation that results when the half-cell reactions are added.

Questions and Problems

1. Briefly define the terms oxidation and reduction.
2. What happens to an oxidizing agent when it oxidizes a substance?
3. If a neutral atom becomes positively charged, has it been oxidized or reduced? Write a general equation for such a change, using M to stand for the neutral atom.
4. If an ion X^- acquires a 2− charge, has it been oxidized or reduced? Write a general equation showing the change.
5. One method of obtaining copper metal is to let a solution containing Cu^{2+} ions trickle over iron. Write equations for the two half-cell reactions involved. Assume the iron becomes Fe^{2+}. In which half-cell reaction does oxidation take place?
6. Nickel metal reacts with cupric ions, Cu^{2+}, and hydrogen ions, H^+, but not with zinc ions, Zn^{2+}. Magnesium metal reacts with Zn^{2+}. In each reaction, ions of 2+ charge are formed. Use these data to expand the last list in Section 24-6.
7. What temperature, concentration, and pressure must be maintained if the standard reduction potentials of half-cell reactions are to be used?
8. Imagine an electrochemical cell consisting of a beaker of copper sulfate solution with a copper electrode, connected to another beaker containing a silver nitrate solution with a silver electrode. Now water is added to the beaker of copper sulfate solution. What change will take place in the current of the cell after the water is added?
9. If a piece of copper metal is dipped into a solution containing Cr^{3+} ions, what will happen? Explain, using half-cell potentials in Appendix 2.
10. What would happen if an aluminum spoon were used to stir an $Fe(NO_3)_2$ solution? What would happen if an iron spoon were used to stir an $AlCl_3$ solution? Refer to Appendix 2.
11. Can $1\,M$ $Fe_2(SO_4)_3$ solution be stored in a container made of nickel? Explain your answer, using Appendix 2.
12. Complete the following equations. Using Appendix 2, determine the net potential of each cell and decide whether a reaction can occur.
    (a) $Zn(s) + Ag^+ \longrightarrow$
    (b) $Cu(s) + Ag^+ \longrightarrow$
    (c) $Sn(s) + Fe^{2+} \longrightarrow$
    (d) $Hg(l) + H^+ \longrightarrow$
13. For each of the following, write the half-cell reactions and determine the net reaction. Using Appendix 2, predict whether the reaction can occur. Give the basis for your prediction.
    (a) $Mg(s) + Sn^{2+} \longrightarrow$
    (b) $Mn(s) + Cs^+ \longrightarrow$
    (c) $Cu(s) + Cl_2 \longrightarrow$
    (d) $Zn(s) + Fe^{2+} \longrightarrow$
    (e) $Fe(s) + Fe^{3+} \longrightarrow$

14. Suppose chemists had assigned the value of zero to the following half-cell reaction.

$$I_2(s) + 2e^- \longrightarrow 2I^-(aq)$$

   (a) What would be $E°$ for the following half-cell reaction? (Use Appendix 2.)

$$Na^+(aq) + e^- \longrightarrow Na(s)$$

   (b) How much would the net potential for the following reaction change?

$$2Na(s) + I_2(s) \longrightarrow 2Na^+(aq) + 2I^-(aq)$$

15. What is the oxidation number of carbon in carbon monoxide? carbon dioxide? diamond?
16. Determine the oxidation number of uranium in UO_3, U_3O_8, UO_2, and UO.
17. When copper is placed in concentrated nitric acid, HNO_3, vigorous bubbling takes place as a brown gas, nitrogen dioxide, NO_2, is evolved. The copper disappears, and the colorless solution changes to greenish-blue due to the formation of cupric ion, Cu^{2+}. Using half-cell reactions, write the net ionic equation for this reaction.
18. Use oxidation numbers to balance the reaction in acid solution between ferrous ion, Fe^{2+}, and permanganate ion, MnO_4^-, to produce ferric ion, Fe^{3+}, and manganous ion, Mn^{2+}.
19. Using oxidation numbers, write a balanced equation for each of the following reactions.
   (a) $HBr + H_2SO_4$ gives $SO_2 + Br_2 + H_2O$
   (b) $NO_3^- + Cl^- + H^+$ gives $NO + Cl_2 + H_2O$
   (c) $Zn + NO_3^- + H^+$ gives $Zn^{2+} + NO_2 + H_2O$
   (d) BrO^- gives $Br^- + BrO_3^-$
20. Show that numbers other than the conventional oxidation numbers can be used to balance the reaction in Question 18. To do this assume that the oxidation number of manganese in MnO_4^- is 2+. Compare your answer to that for Question 18.
21. Using half-cell reactions in Appendix 2, give a balanced equation for each of the following reactions.
   (a) $H_2O_2 + I^- + H^+$ gives $H_2O + I_2$
   (b) $Cr_2O_7^{2-} + Fe^{2+} + H^+$ gives $Cr^{3+} + Fe^{3+} + H_2O$
   (c) $Cu + NO_3^- + H^+$ gives $Cu^{2+} + NO + H_2O$
   (d) $MnO_4^- + Sn^{2+} + H^+$ gives $Mn^{2+} + Sn^{4+} + H_2O$
22. If you wish to plate a silver spoon again, would you make it the anode or the cathode in a cell? Use half-cell reactions in your explanation. How many moles of electrons are needed to plate out one gram of Ag?
23. In the electrolysis of aqueous cupric bromide, $CuBr_2$, 0.500 gram of copper is deposited at one electrode. How many grams of bromine are formed at the other electrode? Write the anode and the cathode half-cell reactions.

APPENDIX 1

Relative Strengths of Acids

In Aqueous Solution at Room Temperature

$$HB \rightleftharpoons H^+(aq) + B^-(aq) \qquad K_A = \frac{[H^+][B^-]}{[HB]}$$

Acid	Strength	Reaction	K_A
perchloric acid	very strong	$HClO_4 \longrightarrow H^+ + ClO_4^-$	very large
hydriodic acid		$HI \longrightarrow H^+ + I^-$	very large
hydrobromic acid		$HBr \longrightarrow H^+ + Br^-$	very large
hydrochloric acid		$HCl \longrightarrow H^+ + Cl^-$	very large
nitric acid		$HNO_3 \longrightarrow H^+ + NO_3^-$	very large
sulfuric acid	very strong	$H_2SO_4 \longrightarrow H^+ + HSO_4^-$	large
oxalic acid	strong	$HOOCCOOH \longrightarrow H^+ + HOOCCOO^-$	5.4×10^{-2}
sulfurous acid ($SO_2 + H_2O$)		$H_2SO_3 \longrightarrow H^+ + HSO_3^-$	1.7×10^{-2}
hydrogen sulfate ion		$HSO_4^- \longrightarrow H^+ + SO_4^{2-}$	1.3×10^{-2}
phosphoric acid		$H_3PO_4 \longrightarrow H^+ + H_2PO_4^-$	7.1×10^{-3}
hydrogen telluride	strong	$H_2Te \longrightarrow H^+ + HTe^-$	2.3×10^{-3}
hydrofluoric acid	weak	$HF \longrightarrow H^+ + F^-$	6.7×10^{-4}
nitrous acid		$HNO_2 \longrightarrow H^+ + NO_2^-$	5.1×10^{-4}
hydrogen selenide		$H_2Se \longrightarrow H^+ + HSe^-$	1.7×10^{-4}
benzoic acid		$C_6H_5COOH \longrightarrow H^+ + C_6H_5COO^-$	6.6×10^{-5}
hydrogen oxalate ion		$HOOCCOO^- \longrightarrow H^+ + OOCCOO^{2-}$	5.4×10^{-5}
acetic acid		$CH_3COOH \longrightarrow H^+ + CH_3COO^-$	1.8×10^{-5}
hydrogen telluride ion		$HTe^- \longrightarrow H^+ + Te^{2-}$	10^{-5}
carbonic acid ($CO_2 + H_2O$)		$H_2CO_3 \longrightarrow H^+ + HCO_3^-$	4.4×10^{-7}
hydrogen sulfide		$H_2S \longrightarrow H^+ + HS^-$	1.0×10^{-7}
dihydrogen phosphate ion		$H_2PO_4^- \longrightarrow H^+ + HPO_4^{2-}$	6.3×10^{-8}
hydrogen sulfite ion		$HSO_3^- \longrightarrow H^+ + SO_3^{2-}$	6.2×10^{-8}
ammonium ion		$NH_4^+ \longrightarrow H^+ + NH_3$	5.7×10^{-10}
hydrogen carbonate ion	weak	$HCO_3^- \longrightarrow H^+ + CO_3^{2-}$	4.7×10^{-11}
hydrogen peroxide	very weak	$H_2O_2 \longrightarrow H^+ + HO_2^-$	2.4×10^{-12}
monohydrogen phosphate ion		$HPO_4^{2-} \longrightarrow H^+ + PO_4^{3-}$	4.4×10^{-13}
hydrogen sulfide ion		$HS^- \longrightarrow H^+ + S^{2-}$	1.3×10^{-13}
water		$H_2O \longrightarrow H^+ + OH^-$	1.0×10^{-14}
hydroxide ion		$OH^- \longrightarrow H^+ + O^{2-}$	$<10^{-30}$
ammonia	very weak	$NH_3 \longrightarrow H^+ + NH^-$	very small

APPENDIX 2 Standard Reduction Potentials for Half-Reactions

Ionic Concentrations = 1 Molar in Water, Temperature = 25°C

Strength As Oxidizing Agent	Half-Reaction	E^0 (volts)	Strength As Reducing Agent
very strong oxidizing agents ↑ increasing strength as oxidizing agent ↑ weak oxidizing agents ↑ very weak oxidizing agents	$F_2(g) + 2e^- \longrightarrow 2F^-$	+2.87	very weak reducing agents ↓ increasing strength as reducing ↓ very strong reducing agents
	$H_2O_2 + 2H^+ + 2e^- \longrightarrow 2H_2O$	+1.77	
	$MnO_4^- + 8H^+ + 5e^- \longrightarrow Mn^{2+} + 4H_2O$	+1.52	
	$Au^{3+} + 3e^- \longrightarrow Au(s)$	+1.50	
	$Cl_2(g) + 2e^- \longrightarrow 2Cl^-$	+1.36	
	$Cr_2O_7^{2-} + 14H^+ + 6e^- \longrightarrow 2Cr^{3+} + 7H_2O$	+1.33	
	$MnO_2(s) + 4H^+ + 2e^- \longrightarrow Mn^{2+} + 2H_2O$	+1.28	
	$\frac{1}{2}O_2(g) + 2H^+ + 2e^- \longrightarrow H_2O$	+1.23	
	$Br_2(l) + 2e^- \longrightarrow 2Br^-$	+1.06	
	$NO_3^- + 4H^+ + 3e^- \longrightarrow NO(g) + 2H_2O$	+0.96	
	$\frac{1}{2}O_2(g) + 2H^+(10^{-7} M) + 2e^- \longrightarrow H_2O$	+0.82	
	$Ag^+ + e^- \longrightarrow Ag(s)$	+0.80	
	$\frac{1}{2}Hg_2^{2+} + e^- \longrightarrow Hg(l)$	+0.79	
	$Hg^{2+} + 2e^- \longrightarrow Hg(l)$	+0.78	
	$NO_3^- + 2H^+ + e^- \longrightarrow NO_2(g) + H_2O$	+0.78	
	$Fe^{3+} + e^- \longrightarrow Fe^{2+}$	+0.77	
	$O_2(g) + 2H^+ + 2e^- \longrightarrow H_2O_2$	+0.68	
	$I_2(s) + 2e^- \longrightarrow 2I^-$	+0.53	
	$Cu^+ + e^- \longrightarrow Cu(s)$	+0.52	
	$Cu^{2+} + 2e^- \longrightarrow Cu(s)$	+0.34	
	$SO_4^{2-} + 4H^+ + 2e^- \longrightarrow SO_2(g) + 2H_2O$	+0.17	
	$Cu^{2+} + e^- \longrightarrow Cu^+$	+0.15	
	$Sn^{4+} + 2e^- \longrightarrow Sn^{2+}$	+0.15	
	$S + 2H^+ + 2e^- \longrightarrow H_2S(g)$	+0.14	
	$2H^+ + 2e^- \longrightarrow H_2(g)$	0.00	
	$Pb^{2+} + 2e^- \longrightarrow Pb(s)$	−0.13	
	$Sn^{2+} + 2e^- \longrightarrow Sn(s)$	−0.14	
	$Ni^{2+} + 2e^- \longrightarrow Ni(s)$	−0.25	
	$Co^{2+} + 2e^- \longrightarrow Co(s)$	−0.28	
	$Se + 2H^+ + 2e^- \longrightarrow H_2Se(g)$	−0.40	
	$Cr^{3+} + e^- \longrightarrow Cr^{2+}$	−0.41	
	$2H^+(10^{-7} M) + 2e^- \longrightarrow H_2(g)$	−0.41	
	$Fe^{2+} + 2e^- \longrightarrow Fe(s)$	−0.44	
	$Ag_2S + 2e^- \longrightarrow 2Ag(s) + S^{2-}$	−0.69	
	$Te + 2H^+ + 2e^- \longrightarrow H_2Te(g)$	−0.72	
	$Cr^{3+} + 2e^- \longrightarrow Cr(s)$	−0.74	
	$Zn^{2+} + 2e^- \longrightarrow Zn(s)$	−0.76	
	$2H_2O + 2e^- \longrightarrow 2OH^- + H_2(g)$	−0.83	
	$Mn^{2+} + 2e^- \longrightarrow Mn(s)$	−1.18	
	$Al^{3+} + 3e^- \longrightarrow Al(s)$	−1.66	
	$Mg^{2+} + 2e^- \longrightarrow Mg(s)$	−2.37	
	$Na^+ + e^- \longrightarrow Na(s)$	−2.71	
	$Ca^{2+} + 2e^- \longrightarrow Ca(s)$	−2.87	
	$Sr^{2+} + 2e^- \longrightarrow Sr(s)$	−2.89	
	$Ba^{2+} + 2e^- \longrightarrow Ba(s)$	−2.90	
	$Cs^+ + e^- \longrightarrow Cs(s)$	−2.92	
	$K^+ + e^- \longrightarrow K(s)$	−2.92	
	$Rb^+ + e^- \longrightarrow Rb(s)$	−2.92	
	$Li^+ + e^- \longrightarrow Li(s)$	−3.00	

APPENDIX 3 Molar Masses of the Elements

Element	Symbol	Atomic Number	Molar Mass
Actinium	Ac	89	(227)
Aluminum	Al	13	26.98
Americium	Am	95	(243)
Antimony	Sb	51	121.75
Argon	Ar	18	39.948
Arsenic	As	33	74.92
Astatine	At	85	(210)
Barium	Ba	56	137.34
Berkelium	Bk	97	(249)
Beryllium	Be	4	9.012
Bismuth	Bi	83	208.98
Boron	B	5	10.81
Bromine	Br	35	79.909
Cadmium	Cd	48	112.40
Calcium	Ca	20	40.08
Californium	Cf	98	(251)
Carbon	C	6	12.011
Cerium	Ce	58	140.12
Cesium	Cs	55	132.91
Chlorine	Cl	17	35.453
Chromium	Cr	24	52.00
Cobalt	Co	27	58.93
Copper	Cu	29	63.54
Curium	Cm	96	(247)
Dysprosium	Dy	66	162.50
Einsteinium	Es	99	(254)
Erbium	Er	68	167.26
Europium	Eu	63	151.96
Fermium	Fm	100	(253)
Fluorine	F	9	19.00
Francium	Fr	87	(223)
Gadolinium	Gd	64	157.25
Gallium	Ga	31	69.72
Germanium	Ge	32	72.59
Gold	Au	79	196.97
Hafnium	Hf	72	178.49
Helium	He	2	4.003
Holmium	Ho	67	164.93
Hydrogen	H	1	1.0080
Indium	In	49	114.82
Iodine	I	53	126.90
Iridium	Ir	77	192.2
Iron	Fe	26	55.85
Krypton	Kr	36	83.80
Lanthanum	La	57	138.91
Lawrencium	Lr	103	(257)
Lead	Pb	82	207.19
Lithium	Li	3	6.939

Based on mass of C^{12} at 12.000. Values in parentheses represent the most stable known isotopes for elements that do not occur naturally.

Element	Symbol	Atomic Number	Molar Mass
Lutetium	Lu	71	174.97
Magnesium	Mg	12	24.312
Manganese	Mn	25	54.94
Mendelevium	Md	101	(256)
Mercury	Hg	80	200.59
Molybdenum	Mo	42	95.94
Neodymium	Nd	60	144.24
Neon	Ne	10	20.183
Neptunium	Np	93	(237)
Nickel	Ni	28	58.71
Niobium	Nb	41	92.91
Nitrogen	N	7	14.007
Nobelium	No	102	(253)
Osmium	Os	76	190.2
Oxygen	O	8	15.9994
Palladium	Pd	46	106.4
Phosphorus	P	15	30.974
Platinum	Pt	78	195.09
Plutonium	Pu	94	(242)
Polonium	Po	84	(210)
Potassium	K	19	39.102
Praseodymium	Pr	59	140.91
Promethium	Pm	61	(147)
Protactinium	Pa	91	(231)
Radium	Ra	88	(226)
Radon	Rn	86	(222)
Rhenium	Re	75	186.23
Rhodium	Rh	45	102.91
Rubidium	Rb	37	85.47
Ruthenium	Ru	44	101.1
Samarium	Sm	62	150.35
Scandium	Sc	21	44.96
Selenium	Se	34	78.96
Silicon	Si	14	28.09
Silver	Ag	47	107.870
Sodium	Na	11	22.9898
Strontium	Sr	38	87.62
Sulfur	S	16	32.064
Tantalum	Ta	73	180.95
Technetium	Tc	43	(99)
Tellurium	Te	52	127.60
Terbium	Tb	65	158.92
Thallium	Tl	81	204.37
Thorium	Th	90	232.04
Thulium	Tm	69	168.93
Tin	Sn	50	118.69
Titanium	Ti	22	47.90
Tungsten	W	74	183.85
Uranium	U	92	238.03
Vanadium	V	23	50.94
Xenon	Xe	54	131.30
Ytterbium	Yb	70	173.04
Yttrium	Y	39	88.91
Zinc	Zn	30	65.37
Zirconium	Zr	40	91.22

APPENDIX 4 Electron Configurations of the Elements

	Orbitals	1s	2s	2p	3s	3p	3d	4s	4p	4d	4f	5s	5p	5d	5f	6s	6p	6d	6f	7s
1	hydrogen	1																		
2	helium	2																		
3	lithium	2	1																	
4	beryllium	2	2																	
5	boron	2	2	1																
6	carbon	2	2	2																
7	nitrogen	2	2	3																
8	oxygen	2	2	4																
9	fluorine	2	2	5																
10	neon	2	2	6																
11	sodium	2	2	6	1															
12	magnesium	2	2	6	2															
13	aluminum	2	2	6	2	1														
14	silicon	2	2	6	2	2														
15	phosphorus	2	2	6	2	3														
16	sulfur	2	2	6	2	4														
17	chlorine	2	2	6	2	5														
18	argon	2	2	6	2	6														
19	potassium	2	2	6	2	6		1												
20	calcium	2	2	6	2	6		2												
21	scandium	2	2	6	2	6	1	2												
22	titanium	2	2	6	2	6	2	2												
23	vanadium	2	2	6	2	6	3	2												
24	chromium	2	2	6	2	6	5	1												
25	manganese	2	2	6	2	6	5	2												
26	iron	2	2	6	2	6	6	2												
27	cobalt	2	2	6	2	6	7	2												
28	nickel	2	2	6	2	6	8	2												
29	copper	2	2	6	2	6	10	1												
30	zinc	2	2	6	2	6	10	2												
31	gallium	2	2	6	2	6	10	2	1											
32	germanium	2	2	6	2	6	10	2	2											
33	arsenic	2	2	6	2	6	10	2	3											
34	selenium	2	2	6	2	6	10	2	4											
35	bromine	2	2	6	2	6	10	2	5											
36	krypton	2	2	6	2	6	10	2	6											
37	rubidium	2	2	6	2	6	10	2	6			1								
38	strontium	2	2	6	2	6	10	2	6			2								
39	yttrium	2	2	6	2	6	10	2	6	1		2								
40	zirconium	2	2	6	2	6	10	2	6	2		2								
41	niobium	2	2	6	2	6	10	2	6	4		1								
42	molybdenum	2	2	6	2	6	10	2	6	5		1								
43	technetium	2	2	6	2	6	10	2	6	6		1?								
44	ruthenium	2	2	6	2	6	10	2	6	7		1								
45	rhodium	2	2	6	2	6	10	2	6	8		1								
46	palladium	2	2	6	2	6	10	2	6	10										
47	silver	2	2	6	2	6	10	2	6	10		1								
48	cadmium	2	2	6	2	6	10	2	6	10		2								
49	indium	2	2	6	2	6	10	2	6	10		2	1							

	Orbitals	1s	2s	2p	3s	3p	3d	4s	4p	4d	4f	5s	5p	5d	5f	6s	6p	6d	6f	7s
50	tin	2	2	6	2	6	10	2	6	10		2	2							
51	antimony	2	2	6	2	6	10	2	6	10		2	3							
52	tellurium	2	2	6	2	6	10	2	6	10		2	4							
53	iodine	2	2	6	2	6	10	2	6	10		2	5							
54	xenon	2	2	6	2	6	10	2	6	10		2	6							
55	cesium	2	2	6	2	6	10	2	6	10		2	6			1				
56	barium	2	2	6	2	6	10	2	6	10		2	6			2				
57	lanthanum	2	2	6	2	6	10	2	6	10		2	6	1		2				
58	cerium	2	2	6	2	6	10	2	6	10	2	2	6			2				
59	praseodymium	2	2	6	2	6	10	2	6	10	3	2	6			2				
60	neodymium	2	2	6	2	6	10	2	6	10	4	2	6			2				
61	promethium	2	2	6	2	6	10	2	6	10	5	2	6			2				
62	samarium	2	2	6	2	6	10	2	6	10	6	2	6			2				
63	europium	2	2	6	2	6	10	2	6	10	7	2	6			2				
64	gadolinium	2	2	6	2	6	10	2	6	10	7	2	6	1		2				
65	terbium	2	2	6	2	6	10	2	6	10	9	2	6			2				
66	dysprosium	2	2	6	2	6	10	2	6	10	10	2	6			2				
67	holmium	2	2	6	2	6	10	2	6	10	11	2	6			2				
68	erbium	2	2	6	2	6	10	2	6	10	12	2	6			2				
69	thulium	2	2	6	2	6	10	2	6	10	13	2	6			2				
70	ytterbium	2	2	6	2	6	10	2	6	10	14	2	6			2				
71	lutetium	2	2	6	2	6	10	2	6	10	14	2	6	1		2				
72	hafnium	2	2	6	2	6	10	2	6	10	14	2	6	2		2				
73	tantalum	2	2	6	2	6	10	2	6	10	14	2	6	3		2				
74	tungsten	2	2	6	2	6	10	2	6	10	14	2	6	4		2				
75	rhenium	2	2	6	2	6	10	2	6	10	14	2	6	5		2				
76	osmium	2	2	6	2	6	10	2	6	10	14	2	6	6		2				
77	iridium	2	2	6	2	6	10	2	6	10	14	2	6	7		2				
78	platinum	2	2	6	2	6	10	2	6	10	14	2	6	9		1				
79	gold	2	2	6	2	6	10	2	6	10	14	2	6	10		1				
80	mercury	2	2	6	2	6	10	2	6	10	14	2	6	10		2				
81	thallium	2	2	6	2	6	10	2	6	10	14	2	6	10		2	1			
82	lead	2	2	6	2	6	10	2	6	10	14	2	6	10		2	2			
83	bismuth	2	2	6	2	6	10	2	6	10	14	2	6	10		2	3			
84	polonium	2	2	6	2	6	10	2	6	10	14	2	6	10		2	4			
85	astatine	2	2	6	2	6	10	2	6	10	14	2	6	10		2	5			
86	radon	2	2	6	2	6	10	2	6	10	14	2	6	10		2	6			
87	francium	2	2	6	2	6	10	2	6	10	14	2	6	10		2	6			1
88	radium	2	2	6	2	6	10	2	6	10	14	2	6	10		2	6			2
89	actinium	2	2	6	2	6	10	2	6	10	14	2	6	10		2	6	1		2
90	thorium	2	2	6	2	6	10	2	6	10	14	2	6	10		2	6	2		2
91	protactinum	2	2	6	2	6	10	2	6	10	14	2	6	10	2	2	6	1		2
92	uranium	2	2	6	2	6	10	2	6	10	14	2	6	10	3	2	6	1		2
93	neptunium	2	2	6	2	6	10	2	6	10	14	2	6	10	4	2	6	1		2
94	plutonium	2	2	6	2	6	10	2	6	10	14	2	6	10	6	2	6			2?
95	americium	2	2	6	2	6	10	2	6	10	14	2	6	10	7	2	6			2
96	curium	2	2	6	2	6	10	2	6	10	14	2	6	10	7	2	6	1		2
97	berkelium	2	2	6	2	6	10	2	6	10	14	2	6	10	8	2	6	1		2
98	californium	2	2	6	2	6	10	2	6	10	14	2	6	10	10	2	6			2?
99	einsteinium	2	2	6	2	6	10	2	6	10	14	2	6	10	11	2	6			2?
100	fermium	2	2	6	2	6	10	2	6	10	14	2	6	10	12	2	6			2?
101	mendelevium	2	2	6	2	6	10	2	6	10	14	2	6	10	13	2	6			2?
102	nobelium	2	2	6	2	6	10	2	6	10	14	2	6	10	13	2	6	1		2?
103	lawrencium	2	2	6	2	6	10	2	6	10	14	2	6	10	14	2	6	1		2?

APPENDIX 5

Names, Formulas, and Charges of Some Common Ions

Positive Ions (Cations)		Negative Ions (Anions)	
aluminum	Al^{3+}	acetate	CH_3COO^-
ammonium	NH_4^+	bromide	Br^-
barium	Ba^{2+}	carbonate	CO_3^{2-}
calcium	Ca^{2+}	hydrogen carbonate ion, bicarbonate	HCO_3^-
chromium(II), chromous	Cr^{2+}	chlorate	ClO_3^-
chromium(III), chromic	Cr^{3+}	chloride	Cl^-
copper(I),* cuprous	Cu^+	chlorite	ClO_2^-
copper(II), cupric	Cu^{2+}	chromate	CrO_4^{2-}
hydrogen, hydronium	H^+, H_3O^+	dichromate	$Cr_2O_7^{2-}$
iron(II),* ferrous	Fe^{2+}	fluoride	F^-
iron(III), ferric	Fe^{3-}	hydroxide	OH^-
lead	Pb^{2+}	hypochlorite	ClO^-
lithium	Li^+	iodide	I^-
magnesium	Mg^{2+}	nitrate	NO_3^-
manganese(II), manganous	Mn^{2+}	nitrite	NO_2^-
mercury(I),* mercurous	Hg_2^{2+}	oxalate	$C_2O_4^{2-}$
mercury(II), mercuric	Hg^{2+}	hydrogen oxalate ion	$HC_2O_4^-$
potassium	K^-	perchlorate	ClO_4^-
silver	Ag^+	permanganate	MnO_4^-
sodium	Na^+	phosphate	PO_4^{3-}
tin(II),* stannous	Sn^{2+}	monohydrogen phosphate	HPO_4^{2-}
tin(IV), stannic	Sn^{4+}	dihydrogen phosphate	$H_2PO_4^-$
zinc	Zn^{2+}	sulfate	SO_4^{2-}
		hydrogen sulfate ion, bisulfate	HSO_4^-
		sulfide	S^{2-}
		hydrogen sulfide ion, bisulfide	HS^-
		sulfite	SO_3^{2-}
		hydrogen sulfite ion, bisulfite	HSO_3^-

*Aqueous solutions are readily oxidized by air.

Note: In ionic compounds the relative number of positive and negative ions is such that the sum of their electric charges is zero.

APPENDIX 6

Solubility of Common Compounds in Water

Negative Ions (Anions)	+	Positive Ions (Cations)	Form	Compounds with Solubility:
All		Alkali ions Li$^+$, Na$^+$, K$^+$, Rb$^+$, Cs$^+$, Fr$^+$		Soluble
All		Hydrogen ion, H$^+$(aq)		Soluble
All		Ammonium ion, NH$_4^+$		Soluble
Nitrate, NO$_3^-$		All		Soluble
Acetate, CH$_3$COO$^-$		All		Soluble
Chloride, Cl$^-$ Bromide, Br$^-$ Iodide, I$^-$		Ag$^+$, Pb^{2+}, Hg$_2^{2+}$, Cu$^+$		Low Solubility
		All others		Soluble
Sulfate, SO$_4^{2-}$		Ag$^+$, Ca^{2+}, Sr^{2+}, Ba^{2+}, Pb^{2+}		Low Solubility
		All others		Soluble
Sulfide, S^{2-}		Alkali ions, H$^+$(aq), NH$_4^+$, Be^{2+}, Mg^{2+}, Ca^{2+}, Sr^{2+}, Ba^{2+}		Soluble Soluble
		All others		Low Solubility
Hydroxide, OH$^-$		Alkali ions, H$^+$(aq), NH$_4^+$, Sr^{2+}, Ba^{2+}		Soluble
		All others		Low Solubility
Phosphate, PO$_4^{3-}$ Carbonate, CO$_3^{2-}$ Sulfite, SO$_3^{2-}$		Alkali ions, H$^+$(aq), NH$_4^+$		Soluble
		All others		Low Solubility

APPENDIX 7 Units

Table of Metric Units

	Unit	Value	Symbol	Approximate magnitude
Length	kilometer	1000 m	km	8 city blocks
	meter	unit	m	height of a drinking fountain
	centimeter	0.01 m	cm	radius of a nickel
	millimeter	0.001 m	mm	thickness of a dime
Mass	kilogram	unit	kg	mass of a full can of motor oil
	gram	0.001 kg	g	mass of 3 aspirin tablets
	milligram	0.001 g	mg	mass of 10 crystals of table salt
Volume	cubic meter	unit	m^3	volume of 3 bathtubs
	liter	0.001 m^3	l	volume of a can of motor oil
	cubic centimeter	0.000 001 m^3	cm^3	volume of 2 nickels
	milliliter	0.001 l	ml	
		cm^3 = ml		
Energy	joule	unit	J	energy to raise the temperature of a raindrop 5°C

Definitions of Some Fundamental Units

Length: meter (or metre)
The meter (m) is the length that is equal to 1,650,763.73 wavelengths in vacuum of the orange-red light in krypton-86.

Mass: kilogram
The kilogram (kg) is the unit of mass equal to the mass of the international prototype of the kilogram (located in Sèvres, France).

Thermodynamic temperature: kelvin
The kelvin (K) is the fraction 1/273.16 of the thermodynamic temperature of the triple point of water.

Time: second
The second (s) is equal to the duration of 9,192,631,770 periods of the radiation corresponding to the transition between two hyperfine levels of the ground state of the cesium-133 atom.

Amount of substance: mole
The mole (mol) is the amount of substance of a system that contains as many elementary particles as there are atoms in 0.012 kilograms of carbon-12.

APPENDIX 8

Laboratory Introduction

You should be thoroughly familiar with the following Laboratory Instructions, in order to make the best and safest possible use of time spent in the laboratory. Following the laboratory instructions are photos that identify much of the equipment that you will be using.

Laboratory Instructions

1. It is important for you to keep in mind at all times that the laboratory is a place for serious investigations.

2. No form of horseplay or practical jokes can be tolerated in the laboratory. What may seem like a casual prank can easily result in someone's getting cut or burned.

3. Wear protective eye glasses whenever there is a chance that some hot or corrosive chemical may splash into your eyes. Be aware of what your neighbors are doing, even though you are not handling anything dangerous. Sometimes an overheated test tube can eject hot material ten or fifteen feet. Some state laws require that you wear glasses at all times in the laboratory.

4. Always prepare for your work in the laboratory by reading the directions for the experiment before coming to class. Save valuable time in the laboratory by preparing forms for recording data in your notebook ahead of time.

5. Be sure that you understand the purpose of the experiment and any special precautions that must be observed before starting your work. Follow the directions very carefully. If you want to modify a procedure, be sure to check with the teacher before you try it.

6. Unauthorized experiments are forbidden. Never just mix a variety of chemicals together to "see what will happen." The results may be tragic. You are encouraged to suggest and to try new experiments, but you must clear your plans with the teacher before proceeding.

7. It is a good idea to wear a rubber or plastic apron in the laboratory to protect your clothing.

8. If you spill an acid or other corrosive chemical on your skin, do not panic. Set down whatever you are doing, and wash your skin at once with large quantities of water. Move quickly, but do not run, to the water supply.

9. If you wear your glasses in the laboratory, there is little chance of getting chemicals in your eyes. If any chemical does get in your eye, wash the chemical out at once with large amounts of water. Either place your eye in contact with running water or splash water repeatedly into your eye.

10. Never taste or touch chemicals or solutions unless you are directed to do so. Some apparently harmless chemicals may be extremely irritating to certain individuals.

11. If you want to observe the odor of a substance, do not inhale its vapors directly. Open the container cautiously and fan some of the vapor toward your nose by passing your hand over the top of the container.

12. If you have been heating glassware, allow plenty of time for it to cool before handling it. The same precaution applies to burners and iron rings on ringstands. Always wait for a ring to cool before trying to adjust its position on the ringstand.

13. Do not handle flammable liquids near an open flame. If a small quantity of a flammable substance does catch fire in a test tube or beaker, either let it burn out or cover the container with a glass plate or a piece of cardboard. If it looks like more than you can handle, immediately get out of the way and get a fire extinguisher.

14. If any substances produce bad odors, or poisonous or corrosive vapors, be sure to handle these substances in a fume hood.

15. Be sure to read the labels on reagent bottles carefully before removing anything from the bottles.

16. Never pour any unused chemicals back into the stock bottles. If you accidentally pour out more than you need, use what you need and dispose of the remainder by pouring it down the sink or placing it in the waste jar. Even though you are sure you would be pouring your excess chemical back into the proper container, your own receptacle may not be clean. It is much cheaper and safer to throw out any excess rather than take the chance of contaminating the entire supply.

17. Never throw any solid substances, including paper, matches, and pieces of metal or glass, or any insoluble or slightly soluble substances into the sink. They can only plug up the drain and somebody will have to clean them out.

18. A clean work area is a safer one than an area that has dirty and excess equipment scattered about. If you spill something at your desk or at the supply desk, clean it up immediately. If it is a corrosive substance, ask the teacher for special instructions.

19. When you start work at the beginning of a period, you will expect to have clean equipment to work with. Be sure that you return all equipment to its proper place in as clean a condition as possible at the end of the period, so that the person following you can also have clean equipment to work with. Clean out your sink and leave your work area ready for the next class.

20. Use leather gloves or several thicknesses of toweling to protect your hands if you are cutting glass tubing or inserting glass tubing into a stopper. When assembling glass-to-rubber connections, wet the glass and the rubber first to prevent them from sticking.

21. A thermometer is one of the most easily broken pieces of equipment in the laboratory. Use a stirring rod instead of a thermometer for stirring. Do not leave a thermometer standing by itself in a beaker or flask. It may make the container top-heavy and is very easy to knock over.

22. Always report any injury to your teacher at once, regardless of how slight it may be.

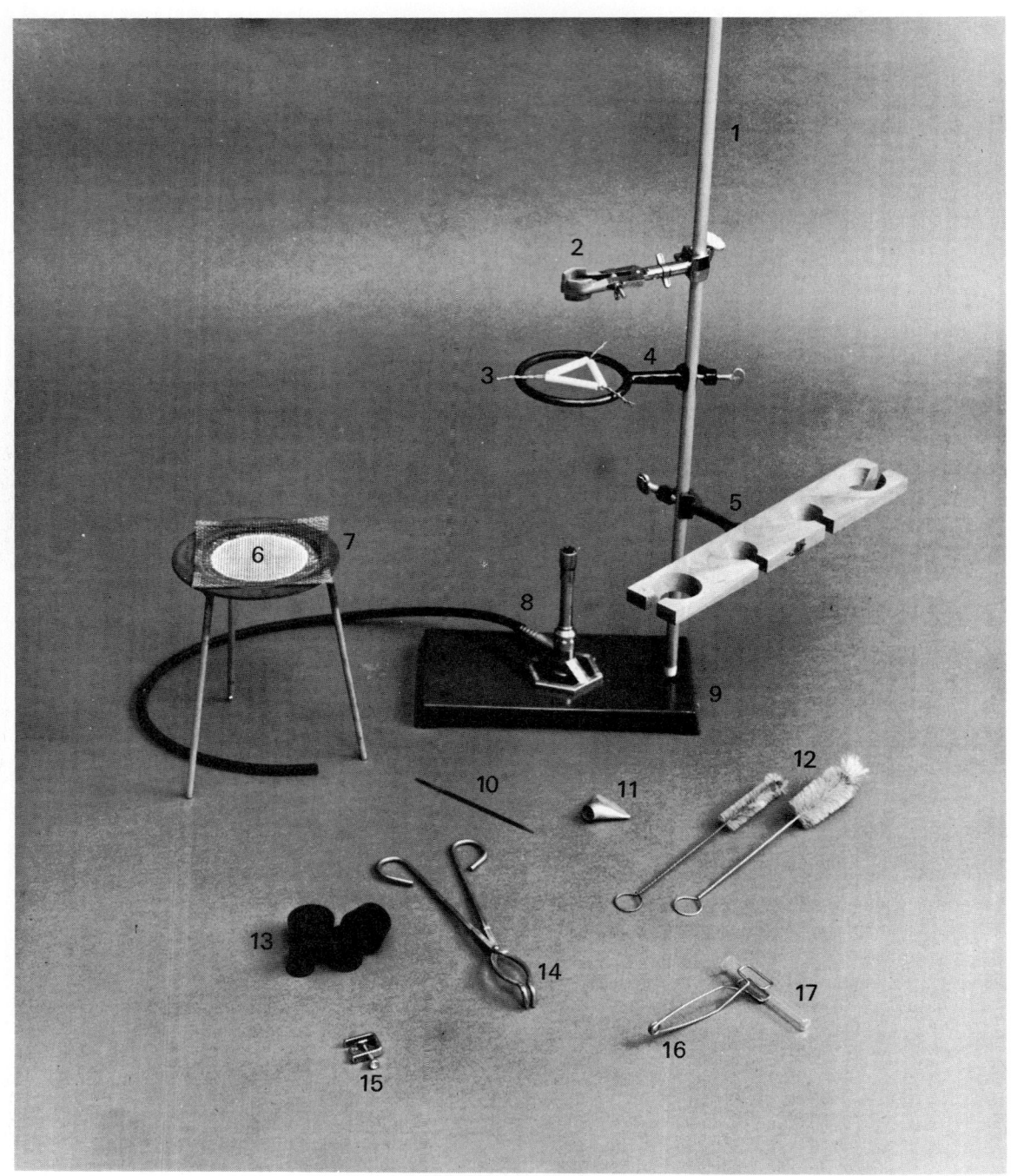

Names of numbered items: 1. ring stand, 2. utility (or buret) clamp, 3. clay triangle, 4. ring, 5. filter rack, 6. wire gauze (with asbestos center), 7. tripod, 8. burner, 9. ring stand base, 10. triangular file, 11. flame spreader, 12. test tube brushes, 13. rubber stoppers, 14. crucible tongs, 15. screw clamp, 16. test tube holder, 17. test tube.

Names of numbered items: 1. graduated cylinders, 2. glass-stoppered flask, 3. Erlenmeyer flasks, 4. plastic wash bottle, 5. buret, 6. stirring rod, 7. thistle tube, 8. thermometer, 9. funnels, 10. round-bottomed flask, 11. dropping bottle, 12. petri dish, 13. watch glass, 14. beakers, 15. glass plate, 16. evaporating dish, 17. medicine dropper, 18. U-tube, 19. test tubes, 20. crucible, 21. mortar and pestle.

GLOSSARY

absolute zero: $-273.16°C$ ($-459.49°F$). The temperature at which all gases would, theoretically, contract to zero volume, if Charles' law were obeyed exactly.

acid: A substance that dissolves in water to release hydrogen ions.

activated complex: The least stable intermediate arrangement of atoms that must be passed through in the course of a reaction.

activation energy: The minimum or "threshold" energy necessary for a reaction to occur. The energy of the activated complex minus the energy of the reactant molecules.

alcohols: Organic compounds containing the —OH (hydroxyl) group. Ethanol, C_2H_5OH, is commonly called "alcohol," but it is only one of many members of the class.

aldehydes: Organic compounds containing the —CHO group.

alkali metals: The group I elements Li, Na, K, Rb, Cs. (Also called *alkalis*.)

alkaline earth metals: The group II elements: Be, Mg, Ca, Sr, Ba, Ra.

alkanes: Saturated hydrocarbons, also called *paraffins*.

alkenes: Hydrocarbons in which two carbon atoms are connected by a double bond.

alkynes: Hydrocarbons in which two carbon atoms are connected by a triple bond.

alloy: A solution of two or more metals in one another.

alpha (α) particle: A particle made up of two protons and two neutrons that is ejected by some radioactive nuclei.

amides: Compounds of the type $RCONH_2$, where R is a hydrocarbon group.

amines: Compounds of the type RNH_2, where R is a hydrocarbon group.

amino acid: Compounds of the type
$$H_2N\underset{H}{\overset{R}{C}}COOH,$$
which polymerize to form proteins.

ångström (Å): A unit of length equal to 10^{-8} cm.

anion: A negative ion.

anode: The electrode at which oxidation occurs in an electrochemical cell. Anions move toward the anode when an electrolyte solution is electrolyzed.

aqueous solution: A solution in which water is the solvent.

atmosphere: The pressure exerted by a column of mercury exactly 760 mm high, at $0°C$.

atom: The smallest particle of an element that can exist alone or in combination with other atoms.

atomic number: The number of protons in the nucleus of an atom.

Avogadro's hypothesis: Equal volumes of gases at the same temperature and pressure contain equal numbers of molecules.

Avogadro's number: The number of atoms, molecules, or other objects in a mole (6.02252×10^{23}).

barometer: A type of manometer for measuring the pressure of the earth's atmosphere.

base: A substance that can combine with hydrogen ions.

beta (β) particle: An electron ejected by a radioactive nucleus.

binary compound: One which contains only two kinds of elements. Examples: HCl, $CaCl_2$, SO_3.

boiling point: The temperature at which the vapor pressure of a liquid is equal to the pressure exerted by the atmosphere on the liquid. The *normal* boiling point is the temperature at which the vapor pressure is 760 Torr.

bond energy: The amount of energy necessary to break a mole of chemical bonds.

Boyle's law: The pressure of a fixed mass of gas is inversely proportional to its volume if the temperature is held constant. (PV = a constant.)

calibrate: To systematically standardize the graduations on a quantitative measuring instrument.

calorie: A unit of heat or energy. Its precise definition is in terms of electrical units, but it is approximately the quantity of heat required to raise the temperature of one gram of water one degree Celsius at about 15°C.

calorimeter: A device for measuring the amount of heat evolved or absorbed in a chemical reaction or phase change.

calorimetry: The measurement of the heats of chemical reactions.

carbohydrates: Molecules consisting solely of C, H, and O atoms, generally with the H and O atoms in the proportion, 2:1.

carboxylic acids: The so-called "organic acids," which contain the carboxyl group, —COOH.

catalyst: A substance added to a chemical reaction to alter (usually to increase) the rate of reaction. The catalyst is not itself changed in the reaction nor does it change the nature of the products. It only changes the rate.

cathode: The electrode at which reduction occurs in an electrochemical cell. Cations move toward the cathode when an electrolyte solution is electrolyzed.

cation: A positive ion.

cellulose: A polymer of glucose with a very high molar mass, containing a type of linkage not easily split by hydrolysis.

Celsius temperature (°C): The temperature on a scale defined so that 0°C is the freezing point of water and 100°C is the boiling point of water.

centi-: A prefix meaning a hundredth, as in centimeter (0.01 meter).

Charles' law: All gases expand by the same fraction of their original volumes when they are heated over the same range of temperature.

chemical equation: A group of formulas and symbols that show what reactants combine, in what proportions, to produce what products, in what proportions. (See **chemical reaction.**)

chemical reaction: The transformation of one or more substances (the reactants) into one or more different substances (the products).

cis (adj.): On the same side. Used in identifying geometric isomers. (See **trans.**)

combustion: The process of burning; the vigorous reaction of a substance with oxygen, giving off heat and light.

compound: A pure substance composed of two or more different kinds of atoms (different elements).

condensed phase: A liquid, solid, or solution.

configuration (of electrons): A specification of how many electrons occupy each orbital. A vacant orbital is not listed; an orbital singly occupied is listed by quantum number and orbital symbol; a doubly occupied orbital is similarly listed, followed by the superscript two. Example: $1s^2 2p$.

control: An experiment identical to the one of interest in all respects except one.

Coulomb's law: The force between two charged bodies is proportional to the product of their charges and inversely proportional to the square of the distance between them.

covalent bond: A link between two atoms, due mainly to the more or less equal sharing of electrons between them.

crystalline state (crystal): The solid form of matter in which the atoms, molecules, or ions occur in a regular, repeated arrangement.

density: Mass per unit volume, usually expressed as grams per milliliter. g/ml.

deuterium: The isotope of hydrogen with a mass number of 2.

diatomic molecule: A molecule consisting of two atoms.

diffraction: The bending of waves through a small opening; a straight beam of parallel waves emerges as a rounded pattern that spreads out.

dipole moment: A quantitative measure of the degree of charge separation, or polarity, in a body with no net charge; defined as the magnitude of the separated charge, q, times the distance between charges, d; that is, dipole moment $= qd$.

divalent: Having the capacity to form two bonds (said of an atom).

elastic collision: a collision in which the colliding objects may change direction, but there is no change in the total kinetic energy.

electrochemical cell: A device in which chemical oxidation and reduction take place, producing or utilizing electric current.

electrolyte: A compound that dissolves in water, or another solvent, to give a conducting solution.

electron: The fundamental negative particle occurring in atoms. Electrons surround the nucleus of an atom and account for nearly all its volume.

element: A substance consisting only of atoms having the same atomic number.

empirical formula: The simplest formula that shows the correct proportions of the various kinds of atoms in a molecule. It may be only a fraction of the molecular formula. Example: Glucose has the molecular formula $C_6H_{12}O_6$; its empirical formula is CH_2O.

endothermic reaction: One that absorbs heat.

energy level: A permitted value of energy for an atom or molecule, or for an electron in an atom or molecule.

enzyme: A protein molecule that has the ability (usually very pronounced and very specific) to catalyze a particular biologically important reaction.

equilibrium: A state in which no net or macroscopic change is occurring in a closed chemical system at uniform temperature, even though there is activity on the molecular scale.

esters: Compounds of the type $RCOOR'$, where R and R' are hydrocarbon groups.

excited state: A state of an atom or molecule in which the electron distribution is less stable (more energetic) than in the normal (ground) state.

exothermic reaction: One that gives out (evolves) heat.

fats: Esters of glycerol.

field: A region of space in which a given effect (magnetic or electric, for example) exists.

fission: The process in which a nucleus splits into two heavy fragments plus a few neutrons.

formula: The code word for a molecule, consisting of symbols for the elements and numerical subscripts for the number of atoms of each element.

frequency (of light): The number of waves passing a point per second; also stated as the number of cycles per second. Unit: sec^{-1}, or hertz, Hz.

functional group: A relatively reactive group (such as —COOH, —OH, and —NH_2) that may be attached to a relatively inert hydrocarbon group. Reactions then occur mainly at or in the functional groups.

gamma (γ) ray: A form of short-wave radiation, similar to but more energetic than X rays, emitted from many radioactive nuclei.

geometric isomers: Molecules having the same types of bonds, in the same order, but having atoms with different spatial relationships. (Also known as *cis-trans* isomers.)

gram (g): A metric unit of mass equal to one thousandth of a kilogram.

ground state: The most stable (normal) state of an atom or molecule in terms of the electron distribution.

half-cell: A cell, containing a single electrode, in which only an oxidation or only a reduction reaction can occur, when it is combined with another half-cell.

halogens: The group VII elements: F, Cl, Br, I.

heat: A form of energy.

heat content: The heat energy stored in a substance during its formation.

heat of formation: The heat released or consumed when one mole of a compound is formed from its constituent elements.

heat of reaction: The heat released or consumed when a reaction occurs.

heat of solution: The heat released or absorbed when a solute dissolves.

heterogeneous system: One in which the components are in two or more different phases.

homogeneous system: One in which all components are in the same phase.

hybrid orbitals: Orbitals produced by the mixing of various pure orbitals (such as s and p orbitals).

hydration: The grouping of water molecules around ions, thus keeping the ions in solution.

hydrocarbon: A compound that contains only hydrogen and carbon atoms. It may be saturated (no double or triple bonds), unsaturated (containing double and/or triple bonds), aromatic (containing benzene or benzene-like rings) or cyclic (saturated or unsaturated but containing a ring of carbon atoms).

hydrogen bond: The additional attraction (2 to 10 kcal/mole) often found between the H atoms in O—H, N—H, and F—H groups and other O, N, or F atoms.

hydronium ion: A hydrated hydrogen ion, $[H(H_2O)_n]^+$.

ideal gas: A gas that follows the law $PV = nRT$ strictly at all temperatures and pressures.

indicator (acid-base): A substance, such as litmus or phenolphthalein, the color of which depends on the H^+ concentration in the solution to which it is added.

infrared light: Light with lower frequencies than red light (less than about 4.3×10^{14} sec^{-1}).

interference: An effect that results when two series of waves merge into each other. Interference effects produced by merging light beams indicate that light has wave-like properties.

ion: An atom that has acquired a positive charge by losing one or more electrons (cation) or has acquired a negative charge by gaining one or more electrons (anion).

ionic bond: A bond in which the attractive force is due entirely, or almost entirely, to electrostatic attraction between two ions.

ionization energy: The energy required to remove the least stable electron from an atom, molecule, or ion.

isomers: Two or more substances having the same molecular formula but different structural formulas.

isotopes: Atoms having the same atomic number (same number of protons) but different mass numbers (different numbers of neutrons).

joule (J): The standard international unit of energy. One calorie equals 4.184 joules.

Kelvin temperature (kelvins, K): Temperature on a scale that is zero at the absolute zero point ($-273.16°C$) and is graduated in degrees equal to Celsius degrees in size.

ketones: Compounds of the general formula $R-\overset{\overset{O}{\|}}{C}-R'$, where R and R' are hydrocarbon groups.

kilo-: A prefix meaning one thousand, as in kilogram (one thousand grams).

kilogram (kg): The standard international scientific unit of mass.

kinetic energy: Energy of motion. Defined as one-half the product of the mass multiplied by the square of the velocity of a particle, $\frac{1}{2}mv^2$.

Law of Additivity of Reaction Heats: When a reaction can be expressed as the algebraic sum of a sequence of two or more other reactions, the heat of reaction is the algebraic sum of the heats of these reactions. (Also known as Hess's law.)

Law of Conservation of Energy: In a chemical reaction, energy is neither created nor destroyed.

Law of Definite Proportions: The masses of the elements in a pure compound always occur in a definite proportion.

Le Chatelier's principle: A system at equilibrium responds to an imposed change in conditions (temperature, pressure, concentration) so as to oppose the change.

liquid: The condensed phase in which molecules are very close together but still have enough energy to move over one another. Thus liquids can flow, and they lack the near perfect orderliness of crystalline solids.

liter (l): The metric unit of volume (equal to 1.057 quarts).

manometer: A device for measuring pressure. It often consists of a column of mercury or other liquid.

mass: The amount of matter that an object contains.

mass number (of an atom): The total number of particles—neutrons plus protons—in a nucleus.

mass spectrograph: An instrument used to measure the masses of positive ions of gases.

mechanism (of a reaction): The sequence of steps, each involving the collision of only a few (two or three) molecules, by which an overall reaction takes place.

meniscus: The curved surface of a liquid in a cylinder, caused by the attraction between the liquid and the walls of the cylinder.

metal: A substance with the characteristic properties of luster and good conductivity of heat and electricity. The bonding is provided by a sea of mobile electrons.

meter (m): The metric unit of length (equal to 39.37 inches).

milli-: A prefix meaning a thousandth, as in milligram (0.001 gram).

molar (adj.): Of or pertaining to one mole. (See also **molarity**.) Example: Molar heat content is the heat content of one mole.

molarity (noun): A unit of concentration of solutions. A one-molar solution (molarity of one) contains one mole of solute in one liter of solution.

molar mass: The mass of 6.02252×10^{23} atoms, molecules, or other objects.

mole: 6.02252×10^{23} atoms, molecules, or other objects.

molecular crystals: Low-melting crystals; the smallest units in the space lattice are molecules, attracted to one another by van der Waals interactions.

molecular formula: A formula that shows how many of each kind of atom are present in a molecule but does not indicate how the atoms are connected. Example: C_2H_6 for ethane.

molecular model: A representation of a molecule (made with balls, sticks, springs, etc.) showing how a molecule might look if enlarged by a factor of about 10^{10}.

molecule: The smallest unit of a substance that can exist independently.

monomer: A small unit that, repeated many times, generates a polymer.

network solid: An element (such as Si, C) or compound (such as SiO_2, As_2O_3) in which covalent bonds connect the atoms into an infinite 2- or 3-dimensional array.

neutralization: The reaction in which OH^- and H^+ ions combine, forming H_2O. Acids can neutralize bases and vice versa.

neutron: A neutral particle with about the same mass as a proton; like the proton, it is found in atomic nuclei.

noble gases: A group of gaseous monatomic elements (He, Ne, Ar, Kr, Xe, Rn) that have either no chemical reactivity (He, Ne, Ar) or rather low reactivity (Kr, Xe, Rn).

nucleus: The positively charged central part of an atom; it contains most of the mass of the atom, but only a very small portion of the atomic volume.

orbital: A region of space around the atomic nucleus in which an electron has significant probability of being found. The main types of orbitals are s (spherical), p (lying along one of the three perpendicular Cartesian axes), d, and f.

organic chemistry: The chemistry of the compounds of carbon, except for the very simple ones, such as CO_2 and Na_2CO_3.

oxidation: 1. In general, the loss of electrons. 2. Reaction of a substance with oxygen or another oxidizing agent.

oxidizing agent: A compound or ion that tends to accept or acquire electrons.

oxidation number: A number assigned to represent the number of electrons lost $(+)$ or gained $(-)$ by an atom in a compound compared to the free, neutral atoms. These numbers are *arbitrary* and do not represent *true* changes.

partial pressure: The contribution made by one gas to the total pressure of a mixture of gases.

Pauli's exclusion principle: The postulate, required by experimental facts, that no more than two electrons can occupy each atomic orbital.

petroleum: A naturally occurring mixture of organic compounds (mainly hydrocarbons) ranging from fairly volatile liquids to heavy, viscous tars. It is also called *crude oil*.

phase: A part of a system that has a uniform set of properties, such as solid, liquid, or gas.

photoelectric effect: The ejection of electrons from a substance by incident electromagnetic radiation.

photon: An indivisible unit of energy making up electromagnetic radiation (also called a quantum). The energy, E, of each photon in radiation of frequency ν is given by $E = h\nu$, where h is Planck's constant.

photosynthesis: The process by which plants synthesize glucose, starches, and cellulose from CO_2 and H_2O, using sunlight as the source of the required energy.

Planck's constant: The number, h, by which the frequency, ν, of light is multiplied to calculate the energy of that light.

polar molecule: One that has an electric dipole moment due to uneven distribution of the electrons over the various nuclei.

polymer: A very large molecule made up of a small unit (the monomer) repeated many times in an extended sequence.

polymerization: The process or reaction in which monomers combine to form polymers.

potential energy: Energy stored in a system (a molecule, for instance) to be released as kinetic energy, electrical energy, heat, light, etc.

precipitate: 1. (verb) To fall out of solution. 2. (noun) A compound formed in a reaction in solution that, because it is insoluble, falls out of solution.

pressure: The force per unit area (units: atmospheres or Torr).

products: Substances produced in a chemical reaction; normally they are written on the right side of a chemical equation.

protein: A type of molecule, vital to life, which consists of one or more long chains of amino acids connected to one another by amide linkages.

proton: The positively charged particle of atomic nuclei.

quanta: The indivisible units of energy (also called photons) in electromagnetic radiation.

quantum mechanics: A set of physical laws and postulates applicable to the behavior of electrons, atoms, and molecules.

quantum number: A number used to designate a particular energy level of an atom; it is also used in algebraic expressions to calculate the magnitude of that energy.

radical: A group of atoms that stays together as a unit through many (though not all) chemical reactions. Examples: SO_4 (sulfate); NO_3 (nitrate). Usually these are present as distinct anions, such as SO_4^{2-} (sulfate ion), NO_3^- (nitrate ion).

radioactivity: The giving off of radiation (electrons, helium nuclei, electromagnetic energy) by unstable atomic nuclei.

rate (of a reaction): The quantity (usually in moles) of a reactant consumed or a product produced per unit of time.

rate-determining step: The slowest reaction in a reaction mechanism.

reactants: Substances that enter into a chemical reaction; normally they appear on the left side of a chemical equation.

reaction mechanism: See **mechanism**.

redox (adj.): A contraction meaning oxidation-reduction.

reducing agent: A compound or ion that tends to release electrons.

reduction: 1. In general, electron gain. 2. A process in which the oxidation state becomes lower (less positive).

reduction potential: The half-cell potential that is a measure of the tendency to gain electrons.

refraction: The bending of light, as by an angular prism; different frequencies are bent through different angles.

regularity: A trend, or pattern, noted in experimental data.

resonance: A phenomenon exhibited by a molecule or ion to which two or more structures differing only in the distribution of electrons can be assigned. These structures are said to be in resonance.

resonance hybrid: The average or mixture of differing structures for a molecule or ion that represents the actual electron distribution.

reversible reaction: A reaction that can go in either direction, depending on conditions such as concentration, temperature, or pressure.

rusting: A particular type of corrosion in which iron is converted to hydrated Fe_2O_3 by the combined action of atmospheric oxygen and water.

salt: 1. The common name of NaCl. 2. The general name for any ionic compound formed when an acid and a base neutralize each other.

saturated solution: A solution containing the maximum amount of solute that will dissolve at the temperature of the experiment. Such a solution can remain in equilibrium with an excess of pure solute.

shell (electron): A particular energy level, as represented by a certain principal quantum number and a letter. Examples: the $2p$ shell or the $3d$ shell.

significant figures: Those digits in a measurement that are completely certain as well as one more, in which there is some uncertainty.

soaps: Metal (usually sodium) salts of long-chain (C_{17}) carboxylic acids.

solubility product: The type of equilibrium constant applicable to the solubility of electrolytes. It consists of the products of the concentrations of all the ions formed, each raised to the appropriate power.

solute: The component of a solution present in a smaller amount and considered to be dissolved in the component present in a larger amount (the solvent).

solution: A mixture, on the molecular level, of two or more substances.

solvent: That component of a solution (usually a liquid) present in the larger amount.

spectroscope (or spectrograph): An optical device that separates light into a spectrum.

spectrum: See **visible light**.

standard temperature and pressure (STP): Zero degrees Celsius (273 K) and one atmosphere (760 Torr) pressure.

starch: A mixture of polymers made of the sugar glucose.

stoichiometry: The quantitative relationships in a chemical reaction.

strong electrolyte: A substance that exclusively gives ions when it dissolves.

structural formula: A formula that shows how the atoms are connected (and perhaps even how they are arranged in space) in a molecule.

structural isomers: Molecules that contain the same number and types of atoms, but have a different arrangement of bonds.

sugars: A class of carbohydrates.

system: Any region and the material in it that you wish to consider.

tetrahedral angle: 109°28'. This is the angle between any two lines drawn from the corners of a tetrahedron to its center.

tetrahedron: A polyhedron having faces that are four identical equilateral triangles.

titration: The process of adding one reactant to another, a drop at a time, until the two are present in exactly equivalent amounts.

Torr: A small unit of pressure equal to 1/760 of an atmosphere, or approximately the pressure exerted by a column of mercury one millimeter high at sea level.

trans (adj.): Across, or on opposite sides. Used in identifying geometric isomers. (See **cis**.)

trivalent (adj.): Having the capacity to form three bonds (said of an atom).

ultraviolet light: Light with frequency greater than that of the light at the violet end of the visible spectrum (greater than 7.5×10^{14} sec^{-1}).

univalent (adj.): Having the capacity to form only one bond (said of an atom).

unsaturated: An adjective used for compounds containing double or triple bonds.

valence (noun): The capacity of an atom to form chemical bonds; also, the number of bonds an atom can form. Example: Oxygen has a valence of two.

valence electrons: The electrons of an atom that are most loosely bound and, therefore, participate in forming chemical bonds; electrons that occupy the valence orbitals.

valence orbitals: Orbitals in the energy range that contains partially-filled orbitals.

van der Waals forces: Very weak forces (about one kcal/mole) between atoms or molecules having all valence orbitals filled.

vapor: The gaseous form of a substance that is a liquid at ordinary temperature and pressure.

vapor pressure: The equilibrium pressure exerted by a vapor over a liquid at a given temperature.

visible light (or spectrum): Light that can be seen by the human eye. It ranges from red (frequency: 4.3×10^{14} sec^{-1}) through orange, yellow, green, blue, to violet (frequency: 7.5×10^{14} sec^{-1}).

volatile (adj.): Easily converted (i.e., at a relatively low temperature) to a gas.

volt: The unit that measures electrical potential.

wavelength (of light): The distance (measured in cm or ångströms) between maxima (crests) in the electromagnetic disturbance.

weak acid: One that is not fully dissociated in aqueous solution.

weak base: A base that does not dissociate completely into OH$^-$ ions and cations.

weak electrolyte: One that is only partially dissociated into ions in aqueous solution.

weight: The gravitational force on a body. It is proportional to the amount of matter in the body (its mass) and, under certain conditions, is equal to the mass. On the earth's surface, mass and weight are practically identical in magnitude.

work: Effort put forth to do a task; the force applied to a body times the distance through which it is moved, expressed in calories or joules.

X rays: A form of high-energy radiation. The wavelengths of X rays are in the range of a few ångströms; hence the diffraction of X rays by molecules (in which the atoms are separated by such distances) is a means of investigating molecular structure.

ACKNOWLEDGEMENTS

p. 2 (top left) American Cancer Society, (top right) Du Pont Company, (bottom) Frank Hoffman, USAEC, Oak Ridge Operations Office
p. 3 Courtesy of Edwards Laboratories, Santa Ana, California
p. 4 Courtesy of Bell Telephone Laboratories
p. 5 National Accelerator Laboratory
p. 18 New York Daily News (EPA)
p. 19 Wide World Photos
p. 20 (top) Charles Pfizer and Company, Inc.
p. 28 NASA
p. 32 Courtesy of National Bureau of Standards
p. 36 (right) Sargent-Welch Scientific Company, Skokie, Illinois
p. 40 University of Maryland
p. 41 NASA
p. 47 NASA
p. 52 Du Pont Company
p. 54 NASA
p. 59 Historical Pictures Service, Chicago
p. 61 Radio Times Hulton Picture Library
p. 69 Bechtel Corporation
p. 76 Courtesy of National Bureau of Standards
p. 77 (bottom left and right) Peter Travers
p. 89 Radio Times Hulton Picture Library
p. 98 (top) Photograph courtesy of the Hale Observatories, (bottom) Courtesy of Cabot Corporation, Boston, Massachusetts
p. 106 Photograph courtesy of the Hale Observatories
p. 108 (bottom) Photo courtesy of Brunswick Bowling and Billiard Division
p. 117 Historical Pictures Service, Chicago
p. 129 Peter Travers
p. 138 U.S. Air Force Photo
p. 143 Environmental Protection Agency
p. 144 Radio Times Hulton Picture Library
p. 145 G. Lotti, The Italian Art and Landscape Foundation, Inc.
p. 162 Prof. Erwin W. Mueller, Pennsylvania State University
p. 164 B. J. Alder and T. E. Wainwright, Lawrence Livermore Laboratory
p. 176 Historical Pictures Service, Chicago
p. 177 Radio Times Hulton Picture Library
p. 178 (top) Bureau of Mines, U.S. Department of the Interior, (bottom) University of California Lawrence Berkeley Laboratory
p. 194 L. Sibaiya, *Types of Graphic Representation of the Periodic System of Chemical Elements*, by Edward G. Mazurs, 1957, figure 94 on page 157
p. 198 Courtesy of Aluminum Company of America
p. 209 Historical Pictures Service, Chicago
p. 214 Ida Noddack-Tacke
p. 215 Argonne National Laboratory
p. 218 H. Armstrong Roberts
p. 224 (top) High Voltage Engineering Corporation
p. 227 General Electric Company
p. 229 (bottom) Fundamental Photographs
p. 231 Culver Pictures, Inc.
p. 240 Radio Times Hulton Picture Library
p. 250 Paul Harteck
p. 256 Radio Times Hulton Picture Library
p. 266 Wide World Photos
p. 268 Floyd Benson, Idaho Nuclear Corporation
p. 269 Los Alamos Scientific Laboratory
p. 271 USAEC Technical Information Center, Oak Ridge, Tennessee
p. 273 Historical Pictures Service, Chicago
p. 274 (top) University of California Lawrence Berkeley Laboratory, (bottom) Cavendish Laboratory
p. 275 (top) French Embassy Press and Information Division, (bottom) Courtesy of Niels Bohr Library, from "A Scientific Autobiography" by Otto Hahn, Charles Scribner's Sons
p. 278 Sargent-Welch Scientific Company
p. 283 By permission from Webster, Farwell, and Drew's *General Physics for Colleges*. Copyright 1923 and 1951 by Appleton-Century-Crofts, Educational Division, Meredith Corporation.
p. 284 *PSSC Physics*, D. C. Heath and Company, Lexington, Massachusetts, 1965
p. 285 *PSSC Physics*, D. C. Heath and Company, Lexington, Massachusetts, 1965
p. 289 (top) Photo by Charles L. Finance, reproduced by permission from *Chemistry—An Experimental Science*, prepared by Chemical Education Material Study, copyright 1963, The Regents of the University of California, W. H. Freeman and Company, Plate III
p. 290 Reproduced by permission from *Chemistry—An Experimental Science*, prepared by Chemical Education Material Study, copyright 1963, The Regents of the University of California, W. H. Freeman and Company, Plate IV
p. 292 Culver Pictures, Inc.
p. 293 Coherent Radiation Laboratories
p. 295 (left) Aden and Marjorie Meinel, (right) Artist: Don Cowen
p. 299 (bottom) Sargent-Welch Scientific Company

Acknowledgements

p. 308 Wide World Photos
p. 318 Jay Orear, *Fundamental Physics*, John Wiley and Sons, Inc., 1961
p. 320 *PSSC Physics*, D. C. Heath Company, Lexington, Massachusetts, 1965
p. 321 *PSSC Physics*, D. C. Heath Company, Lexington, Massachusetts, 1965
p. 322 United Press International Photo
p. 329 (top) E. R. Toon, et al., *Foundations of Chemistry*, Holt, Rinehart and Winston, Inc., 1968
p. 333 Science Related Materials, Inc.
p. 338 Art—*Physical Chemistry* by Gordon Barrow, copyright 1961 by the McGraw-Hill Book Company and used with their permission, Table—Edward G. Mazurs
p. 354 The Bettmann Archive, Inc.
p. 357 H. Gray and G. Haight, *Basic Principles of Chemistry*, Copyright, © 1967, W. A. Benjamin, Inc., Menlo Park, California
p. 381 Underwood and Underwood
p. 387 Westinghouse Photo
p. 389 (bottom) Arnold C. Wahl. Copyright © 1970 by McGraw-Hill, Inc. Used with permission of McGraw-Hill Book Company
p. 390 Arnold C. Wahl, "Molecular Orbital Densities Pictorial Studies", *Science*, pp. 961-967, Figs, 1, 6, 25, February 1966, Copyright 1966 by the American Association for the Advancement of Science
p. 391 Arnold C. Wahl. Copyright © 1970 by McGraw-Hill, Inc. Used with permission of McGraw-Hill Book Company
p. 395 James Pierce, *The Chemistry of Matter*, 1970, Houghton Mifflin Company, p. 279
p. 396 Courtesy of the Charles Supper Company, Natick, Massachusetts
p. 399 *PSSC Physics*, D. C. Heath and Company, Lexington, Massachusetts, 1965
p. 413 Wide World Photos
p. 415 James Pierce, *The Chemistry of Matter*, 1970, Houghton Mifflin Company, p. 154
p. 420 Wide World Photos
p. 432 (top) G. S. Hammond, et al., *Models in Chemical Science*, Copyright, © 1971, W. A. Benjamin, Inc., Menlo Park, California, (bottom) James Pierce, *The Chemistry of Matter*, 1970, Houghton Mifflin Company, p. 126
p. 436 Eastman Kodak Company—This picture was made by Maurice L. Huggins by photographic summation of x-ray diffraction data by Brockway and Robertson

p. 438 Field Museum of Natural History
p. 439 Exxon Company, U.S.A.
p. 454 The Bettmann Archive, Inc.
p. 458 Wide World Photos
p. 470 Wide World Photos
p. 477 Hart, H. and Schuetz, D., *Organic Chemistry: A Short Course*, Houghton Mifflin Co., 4th Edition (1972), figure 11, p. 255
p. 480 Courtesy of Rice Council
p. 484 Collection, The Museum of Modern Art, New York
p. 486 Photo by Charles L. Finance, reproduced from *Chemistry—An Experimental Science*, prepared by Chemical Education Material Study, copyright 1963, The Regents of the University of California, W. H. Freeman and Company, p. 310, courtesy of the publishers
p. 495 Historical Pictures Service, Chicago
p. 497 Jerry Frank—Design Photographers International, Inc.
p. 505 Photo by Charles L. Finance, reproduced from *Chemistry—An Experimental Science*, prepared by Chemical Education Material Study, copyright 1963, The Regents of the University of California, W. H. Freeman and Company, p. 310, courtesy of the publishers
p. 506 (top and bottom) Photos by Charles L. Finance, reproduced from *Chemistry—An Experimental Science*, prepared by Chemical Education Material Study, copyright 1963, The Regents of the University of California, W. H. Freeman and Company, p. 310, courtesy of the publishers
p. 509 H. Gray and G. Haight, *Basic Principles of Chemistry*, Copyright, © 1967, W. A. Benjamin, Inc., Menlo Park, California
p. 510 Smithsonian Institute
p. 521 An X-Ray diffraction pattern of a single ice crystal made by Professor I. Fankuchen at the Polytechnic Institute of Brooklyn
p. 528 Museum of Science, Boston, Massachusetts
p. 539 American Cancer Society
p. 541 H. Embree and H. DeBey, *Introduction to the Chemistry of Life*, 1968, Addison-Wesley, Reading, Massachusetts
p. 543 Courtesy of International Paper Company
p. 545 Du Pont Company
p. 549 Museum of Science, Boston, Massachusetts
p. 558 Du Pont Company
p. 583 Tony Spina, Courtesy Environmental Protection Agency
p. 588 Harold M. Lambert

Acknowledgements

p. *600* C. L. Hosler, Penn State
p. *608* Union Carbide Corporation
p. *629* University of Utah
p. *636* Culver Pictures, Inc.
p. *639* (left and right) Photography by Eastman Kodak for Dr. John Rupley
p. *644* Abram G. Schoenfeld, Photo Researchers, Inc.
p. *648* Culver Pictures, Inc.
p. *662* University of California Lawrence Berkeley Laboratory
p. *672* Historical Pictures Service, Chicago
p. *695* Fisher Scientific Company
p. *701* Photo by Charles L. Finance, reproduced with permission from *Chemistry—An Experimental Science*, prepared by Chemical Education Material Study, copyright 1963, The Regents of the University of California, W. H. Freeman and Company, Plate II
p. *706 Environment* (March 72), Robert Charles Smith
p. *720* Wide World Photos
p. *726* Argonne National Laboratory Photo
p. *750* The Bettmann Archive, Inc.

Photographs by John Urban, Arlington, Massachusetts, on pages—Title page, xviii, 7(bottom), 9(top left and right), 12(left), 17, 43, 46, 50, 63, 74, 77(top), 80, 104, 108(top), 113, 114, 115, 120, 151, 170, 171, 174, 201, 203(bottom), 248, 260, 349, 360, 383, 384, 427, 448, 465, 535, 556, 564(bottom), 593, 611, 613, 646, 690, 694, 709 and 710

Photographs by James Hubbard, Vaughan's Studio, on pages—7(top), 9(bottom), 10, 11, 12(right), 20(bottom), 22, 36(left), 57, 86, 91, 92, 93, 121, 125, 152, 153, 195, 196, 203(top), 219, 220, 221, 228, 229(top), 291, 464, 485, 487, 489, 490, 514, 518, 534, 561, 563, 564(top), 567, 570, 571, 617, 657, 665, 666, 674, 677, 697, 698, 702, 704, 711, 712, 713, 728, 729, 732 and 733

Illustrations by Universal Graphics, Inc., Maple Shade, New Jersey, ANco Technical Services, Inc., Boston, Massachusetts, and Mr. Henri A. Fluchère, Irvington-on-Hudson, New York

We would like to thank Miss Maureen Murphy of The Cambridge High and Latin School, Cambridge, Massachusetts, for allowing us to photograph student investigations.

INDEX

Numbers in boldface (**150**) refer to pages on which the term is defined or explained.

A

Absolute reaction rates, theory of, 629
Absolute temperature scale, 149–150
Absolute zero, **150**
Absorption band, 312–313
Acetaldehyde, model and formula, 466
Acetic acid:
  dissociation of, 715, 721
  hydrogen ion concentration, 704
  model and formula, 468
Acetone:
  absorption bands, 311
  model and formula, 515
  solvent, 515
Acetylene, 411, 451
Acid(s), 690, 750
  amino, 479–480, 537–538, 546, 548
  carboxylic, **467**–468
  metals, reactions with, 151–154
  nucleic, 546–550
  neutralization, 707–**709**
  properties, 696–697
  proton transfer, 719–721
  relative strengths, 715–716, 754
  strong, 715
  titration of, **709**–714
  weak, 715–716, 718–719
Acid-base indicators, 697–698, 704–705, 709
Acid equilibrium constant, 715–716
  determination of, 716–717
Acidic solutions, **694**–695
Acid rain, 706
Activated complex, 627–**628**, 631–635
Activation energy, 625–**628**, 631–635
  and heats of reaction, 628–631

Active site, 638–639
Addition polymerization, **531**–533
Addition reactions, 452
Additivity of reaction heats, law of, **596**
Adenine, 547
Adipic acid, condensation of, 530
Adrenal gland, 636
Adrenalin, 636
Afterburners, 583
Air, composition of, 146
  See also Atmosphere
Air pollution, 144–146, 542, 706
Alchemists, 176
Alcohols, 459–462
  investigating, 462–467
  oxidation of, 462
Aldehydes, **467**
Alizarin yellow R, 700–701
Alkali metals, 202–**203**, 351–352
  properties of, 212
Alkaline earth metals, **352**
Alkanes, 443–446
  models and formulas, 444
  properties of, 446
Alkenes, **450**–453
Alkynes, **450**–453
Alloy, **488**
Alpha (α)-decay, 267–268
Alpha (α) particles, 251, **267**
  deflection by atoms, 252–255
Aluminum, 198
Aluminum oxide, 198
Amide-bond linkage, 530
Amides, **473**
Amines, **472**–473
Amino acids, 479–480, **537**–538
  code for, 546, 548
Amino group, **472**

Ammeter, 728, 736
Ammonia, 125-126
  formation of, 659, 663
  molecular shape, 397, 402
  structure, 373
Ampere, **238**
Analysis, 172
  colorimetric, 665, 668
  gravimetric, 711
  volumetric, 711
  qualitative, 172
  quantitative, 172
Ångström, **258**
Anions, 245
  names and formulas, 760
  of silica, 552-553
Anode, 227, 239, **749**
Aqueous solution, **566**-567
Archimedes' principle, 79
Argon, 215
  discovery of, 210
  electron configuration, 370-371
  properties, 212, 502
Aromatic hydrocarbons, 455-457
Arrhenius, Svante August, 420, 720, 750
Artificial heart, 3
Asbestos, 552-553
  network solid, 505
Aston, F. W., 266
Asymmetric carbon atom, 478
-*ate*, meaning of suffix, 562
Atmosphere (atm), **110**
Atmosphere, water vapor in, 143
Atmospheric pressure, 110-111, 143
Atom(s), 3, 105, 107, **127**
  Bohr's model, 292, 307-310, 319, 322-323
  decay of, 267-268
  diameters, 257-258
  packing of, 490
  Rutherford's model, 254
  structure, 250-258, 262-263, 307-3310, 325-327
  Thomson's model, 251, 253
Atomic bombs, 268
Atomic mass. *See* Molar mass
Atomic numbers, 211, **257**, 756-757
  basis for periodic table, 345-346
  determination of, 355-356
Atomic spectra, 298-301
Atomic Theory, 175, 177, 308
Atomic velocities, distribution of, 622-625
Atom-smasher, 5
Attraction, 163-167
  dipole-dipole, 493
  in condensed phases, 485-486
  of charged particles, 220-222
Automotive smog, 582-583
Avogadro's hypothesis, 126-**127**, 129-131, 133, 155, 158
Avogadro's number, 129

B

Backbone, of nucleic acids, 546-547
Balanced equation, **185**-188
  with half reactions, 744-745
  with oxidation numbers, 743-744
Balances, 32-**33**, 34-39
  care of, 36-37
Balmer, 274
Barkla, 354
Barometer, 111-**112**
Bartlett, Neil, 210
Bases, 690, 750
  formation of, 199-202
  in nucleic acids, 546-548, 550
  neutralization, 707-**709**
  properties of, 697-699
  proton transfer, 719-721
  standardization of solution, 712-713
  titration, **709**-714
  See also Hydroxide ions
Basic solutions, **694**-695
  standardization of, 712-713
Batteries, **746**-747
Becquerel, 273
Bends(caisson disease), 213
Benzene, 454-**455**
  derivatives, 457
  ring, 454
  substitution reactions, 456
Benzoic acid, determining

equilibrium constant, 716-717
Beryllium, bonding capacity, 375-376
Beta (β)-decay, 267-268
Beta (β) particles, 267
Billiard ball analogy, 108-109
Binary compounds, **562**
Biodegradable, 386
Body-centered cubic packing, 490
Bohr, Niels, 274, 300, 307-309
Bohr's atomic model, 292, 307-310, 319, 322-323
Boiling point, **647**
  of hydrogen compounds, 520
  trends in, 213
Boiling point elevation, 516, 519
Bond angles, predicting, 396-398, 400-407
Bond energy, **410**
Bonding:
  double bonds, 408-411, 429-431
  covalent, 363-368, 378-379
  hydrogen, **522**
  ionic, 377-382
  pi (π) bonds, 429-432
  representations of, 366, 368-371
  sigma (σ) bonds, 429-432
  triple bonds, 410-411, 431-432
Bonding capacity, 371-377
Bond length, 364
Boron, bonding capacity, 374-375
Boron trifluoride, 375
Boyle, Robert, 117
Boyle's law, 112-**116**, 124-125, 147, 155, 158
Bravais, 488
Bromination, 452
Bronowski, Jacob, 20
Brønsted, J. M., 720
Buoyancy, 41-47
Buoyancy correction, 41, 45-47, 101, 122
Buoyancy effect, 81-82
  investigating, 42-45
Burets, care and use of, 710-711
Burner, adjustment of, 9
  flame, 650

Burning, 6, 21-24
  See also Combustion

C

Caisson disease, 213
Calcium carbonate, 145
  decomposition of, 649
Calculations. *See* Stoichiometry
Calibration, **34**
Calorie, **58**, 60
Calorimeter, 592, 599
Calorimetry, **599**
Candle, investigating, 5-8, 21-24
Carbohydrates, **197**, 294
  combustion of, 197
Carbon atoms:
  asymmetric, 478
  bonding capacity, 373-374
  layering, 98
  unique property, 440
Carbon compounds, 436-481
  nomenclature, 442-443, 453, 455, 476
  sources of, 437-439
  structure of, 440-442
Carbon dioxide, 409-410
  in the atmosphere, 600-601
Carbonic anhydrase, 638
Carbon monoxide, 143, 582
Carbon tetrachloride, as a solvent, 515
Carbon tetrafluoride, structure of, 374
Carbon-12, 128
Carbonyl group, **469**
Carboxyl group, **467**
Carboxylic acids, **467**-468
Carcinogens, 145
Carothers, Wallace Hume, 545
Carson, Rachel, 474
Catalysis, **631**-635
Catalysts, 576, **578**-579, 630-639, 657, 662
  enzymes, 635, **637**-639
  for afterburners, 583

hydrogen ions, 633-635, 694
hydroxide ions, 694
nitrogen oxides, 575, 579, 660-661
platinum, 579
Cathode, 227, 239, **749**
Cathode-heating coil, 226-227
Cathode-ray tube, 230, 232
Cations, 245
names and formulas, 760
Cavendish, Henry, 210
Cell potentials, 736
measuring, 736-738
Cellulose, 540-542
Celsius, Anders, 61
Celsius temperature scale, 58, 60-61, 150
Center of mass, **30**
Cesium chloride, geometry of, 512
Chadwick, 275
Charge/mass ratio:
of electrons, 232-235
of positive ions, 236
Charles, Jacques, 147
Charles' law, **147**-149, 155, 158
Chemical change, 90
Chemical equations. *See* Equations
Chemical kinetics, 609-639
Chemical reactions. *See* Reactions
Chemistry, 1-4
and environment, 18-19
inorganic, **439**
organic, **437**-481
Chiral molecules, 477-480
Chloride ions, test for, 383
Chloroform, 514
Chromium plating, 751
Chymotrypsin, 637
Circuits, **226**-227
cis-trans isomerism, 423-424
Climatic change, 600-601
Clock reaction, 612
Closed shell configurations, **342**
recurrence of, 344-345
Closed system, **650**-656
Cloud chamber, 274-275
Coal, 437-438
Coke, 438

Collision theory, 57-58, **615**-621, 625
Colorimetric analysis, 665, 668
Combustion, 21-24, 195-197, 270
investigating, 90-92
measuring heat of, 90-92
of carbohydrates, 197
Commoner, Barry, 19
Communication satellite, 4
Compounds, **168**
binary, **562**
carbon, 436-481
conductivity of, 382-385, 510
nomenclature, 561-562
solubility in water, 761
Concentration, 447, **519**
determining colorimetrically, 664-668
effect on equilibrium, 656-657
effect on reaction rates, 613-617, 619-620
square bracket notation, 667-668
Condensation, of gases, 164-167, 485
Condensation polymerization, 529-**531**
of amino acids, 538
of glucose, 541
of isoprene, 544
Conductivity:
of compounds, 382-385, 510
of metals, 497
Conductors, 226
Conservation of energy, law of, 66-67, 70-71, 596
Conservation of mass, law of, 49-51
Conservation of mass and energy, law of, 70-71
Constants:
dissociation, **692**-696
equilibrium, 668-671, 673, 675
Planck's 292-**293**
rate, 620
Contact process, 575-576, 578-579
Continuous spectra, 310-311
Contour lines, 388-391
of hydrogen molecule, 390
of lithium molecule, 390
of two helium atoms, 391

Control, **23**
Conversion factors, 39, **573**-574
Coordination number, 488
Copper, oxidation of, 727-733
Copper oxide, synthesis of, 173-174
Copper plating, 749-751
Copper sulfate, reaction with iron, 562-565
Cortisone, 636
Coulomb's law, **222**
Covalent bond, **365**
  directional character, 498
Covalent bonding, 363-368, 378-379, 720
  and network solids, 494-496, 505-507
Crookes tube, 228-229
Cross-linkage, **535**
Cryostat, 166
Crystalline solids, 486, 507-513
Crystals, 486
  packing of, 510-513
  molecular, **493**
Cubic closest packing, 489-490
Curie, Irène, 275
Curie, Marie and Pierre, 273
Cyclo-, meaning of prefix, 445
Cyclotron, 274-275
Cytosine, 547

D

Dacron, 531-532
Dali, Salvador, 484
Dalton, John, 89, 175, 177, 308
da Vinci, Leonardo, 4
Davisson, Clinton J., 320, 322
Davy, Sir Humphry, 197, 238, 240, 727
DDT, 474-475
de Broglie, Louis Victor, 320-322
Debye, Peter J. W., 420
Debye unit, 417, 420
Decanting a liquid, 564
Defined unit, 32

Definite proportions, law of, 175
Degradability, 386
Degrees of freedom, 672
de Laplace, P. S., 596
Delta, Δ, 65
ΔH, 591-592
$\Delta H^{\ddagger}$, 630
Δt, 86
Density, **78**
  measurement of, 79-83
Deoxy-, meaning of prefix, 550
Deoxyribose sugar, 550
Derivatives, **460**
Derived unit, **78**, 86
  uncertainty in, 86-88, 95-97
Destructive distillation, 438
Detergents, 472
Deuterium, 263-266
Diamond, 494
Diatomic molecules, 180
Diborane, 375
Diffraction, 281-**282**, 284
  of light, 283
  of water waves, 283
Dilute, 519
Dimethyl ether, 182, 184, 421-422
Dipole(s), 416-419, 515-516
  fluctuating, 493
Dipole-dipole attraction, 493
Dipole moment, **416**-418, 420
Discharge tubes, 227-230, 235, 300
Disorder. *See* Randomness
Dissociation, 566
  of solutes, 750
Dissociation constant, **692**
  effect of temperature, 695-696
  of water, 692-694
Divalent, 372
DNA, 528, 546-550
Döbereiner, Johann, 205-206
d orbitals, 330-331
Double bonds, 408-411, 429-431
Double helix, in DNA, 548-549
Dry cells, 746-747
Dust particles, in the atmosphere, 600-601

Dynamic equilibrium, 650-654
  proof of, 653-654

E

"Edison" effect, 274
Einstein, Albert, 70-71, 292, 298
Elastic collisions, **57**, 156-157
Electrical charges, 219-224, 257
Electric currents, 224
Electric dipole, 379-380
Electrochemical cells, 726, 730-**731**, 746-747
  investigating, 727-730
  potentials, **736**-738
Electrolysis, **747**-749
  of ions in solution, 238-243
  of water, 169
Electrolytes, **510**, 675
  solubility in water, 516
  strong and weak, 691-692
Electromagnetic radiation, 279, 281, 290
Electromagnetic spectrum, 290-291
Electron(s), **255**, 257
  behavior in metals, 497-499
  beta-particles, 267
  charge, 231, 233
  charge/mass ratio, 232-235
  competition for, 734-735
  configurations, **341**-345, 758-759
  energy levels, 302-307
  evidence for, 225-232
  magnetic field, 398-400
  mass, 233, 262-263
  spin, 398-400
  tendency to form pairs, 398-400
  wavelike nature, 320-323
Electron cloud, 326
  size of, 403
Electron density maps, 388-391
  of naphthalene, 394
Electron dot representation, 369
Electronic spectra, 311
Electron-pair repulsion theory, 400-404
Electroplating, 749-751
Elements, **167**, 176

  alkaline earth, **352**
  atomic numbers, 756-757
  electron configurations, 758-759
  metallic, 497-498
  molar masses, 756-757
  periodic table, 211
  symbols, 179-180, 756-757
  synthetic, 178
  that form molecular solids, 494
  that form network solids, 496
  See also Halogens and Noble gases
Empirical formulas, determining, 440-441
Endothermic reactions, **592**, 630-631
End point, 709, 711-712
-*ene*, meaning of suffix, 450
Energy, 54-73
  abundance of, 3
  activation, 625-635
  conservation of, 66-67, 70-71
  dissipation of, 67-68
  forms of, 55-58
  ionization, 308, **351**-352, 380, 496
  kinetic, **55**-58, 67
  molecular, 602-604
  nuclear, 268-272
  potential, **56**-58, 67
  promotional, 374
  solar, 294-296
  thermal, 55, 58
  units of, 58, 60
  See also Heat and Light
Energy levels, 302, 305-307
Enzymes, 635, **637**-639
Equations, **132**, 184
  balancing, 185-188, 742-745
  writing, 565-567
Equilibrium, 644-**650**, 651-687
  acid-base, 693-694, 707-708
  altering, 655-659, 663
  concentrations, 668-671, 673-675
  dynamic nature, 650-654
  effect of concentration on, 656-657
  effect of temperature on, 656-657
  factors that determine, 681-686
  in chemical reactions, 645, 650-652
  in phase changes, 645-647, 652-653

in solutions, 647, 653, 675-678
investigating, 664-668, 673-675
Le Chatelier's principle, **657**-659
recognizing, 647-650
Equilibrium constant, 668-671, 673, 675
acid, **715**-717
See also Dissociation constant
Equilibrium law relation, 670-671, 675
derived from reaction rates, 678-681
Errors, **99**-101
Esterases, 637
Esters, 466, **469**, 471
hydrolysis of, 637
Ethane, determining formula, 440-441
Ethanol, 182, 184
physical properties, 421
structure, 419, 421
Ethylene, 408-409, 429-430
model and formula, 451
structure of related molecules, 422-424
Ethyl group, **459**
Evaporator, 648
Exothermic reactions, **592**, 628, 630
Expansion, against a piston, 166-167
Experiments, importance of, 4-5
Explosions, 608
Exponential numbers, 97-98
Eyring, Henry, 629

F

Fahrenheit, Gabriel D., 60-61
Families, 202-205
Faraday, Michael, 238, 240
Faraday's law, 240
investigating, 238-243
Fats, 469, 471-472
Ferric chloride, test for hydroxide ions, 383
Field, **229**
Field ion microscope, 162
Filtrate, 571
Filtration, 570-571
First ionization energy, **351**
First Law of Thermodynamics, **66**-67
Fission, 268-270
Flame, hottest part, 9
Fluctuating dipoles, 493
Fluorides, properties of, 376
Fluorine, bonding of, 369-371, 380
Food chain, 294, 475
f orbitals, 331
Formaldehyde, 409, 466
Formic acid, decomposition of, 633-635
Formulas, **132**, 180-184
determining, 440-442
writing, 560-561
Fossil fuels, 270
Fractional distillation, 439
Freezing point depression, 516
uses of, 517-518
Frequency, **279**
of light, 288
of water waves, 280
Functional groups, 446, **459**-473
nitrogen-containing, 472-473
transfer of, 637
Fusion, 269-270

G

Galileo, 60-61
Gamma (γ) rays, 267
Gaseous nebula, 106
Gases:
investigating masses of equal volumes, 119-124
liquefying, 164-167, 485
particle model, 116-118
pressure, 107-118, 140-143
See also Noble gases
Gas laws, 112-116, 124-125, 147-149, 155-158
deviations from, 164-167
Geiger counter, 274
Genetic code, 546, 548

Geometric isomers, 422-**424**
Gerlach, Walter, 400
Germanium, properties of, 208
Germer, Lester H., 320, 322
Giant molecules, 528-554
Gibbs, Josiah Willard, 672
Glass, 553
Glassy substance, 487
Globular proteins, 638
Glucose, 541
  structure of, 480
Glycerates, **469**, 471-472
Glycerine, 462
Gold foil, bombardment of, 251
Goldstein, E., 236
Golf ball analogy, 681-686
Graduated cylinder, 63-64
Gram, **32**
Graph, **13**
Graphing, 13-15
Graphite, structure, 494, 496
Gravimetric analysis, 711
Gravity, 29-31, 40
Ground level, 305
Ground state, **341**
Groups, **211**
Guanine, 547

H

Haber process, 659, 663
Hahn, Otto, 275
Half-cell reactions, **731**
  potentials, 738-741, 755
  tendency to proceed, 734-735
  using to balance equations, 744-745
Halides, 353
  investigating, 348-350
  of alkali metals, 507-508, 512
Halogens, 203-204, 351-353
  properties, 212, 502
Heat, 55, 58
Heat content, **591**
  molecular, 603-604
Heats of formation, **597**

Heats of reaction, 589-**591**, 597-599
  calculating, 598-599
  measuring, 91-94, 592-595
Heavy water, 265
  synthesis of, 720
Helium, 213, 215
  atom, 339, 341
  interaction between atoms, 367-368
Helix, 540
Hench, P. S., 636
Hertz, **291**
Hess, Germain Henri, 596, 672
Hess's Law of Constant Heat Summation, 596
Heterogeneous system, 616
Hexagonal closest packing, 489-490
Homogeneous system, 616
Hormones, 539, 636
Hybrid orbitals, **404**-407
Hybridization, **405**
Hydration, **516**
Hydrocarbons, **197**, **443**
  aromatic, 455-457
  combustion of, 197
  investigating, 446-449
  pollutants, 582-583
  saturated, **443**-446
  unsaturated, **450**-453
Hydrochloric acid, neutralization, 707-710
Hydrofluoric acid, dissociation of, 715
Hydrogen:
  bonding of molecule, 363-365
  energy levels, 305-310
  ionization energy, 380, 382
  isotopes, 263-265, 269
  preparation of, 151-154, 199-202
  spectrum, 299-305
Hydrogen bomb, 269
Hydrogen bonding, **522**
  in nylon, 531
  in proteins, 540-541
  of hydrogen fluoride, 523
  of water, 523
Hydrogen chloride, 125-126
  reaction with magnesium, 151-154

Hydrogen ion, 692
    as a catalyst, 633-635, 694
    concentration of, 692-694
    calculating concentration, 718
    determining concentration, 700-705
    in acids, 697, 718-719
Hydrogen peroxide, structure, 373
Hydrolysis, 637
Hydrometer, **80**, 82-83, 747
Hydronium ion, **721**
Hydroxide ions, 692
    as a catalyst, 694
    concentration of, 692-694
    in bases, 698-699
    test for, 383
Hydroxides, formation of, 199-202
Hydroxyl group, **459**
Hypothesis, 127

I

-*ic*, meaning of suffix, 562
Ice, crystal structure, 520-521
-*ide*, meaning of suffix, 562
Ideal gas law, **156**-158, 164-167
Indicators, 700-702
Indigo carmine, 700-701
Inelastic collision, **58**
Infrared radiation, **291**
    spectrum, 314
Inorganic chemistry, **439**
Insecticides, 474-475
Interatomic forces, 493-500
Interference patterns, 281-282
    of light waves, 285
    of water waves, 284
International System of Units, 60
Inverse proportion, 116
Ionic bonding, 377-382
Ionic solids, 507-513
    properties, 509-510
    structure, 508-509
Ionization energy, 308, **351**-352, 380
    first, **351**
    of metals 496

Ions, **235**, 244
    anions, 245
    cations, 245
    common, 761
    determining number in solution, 517-518
    formation of, 351
    in solution, 237-243
    of gases, 235-237
    packing in crystals, 510-513
    relative sizes, 511
    valences, 560
Iron, reaction with copper sulfate, 562-565
Iron oxide, 198
Isomers, **182**
    cis-trans, 423-424
    geometric, 422-**424**
    iso, 443
    normal, 443
    optical, 477-480
    structural, 418-**422**
Isotopes, 128, 237, **263**-266
Isotopic tracers, 653-654

J

Johnston, Harold S., 660-662
Joliot, Frédéric, 275
Joule, **60**
Joule, James Prescott, 89
Joule-Thomson effect, 166-167
Julian, Percy Lavon, 458, 636

K

Kekulé, Friedrich A., 454
Kelvin, **60, 150**
Kelvin, Lord (William Thomson), 61
Kendall, Edward Calvin, 636
Ketones, 468-**469**
Kilogram, **32**
Kinetic energy, **55**-58, 67, **156**-157
    distribution of, 622-625
    kinds, 603
Krypton, 76

L

Lambda, λ, 281
Laser beams, 293
Lavoisier, Antoine, 59, 596
Lawrence, Ernest O., 274, 275
Laws:
    Additivity of Reaction Heats, **596**
    Boyle's, 112-**116**, 124-125, 147, 155, 158
    Charles', **147**-149, 155, 158
    Chemical Equilibrium, 670-671, 675, 678-681
    Constant Heat Summation, 596
    Coulomb's, 222
    Conservation of Energy, **66**-67, 596
    Conservation of Mass, 49-51
    Conservation of Mass and Energy, 67, 70-71
    Definite Proportions, **175**
    Gravitation, 29-31
Lead chamber process, 575-576, 578-579
Lead-sulfuric acid battery, 747
Le Chatelier, Henry Louis, 657
Le Chatelier's principle, **657**-659
Lewis, Gilbert Newton, 720
Li, Choh Hao, 539
Light, 278-298
    diffraction, 283
    dual nature, 293, 296
    evidence for particles, 296-298
    frequency, 288
    interference pattern, 285
    investigating wave motion, 285-288
    polarized, 478-479
    speed, 288
    wavelength, **281**
Lignin, 542
Limiting reactant, 572
Linear molecules, 404
Line representation of bonds, 369
Liquefying gases, 164-167
Liquids, 487
    decanting, 564
    measuring density, 82-83
    nitrogen, 485
    See also Solutions
Liter, 78
Lithium, 341
    bonding capacity, 376
Lithium fluoride, bonding in, 377-378
Litmus, 697-698, 709
Livingston, M. S., 274-275
Lowry, T. M., 720
Lysozyme, 638-639

M

Magnesium, 151-154
    combustion of, 195-196
Magnetic field, of electrons, 398-400
Malleability, 497
Manometer, **111**-112
Maps, of electron density, 388-391
Mass, **30**
    and weight, 30-31, 39-48
    center of, **30**
    conservation of, 49-51
    measuring, 33-35, 37-39
    of electron, 233
    of positive ions, 236-237
    units, 31-32
Mass-mass calculations, 574-577
Mass number, **262**
Mass relationships, investigating, 568-572
Mass spectrograph, **236**
Mass-volume calculations, 577-578
Matter, 3
    condensed phases, 485-519
    fundamental properties, 29, 223
    wave properties, 319-323
Measurement, units of, 75-78
Meitner, Lise, 275
Melting points:
    effect of molecular shape, 503
    investigating, 10-15
    trends in, 213
Mendeleev, Dimitri, 206-210, 213-214
Meniscus, **64**

Mercury, toxic compounds, 388
Metals:
  bonding, 496-**500**
  competition for electrons, 734-735
  heats of vaporization, 499
  location in periodic table, 497
  properties, 496-497, 499-500
  reaction with water, 197-202
  synthesis of oxide, 173-174
Meter, 75-77
Methane, 439
  molecular shape, 402
  structure, 374
Methanol, model and formula, 460
Methyl group, **460**
Methyl orange, 700-702
Metric system, 75-78
Mica, 553
  network solid, 505-506
Microscope, field ion, 162
Milky Way, 98
Millikan, Robert, 231, 233, 274, 298, 322
Models, **107**-109, 125
  molecular, 183-185
  particle, 116-118, 163-164, 223-224
  styrofoam, 425-429
Molarity, **519**
Molar masses, **139**-140, 207, 441
  determination of, 517-518
  of elements, 756-757
Mole, **128**-129, 129-130
  method for calculations, 573
Molecular crystals, **493**
Molecular orbitals, 412-415
  of benzene, 455
Molecular model of gases, 139, 141, 156-158
  See also Particle model
Molecular models, 183-185
  Styrofoam, 425-429
Molecular solids, 493-494, 501-504
Molecules, 3, 105, 107, **127**, 167
  chiral, 477-480
  diatomic, 180
  dipole moments, 416-419
  energy, 602-604
  formulas, 180-182, 441
  giant, 528-554
  linear, 404
  mass. See Molar mass
  planar, 403-404
  polarity, **379**-380
  polyatomic, 180-182, 743
  shape, 396-398, 400-407, 416-419
  size, 259-262
  spectra, 311-314
  structure, 395
  vibrations, 312
Moment, **416**
Momentum, **156**-157
Monomer, **529**
Moseley, Henry G. J., 346, 354-356
Moseley diagram, 357
Moth repellent, 11-15
Muller, Paul Hermann, 474
Mulliken, Robert S., 413
Multiple bonds, 410-411, 429-432

N

Naphthalene, 394
Natural gas, 438-439
Nature of the Chemical Bond, The, 381
Neon, 215
Neon sign, 227
Network solids, **494**, 496, 505-507
Neutralization, 707-**709**
  of charges, 222
Neutral solutions, 694-695
Neutrons, **262**-263
  discovery, 275
  mass, 263, 269
Newlands, John, 206
Newton, Isaac, 29-30
Nieuwland, Julius A., 545
Nitrate ion, 508
  molecular orbitals, 414-415
  resonance hybrid, 411-412
Nitration, 452
Nitric acid, 569, 706
Nitrogen:
  bonding capacity, 373

bonding in molecule, 410
liquid, 485
production of compounds, 659, 663
Nitrogen oxides, 143
catalysts, 575, 579, 660-661
equilibrium of, 650-652, 659
pollutants, 582
precautions, 134, 569
Noble gases, **127**, 210, 351-353
configuration, 370-371
properties, 212, 502
uses, 213, 215
Noddack, Ida Tacke, 214
Nomenclature:
inorganic compounds, 561-562
organic compounds, 442-443, 453-455, 476
Normal isomers, 443
Normal boiling point, **647**
Nu, ν, 279
Nuclear energy, 268-272
Nuclear reactions, 267-272
fission, 268-270
fusion, 269-270
Nucleic acids, 546-550
Nucleus, 254-255
diameter, 258
stability, 262-263
Nylon, 529-531, 545

O

Observation, 5-6, 16-17
Octaves, law of, 206
Oil drop experiment, 231-233
Oleic acid, size of molecule, 259-262
Open system, 650, 656
Optical isomers, 477-480
Orange IV, 700-701
Orbital(s), **325**-327
d, 330-331
f, 331
hybrid, **404**-407
molecular, 412-415
p, 330-335
relative energies, 340, 343
s, 328-335
shapes, 328-333
Orbital-occupancy diagrams, 368
Orbital representation of bonds, 368
Organic compounds, **437**-481
nomenclature, 442-443, 453, 455, 476
sources of, 437-439
structure of, 440-442
See also Carbon atoms
Organic reactions, stoichiometry in, 580-581
-*ous*, meaning of suffix, 562
Oxidation, **731**, 734
Oxidation number, **742**-745
Oxidation-reduction, 726-**732**, 733-751
balancing equations for, 742-745
in a beaker, 732
of metals and acids, 733-734
prediction of, 736-741
Oxides, formation of, 195-198
Oxidizing agents, **732**, 755
Oxygen, 125-126
bonding capacity, 371-373
molecular formula, 132
Ozone, 660-662
dangers of, 582-583

P

Packing in crystals, 510-513
investigating, 488-492
Paper, 542-543
Paper mill, 543
*Para*dichlorobenzene, 11-15
Paraffin series. See Alkanes
Partial pressure, **141**-143
Particle accelerator, 178
Particle model, 116-118, 163-164, 223-224
Pauling, Linus C., 381
Pauli's exclusion principle, **341**
Pentane isomers, physical properties, 504
Peptidases, 637
Percentage uncertainty, **95**-96

Periodic table, 206-211
   based on atomic numbers, 345-346
   modern version, 211
Perkin, William, 470
Persistence of Memory, 484
Persistent insecticides, 474-475
Petroleum, 438-439
pH, **695**
   See also Hydrogen ion
pH meter, 695, 700
Phase change, 90
   equilibrium, 645-647, 652-653
Phases, 163-164
Phenolphthalein, 704-705
Phosphorus, combustion of, 196
Photoelectric effect, 292, **296**-298
Photon, **293**, 297
Photosynthesis, 294
Physostigmine, synthesis of, 458
Pi bonds, 429-432
Pioneer 10, 54
Pituitary hormones, 539
Planar molecules, 403-404
Planck, Max, 292, 308
Planck's constant, 292-**293**
Plane-polarized light, 478-479
Platinum, finely divided, 579
Pleated sheet, 540-541
Plunkett, Roy J., 52
Point of equivalence, 709, 711-712
Poisons, 388
Polarity, effect on solubility, 515-516
Polar molecule, **379**-380
Pollutants, 143-146, 582-583
Pollution, 542-543
   air, 144-146, 542, 706
   thermal, 68-69
   water, 388, 542-543
Polonium, discovery of, 273
Polyamides, formation of, 530-531
Polyatomic ions, 508
   oxidation numbers, 742
   valences, 530
Polyatomic molecules, 180-182
   oxidation numbers, 743

Polyester, 531
Polyethylene, 531-532
Polymers, **529**-533, 537-546, 553
   preparation of, 533-537
p orbitals, 330-335
Potential(E), **736**
   of a cell, 736-738
   of a half-cell, 738-741
Potential energy, **56**-58, 67
Powers of ten, 97-98
Precipitate, **566**-567
   predicting, 676-678
Pressure, 109-112, 157
   effect on equilibrium, 658-659
   partial, **141**-143
Pressure-volume relationship, 112-116, 124-125
Principal quantum number(n), **325**-327
Probability, investigating, 323-325
Products, **184**, 566
Promotional energy, 374
Propanol, models and formulas, 461
Proportionality constant, 620
Proteins, **537**-538, 540-541, 635, 637-639
   globular, 638
   hydrolysis of, 637
Protons, 223, **255**, 257
   mass of, 262-263
Proton transfer, 719-721
Pulp, 542
Pure substance, 168

Q

Qualitative analysis, 172
Quanta, 292, 297
Quantitative analysis, 172
Quantum mechanics, 249, 306, 318, 325-327, 333-335
Quantum numbers, 325-327
Quantum Theory, 292, 308
Quartz, network solid, 505-506
Quinine, synthesis of, 470

R

Radiation:
  gamma, 267
  infrared, **291**
  ultraviolet, **291**, 660
  X rays, 273, 355-356
Radioactivity, 265, **267**-268
  dangers of, 272
  discovery of, 273
Radium, discovery of, 273
Radon, 215
Ramsay, William, 210
Random errors, **99**-101
Randomness, 68-70, 684-686
Rate constant, **620**
Rate-determining step, **618**
Rayleigh, J. W. S., 210
Reactants, **184**, 566
  limiting, 572
Reaction coordinate, 627
Reaction mechanism, **618**-620, 631-635, 662
Reaction rates, 608-639
  dependence on reactants, 610-611
  effect of concentration, 613-617, 619-620
  effect of temperature, 614, 620-621
  investigating, 611-615
  See also Catalysts
Reactions, **184**
  addition, 452
  combustion, 21-24, 90-92, 195-197, 270
  completion of, 656
  condensation polymerization, 529-**531**
  endothermic, **592**, 628, 630
  exothermic, **592**, 628, 630
  half-cell, **731**, 734-735, 738-741
  metals with acid, 151-154
  metals and water, 197-202
  of gases, 130-134
  organic, 580-581
  oxidation-reduction, 726-**732**, 733-751
  rates. *See* Reaction rates
  spontaneous, 67-70, 740-741
  substitution, 452
  synthesis of metal oxides, 173-174
Reactors, nuclear, 268, 270-272
  locations of, 271
Recharging, 747
Recycling, 387
  of paper, 543
Redox. *See* Oxidation-reduction
Reducing agents, **732**, 755
Reduction, **731**, 734
  See also Oxidation-reduction
Reduction potentials, **740**-741, 755
Reference half-cell, **738**-740
Refraction, **291**
Refractory, 293
Refrigerators, 167
Regularity, **17**, 19-20
Reichstein, T., 636
Relativity, 70-71
Repulsion, of charged particles, 220-222
Resonance, 411-**412**
  of benzene, 455
Resonance hybrid, **412**
Respiration, 638
Rhenium, discovery of, 214
Ribose sugar, 550
Rillieux, Norbert, 648
RNA, 546, 550
Roentgen, William, 273, 354
Rotational kinetic energy, 603
Rotational spectra, 314
Rubber, 544, 546
Rumford, Count (Benjamin Thompson), 59, 89
Rust, 198
Rutherford, Ernest, 250, 256, 273-275, 308, 354
  atomic model, 254
  scattering experiments, 251-256, 275

S

Salmon, effect of acid rains, 706
Salt bridge, 728-729, 732-733

Salt deposit, 178
Salts, **708**
   formation of, 708-709
   investigating solubility, 348-350
Sanger, Frederick, 539
Saturated hydrocarbons, 443-446
Saturated solution, **676**, 678
Scales, **34**-35
Scientific method, 16-17, 19-21
Scintillation screen, 252
SI, 60
Sigma bonds, 429-432
Significant figures, 86, **96**-98
Silent Spring, 474
Silicates, 505, 551-553
Silica tetrahedra, 551-553
   form band anions, 552
   form chain anions, 552
   form sheet anions, 553
Silver, reduction of, 727-733
Silver acetate, solubility product, 673-675
Silver chloride, precipitation of 570-571
Silver nitrate, 333
   preparation of, 569-570
Smog, 144
   automotive, 582-583
Soaps, **471**-472
Soddy, Frederick, 256, 266
Sodium bicarbonate, decomposition of, 169-172
Sodium chloride:
   crystal, 486-487
   geometry, 513
   lattice, 490-491
Sodium hydroxide, 593
   heat of solution, 593
   neutralization, 707-710
Soil, effect of acid rain, 706
Solar energy, 294-296
Solids:
   crystalline, 486
   density of, 81-82
   heat of solidification, 92-94
   ionic, 507-513
   melting of, 7-10
   metallic, 496-500
   molecular, 493-494, 501-504
   network, **494**, 496, 505-507
   techniques for transferring, 563
Solid wastes, 386-387
Solubility, 348
   calculating, 676
   equilibrium, 647, 653, 675-678
   of compounds in water, 761
   of gases, 126
   of salts, 348-350
   See also Solutions
Solubility product, **676**-678
   determining, 673-675
Solute, **514**
Solutions, **488**, 514-519, 750
   acidic, 696-697
   aqueous, **566**-567
   basic, 697-699
   neutral, 694-695
   pH, 694-695
   saturated, **676**, 678
   standard, 702-703
   standardization, of base, 712-713
Solvents, **514**-516
Soot, 145
s orbitals, 328-335
Sörenson, 695
Space lattice, 488
Specificity, of enzymes, 638-639
Specific gravity, **80**
Specific heat, **62**-65
   of water, 90
Spectra, 278, 291, 298, 310-314
   atomic, 298-301
   continuous, 310-311
   electronic, 311
   hydrogen, 299-305
   rotational, 314
Spectrograph, **288**-289, 291, 299
Spectrometer, 312
Speed of light, 288
sp hybrid orbitals, 405
sp^2 hybrids, 405-406
sp^3 hybrids, 405-407
Spin, 398-400
Spontaneous reactions, 67-70
   predicting, 740-741
Spring scales, **34**-35
Staircase analogy, 301-305

Standard half-cell, **738**-740
Standard solutions, preparation of, 702-703
Standard state, 739
Standardization, of basic solution, 712-713
Standing waves, 321
Starch, 540-541
Stark, Johannes, 400
Stern, Otto, 400
Stoichiometry, 559, **572**-581
Storage cell, 747
STP, 147
Strong acids, 715
Strong electrolyte, **691**
Structural formulas, **181**-182
Structural isomers, 418-**422**
Structure, 127-128, **395**
Styrofoam models, of molecules, 425-429
Sublimation, 451
Substitution reactions, 452
Substrate binding, 638-639
Sugar cane, conversion of, 648
Sulfur dioxide, 143-144
Sulfuric acid, 144-145, 173
   batteries, 747
   in the air, 706
   manufacture of, 574-579
Supersonic transports, 661
Symbols, 179-180
   complete list, 756-757
Synthesis, 172-173
   of a metal oxide, 173-174
Synthetases, 637
Synthetic elements, 178
Synthetic fibers, 529-532
Synthetic polymers, 545
Synthetic rubber, 546
System, **650**
Systematic error, **99**-101

T

Teflon, 52, 533
Tellurium, naturally-occurring
   isotopes, 347
Temperature:
   absolute zero, **150**
   effect on atomic velocities, 624-625
   effect on equilibrium, 656-657
   effect on reaction rates, 614, 620-621
   effect on vapor pressure, 647
   inversion, 600
   scales, 149-150
   units, 58, 60-61
Template, 548
Tetrahedron, **401**-403
Tetravalent, 374
Thermal energy, 55, 58
Thermal pollution, 68-69
Thermodynamics, 672
Thermometers, 60-61
Thermonuclear research device, 269
Thomson, J. J., 231, 233, 251, 256, 274, 308, 322
   atomic model, 251, 253
Threshold energy. *See* Activation energy
Thymine, 547, 550
Thyroxine, 636
Titration, 690, **709**-714
Torr, **111**
Torricelli, Evangelista, 111
Transferases, 637
Translational kinetic energy, 603
Triads, 205-206
Triatomic molecules, 180
Triple bonds, 410-411, 431-432
Tritium, 269
Trivalent, 373

U

Ultraviolet radiation, **291**, 660
Uncertainty:
   in derived quantities, 86-88, 95-97
   in measurements, 85-86
   percentage, **95**-96
Uncertainty Principle, 327-329

Units, **31**
 defined, 32
 derived, **78**, 86
 of energy, 58, 60
 of mass, 31-32
 of measurement, 75-78
 of temperature, 58, 60-61
Univalent, 370
Unsaturated hydrocarbons, **450**-453
Uracil, 550
Uranium isotopes, 264
Urey, Harold, 263, 266

V

Vacuum, 41
Vacuum tube, **227**
Valence, 376-377, 559-561
 electrons, **370**-377
 orbitals, **370**-377
 table, 560
Vanadium pentoxide, as a catalyst, 579
van dan Broek, 354
Van de Graaff generator, 223-224, 275
van der Waals, Johannes D., 495
van der Waals forces, 493, 495, 501-504
 and molecular shape, 503-504
 and molecular size, 502-503
 and number of electrons, 501-502
Vapor pressure, 652-653
 of ethyl alcohol, 647
 of water, 154, 646-647
Vector, 399
Vibrational kinetic energy, 603
Vinegar, 468
Viscous liquids, 486
Voltmeter, 736
Volts, 736
Volumetric analysis, 711
Volume-volume calculations, 578-580
von Guericke, Otto, 117
Vulcanization, 544

W

Water:
 dissociation constant, 692-696
 electrolysis of, 169
 heavy, 265
 hydrogen-bonding, 523
 in the atmosphere, 143
 ionization of, 691-692
 modes of vibration, 312
 molecular formula, 132-133
 molecular shape, 396, 402
 pollution, 388, 542-543
 properties, 168-169
 rotational motions, 314
 solvent, 516, 761
 specific heat, 90
 vapor pressure, 154
 vibrational spectrum, 313
Water gas, 589-590
Wave functions, 333-335
Wavelength, 279
 of light, **281**
 of water waves, 280
Weak acids, 715-716
 competition for hydrogen ions, 718-719
Weak electrolytes, **691**-692
Weather balloon, 138
Weber, Joseph, 40
Weighing, sources of error, 41-48
Weight, 29
 and mass, 30-31, 39, 41-48
Wilson, C. T. R., 274
Woodward, Robert Burns, 470
Work, **67**
Wurtzite lattice, 491

X, Y, Z

Xenon, compounds of, 215-216
X-ray diffraction, 395-396, 420
 by ice, 521
X rays, 355-356
 discovery of, 273
-*yne*, meaning of suffix, 450
Zeeman, Pieter, 400
Zinc sulfide lattice, 491

Molar Masses of the Elements

Element	Symbol	Atomic Number	Molar Mass	Element	Symbol	Atomic Number	Molar Mass
Actinium	Ac	89	(227)	Mercury	Hg	80	200.59
Aluminum	Al	13	26.98	Molybdenum	Mo	42	95.94
Americium	Am	95	(243)	Neodymium	Nd	60	144.24
Antimony	Sb	51	121.75	Neon	Ne	10	20.183
Argon	Ar	18	39.948	Neptunium	Np	93	(237)
Arsenic	As	33	74.92	Nickel	Ni	28	58.71
Astatine	At	85	(210)	Niobium	Nb	41	92.91
Barium	Ba	56	137.34	Nitrogen	N	7	14.007
Berkelium	Bk	97	(249)	Nobelium	No	102	(253)
Beryllium	Be	4	9.012	Osmium	Os	76	190.2
Bismuth	Bi	83	208.98	Oxygen	O	8	15.9994
Boron	B	5	10.81	Palladium	Pd	46	106.4
Bromine	Br	35	79.909	Phosphorus	P	15	30.974
Cadmium	Cd	48	112.40	Platinum	Pt	78	195.09
Calcium	Ca	20	40.08	Plutonium	Pu	94	(242)
Californium	Cf	98	(251)	Polonium	Po	84	(210)
Carbon	C	6	12.011	Potassium	K	19	39.102
Cerium	Ce	58	140.12	Praseodymium	Pr	59	140.91
Cesium	Cs	55	132.91	Promethium	Pm	61	(147)
Chlorine	Cl	17	35.453	Protactinium	Pa	91	(231)
Chromium	Cr	24	52.00	Radium	Ra	88	(226)
Cobalt	Co	27	58.93	Radon	Rn	86	(222)
Copper	Cu	29	63.54	Rhenium	Re	75	186.23
Curium	Cm	96	(247)	Rhodium	Rh	45	102.91
Dysprosium	Dy	66	162.50	Rubidium	Rb	37	85.47
Einsteinium	Es	99	(254)	Ruthenium	Ru	44	101.1
Erbium	Er	68	167.26	Samarium	Sm	62	150.35
Europium	Eu	63	151.96	Scandium	Sc	21	44.96
Fermium	Fm	100	(253)	Selenium	Se	34	78.96
Fluorine	F	9	19.00	Silicon	Si	14	28.09
Francium	Fr	87	(223)	Silver	Ag	47	107.870
Gadolinium	Gd	64	157.25	Sodium	Na	11	22.9898
Gallium	Ga	31	69.72	Strontium	Sr	38	87.62
Germanium	Ge	32	72.59	Sulfur	S	16	32.064
Gold	Au	79	196.97	Tantalum	Ta	73	180.95
Hafnium	Hf	72	178.49	Technetium	Tc	43	(99)
Helium	He	2	4.003	Tellurium	Te	52	127.60
Holmium	Ho	67	164.93	Terbium	Tb	65	158.92
Hydrogen	H	1	1.0080	Thallium	Tl	81	204.37
Indium	In	49	114.82	Thorium	Th	90	232.04
Iodine	I	53	126.90	Thulium	Tm	69	168.93
Iridium	Ir	77	192.2	Tin	Sn	50	118.69
Iron	Fe	26	55.85	Titanium	Ti	22	47.90
Krypton	Kr	36	83.80	Tungsten	W	74	183.85
Lanthanum	La	57	138.91	Uranium	U	92	238.03
Lawrencium	Lr	103	(257)	Vanadium	V	23	50.94
Lead	Pb	82	207.19	Xenon	Xe	54	131.30
Lithium	Li	3	6.939	Ytterbium	Yb	70	173.04
Lutetium	Lu	71	174.97	Yttrium	Y	39	88.91
Magnesium	Mg	12	24.312	Zinc	Zn	30	65.37
Manganese	Mn	25	54.94	Zirconium	Zr	40	91.22
Mendelevium	Md	101	(256)				

Based on mass of C^{12} at 12.000. Values in parentheses represent the most stable known isotopes for elements that do not occur naturally.

PERIODIC TABLE

PERIOD COLUMN	I	II					TRANSITION ELEMENTS							III	IV	V	VI	VII	0	ORBITALS BEING FILLED
$n=1$	1 H 1.00																		2 He 4.00	1s
$n=2$	3 Li 6.94	4 Be 9.01												5 B 10.8	6 C 12.01	7 N 14.01	8 O 16.00	9 F 19.0	10 Ne 20.2	2s2p
$n=3$	11 Na 23.0	12 Mg 24.3												13 Al 27.0	14 Si 28.1	15 P 31.0	16 S 32.1	17 Cl 35.5	18 Ar 39.9	3s3p
$n=4$	19 K 39.1	20 Ca 40.1	21 Sc 45.0	22 Ti 47.9	23 V 50.9	24 Cr 52.0	25 Mn 54.9	26 Fe 55.8	27 Co 58.9	28 Ni 58.7	29 Cu 63.5	30 Zn 65.4		31 Ga 69.7	32 Ge 72.6	33 As 74.9	34 Se 79.0	35 Br 79.9	36 Kr 83.8	4s3d4p
$n=5$	37 Rb 85.5	38 Sr 87.6	39 Y 88.9	40 Zr 91.2	41 Nb 92.9	42 Mo 95.9	43 Tc (99)	44 Ru 101.1	45 Rh 102.9	46 Pd 106.4	47 Ag 107.9	48 Cd 112.4		49 In 114.8	50 Sn 118.7	51 Sb 121.8	52 Te 127.6	53 I 126.9	54 Xe 131.3	5s4d5p
$n=6$	55 Cs 132.9	56 Ba 137.3	57-71 See Below	72 Hf 178.5	73 Ta 180.9	74 W 183.9	75 Re 186.2	76 Os 190.2	77 Ir 192.2	78 Pt 195.1	79 Au 197.0	80 Hg 200.6		81 Tl 204.4	82 Pb 207.2	83 Bi 209.0	84 Po (209)	85 At (210)	86 Rn (222)	6s4f5d6p
$n=7$	87 Fr (223)	88 Ra (226)	89-103 See Below	104 *	105 *															7s5f6d7p

*There Are No Internationally-accepted Names For These Elements.

$n=6$	57 La 138.9	58 Ce 140.1	59 Pr 140.9	60 Nd 144.2	61 Pm (147)	62 Sm 150.4	63 Eu 152.0	64 Gd 157.3	65 Tb 158.9	66 Dy 162.5	67 Ho 164.9	68 Er 167.3	69 Tm 168.9	70 Yb 173.0	71 Lu 175.0	4f
$n=7$	89 Ac (227)	90 Th (232.0)	91 Pa (231)	92 U 238.0	93 Np (237)	94 Pu (242)	95 Am (243)	96 Cm (247)	97 Bk (249)	98 Cf (251)	99 Es (254)	100 Fm (253)	101 Md (256)	102 No (253)	103 Lr (257)	5f

Parenthetical Values Are Mass Numbers of the Most Stable Isotopes